PERIODIC TABLE OF THE ELEMENTS

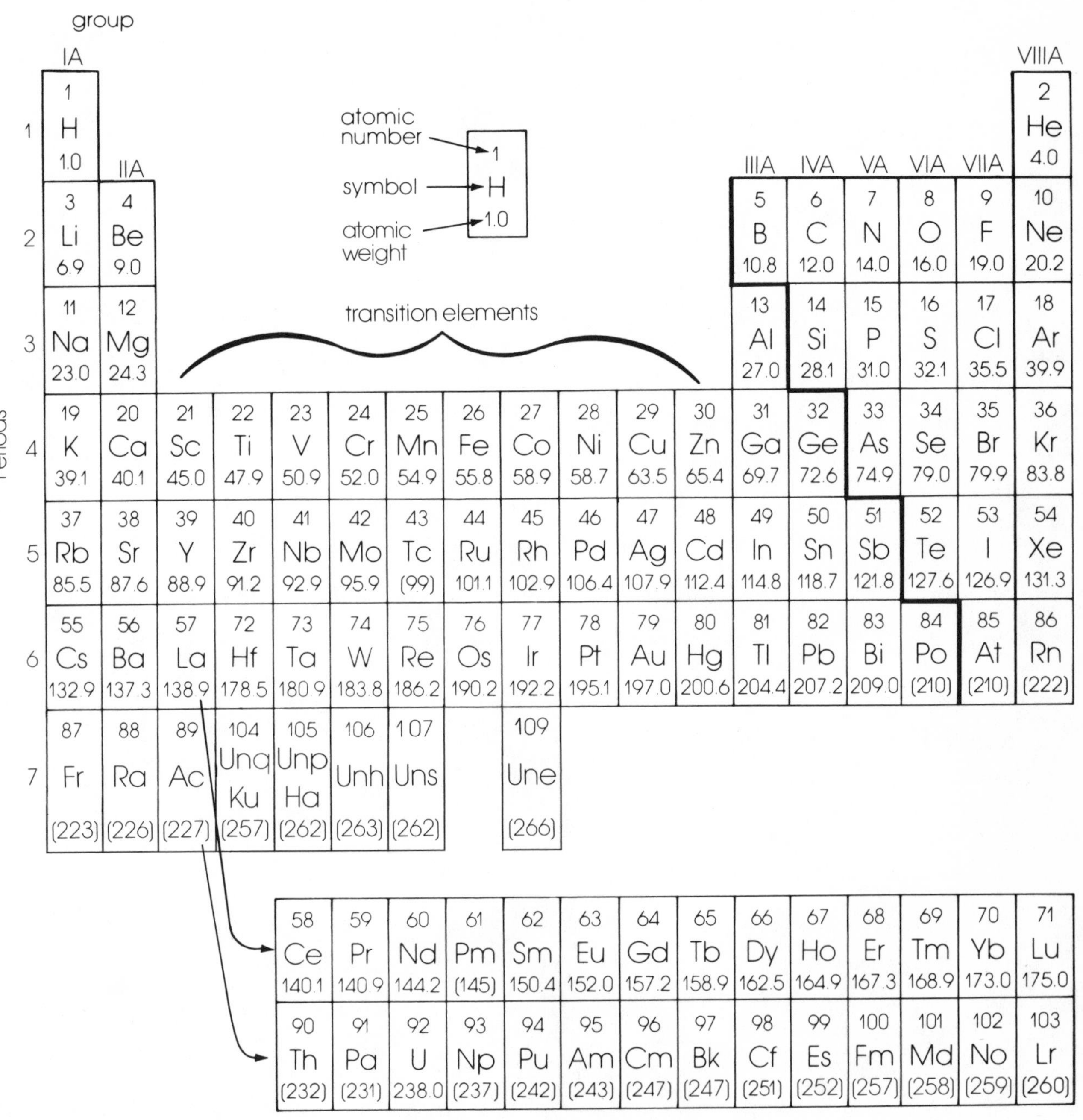

Period	IA	IIA											IIIA	IVA	VA	VIA	VIIA	VIIIA
1	1 H 1.0																	2 He 4.0
2	3 Li 6.9	4 Be 9.0											5 B 10.8	6 C 12.0	7 N 14.0	8 O 16.0	9 F 19.0	10 Ne 20.2
3	11 Na 23.0	12 Mg 24.3											13 Al 27.0	14 Si 28.1	15 P 31.0	16 S 32.1	17 Cl 35.5	18 Ar 39.9
4	19 K 39.1	20 Ca 40.1	21 Sc 45.0	22 Ti 47.9	23 V 50.9	24 Cr 52.0	25 Mn 54.9	26 Fe 55.8	27 Co 58.9	28 Ni 58.7	29 Cu 63.5	30 Zn 65.4	31 Ga 69.7	32 Ge 72.6	33 As 74.9	34 Se 79.0	35 Br 79.9	36 Kr 83.8
5	37 Rb 85.5	38 Sr 87.6	39 Y 88.9	40 Zr 91.2	41 Nb 92.9	42 Mo 95.9	43 Tc (99)	44 Ru 101.1	45 Rh 102.9	46 Pd 106.4	47 Ag 107.9	48 Cd 112.4	49 In 114.8	50 Sn 118.7	51 Sb 121.8	52 Te 127.6	53 I 126.9	54 Xe 131.3
6	55 Cs 132.9	56 Ba 137.3	57 La 138.9	72 Hf 178.5	73 Ta 180.9	74 W 183.8	75 Re 186.2	76 Os 190.2	77 Ir 192.2	78 Pt 195.1	79 Au 197.0	80 Hg 200.6	81 Tl 204.4	82 Pb 207.2	83 Bi 209.0	84 Po (210)	85 At (210)	86 Rn (222)
7	87 Fr (223)	88 Ra (226)	89 Ac (227)	104 Unq Ku (257)	105 Unp Ha (262)	106 Unh (263)	107 Uns (262)		109 Une (266)									

58 Ce 140.1	59 Pr 140.9	60 Nd 144.2	61 Pm (145)	62 Sm 150.4	63 Eu 152.0	64 Gd 157.2	65 Tb 158.9	66 Dy 162.5	67 Ho 164.9	68 Er 167.3	69 Tm 168.9	70 Yb 173.0	71 Lu 175.0
90 Th (232)	91 Pa (231)	92 U 238.0	93 Np (237)	94 Pu (242)	95 Am (243)	96 Cm (247)	97 Bk (247)	98 Cf (251)	99 Es (252)	100 Fm (257)	101 Md (258)	102 No (259)	103 Lr (260)

Note: numbers in parentheses indicate the mass number of the most stable or best-known isotope

Chemistry

FOURTH EDITION

Chemistry

Karen Timberlake
Los Angeles Valley College

HarperCollins*Publishers*

Sponsoring Editor: Lisa Berger
Project Editor: Thomas R. Farrell
Text Design: Caliber Design Planning, Inc.
Cover Design: José R. Fonfrias
Cover Illustration: Baby Shampoo Film, Photo Researchers, Inc., © Tom Branch/
Science Source
Text Art: Fine Line Illustrations, Inc.
Production: Willie Lane
Compositor: Composition House Limited
Printer/Binder: R. R. Donnelley & Sons Company
Cover Printer: Lehigh Press

Chemistry, Fourth Edition

Library of Congress Cataloging-in-Publication Data

Timberlake, Karen.
Chemistry.

Includes index.
1. Chemistry. I. Title.
QD31.2.T55 1988 540 87-2879
ISBN 0-06-046659-6

91 9 8 7

TO MY FAMILY, FRIENDS, AND STUDENTS

The whole art of teaching is only the art of awakening the natural curiosity of young minds.

Anatole France

Happiness is neither virtue nor pleasure nor this thing nor that but simply growth. We are happy when we are growing.

W. B. Yeats

Contents

19 Metabolism: Energy Production in the Living Cell 559

20 Chemistry of Heredity: DNA and RNA and Hormonal Action 592

Preface

Welcome to the fourth edition of *Chemistry*. It is my hope that the reshaping of this text over four editions has resulted in a book that makes teaching and learning chemistry a positive experience for both the teacher and the student. It remains my goal to assist students in their development of critical thinking and to establish a science framework so that students have the concepts and problem-solving techniques to make decisions about issues concerning our environment, medicine, and health that affect all our lives.

Over the years I have found that chemistry is a formidable subject to many students, and thus I have made a practice of associating chemistry concepts with applications to the allied health fields. This bridge between chemistry and the world of the students is designed to help students incorporate the ideas of chemistry into their lives and future careers. I have found discussions of applications of chemistry valuable in increasing student interest, motivation, concentration, and performance in class.

New in This Edition

In response to the needs to my students as well as to suggestions of teachers and reviewers, several changes have been made in the fourth edition. More examples have been added to every chapter. All examples have been rewritten so that they now include fully worked-out solutions. This is intended to assist the student with the patterns of problem solving that lead to correct answers. All the applications to health, including many new ones, are now set off as *Health Notes* for easier identification. Many of the problem sets have been rewritten with more problems to engage the student in more problem solving at various levels of difficulty. A new appendix has been written to deal with calculations and measured numbers.

Pedagogical changes include a fuller use of a second color to highlight major ideas, as well as a change of type size and typeface for easier reading. The wider margins allows for marginal notes for key words and key points. The problem sets have been expanded to include more problems for practice problem solving. The problems are listed by section and objective number for easy reference to the chapter. The appendixes include math and the calculator, scientific notation, and answers to all the problems in the text.

To Accompany the Text

The three supplements to the fourth edition are the *Study Guide*, the *Laboratory Manual*, and the *Instructor's Manual*. The *Study Guide* reviews the basic concepts, provides learning drills, and gives a practice exam, all with answers, for each of the 20 chapters. Students can grade their own practice exams and then check to determine if they have mastered the material. In this way, they can assess areas of difficulty and review the material again.

The *Laboratory Manual* contains some new experiments, including percent water in a hydrate. Each laboratory experiment includes a report page and questions that relate the experiment to the corresponding information in the text. Some questions require essays, which promotes writing skills in the content area. The overall thesis of the *Laboratory Manual* is first to introduce students to basic laboratory skills and then to present some investigative problems to develop the skills of gathering and reporting data, problem solving, calculating, and drawing conclusions. In this edition, as in the previous ones, there is an emphasis on safety in the laboratory. I have attempted to remove those chemicals and procedures that are known or suspected to be dangerous. I have also attempted to reduce the amounts of chemicals needed, in view of the austerity programs that many of our schools now face.

Acknowledgments

I wish to thank my husband, Bill, for his invaluable assistance, support, and advice in the preparation of the manuscript. Thanks also go to my son, John, for his cooperation during the time we were writing this text.

The following people provided outstanding reviews that contained constructive and helpful criticism and support, for which I am very grateful: Thomas Berke, Brookdale Community College; Lew Milner, North Central Technical College, Mansfield, Ohio; John Macklin, University of Washington, Seattle; Thomas Berke, Brookdale Community College, Lincroft, New Jersey;

Barbara Frohardt, Oakdale Community College, Bloomfield Hills, Michigan; and Stanley Mehlman, State University of New York at Farmingdale.

Many thanks are due to the people at Harper & Row, who believe in my approach to teaching chemistry and who have worked with me on this new edition, especially my editor, Lisa Berger, who continually encouraged and supported me.

Writing a chemistry text is an ongoing process as students, teachers, and concepts change. I believe that teaching chemistry involves more than a transmission of chemical facts; teaching also means developing positive attitudes toward science, encouraging students to use new thinking patterns and problem-solving techniques, and developing their reasoning powers. With this aim, I have revised *Chemistry*. I look forward to your use of this text and to hearing from both you and your students. I welcome any suggestions, criticism, or overall comments on this revision.

Karen Timberlake

To the Student

Here you are in chemistry, perhaps because you need a science course or perhaps just because you want to find out something about chemistry. Maybe you want to be a nurse or respiratory therapist or to enter some profession in the health sciences. If so, as you progress through this text, you will discover that chemistry is indeed exciting to learn and that it has an important relation to the world around you. Every chapter in this fourth edition includes application of chemistry to health, medicine, and the environment. Your interest in the sciences will help you learn chemistry, and by learning chemistry you will gain a deeper understanding of physiology, medical care, and major issues of today, including pollution, nuclear energy, and recombinant DNA.

I have designed this text with you in mind. To aid your learning process, each chapter begins with a set of objectives that tell you precisely what to expect in the chapter and what you need to accomplish. Each chapter then has a section called "Scope." This section relates those experiences that involve chemistry or the text material to specific health science areas. Each "Scope" section sets the stage for the chemical concepts discussed throughout the chapter.

As you work through each chapter, take time to consider the objectives for each section. To see if you have mastered the objectives, do the example problems in the section. If you have difficulty with an example, study that part of the unit again before you proceed. For further self-testing, work the problems at the end of the chapter. These problems are grouped by chapter objectives. You can check your answers by referring to Appendix C. It is not necessary to study a chapter all the way through at one time. Instead, you may wish to cover only a few objectives as you study. Also, if you know what your teacher is going to cover in class, you can be prepared for lecture by reading ahead in the text.

To review your knowledge of the important ideas in a chapter, read over the glossary at the end of the chapter. Study the tables and figures, which highlight and summarize important concepts. If you want to go back to a certain topic, look for the marginal notes in color that have been provided for quick reference; those notes are placed alongside the term or terms being discussed.

The study of chemistry involves some hard work, but I hope that you will find the effort rewarding when you see and understand the role of chemistry in many related fields. If you would like to share your feelings about chemistry or offer comments about this text, I should appreciate hearing from you.

Karen Timberlake
Los Angeles Valley College
Van Nuys, CA 91401

1 Measurement in the Health Sciences

Objectives

1.1 Write the units used by scientists to measure length, volume, and mass, then give their abbreviations.

1.2 Write the numerical value of a metric (SI) prefix.

1.3 Given two metric units for the same measurement, write the numerical relationship between them.

1.4 Write a conversion factor for two units that describe the same quantity.

1.5 Use a conversion factor to change from a given unit to another unit.

1.6 Calculate the density or specific gravity of a substance, or use the density or specific gravity to calculate the mass or volume of a substance.

1.7 Given a temperature in degrees Celsius, calculate the corresponding temperature in degrees Fahrenheit or in kelvins.

Scope

What kinds of measurement did you make today? Perhaps you checked your weight by stepping on the scale this morning. Two eggs, a cup of flour, and two cups of milk might have made a pancake batter. You may have filled your car with 10 gallons of gasoline on your way to school. If you were not feeling well, you might have taken your temperature. Try to recall some of the measurements you have made today.

Men and women in the health sciences use measurement every day to evaluate the health of a patient. Temperatures are taken, weights and heights are measured, and certain volumes of blood are drawn for laboratory testing. Medicines may be given whose dosages are stated by the doctor in terms of tablets or injections. In the dental office, hygienists measure out solutions used in fluoride treatments and assist in the preparation of dental materials.

1.1 Units of Measurement

OBJECTIVE Write the units used by scientists to measure length, volume, and mass, then give their abbreviations.

Suppose on a recent visit to the doctor, the nurse charted your mass (weight) as 70.0 kilograms (70.0 kg), your height as 1.78 meters (1.78 m), and your temperature as 37 degrees Celsius (37°C). You might wonder what had happened to you. The system of measurement being used in clinical evaluations is the metric system. You would probably recognize the measurements if they were given in the familiar U.S. customary system. Then your weight would be 154 pounds (154 lb), your height 5 feet 10 inches (5 ft 10 in.), and your temperature 98.6 degrees Fahrenheit (98.6°F).

The metric system is used by scientists everywhere. It is also the common measuring system in all but four countries in the world. The United States, the largest country still using a system based on the English system of measurement, is currently introducing the metric system. The metric system has the advantage of being used and understood by everyone in science and in the health professions. In 1960, a modified form of the metric system called the *International System of Units*, or *Système International* (SI), was adopted to provide additional uniformity in scientific measurements. However, most of the measurements in elementary chemistry classes and in the clinical field are reported using metric units. In this text, we will use metric units and introduce some of the corresponding SI units.

Length

meter

In metric (SI) measurement, a single unit is used for each type of measurement. In measuring length or your height, the unit *meter* (m) is used. You may also see it spelled *metre*. The meter is about 39.4 in. in length, which makes the meter slightly longer than a yard:

$$1 \text{ m} = 39.4 \text{ in.}$$

A comparison of metric (SI) and U.S. customary units for length can be seen in Figure 1.1.

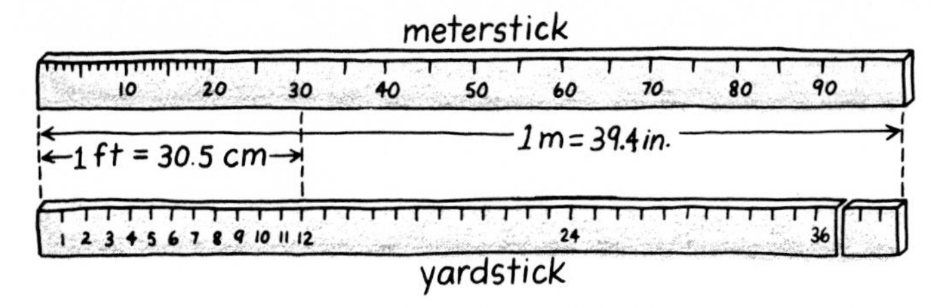

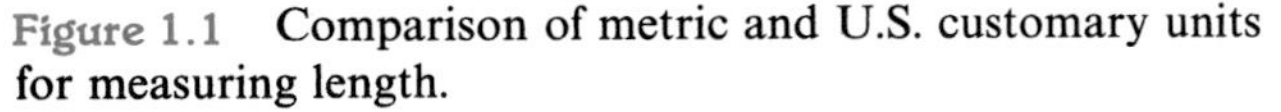
Figure 1.1 Comparison of metric and U.S. customary units for measuring length.

Volume

liter

Volume is the amount of space occupied by any substance. The unit for volume measurement in the metric system is the *liter.* In this text we will use the abbreviation L for liter. The spelling *litre* may also be used. The liter is slightly larger than the one-quart (1-qt) volume used in the U.S. customary system:

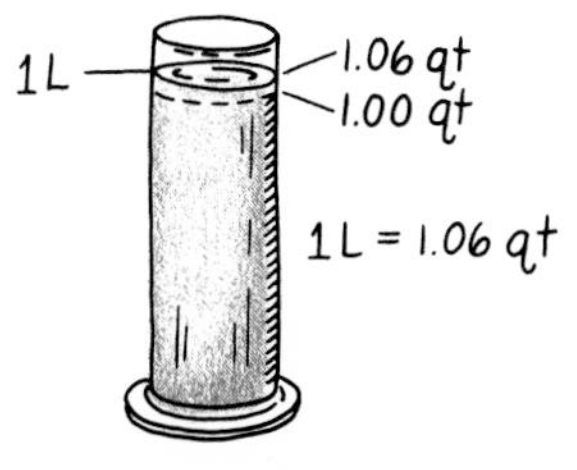

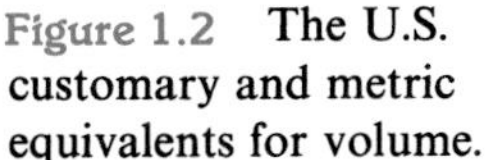
Figure 1.2 The U.S. customary and metric equivalents for volume.

$$1 \text{ L} = 1.06 \text{ qt}$$

Figure 1.2 compares the volumes of a liter and a quart.

Mass

The *mass* of an object is the quantity of material it contains. Everything has mass. Mass is related to the amount of pulling or pushing that must be done to move the object when it is at rest. An object with a large mass requires a stronger push to make it move than an object with a small mass.

In the U.S. customary system, we typically measure the weight of an object rather than its mass. We often use the word *weight* when we really mean mass. However, mass and weight are not the same thing. The weight of an object depends upon the gravitational pull on its mass. Thus, the weight of an object varies with its location. For example, an astronaut with a mass of 75.0 kg has weight of 165 lb on Earth. On the moon, where the gravitational pull is one-sixth of the Earth's, the astronaut will have a weight of 27.5 lb. However, the mass remains the same, 75.0 kg.

kilogram

In the metric system, the unit for mass is the *gram* (g). In comparison to a pound in the U.S. customary system, the gram seems quite small. There are 454 g in 1 lb. In the SI system, the standard unit of mass is the *kilogram.* One kilogram is the same mass as 1000 g. One kilogram is 2.20 lb:

$$1 \text{ kg} = 2.20 \text{ lb}$$

$$1 \text{ lb} = 454 \text{ g}$$

Thus, if you weigh 154 lb, you have a mass of 70.0 kg. Later in this chapter, you will learn how to convert one unit into another; Figure 1.3 illustrates the relationship between kilograms and pounds.

In the laboratory, we use a balance to determine the mass of an unknown object. Its mass is obtained by comparing the unknown mass to a set of known

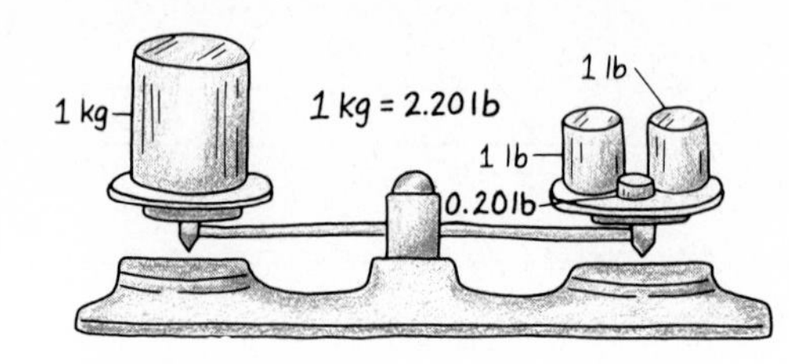

Figure 1.3 Comparison of the pound and the kilogram.

masses. Even though we say we are "weighing" the object, we are really measuring its mass.

Type of Measurement	*Unit*	*Abbreviation*
Distance	meter	m
Volume	liter	L
Mass	gram	g
	kilogram (SI)	kg

Example 1.1 Complete the following table:

Type of Measurement	Metric Base Unit	Abbreviation
Distance	____________	____________
____________	liter	____________
____________	____________	g

Solution

Distance	meter	m
Volume	liter	L
Mass	gram	g

1.2 Metric (SI) Prefixes

OBJECTIVE Write the numerical value of a metric (SI) prefix.

The following list includes various nutrients as well as the National Research Council (NRC) recommended daily allowances for a 25- to 30-year-old woman:

Protein	44 g
Vitamin C	60 mg
Calcium	800 mg
Iron	18 mg
Iodine	150 μg
Potassium	1875–5625 mg

Typical Admission Sheet

Pt: P. Patient
Dr. I. Care
Date: June 17, 1987
No. 00800

Pt admitted: via wheelchair Time: 17:30 *
TPR: 37.0°C-74-14 B/P: 120/90 Height: 1.75 m Weight: 55 kg
Any allergies: penicillin Hearing aid: no Eyeglasses: yes
Previous hospital admissions: no
Special nursing care: none
Medication: Empirin ć codeine #3 (30 mg)
Last dose: Empirin #3 @ 2:00 6/18/87

Disposition of valuables: none
Diet: no special preference
Urine specimen: in lab, sp gr 1.012 6/17/87
Call cord: ✓ TV ✓ Bed ✓
Visiting hours: 1400 - 20:00
Special comments: requested side rail release

Nursing signature: SF Date: 6/18/87

*Hospitals now use a 24-hour clock.

Abbreviations

TPR temperature, pulse, respiration
B/P blood pressure
ć with
Pt patient
sp gr specific gravity

You may notice that the measurement for protein is given in grams, the metric unit for mass. However, for the other recommended allowances, a smaller, more convenient unit such as a milligram or microgram is used. These smaller units of mass were formed by placing prefixes such as *milli-* or *micro-* in front of the unit gram.

prefixes

In the metric (SI) system, prefixes are used to increase or decrease the size of the unit by factors of 10. For example, the NRC recommended allowance for vitamin C is 60 mg. When the prefix milli- is combined with the unit gram, a new mass unit, *milligram*, is formed which is equal to one-thousandth of a gram. Table 1.1 lists some of the metric (SI) prefixes. Since the meaning of the prefixes may also be expressed in scientific notation, a review of this topic is given in Appendix B.

The meaning of a metric measurement can be determined by replacing the prefix term with its numerical value. For example, replacing the prefix *kilo* in

Table 1.1 Some Metric (SI) Prefixes

Prefix	Symbol	Meaning		
Multiples				
mega	M	One million times	1,000,000	10^{6}
kilo	k	One thousand times	1,000	10^{3}
hecto[a]	h	One hundred times	100	10^{2}
deka[a]	da	Ten times	10	10^{1}
Fractions				
deci	d	One tenth of	0.1	10^{-1}
centi	c	One hundredth of	0.01	10^{-2}
milli	m	One thousandth of	0.001	10^{-3}
micro	μ	One millionth of	0.000001	10^{-6}
nano	n	One billionth of	0.0000000001	10^{-9}

[a] These prefixes are included to illustrate the decimal relationship of the prefixes in the metric system, but are seldom used.

kilometer with the numerical value 1000 indicates that one kilometer is the same length as 1000 meters:

Prefix kilo indicates 1000

1 **kilo**meter (1 km) = **1000** meters (1000 m)
1 **kilo**liter (1 kL) = **1000** liters (1000 L)
1 **kilo**gram (1 kg) = **1000** grams (1000 g)

Prefix deci indicates 0.1

1 **deci**meter (1 dm) = **0.1** meter (0.1 m)
1 **deci**liter (1 dL) = **0.1** liter (0.1 L)
1 **deci**gram (1 dg) = **0.1** gram (0.1 g)

Figure 1.4 illustrates some examples of metric measurements for length.

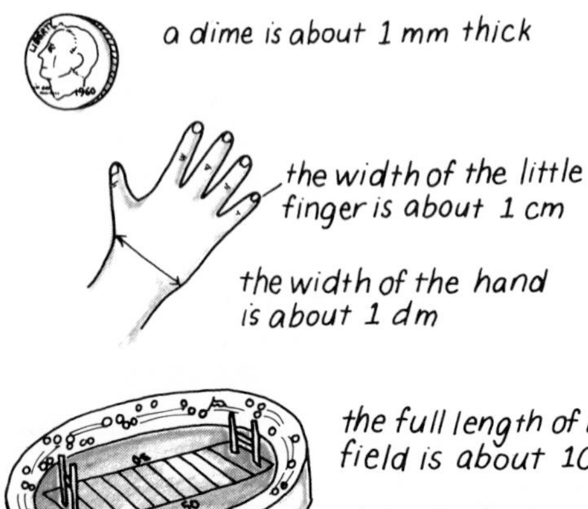

Figure 1.4 Examples of some metric measurements for length.

Example 1.2 Complete the blanks with the correct prefix or numerical value:

1. kilogram = __________ g
2. __________ liter = 0.001 L or 1/1000 L
3. centimeter = __________ m

Solution

1. 1000 g
2. *milli*liter
3. 0.01 m or 1/100 m

1.3 Writing Numerical Relationships for Metric Units

OBJECTIVE Given two metric units for the same measurement, write the numerical relationship between them.

Measurements of Length

Many of the length measurements in the health sciences are much smaller than a meter. An ophthalmologist may measure the diameter of the retina of the eye in centimeters (cm), and the microsurgeon may need to know the length of a nerve in micrometers (μm).

centimeter

A *centimeter* is a unit of length about the width of your little finger. We already know that a centimeter is 1/100 of a meter. If this is so, there must be 100 cm in a meter. See Figure 1.5.

millimeter

A *millimeter* is about the same size as the thickness of a dime. One millimeter is 1/1000 of a meter; there are 1000 mm in a meter. If we compare the millimeter and the centimeter, we find that 1 mm is 0.1 or 1/10 of a centimeter; there are 10 mm in 1 cm (see Figure 1.5):

1 m = 100 cm

1 m = 1000 mm

1 cm = 10 mm

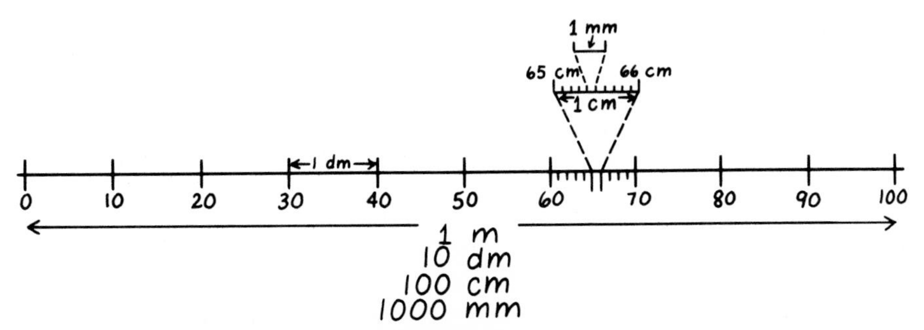

Figure 1.5 Comparison of some metric measurements for length.

Example 1.3

1. Place the following units of length in order, from smallest to largest: centimeter, kilometer, millimeter.
2. Complete the following metric relationships for length:

   a. 1 m = ____________ mm

   b. 1 m = ____________ cm

   c. 1 km = ____________ m

Solution

1. millimeter, centimeter, kilometer
2. a. 1000 mm
   b. 100 cm
   c. 1000 m

Measurements of Volume

In the health sciences, volumes of 1 L or smaller are commonly used. An intravenous (IV) solution usually is made up in a 1-L quantity. Bottles of 500 mL and 250 mL are also used. Small amounts of liquids to be added to the IV or given as injections are measured in milliliters. Figure 1.6 illustrates some laboratory and hospital equipment used for measuring volume.

deciliter

If 1 L is divided into 10 equal portions, each portion would be a *deciliter* (dL). There are 10 dL in 1 L. Dividing a liter into a thousand smaller parts gives

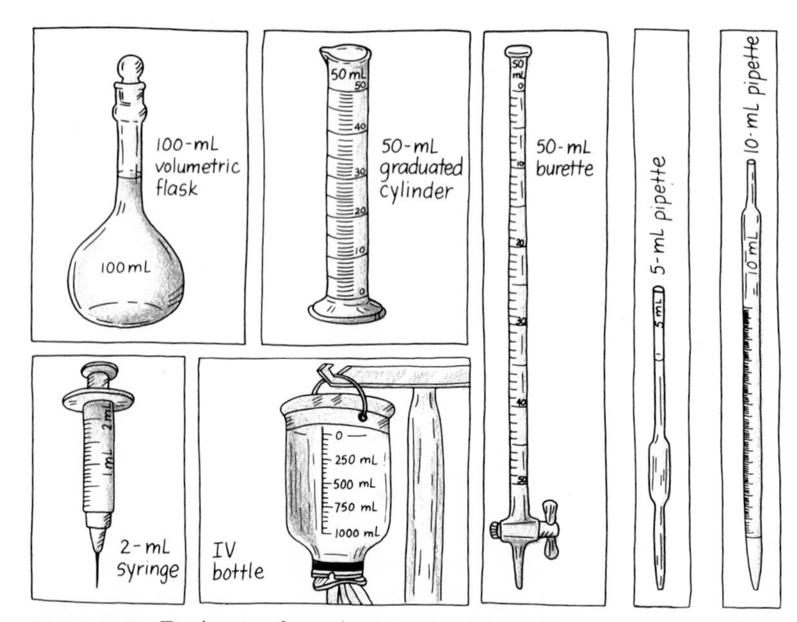

Figure 1.6 Equipment for volume measurement.

milliliter

a volume called the *milliliter* (mL). A 1-L bottle of physiological saline contains 1000 mL of the solution:

1 L = 10 dL

1 L = 1000 mL

1 dL = 100 mL

Laboratory work is often reported in deciliters, which is more convenient to use than its equivalent 100 mL. Observe some of the values reported below for the normal ranges of some substances in the blood.

Some Typical Laboratory Tests

Substance in Blood	*Normal Range Expected*
Albumin	3.5–5.0 g/dL
Ammonia	20–150 μg/dL
Calcium	8.5–10.5 mg/dL
Cholesterol	105–250 mg/dL
Iron (male)	80–160 μg/dL
Protein (total)	6.0–8.0 g/dL

cubic centimeter

The unit *cubic centimeter* (abbreviated cm^3 or cc) refers to a volume that would be contained in a cube whose dimensions are 1 cm on each side. A cubic centimeter has the same volume as a milliliter; the units are used interchangeably:

$$1\ cm^3 = 1\ cc = 1\ mL$$

When you see "1 cm" you are reading about length; when you see "1 cc" or "1 cm^3" you are reading about volume. Figure 1.7 compares the cubic centimeter, milliliter, and liter volumes.

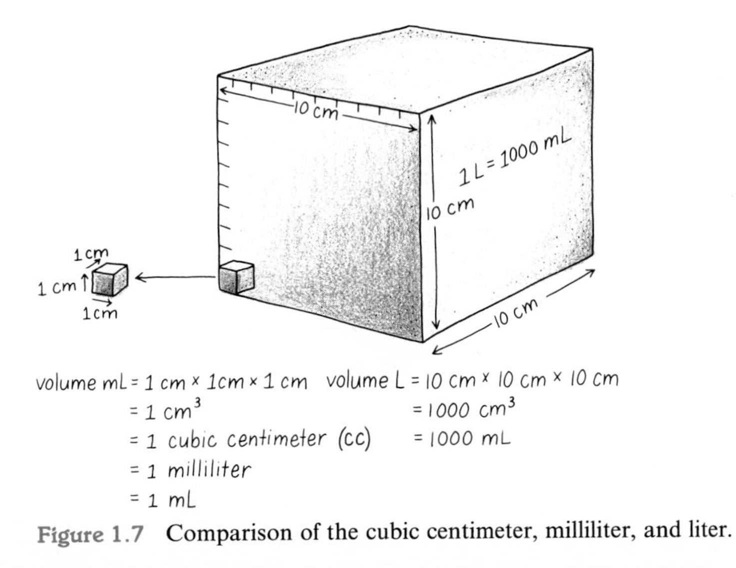

Figure 1.7 Comparison of the cubic centimeter, milliliter, and liter.

Measurements of Mass

kilogram

Several metric mass units may be used in the health sciences. A patient's mass is recorded in kilograms, whereas a laboratory test is reported in g, mg, or μg. A *kilogram* measures 1000 g. One gram is the same mass as 1000 mg:

$$1 \text{ kg} = 1000 \text{ g}$$

$$1 \text{ g} = 1000 \text{ mg}$$

$$1 \text{ mg} = 1000\ \mu\text{g}$$

Example 1.4

1. Place the following units in order, from smallest to largest:
   a. L, mL, kL, dL
   b. cg, kg, mg, g
2. Complete the following list of metric relationships:

   a. 1 L = __________ dL

   b. 1 L = __________ ml

   c. 1 g = __________ mg

   d. 1 kg = __________ g

Solution

1. a. mL, dL, L, kL
   b. mg, cg, g, kg
2. a. 10 dL
   b. 1000 mL
   c. 1000 mg
   d. 1000 g

1.4 Conversion Factors

OBJECTIVE Write a conversion factor for two units that describe the same quantity.

Many problems in chemistry and the health sciences require a change of units. The process of changing one unit to another is a familiar one. Suppose you obtain a piece of plastic tubing that is 2 ft in length, but you want to record its length in inches. You write down 24 in. You used a relationship you already knew to change from one unit (feet) to another (inches). The relationship you used was that 1 ft equals 12 in.:

$$1 \text{ ft} = 12 \text{ in.}$$

Table 1.2 Some "Everyday" Relationships and Factors

Relationship	Conversion Factors
1 yd = 3 ft	$\frac{1 \text{ yd}}{3 \text{ ft}}$ and $\frac{3 \text{ ft}}{1 \text{ yd}}$
1 dollar = 100 cents	$\frac{1 \text{ dollar}}{100 \text{ cents}}$ and $\frac{100 \text{ cents}}{1 \text{ dollar}}$
1 h = 60 min	$\frac{1 \text{ h}}{60 \text{ min}}$ and $\frac{60 \text{ min}}{1 \text{ h}}$
1 gal = 4 qt	$\frac{1 \text{ gal}}{4 \text{ qt}}$ and $\frac{4 \text{ qt}}{1 \text{ gal}}$

conversion factors

This kind of numerical relationship can be written in a fraction form. These fractions are called *conversion factors.* Two factors are always possible from any relationship because either quantity can become the numerator and the other becomes the denominator:

Conversion Factors

$$\frac{12 \text{ in.}}{1 \text{ ft}} \quad \text{and} \quad \frac{1 \text{ ft}}{12 \text{ in.}}$$

These factors are read as "12 in. per 1 ft," or "1 ft per 12 in." The term *per* means to divide. Some useful "everyday" relationships and their corresponding factors are given in Table 1.2.

Example 1.5 Write conversion factors that express the relationship in each of the following statements:

1. There are 16 oz in 1 lb.
2. One day has 24 h.
3. There are 12 eggs in 1 doz eggs.

Solution

Numerical Relationship	*Conversion Factors*
1. 16 oz = 1 lb	$\frac{16 \text{ oz}}{1 \text{ lb}}$ and $\frac{1 \text{ lb}}{16 \text{ oz}}$
2. 1 day = 24 h	$\frac{1 \text{ day}}{24 \text{ h}}$ and $\frac{24 \text{ h}}{1 \text{ day}}$
3. 12 eggs = 1 doz eggs	$\frac{12 \text{ eggs}}{1 \text{ doz eggs}}$ and $\frac{1 \text{ doz eggs}}{12 \text{ eggs}}$

Metric–Metric Conversion Factors

metric factors

In our problem-solving exercises, it will be necessary for you to write conversion factors for any of the metric (SI)–metric relationships. For example, you have learned that there are 100 cm in 1 m. The relationship between meters and centimeters is

$$1 \text{ m} = 100 \text{ cm}$$

The corresponding conversion factors for this relationship are

$$\frac{100 \text{ cm}}{1 \text{ m}} \quad \text{and} \quad \frac{1 \text{ m}}{100 \text{ cm}}$$

Both forms are proper conversion factors for the above relationship; one is just the inverse of the other. The usefulness of conversion factors is enhanced by the fact that we can turn a conversion factor over and use its inverse. Table 1.3 lists some conversion factors that can be derived from your knowledge of prefixes and metric relationships.

U.S. Customary–Metric Conversion Factors

Sometimes it is necessary to change from the U.S. customary system to the metric system. Suppose you have a patient's weight in pounds. In order to determine the quantity of a certain medication, generally given per kilogram of

Table 1.3 Some Useful Metric Conversion Factors

Metric Relationship	Conversion Factors
Length	
1 m = 1000 mm	$\frac{1 \text{ m}}{1000 \text{ mm}}$ and $\frac{1000 \text{ mm}}{1 \text{ m}}$
1 cm = 10 mm	$\frac{1 \text{ cm}}{10 \text{ mm}}$ and $\frac{10 \text{ mm}}{1 \text{ cm}}$
Volume	
1 L = 1000 mL	$\frac{1 \text{ L}}{1000 \text{ mL}}$ and $\frac{1000 \text{ mL}}{1 \text{ L}}$
1 dL = 100 mL	$\frac{1 \text{ dL}}{100 \text{ mL}}$ and $\frac{100 \text{ mL}}{1 \text{ dL}}$
Mass	
1 kg = 1000 g	$\frac{1 \text{ kg}}{1000 \text{ g}}$ and $\frac{1000 \text{ g}}{1 \text{ kg}}$
1 g = 1000 mg	$\frac{1 \text{ g}}{1000 \text{ mg}}$ and $\frac{1000 \text{ mg}}{1 \text{ g}}$

Table 1.4 Some U.S. Customary–Metric Relationships and Their Conversion Factors

Relationship	Conversion Factors
Distance	
2.54 cm = 1 in.	$\frac{2.54\text{ cm}}{1\text{ in.}}$ and $\frac{1\text{ in.}}{2.54\text{ cm}}$
1 m = 39.4 in.	$\frac{1\text{ m}}{39.4\text{ in.}}$ and $\frac{39.4\text{ in.}}{1\text{ m}}$
Volume	
946 mL = 1 qt	$\frac{946\text{ mL}}{1\text{ qt}}$ and $\frac{1\text{ qt}}{946\text{ mL}}$
1 L = 1.06 qt	$\frac{1\text{ L}}{1.06\text{ qt}}$ and $\frac{1.06\text{ qt}}{1\text{ L}}$
Mass	
454 g = 1 lb	$\frac{454\text{ g}}{1\text{ lb}}$ and $\frac{1\text{ lb}}{454\text{ g}}$
1 kg = 2.20 lb	$\frac{1\text{ kg}}{2.20\text{ lb}}$ and $\frac{2.20\text{ lb}}{1\text{ kg}}$

body weight, you need to calculate the mass of your patient in kilograms. In order to solve this problem, you will need a conversion factor that has one unit in the metric system (kilograms) and the other unit in the U.S. customary system (1b). The metric–U.S. customary relationship you might decide to use is

$$1\text{ kg} = 2.20\text{ lb}$$

The corresponding conversion factors would be

$$\frac{1\text{ kg}}{2.20\text{ lb}} \quad \text{and} \quad \frac{2.20\text{ lb}}{1\text{ kg}}$$

In the earlier sections, we mentioned some other U.S. customary–metric relationships. Table 1.4 summarizes these and provides the corresponding conversion factors.

Example 1.6 Write the numerical relationship and corresponding conversion factors for the following:

1. inches and centimeters (U.S. customary–metric)
2. meters and millimeters (metric)

Solution

Relationship	*Conversion Factors*
1. 1 in. = 2.54 cm	$\frac{1\text{ in.}}{2.54\text{ cm}}$ and $\frac{2.54\text{ cm}}{1\text{ in.}}$
2. 1 m = 1000 mm	$\frac{1\text{ m}}{1000\text{ mm}}$ and $\frac{1000\text{ mm}}{1\text{ m}}$

1.5 Calculations Using Conversion Factors

OBJECTIVE Use a conversion factor to change from a given unit to another unit.

The process of problem solving in chemistry and the health sciences is done in the same way you work out everyday problems such as changing feet to inches, dollars to cents, or hours to minutes. To illustrate this problem-solving process, let us take a look at the change of units in an everyday situation.

Suppose you decide to buy 3 lb of apples. The sign at the fruit stand states that 1 lb of apples costs 25 cents. You determine that 3 lb of apples will cost 75 cents. Let us take a close look at your reasoning:

Step 1 The quantity, 3 lb, is the given or initial amount.

Given: 3 lb apples

Step 2 The relationship of 25 cents per pound can be stated in the form of conversion factors:

Relationship: 1 lb apples = 25 cents

Conversion factors: $\frac{1\text{ lb apples}}{25\text{ cents}}$ or $\frac{25\text{ cents}}{1\text{ lb apples}}$

Step 3 By using the given (3 lb in this case) and the conversion factor (25 cents per pound), we can determine the cost of the purchase in cents. Our problem-solving plan is to change from pounds to cents:

Unit Plan: lb → cents

Step 4 To set up the problem, write the given first and multiply by the appropriate conversion factor. Use the conversion factor in such a way that the desired unit (cents) is in the numerator. This will place the unit of lb in the denominator, thus canceling out the given unit of lb:

$$3\,\not{lb} \times \frac{25\text{ cents}}{1\,\not{lb}} = 75\text{ cents}$$

Given (stated unit)	Conversion factor (cancels out given and provides desired unit)	Answer (desired unit)

Take a look at what happened to the units alone. This is a helpful way to check a problem. The unit that is desired in the answer should be the one that is left when the other units have canceled out.

Numerator:
Denominator: $\quad \not{lb} \times \frac{\text{cents}}{\not{lb}} = \text{cents (desired unit)}$

Numerical calculation: $\quad 3 \times \frac{25}{1} = 75$

Final answer: 75 cents

The numerical value of 75 is combined with the desired unit of cents to give the final answer of 75 cents. *With few exceptions, answers to numerical problems must be accompanied by a unit.* (See Figure 1.8.)

Problem Solving with Metric Factors

From the relationships between metric units, we have developed several conversion factors. Metric conversion factors are used when you need to change from one metric unit to another.

Figure 1.8 Conversion of units.

Example 1.7 The width of a room is 5 m. What is the width of the room in centimeters?

Solution: The given unit m must be changed to another metric unit, namely, cm:

Unit Plan: m → cm

The metric relationship and conversion factors for meter and centimeter are

$$1\text{ m} = 100\text{ cm} \qquad \frac{1\text{ m}}{100\text{ cm}} \quad \text{and} \quad \frac{100\text{ cm}}{1\text{ m}}$$

In the problem setup, the given (5 m) is written first, followed by the conversion factor that has the unit cm in the numerator. This places the unit m in the denominator, thus canceling out the units m in the given:

$$\underset{\text{Given}}{5\cancel{\text{ m}}} \times \underset{\text{Metric factor}}{\frac{100\text{ cm}}{1\cancel{\text{ m}}}} = \underset{\text{Answer}}{500\text{ cm}}$$

Example 1.8 A can of frozen orange juice contains 473 mL of juice. How many liters of juice are in the can?

Solution: The given unit is mL which must be changed to the desired unit of L:

Unit Plan: mL → L

The metric relationship and its conversion factors are

$$1\text{ L} = 1000\text{ mL} \qquad \frac{1\text{ L}}{1000\text{ mL}} \quad \text{and} \quad \frac{1000\text{ mL}}{1\text{ L}}$$

The problem setup begins with the given 473 mL, followed by the conversion factor that has the unit mL in the denominator:

$$473\cancel{\text{ mL}} \times \frac{1\text{ L}}{1000\cancel{\text{ mL}}} = 0.473\text{ L}$$

(mL units cancel out)

Mathematically, you need to divide 473 by 1000:

$$\frac{(473)(1)}{1000} = 473 \div 1000 = 0.473$$

Problem Solving with U.S. Customary–Metric Factors

The same problem-solving process can also be used to convert from a U.S. customary to a metric unit, or vice versa. In this case, a conversion factor that links the U.S. customary and metric systems of measurement will be used. Figure 1.9 illustrates some "everyday" measurements in metric and U.S. customary units.

As we continue solving more problems that include measurement, it is necessary to mention that there are some special considerations for making

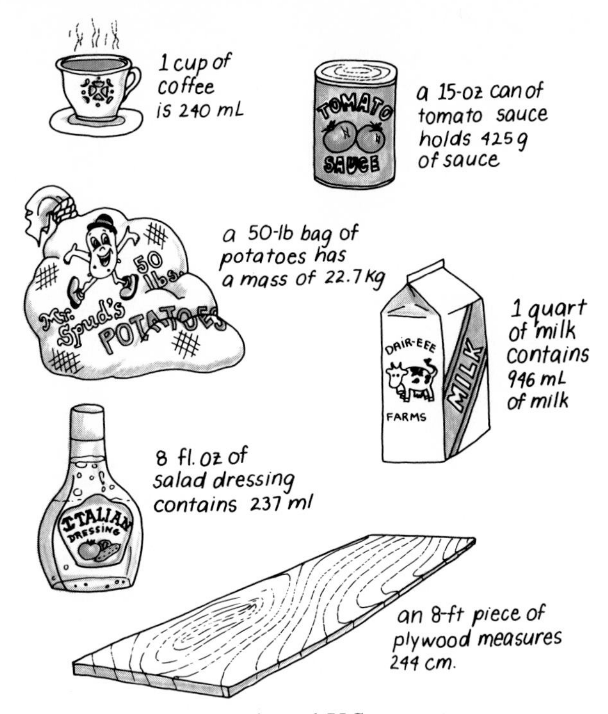

Figure 1.9 The metric and U.S. customary quantities for some "everyday" items.

calculations with measured numbers. You may wish to refer to Appendix A for a review on the following topics: (a) significant figures, (b) rounding-off numbers, (c) calculations with significant figures, (d) calculator use.

Example 1.9 A newborn infant measures 19.0 in. What is the length of the baby in centimeters?

Solution: The initial measurement of 19.0 in. must be changed to a corresponding measurement in centimeters:

Unit Plan: in. → cm

We can state that the relationship of inches and centimeters along with the corresponding conversion factors is

$$1 \text{ in.} = 2.54 \text{ cm} \qquad \frac{1 \text{ in.}}{2.54 \text{ cm}} \quad \text{and} \quad \frac{2.54 \text{ cm}}{1 \text{ in.}}$$

The problem setup shows the given quantity of 19.0 in. followed by the conversion factor that has the unit in. in its denominator:

$$19.0 \cancel{\text{in.}} \times \frac{2.54 \text{ cm}}{1 \cancel{\text{in.}}} = 48.26 \text{ cm (calculator answer)}$$

Final answer = 48.3 cm (rounded to three figures)
See appendix A

Example 1.10 A patient has a weight of 154 lb. You need to prepare a medication based on the patient's mass in kilograms. What is the mass of your patient in kilograms?

Solution: This problem requires a change from the given unit lb to the desired unit of kg:

Unit Plan: lb → kg

The relationship between the units lb and kg require a metric–U.S. customary relationship and conversion factors:

Relationship	*Conversion Factors*
1 kg = 2.20 lb	$\frac{1\text{ kg}}{2.20\text{ lb}}$ and $\frac{2.20\text{ lb}}{1\text{ kg}}$

In the problem setup, the given unit lb is written first, followed by the conversion factor that will cancel out the given unit (see Figure 1.10):

$$\underset{\text{Given}}{154\ \cancel{\text{lb}}} \times \underset{\text{Factor}}{\frac{1\text{ kg}}{2.20\ \cancel{\text{lb}}}} = \underset{\text{Answer}}{70.0\text{ kg}}$$

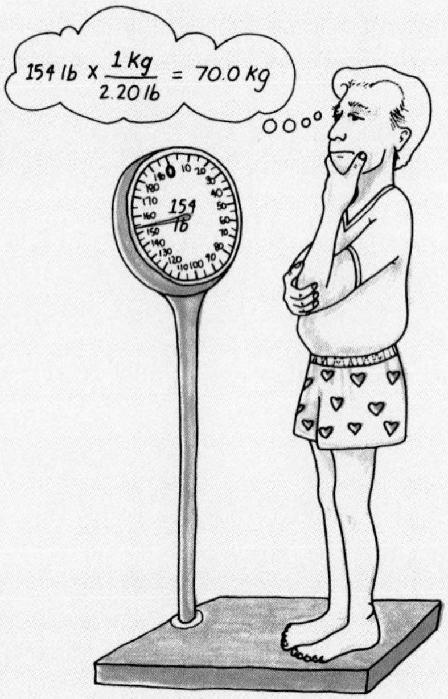

Figure 1.10 Conversion from a U.S. customary unit to a metric unit.

Using Two or More Conversion Factors in Sequence

Suppose you need a 4.00-ft piece of plastic tubing, but you have to record its length in centimeters. If you look over the list of U.S. customary–metric conversion factors, you may notice that there is no single conversion factor that relates centimeters and feet. You could find one in another book perhaps, but that means you have to memorize a lot of relationships, and that really isn't necessary. The problem can be solved in two steps using two conversion factors you already know. First, feet are changed to inches, and then inches are changed to centimeters:

Feet to inches

$$4.00\ \cancel{\text{ft}} \times \frac{12\ \text{in.}}{1\ \cancel{\text{ft}}} = 48.0\ \text{in. (not the desired unit)}$$

U.S. customary factor

Inches to centimeters

$$48.00\ \cancel{\text{in.}} \times \frac{2.54\ \text{cm}}{1\ \cancel{\text{in.}}} = 122\ \text{cm} \quad \text{(desired unit)}$$

Answer

The same problem can be set up using a sequence of conversion factors. One factor follows the other in the problem setup. Arrange each conversion factor to cancel out the given or previous unit until you are left with the desired unit:

Unit Plan: ft → in. → cm

$$4.00\ \text{ft} \times \frac{12\ \text{in.}}{1\ \text{ft}} \times \frac{2.54\ \text{cm}}{1\ \text{in.}} = 122\ \text{cm}$$

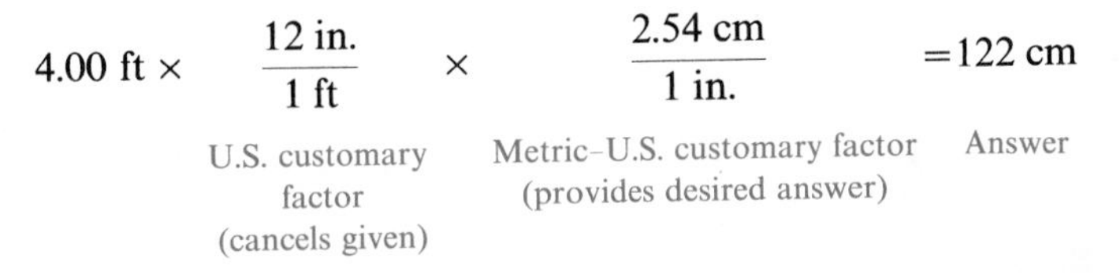

Observe how the units cancel:

$$\cancel{\text{ft}} \times \frac{\cancel{\text{in.}}}{\cancel{\text{ft}}} \times \frac{\text{cm}}{\cancel{\text{in.}}} = \text{cm}$$

The numerical calculation is a series of multiplications:

$$4.00 \times \frac{12}{1} \times \frac{2.54}{1} = 121.92 \quad \text{(calculator answer)}$$

$$= 122 \quad \text{(rounded to three figures)}$$

The final answer is 122 cm.

Using a sequence of two or more conversion factors is a very efficient way to correctly set up and solve problems when you are using a calculator. Once you have the problem set up, the calculations can be done without having to write out all of the in-between values. This process is worth practicing until you understand the purpose of unit cancellation and the mathematical calculations.

Example 1.11 A spaghetti sauce recipe calls for 3.00 cups of tomato sauce. If only metric measures are available, how many milliliters of tomato sauce are needed? There are 4 cups in 1 qt.

Solution: When you think about the relationship between cups and milliliters, you may find that you can't think of one. However, you do know a relationship that could change cups to quarts, and you know another relationship that will change quarts to milliliters:

Unit Plan: cups → quarts → milliliters

For each relationship, there are two conversion factors:

Relationships	*Conversion Factors*
1 qt = 4 cups	$\frac{1\text{ qt}}{4\text{ cups}}$ and $\frac{4\text{ cups}}{1\text{ qt}}$
1 qt = 946 mL	$\frac{1\text{ qt}}{946\text{ mL}}$ and $\frac{946\text{ mL}}{1\text{ qt}}$

Problem Setup:

$$3.00\ \cancel{\text{cups}} \times \frac{1\ \cancel{\text{qt}}}{4\ \cancel{\text{cups}}} \times \frac{946\text{ mL}}{1\ \cancel{\text{qt}}} = 709.5\text{ mL (calculator)}$$

Given — U.S. customary factor (cancels given) — Metric–U.S. customary factor — Desired unit

Final answer = 710 mL (rounded off)

Using Conversion Factors in Clinical Calculations

Conversion factors are widely used in the hospital environment. For example, the dosage in a tablet of medication can be used as a conversion factor. If you are giving an antibiotic that is available in 5-mg tablets, the dosage is written as 5 mg/1 tablet. In many hospitals, the apothecary unit *grain* (gr) is still in use; there are 60 mg in 1 gr.

Some Clinical Conversion Factors

$$\frac{5\text{ mg antibiotic}}{1\text{ tablet}} \qquad \frac{60\text{ mg}}{1\text{ gr}}$$

When you are working with clinical calculations, you often have a doctor's order. The doctor's order contains the given, and the dosage is the conversion factor.

Example 1.12 A doctor orders 0.050 g of a medication. On hand, you have 10-mg tablets available from the drug cabinet. How many tablets are needed?

Solution: The problem requires a change in units from g to tablets:

Unit Plan: g → mg → tables

Relationships	*Conversion Factors*
1 g = 1000 mg	$\frac{1\text{ g}}{1000\text{ mg}}$ and $\frac{1000\text{ mg}}{1\text{ g}}$
1 tablet = 10 mg	$\frac{1\text{ tablet}}{10\text{ mg}}$ and $\frac{10\text{ mg}}{1\text{ tablet}}$

Problem Setup:

$$\underset{\text{(doctor's order)}}{\underset{\text{Given}}{0.050\text{ g}}} \times \underset{\text{(metric)}}{\underset{\text{Factor}}{\frac{1000\text{ mg}}{1\text{ g}}}} \times \underset{\text{(dosage)}}{\underset{\text{Factor}}{\frac{1\text{ tablet}}{10\text{ mg}}}} = \underset{\text{Answer}}{5\text{ tablets}}$$

1.6 Density and Specific Gravity

OBJECTIVE Calculate the density or specific gravity of a substance, or use the density or specific gravity to calculate the mass or volume of a substance.

Density

We can measure the mass of a substance on a balance, and we can determine its volume. However, the separate measurements do not tell us how tightly packed the substance might be or whether its mass is spread out over a large volume or a small one. However, if we relate its mass to its volume, we can make this determination. This relationship is called the *density* of the substance. Density is defined as mass divided by volume:

$$\text{Density} = \frac{\text{Mass}}{\text{Volume}}$$

Table 1.5 Densities of Some Common Substances at 25°C

Substance	Density (g/mL)	Substance	Density (g/mL)
Alcohol (ethyl)	0.79	Lead	11.30
Bone	1.80	Mercury	13.60
Butter	0.86	Olive oil	0.92
Cement	3.00	Seawater	1.03
Diamond	3.52	Water	1.00
Gasoline	0.66	Wood, balsa	0.14
Ice	0.92	Wood, oak	0.90

In the metric system, the density of liquids and solids is often expressed as grams per 1 mL (g/mL), whereas the density of a gas is stated as g/L. Table 1.5 gives the densities of some common substances at 25°C.

Example 1.13 A 50.0-mL sample of buttermilk has a mass of 56.0 g. What is the density of the buttermilk?

Solution

$$\text{Density} = \frac{\text{Mass}}{\text{Volume}} = \frac{56.0\ \text{g}}{50.0\ \text{mL}} = \frac{1.12\ \text{g}}{1\ \text{mL}} = 1.12\ \text{g/mL}$$

Density of Water

The density of water varies slightly with temperature, but the value of 1.00 g/mL is usually used. Suppose you have three water samples. Sample 1 has a mass of 10.0 g and a volume of 10.0 mL. Sample 2 has a mass of 25.0 g and a volume of 25.0 mL. Sample 3 has a mass of 114 g and a volume of 114 mL. What is the density of water in each of the samples?

	Sample 1	*Sample 2*	*Sample 3*
Mass	10.0 g	25.0 g	114 g
Volume	10.0 mL	25.0 mL	114 mL
Density	1.00 g/mL	1.00 g/mL	1.00 g/mL

The densities are all the same because all the samples are the same substance (namely, water), and there is just one density for water. This means that the ratio of mass to volume in all three samples is the same.

HEALTH NOTE

The Density of Urine

The density of urine is often determined as part of a laboratory evaluation of the health of an individual. The density of urine is normally about 1.020 g/mL. This is somewhat greater than the density of water because solids such as urea are dissolved in water in the kidney to form urine. This means that 1 mL of urine has a mass of 1.020 g. If the density of a person's urine is too low or too high, a doctor might suspect a problem with the kidneys. For example, if a urine sample shows a density of 1.001 g/mL, which is significantly lower than normal, malfunctioning of the kidneys is a possibility.

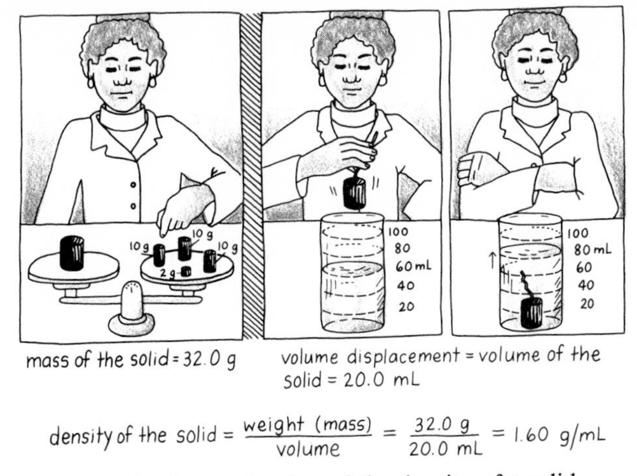

Figure 1.11 Determination of the density of a solid.

Density of Solids

The density of a solid can be determined by measuring the mass and the volume of the object. The mass is obtained by weighing the object on a balance. The volume may be determined by measuring the side lengths if it has a regular shape, or it may be determined by *volume displacement*. A solid that is completely submerged will displace a volume of water equal to the volume of that object. In Figure 1.11, the water level rises from 60 mL to 80 mL after the object is submerged. The difference in the volume of water of 20 mL is equal to the volume of water displaced by the object. The object has a volume of 20 mL. If the mass of the object was measured as 32.0 g, the density of the object is calculated to be 1.60 g/mL.

Example 1.14 A lead weight used in the belt of a scuba diver has a mass of 226 g. When it is carefully dropped into a graduated cylinder containing 200.0 mL of water, the water level rises to 220.0 mL. What is the density of the solid lead weight?

Solution: To calculate the density of the lead weight, we need both the mass and volume of the lead weight. The mass is given as 226 g. The volume will be the displaced volume, which is calculated as follows:

$$\text{Volume displaced} = 220.0\text{ mL} - 200.0\text{ mL} = 20.0\text{ mL}$$

$$\text{Volume of the lead weight} = 20.0\text{ mL}$$

$$\text{Density of lead} = \frac{\text{Mass}}{\text{Volume of lead}} = \frac{226\text{ g}}{20.0\text{ mL}} = 11.3\text{ g/mL}$$

(Be sure to use the volume of the object, and not the volume of water.)

Using Density as a Conversion Factor

In the density expression, both grams and milliliters must appear. Because of the comparison of two units, we can use density as a conversion factor. For example, if the volume and the density of a sample are known, the mass of the sample can be calculated.

Example 1.15 The cast on Suzie's leg has a density of 2.32 g/mL. What is the mass in g of the cast if its volume is 750 mL?

Solution

Relationship (Density)	*Conversion Factors*
2.32 g = 1 mL	$\frac{2.32\text{ g}}{1\text{ mL}}$ and $\frac{1\text{ mL}}{2.32\text{ g}}$

Unit Plan: mL → g

Problem Setup:

$$750\text{ mL} \times \frac{2.32\text{ g}}{1\text{ mL}} = 1740\text{ g}$$

Given — Factor from density

Why Do Objects Sink or Float?

You have probably observed that some things sink, whereas others float. The tendency of an object to sink or float is determined by the relative densities of the object and the fluid. Take a look at Figure 1.12, in which samples of lead and wood are being placed in water. If we compare the densities of lead and wood with the density of water, we find that the density of lead is greater than the density of water, and the density of wood is less. Which sample sinks in the water? Objects that have a greater density than water, such as the lead sample, sink; conversely, objects that are less dense than water, such as oils and most woods, float.

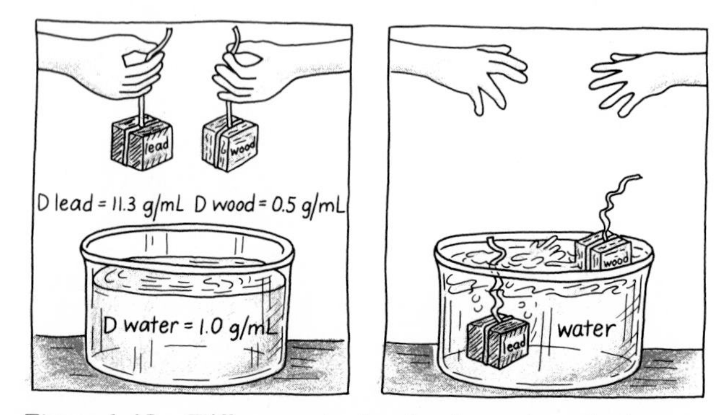

Figure 1.12 Differences in density determine whether an object will sink or float.

HEALTH NOTE

Determination of Percentage Body Fat from Body Density

The mass of the body is made up of protoplasm, extracellular fluid, bone, and adipose tissue (body fat). All but the body fat are lean body mass. A healthy, 70-kg adult should contain 50 percent protoplasm, 24 percent extracellular fluid, 7 percent bone, and 19 percent body fat.

One way to determine the amount of adipose tissue is to measure the whole-body density. The *on-land* mass of the body is determined, followed by a total submersion of the person underwater, during which the *underwater* body mass is determined. Since water helps support the body by giving it buoyancy, the underwater body mass is less than on-land mass. A higher percentage of body fat will make a person more buoyant, causing the underwater mass to be even lower. This occurs because fat has a lower density than the rest of the body materials.

The mass difference between the on-land mass and the underwater mass is known as the *buoyant force*. This buoyant force is the mass of water displaced by the person's body while under water. Using the density of water, the *body's volume* is calculated. Several adjustments, such as subtracting the residual volume of the air trapped in the lungs and the intestine, are made. When the body's volume is determined, it is used along with the body's mass to calculate body density. Suppose a 70.0-kg person has a body volume of 66.7 L:

$$\text{Body density} = \frac{\text{Body mass}}{\text{Body volume}} = \frac{70.0\text{ kg}}{66.7\text{ L}} = 1.05\text{ kg/L or } 1.05\text{ g/mL}$$

When the body density is known, it is compared to a chart that correlates the percentage of adipose tissue with body density. A person with a body density of 1.05 g/mL will have a body fat percentage of 21 percent. This procedure is popular among athletes and runners in determining exercise and diet programs.

Example 1.16 Using Table 1.5, determine whether a lead object will sink or float in mercury, which is liquid at room temperature.

Solution: Table 1.5 gives the density of lead as 11.30 g/mL and the density of mercury as 13.60 g/mL. Since lead is less dense than mercury, the lead object will float on the mercury.

Specific Gravity

Specific gravity (sp gr) is the relationship between the density of a substance and the density of water. Specific gravity is calculated by dividing the density of a sample by the density of water. In this text, we will use a value of 1.00 g/mL for the density of water. In the calculations for specific gravity, all units cancel, and only a number remains. This is one of the few unitless values you will encounter in chemistry:

$$\text{Specific gravity} = \frac{\text{Density of sample (g/mL)}}{\text{Density of water (g/mL)}}$$

Example 1.17 What is the specific gravity of a urine sample that has a density of 1.020 g/mL?

Solution

$$\text{sp gr}_{\text{urine}} = \frac{\text{Density of urine}}{\text{Density of water}} = \frac{1.020\ \cancel{\text{g/mL}}}{1.000\ \cancel{\text{g/mL}}} = \underset{\text{No units}}{1.020}$$

An instrument called a *hydrometer* is used to measure the specific gravity of fluids. You may have used a hydrometer to measure the specific gravity of battery fluid or a sample of urine. Figure 1.13 shows a hydrometer being used to measure the specific gravity of a urine sample.

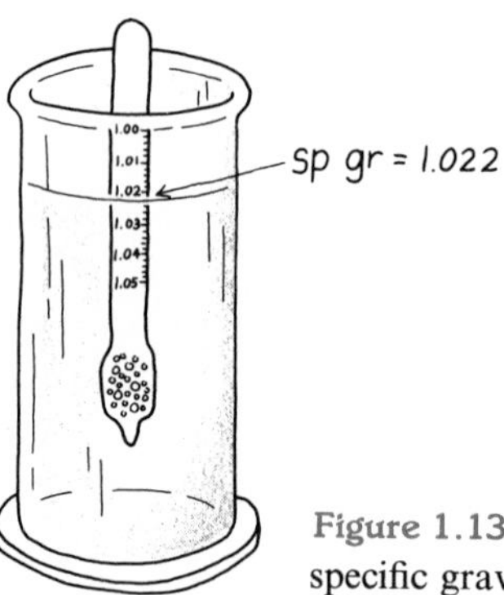

Figure 1.13 Using a hydrometer to measure the specific gravity of a urine sample.

Example 1.18 John took 1 teaspoon (tsp) of cough syrup (sp gr 1.20) for a cough. If there are 5 mL in 1 tsp, what was the mass (g) of the cough syrup?

Solution: First, change from specific gravity to density of the syrup:

$$\underset{\text{sp gr}}{1.20} \times \underset{\text{Density of water}}{\frac{1.00\text{ g}}{1\text{ mL}}} = \underset{\text{Density of syrup}}{1.20\text{ g/mL}}$$

Problem Setup: $5\ \cancel{\text{mL}} \times \frac{1.20\text{ g}}{1\ \cancel{\text{mL}}} = 6\text{ g syrup}$

1.7 Temperature Measurement

OBJECTIVE Given a temperature in degrees Celsius, calculate the corresponding temperature in degrees Fahrenheit or in kelvins.

The temperature of a substance tells you how hot or cold that material is. The temperature of the outside air tells you how hot or cold it is outdoors. Your body temperature tells you if you have a fever or not. Suppose a thermometer indicated that the room temperature was 22.2°, and that your body temperature was 37°. You might wonder what was wrong. It is important to state the temperature units, which in this example are 22.2°C and 37°C, typical room and body temperatures in the metric system. The metric unit, °C, means degrees *Celsius.* Temperature in the U.S. customary system is measured in degrees *Fahrenheit,* °F. The temperature of 22.2°C is the same as 72°F, and 37°C is the same as 98.6°F, normal body temperature.

Units of temperature were determined by assigning certain values to the freezing and boiling points of water. On the Celsius scale, the freezing point of water was assigned a value of 0°C, the boiling point was assigned a value of 100°C. On the Fahrenheit scale, water freezes at 32°F and boils at 212°F. Figure 1.14 compares the Celsius and Fahrenheit temperature scales.

Between the freezing and boiling points of water, there are 180 Fahrenheit units to 100 Celsius units. Or we say that there is 1.8 Fahrenheit unit for every 1 Celsius unit:

$$180 \text{ Fahrenheit units} = 100 \text{ Celsius units}$$

$$\frac{180 \text{ Fahrenheit units}}{100 \text{ Celsius units}} \quad \text{or} \quad \frac{1.8 \text{ Fahrenheit units}}{1 \text{ Celsius unit}}$$

Since the freezing points are different on the two temperature scales, an adjustment must be made. When changing from Celsius temperature to Fahrenheit, a value of 32°F must be added, which adjusts the freezing point. We can use

Figure 1.14 Comparison of Celsius and Fahrenheit temperature scales.

the following equation to change a given Celsius temperature to its corresponding Fahrenheit temperature:

$$T°\text{C} = 1.8(T°\text{C}) + 32$$

Example 1.19 While traveling in Europe, Andrew checks his temperature, which is 38.2°C. What is his body temperature in degrees Fahrenheit?

Solution: To convert from a Celsius temperature to a Fahrenheit temperature, we use the following equation for calculating the Fahrenheit temperature when the Celsius temperature is given:

$$T°\text{F} = 1.8(T°\text{C}) + 32$$

$$T°\text{F} = 1.8(\underbrace{38.2}_{T°\text{C given}}) + 32$$

The multiplication of 1.8(38.2) must be done first, followed by the addition of 32:

$$T°\text{F} = 68.8 + 32$$

$$T°\text{F} = 101°\text{F}$$

In another problem you may be asked to convert from a Fahrenheit temperature to a Celsius temperature. To do this, the temperature equation must be rearranged:

$$T°\text{C} = \frac{(T°\text{F} - 32)}{1.8}$$

Example 1.20 Donyelle is going to cook a turkey at 325°F. If she is using an oven thermometer with Celsius units, at what Celsius temperature should she set the oven?

Solution

$$T°\text{C} = \frac{325 - 32}{1.8} = \frac{293}{1.8} = 163°\text{C}$$

In solving this mathematically, the subtraction 325 − 32 must be done first. The result is then divided by 1.8.

Kelvin Temperature Scale

absolute zero

Another temperature scale that is very important to scientists is the Kelvin (K) scale, the SI unit of temperature. A value of zero is assigned to the lowest possible temperature, called *absolute zero*. Units on the Kelvin scale are called *kelvins*; no degree symbol is used. On the Celsius temperature scale, absolute zero corresponds to −273.15°C. Since the size of the Kelvin and the Celsius

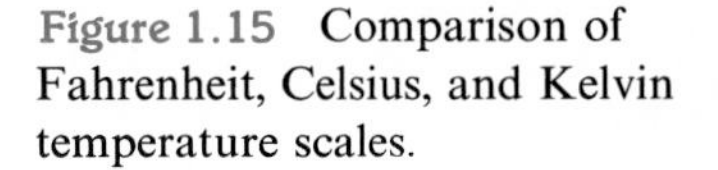

Figure 1.15 Comparison of Fahrenheit, Celsius, and Kelvin temperature scales.

temperature unit is the same, we can calculate a Kelvin temperature by adding 273 (rounded off) to the Celsius temperature:

$$TK = T°C + 273$$

Figure 1.15 shows the relationship of the Fahrenheit, Celsius, and Kelvin temperature scales.

Example 1.21 What is normal body temperature (37.0°C) on the Kelvin scale?

Solution: We can use the equation that states the relationship between Kelvin temperature and Celsius temperature:

$$TK = T°C + 273$$

$$TK = 37.0 + 273$$

$$= 310 \text{ K}$$

Example 1.22 David is running a reaction cooled to 200 K by liquid nitrogen. What is the Celsius temperature of the reaction mixture?

Solution: To find the Celsius temperature, we must rearrange the equation of Kelvin and Celsius to solve for the Celsius temperature:

$$TK = T°C + 273$$

$$T°C = TK - 273$$

$$= 200 - 273$$

$$T°C = -73°C$$

HEALTH NOTE

Variations in Body Temperature

Normal body temperature is considered to be 37.0°C. However, this can vary throughout the day. Generally, oral temperatures of 36.1°C are common when awakening in the morning and climb to a high of 37.2°C between 6 P.M. and 10 P.M. Elevations of temperature above 37.2°C for a person at bed rest are usually an indication of disease.

Elevations in body temperature can occur in individuals involved in prolonged exercise. Body temperatures of marathon runners can range from 39 to 41°C. This occurs because the body's heat production during exercise exceeds the body's ability to lose heat.

Changes of more than 3.5°C from the normal body temperature begin to interfere with bodily functions. Temperatures above 41.1°C can lead to convulsions, particularly in children, and cause permanent brain damage. Temperatures above 42.2°C are rarely observed in humans because of the destruction of cellular enzymes.

Heat stroke (hyperpyrexia) occurs at a body temperature above 41.1°C. Initially, sweat production stops, and the skin becomes hot and dry. The pulse rate is elevated, and respiration becomes weak and rapid. The person generally becomes lethargic and lapses into a coma. Damage to internal organs is a major concern, and treatment must be immediate. The most effective method of removing heat is immersion in an ice-water bath.

In hypothermia, body temperatures can drop as low as 28.5°C. The person may appear cold and pale and have an irregular heart beat. Unconsciousness can occur if body temperature drops below 26.7°C. Respiration is slow and shallow, and oxygenation of the tissues decreases. Treatment involves providing oxygen and increasing blood volume and glucose and saline fluids. Internal temperature may be restored by warming fluids to 37°C and injecting them into the peritoneal cavity.

Glossary

base unit The unit that forms the basis for all other units of measurement. In the metric system, the base units are the meter for length, the liter for volume, and the gram for mass.

Celsius A temperature scale assigning a value of 0°C to the freezing point of water and 100°C to the boiling point of water:

$$T°\text{C} = \frac{T°\text{F} - 32}{1.8}$$

centimeter A unit of length in the metric system; there are 2.54 cm in 1 in.

conversion factor A comparison of two units that measure the same quantity. The factor is written as a fraction. For example, the quantity of 1 kg is also 2.20 lb. Written as a fraction:

$$\frac{2.20 \text{ lb}}{1 \text{ kg}} \quad \text{or} \quad \frac{1 \text{ kg}}{2.20 \text{ lb}}$$

cubic centimeter (cc, cm^3) The volume of a cube with 1-cm sides; equal to 1 mL.

density The relationship of the mass of an object to a volume unit of that object: $D = m/V$. Density is often expressed in g/mL.

gram (g) The metric base unit for mass.

Kelvin (K) A temperature scale based on assignment of zero (absolute) to the lowest possible temperature. One kelvin also represents a unit of temperature measurement on the Kelvin scale:

$$K = T°C + 273$$

kilogram (kg) A metric unit of mass (1000 g) equal to 2.20 lb.

liter (L) The metric base unit for volume, which is slightly larger than a quart.

mass A measure of the quantity of material in an object.

meter (m) The metric base unit for length, which is slightly longer than a yard.

metric system A decimal system of measurement used by scientists and in most countries of the world.

milliliter (mL) A unit of volume in the metric system equal to $\frac{1}{1000}$ L.

prefixes Words preceding a base unit that indicate the size of the measurement. All prefixes are related on a decimal scale. Some important metric prefixes are kilo-, centi-, and milli-.

specific gravity A relationship (unitless) of the density of a material to the density of water:

$$\text{sp gr} = \frac{\text{Density of sample}}{\text{Density of water}}$$

weight A measure of the effect of gravity on the mass of an object.

Problems

Units of Measurement (Objective 1.1)

1.1 Write the name of the metric unit and the type of measurement (mass, volume, or length) indicated in each of the following:
a. 4 m b. 325 g c. 1.5 L d. 5000 m

Prefixes (Objective 1.2)

1.2 Give the abbreviations for the following metric units:
a. milligram b. deciliter c. kilometer d. kilogram

1.3 Write the name of the metric unit for the following abbreviations:
a. cm b. mm c. dL d. kg

1.4 Write the numerical values for the following prefixes:
a. centi b. kilo c. milli d. deci

1.5 Write the correct prefix term for the following numerical values:
a. 0.10 b. 10 c. 1000 d. 1/100

Finding Relationships Between Metric Units (Objective 1.3)

1.6 Place the units or prefixes in order for each set, from smallest to largest:
a. milli, kilo, centi
b. milli, centi, micro
c. deci, milli, mega, deka
d. cg, kg, mg, g, dg
e. hm, mm, dm, m, km
f. kL, L, mL, cL

1.7 Complete the following metric relationships:

a. 1 m = ____________ cm
b. 1 km = ____________ m
c. 1 mm = ____________ m
d. 1 L = ____________ mL
e. 1 L = ____________ dL
f. 1 dL = ____________ L
g. 1 g = ____________ kg
h. 1 g = ____________ mg

Conversion Factors (Objective 1.4)

1.8 Write a numerical relationship and conversion factors for each of the following statements:

a. There are 3 ft in 1 yd.
b. One minute is 60 s.
c. One dollar has 4 quarters.
d. There are 4 qt in 1 gal.
e. One mile is 5280 ft.
f. There are 7 days in 1 week.

1.9 Write the numerical relationship and conversion factors for the following metric units:

a. centimeters and meters
b. milligrams and grams
c. centimeters and millimeters
d. liters and milliliters
e. deciliters and milliliters
f. grams and kilograms

1.10 Write the U.S. customary–metric relationship and conversion factors for the following:

a. centimeters and inches
b. pounds and kilograms
c. quarts and milliliters
d. pounds and grams
e. liters and quarts

Problem Solving (Objective 1.5)

1.11 Use U.S. customary conversion factors to solve the following problems:

a. How many yards are in 24 ft?
b. How many seconds are in 15 min?
c. You need 3 qt of oil for your car. How many gallons is that?
d. One game at the arcade requires one quarter. You have $3.50 in your pocket. How many games can you play?
e. You ran a total of 2.4 miles today. How far is that in feet?

1.12 Use metric conversion factors to solve the following problems:

a. A student's height is 175 cm. How tall is the student in meters?
b. A cooler has a volume of 5500 mL. How many liters are in the cooler?
c. A hummingbird has a mass of 0.050 kg. What is the mass of the hummingbird in grams?
d. The recommended daily allowance of phosphorus is 800 mg. How many grams of phosphorus are recommended?
e. A glass of orange juice contains 0.85 dL of juice. How many milliliters of orange juice is that?
f. A package of chocolate instant pudding contains 2840 mg of sodium. How many grams of sodium is that?

1.13 Solve the following problems using one or more conversion factors:

a. A container of lemonade holds 0.750 qt. How many milliliters of juice can be held in the container?
b. What is the mass in kilograms of a person who weighs 165 lb?

c. The femur, or thigh bone, is the longest bone in the body. In a 6-ft person, the femur might be 19.5 in. long. What is the length of that person's femur in millimeters?
d. A dialysis unit requires 75,000 mL of distilled water. How many gallons of water must be used? (*Hint*: 1 gal = 4 qt.)
e. You need 4.0 oz of a steroid ointment. If there are 16 oz in 1 lb, how many grams does the pharmacist need to prepare?
f. An injured person is losing blood from an arterial wound at the rate of 0.5 mL/s. If it takes 40 min to get to the hospital, and the loss of blood continues at the same rate, how many pints of blood will be lost by the time the person gets to the hospital? (*Hint*: 1 qt = 2 pt.)
g. You are setting up a fish tank and need a 5.0-ft piece of plastic tubing. How many millimeters of tubing will you need to buy?
h. A person on a diet has been losing weight at the rate of 4 lb/week. If the person has been on the diet for 6 weeks, how many kilograms were lost?

1.14 Use clinical and other conversion factors to solve the following health-related problems:
a. The daily dose of ampicillin for the treatment of an ear infection is 100 mg/kg body weight. What is the daily dose for a 34-lb toddler?
b. A patient needs 0.024 g of a sulfa drug. There are 8-mg tablets in stock. How many tablets should be given?
c. You have used 250 L of distilled water for a dialysis patient. How many gallons of water is that?
d. An intramuscular medication is given at 5 mg/kg body weight. How much medication do you need to give an 80-lb patient?
e. The doctor has ordered 1.0 g tetracycline to be given every 6 h to a patient. If your stock on hand consists of 500-mg tablets, how many will you need for 1 day's treatment?
f. A doctor has ordered 300 mg of atropine IM. Atropine is available at 0.5 g/mL. How many milliliters would you give?
g. A patient has received 4 pt of blood. How many deciliters would that be? (*Hint*: 1 qt = 2 pt.)

Density and Specific Gravity (Objective 1.6)

1.15 Calculate the density of the following samples:
a. Twenty milliliters of a salt solution that has a mass of 24 g.
b. A solid object with a mass of 75 g and a volume of 17 mL.
c. A gem with a mass of 22.5 g. When placed in a graduated cylinder containing 20.0 mL of water, the level of water rises to 34.5 mL.
d. A medication, if the contents of a syringe filled to 3.00 mL have a mass of 3.85 g.
e. Thirty milliliters of a urine sample that has a mass of 31.5 g.

1.16 Use density values to solve the following problems:
a. What is the mass, in grams, of 150 mL of a liquid that has a density of 1.4 g/mL?
b. What is the mass of a sucrose solution that fills a 500-mL IV bottle if the density of the sucrose solution is 1.15 g/mL?
c. Beverly, a sculptor, has prepared a mold for casting a bronze figure. The figure has a volume of 220 mL. If bronze has a density of 7.8 g/mL, how many grams of bronze does Beverly need to prepare the figure?
d. A solid material has a density of 5.0 g/mL. What is the volume (mL) of 100 g of the material?

e. A fish tank holds 30 gal of water. Using the density of 1.0 g/mL for water, determine the number of pounds of water in the fish tank.
f. Copper has a density of 8.9 g/mL, and aluminum has a density of 2.70 g/mL. A graduated cylinder contains 20.0 mL of water. To what level will the water rise in the cylinder when 32.4 g of copper and 8.6 g of aluminum are added?

1.17 Solve the following specific gravity problems:
a. A urine sample has a density of 1.030 g/mL. What is the specific gravity of the sample?
b. The specific gravity of an oil is 0.85. What is its density?
c. A solution has a specific gravity of 1.72. What is the mass, in grams, of a 250-mL sample of the solution?
d. A liquid has a volume of 400 mL and a mass of 450 g. What is the specific gravity of the liquid?
e. A bottle containing 300 g of cleaning solution has fallen and broken on the floor. If the solution in the bottle has a specific gravity of 0.850, what volume of solution needs to be cleaned up?
f. Butter has a specific gravity of 0.86. What is the volume, in liters, of 0.25 lbs of butter?

Measurement of Temperature (Objective 1.7)

1.18 Solve the following conversions between Celsius and Fahrenheit temperatures:

a. 36°C = ____________°F   d. 60°F = ____________°C

b. −20°C = ____________°F   e. 110°F = ____________°C

c. 150°C = ____________°F   f. −25°F = ____________°C

1.19 Solve the following conversions between Kelvin, Celsius, and Fahrenheit temperatures:

a. 60°C = ____________K   d. 540 K = ____________°C

b. −20°C = ____________K   e. 220 K = ____________°C

c. 20°F = ____________K   f. 1000 K = ____________°F

1.20 A patient has a high fever of 106°F. What does this read on a Celsius thermometer?

1.21 A 4-year-old child has a temperature of 38.7°C. Since high fevers cause convulsions in children, it is recommended that phenobarbital be given if the temperature exceeds 101°F. Should phenobarbital be given now?

1.22 Hot compresses are being prepared for the patient in room 32B. The water is heated to 145°F. What is the temperature of the hot water in °C?

1.23 On the planet Mercury, the average nighttime temperature is 13 K, and the average daytime temperature is 683 K. What are the corresponding Fahrenheit temperatures?

Atoms and Elements

Objectives

2.1 Given the name of an element, write its correct symbol; given the symbol, write the correct name.

2.2 Use the periodic table to state the elements in a group or period and identify them as metals or nonmetals.

2.3 State the names of the three subatomic particles in the atom. Describe their electrical charges, relative masses, and locations within the atom.

2.4 Given the atomic number and mass number of a neutral atom, state the number of protons, neutrons, and electrons.

2.5 Using the periodic table, state the atomic weight of an element.

2.6 Given the name or symbol of one of the first 20 elements on the periodic table, write its electron arrangement.

2.7 Use the electron arrangement of an element to state its group number and to explain periodic law.

Scope

Everything around you that occupies space is *matter*. Atoms are minute bits of matter. Billions of atoms are packed together to build you and all the materials you can see around you. The paper in this book contains atoms of carbon, hydrogen, and oxygen. The ink on this paper, even the dot over the letter "i," contains huge numbers of atoms. There are as many atoms in that dot as there are seconds in 10 billion years.

The primary substances from which all things are made are called *elements*. There is a different kind of atom for each kind of element. Atoms of the elements calcium and phosphorus build your teeth and bones. The hemoglobin that transports oxygen in your blood contains atoms of iron. Atoms of carbon, hydrogen, oxygen, and nitrogen derived from the digestion of food are used by the cells of your body to build proteins. There are about 109 different elements known today. Of these, 91 elements occur naturally and are found in the compounds that make up our world. The remaining elements have been produced artificially by scientists and are not found in nature. The variation in the many different materials around you is a result of the types of elements and numbers of atoms that make up the different substances.

Doctors, nurses, and laboratory technicians know which elements should be in your body. Laboratory tests of body fluids such as blood and urine determine whether the amounts present are normal or abnormal. The quantities of sodium and potassium in your blood warn a doctor of the possibility of disease or metabolic malfunctioning.

2.1 Elements and Symbols

OBJECTIVE Given the name of an element, write its correct symbol; given the symbol, write the correct name.

element

atom

Elements are the primary substances that build all other things. An *atom* is the smallest particle of an element that still retains the characteristics of that element. You cannot see an atom or even a hundred atoms with the naked eye. However, when billions and billions of atoms are packed together, the characteristic of each atom is added to the next until we can see the characteristics we associate with the element. For example, a small piece of the shiny, copper-colored element we call copper consists of many, many copper atoms.

Symbols

symbol

Scientists have a shorthand system for writing the names of the elements. The name of an element is represented by a symbol that contains one or two letters from the name. A few symbols are derived from the ancient Latin or Greek names of these elements. For example, the Latin word for sodium is *natrium*; the symbol is Na. The symbol for potassium, K, comes from the Latin name *kalium*. See Table 2.1 for the names and symbols of the commonly used elements. Learning the names and symbols of the elements listed in Table 2.1 will greatly help your learning of chemistry. A complete list of all the elements (both naturally occurring and artificially produced) and their symbols appears at the front of the book.

Traditionally, the country in which a new element is discovered is given the honor of providing a name for that element. In the case of element 104, both

Table 2.1 Names and Symbols of Commonly Used Elements

Name	Symbol	Name	Symbol
Aluminum	Al	Lead	Pb (plumbum)
Argon	Ar	Lithium	Li
Barium	Ba	Magnesium	Mg
Boron	B	Mercury	Hg (hydrargyrum)
Bromine	Br	Neon	Ne
Cadmium	Cd	Nickel	Ni
Calcium	Ca	Nitrogen	N
Carbon	C	Oxygen	O
Chlorine	Cl	Phosphorus	P
Cobalt	Co	Potassium	K (kalium)
Copper	Cu (cuprum)	Silicon	Si
Fluorine	F	Silver	Ag (argentum)
Gold	Au (aurum)	Sodium	Na (natrium)
Helium	He	Strontium	Sr
Hydrogen	H	Sulfur	S
Iodine	I	Tin	Sn (stannum)
Iron	Fe (ferrum)	Zinc	Zn

the United States and the Soviet Union claim that their scientists were the first to discover this element. In some texts, you will see the name Kurchatovium, the Russian name, and in other texts, you will see the name Rutherfordium, the American name. A new system of naming has been proposed to avoid this problem for elements 104 and up. It consists of naming each part of the number:

0: nil	1: un	2: bi	3: tri	4: quad
5: pent	6: hex	7: sept	8: oct	9: en

Thus, the name for element 104 would be unnilquadium. The name for element 105 would be unnilpentium:

1	0	4		1	0	5	
un	nil	quad	ium	un	nil	pent	ium

HEALTH NOTE

Latin Names for Elements in Clinical Usage

In medicine, the Latin names for sodium and potassium are often used. The condition of excess sodium in the body is called *hypernatremia*, and a low sodium level is called *hyponatremia*. In the case of potassium, both the modern name and the Latin name are used. For example, a high potassium level may be called *hyperpotassemia* or *hyperkalemia*; a potassium level that is below normal may be called *hypopotassemia* or *hypokalemia*.

Example 2.1 Give the names for the following chemical symbols:

1. Zn
2. K
3. H
4. Fe

Solution

1. zinc
2. potassium
3. hydrogen
4. iron

Example 2.2 Give the chemical symbol for each of the following elements:

1. carbon
2. nitrogen
3. sodium
4. copper

Solution

1. C
2. N
3. Na
4 Cu

There is a small group of elements that is important to the human body. These elements are always found in combination with other elements to form

Table 2.2 Elements Important in the Human Body

Name	Symbol	Amount (Approx.) in a 60-kg Person (kg)	Found in
Oxygen	O	39	Water, carbohydrates, fats, and proteins
Carbon	C	11	Carbohydrates, fats, and proteins
Hydrogen	H	6	Water, carbohydrates, fats, and proteins
Nitrogen	N	2	Proteins, DNA, and RNA
Calcium	Ca	1	Bones and teeth
Phosphorus	P	0.6	Bones and teeth, DNA, and RNA
Potassium	K	0.2	Inside cells; important in conduction of nerve impulses
Sulfur	S	0.2	Some amino acids
Sodium	Na	0.1	Body fluids; important in nerve conduction and fluid balance
Magnesium	Mg	0.1	Bone; important in enzyme function
Chlorine	Cl	0.1	Outside cells; major electrolyte

compounds such as amino acids, carbohydrates, or vitamins. Table 2.2 lists some elements that are important in the human body.

2.2 The Periodic Table

OBJECTIVE Use the periodic table to state the elements in a group or period and identify them as metals or nonmetals.

All of the elements are listed by their symbols in the *periodic table of elements.* A complete periodic table is located at the front of this book. A single horizontal row in the periodic table is called a *period.* For example, the row of elements Li, Be, B, C, N, O, F, and Ne represents a period.

period

family

group

Each vertical column contains elements belonging to a *family* or group of elements. For example, the elements Li, Na, K, Rb, Cs, and Fr are in vertical column IA; they compose the family of elements known as the *alkali metals.* (See Figure 2.1.) Most elements within a family exhibit similar properties.

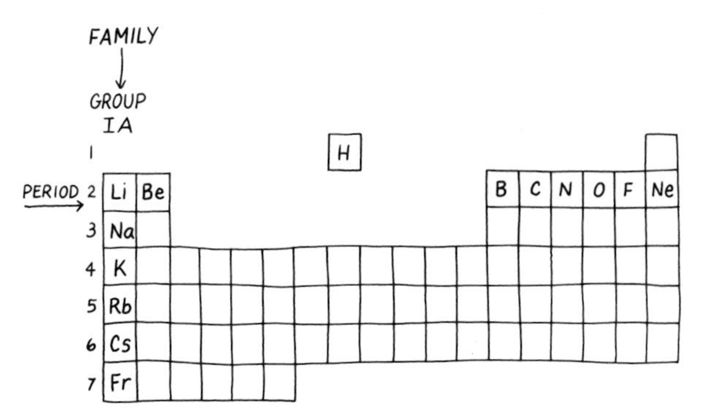

Figure 2.1 Families and periods of elements in the periodic table.

Example 2.3 State if each set of elements represents elements in a family, a period, or neither:

1. F, Cl, Br
2. Na, Al, P
3. K, Al, O

Solution

1. Family. The elements F, Cl, and Br are part of a family of elements; they all appear in the vertical column VIIA.
2. Period. The elements Na, Al, and P all appear in the third row or third period on the periodic table.
3. Neither. The elements K, Al, and O are neither in the same family nor in the same period.

HEALTH NOTE

Calcium and Strontium

Calcium (Ca) and strontium (Sr) in the second vertical column (IIA) are elements of the same family. The chemical behavior of strontium is so similar to that of calcium that if strontium is ingested, it replaces some of the calcium in the bones and teeth. This similarity in behavior caused great concern among scientists during nuclear testing, since radioactive strontium is a product of a nuclear detonation. If the radioactive strontium were to drift to cattle-grazing lands, the radioactive strontium could become a part of cow's milk and eventually find its way to the bones of young children. Once there, the effects of the radioactivity are detrimental to proper growth and development.

Metals and Nonmetals

Another feature of the periodic table is the heavy zigzag line that separates the elements into *metals* and *nonmetals*. The metals are those elements to the left of the line except hydrogen, and the nonmetals are the elements to the right.

metals

In general, most metals are shiny after they are polished. They can be shaped into wires (ductile) or hammered into a flat shape (malleable). Metals are often good conductors of heat and electricity. They usually melt at higher temperatures than nonmetals. Some typical metals are copper (Cu), aluminum (Al), gold (Au), silver (Ag), iron (Fe), tin (Sn), and mercury (Hg). (See Figure 2.2.)

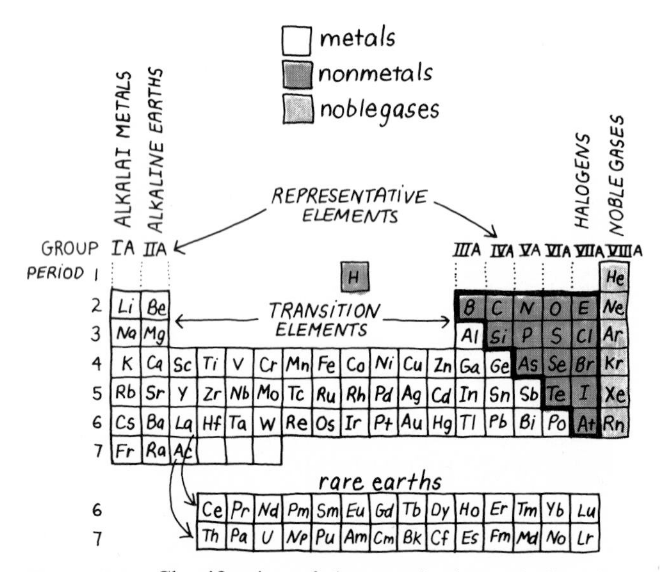

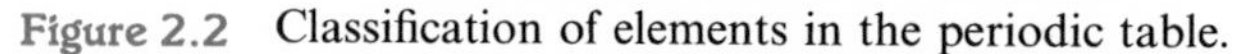
Figure 2.2 Classification of elements in the periodic table.

Table 2.3 A Comparison of Silver, a Metal, with Sulfur, a Nonmetal

Silver, a Metal	Sulfur, a Nonmetal
A pure, white solid	A pale yellow solid
Brilliant luster	Dull
Extremely ductile	Brittle
Can be hammered into sheets (malleable)	Shatters
Good conductor of heat and electricity	Poor conductor, good insulator
Used in coins, jewelry, and tableware	Used in gun powder, rubber, and fungicides
Density 10.5 g/mL	Density 2.1 g/mL
Melting point 962°C	Melting point 113°C

nonmetals

Nonmetals, which appear on the right side of the zigzag line, are not very shiny or ductile and are often poor conductors of heat and electricity. They typically have low melting points and low densities. You may have heard of nonmetals such as hydrogen, carbon, nitrogen, oxygen, chlorine, and sulfur. Table 2.3 compares some characteristics of silver, a metal, with those of sulfur, a nonmetal.

Example 2.4 Using a periodic table, classify the following elements as metals or nonmetals:

1. Na
2. Si
3. Cl
4. Cu

Solution

1. Metal; sodium is located to the left of the heavy zig-zag line that separates metals and nonmetals.
2. Nonmetal; silicon is to the right of the zig-zag line.
3. Nonmetal; chlorine is located to the right of the heavy line, which is the nonmetal region of the periodic chart.
4. Metal; copper is located to the left of the heavy line.

Representative Elements, Transition Elements, and Noble Gases

There are other ways by which elements are also classified as shown in Figure 2.2. All of the "A" group elements are known as *representative* elements. The elements in Group IA are called the *alkali metals.* The Group IIA elements are

HEALTH NOTE

Some Important Trace Elements in the Body

Some of the metals and nonmetals are important as trace elements in the body. When the body is deficient in certain trace elements, biological processes are disrupted. Some of the trace elements and their recommended daily allowances (RDA), functions, deficiency symptoms, and dietary sources are listed in Table 2.4.

Table 2.4 Some Important Trace Elements in Your Body

Element	Adult RDA[a]	Biological Function	Deficiency Symptoms	Dietary Sources
Iron (Fe)	M 10 mg F 18 mg	Formation of hemoglobin; enzymes	Dry skin, spoon nails, decreased hemoglobin count, anemia	Beef, kidneys, liver; egg yolk, oysters, spinach, beans, apricots, raisins, whole wheat bread
Cobalt (Co)		Formation of vitamin B_{12} structure	Pernicious anemia	Plentiful in most foods as vitamin B_{12}
Copper (Cu)	2.0–3.0 mg	Necessary in many enzyme systems; growth; aids formation of red blood cells and collagen	Uncommon; anemia, decreased white cell count, and bone demineralization	Nuts, organ meats, whole wheat grains, shellfish, eggs, poultry, leafy green vegetables
Zinc (Zn)	15 mg	Amino acid metabolism; enzyme systems, energy production; collagen	Retarded growth and bone formation; skin inflammation, loss of taste and smell, poor healing	Wheat germ, shellfish, milk, lima beans, fish, eggs, whole grains, turkey (dark meat), cheddar cheese
Manganese (Mn)	2.5–5.0 mg	Necessary for some enzyme systems; collagen formation; bone formation; central nervous system; fat and carbohydrate metabolism; blood clotting	Abnormal skeletal growth; impairment of central nervous system	Cereals, peas, beans, lettuce, wheat bran, meat, poultry, fish
Iodine (I)	150 μg	Necessary for activity of thyroid gland	Hypothyroidism; goiter; cretinism	Iodized table salt, seafood
Fluorine (F)	1.5–4.0 mg	Necessary for solid teeth formation and retention of calcium in bones with aging	Dental cavities	Tea, fish, milk, eggs, water in some areas, supplementary drops, toothpaste
Selenium (Se)	0.05–0.2 mg	Protection of cellular membranes; proper function of heart muscle		Fish, meat, bread, and cereals

[a] M, male; F, female.

referred to as the *alkaline earth metals.* Group VIIA elements are called the *halogens*, and elements in Group VIIIA are known as the *noble* (or *rare*) *gases.* The elements that are located between Group IIA and Group IIIA are called the *transition elements.*

2.3 The Atom

OBJECTIVE State the names of the three subatomic particles in the atom. Describe their electrical charges, relative masses, and locations within the atom.

Until recently, no one had even seen an atom, but scientists think they have a good idea of what an atom is like. At the end of the nineteenth century, an English chemist named John Dalton did experiments that suggested that elements were composed of tiny particles called atoms. By the early part of the twentieth century, experiments had shown that the atom was made up of even smaller bits of matter called *subatomic particles.* Much of the chemistry of an element depends upon the subatomic particles that make up the atoms of that element.

subatomic particles

There are three subatomic particles that are of interest to us: *protons*, *neutrons*, and *electrons.* Two of these, the proton and the neutron, form a dense *nucleus* at the center of the atom. The nucleus is extremely small compared to the size of the atom. If we consider the atom as a cloud filling a football stadium, then the nucleus would be about the size of a golf ball located at the center of the stadium. The large space outside of the nucleus would be occupied only by electrons. (See Figure 2.3.)

nucleus

Electrical Charges in an Atom

An important characteristic of some of the subatomic particles is their electrical charge. An electrical charge may be positive or negative. *Like charges repel*; they move away from each other. When you brush your hair on a dry day, the like charges that build up on the brush and your hair cause the hair to fly away from the brush.

like charges repel

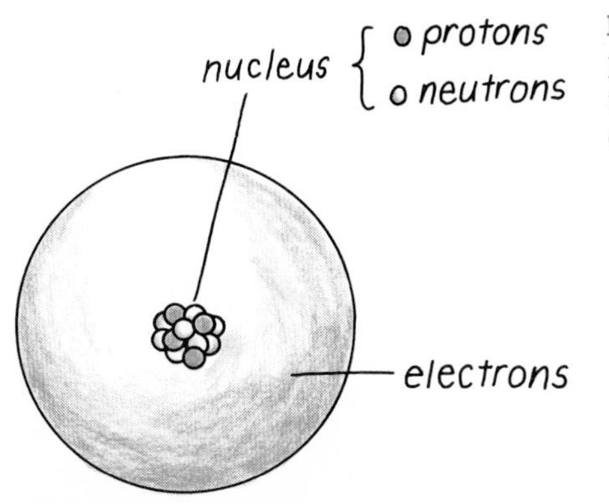

Figure 2.3 Parts of an atom. The protons and neutrons are located in the nucleus at the center of the atom; the electrons are located outside the nucleus.

opposite charges attract

Opposite or Unlike Charges Attract. The crackle of clothes taken from the clothes dryer indicates the presence of electrical charges. The clinginess of the clothing is caused by the attraction of opposite, unlike charges. (See Figure 2.4.)

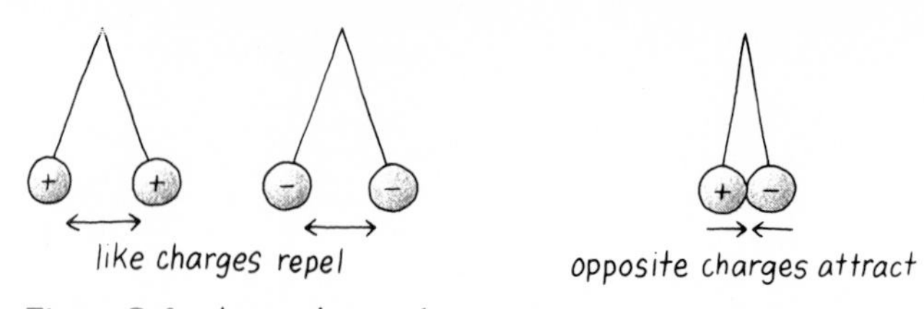

Figure 2.4 Attraction and repulsion of electrical charges.

proton p^+, electron e^-, neutron, n^0

Two of the subatomic particles carry electrical charges. The *proton* has a positive (+) electrical charge and can be written as p^+. An *electron* has an electrical charge opposite to that of the proton; the *electron* is negative (−). Its symbol is e^-. The *neutron* is not electrically charged; it is a neutral particle which has a symbol of n^0.

A Neutral Atom Always Has an Equal Number of Protons and Electrons. When the number of positive charges is the same as the number of negative charges, the overall charge on the atom is zero (0); this means that the atom is neutral. For example, an atom of fluorine with nine protons and nine electrons is neutral:

$$9p^+ + 9e^- = 0 \text{ electrical charge} = \text{neutral atom}$$

Masses of the Subatomic Particles

amu

All of the subatomic particles are extremely small compared to the things you see around you. Instead of using grams for mass, chemists find it more convenient to use a unit called an *atomic mass unit* (*amu*) to represent the mass of very small particles:

$$1 \text{ amu} = 1.66 \times 10^{-24} \text{ g}$$

The proton has a mass of about 1 amu. The neutron, which is almost the same size as the proton, also has a mass of about 1 amu. The electron, which is about 1/1836 the mass of a proton, has such a small mass that we usually consider its mass as negligible or zero (0). For a summary of these three subatomic particles, see Table 2.5.

Table 2.5 Particles in the Atom

Particle	Symbol	Mass (Approx.)	Charge	Location
Proton	p^+	1 amu	1+	Nucleus
Neutron	n^0	1 amu	0	Nucleus
Electron	e^-	"0"	1−	Outside of nucleus

Example 2.5 Complete the table for subatomic particles:

Name	Symbol	Mass	Charge	Location in the atom
Electron	______	______	______	______________
__________	______	1 amu	0	______________

Solution

Name	Symbol	Mass	Charge	Location in the atom
Electron	e^-	0	1–	Outside of nucleus
Neutron	n^0	1 amu	0	Nucleus

2.4 Atomic Number and Mass Number

OBJECTIVE Given the atomic number and mass number of a neutral atom, state the number of protons, neutrons, and electrons.

atomic number

The *atomic number* represents the number of protons contained in an atom of any element:

Atomic number = Number of protons

The elements in the periodic table are arranged in order of increasing atomic number and, therefore, by increasing number of protons. For example, a hydrogen atom (atomic number 1) has one proton, a helium atom (atomic number 2) has two protons, an atom of lithium (atomic number 3) has three protons, and so on, for all the elements in the periodic table.

For neutral atoms, the atomic number also gives the number of electrons. Table 2.6 lists the atomic number and corresponding number of protons and electrons for some neutral atoms.

Table 2.6 Atomic Numbers and Number of Protons and Electrons for Some Neutral Atoms

Element	Symbol	Atomic Number	Protons	Electrons
Hydrogen	H	1	1	1
Helium	He	2	2	2
Lithium	Li	3	3	3
Oxygen	O	8	8	8
Sodium	Na	11	11	11
Iron	Fe	26	26	26

Example 2.6 Provide the atomic number, number of protons, and number of electrons for neutral atoms of the following elements:

1. nitrogen
2. magnesium
3. copper

Solution

1. nitrogen (N): atomic number 7; 7 protons and 7 electrons
2. magnesium (Mg): atomic number 12; 12 protons and 12 electrons
3. copper (Cu): atomic number 29; 29 protons and 29 electrons

Mass Number

The *mass number* is the sum of all the protons and neutrons in the nucleus of an atom. It is always a whole number:

$$\text{Mass number} = \text{Number of protons} + \text{Number of neutrons}$$

For example, an atom of potassium with 19 protons and 20 neutrons has a mass number of 39. If you are given the number of protons and the number of neutrons of an atom, you can calculate the mass number by addition:

Element	K	O	Al	Fe
Number of protons	19	8	13	26
+Number of neutrons	+20	+8	+14	+30
Mass number (total)	39	16	27	56

Example 2.7 Calculate the mass number for atoms with the following number of protons and neutrons:

1. 5 protons and 6 neutrons
2. 18 protons and 22 neutrons
3. 48 protons and 64 neutrons

Solution

1. $5 + 6 = \text{mass number} = 11$
2. $18 + 22 = \text{mass number} = 40$
3. $48 + 64 = \text{mass number} = 112$

Table 2.7 Atomic Data for Several Elements

Element	Symbol	Atomic Number	Mass Number	Number of Protons	Number of Neutrons	Number of Electrons
Hydrogen	H	1	1	1	0	1
Nitrogen	N	7	14	7	7	7
Phosphorous	P	15	31	15	16	15
Chlorine	Cl	17	35	17	18	17
Iron	Fe	26	56	26	30	26

If you already know the atomic number and the mass number, you can calculate the number of neutrons. The number of neutrons is the *difference* between the mass number and the atomic number:

Number of neutrons = Mass number − Number of protons

Number of neutrons = Mass number − Atomic number

Element	C	Cl	Fe
Mass number	12	37	58
− Atomic number	−6	−17	−26
Number of neutrons	6	20	32

Table 2.7 illustrates the atomic number, mass number, and number of protons, neutrons, and electrons for some neutral atoms.

Example 2.8 A neutral atom of phosphorus has an atomic number of 15 and a mass number of 31. Determine:

1. the number of protons in the atoms
2. the number of neutrons in the atom
3. the number of electrons

Solution

1. There are 15 protons in a neutral atom of phosphorus. The atomic number 15 is equal to the number of protons.
2. There are 16 neutrons in the phosphorus atom. The number of neutrons is obtained by subtracting the atomic number from the mass number:

   Mass number (31) − Atomic number (15) = Number of neutrons (16)

3. There are 15 electrons in a neutral phosphorus atom. In a neutral atom, the number of electrons is the same as the number of protons.

2.5 Isotopes and Atomic Weight

OBJECTIVE Using the periodic table, state the atomic weight of an element.

Isotopes

isotopes

Isotopes are the atoms of the same element that have the same atomic number but different mass numbers. Scientists once thought that all atoms of an element were exactly alike, but then it was discovered that atoms of the same element could have different numbers of neutrons contained in the nucleus. These different forms of atoms of the same element are called isotopes. For example, all magnesium atoms have 12 protons. However, one type of atom has 12 neutrons, another has 13 neutrons, and yet another has 14 neutrons. The differences in neutrons among these atoms affects the mass number. The three kinds of isotopes of magnesium have the same atomic number, but different mass numbers.

nuclear symbol

In order to distinguish between the isotopes of an element, the nucleus of each isotope can be represented by a *nuclear symbol*. This is done by writing the symbol of the element and placing the mass number in the upper left corner and the atomic number in the lower left corner. The three isotopes of magnesium and their nuclear symbols are illustrated in Figure 2.5. The percent (%) natural abundance indicates the relative amounts of each isotope that occur in a naturally occurring sample of that element.

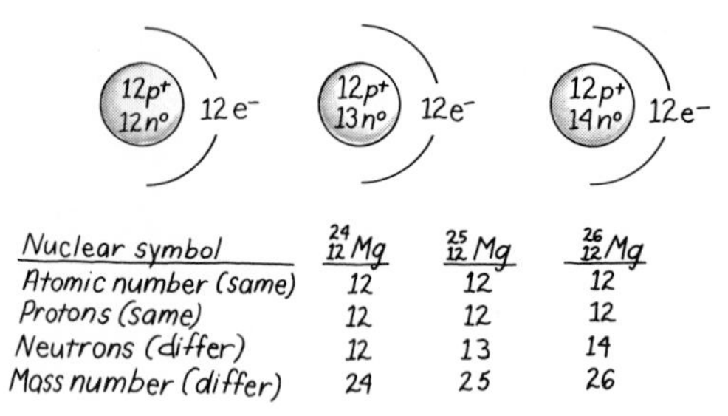

Figure 2.5 Isotopes of magnesium.

Table 2.8 Atomic Numbers and Mass Numbers of Isotopes

Element	Atomic Number	Mass Number	Nuclear Symbol	% Natural Abundance
S	16	32	$^{32}_{16}S$	95.0
		33	$^{33}_{16}S$	0.8
		34	$^{34}_{16}S$	4.2
Cl	17	35	$^{35}_{17}Cl$	75.5
		37	$^{37}_{17}Cl$	24.5

Table 2.8 lists the atomic number and the mass number for the isotopes of some selected elements. Most elements consist of more than one isotope.

Example 2.9 State the number of protons and neutrons in the following naturally occurring isotopes of oxygen:

1. $^{16}_{8}O$
2. $^{17}_{8}O$
3. $^{18}_{8}O$

Solution: Since all the isotopes have atomic number 8, they each have 8 protons. The number of neutrons is found by subtracting the atomic number (8) from each of their mass numbers:

1. 8 protons; 8 neutrons
2. 8 protons; 9 neutrons
3. 8 protons; 10 neutrons

Atomic Weight

Most of the time, scientists work with a sample of an element that contains a great many atoms. Since this sample includes all of the isotopes of that element, it is necessary to consider an "average atom" of the element. This average atom would have an "average atomic mass" based on the atomic masses of all of its isotopes. To do this, a weighted average is calculated using the percent abundance of each naturally occurring isotope.

For example, if you had 100 atoms in a sample of chlorine, approximately 75 of the atoms would have an atomic mass of 35.0, and the other 25 atoms would have atomic mass of 37.0. This assortment of isotopes is based on the naturally occurring abundance of chlorine isotopes (Table 2.8). Using the decimal form of the percent abundance of each isotope and the atomic masses, we can calculate a weighted average:

Isotope	*Abundance* ×	*Atomic mass*	
$^{35}_{17}Cl$	0.755	× 35	= 26.4
$^{37}_{17}Cl$	0.245	× 37	= 9.1
		Atomic weight (average mass)	= 35.5

We have illustrated here the effect of the various isotopes on the overall atomic weight. However, you do not need to know how to calculate the atomic weight of an average atom of an element. This example is to illustrate that atomic weights will not necessarily be whole numbers, since they represent the average atomic mass for all the isotopes of each element. In the periodic table at the front of this book, the atomic weight appears below the symbol of each element in the periodic table. Although atomic weights are known to several decimal places, we have rounded off their values in this text to the tenths (0.1) place.

Example 2.10 Using the periodic table, write the atomic weight for each of the following elements:

1. hydrogen
2. iron
3. sulfur
4. potassium

Solution

1. 1.0
2. 55.8
3. 32.1
4. 39.1

2.6 Energy Levels in the Atom

OBJECTIVE Given the name or symbol of one of the first 20 elements in the periodic table, write its electron arrangement.

The chemistry of an element is primarily determined by the number of electrons in the atoms of that element and the way in which the electrons are arranged about the nucleus. Every electron is known to have a specific amount of energy, and electrons of similar energy occupy the same energy levels. The first energy level (level 1), which is closest to the nucleus, is the lowest energy level. It holds electrons with the lowest possible energy. There are energy levels further from the nucleus and higher in energy such as energy level 2, followed by energy level 3, and so on. The energy levels actually describe the energies of electrons rather than their locations in the atom. However, the electrons in the lower energy levels are usually closer to the nucleus, whereas electrons in the higher energy levels are usually further away.

Table 2.9 Main Energy Levels

Energy Level	Maximum Number of Electrons
1	2
2	8
3	18
4	32

There is a limit to the number of electrons possible in each energy level. The first energy level (1) can hold two electrons; the second level (2) can hold up to eight electrons; the third energy level (3) can take a maximum of 18 electrons. Table 2.9 summarizes the designations for the first four energy levels and the maximum number of electrons that each can hold. There are additional energy levels in the atom, but they are beyond our consideration in this text.

Electron Arrangements for the First Twenty Elements

If we consider the first 20 elements in the periodic table, we can determine the number of electrons in each energy level. This is called the *electron arrangement* for the atom. Electrons are placed in order, first filling the lowest energy level, then filling the next-lowest energy level, and so on. Thus, there is a buildup in energy as the energy levels are filled.

We might think of energy levels as floors in a hotel. The ground floor fills first, then the second floor, and so on. In Figure 2.6, two electrons are placed in the first (1), or lowest, energy level. The next eight electrons can be placed in the second (2) energy level. Both of these levels are now completely filled. In the third (3) energy level, we have placed just eight electrons. This energy level (3) stops filling for a while even though it can hold additional electrons. This break in filling is due to the stability of some subdivisions within the third energy level. Thus, the third energy level remains stable with eight electrons, whereas the next two electrons go into the fourth (4) energy level.

Now we can proceed writing the electron arrangements or electron configurations for the first 20 elements. The single electron of hydrogen and the two electrons of helium are placed in the first (1) energy level. Sometimes, when we wish to draw a simple diagram of the electron arrangement, we indicate the nucleus of the atom and draw curved lines to represent each of the occupied energy levels. We will use the most common isotopes in our illustrations of

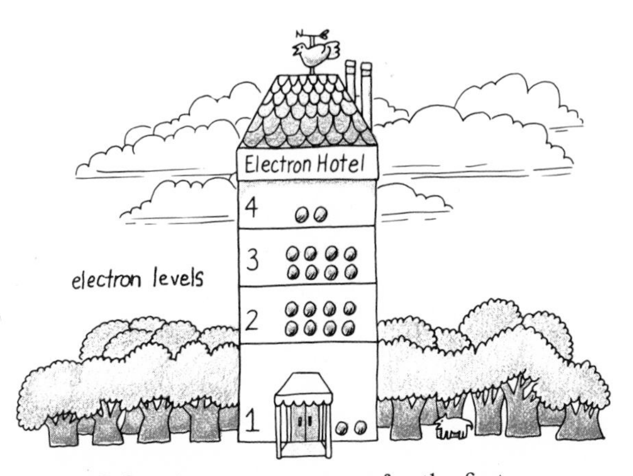

Figure 2.6 Electron occupancy for the first 20 electrons.

electron configurations. The electron configurations for hydrogen and helium would appear as follows:

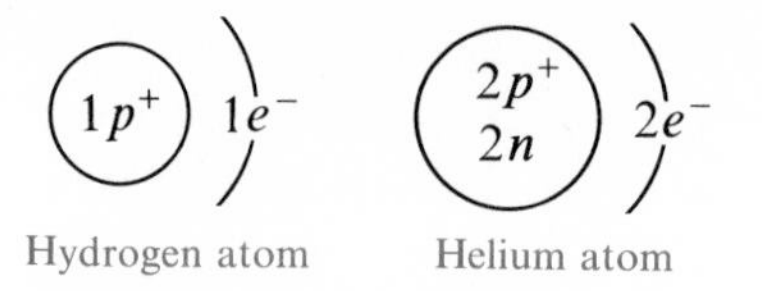

Hydrogen atom Helium atom

Elements of the second period (Li to Ne) have electrons that complete the first energy level and go into the second energy level. For example, lithium, with three electrons, places two electrons in the first level. The third remaining electron must go into the second energy level. An atom of carbon, with a total of six electrons, places two electrons in the first level. The remaining four electrons in carbon go into the second level. An atom of neon completely fills the first and second energy levels with two electrons going into the first level and the remaining eight electrons going into the second level:

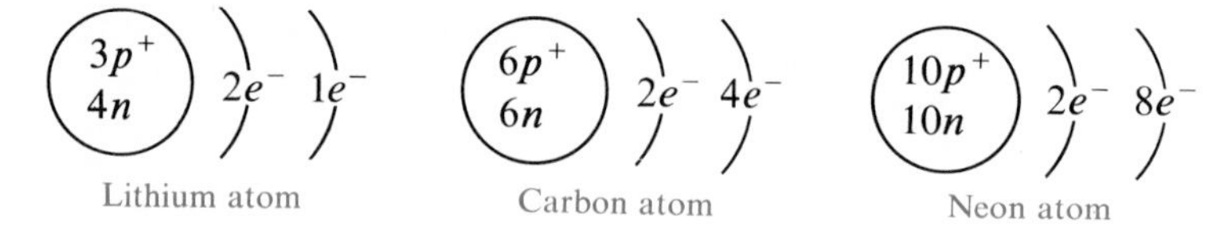

Lithium atom Carbon atom Neon atom

The last electron of sodium (11) goes into the third energy level since the first and second levels are already filled. The rest of the elements in the third period continue to place their remaining electrons in the third energy level. For example, in a sulfur atom, which contains 16 electrons, two electrons are placed in the first level, eight electrons go into the second level, and the remaining six electrons go into the third energy level. Argon, which contains 18 electrons, has eight electrons in the third level:

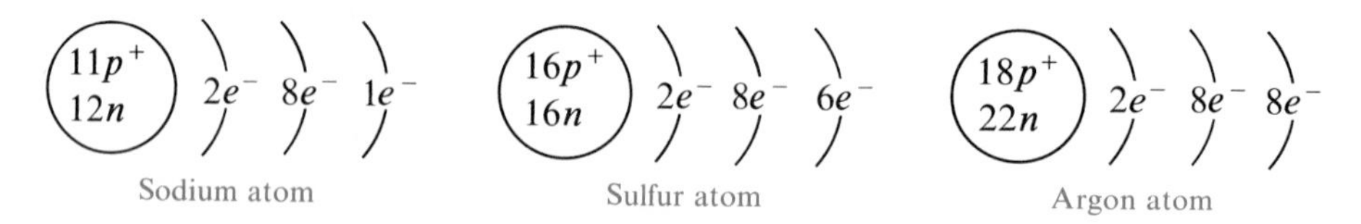

Sodium atom Sulfur atom Argon atom

As noted earlier, the third shell takes up to eight electrons and stops filling for a while. This third level will begin filling again with elements of higher atomic numbers. For our purposes, when the third level has accumulated eight electrons, it has sufficient stability so that the remaining electrons in potassium and calcium actually go into the fourth energy level:

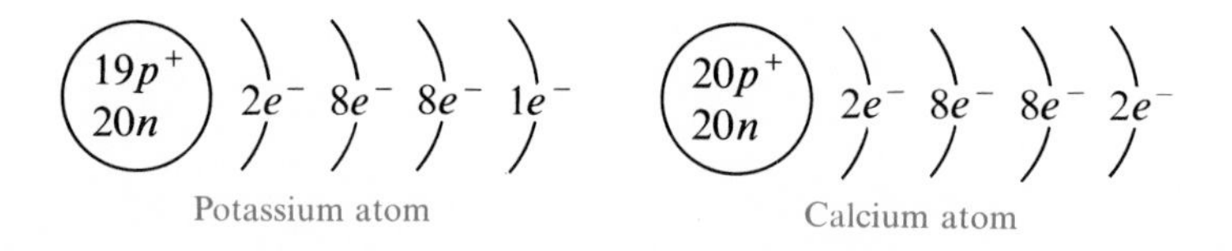

Potassium atom Calcium atom

The electron arrangements for the first 20 elements are summarized in Table 2.10.

Table 2.10 Electron Arrangements for the First 20 Elements

Atomic Number	Element	Energy Level 1	2	3	4
1	H	1			
2	He	2			
3	Li	2	1		
4	Be	2	2		
5	B	2	3		
6	C	2	4		
7	N	2	5		
8	O	2	6		
9	F	2	7		
10	Ne	2	8		
11	Na	2	8	1	
12	Mg	2	8	2	
13	Al	2	8	3	
14	Si	2	8	4	
15	P	2	8	5	
16	S	2	8	6	
17	Cl	2	8	7	
18	Ar	2	8	8	
19	K	2	8	8	1
20	Ca	2	8	8	2

Example 2.11 Write the electron arrangements for each of following neutral atoms:

1. An oxygen atom with atomic number 8 and mass number 16.
2. A chlorine atom with atomic number 17 and mass number 37.

Solution

1. The nucleus of this oxygen atom has eight protons and eight neutrons. In a neutral oxygen atom, there are eight electrons. In the electron arrangement, the first two electrons go into the first energy level, and the remaining six electrons go into the second energy level:

$$\begin{pmatrix} 8p^+ \\ 8n \end{pmatrix} \; 2e^- \; 6e^-$$

2. This chlorine nucleus has 17 protons and 20 neutrons. Since it is a neutral atom, there are 17 electrons. In its electron arrangement, the first two electrons go into the first energy level, the next eight electrons go into the second energy level, and the remaining seven electrons are placed in the third energy level:

$$\begin{pmatrix} 17p^+ \\ 20n \end{pmatrix} \; 2e^- \; 8e^- \; 7e^-$$

Energy-Level Changes

Whenever possible, electrons occupy the lowest and most stable energy levels. However, if some outside source of energy such as heat or light is made available, an electron can absorb a certain amount of energy, then jump to a vacancy in a higher energy level. This high-energy state, however, is an unstable situation, and the electron will drop to a vacant space in a lower, more stable energy level. When the electron drops to a lower energy level, energy is released. (See Figure 2.7.)

Consider a staircase, such as the one in Figure 2.8. You are standing at the bottom of the stairs with a ball in your hand. When you throw the ball up the stairs, you provide the outside source of energy. After the ball reaches its highest point, it will come down the steps; it might skip a step or two, but each time it bounces on a lower step. Note that the ball cannot land anywhere between the steps. When an electron drops, it too must land in lower energy levels and never between them.

When electrons drop to different energy levels, different amounts of energy are emitted. High-energy emissions, which include x-rays, are more dangerous

HEALTH NOTE

Biological Reactions to Sunlight

Our everyday life depends on sunlight, but exposure to sunlight can have damaging effects on living cells, and too much exposure can even cause their death. The list of damaging effects of sunlight includes sunburn, skin cancer, premature aging of the skin, changes in the DNA of the cells, inflammation of the eyes, and perhaps cataracts. Some drugs, like the acne medications Accutane and Retinin A, as well as some antibiotics and diuretics, cause undesirable changes in the skin's reaction to sunlight.

About 10 percent of all the sunlight is in the ultraviolet (UV), high-energy range. About 50 percent of the sunlight is in the visible, moderate-energy range, and another 40 percent is in the low-energy, infrared (IR) range.

High-energy radiation is the most damaging, biologically speaking. Most of the radiation in this range is absorbed in the epidermis of the skin. The degree to which radiation is absorbed depends on the thickness of the epidermis, the hydration of the skin, the amount of coloring pigments and proteins of the skin, and the arrangement of the blood vessels. In light-skinned people, 85–90 percent of the radiation is absorbed by the epidermis, with the rest reaching the dermis layer. In dark-skinned people, 90–95 percent of the radiation is absorbed by the epidermis, with a smaller percentage reaching the dermis.

Radiation from the sun can cause sunburn, tanning, and skin aging. The same radiation can alter the cells, which leads to skin cancers. However, medicine can take advantage of the beneficial effect of sunlight. Phototherapy can be used to treat certain skin conditions, including psoriasis, eczema, and dermatitis. Low-energy radiation is used to break down bilirubin in neonatal jaundice. When radiation is combined with certain chemicals during photochemotherapy, the skin becomes more photosensitive.

Figure 2.7 Change of energy levels as energy is absorbed and emitted.

Figure 2.8 Model for electron energy-level changes.

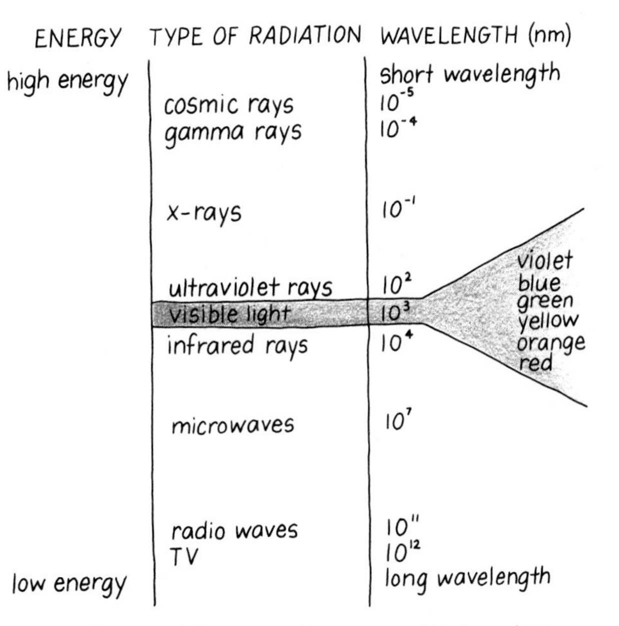

Figure 2.9 Various types of energy emitted as electrons change energy levels.

to the body because they can penetrate the body tissue and cause disruption within the cells. Low-energy emissions include infrared rays (heat), radio waves, and microwaves. Figure 2.9 illustrates some forms of energy emitted when electrons change energy levels in atoms.

2.7 Periodic Law

OBJECTIVE Use the electron arrangement of an element to state its group number and to explain the periodic law.

When elements were grouped together on the periodic table, it was discovered that each vertical column (family) contained elements with similar properties. This repetition of the physical and chemical properties with increasing atomic number is known as the *periodic law*. Later, when the patterns of electron arrangement became known, each element in a family was found to have another similarity. All the elements in a family have the same number of electrons in each of their outermost energy levels.

For example, the elements lithium, sodium, and potassium are part of a family of elements. If we look at their electron arrangement, we find that each of these elements has one electron in its outermost energy level. See Table 2.11. Thus, the similarity of chemical and physical properties among a family of elements could now be associated with a repetition in the number of electrons in their outermost energy levels.

Group Number

The group numbers IA–VIIIA can be used to identify the number of electrons in the outermost electron level of a representative element. The group number, which is a roman numeral, is located at the top of each vertical column in the periodic table. All elements in Group IA have one electron in their outer energy level, elements in Group IIA have two electrons in their outer energy level, elements in Group IIIA have three electrons in their outer energy level, and so on. We are most interested in the number of electrons in the outer energy level because these electrons have the greatest effect on the way that atoms form compounds. Thus, the electron arrangement and particularly the outermost electrons determine the chemistry of each element.

Table 2.11 A Comparison of Electron Arrangements of Some Group IA Elements

		Energy Levels				
Element	Atomic Number	1	2	3	4	
Lithium	3	$2e^-$	$1e^-$			one electron in each outer energy level
Sodium	11	$2e^-$	$8e^-$	$1e^-$		
Potassium	19	$2e^-$	$8e^-$	$8e^-$	$1e^-$	

Table 2.12 Electron Arrangements, by Family, for the First 20 Elements

	Energy Level					Energy Level					Energy Level			
Element	1	2	3	4	Element	1	2	3	4	Element	1	2	3	4
	Group I					Group II					Group III			
H	①[a]													
Li	2	①			Be	2	②			B	2	③		
Na	2	8	①		Mg	2	8	②		Al	2	8	③	
K	2	8	8	①	Ca	2	8	8	②					
	Group IV					Group V					Group VI			
C	2	④			N	2	⑤			O	2	⑥		
Si	2	8	④		P	2	8	⑤		S	2	8	⑥	
	Group VII					Group VII								
					He	②								
F	2	⑦			Ne	2	⑧							
Cl	2	8	⑦		Ar	2	8	⑧						

[a] ○ = Outer-shell electrons.

All of the noble gases (Group VIIIA) except helium have eight electrons in their outer energy levels. As a family, the noble gases are very stable and quite unreactive. At ordinary temperatures and experimental conditions, they will not enter into chemical reactions, that is, they do not easily combine with other elements to form compounds. Table 2.12 shows the electron arrangements, by families, of the first 20 elements.

HEALTH NOTE

Uses of Some Noble Gases

Inert gases are used when it is necessary to have a substance that does not react. Scuba divers normally use pressurized air, a mixture of nitrogen and oxygen gases, for breathing under water. If the air mixture is used at depths where pressure is high, the nitrogen is absorbed into the blood.

High levels of nitrogen in the blood can cause a dangerous mental disorientation. To avoid this problem, a mixture of oxygen and helium can be substituted. The diver still obtains the necessary oxygen, but the helium does not cause a mental disorientation. However, the helium does change the vibrations of the vocal cords, making the voice sound like Donald Duck.

Many lighting fixtures use a noble gas such as neon or argon to fill the lighting tubes. The heating elements that produce light with the tubes get very hot, but the noble gas does not react. The gas used to fill a blimp is the noble gas helium. When airships were first invented, a very light gas, hydrogen, was used to fill them and lift them into the air. However, hydrogen-filled dirigibles exploded because of the extreme reactivity of hydrogen gas. The unreactive helium gas presents no danger of explosion.

Example 2.12 Using the periodic table, write the group number and the number of electrons in the outer electron level of the following elements:

1. sodium
2. sulfur
3. aluminum

Solution

1. Sodium is in Group IA; sodium has one electron in the outermost electron level.
2. Sulfur is in Group VIA; sulfur has six electrons in its highest electron level.
3. Aluminum is in Group IIIA. There are three electrons in the highest electron level of aluminum.

Problems

Elements and Symbols (Objective 2.1)

2.1 Write the symbols of the following elements:

a. copper	f. barium	k. sulfur
b. silicon	g. lead	l. aluminum
c. potassium	h. neon	m. helium
d. cobalt	i. oxygen	n. boron
e. iron	j. lithium	o. hydrogen

2.2 Write the correct name of the element for each symbol:

a. C	f. F	k. K
b. Cl	g. Ar	l. Ni
c. I	h. Zn	m. Hg
d. P	i. Mg	n. Ca
e. Ag	j. Na	o. Br

The Periodic Table (Objective 2.2)

2.3 Classify the following elements as metals or nonmetals:

a. phosphorus	f. sulfur
b. magnesium	g. silicon
c. silver	h. nitrogen
d. fluorine	i. aluminum
e. nickel	j. sodium

2.4 Use the periodic table to identify each set of elements as part of a *family* or part of a *period* of elements. If the set does *not* represent a family or a period, write *none*:

a. B, Al, Ga	d. Si, P, S
b. Na, Mg, Al	e. Cu, Ag, Au
c. Cl, Br, I	f. O and S

2.5 The following are trace elements that have been found to be crucial to the biochemical and physiological processes in the body. Write *metal* or *nonmetal* for each element:

a. zinc
b. cobalt
c. manganese (Mn)
d. iodine
e. copper
f. selenium (Se)
g. nickel
h. iron

Parts of the Atom (Objective 2.3)

2.6 Use *proton, neutron,* or *electron* to identify the subatomic particle that each of the following statements describe:
a. has the smallest mass
b. carries a positive charge
c. located outside the nucleus
d. is electrically neutral
e. carries a negative charge
f. has a mass about the same as a proton

Atomic Number and Mass Number (Objective 2.4)

2.7 Write the atomic number and the mass number for the following neutral atoms:
a. an atom with 15 protons and 16 neutrons
b. an atom with 35 protons and 45 neutrons
c. an atom with 11 electrons and 12 neutrons
d. an atom with 26 electrons and 30 neutrons

2.8 Complete the following table for neutral atoms:

	Atomic Number	Mass Number	Protons	Neutrons	Electrons	Name	Symbol
a.	______	27	______	______	______	______	Al
b.	12	______	______	12	______	______	______
c.	______	______	6	7	______	______	______
d.	______	______	16	15	______	______	______
e.	______	34	______	______	16	______	______
f.	20	______	______	22	______	______	______

Atomic Weight and Isotopes (Objective 2.5)

2.9 State the number of protons, neutrons, and electrons in isotopes with the following symbols:
a. $^{27}_{13}Al$ b. $^{52}_{24}Cr$ c. $^{34}_{16}S$ d. $^{56}_{26}Fe$

2.10 Write a nuclear symbol for an atom with:
a. mass number 44 and atomic number 20
b. 28 protons and 31 neutrons
c. mass number 24 and 13 neutrons
d. 35 electrons and 45 neutrons

2.11 There are four isotopes of sulfur with mass numbers 32, 33, 34, and 36.
a. Write the nuclear symbols for these isotopes of sulfur.
b. How are these isotopes alike?
c. How are they different?

2.12 Using the periodic table, write the atomic weight of each of the following elements (state the atomic weight to the first decimal place; for example, calcium 40.1):
a. oxygen c. iron e. magnesium g. sodium
b. nitrogen d. hydrogen f. chlorine h. phosphorus

Energy Levels in the Atom (Objective 2.6)

2.13 Write the electron arrangements for the following elements (*example*: sodium 2, 8, 1):
a. carbon c. aluminum e. potassium g. nitrogen
b. argon d. sulfur f. phosphorus h. neon

2.14 Identify the elements with the following electron arrangements:

	Energy Level 1	2	3
a.	$2e^-$	$1e^-$	
b.	$2e^-$	$8e^-$	$2e^-$
c.	$1e^-$		
d.	$2e^-$	$8e^-$	$7e^-$
e.	$2e^-$	$6e^-$	

2.15 Use the following words to complete the statements below:
emit absorb
high-energy low-energy
a. Electrons can jump to higher energy levels when they ____________ a specific amount of energy.
b. When electrons drop to lower energy levels, they ____________ a certain amount of energy.
c. ____________ radiation is considered to be more dangerous to living cells than ____________ radiation.

Periodic Law (Objective 2.7)

2.16 The elements boron and aluminum are in the same group in the periodic table.
a. Write the electron arrangements for B and Al.
b. How many electrons are in the outer energy level for each?
c. What is their group number?
d. What part of the electron arrangement tells you the group number?

2.17 Write the number of electrons in the outer energy level and the group number for each of the following elements (*example*: Fluorine $7e^-$; Group VIIA):
a. magnesium c. oxygen e. lithium g. silicon
b. chlorine d. nitrogen f. neon h. neon

Compounds and Their Bonds

Objectives

3.1 Use the periodic table to write electron dot structures for the first 20 elements.

3.2 Use the octet rule to predict the loss or gain of electrons by a Group A element.

3.3 Predict the ionic charges for the Group A elements. State the ions and charges for the Group B elements Fe, Cu, Ag, and Zn.

3.4 Using the ionic charges, write the correct formula of an ionic compound.

3.5 Given the formula of an ionic compound, write the correct name; given the name, write the correct formula.

3.6 Use electron dot structures to diagram the formulation of covalent bonds in a covalent compound.

3.7 Given the formula of a covalent compound, write its correct name; given the name of a covalent compound, write its formula.

3.8 Identify a bond as polar covalent, nonpolar covalent, or ionic.

Scope

Most of the things you see around you are made of substances in which atoms are combined with other atoms. These combinations are called *compounds*. Although there are 109 elements now known, there are millions of different compounds because of the many different combinations of atoms.

Most substances necessary for life are compounds. For example, the human body is about 60–65 percent water by weight. Water is a compound composed of the elements hydrogen and oxygen. Some other compounds necessary for life are carbohydrates, fats, and proteins. They all are made of the elements carbon, hydrogen, and oxygen; proteins also contain nitrogen and sulfur. These compounds, obtained from your diet, build cells, provide energy, and ensure proper function.

It is important to understand compounds, their atoms, and the ways in which the atoms are bonded together. The concept of the bonds in compounds will enable you to understand better the chemical behavior and function of compounds in your body that are part of the life process.

3.1 Valence Electrons

OBJECTIVE Use the periodic table to write electron dot structures for the first 20 elements.

The outermost electrons in the electron arrangement of an atom are a major factor in determining the chemical behavior of that element. These outer electrons are called the *valence electrons*. As we saw in Chapter 2, the number of valence electrons for any Group A element is equal to the group number:

Group A number = Valence electrons

For example, potassium is in Group IA. We know from the group number that potassium must have one valence electron. We can check this by reviewing its electron arrangement. An atom of potassium has a total of 19 electrons, which are arranged in four energy levels as 2, 8, 8, 1. The one electron in the fourth energy level is the valence electron, as indicated by the group number IA.

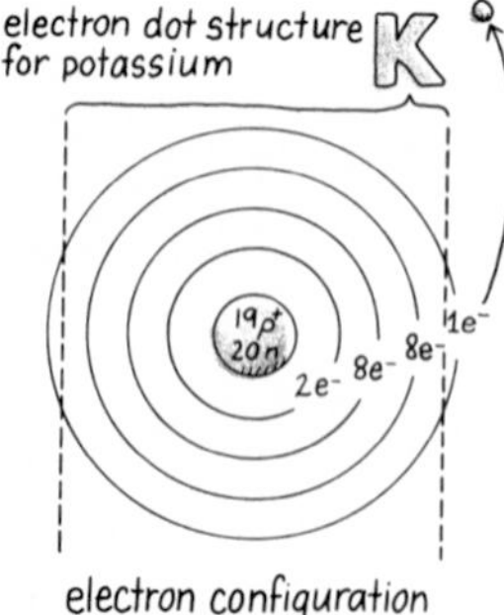

Figure 3.1 Relationship between electron dot structure and electron arrangement for potassium.

Electron Dot Structures

An *electron dot structure* represents the valence electrons of an atom. In an electron dot structure, the valence electrons are indicated by dots surrounding the symbol of the element. Figure 3.1 illustrates the relationship between the electron arrangement and the electron dot structure of potassium. We can think of the symbol as having four sides like a square. It does not matter on which side you start to place the dots. Any of the following would be acceptable for the electron dot structure of magnesium:

$\dot{\text{Mg}}\cdot \quad \cdot\text{Mg}\cdot \quad \cdot\underset{\cdot}{\text{Mg}} \quad \cdot\dot{\text{Mg}}$

Table 3.1 compares the electron arrangements, group numbers, and electron dot structures for some elements. Note that there are, at most, two dots on each side of the symbol. The maximum number of electron dots that can be placed around a symbol is eight. Table 3.2 lists the electron dot structures of the first 20 elements of the periodic table.

Table 3.1 Relationship of Electron Arrangement and Electron Dot Structure

Element	Energy Level 1	2	3	4	Group Number	Electron Dot Structure
H	1e				IA	H·
C	2e	4e			IVA	·Ċ·
Mg	2e	8e	2e		IIA	Ṁg·
N	2e	5e			VA	·N̈·
Cl	2e	8e	7e		VIIA	:C̈l·

Table 3.2 Electron Dot Structures for the First 20 Elements in the Periodic Table[a]

IA	IIA	IIIA	IVA	VA	VIA	VIIA	VIIIA
H·							He:
Li·	Ḃe·	·Ḃ·	·Ċ·	·N̈·	:Ö·	:F̈·	:N̈e:
Na·	Ṁg·	·Ȧl·	·Ṡi·	·P̈·	:S̈·	:C̈l·	:Är:
K·	Ċa·						

[a] Note that helium appears with the other noble gases in Group VIIIA even though it has only two electrons: These two electrons fill the first energy level, thus making it stable.

Example 3.1 Write the electron dot structure for an atom of

1. chlorine
2. magnesium
3. argon

Solution

1. The group number of each element can be used to determine the number of valence electrons. Chlorine in Group VIIA has seven valence electrons:

   :C̈l·

2. Magnesium, Group IIA, has two valence electrons:

   Ṁg·

3. Argon, Group VIIIA, has eight valence electrons:

   :Är:

3.2 Octet Rule

OBJECTIVE Use the octet rule to predict the loss or gain of electrons by a Group A element.

All the elements, except the noble gases, show a strong tendency to join with other elements and form compounds. The noble gases do not usually combine with other elements because the electrons in the atoms of Group VIIIA are already in a stable arrangement. All of the noble gases except helium have eight valence electrons, called an *octet*, as seen in Figure 3.2. Helium is an exception because it is stable when the first energy level is complete with two electrons.

octet consists of eight valence electrons

	energy level 1	2	3	4	electron dot structure
He	2				$\overset{..}{He}$
Ne	2	8			$:\overset{..}{\underset{..}{Ne}}:$
Ar	2	8	8		$:\overset{..}{\underset{..}{Ar}}:$
Kr	2	8	18	8	$:\overset{..}{\underset{..}{Kr}}:$

(shaded: valence electrons)

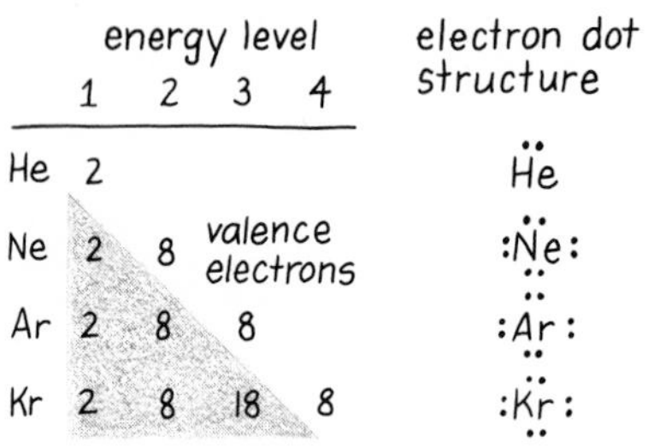

Figure 3.2 Electron arrangements for some noble gases.

An octet of valence electrons makes the noble gases particularly stable. Other elements achieve the stable arrangement of the noble gases by forming compounds. By losing, gaining, or sharing electrons, atoms in Groups IA–VIIA will acquire the electron configuration (an octet) and therefore the stability of one of the noble gases. This tendency for atoms to adjust their valence shells to eight valence electrons is called the *octet rule*. We must point out here that there are compounds with atoms that do not have octets. However, they will not be discussed in this text.

octet rule

Example 3.2 Write the electron dot structure for neon (Ne), sodium (Na), and fluorine (F). Which atoms have octets and which do not? Which atoms will form compounds and which will not? Why?

Solution

$:\overset{..}{\underset{..}{Ne}}:$ $Na\cdot$ $:\overset{..}{\underset{..}{F}}\cdot$

Neon has eight valence electrons. It is already stable and will not form compounds. However, sodium (one valence electron) and fluorine (seven valence electrons) are not as stable as neon. (They do not have eight valence electrons.) Atoms of sodium and fluorine will form those compounds that provide each atom with an octet of valence electrons.

3.3 Ions and Their Charges

OBJECTIVE Predict the ionic charges for the Group A elements. State the ions and charges for the Group B elements Fe, Cu, Ag, and Zn.

Formation of Positive Ions

Atoms of metals form positive ions when they lose electrons. For example, a potassium atom has one valence electron. By losing that valence electron, the third energy level becomes the outermost level. Since it now has eight electrons, an octet, potassium is stable. It has the electron configuration of the noble gas argon. (See Figure 3.3.)

metals lose electrons

The loss of the valence electron by a potassium atom has an effect on the electrical balance of the atom. With the loss of one electron, the number of electrons decreases from 19 to 18. Since there are 19 protons in the nucleus (the number of protons has not changed), the atom is no longer neutral. We can calculate the electrical charge as follows:

Protons in a potassium ion	= 19+	
Electrons in a potassium ion	= 18−	
Electrical charge	1+	K^+

When an atom acquires an electrical charge, it is called an *ion.* In this example, the loss of an electron by a potassium atom forms a potassium ion with a positive charge (1+).

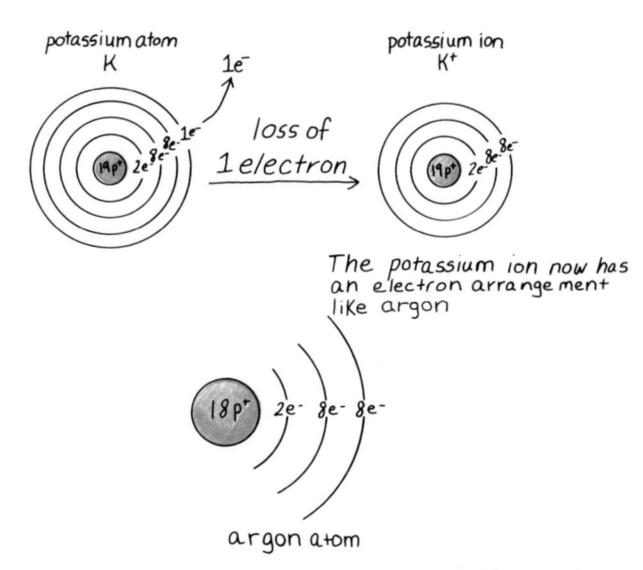

Figure 3.3 The electron configuration of the potassium atom and potassium ion compared.

Figure 3.4 Representation of the formation of a magnesium ion.

metals form positive ions

As a general rule, metals lose their valence electrons to form positively charged ions. Positive ions are also called *cations* (pronounced *cat'-ion*). Let us look at the way that another ion, the magnesium ion, is formed. Magnesium has two valence electrons. By losing two electrons, a magnesium ion forms which has the same electron arrangement as neon, the stable noble gas nearest to it in the periodic table. (See Figure 3.4.)

The loss of two electrons decreases the number of electrons to 10. Since there are still 12 protons in the nucleus, the difference in electrical charge gives the magnesium ion a positive (2+) charge:

Protons in a magnesium ion	= 12+
Electrons in a magnesium ion	= 10−
Electrical charge	= 2+

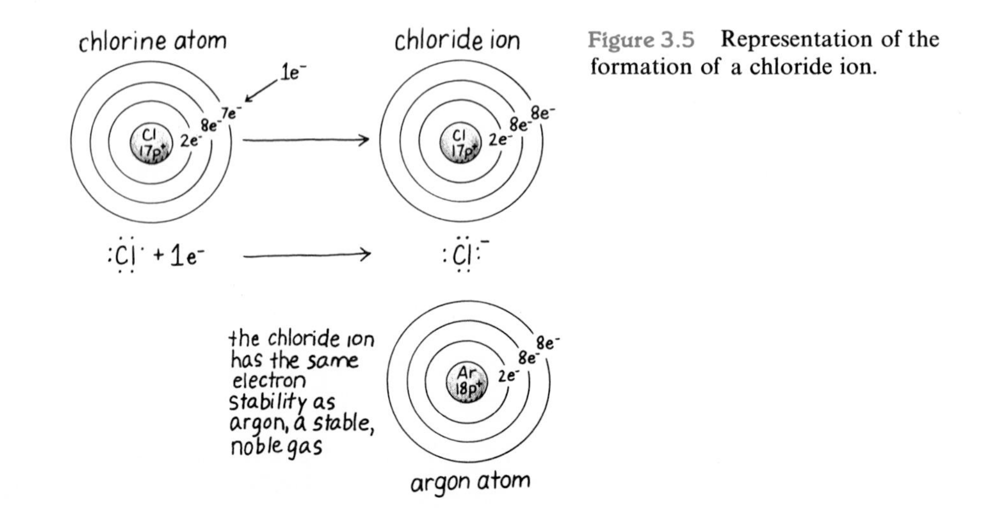

Figure 3.5 Representation of the formation of a chloride ion.

Formation of Negative Ions

nonmetals can form negative ions by gaining electrons

In contrast to the metals, nonmetals from Groups VA, VIA, and VIIA gain electrons to form octets. For example, an atom of chlorine (Group VIIA) has seven valence electrons. By gaining one more electron, chlorine acquires a noble gas configuration and becomes stable. It is now a chloride ion with an ionic charge of 1−:

Protons in a chloride ion	= 17+	
Electrons in a chloride ion	= 18−	
Electrical charge	= 1−	Cl^-

A negatively charged ion is also called an *anion* (pronounced *an′-ion*). The formation of a chloride ion is illustrated in Figure 3.5.

Example 3.3

1. Identify the elements calcium and oxygen as metals or nonmetals.
2. Write the electron arrangement for each.
3. State the number of electrons lost or gained when they form ions.
4. Write the resulting electron arrangement of the ion.
5. Give the symbol of the ions, including ionic charge.

Solution

Element	Metal or Nonmetal	Electron Arrangement	Electrons Lost or Gained	Electron Arrangement (ion)	Symbol of Ion
Ca	Metal	2, 8, 8, 2	Loss of $2e^-$	2, 8, 8	Ca^{2+}
O	Nonmetal	2, 6	Gains $2e^-$	2, 8	O^{2-}

Ionic Charges for Group A Elements

All elements within a group will gain or lose the same number of electrons and have identical ionic charges. Therefore, we can assign an ionic charge to several Group A elements by using the periodic table. The metals in Groups IA, IIA, and IIIA form positive ions. Group IA ions all have an ionic charge of 1+, Group IIA ions have an ionic charge of 2+, and Group IIIA ions have an ionic charge of 3+.

When nonmetals form ions, they have negative charges. Group VA ions have a 3− charge, Group VIA ions have a 2− charge, and Group VIIA ions have a 1− charge. Note that Group IVA is omitted from our discussion of ions. Elements in Group IVA do not typically form simple ions. Table 3.3 lists the ionic charges of Group A elements that typically form ions. Table 3.4 lists the ions of some Group A elements.

Table 3.3 Ionic Charges for Group A Elements

Group Number	Valence Electrons	Electron Activity to Become Stable	Ionic Charge	Example
IA	$1e^-$	Lose $1e^-$	1+	$Na\cdot \xrightarrow{\text{lose } 1e^-} Na^+$
IIA	$2e^-$	Lose $2e^-$	2+	$\dot{Ca}\cdot \xrightarrow{\text{lose } 2e^-} Ca^{2+}$
IIIA	$3e^-$	Lose $3e^-$	3+	$\cdot\dot{Al}\cdot \xrightarrow{\text{lose } 3e^-} Al^{3+}$
IVA	$4e^-$	Ion formation does not typically occur.		
VA	$5e^-$	Gain $3e^-$	3−	$\cdot\ddot{N}\cdot \xrightarrow{\text{gain } 3e^-} :\ddot{N}:^{3-}$
VIA	$6e^-$	Gain $2e^-$	2−	$:\ddot{O}\cdot \xrightarrow{\text{gain } 2e^-} :\ddot{O}:^{2-}$
VIIA	$7e^-$	Gain $1e^-$	1−	$:\ddot{Cl}\cdot \xrightarrow{\text{gain } 1e^-} :\ddot{Cl}:^-$
VIIIA	$8e^-$	Ion formation does not occur. Already stable.		

Table 3.4 Ions of Some Group A Elements

Metals			Nonmetals		
IA	IIA	IIIA	VA	VIA	VIIA
Li^+		Al^{3+}	N^{3-}	O^{2-}	F^-
Na^+	Mg^{2+}		P^{3-}	S^{2-}	Cl^-
K^+	Ca^{2+}				Br^-
	Ba^{2+}				I^-

Example 3.4 Using a periodic table and the group number, predict the ion for the following:

1. Al
2. Br

Solution

1. The element aluminum (Al) is found in Group IIIA in the periodic table. Since Group IIIA elements have three electrons, we would predict that aluminum would lose three electrons to form an ion with a 3+ ionic charge:

   Al^{3+}

2. The element bromine (Br) is found in Group VIIA. Since bromine has seven valence electrons, we would predict that it needs to gain one electron to have an octet. This would form an ion with a 1− ionic charge:

   Br^-

Names of Group A Ions

The positive ions in Groups IA, IIA, and IIIA are named by using their elemental names. However, the negative ions of the nonmetals replace the end of their elemental names with *-ide*. Table 3.5 lists the symbols and names of some important ions.

Table 3.5 Symbols and Names of Some Important Ions

	Ion Symbol	Name of Ion
Group IA	Li^+	Lithium ion
	Na^+	Sodium ion
	K^+	Potassium ion
Group IIA	Mg^{2+}	Magnesium ion
	Ca^{2+}	Calcium ion
	Ba^{2+}	Barium ion
Group IIIA	Al^{3+}	Aluminum ion
Group VA	N^{3-}	Nitride ion
	P^{3-}	Phosphide ion
Group VIA	O^{2-}	Oxide ion
	S^{2-}	Sulfide ion
Group VIIA	F^-	Fluoride ion
	Cl^-	Chloride ion
	Br^-	Bromide ion
	I^-	Iodide ion

Positive Ions from Some Transition Elements (Group B)

The transition elements (Group B) are metals and form positive ions. However, we cannot predict their ionic charges because for the same element there may be more than one type of positive ion. However, there are only a few Group B ions that you need to know.

iron forms two ions, Fe^{2+} and Fe^{3+}

Iron is a transition metal that forms two kinds of positive ions. If two electrons are lost, an iron ion with a 2+ charge results; if three electrons are lost, an iron ion with a 3+ charge is formed:

Fe^{2+}	Fe^{3+}
Iron(II) ion	Iron(III) ion
Ferrous ion	Ferric ion

When there is the possibility of two different ions for the same element, a naming system is needed that will distinguish between the ions. In the modern naming system, the charge of the ion is shown as a roman numeral following the name of the element. The iron ion with a 2+ charge is named iron(II); the ion with the 3+ charge is named iron(III). An older naming system still in use adds endings to the root of the Latin name, *-ous* for the lower charge, and *-ic* for the higher charge. In this system, the iron ion with the 2+ charge is called the *ferrous ion*; the iron ion with the 3+ charge is called the *ferric ion.*

copper forms two ions, Cu^{+} and Cu^{2+}

The transition element copper also forms two different ions. There is a copper ion with a charge of 1+; the other has a charge of 2+. The copper ion with the 1+ charge is named the *copper(I) ion*, or the *cuprous ion*; the copper ion with the 2+ charge is called the *copper(II) ion*, or the *cupric ion*:

Cu^{+}	Cu^{2+}
Copper(I) ion	Copper(II) ion
Cuprous ion	Cupric ion

zinc ion, Zn^{2+}
silver ion, Ag^{+}

Two other ions from the transition elements whose charge you need to know are the zinc ion, which has a charge of 2+, and the silver ion, which has a charge of 1+. Although zinc and silver are transition elements, they only form one kind of ion, so their elemental names are sufficient for these ions. Table 3.6 lists the ions of the transition element we have discussed.

Table 3.6 Ions of Some Transition Elements

Ionic Charge		
1+	2+	3+
Cu^{+} (copper(I) ion, cuprous ion)	Cu^{2+} (copper(II) ion, cupric ion)	
	Fe^{2+} (iron(II) ion, ferrous ion)	Fe^{3+} (iron(III) ion, ferric ion)
Ag^{+} (silver ion)	Zn^{2+} (zinc ion)	

Example 3.5 Write the ionic symbols and the ionic names for the following elements:

1. magnesium
2. chlorine
3. iron

Solution

1. Mg^{2+} magnesium ion
2. Cl^- chloride ion
3. Fe^{2+} iron(II) ion, or ferrous ion
   Fe^{3+} iron(III) ion, or ferric ion

HEALTH NOTE

Some Important Ions in the Body

Table 3.7 lists several important ions in the body.

Table 3.7 Ions in the Body

Ion	Occurrence	Function	Source	Result of Too Little	Result of Too Much
Na^+	Principal cation outside the cell	Regulation and control of body fluids	Salt, seafoods, meats	Hyponatremia, anxiety, diarrhea, circulatory failure, decrease in body fluid	Hypernatremia, little urine, thirst, edema
K^+	Principal cation inside the cell	Regulation of body fluids and cellular functions	Bananas, orange juice, skim milk, prunes, meats	Hypokalemia (hypopotassemia), lethargy, muscle weakness, failure of neurological impulses	Hyperkalemia (hyperpotassemia), irritability, nausea, little urine, cardiac arrest
Ca^{2+}	Cation outside the cell; 90% of calcium in the body in bone as $Ca_3(PO_4)_2$ or $CaCO_3$	Major cation of bone; muscle smoothant	Milk, cheese, butter, meats, some vegetables	Hypocalcemia, tingling finger tips, muscle cramps, tetany	Hypercalcemia, relaxed muscles, kidney stones, deep bone pain, nausea
Mg^{2+}	Cation outside the cell; 70% of magnesium in the body in bone structure	Essential for certain enzymes, muscles, and nerve control	Widely distributed (part of chlorophyll of all green plants), nuts, grains	Disorientation, hypertension, tremors, slow pulse	Drowsiness
Cl^-	Principal anion outside the cell	Gastric juice, regulation of body fluids	Salt, seafoods, meats	Same as for Na^+	Same as for Na^+

3.4 Ionic Compounds

OBJECTIVE Using the ionic charges, write the correct formula of an ionic compound.

oxidation

reduction

In the formation of an ionic compound, electrons are transferred from the metal atoms to the nonmetal atoms. There are actually two events that occur in the formation of an ionic compound. The metal loses electrons; this is called *oxidation.* At the same time, the nonmetal gains electrons; this is called *reduction.* Since both reduction and oxidation must occur in the formation of an ionic compound, the whole process is also referred to as a *redox reaction.*

We can use diagrams to illustrate the formation of ionic compounds using a metal and a nonmetal such as sodium and chlorine. The loss of an electron (oxidation) from the sodium atom and the gain of an electron (reduction) by the chlorine atom results in the formation of oppositely charged ions:

$$Na \longrightarrow Na^+ + 1e^- \quad \text{(Oxidation)}$$

$$Cl + 1e^- \longrightarrow Cl^- \quad \text{(Reduction)}$$

ionic bond

An attraction called an *ionic bond* occurs between the positive and negative ions. The compound is called an ionic compound. See Figure 3.6.

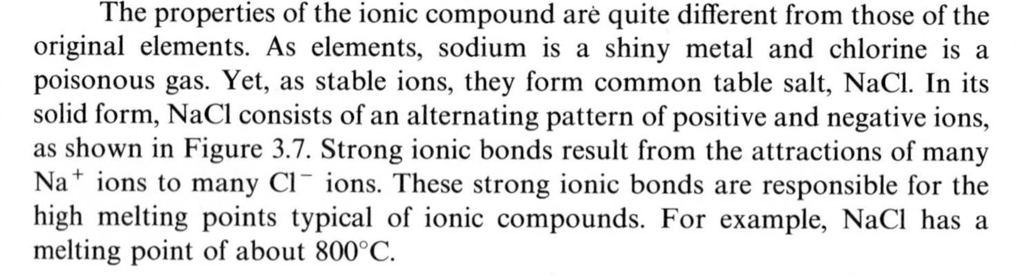

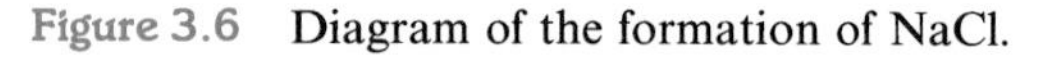

Figure 3.6 Diagram of the formation of NaCl.

NaCl is an ionic compound

The properties of the ionic compound are quite different from those of the original elements. As elements, sodium is a shiny metal and chlorine is a poisonous gas. Yet, as stable ions, they form common table salt, NaCl. In its solid form, NaCl consists of an alternating pattern of positive and negative ions, as shown in Figure 3.7. Strong ionic bonds result from the attractions of many Na^+ ions to many Cl^- ions. These strong ionic bonds are responsible for the high melting points typical of ionic compounds. For example, NaCl has a melting point of about 800°C.

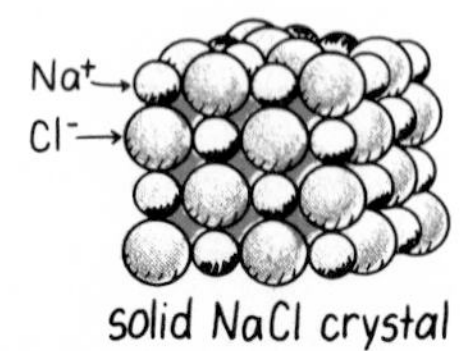

Figure 3.7 Sodium chloride and its solid structure.

Charge Balance in Ionic Compounds

The formula of an ionic compound indicates the number and kinds of ions that make up a unit of the ionic compound. In the diagram of the formation of NaCl, we see that the total positive charge is equal to the total negative charge. This means that the sum of the ionic charges is zero. In the NaCl formula, there is one sodium ion (Na^+) and one chloride ion (Cl^-). The sum of their ionic charges in the formula is zero:

$$\begin{aligned} NaCl &= \\ Na^{+} &= 1+ \\ Cl^{-} &= \underline{1-} \\ \text{Sum of charges} &= 0 \end{aligned}$$

Subscripts in Formulas

For many ionic compounds, the formula requires two or three positive or negative ions. For example, we can diagram the formation of an ionic compound of lithium and oxygen. First, we write the electron dot structures of each element:

$$\text{Li}\cdot \qquad :\ddot{\text{O}}\cdot$$

Then, we can determine the loss and gain of electrons that must take place for each to form octets. The electron dot structures indicate that lithium needs to lose one electron and that oxygen needs to gain two electrons to reach an octet:

$$(\text{Li}\cdot \longrightarrow \text{Li}^{+} + 1e^{-}) \times 2$$

$$:\ddot{\text{O}}\cdot + 2e^{-} = :\ddot{\underset{..}{\text{O}}}:^{2-}$$

In the diagram, we need to use two lithium atoms, which provide the two electrons needed by the oxygen atom. (See Figure 3.8.)

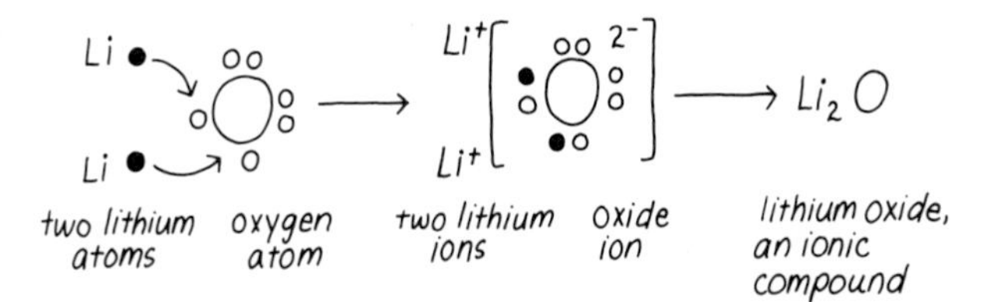

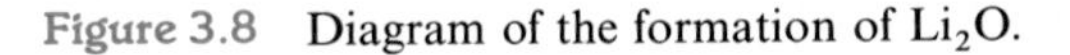
Figure 3.8 Diagram of the formation of Li_2O.

When more than one atom of a particular element is needed in the formation of an ionic compound, that number is shown as a *subscript* following the symbol of that element in the formula. The number 1 does not appear as a subscript; it is understood. Make sure that you understand the reason for each subscript in the ionic formula. Thus, we can write the final formula as Li_2O.

subscript

Additional diagrams for the formation of some ionic compounds and their formulas are seen in Table 3.8.

Writing Ionic Formulas from Ionic Charges

The diagram helps us visualize the transfer of electrons, but you should now be able to write an ionic formula directly from the ionic charges of the ions in the compound. Remember that an ionic formula never shows the charges of the

Table 3.8 Formation of Some Ionic Compounds

Atoms	Ions	Formula	Name
Ca → 2 ·I:	Ca^{2+}, 2 :I:$^-$	CaI_2	Calcium iodide
	Net charge: $1(+2) + 2(-1) = 0$		
·Al· → 3 ·Cl:	Al^{3+}, 3 :Cl:$^-$	$AlCl_3$	Aluminum chloride
	Net charge: $1(+3) + 3(-1) = 0$		
2 Na· → ·S:	2 Na^+, :S:$^{2-}$	Na_2S	Sodium sulfide
	Net charge: $2(+1) + 1(-2) = 0$		

ions. However, it represents the combination of positive ions and negative ions whose sum is zero.

Example 3.6 Write the formula for the ionic compound formed between magnesium ions (Mg^{2+}) and chloride ions (Cl^-).

Solution

Step 1. Write the ions of magnesium and chloride:

Mg^{2+} and Cl^-

Step 2. Determine the number of these ions needed for charge balance. The charge (2+) on the magnesium ion can be balanced by a negative (2−) charge. Since a chloride ion has only a 1− charge, two chloride ions are needed to provide a total 2− charge:

Ions:	Mg^{2+}	Cl^- Cl^-
Overall charge:	2+	2− = 0

Step 3. Write the correct formula (using subscripts) to indicate the use of two chloride ions:

$$Mg^{2+} + 2Cl^- = MgCl_2$$

Magnesium chloride

Example 3.7 Write the formula for a compound formed by iron(III) ions and oxide ions.

Solution

Step 1. Write the ions for iron(III) and oxide ion:

Fe^{3+} and O^{2-}

Step 2. Determine the number of positive ions and negative ions that will balance the charge. (When charges are balanced, their sum is zero.) In this case, we need two Fe^{3+} ions and three oxide ions:

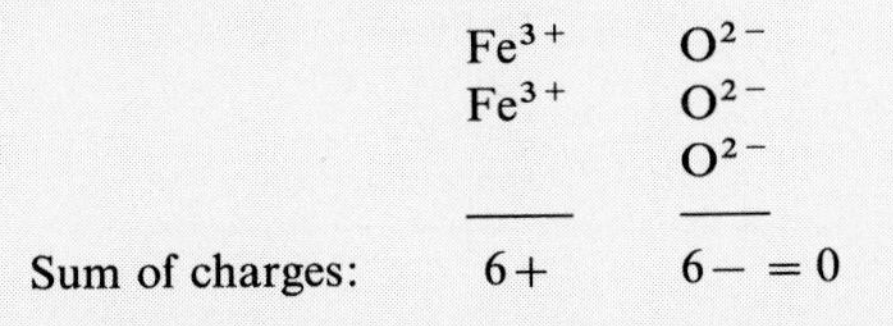

Step 3. Write the formula of the ionic compound. The use of two iron ions is shown by the subscript 2 after the symbol Fe; the three oxide ions are shown by the subscript 3 after the symbol for oxygen:

Fe_2O_3
Iron(III) oxide,
or ferric oxide

Polyatomic Ions

A *polyatomic ion* is a group of atoms with an overall electrical charge. This is a special kind of ion that consists of several atoms with an overall electrical charge. Most polyatomic ions are composed of a nonmetal such as phosphorus, sulfur, carbon, or nitrogen along with several oxygen atoms. All but one that we will discuss here are negatively charged. That means that the group gained one or more electrons to complete the octets of the atoms in the group.

We can diagram the hydroxide ion using its valence electrons as a guide. There is a shared pair of electrons between the oxygen atom and the hydrogen atom. This is called a *covalent bond*, which you will study in the next section. If we draw only the valence electrons, the oxygen atom does not have an octet. The octet for oxygen is completed when the OH group gains an electron. (The electron would come from a metal that is losing electrons.) The gain of one electron gives an ionic charge (1−) to the OH group. The resulting OH^- is called the hydroxide ion, which is one of the polyatomic ions:

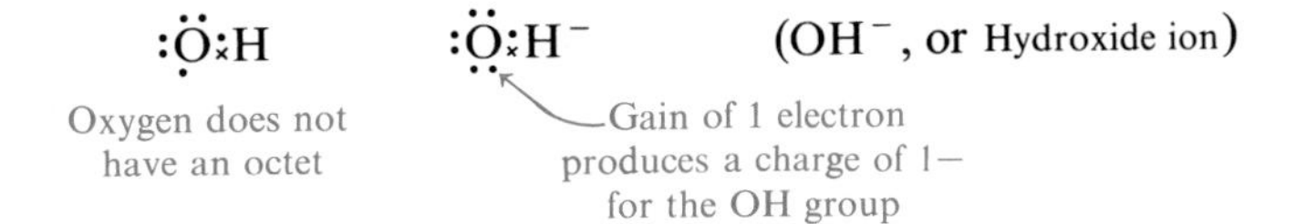

The most common of the negatively charged polyatomic ions have names that end in *-ate*. Those ions having one oxygen less than the common form have

Table 3.9 Some Important Polyatomic Ions

Ionic Charge	Polyatomic Ion
1+	NH_4^+ (ammonium ion)
1−	HCO_3^- (hydrogen carbonate ion, or bicarbonate)
	NO_3^- (nitrate ion)
	OH^- (hydroxide ion)
	$C_2H_3O_2^-$ (acetate ion)
	HSO_4^- (hydrogen sulfate, or bisulfate)
	NO_2^- (nitrite ion)
	CN^- (cyanide ion)
2−	CO_3^{2-} (carbonate ion)
	SO_4^{2-} (sulfate ion)
	SO_3^{2-} (sulfite ion)
3−	PO_4^{3-} (phosphate ion)

names that end in *-ite*. The endings -ate or -ite will help you to recognize a polyatomic ion in a name. The hydroxide ion and cynanide ion are exceptions to this naming pattern. The only positively charged polyatomic ion we will use is the ammonium ion. Several important polyatomic ions are listed in Table 3.9. You will need to memorize the number of oxygen atoms and the charge associated with each polyatomic ion.

We must make it very clear that no polyatomic ion can exist by itself. It must be associated with ions of the opposite charge. The bonding between polyatomic ions and other ions is one of electrical attraction and thus is an ionic bond. For example, the compound sodium hydroxide consists of a sodium ion (Na^+) and a hydroxide ion (OH^-). They are held together by an ionic bond as a result of the attractions between the positive and negative charges:

$$NaOH = Na^+ \underset{\text{bonds}}{\overset{\text{Ionic}}{\longleftrightarrow}} OH^-$$

Writing Formulas with Polyatomic Ions

The writing of a correct formula with polyatomic ions follows the same rules of charge balance as formulas of any ionic compound. The total negative charge must be equal to the total positive charge. For example, consider the formula for sodium nitrate. The ions in this compound are

	Na^+	NO_3^-
	Sodium ion	Nitrate ion
Charge:	1+	1−

Since one ion of each will provide a charge balance, the formula can be written as

$NaNO_3$
Sodium nitrate

When more than one polyatomic ion is needed, parentheses are used to enclose the formula of the ion, and the subscript to balance the charge is written just after the closing parenthesis. Consider the formula for calcium nitrate. The ions in this compound are the calcium ion and the nitrate ion, a polyatomic ion:

Ca^{2+}	NO_3^-
Calcium ion	Nitrate ion

To provide as much negative charge as positive charge, two nitrate ions must be used. The formula including the parentheses around the nitrate ion is

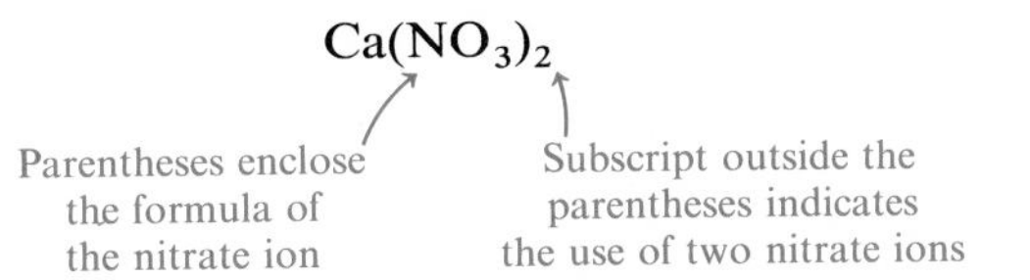

Example 3.8 Write the formula for aluminum hydroxide.

Solution: The name of this compound indicates the presence of the metal aluminum, whose ionic form is Al^{3+}, and the hydroxide ion, which is a polyatomic ion OH^-:

Ions:	Al^{3+}	OH^-	
	Aluminum ion	Hydroxide ion	
Number of ions needed for charge balance:	Al^{3+}	OH^- OH^- OH^-	
Sum of charges:	3+	3−	= 0
Formula:	$Al(OH)_3$		

Table 3.10 lists some ionic compounds with polyatomic ions commonly used in medicine and industry.

Table 3.10 Some Ionic Compounds with Polyatomic Ions

Formula	Compound Name	Use
$BaSO_4$	Barium sulfate	Radiopaque medium
$CaCO_3$	Calcium carbonate	Antacid; calcium supplement
$Ca_3(PO_4)_2$	Calcium phosphate	Calcium replenisher
$CaSO_3$	Calcium sulfite	Preservative in cider and fruit juices
$CaSO_4$	Calcium sulfate	Plaster casts
$AgNO_3$	Silver nitrate	Topical anti-infective
$NaHCO_3$	Sodium bicarbonate	Antacid
$Zn_3(PO_4)_2$	Zinc phosphate	Dental cements
$FePO_4$	Iron(III) phosphate	Food and bread enrichment
K_2CO_3	Potassium carbonate	Alkalizer, diuretic
$Al_2(SO_4)_3$	Aluminum sulfate	Antiperspirant, anti-infective
$AlPO_4$	Aluminum phosphate	Antacid
$MgSO_4$	Magnesium sulfate	Cathartic, "epsom salts"

HEALTH NOTE

Polyatomic Ions in Bone and Teeth

Bone structure consists of two parts: (1) a solid mineral material and (2) a phase consisting primarily of collagen proteins. The mineral substance is a compound called *hydroxyapatite*, a solid formed from calcium ions, phosphate ions, and hydroxide ions. This material is deposited in the web of collagen to form a very durable bone material:

$Ca_{10}(PO_4)_6(OH)_2$
Hydroxyapatite

In most individuals, bone material is continuously being absorbed and re-formed. After age 40, *osteoporosis* may occur, a condition in which more bone material is lost than formed. Bone mass reduction occurs at a faster rate in women than in men and occurs at different rates in different parts of the body skeleton. The reduction in bone mass can be as much as 50 percent over a period of 30 to 40 years. It is recommended that persons over 35, especially women, take a daily calcium supplement in their diet.

3.5 Naming Ionic Compounds

OBJECTIVE Given the formula of an ionic compound, write the correct name; given the name, write the correct formula.

In the name of an ionic compound, the name of the positive ion is written first; the name of the negative ion is written second. For compounds with only two different elements (binary), the positive ion (metal) is named as the element; the negative ion (nonmetal) is named by replacing the end of its elemental name with *-ide*.

To name an ionic compound with three elements (ternary), the polyatomic ion must be recognized and named by its proper name. We looked at the names of the common polyatomic ions in Section 3.4. The metal ion is named first, followed by the name of the polyatomic ion if it is the negative ion. Then the name usually ends in *-ate* or *-ite*.

Subscripts are never mentioned in the names of ionic compounds: they are understood as a result of the charge balance of the ions in the compound:

Ionic formula:	$MgCl_2$ (binary)		Na_2SO_4 (ternary)	
Ions:	Metal ion	Nonmetal ion	Metal ion	Polyatomic ion
	Mg^{2+}	Cl^-	Na^+	SO_4^{2-}
Name:	Magnesium	chloride	Sodium	Sulfate

Table 3.11 lists some examples of the formulas and names of some ionic compounds.

Table 3.11 Formulas and Names of Some Ionic Compounds

Formula (Binary)	Name	Formula (Ternary)	Name
MgS	Magnesium sulfide	$MgSO_4$	Magnesium sulfate
$CaBr_2$	Calcium bromide	$Ca(NO_3)_2$	Calcium nitrate
$AlCl_3$	Aluminum chloride	$Al_2(CO_3)_3$	Aluminum carbonate
Na_3P	Sodium phosphide	Na_3PO_4	Sodium phosphate
Cu_2S	Copper(I) sulfide (cuprous sulfide)	$Fe_2(SO_4)_3$	Iron(III) sulfate (ferric sulfate)

Naming Ionic Compounds with a Transition Metal Ion

When the positive ion is a transition metal such as iron or copper, the ionic charge of the metal ion must be included in the name for the positive ion. Recall that iron and copper can form two possible ions each.

Compound	*Name*
$FeCl_2$	Iron(II) chloride, or ferrous chloride
$FeCl_3$	Iron(III) chloride, or ferric chloride
Cu_2S	Copper(I) sulfide, or cuprous sulfide
$CuCl_2$	Copper(II) chloride, or cupric chloride

Example 3.9 Write the name of the following ionic compounds:

1. Na_2O
2. $Mg(HCO_3)_2$
3. NH_4Cl

Solution

1. The metal ion in the formula is the sodium ion (Na^+); the nonmetal ion is the oxide ion (O^{2-}):

   Sodium oxide

2. The metal ion is the magnesium ion (Mg^{2+}). The remaining part of the formula represents the polyatomic ion HCO_3^-, which is the hydrogen carbonate, or bicarbonate, ion. Note that there are more than three elements in the formula, but that the naming proceeds as it does for ternary compounds:

   Magnesium hydrogen carbonate (magnesium bicarbonate)

3. The positive ion in this formula is the polyatomic ion NH_4^+, known as the *ammonium ion.* The negative ion is the chloride ion (Cl^-):

   Ammonium chloride

Example 3.10 Write the name for the compound whose formula is FeO.

Solution: Since iron is a transition metal with more than one possible charge, it is necessary to determine its particular charge in this compound. To do this, we use the oxide ion. We know that an oxide ion has a charge of 2−. To keep a balance of charge for the formula, the overall positive charge in the compound must be 2+. Since there is only one iron ion in the formula, it must have an ionic charge of 2+:

		Fe	O
Charge of oxide ion:			2−
Charge balance:		2+	2−
Charge of Fe ion:	1 Fe =	2+	

The name of the Fe^{2+} ion is iron(II), or ferrous ion. The compound FeO is named

Iron(II) oxide, or Ferrous oxide

Example 3.11 Write the name of the compound whose formula is Cu_3PO_4.

Solution: Copper is a transition element and can form positive ions with two different charges. We will use the known charge of the phosphate ion, 3−, in order to determine the charge of the copper ion used in this compound. From the charge balance, we see that the total positive charge has to be 3+. This is divided among the three copper ions. Therefore, each copper ion must have a charge of 1+:

		Cu_3	PO_4
Charge of phosphate ion:			3−
Charge balance:		3+	3−
Charge of Cu ion:	3 Cu =	3+	
	1 Cu =	1+	
Positive ion:	=	Cu^+	

The name of Cu^+ is copper(I) ion, or cuprous ion. The name of the compound Cu_3PO_4 is

Copper(I) phosphate, or Cuprous phosphate

Summary of Naming Positive Ions with Variable Valences

1. Assign the known ionic charge to the negative ion.
2. Multiply the negative ion by any subscript to find the total negative charge.
3. State the total positive charge that balances the negative charge.
4. Assign a positive charge to the metal ions. If there is a subscript, divide by the number of positive ions first.
5. Name the compound including the charge of the positive ion as a roman numeral; or use the appropriate (*-ous* or -ic) ending.

Writing an Ionic Formula from the Name of an Ionic Compound

When you are given the name of a compound, you can use it to write the formula. The first part of the name tells you about the metal ion; the second part of the name tells you about the nonmetal ion. The correct formula is written by using subscripts to indicate the number of positive and negative ions that give a net charge of zero.

Example 3.12 Write the formula for lithium sulfate.

Solution

Step 1. Write the positive ion and the negative ion from the name:

Lithium	sulfate
Li^+	SO_4^{2-}

Step 2. Balance the charges to make the overall charge of the formula equal 0. We need two lithium ions (2+ charge) to balance one sulfate ion (2− charge):

Ions needed to balance:	Li^+ Li^+	SO_4^{2-}
Total charge:	$2(+)$	$2-$
	$2(+)$ +	$2(-) = 0$

Step 3. Write the correct formula of the compound:

Li_2SO_4

Example 3.13 Write the correct formula of iron(III) chloride.

Solution

Step 1. Write the ions. The roman numeral III indicates that the iron ion in this compound has a 3+ charge:

	Iron(III)	chloride
Ions:	Fe^{3+}	Cl^-

Step 2. Balance the charges:

Ions needed to balance:	Fe^{3+}	Cl^- Cl^- Cl^-
Total charge:	$1(3+)$	$3(-)$
	$3(+)$ +	$3(-) = 0$

Step 3. Write the formula of the compound:

$FeCl_3$

3.6 Covalent Bonds

OBJECTIVE Use electron dot structures to diagram the formation of covalent bonds in a covalent compound.

covalent bonds are shared electrons

Two nonmetals form a compound through *covalent bonding*, which is the sharing of electrons. In a covalent bond, two atoms share one or more of their valence electrons to form octets. Covalent bonds occur in compounds composed of two nonmetals from Groups IVA, VA, VIA, and VIIA and hydrogen. (Recall that hydrogen is a nonmetal even though it is listed in Group IA.) If we look at the number of valence electrons in atoms of nonmetals, we see that there are four valence electrons or more. That is too many electrons for an atom to lose. (Hydrogen only has one valence electron, but it is held very tightly.) Therefore, ionic bonds are not possible when two nonmetals combine. Nonmetals must bond through a sharing of electrons until an octet for each is achieved.

In order to diagram a covalent compound, we need to know the number of electrons needed by each atom to become stable. Let us consider the formation of a covalent bond between two hydrogen atoms. A hydrogen atom is stable, like its nearest noble gas helium, when the first valence shell is filled with two electrons. By sharing a pair of electrons, a covalent bond is formed:

$$H\cdot + H\cdot \longrightarrow \quad H{:}H \quad H{-}H \quad H_2$$

Shared pair of electrons (H:H); Covalent bond (H—H)

A shared pair of electrons is called a *covalent bond.* It is represented as a pair of electrons shared between two atoms or as a dash between the two atoms. When atoms are bonded together through covalent bonding, the resulting unit is called a *molecule.*

The number of electrons that a nonmetal needs to share can be determined from its group number or number of valence electrons:

Group	*Elements*	*Valence Electrons*	*Electrons Needed*	*Bonds Formed*
IA	H	$1e^-$	$1e^-$	1
IVA	C, Si	$4e^-$	$4e^-$	4
VA	N, P	$5e^-$	$3e^-$	3
VIA	O, S	$6e^-$	$2e^-$	2
VIIA	F, Cl, Br, I	$7e^-$	$1e^-$	1

For example, a chlorine molecule, Cl_2, consists of two chlorine atoms. Each chlorine atom has seven valence electrons. By sharing a pair of electrons, each of the chlorine atoms achieves an octet:

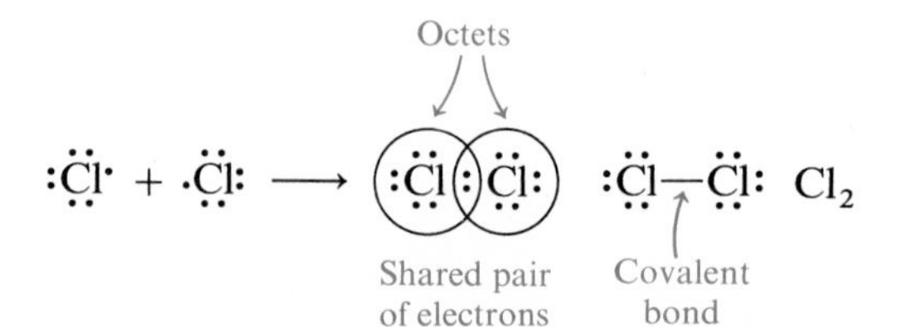

Diatomic Molecules

Hydrogen (H_2) and chlorine (Cl_2) are just two of several nonmetals that form diatomic (two-atomed) molecules by covalent bonding. A summary of elements that form diatomic molecules is listed below:

Element	*Diatomic Molecule*	*Name*
H	H_2	Hydrogen
F	F_2	Fluorine
N	N_2	Nitrogen
O	O_2	Oxygen
Cl	Cl_2	Chlorine
Br	Br_2	Bromine
I	I_2	Iodine

Molecules with Several Covalent Bonds

Most molecules have more than one covalent bond. When there are two or more atoms attached to a central atom, each bonded atom will be sharing electrons with the central atom. Therefore, it is possible to have several covalent bonds in a molecule.

For example, carbon in Group IVA has four valence electrons and requires four more electrons to reach an octet. Therefore carbon must share four electrons. Hydrogen needs to share one electron to reach a stable configuration. Below we show the diagram of the compound formed when four hydrogen atoms share electrons with one carbon atom:

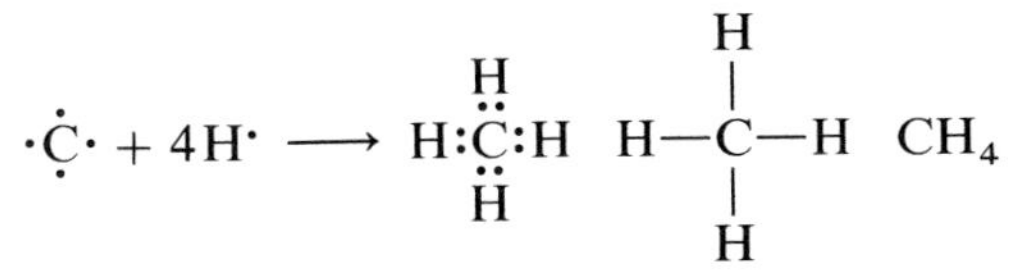

$$\cdot\dot{\underset{\cdot}{\mathrm{C}}}\cdot + 4\mathrm{H}\cdot \longrightarrow \begin{array}{c}\mathrm{H}\\ \cdot\cdot\\ \mathrm{H{:}C{:}H}\\ \cdot\cdot\\ \mathrm{H}\end{array} \quad \begin{array}{c}\mathrm{H}\\ |\\ \mathrm{H—C—H}\\ |\\ \mathrm{H}\end{array} \quad \mathrm{CH_4}$$

Example 3.14 Water has the formula H_2O. Draw the electron dot diagram for the covalent compound.

Solution

Step 1. Write the electron dot structures: $\mathrm{H}\cdot \quad :\ddot{\underset{\cdot}{\mathrm{O}}}\cdot$

Step 2. Determine the number of electrons that each element needs to share. Hydrogen needs one more electron and oxygen needs to share two electrons.

Step 3. Draw the electron dot structure:

$$\begin{array}{c}\mathrm{H{:}\ddot{\underset{\cdot\cdot}{O}}{:}}\\ \mathrm{H}\end{array} \qquad \begin{array}{c}\mathrm{H—\ddot{O}{:}}\\ |\\ \mathrm{H}\end{array}$$

Shared pairs — Covalent bonds

The electron dot diagrams for several molecules are shown in Table 3.12.

Table 3.12 Examples of Some Covalent Compounds

Group Number:	H(IA)	IVA	VA	VIA	VIIIA
Electrons:	$1e^-$	$4e^-$	$5e^-$	$6e^-$	$7e^-$
Dot Structure:	H·	·C·	·N̈·	:S̈·	:C̈l·
Examples:	H:H	H:C̈:H (with H above and below C)	H:N̈:H (with H below N)	:S̈:C̈l: / :C̈l:	:C̈l:C̈l:
	H—H	H—C—H (with H above and below C)	H—N̈—H (with H below N)	:S̈—C̈l: / | / :C̈l:	:C̈l—C̈l:
	H_2	CH_4	NH_3	SCl_2	Cl_2

Example 3.15 Write the electron dot structure for NCl_3.

Solution: The electron dot structures indicate that nitrogen has five valence electrons and chlorine has seven:

·N̈· :C̈l·

To form an octet, nitrogen has to share three electrons. Each atom of chlorine will share one electron to form an octet:

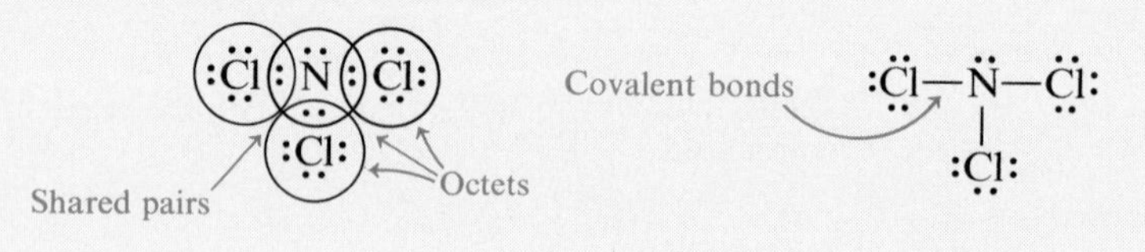

Multiple Bonds

Up to now, we have shown how an octet forms by sharing a single pair of electrons. Sometimes, in covalent compounds, atoms must share two pairs or three pairs of electrons to acquire an octet for each atom. A double pair of electrons is called a *double bond*; a triple pair of electrons is called a *triple bond*.

double bond, triple bond

We can determine the existence and positions of the double or triple bonds when we find that drawing all single covalent bonds does not complete the octets of all the atoms in the covalent compound. For example, we can try to diagram the covalent compound C_2H_4. The skeleton structure of atoms is as follows:

```
H  C  C  H
   H  H
```

If we draw in the electrons as single bonds, we find that the carbon atoms do not have complete octets:

```
H:Ċ:Ċ:H
  ·· ··
  H  H
```

Since the hydrogen atoms are complete with two electrons, the only change we can make is to double up the valence electrons on the carbon atoms.

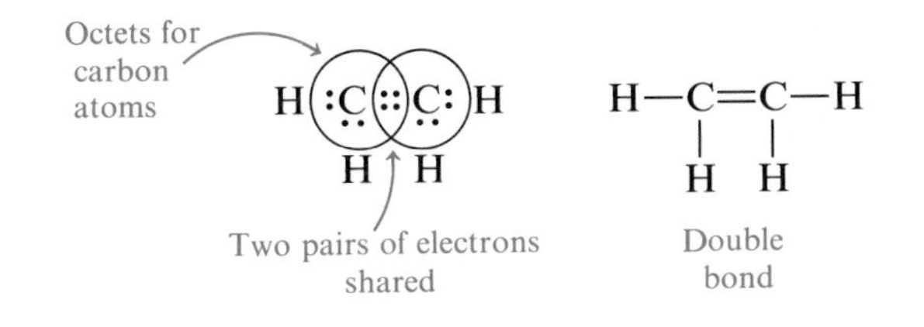

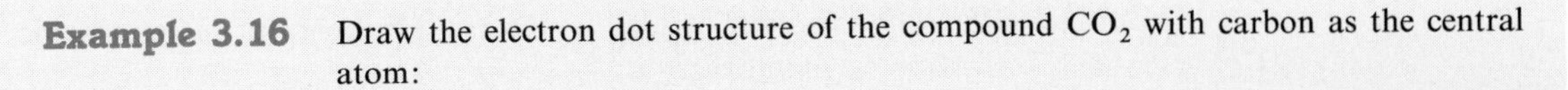

Example 3.16 Draw the electron dot structure of the compound CO_2 with carbon as the central atom:

O C O

Solution: Each oxygen atom needs two electrons, and the carbon atom needs four electrons. Octets are achieved for all the atoms when each oxygen atom shares two pairs of electrons with the carbon atom:

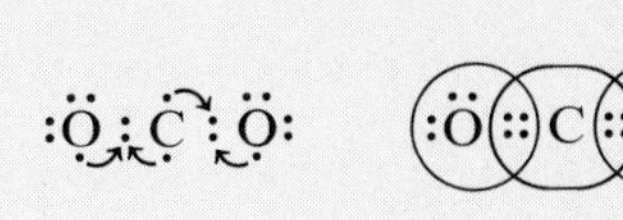

Example 3.17 Write the electron dot structure for nitrogen, N_2.

Solution: Each nitrogen atom has to share three electrons to achieve octets in the valence shells. The sharing of three pairs of electrons between the same two atoms is called a *triple bond*:

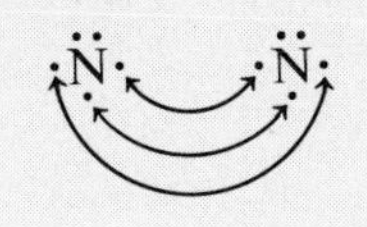

N:::N :N≡N:

Three shared pairs = Triple bond

3.7 Naming Covalent Compounds

OBJECTIVE Given the formula of a covalent compound, write its correct name; given the name of a covalent compound, write its formula.

prefixes In the names of covalent compounds, *prefixes* indicate the number of atoms of each element in the compound. The prefixes refer to the subscripts that are shown in the formula of the compound.

Some Prefixes Used in Naming Covalent Compounds

mono-	one
di-	two
tri-	three
tetra-	four
penta-	five

To name a covalent compound, the element that appears first in the formula uses its elemental name; the second element uses the root of its elemental name followed by *-ide*. In the naming of covalent compounds, prefixes are used to indicate the number of atoms of each element. The prefix to use is determined by the subscripts in the formula:

Naming Some Covalent Compounds

CO	Carbon monoxide*	CO_2	Carbon dioxide
PCl_3	Phosphorous trichloride	P_2O_5	Diphosphorous pentoxide
CCl_4	Carbon tetrachloride	N_2O	Dinitrogen monoxide

Example 3.18 Write the names of the following covalent compounds:

1. NCl_3
2. N_2O_4

Solution

1. The first element is nitrogen, a nonmetal. Since there is one atom of nitrogen, the name is given as nitrogen. (It could be mononitrogen, but the prefix mono- is usually dropped from the name of the first element.) The three chlorine atoms are indicated by the prefix tri- in the name of the second element trichloride.

   Nitrogen trichloride

2. The first element is nitrogen. Since there are two nitrogen atoms, it is named dinitrogen. The four oxygen atoms are named tetroxide. When the vowels o-o or a-o appear together, the vowel ending of the prefix is dropped:

   Dinitrogen tetroxide

* The prefix *mono-* is understood for the first element and may be omitted.

Example 3.19 Write the formulas of the following covalent compounds from their names:

1. Sulfur dichloride
2. Iodine pentafluoride

Solution

1. The first element is sulfur. Since there is no prefix, there is one atom of sulfur. The second element in the formula is chlorine. The prefix *di-* indicates two atoms of chlorine, so a subscript 2 appears in the formulas:

   SCl_2

2. The first element in the formula is iodine. The second element is fluorine. The prefix *penta-* indicates that there are five atoms of fluorine. A subscript 5 follows the symbol for fluorine:

   IF_5

Table 3.13 lists the formulas, names, and commercial uses of some covalent compounds.

Table 3.13 Some Common Covalent Compounds

Formula	Name	Uses
CS_2	Carbon disulfide	Manufacture of rayon
CO_2	Carbon dioxide	Carbonation of beverages, fire extinguishers, propellant in aerosols, dry ice
SiO_2	Silicon dioxide	Manufacture of glass
NCl_3	Nitrogen trichloride	Bleaching of flour in some countries (prohibited in United States)
SO_2	Sulfur dioxide	Preserving fruits, vegetables; disinfectant in breweries, bleaching textiles
SO_3	Sulfur trioxide	Manufacture of explosives
SF_6	Sulfur hexafluoride	Electrical circuits
P_2S_5	Diphosphorus pentasulfide	Manufacture of safety matches
ClO_2	Chlorine dioxide	Bleaching paper pulp, flour, leather
ClF_3	Chlorine trifluoride	Rocket propellant

3.8 Polarity

OBJECTIVE Identify a bond as polar covalent, nonpolar covalent, or ionic.

The *polarity* of a covalent bond is determined by the attraction of each atom in the bond for a shared pair of electrons. The tendency of an element to attract the shared electrons in a bond is called *electronegativity*. The element fluorine, which has the greatest attraction for a shared pair of electrons, is the most

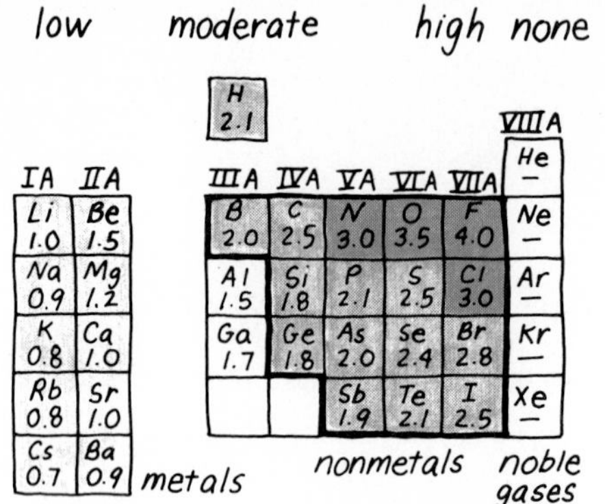

Figure 3.9 Regions of electronegativity in the periodic table.

electronegative element. It has the highest value (4.0) on the *Pauling electronegativity scale.* This is a relative scale derived from some measured properties of atoms and is calculated slightly differently, by different people. Some typical electronegativity values are given in Figure 3.9.

nonpolar covalent bonding

:Cl : Cl: H : H

shared equally shared equally

Figure 3.10 Equal sharing of electrons between identical atoms.

When a covalent bond occurs between two different nonmetals, there is an unequal sharing of the electron pair. The shared electron pair will be pulled closer to the atom with the higher electronegativity. For example, the pair of electrons of the covalent bond between hydrogen (H) and chlorine (Cl) is pulled closer to the chlorine. This gives the chlorine atom a slight increase in negative charge; we say it has a partial negative charge. The shared pair of electrons is slightly removed from the hydrogen atom to give hydrogen a partial positive charge. The Greek letter delta, δ, is used to indicate the partial charges as delta positive, δ^+, and delta negative, δ^-:

$$\overset{\delta^+}{\text{H}}\ \overset{\delta^-}{\text{Cl}}$$

electrons in a polar covalent bond are shared unequally

dipole

This type of covalent bond where electrons are shared unequally is called a *polar covalent bond.* A separation of charge in a covalent bond that gives a positive and negative pole is also called a *dipole.* The partial charges may also be indicated by an arrow pointing toward the more electronegative atom:

$$\overset{+\!\longrightarrow}{\text{H—Cl}}$$

electrons in a nonpolar covalent bond are shared equally

If the two atoms in a covalent bond are identical, there is no electronegativity difference. Both atoms have the same attraction for electrons, so the shared electron pair is shared equally. Such a covalent bond is called a *nonpolar covalent bond.* For example, the covalent bonds in H_2 and Cl_2 are nonpolar bonds (see Figure 3.10).

	Electronegativity Difference	*Type of Covalent Bond*
H_2:	$(2.1 - 2.1) = 0$	Nonpolar
Cl_2:	$(3.0 - 3.0) = 0$	Nonpolar

The sharing of electrons in covalent compounds is compared for nonpolar covalent bonds and polar covalent bonds in Figure 3.11.

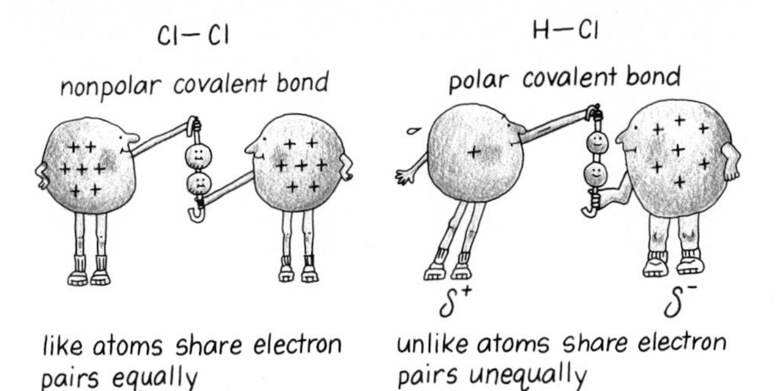

Figure 3.11 Nonpolar covalent bonds compared to polar covalent bonds.

Example 3.20 Indicate the polarity of the bond as polar or nonpolar in the following compounds. If polar, indicate the partial charges with δ^+ and δ^-:

1. HBr
2. F_2

Solution

1. The covalent compound HBr has a covalent bond between two different nonmetals. The bond is polar. It has an electronegativity difference (2.8 – 2.1) of 0.7. The bromine atom has the greater electronegativity, so bromine will be partially negative in the polar bond:

$$\overset{\delta^+}{H}:\overset{\delta^-}{\ddot{\underset{\cdot\cdot}{Br}}}: \qquad \overset{\delta^+}{H}—\overset{\delta^-}{Br}$$

2. The covalent compound F_2 has a covalent bond between identical atoms. The electronegativity difference (4.0 – 4.0) is zero, and the covalent bond is nonpolar:

$$:\ddot{\underset{\cdot\cdot}{F}}:\ddot{\underset{\cdot\cdot}{F}}: \qquad F—F$$

A Review of Types of Bonding

The types of bonds between atoms range from the nonpolar covalent bond, where electrons are shared equally, all the way to the ionic bond, where electrons are transferred from one atom to another. This variation is continuous; there is no definite division where one type of bond stops and the next starts. However, for purposes of discussion, we will use some general rules to help predict the type of bond that would be expected between two elements. (See Figure 3.12.) A summary of general rules for predicting bond type is given in the list on the following page.

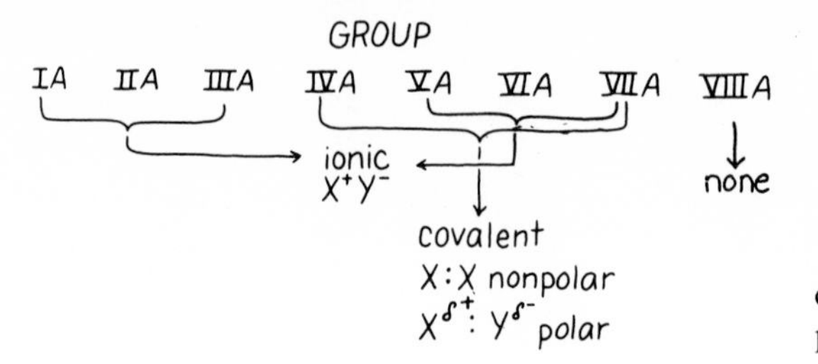

Figure 3.12 Types of bonding expected between groups in the periodic table.

Some General Rules for Predicting Bond Type

1. A bond between a metal and a nonmetal is usually ionic.
2. A bond between two nonmetals is covalent.
   a. A covalent bond between two atoms of the same element is nonpolar.
   b. A covalent bond between two atoms of different nonmetals is polar.

Examples of the different bonding types are given in Table 3.14.

Table 3.14 Examples of Different Types of Bonding

Elements to Bond	Electron Dot Structure		Types of Elements	Types of Bonding	Molecule or Ionic Unit	Formula
F and F	:F̤̈·	:F̤̈·	Two nonmetals (same element)	Covalent (nonpolar)	:F̤̈:F̤̈:	F_2
Na and F	Na·	:F̤̈·	Metal, nonmetal	Ionic	Na^+:F̤̈:	NaF
P and Cl	·P̤̈·	:C̤̈l·	Two nonmetals (different elements)	Covalent (polar)	:C̤̈l—P̈—C̤̈l: with :C̤̈l: bonded below P	PCl_3
Mg and S	Ṁg·	:S̤̈·	Metal, nonmetal	Ionic	Mg^{2+}:S̤̈:$^{2-}$	MgS

Example 3.21 Predict the type of bond between each of the following pairs of elements:
1. Ca and Cl
2. P and S
3. Br and Br

Solution

1. A bond between a metal and a nonmetal would be ionic.
2. A bond between two different nonmetals is a polar covalent bond.
3. A bond between atoms of the same element gives an electronegativity difference of zero. The Br–Br bond would be a nonpolar covalent bond.

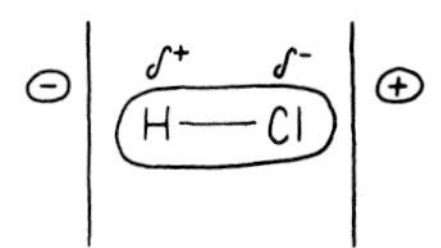

Figure 3.13 Orientation of HCl, a polar molecule, in an electric field.

Polar and Nonpolar Molecules

We have seen that the bond of HCl is a polar covalent bond. Since the HCl molecule has a single dipole, we say that HCl is a *polar molecule*. A polar molecule such as HCl is affected by an electric field; the dipoles of the molecules line up with the opposite charges of the field. (See Figure 3.13.)

However, if a diatomic molecule consists of two atoms of the same element, such as H—H or Cl—Cl, the covalent bond is nonpolar and there is no dipole. A molecule with nonpolar bonds is a *nonpolar molecule* and is not affected by an electric field.

Polarity of Molecules with Two or More Dipoles

In molecules with two or more polar bonds, we refer to the *central atom*, the atom to which all the other atoms in the molecule are attached. Sometimes, there are pairs of electrons on the central atom that are unshared (not bonded to another atom); they are called *lone pairs*. The shared pairs and the lone pairs of electrons about the central atom determine the overall polarity of the molecule.

nonpolar molecule

Pairs of electrons are arranged around the central atom to give the greatest distance between the electron pairs. This reduces the repulsion that exists between the electrons. When all the electrons about the central atom are shared (no lone pairs), the molecule will be *nonpolar*. The dipoles are balanced, and the effect of the dipoles on the molecule cancels out. For example, in the compound CO_2, there are two oxygen atoms bonded to the central carbon atom by double bonds. All of the electron pairs of the central atom are shared; there are no lone pairs on the carbon atom. The dipoles line up in opposite directions, and the molecule is nonpolar:

:Ö::C::Ö: O=C=O

Dipoles in opposite directions

There are also nonpolar molecules with three and four atoms attached to the central atom. Once again, all of the electrons of the central atom are shared, and the dipoles of the individual polar bonds cancel each other out:

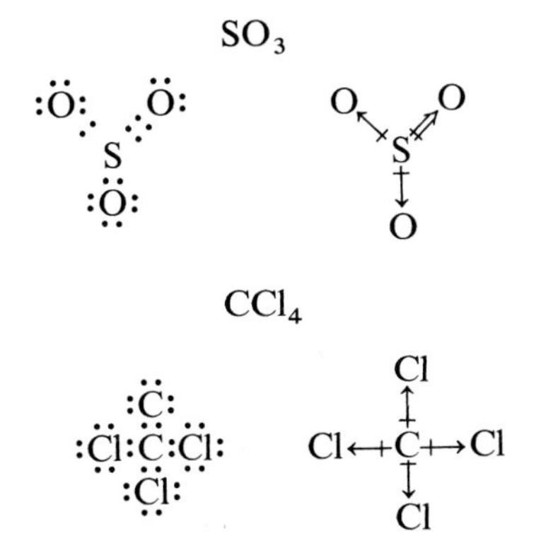

If a central atom consists of a combination of shared pairs and unshared pairs of electrons, the molecule will be *polar*. There is an overall charge separation for the whole molecule because the dipoles do not cancel out. For example, the electron dot diagrams for ammonia (NH_3) and water (H_2O) both indicate lone pairs of electrons:

polar molecule

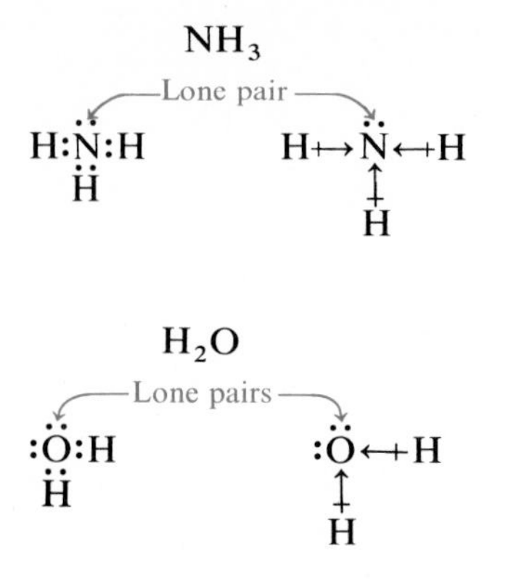

The topic of polar and nonpolar molecules is much more complex than we have discussed. However, in this text we have considered only molecules with one, two, three or four bonds of identical atoms attached to a central atom. We have not indicated the shapes of the molecules but have described a method for determining whether a molecule is polar or nonpolar. Polar molecules are attracted to other polar molecules but not to nonpolar molecules. This has an important effect upon later topics such as solutions and the compatibility of drugs. Table 3.15 summarizes this information.

Table 3.15 Polarity of Molecules

Shared Pairs	Lone Pairs	Type of Molecule	Examples
Nonpolar Molecules with Nonpolar Bonds			
—	—	Nonpolar	H_2, Cl_2, N_2
Nonpolar Molecules with Polar Bonds			
2	0	Nonpolar	CO_2
3	0	Nonpolar	SO_3
4	0	Nonpolar	CH_4, CCl_4
Polar Molecules with Polar Bonds			
1	—	Polar	HCl, HF
2	1	Polar	SO_2
2	2	Polar	H_2O, SCl_2
3	1	Polar	NH_3, PCl_3

Example 3.22 For the following molecules, write the electron dot diagram and state whether the molecule is polar or nonpolar.

1. F_2
2. PCl_3

Solution

1. The diatomic molecule of fluorine consists of two atoms of the same element. The covalent bond is nonpolar; the molecule is nonpolar:

:F:F: F—F

No dipole

2. The electron dot diagram shows that there is a lone pair on the central phosphorus atom. The molecule will be polar.

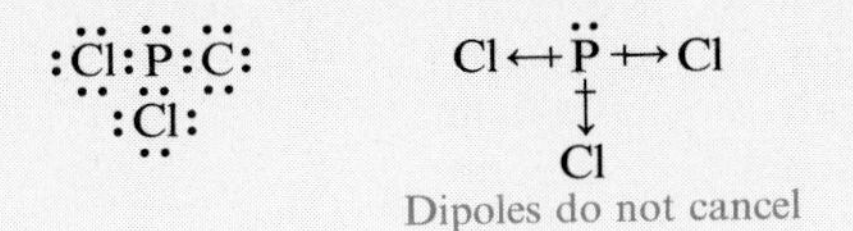

Dipoles do not cancel

Glossary

anion A negatively charged ion such as Cl^- or O^{2-}.

cation A positively charged ion such as Na^+, Mg^{2+}, or Al^{3+}.

charge The difference between the number of protons (positive) and electrons (negative), written to the upper right side of the symbol for the element.

chemical activity The tendency of an element to form a compound.

covalent bond A sharing of valance electrons by atoms to achieve stability.

electron dot structure Representation of an atom that shows each valence electron as a dot.

electronegativity value A number used to indicate a measure of the ability of an element to attract electrons.

group number A roman numeral appearing at the top of each family that indicates the number of electrons in the valence shell.

ion An atom with an electrical charge because of having a different number of protons and electrons.

ionic bond The attraction of oppositely charged ions that holds ionic compounds together.

ionic charge An electrical charge on an ion resulting from a loss or gain of electrons.

ionic compound A compound consisting of ions.

nonpolar bond A covalent bond in which electrons are shared equally by two atoms of identical electronegativity.

nonpolar molecule A molecule consisting of nonpolar bonds or a molecule in which the dipoles cancel out.

octet rule Elements with eight electrons (octet) in their valence shells appear to be especially stable and very unreactive. Elements with one to seven valence electrons seem to react with other elements by forming ionic or covalent bonds to acquire an arrangement of eight electrons in the outer shell.

polar bond A covalent bond that has a positive and negative end because of an unequal sharing of electrons by two atoms.

polar molecule A molecule consisting of dipoles that do not cancel out.

polyatomic ion A group of nonmetals bonded through covalent bonds with an overall electrical charge.

triple bond A sharing of three pairs of electrons by two atoms.

valence electrons The electrons in the outermost shell, which are largely responsible for the chemical properties of elements.

valence shell The outermost energy level of an atom that contains at least one electron.

Problems

Valence Electrons (Objective 3.1)

3.1 Write an electron dot structure for each element:

a. sulfur	f. carbon
b. nitrogen	g. oxygen
c. calcium	h. fluorine
d. sodium	i. lithium
e. potassium	j. chlorine

3.2 State the group number for each of the following electron dot structures for element X:

a. $\cdot\dot{X}\cdot$ b. $:\underset{\cdot}{\ddot{X}}\cdot$ c. $X^{\cdot}$ d. $\cdot\underset{\cdot}{\ddot{X}}\cdot$ e. $:\underset{\cdot\cdot}{\ddot{X}}\cdot$

3.3 State the number of valence electrons in an atom of each of the following elements:

a. N	c. Br	e. S	g. Ba	i Cl
b. O	d. K	f. Na	h. Al	j. I

Octet Rule (Objective 3.2)

3.4 Write the electron arrangement for each of the following elements. Indicate whether it would lose or gain electrons:

a. neon b. oxygen c. lithium d. argon e. phosphorus

3.5 Classify each of the following elements as a metal or nonmetal:

a. N	c. Ca	e. Cl	g. Al	i. K
b. Na	d. S	f. Li	h. Mg	j. P

3.6 State the number of electrons lost or gained when ions are formed by each of the following elements:

a. Mg	c. Cl	e. Al	g. S	i. Li
b. P	d. Na	f. O	h. F	j. N

Ionic Charges (Objective 3.3)

3.7 State the charge on atoms or ions that have the following numbers of subatomic particles:

a. 9 protons, 10 electrons, 9 neutrons
b. 12 protons, 10 electrons, 12 neutrons
c. 15 protons, 18 electrons, 16 neutrons
d. 19 protons, 18 electrons, 20 neutrons
e. 20 protons, 20 electrons, 22 neutrons

3.8 Give the symbol and name for an ion of each element:

a. chlorine
b. magnesium
c. potassium
d. oxygen
e. aluminum
f. fluorine
g. calcium
h. phosphorus
i. sodium
j. lithium

3.9 Write the name of each ion:

a. Na^+
b. Mg^{2+}
c. F^-
d. Cl^-
e. O^{2-}
f. Al^{3+}
g. Ca^{2+}
h. K^+

3.10 Write the electron arrangement for each of the following ions:

a. Na^+ b. S^{2-} c. Ca^{2+} d. Cl^-

3.11 Write the name of the following ions from the transition metals:

a. Fe^{2+}
b. Ag^+
c. Cu^{2+}
d. Cu^+
e. Fe^{3+}
f. Zn^{2+}

3.12 Write the symbol, including charge, for each of the following:

a. potassium ion
b. copper(II) ion
c. iron(III) ion
d. ferrous ion
e. silver ion
f. zinc ion

3.13 Name the following polyatomic ions:

a. SO_4^{2-}
b. CO_3^{2-}
c. PO_4^{3-}
d. NO_3^-
e. OH^-
f. SO_3^{2-}

3.14 Write the formulas for the following polyatomic ions:

a. bicarbonate ion
b. ammonium ion
c. phosphate ion
d. nitrite ion
e. sulfite ion
f. hydroxide ion

Writing Ionic Formulas (Objective 3.4)

3.15 Use electron dot structures to diagram the formation of the following ionic compounds:

a. KCl b. $MgCl_2$ c. Na_3N d. MgS e. $AlCl_3$

3.16 Write the correct ionic formula for compounds formed between the following ions:

a. Na^+ and O^{2-}
b. Fe^{3+} and Cl^-
c. Ba^{2+} and Cl^-
d. Cu^{2+} and O^{2-}
e. K^+ and I^-
f. Zn^{2+} and Cl^-
g. Al^{3+} and S^{2-}
h. Li^+ and S^{2-}
i. Fe^{3+} and O^{2-}
j. Ag^+ and N^{3-}

3.17 Write the correct formula for ionic compounds formed by the following metals and nonmetals:

a. sodium and sulfur
b. aluminum and oxygen
c. iron(II) and chlorine
d. copper(I) and sulfur
e. calcium and chlorine
f. barium and bromine
g. lithium and chlorine
h. zinc and phosphorus
i. silver and nitrogen
j. iron(III) and oxygen

Naming Ionic Compounds (Objective 3.5)

3.18 Write names for the following ionic compounds:
a. Al_2O_3 e. Na_2S
b. $CaCl_2$ f. K_3P
c. Na_2O g. MgO
d. Mg_3N_2 h. $LiBr$

3.19 Indicate the valence of the metal ion in each of the following ionic compounds:
a. $MgCl_2$ b. FeO c. Cu_2S d. Ag_20 e. Fe_2O_3

3.20 Write names for the following ionic compounds:
a. $FeCl_2$ e. Ag_3P
b. CuO f. Na_2S
c. Fe_2S_3 g. ZnF_2
d. $CuCl$ h. AlP

3.21 Write formulas for the following ionic compounds:
a. magnesium chloride f. iron(III) oxide
b. sodium sulfide g. barium fluoride
c. cuprous oxide h. aluminum chloride
d. zinc phosphide i. silver sulfide
e. barium nitride j. copper(II) chloride

3.22 Circle the polyatomic ion in each of the following formulas and write the correct name of the compound:
a. Na_2CO_3 f. KOH
b. NH_4Cl g. $NaNO_3$
c. Li_3PO_4 h. $CuCO_3$
d. $Cu(NO_2)_2$ i. $NaHCO_3$
e. $FeSO_3$ j. $BaSO_4$

3.23 Write the correct formula for the following compounds:
a. barium hydroxide f. aluminum sulfite
b. sodium sulfate g. ammonium oxide
c. iron(II) nitrate h. magnesium bicarbonate
d. zinc phosphate i. sodium nitrite
e. silver carbonate j. copper(I) sulfate

Covalent Bonds (Objective 3.6)

3.24 Write the electron dot structure for the following covalent compounds:
a. Br_2 c. HF e. NCl_3 g. H_2
b. H_2S d. OF_2 f. CCl_4 h. SiF_4

Naming Covalent Compounds (Objective 3.7)

3.25 Name the following covalent compounds:
a. H_2 c. SCl_2 e. NI_3 g. P_2O_5
b. CBr_4 d. HF f. CS_2 h. Cl_2O

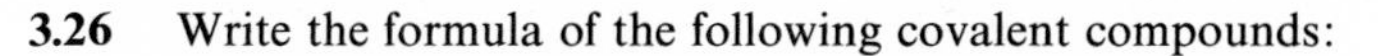

3.26 Write the formula of the following covalent compounds:

a. carbon tetrachloride
b. carbon monoxide
c. phosphorus trichloride
d. dinitrogen tetroxide
e. oxygen difluoride
f. hydrogen monochloride
g. dinitrogen oxide
h. chlorine (diatomic)

Polarity (Objective 3.8)

3.27 Place the symbols for partially positive (δ^+) and partially negative (δ^-) above the appropriate atoms in the following covalent bonds:

a. H—F b. C—Cl c. N—O d. H—O e. S—O

3.28 Identify the bonding between the following pairs of elements as ionic, polar covalent, or nonpolar covalent. If no bond forms, write *none*.

a. Cl and Cl
b. Ne and O
c. F and F
d. Mg and F
e. S and Cl
f. Na and Cl
g. O and F
h. K and S
i. H and H
j. N and H

3.29 Identify the following as polar or nonpolar molecules:

a. Br—Br
b. H—Br
c. S=C=S

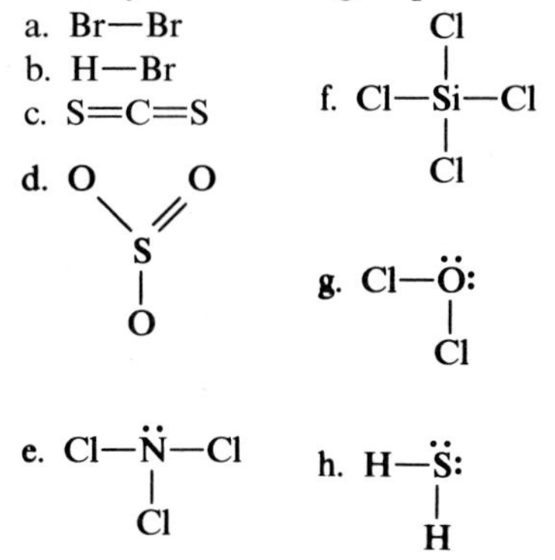

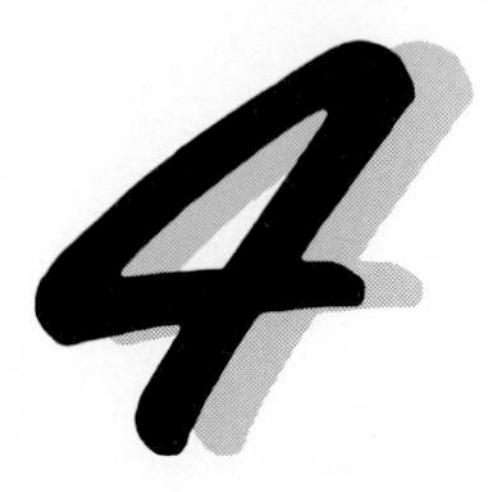

Chemical Quantities and Equations

Objectives

4.1 Given the formula of a compound, calculate its formula weight.

4.2 State the molar mass of an element or a compound.

4.3 Given the number of moles of a substance, calculate its mass in grams; given the mass, calculate the number of moles.

4.4 Identify a description of a reaction as a physical or a chemical change. Given a chemical equation, state its meaning in words; write a chemical equation from a word description of the chemical reaction.

4.5 Write a balanced equation for a chemical reaction.

4.6 Given the amount of reactant and the balanced equation, calculate the quantity of product; calculate the amount of reactant needed to produce a given quantity of product.

Scope

You are now ready to learn something about the way a chemist measures amounts of atoms, molecules, and ions. The formula of a compound tells us the numbers of atoms that make up a compound as well as the amount of each element present. This information allows the scientist to develop new drugs, cosmetics, fabrics, and dyes.

By using chemical equations, the scientist calculates how much reactant to use and how much product it will make. The concept of using balanced chemical equations for predicting the weights of reacting substances is important in our everyday lives. The carburetor of an engine is adjusted to allow the correct amounts of fuel and oxygen from the air to mix together so the engine will run properly. When preparing pancakes, the correct amounts of pancake mix, eggs, and milk must be thoroughly mixed to get good results. In medicine, the correct dosages of medications must be given to ensure the proper reactions within the body. In chemistry, the amount of each reactant is also carefully measured to obtain the correct products in the desired amounts.

4.1 Formula Weight

OBJECTIVE Given the formula of a compound, calculate its formula weight.

The formula of a compound represents the number of atoms in the smallest unit (molecule or formula unit). Its mass, which is called *formula weight*, can be calculated by adding the atomic weights of all the elements in the formula. The atomic weights of the elements are found in the periodic table. In this text, the atomic weight is listed under the symbol of each element. (Recall that the atomic weight of an element is the average atomic mass of all its isotopes.) Since atomic weights are expressed in atomic mass units (amu), the formula weight is also expressed in amu.

Example 4.1 Calculate the formula weight of $CaCO_3$, an antacid.

Solution: We use the subscripts in the formula to determine the number atoms of each element. In the formula of the antacid $CaCO_3$, there is one atom of calcium, one atom of carbon, and three atoms of oxygen. The atomic weight of each element is obtained from the periodic table, and the formula weight is calculated as follows:

Element	Number of atoms in the formula	×	Atomic weight	=	Weight of element in the formula
Ca	1 atom Ca	×	$\frac{40.1 \text{ amu}}{1 \text{ atom Ca}}$	=	40.1 amu
C	1 atom C	×	$\frac{12.0 \text{ amu}}{1 \text{ atom C}}$	=	12.0 amu
O	3 atoms O	×	$\frac{16.0 \text{ amu}}{1 \text{ atom O}}$	=	48.0 amu
	Formula weight of $CaCO_3$			=	100.1 amu

Example 4.2 The compound aluminum nitrate, $Al(NO_3)_3$, is used in antiperspirants. What is its formula weight?

Solution: When you have a formula with a polyatomic ion, remember to consider the subscript outside the parentheses. In the formula for the antiperspirant $Al(NO_3)_3$, there is one atom of aluminum, three atoms of nitrogen, and nine atoms of oxygen. Using the atomic weights for each of these elements as found in the periodic table, we can calculate the formula weight as follows:

Element	Number of atoms in the formula	×	Atomic weight	=	Weight of element in the formula
Al	1 atom Al	×	$\frac{27.0 \text{ amu}}{1 \text{ atom Al}}$	=	27.0 amu
N	3 atoms N	×	$\frac{14.0 \text{ amu}}{1 \text{ atom N}}$	=	42.0 amu
O	9 atoms O	×	$\frac{16.0 \text{ amu}}{1 \text{ atom O}}$	=	144.0 amu
	Formula weight of $Al(NO_3)_3$			=	213.0 amu

4.2 The Mole and Its Mass

OBJECTIVE State the molar mass of an element or a compound.

It is most unlikely that you will ever determine the mass of a single atom or molecule. They are much too small to weigh out, even on the most accurate balance. Instead, chemists work with samples that contains many atoms or molecules. You, too, work with groups or collections of things. Perhaps you buy eggs by the dozen, pencils by the gross, paper by the ream, and soda by the case. The terms *dozen*, *gross*, *ream*, and *case* are names for groups of items. It is common practice to give names to collections of small units to create a larger, more practical unit. Chemists use the unit *mole* (abbreviated mol) to name a collection made up of many small particles such as atoms or molecules. The unit mole can now be added to your list of names for groups of particles. (See Figure 4.1.)

mole

Name of Collection	*Number of Items in Collection*
1 trio	3
1 dozen	12
1 case	24
1 gross	144
1 ream	500
1 mol	6.02×10^{23}

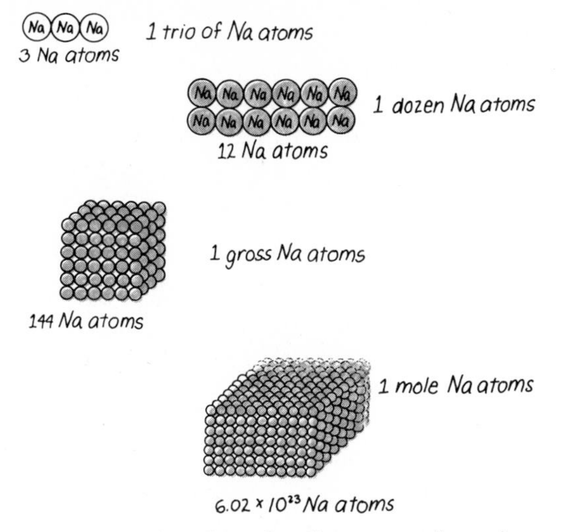

Figure 4.1 Number of sodium atoms in various kinds of collections.

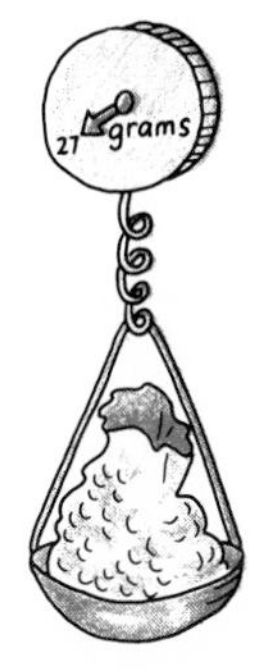

Figure 4.2 Examples of 1-mol quantities of elements.

Avogadro's Number

If you could count all the atoms in a mole of an element you would count out 6.02×10^{23} atoms. This number is called *Avogadro's number.* It is named after Amedeo Avogadro, an eighteenth-century Italian physicist. Avogadro's number is quite a large number. It looks like this written out:

Avogadro's Number

602 000 000 000 000 000 000 000 *or* 6.02×10^{23}

In one mole of any element, there is Avogadro's number of atoms. (See Figure 4.2.) For example, 1 mol of sodium atoms, 1 mol of phosphorus atoms, and 1 mol of iron atoms all contain the same number (Avogadro's) of atoms:

1 mol Na atoms = 6.02×10^{23} Na atoms

1 mol P atoms = 6.02×10^{23} P atoms

1 mol Fe atoms = 6.02×10^{23} Fe atoms

One mole of any compound also contains Avogadro's number of particles:

1 mol KCl = 6.02×10^{23} formula units KCl

1 mol CO_2 = 6.02×10^{23} molecules CO_2

1 mol $C_6H_{12}O_6$ = 6.02×10^{23} molecules $C_6H_{12}O_6$

For compounds, the individual particles are molecules or formula units of that compound. The formula unit gives the number of positive and negative ions in the crystal lattice of an ionic compound that have an overall ionic charge of zero.

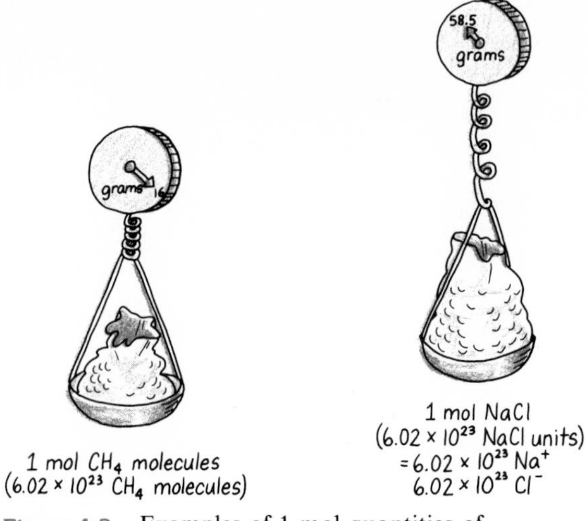

Figure 4.3 Examples of 1-mol quantities of compounds.

Only particles of covalent compounds exist as individual molecules. (See Figure 4.3.)

Molar Mass of an Element

We have seen that one mole of any element or compound contains the same number of particles. However, 1 mol quantities of different elements or compounds are not equal in mass. The mass of one mole of an element or a compound depends upon the mass of its particles. The mass of one mole, called the *molar mass* of a substance, is equal to its atomic weight expressed in grams. For example, the atomic weight of sodium is 23.0 amu. One mole of sodium atoms (6.02×10^{23} atoms) has a molar mass of 23.0 g. (See Figure 4.4.) Using the periodic table, we can express the molar mass of any element. (See Table 4.1.)

the molar mass is the formula weight in grams

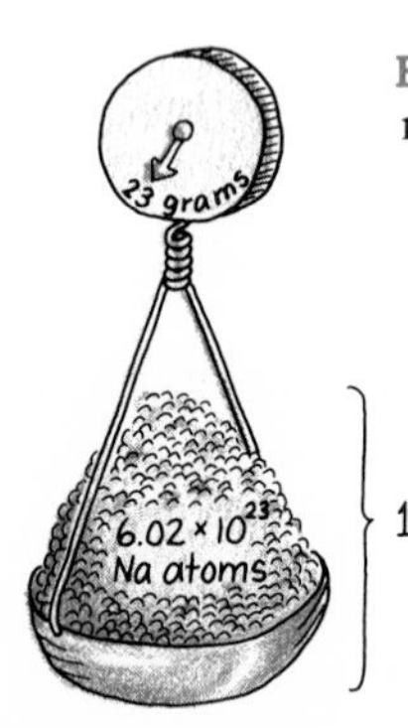

Figure 4.4 One mole of sodium atoms has a mass of 23.0 g.

Table 4.1 Comparison of Atomic Weight and Molar Mass of Some Elements

Element	Atomic Weight[a] of One Atom (amu)	Molar Mass of One Mole (g)	Atoms of Element in Molar Mass
C	12.0	12.0	6.02×10^{23}
Na	23.0	23.0	6.02×10^{23}
Cl	35.5	35.5	6.02×10^{23}
Fe	55.8	55.8	6.02×10^{23}

[a] Atomic weights are rounded to three significant figures.

Example 4.3 Write the molar mass for the following elements:

1. sulfur
2. oxygen
3. hydrogen

Solution: Using the periodic table, we can obtain the atomic weights of each element and express it in grams:

	Atomic weight (amu)	*Molar mass* (g)
1. S	32.1	32.1
2. O	16.0	16.0
3. H	1.0	1.0

Molar Mass of a Compound

Many of the substances we work with are compounds. The molar mass of any compound is its formula weight expressed in grams. Table 4.2 compares the formula weight and the molar mass of several compounds. Recall that one mole of a compound always contains Avogadro's number, 6.02×10^{23}, of molecules or units of that compound.

Table 4.2 Comparison of Formula Weight and Molar Mass of Some Compounds

Compound	Formula Weight (amu)	Molar Mass (g)	Number of Particles in Molar Mass
NaF (preventative for dental caries)	42.0	42.0	6.02×10^{23}
$CaCO_3$ (antacid)	100.1	100.1	6.02×10^{23}
$C_6H_{12}O_6$ (glucose)	180.0	180.0	6.02×10^{23}
$C_8H_{10}N_4O_2$ (caffeine)	194.0	194.0	6.02×10^{23}

Example 4.4 Write the molar mass of the following compounds:

1. NaBr (sedative)
2. $FeSO_4$ (iron supplement)
3. $C_6H_8O_6$ (vitamin C)

Solution: The molar mass of each compound is the number of grams in 1 mol, that is, the formula weight in grams.

1. The formula weight of NaBr is 102.9 amu. Thus,

$$\text{Molar mass NaBr} = 102.9 \text{ g}$$

2. The formula weight of $FeSO_4$ is 151.9 amu. Thus,

$$\text{Molar mass } FeSO_4 = 151.9 \text{ g}$$

3. The formula weight of $C_6H_8O_6$ is 176.0 amu. Thus,

$$\text{Molar mass } C_6H_8O_6 = 176.0 \text{ g}$$

4.3 Conversions Between Moles and Mass

OBJECTIVE Given the number of moles of a substance, calculate its mass in grams; given the mass, calculate the number of moles.

A molar mass is very useful for work in the laboratory when you need to change from moles to grams of a substance, or from grams to moles. To make these calculations, we need to write a molar mass in the form of a conversion factor:

$$\text{Molar mass} = \frac{\text{g}}{1 \text{ mol}}$$

For example, carbon has a molar mass of 12.0 g. The possible conversion factors for the molar mass of carbon are

$$\frac{12.0 \text{ g C}}{1 \text{ mol C}} \quad \text{and} \quad \frac{1 \text{ mol C}}{12.0 \text{ g C}}$$

The compound CO_2 has a molar mass of 44.0 g/mol. The conversion factors from the molar mass of CO_2 are

$$\frac{44.0 \text{ g } CO_2}{1 \text{ mol } CO_2} \quad \text{and} \quad \frac{1 \text{ mol } CO_2}{44.0 \text{ g } CO_2}$$

Problem Solving with Molar Conversion Factors

When the number of moles of an element or compound is given, the number of grams can be calculated. It will be necessary to first calculate the molar mass of the substance in each of these types of problems.

Example 4.5 Calculate the mass (in grams) of 2.00 mol of sodium (Na).

Solution: In the periodic table, we find that the atomic weight of sodium is 23.0 amu. Therefore, the molar mass of sodium is 23.0 g. The molar mass conversion factors for sodium are

$$\frac{23.0 \text{ g Na}}{1 \text{ mol Na}} \quad \text{and} \quad \frac{1 \text{ mol Na}}{23.0 \text{ g Na}}$$

The molar factor we have obtained can now be used to solve the problem. The given quantity of 2.00 mol of sodium is multiplied by the molar mass factor (selected from the above that cancels the mole units)

$$\underset{\text{Given}}{2.00 \text{ mol Na}} \times \underset{\text{Molar mass as a conversion factor}}{\frac{23.0\text{g Na}}{1 \text{ mol Na}}} = \underset{\text{Answer}}{46.0 \text{ g Na}}$$

Example 4.6 Camphor has a formula of $C_{10}H_{16}O$. Calculate the number of grams in 0.150 mol of camphor.

Solution: Using the periodic table, we will need to calculate the formula weight and the molar mass of camphor:

$$10 \text{ atoms C} \times \frac{12.0 \text{ amu C}}{1 \text{ atom C}} = 120 \text{ amu}$$

$$16 \text{ atoms H} \times \frac{1.0 \text{ amu H}}{1 \text{ atom H}} = 16 \text{ amu}$$

$$1 \text{ atom O} \times \frac{16.0 \text{ amu O}}{1 \text{ atom O}} = 16.0 \text{ amu}$$

Formula weight of camphor = 152.0 amu

Molar mass of camphor = 152.0 g

The molar mass of camphor can be used to write two possible molar mass conversion factors:

$$\text{Molar factors} \quad \frac{152.0 \text{ g camphor}}{1 \text{ mol camphor}} \quad \text{or} \quad \frac{1 \text{ mol camphor}}{152.0 \text{ g camphor}}$$

The problem setup using a molar mass conversion factor is:

$$0.150 \text{ mol camphor} \times \underset{\text{Molar mass conversion factor}}{\frac{152.0 \text{ g camphor}}{1 \text{ mol camphor}}} = 22.8 \text{ g camphor}$$

When the number of grams of an element or a compound is known, the number of moles can be calculated. Be sure to obtain the formula weight first.

Example 4.7 The total bone mass of a typical 70-kg person contains 115 g of phosphorus. How many moles of phosphorus are in the bone?

Solution: Using the periodic table, we find that 1 mol of phosphorus has a mass of 31.0 g. As a conversion factor, this molar mass takes the following forms:

$$\frac{31.0 \text{ g P}}{1 \text{ mol P}} \quad \text{or} \quad \frac{1 \text{ mol P}}{31.0 \text{ g P}}$$

The problem setup begins with the given quantity of 115 g phosphorus. The conversion factor selected is the one that cancels the unit gram:

$$115 \cancel{\text{g P}} \times \frac{1 \text{ mol P}}{31.0 \cancel{\text{g P}}} = 3.71 \text{ mol P}$$

Molar mass conversion factor

Example 4.8 Sodium bicarbonate, $NaHCO_3$, is used as an antacid. How many moles of antacid are in 63.0 g?

Solution: The formula weight of $NaHCO_3$ is 84.0 amu (23.0 amu + 1.0 amu + 12.0 amu + 3 × 16.0 amu). Therefore, we can write the molar mass of $NaHCO_3$ as 84.0 g. This can be written in the form of conversion factors:

$$\frac{84.0 \text{ g } NaHCO_3}{1 \text{ mol } NaHCO_3} \quad \text{or} \quad \frac{1 \text{ mol } NaHCO_3}{84.0 \text{ g } NaHCO_3}$$

To set up the calculation for the problem, we write the 63.0 g $NaHCO_3$ and multiply by the factor that cancels the unit gram:

$$63 \cancel{\text{g } NaHCO_3} \times \frac{1 \text{ mol } NaHCO_3}{84.0 \cancel{\text{g } NaHCO_3}} = 0.75 \text{ mol } NaHCO_3$$

Molar mass conversion factor

4.4 Chemical Reactions

OBJECTIVES Identify a description of a reaction as a physical or a chemical change. Given a chemical equation, state its meaning in words; write a chemical equation from a word description of the chemical reaction.

Physical and Chemical Changes

Suppose you are watching a log burn in the fireplace while ice cubes melt in a glass next to you. Both of these events can be described in terms of changes in their physical properties. The log has burned to a grayish-white ash as it produced heat and light. The chemical bonds in the wood of the log were changed as new substances with new physical properties were produced. When a change in physical properties results in the formation of new substances, a

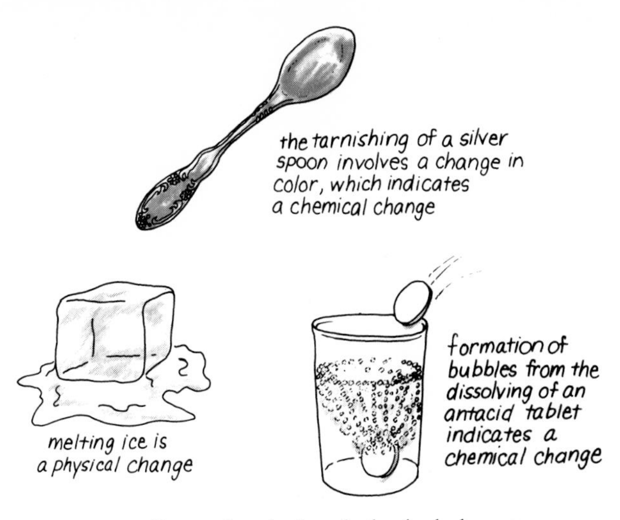

Figure 4.5 Some chemical and physical changes.

chemical change chemical reaction has taken place. We call the burning of wood a *chemical change*.

The ice cubes in the glass went through a change in physical properties, too. However, no new substances have formed. Water was the substance of the ice cubes and water is the substance of the liquid. When physical properties change without the formation of new substances, a *physical change* has taken place. Let us look at some everyday examples of physical and chemical changes (see Figure 4.5):

physical change

Physical Changes	*Chemical Changes*
Melting ice	Rusting nail
Boiling water	Bleaching stain
Tearing paper	Burning log
Rolling ball	Baking cake
Breaking glass	Fermenting grapes

Example 4.9 Identify each of the following as a physical or chemical change:

1. freezing water
2. burning a match
3. breaking a match
4. tarnishing silver

Solution

1. Physical. Freezing water involves only a change in state from liquid water to solid ice, but no change has occurred in the substance itself.

2. Chemical. Burning a match causes the formation of new substances that were not present prior to striking the match.
3. Physical. Breaking a match does not change any of the substances in the match.
4. Chemical. The tarnishing of silver involves the formation of a new substance.

Chemical Equations

Whenever you put together a model, build something from a kit, follow a recipe, prepare medication, or clean a patient's teeth, you follow a set of directions. These directions tell you what to use, how much to use, and the kind and quantity of product you should obtain. The chemist also has a set of directions for every chemical reaction. This set of directions is called a *chemical equation.*

reactants, products

We might compare a chemical equation to a recipe. Suppose you looked up a recipe for making pancakes. The ingredients in the recipe are called the *reactants* by the chemist, and the pancakes you make are the *products.* An arrow pointing from the reactants to the products indicates that a reaction has taken place. The pancakes represent the new substances that result from the reaction:

$$\underset{\text{Reactants}}{\text{Milk + Eggs + Pancake mix}} \xrightarrow{\text{react to form}} \underset{\text{products}}{\text{Pancakes}}$$

coefficients are the numbers in front of the formulas in an equation

The recipe is not yet complete. You need more than just the ingredients before you can follow a recipe. You need to know how much of each reactant to use. The amount of each ingredient is indicated by a number placed in front of the reactant. In a chemical equation, these numbers are called *coefficients* and appear in front of the formulas in the equation. They indicate the amount of each reactant needed for reaction. They also tell us the corresponding quantities of products that will form. Our everyday equation for making pancakes will be complete or *balanced* by writing the following recipe:

$$\text{1 cup milk + 2 eggs + 2 cups pancake mix} \longrightarrow \text{12 pancakes}$$

Stating a Chemical Equation in Words

A chemical equation uses symbols to describe a chemical reaction. It tells us what occurs when a chemical change takes place. For example, the following equation represents the reaction in which the carbon in barbecue briquettes react with oxygen to produce carbon dioxide. The amounts of each reactant and product can be stated in terms of atoms or molecules. The coefficient 1 is understood and is usually omitted:

$$1C + 1O_2 \longrightarrow 1CO_2$$

$$\text{or} \quad C + O_2 \longrightarrow CO_2$$

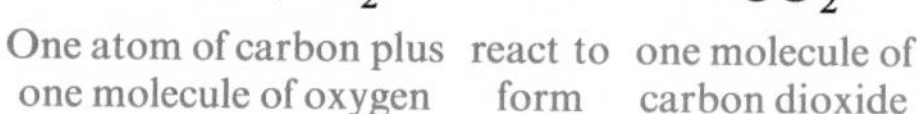

The gas propane is used as a fuel for campfire stoves. Its reaction with oxygen is described by the following equation, written in symbols and in words:

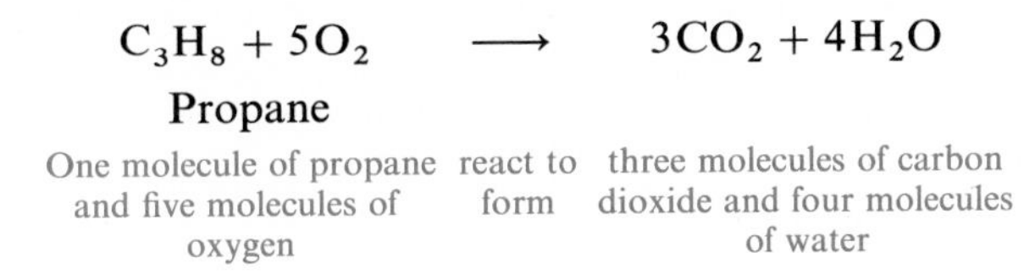

Example 4.10 State the meaning, in words, of the following chemical equations:

1. $H_2 + Cl_2 \longrightarrow 2HCl$
2. $4NH_3 + 3O_2 \longrightarrow 2N_2 + 6H_2O$
   Ammonia

Solution

1. One molecule of hydrogen plus one molecule of chlorine react to form two molecules of hydrogen chloride.
2. Four molecules of ammonia plus three molecules of oxygen react to form two molecules of nitrogen and six molecules of water.

Example 4.11 Write a chemical equation for the following reactions stated in words:

1. One molecule of ethanol (C_2H_6O) plus three molecules of oxygen (O_2) react to form two molecules of carbon dioxide and three molecules of water.
2. Two atoms of aluminum and three atoms of sulfur react to form one formula unit of aluminum sulfide.

Solution

1. $C_2H_6O + 3O_2 \longrightarrow 2CO_2 + 3H_2O$
2. $2Al + 3S \longrightarrow Al_2S_3$

4.5 Balancing Equations

OBJECTIVE Write a balanced equation for a chemical reaction.

The *law of conservation of matter* states that atoms are neither created nor destroyed during a chemical reaction; the atoms of the reactants are just rearranged to form the products. Therefore, in a chemical equation, there must be the same number of atoms for each element on the reactant side as on the product side. For example, consider the reaction when carbon is heated in an atmosphere with a limited quantity of oxygen:

Reactants *Products*

$$2C + O_2 \longrightarrow 2CO$$

In this equation, there are two atoms of carbon on the reactant side and two atoms of carbon on the product side. Two atoms of oxygen start the reaction as a molecule of oxygen. On the product side, the oxygen atoms are part of two carbon monoxide molecules. Note that the coefficient in front of carbon monoxide doubles everything in the formula.

Reactants		*Products*
$2C + O_2$	$\longrightarrow$	$2CO$
$2 \times C = 2$ atoms C	=	$2 \times CO = 2$ atoms C
$1 \times O_2 = 2$ atoms O		2 atoms O

Example 4.12 State the number of atoms of each of the elements on the reactant and product sides for the following balanced equations:

1. $C + 2H_2 \longrightarrow CH_4$
2. $2Fe + 3Cl_2 \longrightarrow 2FeCl_3$
3. $C_4H_8 + 6O_2 \longrightarrow 4CO_2 + 4H_2O$

Solution

	Reactants			*Products*	
1. C:	$1 \times C$	= 1 atom C	=	$1 \times CH_4$	= 1 atom C
H:	$2 \times H_2$	= 4 atoms H			4 atoms H
2. Fe:	$2 \times Fe$	= 2 atoms Fe	=	$2 \times FeCl_3$	= 2 atoms Fe
Cl:	$3 \times Cl_2$	= 6 atoms Cl			6 atoms Cl
3. C:	$1 \times C_4H_8$	= 4 atoms C	=	$4 \times CO_2$	= 4 atoms C
H:		= 8 atoms H		$4 \times H_2O$	= 8 atoms H
				$4 \times CO_2$	= 8 atoms O } Add
				$4 \times H_2O$	= 4 atoms O } Add
O:	$6 \times O_2$	= 12 atoms O			12 atoms O

Balancing a Chemical Equation

Equations are balanced by placing coefficients in front of the formulas in order to equalize the number of each type of atom on both sides of the equation. You may wonder why we must balance an equation. An equation is balanced to conform to the law of conservation of matter. When a reaction occurs, atoms separate and recombine to form new compounds. No new atoms enter the reaction, nor do any of the original atoms disappear. There are as many of each kind of atom in the products as there were in the reactants.

Example 4.13 Balance the following equation:

$$H_2 + Cl_2 \longrightarrow HCl$$

Solution: We can set up a score sheet for the reaction by writing the total of each type of atom on the reactant side and on the product side:

$$H_2 + Cl_2 \longrightarrow HCl$$

Reactants	*Products*
2H	1H
2Cl	1Cl

Both the hydrogen and the chlorine are unbalanced. (There are more hydrogen and chlorine atoms on the reactant side.) We might start balancing the chlorine first. A coefficient of 2 is placed in front of the HCl formula. The coefficient 2 cannot be placed in between the symbols of the fomula nor can it be written as a subscript.

$$H_2 + Cl_2 \longrightarrow 2HCl$$

Reactants	*Products*	
2H	2H	
2Cl	2Cl	(balanced)

Checking the other element, hydrogen, we find that it became balanced when the coefficient of 2 was placed in front of the HCl formula. The equation is now balanced. There are two atoms of H on the reactant side and on the product side. There are also two atoms of Cl on both sides of the equation. (See Figure 4.6.)

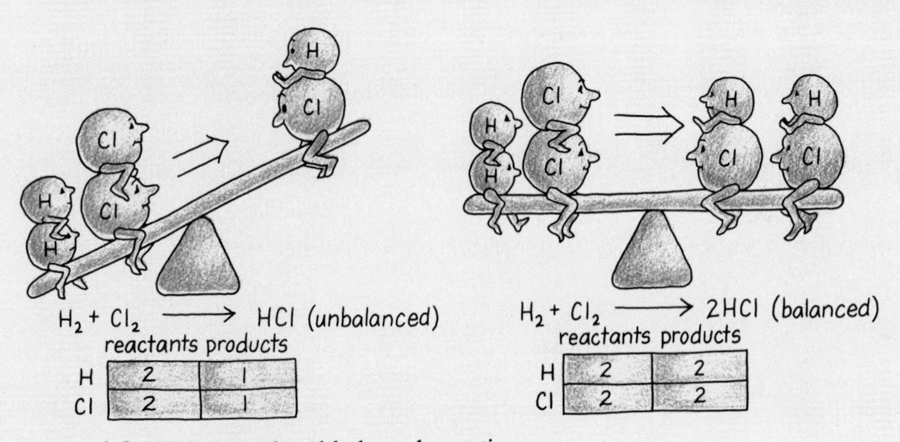

Figure 4.6 Unbalanced and balanced equations.

Some hints for balancing equations are given in Table 4.3. Remember, they are only hints. Balancing equations in this way is largely a matter of trial and error.

Table 4.3 Hints for Balancing Chemical Equations

1. Count the number of atoms for each element or ion on the reactant side and then on the product side.
2. Determine which are unequal.
3. Balance one element at a time. The most likely starting place is with a metal or the elements in a complicated formula that has subscripts. The elements hydrogen and oxygen and polyatomic ions are usually balanced last.
4. Start balancing one of the elements by placing a small whole number (coefficient) in front of the symbol or formula that contains an atom (or atoms) of the element you are trying to balance. Note that this coefficient multiplies the entire formula that follows it. *No changes can be made in the subscripts of the formulas while you balance an equation.*
5. Check to see if the equation is completely balanced. Sometimes, balancing one element will cause another to become unbalanced. Then you must return to the other element and rebalance. The final ratio of coefficients should be as small as possible, that is, not divisible by a whole number.
6. If the equation is still unbalanced, multiply through by 2, and repeat these steps.

Example 4.14 Balance the following equation:

$$C + SO_2 \longrightarrow CS_2 + CO$$

Solution: We start by checking the number of each atom on the reactant and product sides:

$$C + SO_2 \longrightarrow CS_2 + CO$$

Reactants	*Products*
1C	2C
1S	2S
2O	1O

We might start balancing with the element sulfur. We will balance the carbon atoms last because carbon appear in two formulas on the product side. The two sulfur atoms on the product side can be balanced by placing a coefficient 2 in front of the SO_2 formula:

$$C + 2SO_2 \longrightarrow CS_2 + CO$$

Reactants	*Products*
1C	2C
2S	2S
4O	1O

In order to balance the four oxygen atoms in the reactants, we place a coefficient 4 in front of the CO formula and recheck the number of atoms:

$$C + 2SO_2 \longrightarrow CS_2 + 4CO$$

Reactants	*Products*
1C	5C
2S	2S
4O	4O

Now we can balance the carbon atoms. Be sure to count all the carbon atoms on the product side. There are five carbon atoms on the product side (one in CS_2 and four in 4CO). We need a coefficient 5 in front of the symbol for carbon on the reactant side. Our final check of atoms in the reactants and the products is equal, thus the equation is balanced:

$$5C + 2SO_2 \longrightarrow CS_2 + 4CO$$

Reactants	*Products*	
5C	5C	
2S	2S	
4O	4O	(balanced)

Example 4.15 Balance the following equation:

$$Al + CuSO_4 \longrightarrow Al_2(SO_4)_3 + Cu$$

Solution: Counting the number of atoms on the reactant side and on the product side, we find that the equation is unbalanced. When reactions involve the same polyatomic ion on the reactant and product side, the ion can be counted as a unit:

$$Al + CuSO_4 \longrightarrow Al_2(SO_4)_3 + Cu$$

Reactants	*Products*
1 Al	2 Al
1 Cu	1 Cu
1 SO_4 (Polyatomic ion)	3 SO_4

We might start balancing the equation for aluminum by placing a coefficient 2 in front of the aluminum atom on the reactant side:

$$2Al + CuSO_4 \longrightarrow Al_2(SO_4)_3 + Cu$$

Reactants	*Products*
2 Al	2 Al
1 Cu	1 Cu
1 SO_4	3 SO_4

The sulfate ion can be balanced by placing a coefficient 3 in front of the $CuSO_4$ formula:

$$2Al + CuSO_4 \longrightarrow Al_2(SO_4)_3 + Cu$$

Reactants	*Products*
2 Al	2 Al
3 Cu	1 Cu
3 SO_4	3 SO_4

Our balancing of the sulfate ion caused an unbalance of the Cu atoms. This can be remedied by placing a coefficient 3 in front of the Cu on the product side:

$$2Al + 3CuSO_4 \longrightarrow Al_2(SO_4)_3 + 3Cu$$

Reactants	*Products*	
2Al	2Al	
3Cu	3Cu	
$3SO_4$	$3SO_4$	(balanced!)

Our final check of the atoms on the reactant and products sides indicates that the equation is balanced.

Other Symbols Used in Equations

Sometimes other symbols are used to represent certain reaction conditions. Some of these symbols are:

(*s*) solid	$\rightleftarrows$ reversible reaction
(*l*) liquid	$\downarrow$ solid forms
(*g*) gas	$\uparrow$ gas forms
Δ heat	(aq) dissolved in water

Here are four examples of equations that include several of the above symbols:

$$2P(s) + 3H_2(g) \longrightarrow 2PH_3(g)$$

$$2Na(s) + 2H_2O(l) \longrightarrow H_2(g) + 2NaOH(aq)$$

$$Mg(s) + 2HCl(aq) \longrightarrow MgCl_2(aq) + H_2(g)$$

$$2Mg(s) + O_2(g) \longrightarrow 2MgO(s)$$

4.6 Calculations Using Chemical Equations

OBJECTIVE Given the amount of reactant and the balanced equation, calculate the quantity of product; calculate the amount of reactant needed to produce a given quantity of product.

Now we need to look at some more ways to interpret an equation. We have used the coefficients to determine the numbers of atoms and molecules taking part in a chemical reaction. If we consider that there are many atoms and molecules participating in the reaction, perhaps as many as Avogadro's number, we will count the moles of atoms and molecules. The coefficients in the equation can also tell us the numbers of moles of reactants and products.

$$N_2 + O_2 \longrightarrow 2NO$$

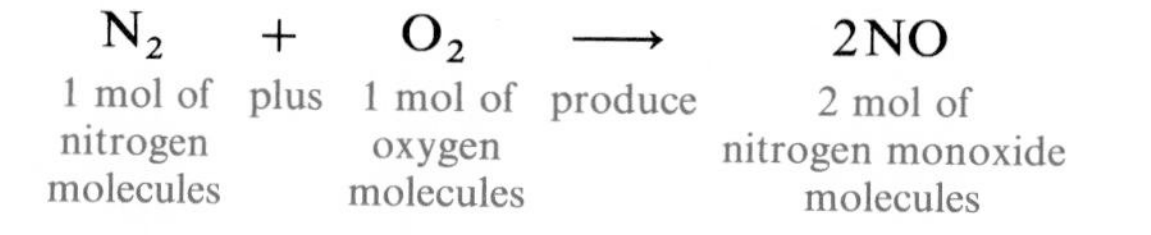

Figure 4.7 Diagram of reactions of matter in a recipe for pancakes.

By knowing that an equation also states the number of moles of reactants and products, we can determine the mass of the reactants and the mass of the products. We find that they are equal. Such a result is expected according to the *law of conservation of mass.* The mass of the reactants and the products in the above equation are calculated as follows:

Mass of Reactants = *Mass of the Products*

$$1\ \cancel{\text{mol}}\ N_2 \times \frac{28.0\ \text{g}}{1\ \cancel{\text{mol}}} = 28.0\ \text{g}\ N_2 \qquad 2\ \cancel{\text{mol}}\ \text{NO} \times \frac{30.0\ \text{g}}{1\ \cancel{\text{mol}}} = 60.0\ \text{g NO}$$

$$1\ \cancel{\text{mol}}\ O_2 \times \frac{32.0\ \text{g}}{1\ \cancel{\text{mol}}} = \frac{32.0\ \text{g}\ O_2}{60.0\ \text{g}} = \frac{}{60.0\ \text{g}}$$

Let us use our pancake recipe here as an analogy. If we list the mass of each ingredient, we find that the mass of the pancakes is the sum of the mass of the ingredients. (Any water loss is included in the mass of the pancakes.) The ingredients are not lost; the milk or the eggs or the pancake mix do not disappear. They only change form in becoming pancakes. They have rearranged their atoms and molecules into pancakes, and the mass of each ingredient is now part of the mass of the pancakes. Mass has been conserved (see Figure 4.7).

1 cup milk + 2 eggs + 2 cups pancake mix ⟶ 12 pancakes
250 g + 150 g + 300 g ⟶ 700 g

Example 4.16 Demonstrate the law of conservation of mass by calculating the mass of the reactants and the mass of the products in the following equation:

$$2\text{Na} + \text{Cl}_2 \longrightarrow 2\text{NaCl}$$

Solution: Multiplying the number of moles (indicated by the coefficients) by their formula weights gives the total mass of the reactants and products:

Reactants		*Products*
$2\text{Na} + \text{Cl}_2$	⟶	2NaCl
2 × 23.0 g + 71.0 g	=	2 × 58.5 g
46.0 g + 71.0 g	=	117 g
117 g reactants	=	117 g products

Conversion Factors from Balanced Equations

A chemical equation is also a source of several conversion factors. Since we can interpret the coefficients in terms of mole relationships among the reactants and products, we can write mole–mole factors for any two compounds in the equation. Let us consider the following reaction:

$$N_2 + 3H_2 \longrightarrow 2NH_3$$

The equation can be put into words to remind you of its meaning in moles:

1 mol N_2 and 3 mol H_2 react to form 2 mol NH_3

mole–mole factors From this statement, six possible mole–mole conversion factors can be derived:

Mole–Mole Factors for Nitrogen and Hydrogen

$$\frac{1 \text{ mol } N_2}{3 \text{ mol } H_2} \quad \text{and} \quad \frac{3 \text{ mol } H_2}{1 \text{ mol } N_2}$$

Mole–Mole Factors for Nitrogen and Ammonia

$$\frac{1 \text{ mol } N_2}{2 \text{ mol } NH_3} \quad \text{and} \quad \frac{2 \text{ mol } NH_3}{1 \text{ mol } N_2}$$

Mole–Mole Factors for Hydrogen and Ammonia

$$\frac{3 \text{ mol } H_2}{2 \text{ mol } NH_3} \quad \text{and} \quad \frac{2 \text{ mol } NH_3}{3 \text{ mol } H_2}$$

Example 4.17 State the mole–mole factors for the substances in the following balanced equation:

$$2Mg + O_2 \longrightarrow 2MgO$$

Solution: There are six mole–mole factors that can be derived from this equation:

$$\frac{2 \text{ mol Mg}}{1 \text{ mol } O_2} \qquad \frac{1 \text{ mol } O_2}{2 \text{ mol Mg}}$$

$$\frac{2 \text{ mol Mg}}{2 \text{ mol MgO}} \qquad \frac{2 \text{ mol MgO}}{2 \text{ mol Mg}}$$

$$\frac{1 \text{ mol } O_2}{2 \text{ mol MgO}} \qquad \frac{2 \text{ mol MgO}}{1 \text{ mol } O_2}$$

Problem Solving with Mole–Mole Factors

Using mole–mole conversion factors, we can now answer some questions about the amounts of reactants and products in a reaction. In this type of problem, you will be given the number of moles for one compound and asked to find the number of moles of one of the other compounds in the reaction. The mole–mole factor is selected from the equation to cancel the moles of the given compound and give the number of moles of the requested one.

Example 4.18 Consider the reaction of hydrogen and oxygen in the formation of water:

$$2H_2 + O_2 \longrightarrow 2H_2O$$

How many moles of hydrogen are needed to react with 5.0 moles of oxygen?

Solution: The given is 5.0 mol O_2. The request is for the number of moles of H_2 that will be needed for the reaction. From the equation, we see that 2 mol of H_2 react with every 1 mol of O_2. The possible conversion factors between H_2 and O_2 are the following:

$$\frac{2 \text{ mol } H_2}{1 \text{ mol } O_2} \quad \text{and} \quad \frac{1 \text{ mol } O_2}{2 \text{ mol } H_2}$$

The factor to use is the one with 1 mol O_2 in the denominator to cancel the given moles of O_2:

$$5.0 \cancel{\text{mol}} \; O_2 \times \underbrace{\frac{2 \text{ mol } H_2}{1 \cancel{\text{mol}} \; O_2}}_{\text{Mole–mole factor}} = 10 \text{ mol } H_2$$

The answer tells us that 10 mol of H_2 are needed to completely react with 5 mol of O_2.

Calculating Mass in Chemical Reactions

In the laboratory, you will be weighing quantities of reactants or products in grams, not moles. However, the mole–mole factors are still the key to solving problems that state the reactants or products in grams. The conversion of moles to grams, or grams to moles, can be accomplished by including the molar mass of the appropriate compound in the problem-solving setup. (See Figure 4.8 for a diagram of this problem-solving process.)

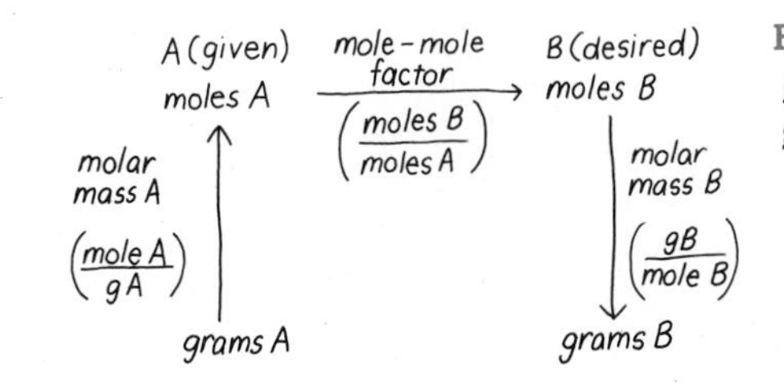

Figure 4.8 A pathway for converting from grams or moles of given quantity A to the grams or moles of the desired quantity B.

Example 4.19 Consider the following equation:

$$\underset{\text{Propane}}{C_3H_8} + 5O_2 \longrightarrow 3CO_2 + 4H_2O$$

1. How many grams of carbon dioxide are produced when 2.00 mol of propane completely react? (You may assume that there is sufficient oxygen for the reaction.)
2. How many grams of water are produced when 80.0 g of oxygen undergo reaction? (Assume enough propane is available for the oxygen to react with.)

Solution

1. First, we need to find out the number of moles of CO_2 produced. This can be done by using the mole–mole ratio. From the equation, we know that 3 mol of CO_2 are produced for every 1 mol of propane that reacts. The molar mass of CO_2 is 44.0 g/mol. This factor will change the moles of CO_2 to grams of CO_2:

   Unit Plan: mol $C_3H_8 \longrightarrow$ mol $CO_2 \longrightarrow$ g CO_2

$$\underset{\text{Given}}{2.00\ \cancel{\text{mol } C_3H_8}} \times \underset{\substack{\text{Mole–mole factor}\\\text{(coefficients)}}}{\frac{3\ \cancel{\text{mol } CO_2}}{1\ \cancel{\text{mol } C_3H_8}}} \times \underset{\substack{\text{Molar mass } CO_2\\\text{(formula weight)}}}{\frac{44.0\text{ g } CO_2}{1\ \cancel{\text{mol } CO_2}}} = 264\text{ g } CO_2$$

2. The given of 80.0 g O_2 is changed to grams of O_2 using the molar mass of O_2 (32.0 g/mol). Then the moles of O_2 can be converted to moles of H_2O. The equation gives the mole–mole relationship of 4 mol of H_2O to every 5 mol of O_2. The moles of H_2O are finally converted to grams of H_2O using the molar mass of water.

   Unit Plan: g $O_2 \longrightarrow$ mol $O_2 \longrightarrow$ mol $H_2O \longrightarrow$ g H_2O

$$\underset{\text{Given}}{80.0\ \cancel{\text{g } O_2}} \times \underset{\substack{\text{Molar}\\\text{mass } O_2}}{\frac{1\ \cancel{\text{mol } O_2}}{32\ \cancel{\text{g } O_2}}} \times \underset{\substack{\text{Mole–mole}\\\text{factor}}}{\frac{4\ \cancel{\text{mol } H_2O}}{5\ \cancel{\text{mol } O_2}}} \times \underset{\substack{\text{Molar}\\\text{mass } H_2O}}{\frac{18.0\text{ g } H_2O}{1\ \cancel{\text{mol } H_2O}}} = 36.0\text{ g } H_2O$$

The answer tells us that 36.0 g of water are produced when 80.0 g of oxygen undergo reaction.

In summary, a balanced chemical equation tells you a great deal about a reaction. Using the balanced reaction equation and a periodic chart, you should be able to determine the following information about the compounds in a reaction:

1. the molar mass of any reactant and product
2. the mole–mole conversion factors for any two compounds in the reaction
3. the number of moles of a reactant used or a product formed when given the number of moles of one of the other compounds
4. the number of grams of a reactant used or a product formed when given the number of moles or grams of one of the other compounds

Glossary

Avogadro's number The number of items in a mole, namely, 6.02×10^{23}.

chemical change The formation of a new substance causing a change in the physical properties.

chemical reaction The process by which a chemical change takes place.

equation A way to represent a chemical reaction using chemical formulas to indicate the reactants and products.

formula weight The sum of the atomic weights of all the atoms in a formula.

law of conservation of matter A law that states that atoms are neither created nor destroyed in a chemical reaction, but may be rearranged.

molar mass The mass of 1 mol of an element or compound equal to the formula weight expressed in grams.

mole A collection of atoms, molecules, or ions that contains 6.02×10^{23} of these items.

physical change A change in which the physical properties are altered but the composition of the substance remains the same.

products The substances produced in a chemical reaction.

reactants The initial substances that undergo change in a chemical reaction.

Problems

Formula Weight (Objective 4.1)

4.1 State the total number of atoms of each element in each of the following formulas:
a. Al_2O_3
b. $Al(OH)_3$
c. $Al_2(SO_4)_3$
d. $(NH_4)_2CO_3$
e. $Mg(HCO_3)_2$

4.2 Calculate the formula weight of the following compounds:
a. NaF
b. KCl
c. $MgCl_2$
d. H_2
e. $CaCO_3$
f. $Al(NO_3)_3$
g. $(NH_4)_2SO_4$
h. $C_6H_{12}O_6$

4.3 Calculate the formula weight of the following compounds:
a. ethanol (C_2H_6O)
b. ascorbic acid, (vitamin C, $C_6H_8O_6$)
c. tetracycline ($C_{22}H_{25}ClN_2O_8$)

The Mole and Its Mass (Objective 4.2)

4.4 Write the molar mass (g/mol) for the following elements:
a. Na
b. Cl
c. Pb
d. Fe
e. Mg
f. I

4.5 Write the molar mass (g/mol) for the following compounds:
a. C_3H_6 (cyclopropane, an inhaled anesthetic)
b. NH_4NO_3 (diuretic)
c. $NaHCO_3$ (baking powder)
d. $C_{20}H_{18}O_4$ (cyclocumarol, anticoagulant)
e. $Al(OH)_3$ (antacid)
f. $C_{29}H_{50}O_2$ (vitamin E)

Conversions Between Moles and Mass (Objective 4.3)

4.6 Calculate the mass in grams of

a. 2.0 mol of calcium
b. 0.12 mol of sulfur
c. 10.0 mol of aluminum
d. 4.5 mol of carbon
e. 0.50 mol of tin
f. 0.0085 mol of copper
g. 1.0 mol of potassium
h. 2.5 mol of phosphorus

4.7 Calculate the mass in grams of

a. 0.50 mol of NH_3
b. 2.0 mol Na_2O
c. 0.40 mol $Ca(NO_3)_2$
d. 5.0 mol HCl
e. 7.5 mol C_2H_6O
f. 0.10 mol $(NH_4)_3PO_4$

4.8 Solve the following problems:

a. The compound $MgSO_4$ is called Epsom salts. How many grams will you need to prepare a bath with 5.0 mol of Epsom salts?
b. In a bottle of soda, there is 0.25 mol of CO_2. How many grams of CO_2 are there?
c. Cyclopropane, C_3H_6, is an anesthetic given by inhalation. How many grams are in a 0.25 mol sample of cyclopropane.
d. The sedative, Demerol has the formula $C_{15}H_{22}ClNO_2$. How many grams are present in 0.025 mol of Demerol?

4.9 Calculate the number of moles in

a. 50.0 g of silver
b. 40.0 g of copper
c. 10.0 g of carbon
d. 150 g of iron
e. 400 g of sulfur
f. 1.50 g of silicon

4.10 Solve the following problems:

a. A nickel has a mass of 5.10 g. If it is 100 percent pure nickel, how many moles of nickel does it contain?
b. A gold nugget weighs 35.0 g. How many moles of gold are in the nugget?
c. You have collected 20.0 lb of aluminum cans. How many moles of aluminum do you have?

4.11 Calculate the number of moles in the following compounds:

a. 1.00 g of H_2
b. 160 g of CH_4
c. 222 g of $CaCl_2$
d. 100 g of SO_2
e. 36.0 g of H_2O
f. 0.200 g of $Cu(NO_3)_2$

4.12 Solve the following problems:

a. A can of Drāno contains 480 g of NaOH. How many moles of NaOH are in the can of Drāno?
b. The human body contains about 60 percent water (H_2O). If a person weighs 70 kg, how many moles of water are present in that person's body?
c. An alcohol blood level of 400 mg alcohol per 100 mL of blood can cause coma and possibly be fatal. How many moles of alcohol (C_2H_6O) are there in 100 mL?

Chemical Reactions (Objective 4.4)

4.13 Identify each of the following as a chemical or a physical change:

a. a piece of chocolate melting in your hand
b. sawing a board in half
c. boiling water
d. neutralizing stomach acid with an antacid tablet

e. evaporation of water from wet clothes
f. gasoline burning in a car engine
g. formation of green leaves in plants

4.14 Describe, in words, the meaning of the following equations:
a. $2NO + O_2 \longrightarrow 2NO_2$
b. $2H_2S + 3O_2 \longrightarrow 2SO_2 + 2H_2O$
c. $C_2H_6O + 3O_2 \longrightarrow 2CO_2 + 3H_2O$
ethanol

4.15 Write an equation using correct symbols for the elements and correct formulas for the compounds:
a. Four molecules of ammonia (NH_3) and three molecules of oxygen react to form two molecules of nitrogen and six molecules of water.
b. Four iron atoms and three oxygen molecules react to form two formula units of iron(III) oxide.
c. Two molecules of propanol (C_3H_8O) and nine molecules of oxygen react to form six molecules of carbon dioxide and eight molecules of H_2O.

Balancing Chemical Equations (Objective 4.5)

4.16 State the number of atoms of each element on the reactant side and on the product side in each of the following balanced equations:
a. $2Na + Cl_2 \longrightarrow 2NaCl$
b. $N_2H_4 + 2H_2O_2 \longrightarrow N_2 + 4H_2O$
c. $P_4O_{10} + 6H_2O \longrightarrow 4H_3PO_4$

4.17 Balance the following equations:
a. $N_2 + O_2 \longrightarrow NO$
b. $HgO \longrightarrow Hg + O_2$
c. $Fe + O_2 \longrightarrow Fe_2O_3$
d. $Na + Cl_2 \longrightarrow NaCl$
e. $Cu_2O + O_2 \longrightarrow CuO$

4.18 Balance the following equations:
a. $Al + Br_2 \longrightarrow AlBr_3$
b. $P_4 + O_2 \longrightarrow P_4O_{10}$
c. $C_3H_8 + O_2 \longrightarrow CO_2 + H_2O$
d. $Sb_2S_3 + HCl \longrightarrow SbCl_3 + H_2S$
e. $Fe_2O_3 + C \longrightarrow Fe + CO$

4.19 Balance the following equations:
a. $Mg + AgNO_3 \longrightarrow Mg(NO_3)_2 + Ag$
b. $CuCO_3 \longrightarrow CuO + CO_2$
c. $CaCO_3 \longrightarrow CaO + CO_2$
d. $Al + CuSO_4 \longrightarrow Cu + Al_2(SO_4)_3$
e. $Pb(NO_3)_2 + NaCl \longrightarrow PbCl_2 + NaNO_3$

4.20 Balance the following equations:
a. $Zn + H_2SO_4 \longrightarrow ZnSO_4 + H_2$
b. $Al_2(SO_4)_3 + KOH \longrightarrow Al(OH)_3 + K_2SO_4$
c. $K_2SO_4 + BaCl_2 \longrightarrow BaSO_4 + KCl$

Calculations Using Chemical Equations (Objective 4.6)

4.21 Give an interpretation of the following equations in terms of moles:
a. $NaCl + AgNO_3 \longrightarrow AgCl + NaNO_3$
b. $4Al + 3O_2 \longrightarrow 2Al_2O_3$

4.22 Demonstrate the law of conservation of mass by calculating the total mass of the reactants and the total mass of the products in each of the following balanced equations:
a. $N_2 + O_2 \longrightarrow 2NO$
b. $CaCO_3 \longrightarrow CaO + CO_2$
c. $2SO_2 + O_2 \longrightarrow 2SO_3$
d. $4Al + 3O_2 \longrightarrow 2Al_2O_3$

4.23 Copper metal reacts with sulfur to form copper(I) sulfide by the following equation:

$$2Cu + S \longrightarrow Cu_2S$$

a. How many moles of S are needed to react with 2.0 mol of Cu?
b. How many moles of Cu_2S will be produced from 4.0 mol of S?
c. How many grams of Cu_2S will be produced when 2.5 mol of Cu react?

4.24 Nitrogen gas will react with hydrogen gas to produce ammonia by the following equation:

$$N_2(g) + 3H_2(g) \longrightarrow 2NH_3(g)$$

a. How many moles of H_2 are needed to react with 1.0 mol of N_2?
b. How many grams of H_2 are needed to react with 2.0 mol of N_2?
c. How many grams of NH_3 will be produced when 12 g of H_2 react?

4.25 Ammonia and oxygen react to form nitrogen and water.

$$\underset{\text{Ammonia}}{4NH_3} + 3O_2 \longrightarrow 2N_2 + 6H_2O$$

a. How many moles of O_2 are needed to react with 8.0 mol of NH_3?
b. How many grams of N_2 will be produced when 170 g of NH_3 react?
c. How many grams of O_2 must react to produce 90 g of H_2O?
d. How many pounds of water are formed when 34.0 g of ammonia undergo reaction?

4.26 Consider the equation for the combustion of propane in a camping stove:

$$\underset{\text{Propane}}{C_3H_8} + 5O_2 \longrightarrow 3CO_2 + 4H_2O$$

a. How many moles of O_2 are needed to completely react with 4.0 mol of propane?
b. How many grams of H_2O will be produced when 24.0 g of C_3H_8 react?
c. If you need to produce 88 g of CO_2, how many grams of O_2 must react?

4.27 At the winery, glucose in grapes undergoes fermentation to produce ethyl alcohol and carbon dioxide by the following equation:

$$C_6H_{12}O_6 \longrightarrow 2C_2H_6O + 2CO_2$$

a. How many moles of CO_2 are produced when 500.0 g of glucose undergo fermentation?
b. How many grams of ethanol would be formed from the reaction of 0.240 kg of glucose?

5 Nuclear Radiation

Objectives

5.1 Given a nuclear symbol, write the mass number, number of protons, and number of neutrons, and vice versa.

5.2 Write the nuclear symbols for an alpha particle, beta particle, and gamma ray.

5.3 Use nuclear symbols to write a nuclear equation for the radioactive decay of a radioisotope.

5.4 Write a nuclear equation for the artificial production of a radioisotope.

5.5 Explain the process by which a radioisotope in the body is detected.

5.6 Given the mass of a radioisotope and its half-life, determine the quantity that remains after one or more half-lives have passed.

5.7 Identify the characteristics of fission and fusion processes.

5.8 Describe the units of radiation measurement.

Scope

Radiation is a tool of nuclear medicine. Its application to the diagnosis and treatment of disease increases every year. Radiation can be used to measure small amounts of specific components in the blood. Because some radioactive elements accumulate in a particular organ of the body, it is possible to locate a tumor in an organ, to determine the size and shape of an organ, and to measure the level of functioning of the cells in the organ. Radioactive iodine, for example, is absorbed by the thyroid gland, and radioactive technetium can be made to accumulate in the brain, liver, bone, or spleen. The radiation emitted from radioactive isotopes can be detected and measured by the radiologist.

When radiation passes through the cells of the body, it may damage them. For this reason, the amount of radiation received by a patient as well as the radiologist is very carefully controlled. In radiation therapy, radioactivity is used to destroy abnormal cells within the body. Gamma rays from radioactive cobalt, x-rays, and electrons from linear accelerators (linacs) are used in the treatment of cancer. The radiation penetrates the abnormal cells, thereby inhibiting their growth. Nearby normal cells, also affected by the radiation, have a greater capacity to recover than do the abnormal cells.

5.1 Nuclear Symbols

OBJECTIVE Given a nuclear symbol, write the mass number, number of protons, and number of neutrons, and vice versa.

nuclear symbol A nuclear symbol identifies the number of protons and neutrons contained in the nucleus of a radioactive atom or particle. Recall from our discussion in Chapter 2 that:

1. The *atomic symbol* identifies the element or particle;
2. The *mass number* gives the number of protons and neutrons;
3. The *atomic number* gives the number of protons.

To write the nuclear symbol for an atom of nitrogen (N) with a mass number of 13 and an atomic number of 7, first write the symbol for the element. Then place the mass number in the upper left-hand corner and the atomic number in the lower left-hand corner:

Mass number →
Atomic number → $^{13}_{7}N$

isotopes All isotopes of nitrogen have the same atomic number, 7, but they have different mass numbers. Recall from Chapter 2 that isotopes of an element have the same number of protons, but differ in numbers of neutrons. This particular isotope has seven protons and six neutrons.

If we were to write an isotope of nitrogen with seven protons and eight neutrons, we would identify it as ^{15}N and write its nuclear symbol as

$$^{15}_{7}N$$

The most common isotope of nitrogen has seven protons and seven neutrons and is written as

$$^{14}_{7}N$$

Nuclear symbols for some of the isotopes of carbon are shown in Figure 5.1.

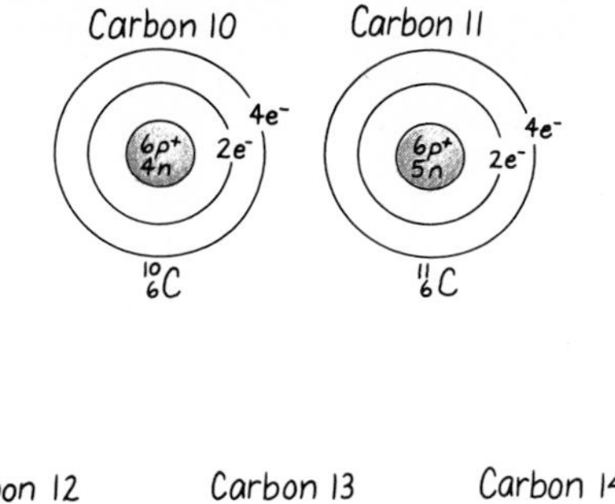

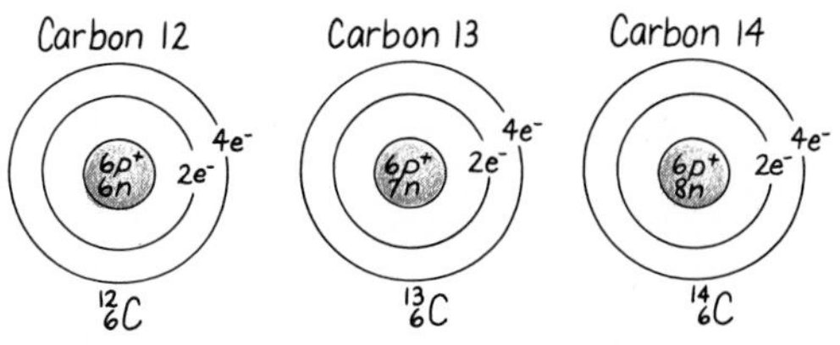

Figure 5.1 Isotopes of carbon and their nuclear symbols. All isotopes of carbon have six protons but vary in mass number.

Example 5.1 Write the name of the element, mass number, atomic number, number of protons, and number of neutrons for the following nuclear symbols:

1. $^{32}_{16}S$
2. $^{131}_{53}I$

Solution

1. The subscript 16 is the atomic number of sulfur, which has 16 protons. The number of neutrons is obtained by subtracting the number of protons from the mass number:

   number of neutrons = 32 − 16 = 16 neutrons

   $^{32}_{16}S$ sulfur; 16 protons; 16 neutrons

2. The atomic number of 53 for iodine indicates that there are 53 protons. The number of neutrons in this isotope of iodine is the difference between the mass number and the number of protons:

   number of neutrons = 131 − 53 = 78 neutrons

   $^{131}_{53}I$ iodine; 53 protons; 78 neutrons

5.2 Radioactivity

OBJECTIVE Write the nuclear symbols for an alpha particle, beta particle, and gamma ray.

Nuclear radioactivity is the process by which the nucleus of an unstable isotope releases radiation in the form of particles or rays. The resulting nucleus is more stable. In general, most naturally occurring isotopes of elements up to atomic number 20 are stable. For the elements with atomic number 20–80, the stable

Table 5.1 Examples of Stable and Unstable Isotopes

Some Stable Isotopes	Unstable and Radioactive Isotopes	
$^{24}_{12}Mg$	$^{23}_{12}Mg$	$^{27}_{12}Mg$
$^{127}_{53}I$	$^{125}_{53}I$	$^{131}_{53}I$
$^{209}_{83}Bi$	$^{208}_{83}Bi$	$^{210}_{83}Bi$
None	$^{209}_{84}Po$	$^{210}_{84}Po$
None	$^{235}_{92}U$	$^{238}_{92}U$

nuclei have more neutrons than protons. Elements of atomic number 84 and above have no stable nuclei; all isotopes are unstable and radioactive. Table 5.1 lists some examples of stable and unstable (radioactive) isotopes.

Types of Radiation

You cannot feel the radiation particles or rays, but their great energy is capable of creating havoc within the cells of your body. The three most common types of radiation are *alpha particles, beta particles,* and *gamma rays.*

alpha particle, $^{4}_{2}\alpha$

The alpha particle is a low-energy particle that contains two protons and two neutrons. It is identical to the nucleus of a helium atom. This gives the alpha particle the same mass number (4) and atomic number (2) as a helium nucleus. Because there are no electrons involved, the alpha particle or helium nucleus has a charge of 2+. This means that an alpha particle is a positive ion. It is represented by the Greek letter alpha (α) or the symbol for helium:

Alpha particle: α $\quad ^{4}_{2}\alpha$ $\quad ^{4}_{2}He$

Alpha particles travel only a few centimeters in the air because they are the heaviest of the radiation particles. They can be stopped by paper, clothing, and the skin. They do not cause major damage unless the source of alpha particles enters the body through ingestion, inhalation, or open wounds. Then, severe internal damage can occur.

beta particle, $^{0}_{-1}\beta$

Beta particles are fast-moving electrons that are formed in an unstable nucleus and emitted at high speed. A beta particle is identical to an electron. It has a charge of −1 and very little mass. The beta particle is represented by the symbol for the Greek letter beta (β) or the symbol for the electron.

Beta particle: β $\quad ^{0}_{-1}\beta$ $\quad ^{0}_{-1}e$

Beta particles move much faster and with greater energy than alpha particles. Beta particles travel several meters through the air and penetrate 1–2 mm into solid material. External exposure to beta particles damages the skin, but the particles are stopped before they can reach the internal organs. Heavy clothing and gloves protect the skin from beta particles.

gamma ray, γ

Gamma rays are not particles, but instead are waves of high-energy radiation that are emitted from the nucleus of a radioactive atom. An unstable

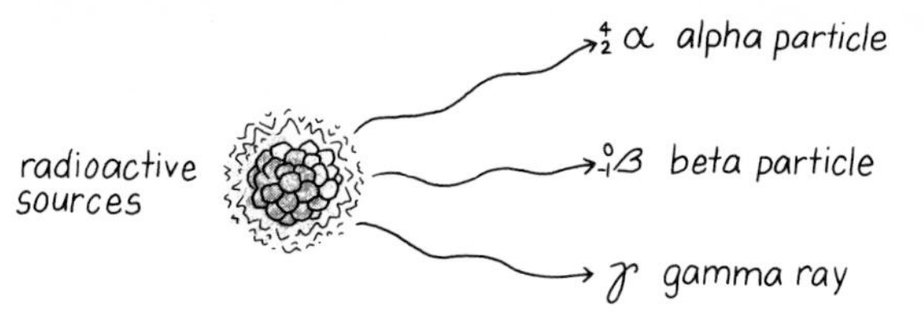

Figure 5.2 Unstable nuclei emitting alpha (α), beta (β), and gamma (γ) radiation.

nucleus releases some energy when its protons and neutrons rearrange to become more stable. Since gamma rays are waves of energy only, there is no mass or charge associated with the symbol for gamma rays (see Figure 5.2):

Gamma ray: γ $^0_0\gamma$

Gamma rays travel great distances through the air and penetrate deeply into solid material such as body tissues. Only heavy shielding, such as lead or concrete, will stop them. Any exposure to gamma rays is extremely hazardous. Radiologists who work with gamma radiation follow a strict set of precautions.

Example 5.2 Write the nuclear symbol for an alpha particle.

Solution: The alpha particle has the same number of protons and neutrons as in the nucleus of a helium atom. It has a mass number of 4 and an atomic number of 2:

$^4_2\alpha$

Radiation Protection

When working in areas containing radioactive isotopes, be sure that proper shielding is available. Lab coats and gloves are sufficient shielding for alpha and beta radiation, but lead or concrete shielding is needed for gamma rays. Even the syringe used to give an injection of radioisotopes should always be placed inside a special lead-glass cover. Table 5.2 summarizes the types of radiation and their required shielding materials.

shielding

When preparing radioactive materials, the radiologist wears special gloves and works behind leaded windows. Long tongs are operated within the work

Table 5.2 Types of Radiation and Shielding Materials

		Distance Traveled		
Type	Symbol	Through Air	Into Solid	Shielding
Alpha	$^4_2\alpha$	2–3 cm	None	Paper, clothing, skin
Beta	$^{0}_{-1}\beta$	Several meters	1–2 mm	Heavy clothing
Gamma	$^0_0\gamma$	Great distance	Deep	Lead, concrete

Figure 5.3 Shielding materials needed to absorb alpha, beta, and gamma radiation.

area to pick up vials of radioactive material, keeping them away from the hands and body. (See Figure 5.3.)

time

Keep the time you must spend in a radioactive area to a minimum. A certain amount of radiation is emitted every minute, so the amount of radiation received by your body accumulates continuously.

distance

Keep your distance! The greater the distance from the source of radiation, the lower the intensity of radiation received. The intensity of radiation decreases with the square of the distance from the radioactive object. If you double your distance from the radiation source, the intensity of radiation drops to $(\frac{1}{2})^2$ or one-fourth of its previous value, as Figure 5.4 shows. For this reason, the dentist or x-ray technician leaves the room or stands behind a shield to take your x-rays. Since these people are exposed to radiation every day, they must minimize the amount of radiation they receive.

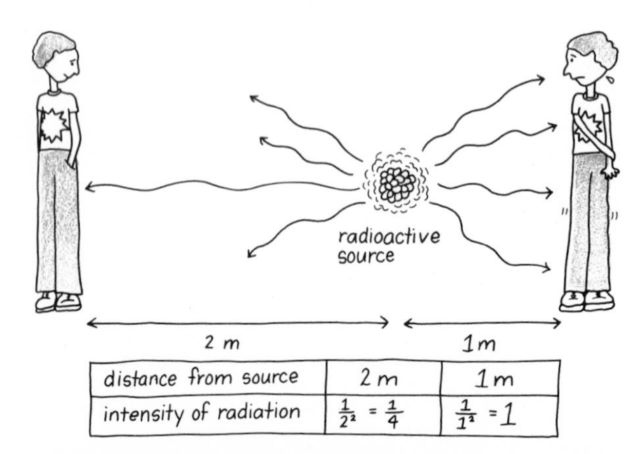

distance from source	2 m	1 m
intensity of radiation	$\frac{1}{2^2} = \frac{1}{4}$	$\frac{1}{1^2} = 1$

Figure 5.4 The intensity of radiation decreases as the distance from radiactive source increases.

HEALTH NOTE

Sensitivity of Cells to Radiation

The cells that are the most sensitive to radiation are young, immature cells undergoing rapid cellular division. Cells of the bone marrow, reproductive organs, and intestinal lining are among the kinds of cells most sensitive to radiation. It is very important to limit the radiation received by pregnant women or growing children because of the large number of rapidly dividing cells in their bodies. Cancer cells are also immature, rapidly dividing cells and are highly sensitive to radiation. For these reasons, radiation treatment is sometimes used to destroy cells of cancerous tissue. The surrounding normal tissue exhibits a greater resistance to radiation and suffers less damage. In addition, normal tissue is able to repair itself more rapidly than cancerous tissue. Cells such as those of the nerves, muscles, and adult bones have low sensitivity to radiation because they undergo little or no cellular division.

Example 5.3 How do shielding, time, and distance affect your exposure to radiation?

Solution: Shielding absorbs the radiation before it reaches your body. The amount of exposure increases with time; so the less time spent near a radioactive source, the lower the exposure. Keeping as far as possible from the source of radiation keeps exposure to a minimum.

5.3 Nuclear Reactions

OBJECTIVE Use nuclear symbols to write a nuclear equation for the radioactive decay of a radioisotope.

radioactive decay

The process whereby a radioactive isotope breaks down to release some type of radiation is called *radioactive decay*. The process occurs in *unstable* nuclei only. The result of radioactive decay is the formation of new isotopes and the emission of radiation particles. If the newly formed isotopes are stable, no further decay occurs.

nuclear equation

The process of radioactive decay can be represented by a *nuclear equation*. A nuclear equation uses the nuclear symbols for the radioactive isotopes that undergo decay as well as the newly formed isotopes and the radiation particles or rays that are emitted. A radioactive isotope can be classified according to the type of radiation it emits. Alpha emitters produce alpha particles during the decay process; beta emitters release beta particles; and gamma emitters give off gamma rays.

Alpha Emitters

Alpha emitters are radioisotopes that emit alpha particles. For example, an isotope of uranium $^{238}_{92}U$, decays to thorium, $^{234}_{90}Th$, by emitting an alpha particle (see Figure 5.5):

$$^{238}_{92}U \longrightarrow \, ^{234}_{90}Th + \, ^{4}_{2}\alpha$$

The alpha particle emitted from the uranium nucleus contains 2 protons, so the newly formed nucleus has just 90 protons and a new atomic number 90. The element with atomic number 90 is thorium, so the nucleus produced in radioactive decay of uranium is a thorium nucleus. The alpha particle emitted also has a mass number of 4, so the new nucleus of thorium has a mass number of 234. Note that in a balanced nuclear reaction, the sums of the mass numbers on each side of the equation are equal and the sums of the atomic numbers on each side of the equation are also equal:

	Uranium	=	thorium +	alpha particle
Mass numbers:	238	=	234 +	4
	U	$\longrightarrow$	Th	α
Atomic numbers:	92	=	90 +	2

Completing a Nuclear Equation

Let us look at another example of radioactive decay. A radioisotope of radium, $^{226}_{88}Ra$, emits an alpha particle to form new isotope X whose mass number (A), atomic number (Z), and identity we do not know:

$$\underset{\text{Radioisotope}}{^{226}_{88}Ra} \longrightarrow \underset{\text{Alpha particle emitted}}{^{4}_{2}\alpha} + \underset{\text{New isotope formed}}{^{A}_{Z}X}$$

To complete the nuclear equation, we need to determine the mass number (A) and atomic number (Z) of the new nucleus. Since the sum of the mass numbers of the reactants and the products must be equal, the mass number (A) of the

finding the mass number

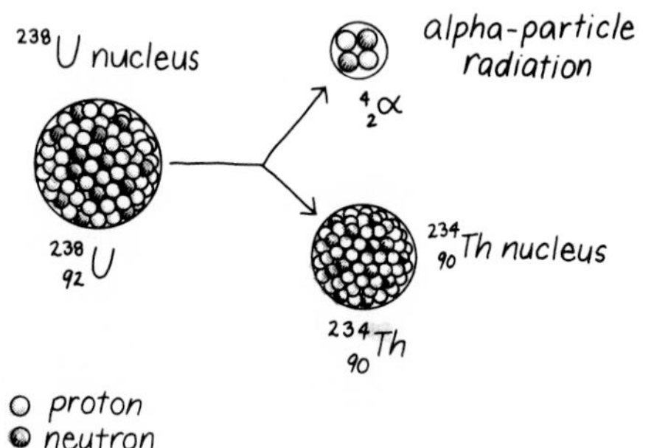

Figure 5.5 An unstable ^{238}U nucleus undergoes radioactive decay, emitting an alpha particle.

unknown isotope would be the difference of the mass numbers of the radium and the alpha particle:

$$226 = 4 + A$$
$$226 - 4 = A$$
$$222 = A$$

finding the atomic number

To calculate the atomic number (Z) of the new nucleus, we subtract the number of protons of the alpha particle (2) from the atomic number of radium:

$$88 = 2 + Z$$
$$88 - 2 = Z$$
$$86 = Z$$

We now know that the new nucleus has a mass number of 222 and an atomic number of 86:

$$^{226}_{88}Ra \longrightarrow ^{4}_{2}\alpha + ^{222}_{86}X$$

identify the unknown element

All that is needed now is the symbol of the element X, which has an atomic number of 86. This can be found by looking at the periodic chart, where you will find that radon (Rn) has atomic number 86. The completed nuclear equation then reads:

$$^{226}_{88}Ra \longrightarrow ^{4}_{2}\alpha + ^{222}_{86}Rn$$

Example 5.4 Write a nuclear equation for the decay of $^{185}_{79}Au$ by the emission of an alpha particle.

Solution: From the description of the nuclear reaction, we know that the nuclear equation begins with an isotope $^{185}_{79}Au$ and results in the formation of an alpha particle $^{4}_{2}\alpha$ and some new isotope:

$$^{185}_{79}Au \longrightarrow ^{4}_{2}\alpha + ^{A}_{Z}X$$

The mass number of X is 181, obtained by subtracting the mass number of the alpha particle (4) from the mass number of the gold isotope (185). The atomic number of X is 77, obtained by subtracting the number of protons in the alpha particle from the atomic number of the gold isotope (79 − 2):

Mass number (A): $185 = 4 + A$
$181 = A$

Atomic number (Z): $79 = 2 + Z$
$77 = Z$

The element with atomic number 77 is iridium, Ir. The completed nuclear equation for the reaction is as follows:

$$^{185}_{79}Au \longrightarrow ^{4}_{2}\alpha + ^{181}_{77}Ir$$

Beta Emitters

beta particle

Some nuclear reactions involve radioisotopes that decay by emitting beta particles. This type of radioisotope is called a *beta emitter*. The formation of a beta particle is a bit more subtle than the formation of the alpha particle. A beta particle is formed in the nucleus of the atom by the transformation of a neutron (^1_0n) into a proton (^1_1p or ^1_1H) and an electron ($^{\ 0}_{-1}e$):

$$^1_0n \longrightarrow {}^1_1p + {}^{\ 0}_{-1}e$$

The newly formed proton remains in the nucleus, but the electron is released with high energy and is called a *beta particle*. The equation for the transformation of a neutron in the nucleus can be rewritten using the nuclear symbol for a beta particle:

$$^1_0n \longrightarrow {}^1_1p + {}^{\ 0}_{-1}\beta$$

Note that the electrical charge of the neutron (0) is equal to the sum of the charge of the proton (+1) and the charge of the electron (−1).

Change in Atomic Number from Beta Radiation

A change occurs in the number of protons found in the nucleus when a neutron breaks apart into a proton and a beta particle. The number of neutrons decreases by 1, while the number of protons increases by 1. The atomic number (number of protons) of the newly formed isotope is now 1 greater than that of the original isotope. However, the mass number of the newly formed nucleus stays the same.

Let us see what this means in some nuclear equations for beta emitters. In the following nuclear reaction, a carbon radioisotope decays to a nitrogen isotope and a beta particle (see Figure 5.6):

	$^{14}_6C$	$\longrightarrow$	$^{14}_7N +$	$^{\ 0}_{-1}\beta$
Mass numbers:	14	=	14 +	0
Atomic numbers or charge:	6	=	7	+ (−1)

When a beta particle is emitted, the mass number does not change, but the atomic number does, representing a change of one element into another. The

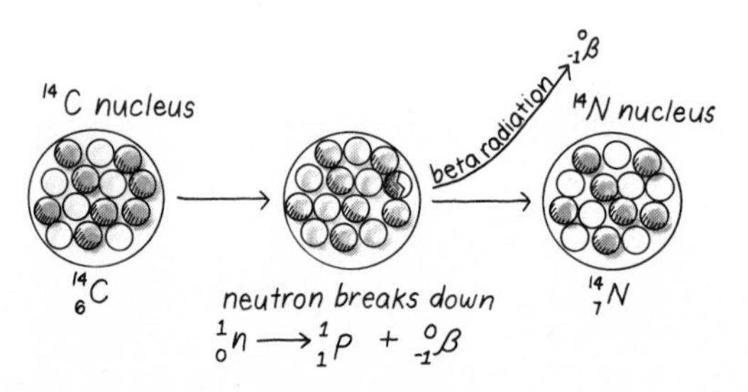

Figure 5.6 A breakdown in a neutron forms the beta particle.

radioactive isotopes of several biologically important elements are beta emitters:

$$^{24}_{11}\text{Na} \longrightarrow\ ^{24}_{12}\text{Mg} +\ ^{0}_{-1}\beta$$

$$^{59}_{26}\text{Fe} \longrightarrow\ ^{59}_{27}\text{Co} +\ ^{0}_{-1}\beta$$

Example 5.5 Write a nuclear equation for the decay of $^{16}_{7}\text{N}$ by the emission of a beta particle.

Solution: The description of the nuclear reaction indicates that the equation reactant is a nitrogen isotope and that the decay products are a beta particle and some unknown decay product:

$$^{16}_{7}\text{N} \longrightarrow\ ^{0}_{-1}\beta +\ ^{A}_{Z}\text{X}$$

The unknown decay product is identified by determining the mass number and atomic number that will provide a balance:

Mass Number of X

$$\begin{aligned} 16 &= 0 + A \\ 16 - 0 &= A \\ 16 &= A \end{aligned}$$

Atomic number of X

$$\begin{aligned} 7 &= -1 + Z \\ 7 - (-1) &= Z \\ 8 &= Z \end{aligned}$$

The element with atomic number 8 is oxygen (O). We can write the finished nuclear equation for this nuclear reaction of a beta emitter:

$$^{16}_{7}\text{N} \longrightarrow\ ^{0}_{-1}\beta +\ ^{16}_{8}\text{O}$$

HEALTH NOTE

Uses of Beta Emitters in Medicine

When a radiologist wants to treat a malignancy within the body, a beta emitter is sometimes used. The short range of penetration into the tissue by beta particles is advantageous. For example, some malignant tumors produce fluid within the body tissues. This is uncomfortable for the patient and leads to edema. A colloidal chromic phosphate compound containing ^{32}P, a beta emitter, is injected into the body cavity containing the tumor. The beta particles have a tissue range of 2–8 mm, so only the tissue and the malignancy within that range are damaged. Such damage to the cells within the tumor slows or stops the growth of the tumor and decreases the production of fluid.

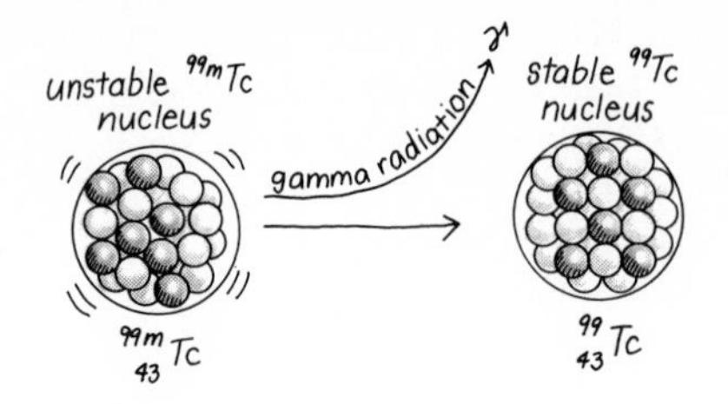

Figure 5.7 Gamma radiation occurs when an unstable nucleus rearranges to a more stable nucleus.

Gamma Emitters

In radiology, one of the most commonly used gamma emitters is an unstable form of technetium (Tc). This high-energy, unstable state is sometimes called the *metastable* form and can be indicated by the letter "m." The unstable nucleus emits a gamma ray and goes to a more stable, less energetic isotope (see Figure 5.7):

$$^{99m}_{43}Tc \longrightarrow {}^{0}_{0}\gamma + {}^{99}_{43}Tc$$

HEALTH NOTE

Cellular Damage Caused by Radiation

ion pair

ionizing radiation

Radiation particles and rays cut a path right through the cells of the body. As the radiation passes through the cells, collisions occur with the molecules within the cells. This interaction may strip an electron from an atom within the cell. The separation of an electron from the atom produces an *ion pair*—a negatively charged electron (e^-) and a positively charged ion (H_2O^+). Figure 5.8 shows how this might happen. Radiation that produces ion pairs is called *ionizing radiation.*

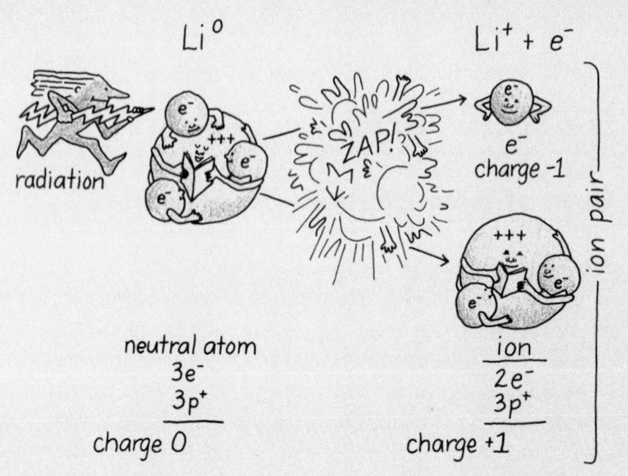

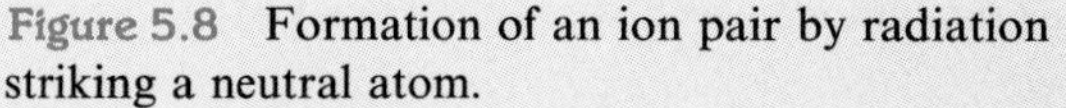

Figure 5.8 Formation of an ion pair by radiation striking a neutral atom.

5.4 Artificial Radioactivity

OBJECTIVE Write a nuclear equation for the artificial production of a radioisotope.

A stable, nonradioactive isotope can be made into a radioactive isotope by causing a collision between the nonradioactive isotope and fast-moving nuclear particles such as protons, neutrons, electrons or alpha particles. When one of these particles is absorbed by the nucleus, the nucleus becomes unstable and therefore radioactive. The newly formed radioisotope emits radiation as it acquires a more stable state.

Making a Stable Nucleus Radioactive

A nuclear reactor contains radioactive isotopes that continuously emit particles at high speeds. When a stable isotope is placed in the reactor, it is bombarded with these fast-moving particles. For example, a stable form of calcium, $^{40}_{20}Ca$, may be placed in a nuclear reactor. If the calcium nucleus absorbs a fast moving neutron, it is changed into an unstable isotope of potassium, $^{40}_{19}K$. A proton is

Radiation particles may also collide with water molecules within the cells, causing the separation of a proton. The resulting ion pair would be H^+ and OH^-. Figure 5.9 illustrates the formation of ion pairs from water molecules.

Ions produced by radiation bombardment of the cells disrupt activity within a cell so that the cell can no longer function properly. If this disruption is great, the cell dies. If not, the cell may repair itself and function again. However, such repair is not always complete. Long-range effects of radiation include a shortened life span, malignant tumors, leukemia, anemia, and genetic mutations.

The disruption of cellular activity occurs in two steps. The initial radiation causes changes in the chemical components of the cells, such as the formation of ion pairs in the cell. As a result, the cell loses its ability to produce necessary materials such as a hormone or red blood cell. For example, the effect of radiation in the bones is damage to some cells of the bone marrow. The damaged bone marrow is no longer able to produce as many red blood cells.

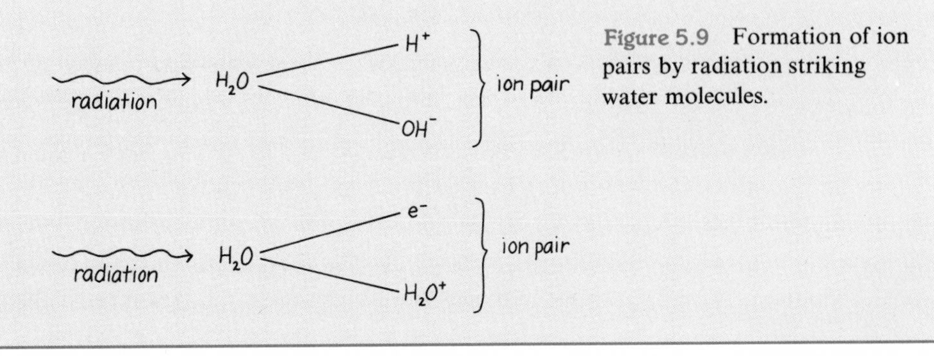

Figure 5.9 Formation of ion pairs by radiation striking water molecules.

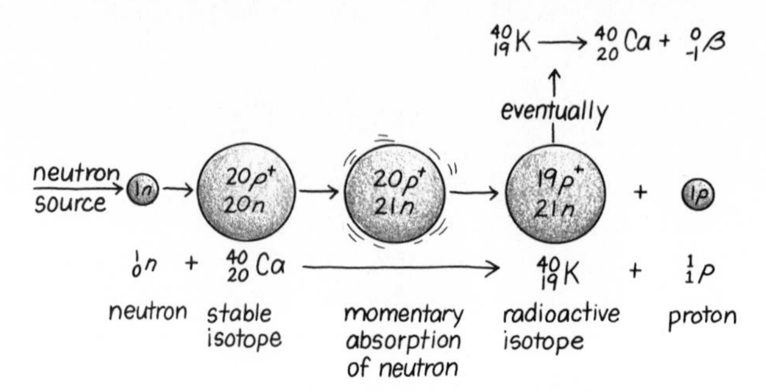

Figure 5.10 Transmutation: Formation of radioactive isotopes by bombardment with accelerated neutrons.

lost during the reaction. The process of changing one element into another is called *transmutation* (see Figure 5.10):

$$^{40}_{20}\text{Ca} + ^{1}_{0}n \longrightarrow ^{40}_{19}\text{K} + ^{1}_{1}\text{H}$$

$^{40}_{20}$Ca	$^{1}_{0}n$	$^{40}_{19}$K	$^{1}_{1}$H
Stable isotope (nonradioactive)	Neutron from reactor	Unstable isotope (radioactive)	Proton emitted

In another example, an alpha particle collides with a stable isotope of nitrogen. The bombardment results in the formation of an unstable oxygen isotope $^{17}_{8}$O and a proton:

$$^{14}_{7}\text{N} + ^{4}_{2}\alpha \longrightarrow ^{17}_{8}\text{O} + ^{1}_{1}\text{H}$$

$^{14}_{7}$N	$^{4}_{2}\alpha$	$^{17}_{8}$O	$^{1}_{1}$H
Stable isotope	Alpha particle	Unstable isotope (radioactive)	Proton emitted

Example 5.6 Write the nuclear equation for the neutron bombardment of $^{13}_{6}$C to produce the radioisotope $^{14}_{6}$C.

Solution: The reactants in this nuclear equation are the $^{13}_{6}$C isotope and a neutron. The product is $^{14}_{6}$C:

$$^{13}_{6}\text{C} + ^{1}_{0}n \longrightarrow ^{14}_{6}\text{C}$$

Since the mass numbers on the reactant and product are equal, there are no other products. This is also true of the atomic numbers (or number of protons).

Example 5.7 Write in words the meaning of the following nuclear equation:

$$^{15}_{7}\text{N} + ^{1}_{1}\text{H} \longrightarrow ^{12}_{6}\text{C} + ^{4}_{2}\alpha$$

Solution: In the nuclear reaction, the nitrogen isotope, ^{15}N, is bombarded by a proton to produce a carbon isotope, ^{12}C, and an alpha particle.

HEALTH NOTE

Producing Radioisotopes for Nuclear Medicine

Of the many artificial radioisotopes (over 1500) produced today, several find use in nuclear medicine. For example, the radioisotope of gold, ^{198}Au, is produced by bombarding a stable isotope of gold with a neutron:

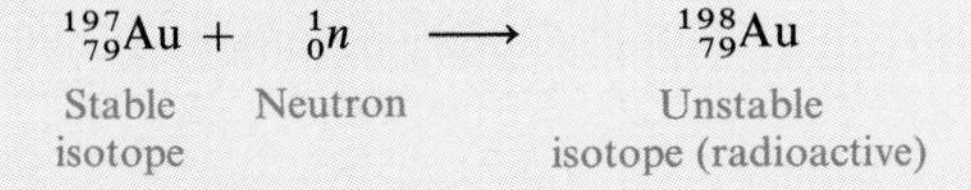

The resulting radioactive isotope is used in studies of the liver and lymph nodes. It decays by emitting a beta particle to form a stable mercury isotope:

$$^{198}_{79}Au \longrightarrow ^{0}_{-1}\beta + ^{198}_{80}Hg$$

Radioisotope — Radiation product — Stable product

The metastable form of technetium is produced by neutron bombardment of molybdenum, $^{98}_{42}Mo$. Many radiation laboratories have small reactors that produce ^{99m}Tc radioisotope every day:

$$^{98}_{42}Mo + ^{1}_{0}n \longrightarrow ^{99m}_{43}Tc + ^{0}_{-1}\beta$$

Stable isotope — Neutron — Unstable isotope (radioactive)

The radioisotope ^{99m}Tc produced from this bombardment reaction is used for detection of brain tumors and for liver and spleen examinations. The radioactive technetium decays by emitting gamma rays.

5.5 Radiation Detection

OBJECTIVE Explain the process by which a radioisotope in the body is detected.

The chemistry of all isotopes of an element is identical even if the isotope is radioactive. The cells in the body cannot differentiate between a nonradioactive atom and a radioactive one. Therefore, a small amount of a radioisotope injected into a patient interacts, metabolizes, and locates in the same organs as the nonradioisotopes of that element. For example, the radioisotopes ^{131}I and ^{125}I are absorbed by the thyroid because the thyroid attracts most of the iodine in the body. The radioisotope ^{75}Se locates in the pancreas; ^{99m}Tc will be absorbed by the brain, liver, bone, and spleen; and ^{67}Ga concentrates in certain malignancies, mainly lymphomas. Table 5.3 lists some radioisotopes used in nuclear medicine and the organs in which they concentrate.

Table 5.3 Some Radioisotopes Used in Nuclear Medicine

Element	Radioisotope	Location in Body
Iodine	${}^{131}_{53}I$	Thyroid, lung, kidney
Chromium	${}^{51}_{24}Cr$	Spleen
Selenium	${}^{75}_{34}Se$	Pancreas
Technetium	${}^{99m}_{43}Tc$	Brain, lung, liver, spleen, bone
Gallium	${}^{67}_{31}Ga$	Lymphomas

Detecting Radioisotopes in the Body

scanner

Suppose a radiologist wants to examine the condition of an organ. How is this done? First, the patient is given a radioisotope that concentrates in that organ. Then an apparatus called a *scanner* is used to produce an image of the organ. The scanner moves slowly across the patient's body above the region where the organ containing the radioisotope is located. The radiation emitted from the radioisotope in the organ passes through the body and strikes a detection tube within the scanner.

In the detection tube, ion pairs are created by the radiation. The charged ions cause a burst of electrical current, which a detector changes to a flash of light. The light exposes a photographic plate, producing an image of the organ. See Figure 5.11.

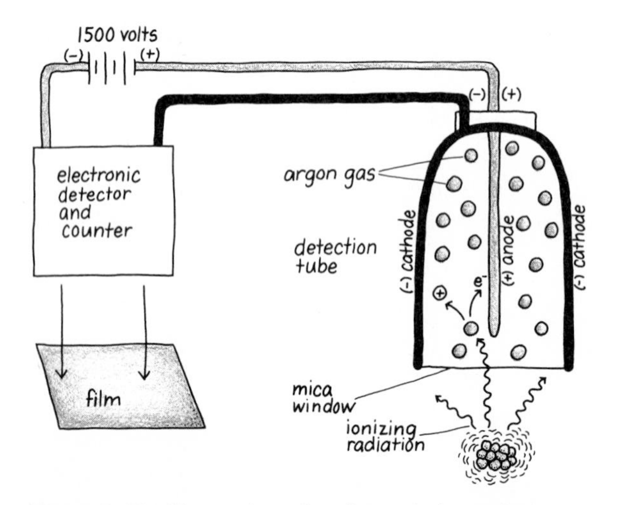

Figure 5.11 Formation of an ion pair by radiation passing through a detection tube, thereby causing a current pulse, which is amplified and counted.

A large detecting camera is also used to view the entire organ. The camera is positioned over the patient, where it picks up radiation from all parts of the organ. The resulting image is called a *scan*. The location and distribution of radioisotope in the organ can tell a radiologist the size, shape, and functioning of the organ.

scan

When there is decreased activity of cells in an organ, there is less uptake of radioisotope relative to the surrounding tissue and therefore less radiation emitted from that part of the organ. When there is increased activity of the cells in the organ, there is increased concentration of radioisotope and more radiation is picked up.

Example 5.8 How does a scanner detect radiation?

Solution: When radiation passes through the detection chamber of the scanner, ion pairs form. These charged ions produce a burst of current, producing a burst of light, which can be recorded.

HEALTH NOTE

Use of Radioactive Iodine in Testing Thyroid Function

A common method of determining thyroid function is by the use of radioactive iodine uptake (RAIU). Although the ^{131}I radioisotope can be used, ^{125}I is preferred because it delivers a lower radiation dosage to the patient. The radioisotope is taken orally and mixes with the iodine already present in the body. The patient returns to the doctor in 24 hours for a measure of iodine uptake. A detection tube held up to the area of the thyroid gland counts the radiation coming from the thyroid gland.

radioactive iodine uptake depends on the activity of the thyroid gland

The iodine uptake is directly proportional to the activity of the thyroid gland. A patient with a hyperactive thyroid will have a high reading of radioactive iodine in the thyroid, whereas a patient with a hypoactive thyroid will record low values.

If the patient has hyperthyroidism, treatment is begun to lower the activity of the thyroid. One treatment involves giving the patient a therapeutic dosage of radioactive iodine which has a higher radiation count than the diagnostic dose. The radioactive iodine locates in the thyroid where its radiation causes damage to a percentage of the thyroid cells. The damaged cells can no longer produce the thyroid hormone. The decreased number of functioning thyroid cells lowers the production of thyroid hormone and brings the condition under control.

HEALTH NOTE

Visualization of the Brain and Skull

brain

In a normal brain scan, radioactive isotopes do not enter brain cells because of the presence of a protective blood–brain barrier, and there is no concentration of radioisotopes within the brain. However, if a blood vessel has broken or there is a brain tumor drawing on the blood supply, radioisotope uptake does occur within the brain and appears on the photoscan. (See Figure 5.12.)

CT scan

Another method used to visualize the brain is known as *computerized tomography* (CT) *scan.* A computer monitors the degree of absorption of 30,000 *x*-ray beams directed at the brain at successive levels. The differences in absorption based upon the densities of the tissues and fluids in the brain provide a series of images of the brain. This technique is successful in the identification of brain hemorrhages, tumors, and atrophy.

magnetic resonance imaging

Magnetic resonance imaging is a new form of visualization of internal organs that shows great promise for the future. It is predicted that magnetic resonance

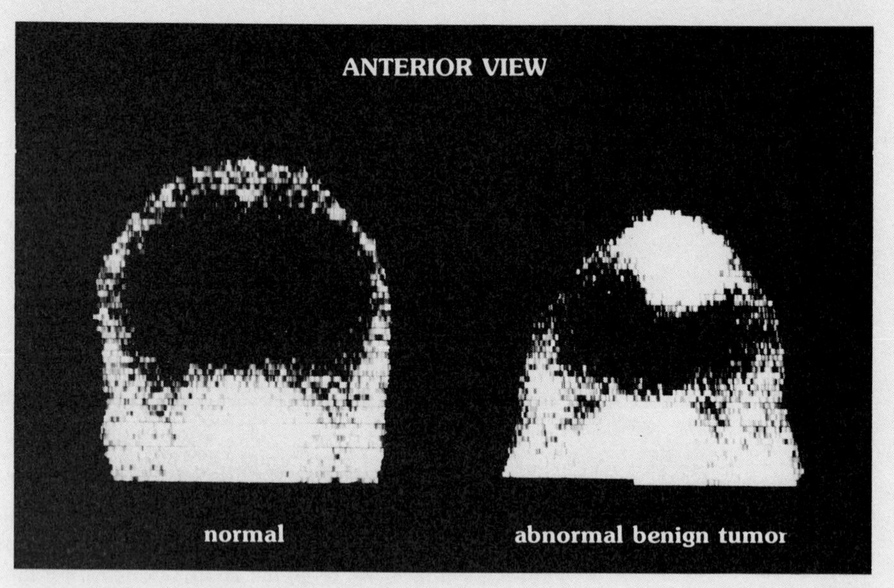

Figure 5.12 Brain scans: four views.

5.6 Half-Life of a Radioisotope

OBJECTIVE Given the mass of a radioisotope and its half-life, determine the quantity that remains after one or more half-lives have passed.

half-life

The *half-life* of a radioactive isotope is the amount of time it takes for one-half of the sample to decay (emit radiation). For example, a radioisotope of iodine, ^{131}I, has a half-life of 8 days. Suppose you have a 40-mg sample. In 8 days, one-half,

imaging may replace the use of x-rays and current scanning techniques including the CT scanner. This particular imaging technique differs from current forms of imaging because it uses no radiation. Strong radio impulses are sent through the body of a patient placed in a magnetic field which alter the energy level of the nuclei in hydrogen atoms. These changes are detected, and computers create an image that can be displayed on a television monitor.

A major factor in the slow appearance of the imagers is the cost of the scanner system and the charge to the patient. Some doctors are also concerned with the ability of the technique to differentiate properly between certain kinds of abnormalities. However, this new imaging technique has already found use in simplifying the detection of multiple sclerosis, abnormalities of the spine and brain, tumors, and birth malformations. You will be able to determine the success of the magnetic resonance imaging technique yourself by observing its impact on the field of imaging techniques over the next few years.

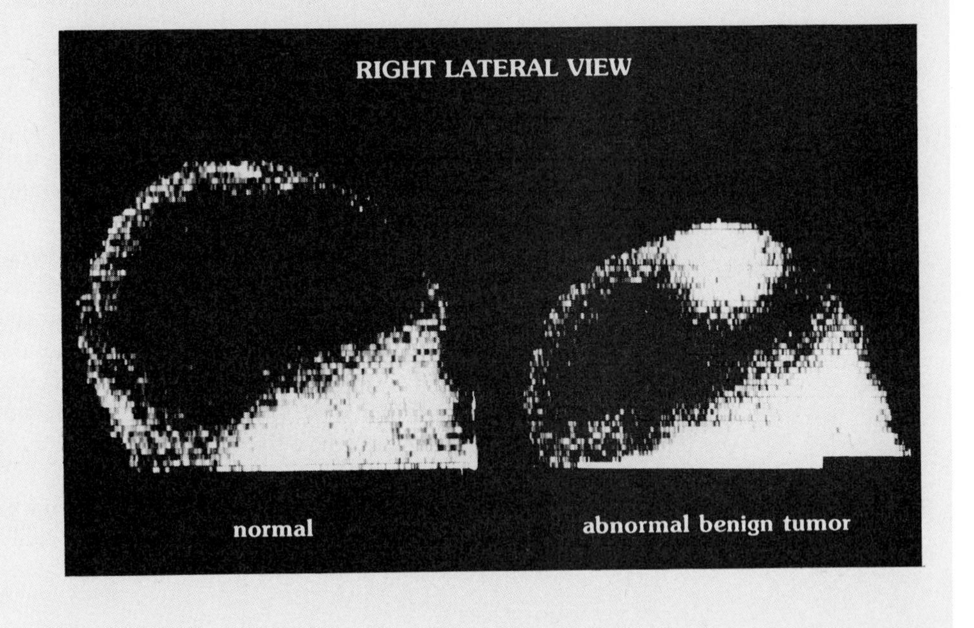

20 mg, of the original sample will emit radiation (decay) changing to a new substance. The other half of the original sample, 20 mg, is left. In the next 8 days, half of the 20 mg sample decays leaving 10 mg still radioactive:

$$40 \text{ mg } {}^{131}\text{I} \xrightarrow[\text{One half-life}]{\text{8 days}} 20 \text{ mg } {}^{131}\text{I} \xrightarrow[\text{Two half-lives}]{\text{8 days}} 10 \text{ mg } {}^{131}\text{I} \xrightarrow[\text{Three half-lives}]{\text{8 days}} 5 \text{ mg } {}^{131}\text{I}$$

Some radioactive isotopes decay very rapidly, with most of the atoms emitting radiation within a few hours or even seconds. Such radioisotopes have

Table 5.4 Half-Lives and Uses of Some Medical Radioisotopes

Radioisotope	Half-Life	Radiation	Uses
^{51}Cr	28 days	γ	Blood volume, spleen imaging
^{198}Au	2.7 days	β, γ	Liver imaging, treatment of abdominal carcinoma
^{131}I	8 days	β, γ	Studies of thyroid; treatment of thyroid and hyperthyroidism
^{125}I	60 days	γ	Thyroid imaging, plasma volume, fat absorption
^{59}Fe	46 days	β, γ	Blood volume determination
^{32}P	14 days	β	Treatment of leukemia, polycythemia vera, and lymphomas; diagnostic for brain and breast tumors
^{42}K	12 h	β, γ	Localization of brain tumors
^{24}Na	15 h	β, γ	Study of vascular disease, extracellular volume, and circulation time
^{85}Sr	64 days	γ	Bone imaging with known malignancies
^{99m}Tc	6 h	γ	Diagnostic brain, thyroid, lung, kidney, and bone marrow scans

short half-lives. Other radioactive isotopes have long half-lives. They take a long time to disintegrate and continue to produce radiation over a very long period of time, even hundreds or millions of years.

Many of the radioisotopes used in diagnostic work and therapy have short half-lives. For example, ^{99m}Tc is often used because it has a short half-life and is a pure gamma emitter. Gamma emission is desirable for diagnostic work because gamma rays pass through the body and are detected. Alpha and beta particles do not travel far enough through the body to be easily detected, but they still cause internal damage. Table 5.4 lists some radioisotopes used in medicine and gives their half-lives and medical use. Such radioisotopes are produced artificially.

Radioisotopes with long half-lives are usually naturally occurring isotopes of the elements. Some radioisotopes with long half-lives are listed in Table 5.5.

Table 5.5 Half-Lives of Some Naturally Occurring Radioisotopes

Element	Radioisotope	Half-Life	Radiation
Carbon	$^{14}_{6}C$	5730 yr	β
Potassium	$^{40}_{19}K$	1.26×10^9 yr	β, γ
Lead	$^{210}_{82}Pb$	20 yr	α, β, γ
Radium	$^{226}_{88}Ra$	1600 yr	α, γ
Uranium	$^{235}_{92}U$	7.1×10^8 yr	α, γ

Decay Curve

one-half of a radioactive sample emits radiation in 1 half-life

Every radioisotope decays at a constant rate. This makes it possible to calculate the quantity of radioactive sample at any given time. For example, ^{99}Tc has a half-life of 6 h. Suppose a patient receives 10,000 atoms of ^{99m}Tc. In 6 h, 5000 atoms of ^{99m}Tc remain undecayed:

$$\begin{aligned}\text{Quantity of radioisotope after one half-life} &= \text{Initial sample} \times \tfrac{1}{2}\\ &= 10{,}000 \text{ atoms} \times \tfrac{1}{2}\\ &= 5{,}000 \text{ atoms}\end{aligned}$$

After 12 h, or two half-lives, half of those 5000 atoms will have decayed, so 2500 atoms of radioactive isotope are left:

$$\begin{aligned}\text{Quantity of radioisotope after two half-lives} &= \text{Initial sample} \times \underbrace{\tfrac{1}{2} \times \tfrac{1}{2}}_{\text{(Two half-lives)}}\\ &= 10{,}000 \text{ atoms} \times \tfrac{1}{2} \times \tfrac{1}{2}\\ &= 2{,}500 \text{ atoms}\end{aligned}$$

After 18 h, or three half-lives, there are only 1250 atoms undecayed. A graph of this rate of decay, called a *decay curve*, is shown in Figure 5.13.

$$\begin{aligned}\text{Quantity of radioisotope after three half-lives} &= \text{Initial sample} \times \underbrace{\tfrac{1}{2} \times \tfrac{1}{2} \times \tfrac{1}{2}}_{\text{(Three half-lives)}}\\ &= 10{,}000 \text{ atoms} \times \tfrac{1}{2} \times \tfrac{1}{2} \times \tfrac{1}{2}\\ &= 1{,}250 \text{ atoms}\end{aligned}$$

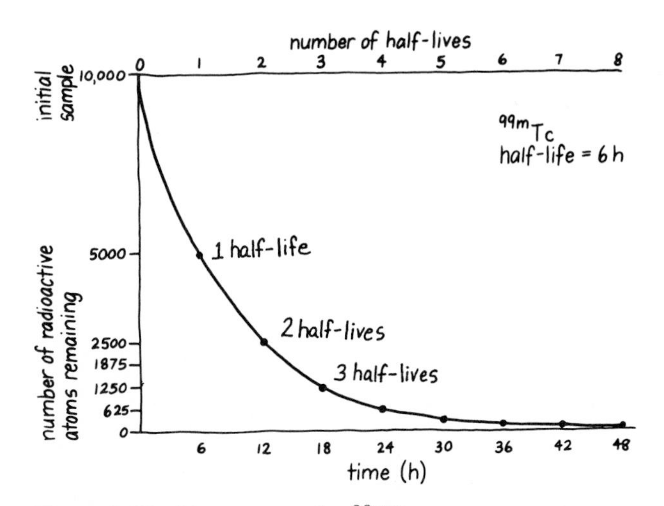

Figure 5.13 Decay curve for ^{99m}Tc.

Another way to calculate the amount of radioactive technetium remaining in the sample is to construct a chart like the one shown below, with columns for the number of half-lives and for time that has passed, as well as for the amount of radioactive isotope that is left in the sample:

Number of Half-Lives	*Elapsed Time*	*Amount of Radioactive Isotope Remaining*
0	0	10,000 atoms (initial)
1	6 h	5,000 atoms
2	12 h	2,500 atoms
3	18 h	1,250 atoms

HEALTH NOTE

Dating Ancient Objects

carbon dating

A technique known as *carbon dating* is widely used by archeologists and historians as a way of measuring the age of an ancient object. Radioactive carbon-14 is continuously produced in the upper atmosphere, where it reacts with oxygen to form radioactive carbon dioxide, $^{14}CO_2$. Since CO_2 is absorbed by plants during the process of photosynthesis and incorporated into plant tissues, some of the radioactive carbon will be incorporated into the carbohydrates produced by the plant. Because plants and plant products are part of animal diets, the radioactive carbon-14 is also incorporated into animal tissues. A constant level of carbon-14 is reached in living plants and animals. After the plant or animal dies, no more carbon-14 is taken up and the level of radioactivity falls as carbon-14 undergoes radioactive decay.

Carbon-14 has a half-life of about 6000 yr. Scientists use its half-life to calculate the quantity of time that has passed since the time that plant or animal died. The smaller the percentage of carbon-14 in the sample, the greater the number of half-lives that have passed. Thus, the approximate age of the sample can be determined. Samples that can be measured include paint from plant pigments, cooking utensils, bone material, cotton or woolen clothing, and wood materials.

potassium argon dating

There are other radiochemical methods that are now being used to establish dates for ancient objects. For example, potassium-40, ^{40}K, has a half-life of 1.3×10^6 yr. It decays to argon-40, ^{40}Ar, in a process called *electron capture*. By measuring the ratio of ^{40}K to ^{40}Ar, dates reaching back as far as 1 million years (1×10^6 yr) can be established:

$$^{40}_{19}K + {}^{\ 0}_{-1}e \longrightarrow {}^{40}_{18}Ar$$

^{238}U dating

Another dating method is based on the radioisotope that decays, through a series of reactions, to ^{206}Pb. The uranium-238, ^{238}U isotope has a half-life of about 4×10^9 (four billion) yr. This radioisotope is measured to date objects with ages in the range of a billion years.

Example 5.9 The radioisotope of phosphorus, $^{32}_{15}P$, has a half-life of 14 days. How much of a 80.0-mg ^{32}P sample is still radioactive after the sample has decayed over a period of 28 days?

Solution: To solve the problem, we need to first determine the number of half-lives that have elapsed over the given time period:

$$\text{Number of half-lives} = 28 \text{ days} \times \frac{1 \text{ half-life}}{14 \text{ days}} = 2 \text{ half-lives}$$

The amount of sample remaining radioactive can now be calculated:

$$\begin{aligned}\text{Quantity of sample still radioactive after 2 half-lives} &= \underset{\text{(Initial sample)}}{80.0 \text{ mg}} \times \underset{\text{(Two half-lives)}}{\tfrac{1}{2} \times \tfrac{1}{2}} \\ &= 80.0 \times \tfrac{1}{4} \\ &= 20.0 \text{ mg}\end{aligned}$$

After 28 days, 20.0 mg of the original 80.0-mg phosphorus-32 sample remain radioactive.

Alternatively, a table can be constructed to give the number of half-lives and the resulting amount of radioactive isotope remaining in the sample:

Number of Half-Lives	*Elapsed Time*	*Amount of Radioactive Isotope Remaining*
0	0	80.0 mg (initial sample)
1	14 days	40.0 mg
2	28 days	20.0 mg (remain)

5.7 Nuclear Fission and Fusion

OBJECTIVE Identify the characteristics of fission and fusion processes.

nuclear fission
atomic energy

In the 1930s, scientists were experimenting with bombarding uranium-235 with neutrons. They discovered that two (and sometimes more) medium-weight nuclei were produced as well as a great amount of energy. This was the discovery of a new kind of nuclear reaction called *nuclear fission*. The energy associated with nuclear fission was called *atomic energy*.

When an isotope of uranium, $^{235}_{92}U$, is bombarded by neutrons, two medium-weight nuclei, several neutrons, and a great amount of energy are produced. There are several fission products that may result. The nuclear equation for one possible nuclear fission reaction is as follows:

$$^{235}_{92}\text{U} + {}^{1}_{0}n \longrightarrow {}^{139}_{56}\text{Ba} + {}^{94}_{36}\text{Kr} + 3\,{}^{1}_{0}n + \text{Energy}$$

If you could weigh these products with great accuracy, you would find that the sum of their masses is slightly less than the total mass of the starting material.

This missing mass has been converted into energy in an amount that can be calculated from an equation derived by Albert Einstein:

$$E = mc^2$$

E is the energy released, m is the mass lost, and c is the speed of light, 3×10^{10} cm/s. Even though the mass change is very small, it is multiplied by the speed of light squared, 9×10^{20}, to give a large value for the energy.

The fission reaction begins when the nucleus of a uranium atom is hit by a neutron. The nucleus undergoes fission by splitting into smaller isotopes of barium ($^{139}_{56}Ba$) and krypton ($^{94}_{36}Kr$), along with some neutrons and a large amount of energy. The neutrons released are capable of bombarding more uranium nuclei, and each resulting fission reaction releases additional neutrons. The number of bombarding neutrons increases rapidly, as does the number of uranium atoms undergoing fission. Once started, the fission reaction can run by itself in a chain reaction, with a rapid production of heat and energy. (See Figure 5.14.)

In the atomic bomb, sufficient quantities of uranium-235 are brought together to provide a critical mass that will sustain a nuclear chain reaction. The buildup of vast amounts of heat and energy of the chain reaction eventually creates an atomic explosion.

Nuclear Power Plants

In a nuclear power plant, the fission reaction is slowed down. The quantity of uranium-235 is held below critical mass, so it cannot sustain a chain reaction.

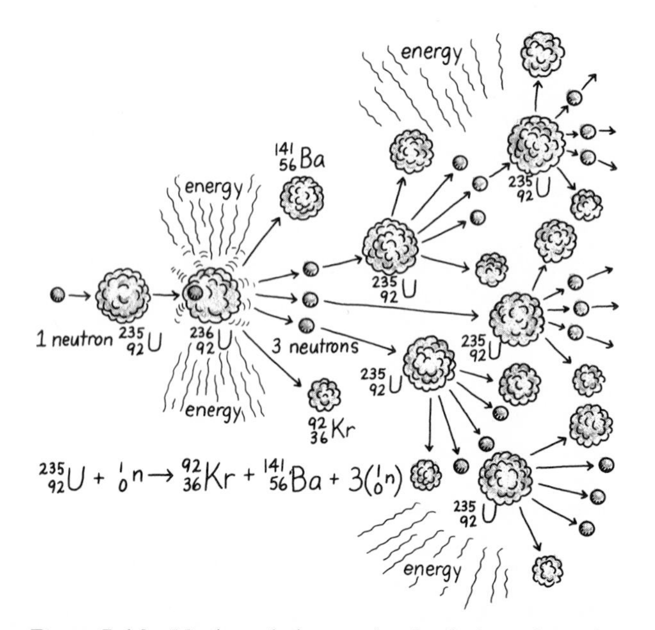

Figure 5.14 Nuclear chain reaction by fission of uranium 235.

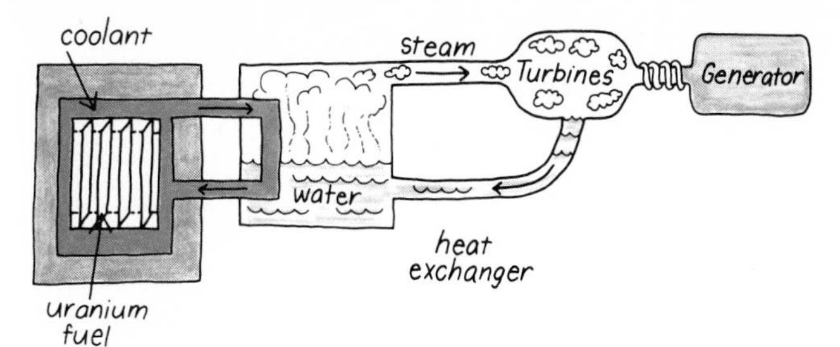

Figure 5.15 The generation of electricity in a nuclear power plant.

Control rods placed among the samples of uranium absorb some of the fast-moving neutrons. In this way, less fission occurs, giving a slower, controlled production of energy. The energy that is released from the fission reactions is given off as heat. The heat is transferred to containers of water, which when heated produce steam. The steam then passes through turbines, which produce electricity. (See Figure 5.15.)

Although nuclear power plants have the potential to solve some of our energy needs, there are major problems. One such problem is the production of nuclear waste. As the fission reaction occurs, there is an accumulation of waste products that are radioactive isotopes. These waste products have long half-lives and must be stored for a very long time where there is no chance of contamination. Unfortunately, we have not yet found a good solution to this problem. The method for handling the accumulation of radioactive waste products is one of much concern.

Nuclear Fusion

nuclear fusion In nuclear fusion, two small nuclei combine to form a larger nucleus. In the fusion process, mass is lost, and a tremendous amount of energy is released, even more than in nuclear fission. A very high temperature (100,000,000°C) is required for the fusion. One place where this temperature occurs is in the sun, where fusion reactions are occurring continuously to provide us with heat and light. One example of a fusion reaction is the following, in which two isotopes of hydrogen combine to form helium (see Figure 5.16):

$$\underset{\text{Deuterium}}{{}^{2}_{1}\text{H}} + {}^{3}_{1}\text{H} \longrightarrow {}^{4}_{2}\text{He} + {}^{1}_{0}n + \text{Energy}$$

The fusion reaction has tremendous potential as a possible source for future energy needs. One of the advantages of fusion as an energy source is that the deuterium isotope of hydrogen is plentiful in the oceans. Although scientists now expect some radioactive waste from fusion reactors, the amounts of radioactive waste are expected to be much less than from fission; also these wastes are expected to have shorter half-lives. However, fusion is still in the

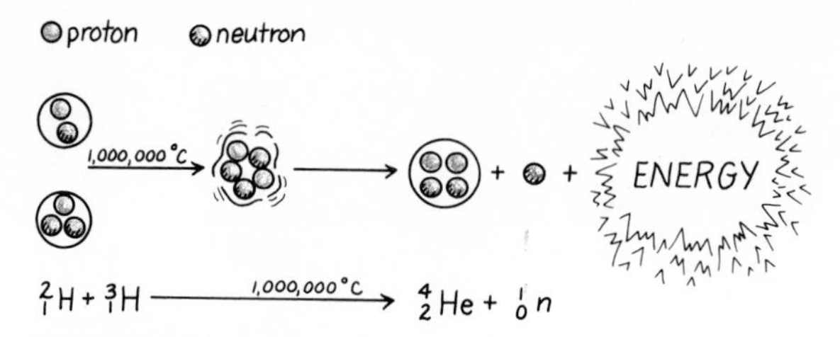

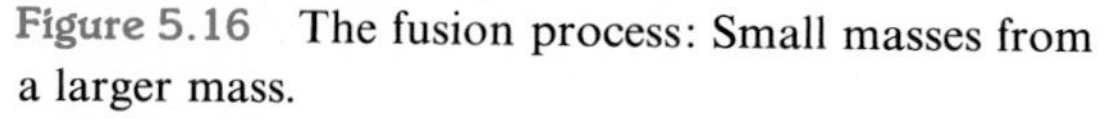

Figure 5.16 The fusion process: Small masses from a larger mass.

experimental stage because the extremely high temperatures needed for fusion have been difficult to reach and even more difficult to maintain. Several research groups around the world are currently attempting to develop the technology needed to make the fusion reaction a reality in our lifetimes.

Example 5.10 Indicate whether each of the following are characteristic of nuclear fission or nuclear fusion processes:

1. Small nuclei combine to form larger nuclei.
2. Large amounts of energy are released.
3. Very high temperatures are needed for reaction.

Solution

1. Fusion
2. Both
3. Fission

5.8 Units of Radiation Measurement

OBJECTIVE Describe the units of radiation measurement.

curie
roentgen
rad, rem

There are several units used to describe the level or quantity of radiation that is emitted by a radioisotope or received by a patient. The *curie* (Ci) measures the number of atoms that break down (disintegrate) during radioactive decay. The *roentgen* (R) measures the amount of ionization (ion pairs) of air molecules caused by radiation. The biological effects of radiation are described in units called the *rad* and the *rem.* These biological units reflect the radiation absorbed by the body and the damage caused by specific kinds of radiation. Table 5.6 explains these various radiation units of measurement.

Exposure to Radiation

You are always being exposed to some radiation every day. Naturally occurring radioactive isotopes are part of the atoms of wood, brick, and concrete in our

Table 5.6 Units of Radiation Measurement

Unit	Symbol	Explanation
Curie	Ci	The amount of radioactive material that causes 3.7×10^{10} (37 billion) disintegrations/second.
Roentgen	R	The ionization ability of x-rays and gamma rays. One roentgen produces 2×10^9 ion pairs of 1 mL of air.
Radiation absorbed dosage	rad	The absorption of radiation by body tissue. In soft tissue, the exposure to 1 R will produce an absorption dosage of about 1 rad.
Roentgen equivalent man	rem	The biological damage produced by the absorption of radiation. For beta and gamma rays, 1 rem is the biological effect of 1 rad of radiation. The term millirem (mrem) is often used in the hospital for radiation measurement (1 rem = 1000 mrem).

homes and the buildings where we work and go to school. Radioactivity is present in the soil, the food we eat, the water we drink, and even in our bodies. For example, a naturally occurring radioactive isotope of potassium, ${}^{40}_{19}K$, is found in the body because ${}^{40}_{19}K$ is always found with all sources of posassium, including food.

The atmosphere is another natural source of radiation. We are constantly being exposed to radiation from cosmic rays. At higher elevations, the amount of radiation from cosmic rays is greater because there are fewer atmospheric molecules to absorb the radiation. People living in Denver or those flying in an

Table 5.7 Radiation Typically Received by U.S. Citizen

Source	Average Dose/Year (mrem)
Environment	
Natural	
Ground	15
Air, water, food	30
Cosmic rays	40
Wood, concrete, brick	50
Global fallout	4
Medical	
Diagnostic	
Chest x-ray	50
Dental x-ray	20
Upper gastrointestinal tract x-ray	200
Occupational	2
Other	
Television	2
Air travel	1

airplane receive more radiation from cosmic rays than do people living in Los Angeles or New York. You are also exposed to global fallout created by nuclear testing. Even though the testing may be far away, some radioactive isotopes get into the atmosphere. Wind currents carry that radioactivity around the world, and it is carried to the earth during rain and weather changes.

Television also emits a low level of radiation. Dental and chest x-rays add to our radiation exposure as well. Table 5.7 lists some sources of radiation that affect us all. All together, the average citizen in the United States receives about 170 millirems (mrem) of radiation annually.

Example 5.11 A treatment for leukemia is 2 mCi of the isotope ^{32}P. If a patient receives six treatments, how many mCi of ^{32}P were given?

Solution

$$6 \text{ \sout{treatments}} \times \frac{2 \text{ mCi}}{1 \text{ \sout{treatment}}} = 12 \text{ mCi}$$

HEALTH NOTE

Radiation Exposure in Diagnostic and Therapeutic Procedures

We can compare the levels of radiation exposure during diagnostic procedures in nuclear medicine. The dosages used for diagnostic procedures are much less than the doses given in radiation therapy. (See Tables 5.8 and 5.9.) For example, in a study of the lungs, only enough radioisotope such as ^{131}I is given to produce a good scan. In diagnostic procedures, the radiologist wants to minimize the possible damage to the organ or tissues. However, when there is a malignant tumor, the lung receives much higher doses of ^{131}I. In therapy, the radiologist wishes to destroy the cells in the tumor. There is still damage to surrounding healthy tissue, but healthy cells are more capable of surviving radiation and repairing themselves than are malignant cells.

Table 5.8 Examples of Radiation From Diagnostic Procedures Using Radioisotopes

Radioisotope	Organ	Dose (rem)
^{99m}Tc	Liver	0.3
^{131}I	Thyroid	50.0
^{131}I	Lung	2.0

Table 5.9 Doses Used in Radiation Therapy

Condition	Dose (rem)
Skin cancer	5000–6000
Lung cancer	6000
Lymphoma	4500
Brain Tumor	6000–7000

HEALTH NOTE

Maximum Permissible Dose

maximum permissible dose

Any person working with radiation, such as the radiologist, radiation technician, or nurse whose patient has received a radioisotope, must wear some type of detection apparatus to measure exposure to radiation. A standard for occupational exposure is the *maximum permissible dose* (MPD). If this dose is not exceeded, the probability of injury is minimized. Exposure is monitored by wearing a badge, ring, or pin containing a small piece of photographic film. The window covering the film absorbs beta rays but lets gamma rays through to hit and expose the film. The darker the film, the more exposure to radiation the person has had. These badges are checked periodically to prevent anyone from being exposed to more than the maximum permissible dose. The MPD for occupational exposure is 5 rem (5000 mrem) per year.

Lethal Dose

The larger the dose of radiation received at one time, the greater the effect on the body. Whole-body dosages under 25 rem usually cannot be detected. Whole-body exposure of 25–50 rem produces a temporary decrease in the white blood cells. If the dosage is 100 rem or higher, the person suffers nausea, vomiting, fatigue, and possible hemorrhaging. A dosage greater than 300 rem lowers the white blood cell count to zero. The patient suffers diarrhea and hair loss. Exposure to dosages over 500 rem is expected to cause death in 50 percent of the people receiving that amount. This amount of radiation is called the *lethal dose* for one-half the population test (LD_{50}), and the value varies considerably for different life forms, as Table 5.10 shows.

Table 5.10 Lethal Dosages of Radiation for Various Life Forms

Life Form	LD_{50} (rem)
Insect	100,000
Bacterium	50,000
Rat	850
Cattle	750
Hamster	700
Mouse	530
Human	500
Dog	325
Guinea pig	200

Glossary

alpha particle A nuclear particle containing two p^+ and two n^0, symbolized as α or ${}^4_2\alpha$.

artificial radioactivity The creation of a radioactive element by bombarding a stable nucleus with nuclear particles. The resulting unstable atom undergoes radioactive decay.

beta particle A nuclear particle that is a fast-moving electron emitted from the nucleus of a radioactive atom. Symbolized by β or e^-.

carbon dating A process used to date archeological specimens that contain carbon. The age is determined by the ratio of the radioactive ^{14}C to the ^{12}C in the sample.

chain reaction A fission reaction that will continue once it has been initiated.

curie A unit of radiation equal to 3.7×10^{10} disintegrations per second.

fission A process in which large radioactive atoms break apart into smaller atoms, thereby releasing energy.

fusion A process in which small nuclei combine to form larger ones, thereby releasing energy.

gamma ray High-energy radiation similar to x-rays, symbolized by γ.

half-life The time it takes for one-half of a radioactive sample to decay.

ion pairs Two oppositely charged ions produced when radiation strikes a molecule.

LD_{50} The amount of radiation that will cause death in half the population exposed to the radiation.

MPD (maximum permissible dose) The most exposure allowed in occupational radiology.

nuclear symbol A symbol that shows the number of protons and neutrons in an atom.

rad A measure of an amount of radiation absorbed by the body.

radiation therapy The use of radiation to destroy harmful tissue in the body.

radioactive decay The breakdown of a radioactive nucleus with the release of radiation.

radioisotope A radioactive form of an element.

rem A measure of the damage caused by radiation.

roentgen The amount of radiation that will produce 2×10^9 ion pairs in 1 mL of air.

scan The image of an organ created by radioactive isotopes that accumulate in the organ.

shielding Material used to protect a person from radiation.

x-rays A form of radiation produced when electrons change levels within an atom.

Problems

Nuclear Symbols (Objective 5.1)

5.1 Supply the missing information.

Medical Use	Nuclear Symbol	Mass Number	Number of Protons	Number of Neutrons
Spleen imaging	${}^{51}_{24}Cr$	______	______	______
Malignancies	______	60	27	______
Blood volume	______	______	26	33
Hyperthyroidism	${}^{131}_{53}I$	______	______	______
Leukemia treatment	______	32	______	17

5.2 Write the nuclear symbols for the radioactive isotopes of bromine that have 40 neutrons, 41 neutrons, 42 neutrons, 45 neutrons, 47 neutrons, and 48 neutrons.

Radioactivity (Objective 5.2)

5.3 Write a nuclear symbol for the following:
a. an alpha particle
b. a neutron
c. a beta particle
d. a proton
e. a gamma ray

5.4 Identify the symbol for X in each of the following nuclear particles:
a. ${}^{0}_{-1}X$ b. ${}^{4}_{2}X$ c. ${}^{1}_{0}X$ d. ${}^{0}_{0}X$ e. ${}^{1}_{1}X$

5.5 Describe each of the following protective methods for limiting radiation exposure:
a. shielding
b. length of time of exposure
c. distance from the radioactive source

Nuclear Equations (Objective 5.3)

5.6 Write a balanced nuclear equation for the alpha decay of the following radioactive isotopes:
a. ${}^{208}_{84}Po$ b. ${}^{232}_{90}Th$ c. ${}^{251}_{102}No$

5.7 Write a balanced nuclear equation for the beta decay of the following radioactive isotopes:
a. ${}^{25}_{11}Na$ b. ${}^{59}_{26}Fe$ c. ${}^{92}_{38}Sr$

5.8 Write a complete nuclear symbol for the unknown particle or isotope in the following nuclear equations:
a. ${}^{28}_{13}Al \longrightarrow X + {}^{0}_{-1}\beta$
b. $X \longrightarrow {}^{86}_{36}Kr + {}^{1}_{0}n$
c. ${}^{66}_{29}Cu \longrightarrow {}^{66}_{30}Zn + X$
d. $X \longrightarrow {}^{4}_{2}\alpha + {}^{234}_{90}Th$
e. ${}^{11}_{6}C \longrightarrow {}^{7}_{4}Be + X$
f. ${}^{211}_{83}Bi \longrightarrow X + {}^{4}_{2}\alpha$
g. ${}^{35}_{16}S \longrightarrow X + {}^{0}_{-1}\beta$

Artificial Radioactivity (Objective 5.4)

5.9 Complete the following nuclear equations for bombardment reactions by writing the nuclear symbol for the unknown component:
a. ${}^{9}_{4}Be + {}^{1}_{0}n \longrightarrow X$
b. ${}^{32}_{16}S + X \longrightarrow {}^{32}_{15}P$
c. $X + {}^{1}_{0}n \longrightarrow {}^{24}_{11}Na + {}^{4}_{2}\alpha$
d. ${}^{40}_{18}Ar + X \longrightarrow {}^{43}_{19}K + {}^{1}_{1}H$
e. ${}^{238}_{92}U + {}^{1}_{0}n \longrightarrow X$
f. $X + {}^{1}_{0}n \longrightarrow {}^{14}_{6}C + {}^{1}_{1}H$

Radiation Detection (Objective 5.5)

5.10 How does a detection tube in a scanner detect radiation from a gamma-emitting isotope?

5.11 a. Bone and bony structures consist primarily of $Ca_3(PO_4)_2$. Why are the radioisotopes ^{47}Ca, ^{32}P, or ^{85}Sr used in the determination of bone lesions and bone tumors?
b. A patient with polycythemia vera (excess production of red blood cells) receives radioactive phosphorus. Why would the production of red blood cells in the bone marrow be reduced?
c. Treatment with ^{131}I decreases the amount of hormone produced by the thyroid gland. Why?

Half-Life (Objective 5.6)

5.12 You have a 400.0-mg sample of ^{59}Fe, which has a half-life of 44 days. What mass of the radioactive sample remains after:
a. one half-life?
b. two half-lives?
c. three half-lives?
d. five half-lives?
e. 88 days?
f. 132 days?

5.13 a. Chromium-51 has a half-life of 28 days. How much of a 50-μg sample will still be radioactive after 84 days?
b. Gallium-67 has a half-life of 78 h. How long will it take for the radiation level of ^{67}Ga to fall to one-fourth of the original level? To one-eighth?
c. Iodine-125 has a 60-day half-life. How much of a 200-mg sample will be left after 1 yr? (*Hint*: Round off to 360 days for ease of computation.)
d. Technetium-99m has a half-life of 6 h. How much of a 500-mg sample remains after 36 h?

Fission and Fusion (Objective 5.7)

5.14 Indicate whether each of following are characteristic of the fission or fusion process or both:
a. Neutrons bombard a nucleus.
b. This process occurs in the sun.
c. A large nucleus splits into smaller nuclei.
d. Small nuclei form larger nuclei.
e. Very high temperatures are required to initiate the reaction.
f. Little radioactive waste results.
g. Hydrogen and helium nuclei are the reactants.
h. Large amounts of energy are released when the reaction occurs.
i. Used in some electrical power plants to produce electricity.

Measurement of Radiation (Objective 5.8)

5.15 a. The recommended dosage of ^{131}I is 4.2 μCi/kg body weight. How many microcuries of ^{131}I are needed for a 70.0-kg patient?
b. The dosage of ^{99m}Tc for a lung scan is 20 μCi/kg body weight. How many millicuries should be given to a 50.0-kg patient? (*Hint*: 1 mCi = 1000 μCi.)

Energy and the States of Matter

Objectives

6.1 Define the following terms: energy, potential energy, kinetic energy, and work.

6.2 Given the mass of a water sample and the change in temperature, calculate the heat energy in calories or joules.

6.3 Identify a substance as a solid, liquid, or gas in terms of the (1) arrangement of particles, (2) motion of particles, (3) distance between particles, (4) attractive forces between particles, (5) shape of the substance, and (6) volume.

6.4 On a heating or cooling curve, identify the parts of the graph that correspond to the solid, liquid, or gas state; the melting or freezing point; and the boiling or condensation point. Given this information for a substance, draw the heating or cooling curve.

6.5 Calculate the number of calories or joules absorbed or released when a given mass of water undergoes a change of state.

6.6 Use calorimetry to calculate the energy of a food in calories, kilocalories, or joules.

Scope

In order to be set in motion, all matter requires energy. This means that energy does work by making things move. The water in the tea kettle does not boil by itself but requires heat from the burner. As more energy is supplied in the form of heat, the molecules of water in the tea kettle move faster. As their motion increases, the water gets hotter.

Matter can take the physical form of a solid, liquid, or gas. An ice cube is solid water, whereas the water running out of a faucet is liquid. When water evaporates from a wet spot or boils in a pan on the stove, it is escaping as a gas. Gases are usually invisible to us, but we can detect them if they have some characteristic odor or color. For example, you know when someone opens a bottle of perfume or ammonia because you can smell those gases as they leave the bottle and fill the room. The characteristic color of smog is caused by the brown color of the gas NO_2.

Matter undergoes a change in state when it is transformed from a solid to a liquid (melting), from a liquid to a gas (vaporization), or from a gas to a liquid (condensation). You see these changes of state when you leave an ice cube on the counter and it melts, or when you boil a pot of water to make tea. All changes of state are accompanied by energy changes. In this chapter we will look at the concept of energy and how much energy is involved in the changes of state and in the food we eat.

6.1 Energy

OBJECTIVE Define the following terms: energy, potential energy, kinetic energy, and work.

Energy is everywhere. Energy is being used to move your muscles when you are playing tennis, studying, working, and even sleeping. You require more energy to walk 5 miles than you do to walk around the block. More energy is needed to move a heavy object than a light object.

Work

When you are running, or moving furniture, or thinking, you are using energy to do work. In fact, energy is defined as *the ability to do work.* For you to do work, you must expend a certain amount of energy. Suppose you are climbing a steep hill. While climbing that hill you are expending energy. Perhaps you become too tired to finish. We could say that you do not have sufficient energy to do any more work. Now, suppose you sit down and have lunch. In a while, you will have obtained some energy from the food and you will be able to do more work and complete the climb.

Potential and Kinetic Energy

Energy may be in an active or inactive state. Inactive energy does no work; it serves as a store of energy and is capable of doing work in the future. We refer to inactive energy as *potential energy.* On the other hand, energy that is doing work

HEALTH NOTE

Biological Work

In the cells of the body, the process of metabolism converts potential energy from nutrients in our diet to kinetic energy. Kinetic energy is needed to do biological work. There are different kinds of biological work to be done. When you contract a muscle, you are doing *mechanical work*. In the movement of an arm or a leg or in the involuntary movement of the heart, mechanical work is done, and energy is expended.

mechanical work

$$\text{Relaxed muscle} \xrightarrow{\text{Energy to do mechanical work}} \text{Contracted muscle}$$

chemical work

Chemical work is also done in the cells of the body. Cells of living organisms are involved in processes of growth and maintenance that depend on the production (biosynthesis) of macromolecules from simpler molecules. For example, proteins (macromolecules built from small amino acid molecules) are necessary for the growth of a cell. The formation of these complex compounds is chemical work and requires energy:

$$\text{Simple molecules} \xrightarrow{\text{Energy to do chemical work}} \text{Macromolecules}$$

transport work

Transport work is done when digested materials are moved from the intestinal membrane to the cells of organs and tissues. Moving these components from a lower to a higher concentration requires energy, and the process is called *active transport*:

$$\text{Digestion products} \xrightarrow{\text{Energy for transport}} \text{Cellular nutrients}$$

is active energy; this is called *kinetic energy*—the energy of motion. A boulder resting on a mountain has potential energy because of its location. When the boulder moves and rolls down the mountain it has kinetic energy. The chemical bonds in the food you eat have potential energy. When the food is digested, the stored energy is released and you have kinetic energy to do work.

Example 6.1 State whether each of the following phrases describes potential or kinetic energy:

1. stored energy
2. a falling rock
3. energy in a candy bar
4. a stretched rubber band

Solution

1. Potential energy is inactive, stored energy.
2. Kinetic energy is the energy of motion.
3. Potential energy is stored in the candy bar.
4. Potential energy is present in a stretched rubber band capable of doing work.

Forms of Energy

chemical energy
light
electrical

Energy can take many forms. The energy available in the bonds of a chemical compound is *chemical energy*. The energy of electromagnetic radiation takes the form of *light*, which may be in the visible range or in the invisible ranges of infrared, ultraviolet, x-rays, or gamma rays. The *electrical energy* in your home is yet another form of energy.

thermal (heat)

You are probably most familiar with *thermal energy*, or *heat*. Heat is associated with the motion of particles in a substance. A frozen pizza feels cold to the touch because the particles in the pizza have low kinetic energy and are moving very slowly. As the pizza is heated, the kinetic energy increases and the pizza becomes warm. Eventually the molecules in the pizza pick up enough energy to make the pizza hot and ready to eat. When you warm food, the food gets hot because thermal energy is provided by the flame or the electrical coils. Heat flows from a hot object (electrical coils) to a cooler object (food).

One form of energy may be converted into another. The burning of wood converts chemical energy into thermal energy. The electrical energy in your home is converted to light energy when you turn on a light switch or to thermal energy when you use a hair dryer.

6.2 Measuring Heat Energy

OBJECTIVE Given the mass of a water sample and the change in temperature, calculate the heat energy in calories or joules.

calorie
joule

In chemistry and the health sciences, heat is measured in units called *calories* (cal) or *joules* (J) (pronounced "jewels"). Using water as a standard, one calorie is defined as the amount of heat needed to raise the temperature of 1 g of water by 1°C (from 14.5°C to 15.5°C) (see Figure 6.1). The SI unit of heat is the joule:

$$1 \text{ cal} = 4.184 \text{ J}$$

Specific Heat

The *specific heat* of a substance is a measure of its ability to absorb heat. It is defined as the number of calories or number of joules needed to raise the temperature of 1 g of substance 1°C:

$$\text{Specific heat of a substance} = \frac{\text{cal}}{\text{g °C}} \quad \text{or} \quad \frac{\text{joules}}{\text{g °C}}$$

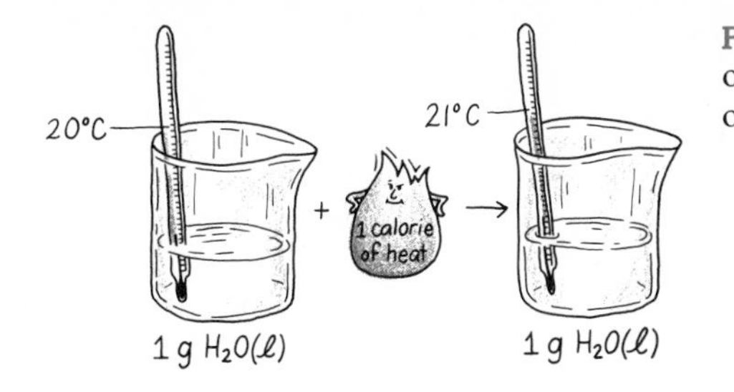

Figure 6.1 One calorie is the amount of heat needed to raise the temperature of 1 g of water 1°C.

Every substance has a specific heat. We have seen that the specific heat of water is 1 cal/g °C or 4.184 J/g °C:

$$\text{Specific heat of water} = \frac{1\ \text{cal}}{\text{g}\ ^\circ\text{C}} = \frac{4.184\ \text{J}}{\text{g}\ ^\circ\text{C}}$$

the large specific heat of water helps maintain body temperature

Water has one of the largest specific heats of any substance, which means that 1 g of water absorbs more heat than most substances. In the body, water serves to absorb body heat without causing changes in body temperature. Table 6.1 lists the specific heats of a variety of materials.

A substance with a large specific heat (such as water) will not change temperature as much as a substance with a small specific heat. For example, on a hot day, the ocean water remains cool to the touch, but the sand can burn your feet.

When materials with small specific heats absorb energy, they reach high temperatures rapidly. For this reason, pans for cooking are often coated with aluminum or copper. These materials with low specific heats cause the pan to heat quickly and to transfer the heat to the food being cooked. The handles on the pans are made of different materials, or they would become too hot to be picked up.

Calculations Using Specific Heat

mass
ΔT
specific heat

We can calculate the amount of heat involved in heating or cooling a certain quantity of a substance that does not change state. To do this, we need to know the mass of the sample, the temperature change, and the specific heat. The mass in grams is multiplied by the change in temperature (ΔT) in degrees Celsius and the specific heat of the substance. Note that the units g and °C cancel to give an

Table 6.1 Specific Heats of a Variety of Materials

Material	Specific Heat (cal/g °C)	Material	Specific Heat (cal/g °C)
Aluminum	0.22	Water (*l*)	1.00
Copper	0.092	Ethanol (*l*)	0.58
Gold	0.031	Water (ice)	0.50
Iron	0.106	Sand	0.19
Silver	0.057	Wood	0.42
Tin	0.051		

answer in units cal or joules:

$$\text{Heat} = \text{Mass} \times \text{Temperature change} \times \text{Specific heat}$$

$$\text{Calories} = \text{g} \times {}^\circ\text{C} \times \frac{\text{cal}}{\text{g}\,{}^\circ\text{C}}$$

or

$$\text{Joules} = \text{g} \times {}^\circ\text{C} \times \frac{\text{joules}}{\text{g}\,{}^\circ\text{C}}$$

Example 6.2 Calculate the calories required to raise the temperature of 100 g of water from:
1. 24°C to 25°C
2. 0°C to 100°C

Solution

1. The specific heat for water is 1 cal/g °C. The temperature change is 1°C (25°C – 24°C). The heat required is calculated by multiplying the mass (100 g) by the temperature change (1°C) and by the specific heat of 1 cal/g °C:

$$\text{Heat} = 100\ \text{g} \times 1^\circ\text{C} \times \frac{1\ \text{cal}}{\text{g}\,{}^\circ\text{C}}$$
$$= 100\ \text{cal}$$

2. In this problem we have the same mass (100 g), but a bigger temperature change of 100°C (100°C – 0°C). Using the specific heat of liquid water again, 1 cal/g °C, we calculate the heat required:

$$\text{Heat} = 100\ \text{g} \times 100^\circ\text{C} \times \frac{1\ \text{cal}}{\text{g}\,{}^\circ\text{C}}$$
$$= 10{,}000\ \text{cal}$$

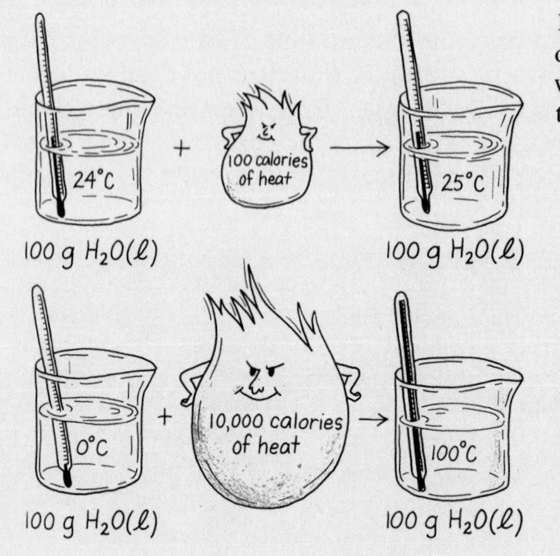

Figure 6.2 For the same quantity of water, more heat is needed when there is a greater change in temperature.

Note that for the same quantity of water, more heat is needed when there is a greater temperature change. (See Figure 6.2.)

Example 6.3 Calculate the number of calories needed to heat 15.0 g of copper from 30°C to 70°C.

Solution: The table of specific heats (see Table 6.1) gives the specific heat of copper as 0.092 cal/g °C. The temperature change ($t_2 - t_1$) is 40°C (70°C − 30°C). The heat needed to heat the copper is then calculated by multiplying the mass (15.0 g) by the temperature change (40°C) and by the specific heat 0.092 cal/g °C.

$$\text{Heat} = 15.0\,\cancel{g} \times 40\,\cancel{°C} \times \frac{0.092}{\cancel{g}\,\cancel{°C}}$$

$$= 55 \text{ cal}$$

Example 6.4 Calculate the number of joules needed to heat 5.0 g of water from 15°C to 60°C.

Solution: We know that the specific heat (in joules) of liquid water is

$$\frac{4.184 \text{ J}}{\text{g °C}}$$

The change in temperature (60°C − 15°C) is 45°C. Using the mass and the specific heat of water, we can calculate the number of joules needed to heat the water:

$$\text{Heat} = 5.0\,\cancel{g} \times 45\,\cancel{°C} \times \frac{4.184 \text{ J}}{\cancel{g}\,\cancel{°C}}$$

$$= 940 \text{ joules}$$

6.3 States of Matter

OBJECTIVE Identify a substance as a solid, liquid, or gas in terms of the (1) arrangement of particles, (2) motion of particles, (3) distance between particles, (4) attractive forces between particles, (5) shape of the substance, and (6) volume.

Solids

regular pattern

definite shape and volume

In solids, the particles (which may be molecules, atoms, or ions) are closely packed together in a very stable arrangement. The particles of a solid are arranged in a regular and predictable pattern. This gives the solid a definite shape and a definite volume. Teeth and bones in your body are solids composed primarily of calcium cations and phosphate anions. As solids, they have definite shapes and volumes. The particles in a solid are held in relatively fixed positions, so their motion is very restricted. However, they are not motionless but vibrate slightly within the solid structure.

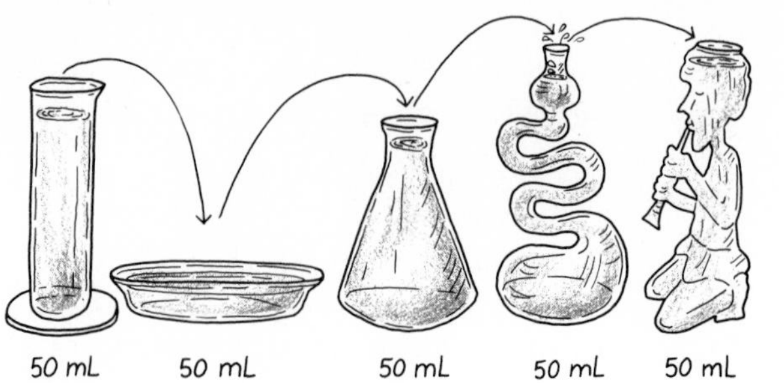

Figure 6.3 A liquid maintains a definite volume but takes the shape of the container.

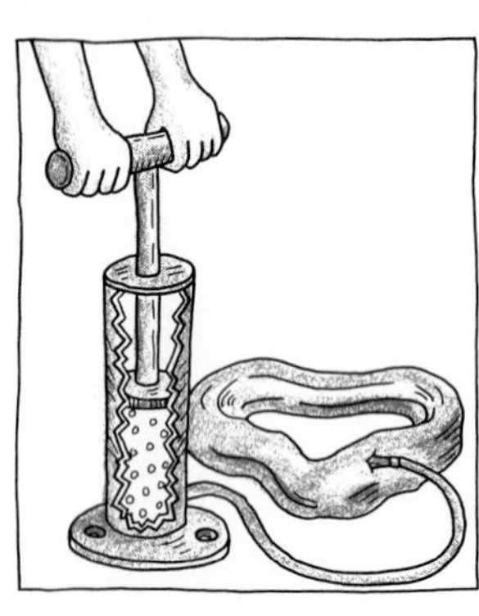

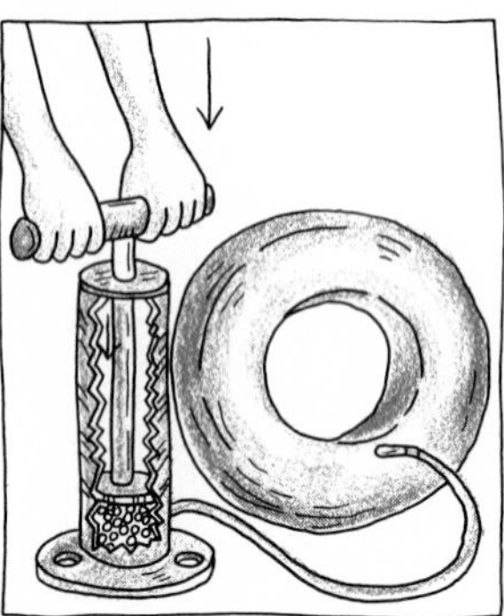
Figure 6.4 A gas takes the shape and volume of the container.

Liquids

In the liquid state, the molecules or ions of a compound conform to the shape of the container. The total volume of liquid material is maintained. Particles in the liquid state have greater freedom of movement than when they are arranged as a solid. In the liquid state, the particles are not in an ordered pattern, but they are still rather strongly attracted. The particles stay close together, which is why it is difficult to compress a liquid. (See Figure 6.3.)

Gases

Molecules in a gaseous state move with great speed in a straight-line path until they collide with other gas molecules or the walls of a container. *Gases have no definite shape or volume; they take the shape and volume of the container they are placed in.* (See Figure 6.4.) Because there are great distances between gas molecules, there is essentially no attraction among the molecules. Table 6.2 lists the characteristics of the three states of matter; Figure 6.5 provides a visual comparison.

Table 6.2 Characteristics of the Three States of Matter

State	Shape	Volume	Arrangement of Molecules	Distance Between Molecules	Attraction Between Molecules	Motion of Molecules
	Visible			**Nonvisible**		
Solid	Has its own shape	Has its own volume	Definite pattern	Very close	Strong	Slow, vibrate
Liquid	Shape of container	Has its own volume	Random	Close	Moderate	Moderate
Gas	Shape of of container	Fills volume of container	Random	Far apart	None	Very fast

Figure 6.5 Comparison of molecular arrangement in the different states of matter.

Example 6.5 Identify each of the following as characteristic of a solid, liquid, or gas:

1. maintains its own volume, but takes the shape of the container
2. has a highly ordered arrangement of particles
3. has little or no attraction between particles

Solution

1. A *liquid* has a definite volume, but takes the shape of the container.
2. In a *solid*, there are strong attractions between particles to give a definite pattern.
3. Particles of a *gas* move so fast and are so far apart that there is little or no attraction between them.

6.4 Changes of State

OBJECTIVE On a heating or cooling curve, identify the parts of the graph that correspond to the solid, liquid, or gas state; the melting or freezing point; and the boiling or condensation point. Given this information for a substance, draw the heating or cooling curve.

As a solid is heated, the movement of the particles increases. A temperature is eventually reached at which the bonds between the particles of the solid are broken. The particles are free to move about in random patterns as the sample changes from a solid to a liquid. This change of state occurs at a specific temperature, referred to as the *melting point* (mp). When a solid melts to form a liquid, a new state of matter appears with new physical properties. We say that a *change of state* has taken place.

melting point

Melting Point

The melting point of a substance gives us an indication of the strength of the attractive forces between the particles. If the melting point is rather low, then the attractive forces that hold the particles in the solid are relatively weak. For the solid to melt, these particles require only a small amount of heat, which can be obtained at relatively low temperatures. On the other hand, a high melting point indicates strong attractive forces between the particles of the solid. Then much more energy is required to separate the particles, thus a higher temperature is needed. The attraction between the positive and negative ions in an ionic solid is much greater than the attraction between covalent molecules in a molecular

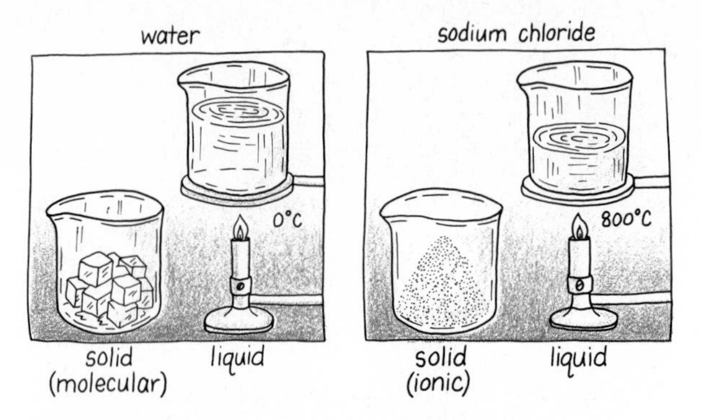

Figure 6.6 Comparison of the melting points of water (molecular solid) and sodium chloride (ionic solid).

solid. (See Figure 6.6.) In general, ionic solids have high melting points, whereas covalent compounds have much lower melting points. (See Table 6.3.)

Evaporation

Within a liquid, the molecules move about in a random, zigzag pattern. At the surface of the liquid, there are occasionally molecules that obtain sufficient energy to escape from the liquid. They break away from the molecules of the liquid as they transform into individual molecules of gas. The escape of liquid molecules below the boiling point of the liquid, which occurs only at the surface, is called *evaporation.*

If you place a container of water (liquid) in the open air, all the water molecules will eventually evaporate. However, if you fill the container with water and place a tight cover on it, the level of water will lower for a time and

Table 6.3 Melting Points of Ionic and Molecular Solids

Type of Solid	Compound	Formula	mp (°C)
Ionic salts	Calcium phosphate	$Ca_3(PO_4)_2$	1670
	Sodium chloride	$NaCl$	800
	Potassium bromide	KBr	730
	Ferrous chloride	$FeCl_2$	670
Molecular solids	Glucose	$C_6H_{12}O_6$	146
	Benzene	C_6H_6	5.5
	Water (ice)	H_2O	0
	Carbon tetrachloride	CCl_4	−73
	Sulfur dioxide	SO_2	−23
	Hydrogen sulfide	H_2S	−83
	Carbon monoxide	CO	−207

Figure 6.7 (a) Evaporation of a liquid from an open container; (b) dynamic equilibrium between liquid and vapor in a closed container; (c) boiling when vapor forms within a liquid.

then remain at a constant level within the closed container. Some of the water molecules evaporate, forming gaseous water called *water vapor*.

water vapor

When some of the water vapor molecules collide and return to the liquid state, they undergo *condensation*. A point is reached at which the number of water molecules evaporating is equal to the number of water molecules condensing. This is an example of *dynamic equilibrium*, and there is no further change in the water level in the closed container. As Figure 6.7 shows, dynamic equilibrium exists when the rates of change occurring in opposite directions are equal:

condensation

$$H_2O(l) \underset{\text{Rate of condensation}}{\overset{\text{Rate of evaporation}}{\rightleftharpoons}} H_2O(g)$$

Boiling Point

As more heat is added to a liquid, there is an increase in the amount of water vapor that forms. Boiling of the liquid begins when vapor molecules form within the liquid as well as at the surface. (See Figure 6.7.) The boiling of a liquid occurs at a definite temperature called the *boiling point* (bp).

Example 6.6 Indicate whether the following statements describe melting, melting point, evaporation, boiling, or boiling point:

1. the escape of liquid molecules from the surface only
2. the temperature at which a solid changes to a liquid
3. the formation of vapor bubbles within the liquid

Solution

1. During *evaporation*, only molecules at the surface of a liquid obtain sufficient energy to evaporate.
2. The *melting point* is the specific temperature at which a solid changes to a liquid.
3. When vapor bubbles form within the liquid, we say that the liquid is *boiling*.

HEALTH NOTE

Beneficial Effects of a Fever

When bacteria or viruses invade the body, they are surrounded by activated white blood cells. In the process, a hormone, EP (endogenous pyrogen), is released into the blood stream and travels to the brain. The effect of the hormone is to make the brain reset the body temperature to a higher level. Chills set in and blood vessels constrict to reduce heat loss; body fat is broken down to produce more heat. The EP hormone also reduces the blood levels of iron.

Some research is now showing that a fever along with reduced iron plays an important role in inhibiting the growth of bacteria. Other studies indicate that antibiotics may be more effective during a fever. A fever also triggers the production of interferon, which the body used to fight viruses. While a high fever can certainly be harmful, it may be that a moderate fever should run for a time before it is reduced.

Patients with high fevers can be cooled by a sponge bath. The water evaporates from the skin and lowers the temperature of the body surface. Sometimes ice packs or ice blankets are used; the body heat is used to convert the ice to liquid water.

For some local surgeries, certain sprays are used to reduce the temperature on an area of skin. The heat from the surface of the skin evaporates these liquids very quickly. A rapid drop in temperature occurs, and the sprayed area becomes so cold that it feels numb. The numbness allows localized surgery to be done on the sprayed portion of the skin.

Heating and Cooling Curves

Changes of state can be represented on *heating and cooling curves*, which diagram the states of matter and their changes as the temperature changes. We begin drawing a heating curve with the solid state at a low temperature. The line increases on a diagonal until the melting point is reached. At the melting point, solid changes to a liquid and there is no change in the temperature. The melting point is represented by a flat line, or plateau, on the heating curve. (See Figure 6.8.)

When all of the solid has been converted to liquid, the temperature begins to rise again. The molecules of the liquid move faster and faster as their kinetic

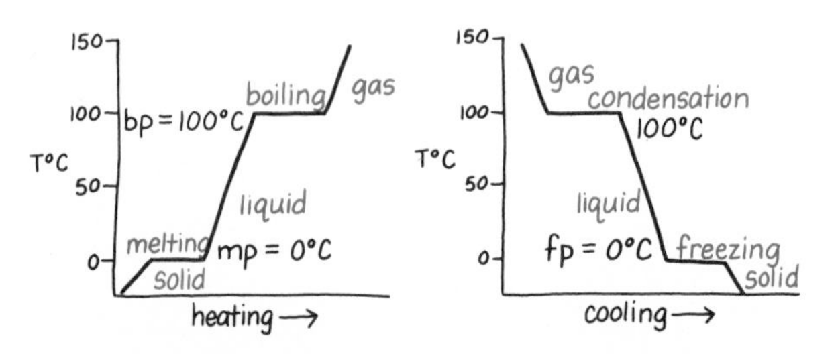

Figure 6.8 Heating and cooling curves for water.

energy increases. When the boiling point is reached, a flat line appears again on the heating curve. The temperature of boiling remains constant as molecules of the liquid use the energy to change to gas. When all of the molecules are in the gaseous state, the temperature rises as the motion of the gas molecules increases. (See Figure 6.8.)

freezing

A cooling curve represents a loss of heat. The line begins at a high temperature in the gaseous state. As the temperature drops, the gas cools. The change from gas to liquid is *condensation*, and the change from liquid to solid is *freezing*. Both are indicated by horizontal lines on the cooling curve. Note that boiling point and condensation occur at exactly the same temperature as do melting and freezing.

Example 6.7 Draw a heating curve for water and write on the proper part of the curve: solid, liquid, gas, melting point, and boiling point.

Solution

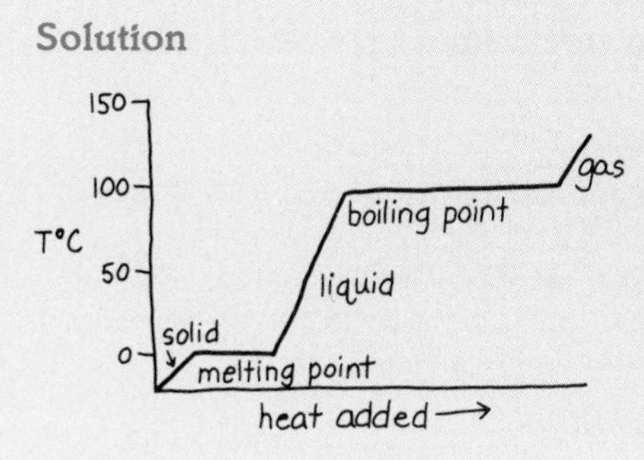

6.5 Energy in Changes of State

OBJECTIVE Calculate the number of calories or joules absorbed or released when a given mass of water undergoes a change of state.

We have seen that on heating and cooling curves constant temperatures are maintained when a substance undergoes a change of state such as melting or boiling. Now we will look at the energy associated with these changes of state.

Heat of Fusion

When a solid, such as ice, is heated, the energy absorbed causes an increase in the motion of the particles in the solid. When the melting point is reached, all the energy is used to break apart the bonds that hold water molecules together in the solid ice. There is no change in temperature; it remains constant. The amount of heat needed to change 1 g of solid to 1 g of liquid at its melting point

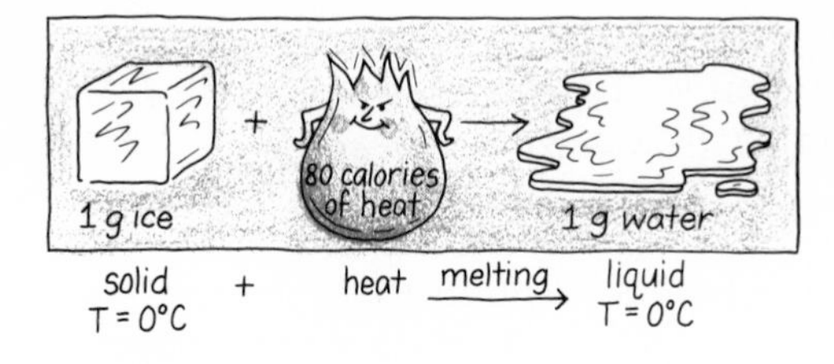

Figure 6.9 The heat of fusion for water; 1 g of ice (solid) requires 80 cal of heat to completely melt.

is called the *heat of fusion.* At the melting point of ice (0°C), 80 cal (334 joules) are required to melt 1 g of ice (solid) (see Figure 6.9):

$$\text{Heat of fusion (0°C)} = \frac{80\text{ cal}}{\text{g ice}} \quad \text{or} \quad \frac{334\text{ J}}{\text{g ice}}$$

The amount of heat needed to melt a certain mass of ice can be calculated by multiplying the mass of the solid by the heat of fusion. Note that there is no temperature change since melting is a constant-temperature process:

$$\text{Heat to melt} = \text{Mass (g)} \times \text{Heat of fusion (cal/g)}$$

Example 6.8 How many calories are needed to melt 50 g of ice at 0°C, the melting point of ice?

Solution: The number of calories of heat required to melt 50 g of ice is calculated by multiplying the mass of the ice (50 g) by the heat of fusion of water (80 cal/g). The unit g cancels out to give an answer in cal:

Water (Solid Ice) to Water (Liquid) at 0°C

$$\text{Heat} = \underbrace{50\,\cancel{\text{g}}}_{\text{Mass of ice}} \times \underbrace{\frac{80\text{ cal}}{\cancel{\text{g}}}}_{\text{Heat of fusion of water}}$$

$$= 4000\text{ cal}$$

Therefore the heat in calories required to melt 50 g of ice is 4000 cal. (See Figure 6.10.)

Figure 6.10 The heat required to melt 50 g of ice:

$$50\text{ g} \times (80\text{ cal/g}) = 4000\text{ cal}$$

The heat of fusion also applies to the freezing of water. When the temperature of liquid water is lowered, the liquid loses heat. Eventually the molecules slow down so much that their attractive forces are sufficient to form

Figure 6.11 Amount of heat released when 1 g of water freezes.

solid crystals. This change is called *freezing*. The freezing point is the same as the melting point; the freezing point of water is 0°C. During the freezing process, also called *crystallization* or *solidification*, heat is released. The heat released during freezing is the same as the heat of fusion, also called *heat of crystallization*. The freezing of 1 g of water releases 80 cal of heat. (See Figure 6.11.)

Example 6.9 In a freezer, heat is removed from water (liquid) that has been placed in an ice cube tray. How many calories are absorbed by the freezer when 150 g of water is frozen at 0°C?

Solution: The heat released at the freezing point of water is calculated by multiplying the mass of the water in the ice cube tray by the heat of fusion of water:

Water (Liquid) to Water (Solid Ice) at 0°C

$$\text{Heat} = 150\,\cancel{g} \times \underbrace{\frac{80\text{ cal}}{\cancel{g}}}_{\text{Heat of fusion of water}}$$

$$= 12{,}000\text{ cal}$$

Heat of Vaporization

The energy required to transform 1 g of liquid into gas or vapor at its boiling point is called the *heat of vaporization*. This change of state is called *boiling*. For water, boiling occurs at 100°C. One gram of water requires 540 cal for complete conversion from the liquid to the gaseous states (see Figure 6.12):

$$\text{Heat of vaporization of water at } 100°\text{C} = \frac{540\text{ cal}}{\text{g } H_2O}$$

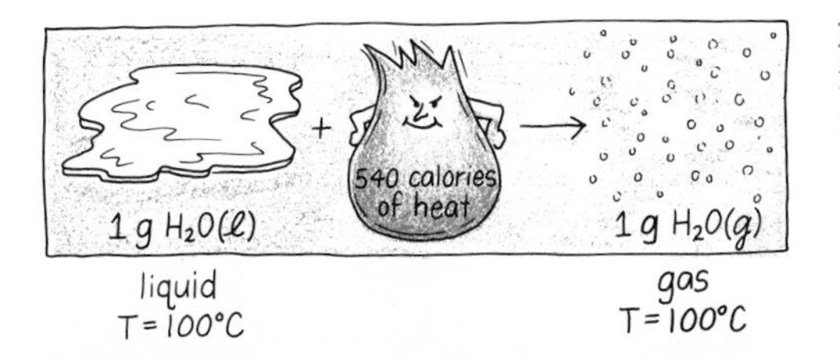

Figure 6.12 Heat of vaporization for water; 1 g water (liquid) requires 540 cal to completely vaporize (gas).

HEALTH NOTE

Homeostasis: Regulation of Body Temperature

The body is an open system through which nutrients, waste products, and heat are constantly being exchanged with the environment. The maintenance of a constant internal temperature is essential to the efficient functioning of our body cells. At low temperatures, essential metabolic reactions proceed too slowly, producing too little of the body's crucial materials. At high temperatures, the structures of the enzymes that regulate our metabolic reactions can change and cease to be active.

Body temperature is regulated by an elaborate series of control systems and usually does not go below 36°C (97°F) or above 40°C (104°F). The ability to maintain a constant body temperature is called *homeostasis.* It is a system by which changes in the external environment are balanced by changes in the internal environment. It is critical to our survival that our body balance heat gain with heat loss. If we do not lose enough heat, our internal temperature rises; if we lose too much heat, our internal temperature drops. Let us look at several of these control mechanisms.

radiation

If you are in an environment with a temperature above body temperature, your body loses heat through radiation, convection, and evaporation. *Radiation* is a process in which heat flows from a warmer body to a cooler one. When the external temperature increases, receptors in the skin send signals to the hypothalamus gland in the brain. A temperature control center in the hypothalamus causes the muscles in the blood vessels of the skin to relax, which causes the blood vessels of the skin to expand. Warm blood flows through the skin, and heat is radiated away. At the same time, the air next to the skin is heated, rises, and is replaced by cooler air. Thus, still more heat is lost by the process of *convection.* In addition, with increased external temperature, three things happen: Muscular activity is slowed, metabolic reactions are reduced, and less internal heat is produced.

convection

An increase in external temperature also stimulates the sweat glands. These glands in the skin produce a fluid that consists mostly of water and some salts. As this fluid evaporates from the skin, heat is lost, and the body temperature is lowered. A gram of sweat (assuming pure water) removes 540 cal of heat from the body surface as

Example 6.10 How many calories would be needed to change 50 g of water (liquid) at 100°C to 50 g of water vapor at 100°C?

Solution: The change of state in this problem occurs at the boiling point of water, 100°C. The heat of vaporization of water is 540 cal/g. We calculate the calories required to change 50 g of liquid water to steam by multiplying the mass of the water by the heat of vaporization for water:

Water (Liquid) to Water (Gas) at 100°C

$$\text{Heat} = 50\ \cancel{\text{g}} \times \frac{540\ \text{cal}}{\cancel{\text{g}}} = 27{,}000\ \text{calories}$$

it evaporates. Each time 1 g of sweat evaporates, the heat content of the skin is lowered by 540 cal. This lowering of the skin's heat content creates the cooling effect of *evaporation*.

evaporation

An increase in humidity means that there is more water in the air. The sweat on the skin will not evaporate as readily, so it is more difficult to keep cool. You feel uncomfortable and "sticky," and you might say, "It's not the heat, it's the humidity."

On the other hand, if the external environment is cold, epinephrine is released, causing an increase in metabolic rate, which increases the production of heat. The change in temperature on the skin also signals the hypothalamus to increase the contraction of the muscles in the blood vessels. As the vessels contract, less blood flows through the skin, conserving the heat that might be lost through the skin. The production of sweat also stops, to lower the heat lost by evaporation. (See Figure 6.13.)

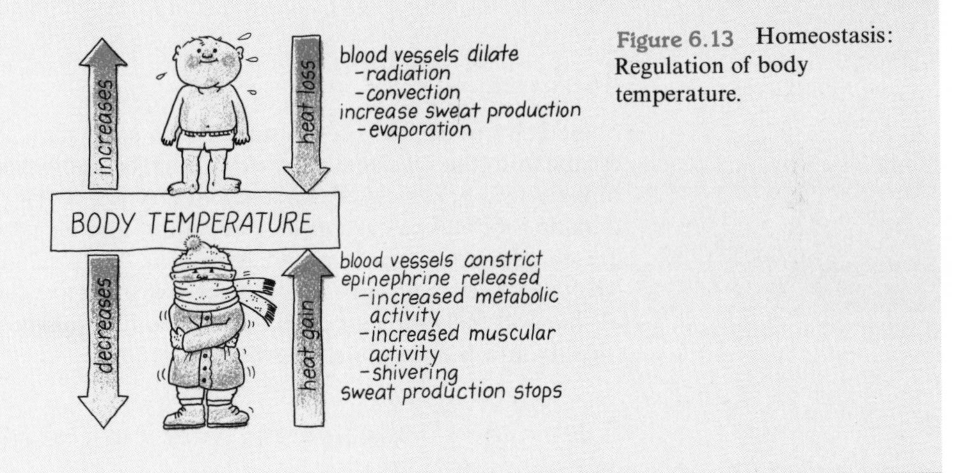

Figure 6.13 Homeostasis: Regulation of body temperature.

The heat of vaporization also applies to the condensation of gas into liquid. If a gas is cooled, the molecules slow down until the attractive forces bring them together, forming drops of liquid. The change from gas to liquid is called *condensation*. For water, steam changes to liquid at 100°C. Upon condensation, energy is released. Thus condensation of 1 g of water (gas) releases 540 cal of heat. (See Figure 6.14.)

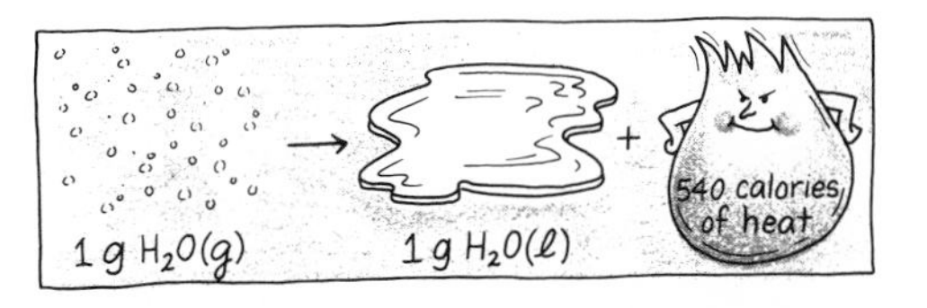

Figure 6.14 Amount of heat released when 1 g of water condenses.

Example 6.11 When steam from a pan of boiling water reaches the window, it condenses and changes into liquid. How much heat (in kilocalories) is released when 25 g of water vapor condenses to liquid water at the condensation point of 100°C?

Solution: The amount of heat released by the condensation of the water vapor at 100°C is calculated by multiplying the mass of the sample by the heat of vaporization.

Water (Gas) to Water (Liquid) at 100°C

$$\text{Heat} = 25\ \cancel{g} \times \frac{540\ \text{cal}}{\cancel{g}}$$

$$= 13{,}500\ \text{cal}$$

$$\text{Kilocalories} = 13{,}500\ \cancel{\text{cal}} \times \frac{1\ \text{cal}}{1000\ \cancel{\text{cal}}}$$

$$= 13.5\ \text{kcal}$$

Sublimation

Sometimes, molecules on the surface of a solid may acquire enough energy to escape from the solid and go directly into the gas state. The transition of a solid directly to a gas is called *sublimation.* Dry ice is dry because the solid carbon dioxide sublimes to carbon dioxide vapor, and no liquid is formed. Mothballs are compounds of carbon and hydrogen that sublime at room temperature. They just seem to disappear. However, we can detect the gas formed by the odor of the mothballs. In very cold areas, snow does not turn to slush, but sublimes directly into the gas state. (See Figure 6.15.)

Calculations Using Specific Heat and Energy of Change of State

Energy problems can combine both the heat of fusion and vaporization with the specific heat of the substance. This type of problem requires several steps.

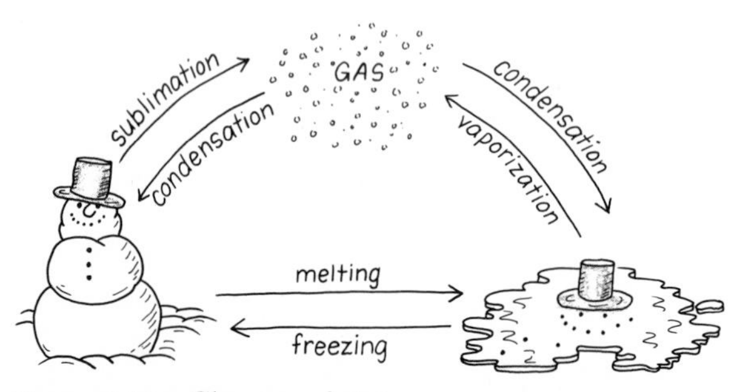

Figure 6.15 Changes of state.

Sometimes it helps to sketch the graph of the heating or cooling curve that fits the problem. Each time you reach a corner on the curve, another energy calculation is needed.

Example 6.12 A sample of 200 g of water at 40°C is heated to boiling at 100°C and then converted to vapor at 100°C. How many calories were needed to heat the sample?

Solution: There are two steps in the calculation of the heat needed. In the first step, the liquid water is heated from 40°C to 100°C, with a temperature change of 60°C.

Step 1. Heating Water

$$\text{Calories} = \text{g} \times {}^\circ\text{C} \times \frac{1\ \text{cal}}{\text{g}\ {}^\circ\text{C}}$$

$$= 200\ \cancel{\text{g}} \times 60\ \cancel{{}^\circ\text{C}} \times \frac{1\ \text{cal}}{\cancel{\text{g}}\ \cancel{{}^\circ\text{C}}}$$

$$= 12{,}000\ \text{cal}$$

In the second step, the liquid water is converted to water vapor at 100°C. This heat calculation requires the use of the heat of vaporization of water, 540 cal/g.

Step 2. Change Liquid to Steam

$$\text{Calories} = \text{Mass} \times \text{Heat of vaporization}$$

$$= 200\ \cancel{\text{g}} \times \frac{540\ \text{cal}}{1\ \cancel{\text{g}}}$$

$$= 108{,}000\ \text{cal}$$

The total heat required for the process is the sum of Step 1 and Step 2:

1. Heating water:	12,000 cal
2. Change liquid to steam:	108,000 cal
Total heat needed:	120,000 cal

Example 6.13 How many kilocalories (kcal) are needed to convert 100 g of ice at 0°C to gas at 100°C?

Solution: This problem requires several steps. It may be helpful to sketch the portion of the heating curve that fits the changes in state. (See Figure 6.16 for the graph.) From the graph, we see that there are three changes in energy, which are melting the ice at 0°C (solid to liquid), heating the water from 0°C to 100°C, and

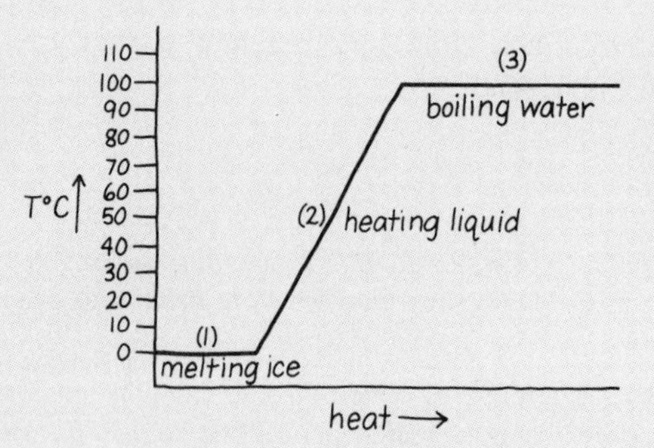

Figure 6.16 Sketch of a graph to aid problem solving:

Step 1: Ice melts; use heat of fusion.
Step 2: Water changes temperature; use specific heat of water.
Step 3: Water boils; use heat of vaporization.

vaporizing the water at 100°C (liquid to gas). The total heat required for all these changes is the sum of the following calculations:

1. Heat to melt solid to liquid at 0°C:

$$100\ \cancel{\text{g ice}} \times \frac{80\ \text{cal}}{\cancel{\text{g ice}}} = 8000\ \text{cal to melt ice}$$

Heat of fusion for ice

2. Heat to raise the temperature of liquid water from 0°C to 100°C using the specific heat of water (1 cal/g°C):

Temperature change = 100°C − 0°C = 100°C

$$100\ \cancel{\text{g}} \times 100\ \cancel{^\circ\text{C}} \times \frac{1\ \text{cal}}{\cancel{\text{g}}\ \cancel{^\circ\text{C}}} = 10{,}000\ \text{cal}$$

3. Heat to vaporize liquid to gas at 100°C:

$$100\ \cancel{\text{g}} \times \frac{540\ \text{cal}}{\cancel{\text{g}}} = 54{,}000\ \text{cal to vaporize}$$

Total Energy

1. Solid to liquid (melt) ice at 0°C:	8,000 cal
2. Heating water from 0°C to 100°C:	10,000 cal
3. Liquid to gas (vaporize) at 100°C:	54,000 cal
	72,000 cal

To change calories to kilocalories, we use the following factor:

1 kcal = 1000 cal

Total Heat (in Kilocalories)

$$72{,}000\ \cancel{\text{cal}} \times \frac{1\ \text{kcal}}{1000\ \cancel{\text{cal}}} = 72\ \text{kcal}$$

HEALTH NOTE

Steam Burns

Whereas hot water (liquid) at 100°C will cause burns and damage to the skin, the effect of steam on the skin is even more dangerous. Let us consider 100 g of hot water at 100°C. If this water falls on a person's skin, the temperature of water will drop to body temperature, 37°C. Heat is lost as the water cools. This heat burns the skin. The temperature change for the water is 63°C:

$$100\ \cancel{\text{g water}} \times 63\cancel{^\circ\text{C}} \times \frac{1\ \text{cal}}{\cancel{\text{g}\ ^\circ\text{C}}} = 6300\ \text{cal heat}$$

Now let us calculate the amount of heat released when 100 g of steam at 100°C hits the body. First, the steam changes to liquid water:

Steam to water (liquid) at 100°C

$$100\ \cancel{\text{g steam}} \times \frac{540\ \text{cal}}{\cancel{\text{g steam}}} = 54{,}000\ \text{cal heat}$$

Now the 100 g of liquid water drops in temperature and releases still more heat to burn the skin. This value is the same as that of the preceding problem, which is 6300 cal. The total amount of heat released from this steam burn would be the total from the condensation of the steam *and* the cooling:

Condensation of steam at 100°C	= 54,000 cal heat
Cooling liquid water (100°C to 37°C)	= 6,300 cal heat
Total heat released from 100 g steam	= 60,300 cal heat

The amount of heat released from steam is almost 10 times that from hot water. This tremendous release of energy from the condensation of steam to liquid water is the reason for the severity of steam burns.

6.6 Calorimetry

OBJECTIVE Use calorimetry to calculate the energy of a food in calories, kilocalories, or joules.

Heat of Reaction

exothermic
endothermic

In a chemical reaction, there is usually a change in thermal energy. A reaction that gives off heat is called an *exothermic reaction*; a reaction that absorbs heat is an *endothermic reaction*. For example, the combustion of methane, CH_4, is an exothermic reaction:

$$CH_4 + 2O_2 \longrightarrow CO_2 + 2H_2O + \underset{\text{Heat given off}}{211\ \text{kcal}}$$

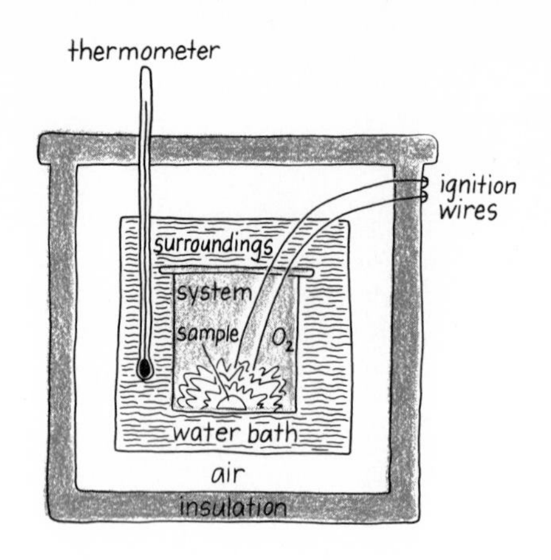

Figure 6.17 A calorimeter: The energy released by a sample undergoing combustion is measured by knowing the mass of water and the change in temperature for the reaction.

The synthesis of hydrogen iodide from hydrogen and iodine is an endothermic reaction:

$$H_2 + I_2 + \underset{\text{Heat absorbed}}{12\text{ kcal}} \longrightarrow 2HI$$

Measuring Energy Change

calorimeter

We can measure the energy change of a reaction by using an apparatus called a *calorimeter*. The reactants are placed in a container within the calorimeter. The reaction chamber is surrounded by water. When a sample is reacted, the heat change in the sample is indicated by a change in the temperature of the surrounding water. By measuring the temperature of the water before and after the reaction, the heat of the reaction can be calculated. (See Figure 6.17.)

1 Cal in nutrition equals 1000 cal

To measure the energy content of a food, a sample of the food is placed in the calorimeter and oxygen is added. Electrical wires inside the reaction container ignite the food, and combustion occurs. The heat that is given off is transferred to the surrounding water. When you are determining your caloric intake, the "calories" you are counting are actually kilocalories. In the field of nutrition, it is common to use Cal with a capital "C" to mean 1000 cal or 1 kcal. This terminology appears on many labels on food packages to indicate the number of Calories in a serving. Nutritional values may also be given in kilojoules (kJ):

$$1\text{ kcal} = 1000\text{ cal}$$

$$1\text{ kJ} = 4184\text{ cal}$$

Example 6.14 A slice of whole-wheat bread was placed in a calorimeter. The mass of the surrounding water was 1000 g, and the initial temperature of the water was 25°C. Following combustion of the bread, the temperature of the water increased to 90°C. How many kilocalories (Cal) were in the slice of bread?

Solution: To solve this problem, we calculate the amount of heat (in kilocalories) that caused a 65°C temperature change for 1000 g of water in the calorimeter. Since the combustion of the slice of bread provided the heat to increase the temperature of the water, we associate the same amount of energy with the slice of bread:

$$\text{Heat} = 1000\ \cancel{\text{g water}} \times 65\cancel{^\circ\text{C}} \times \frac{1\ \text{cal}}{\cancel{\text{g}}\cancel{^\circ\text{C}}}$$

$$= 65{,}000\ \text{cal}$$

$$\text{Kilocalories} = 65{,}000\ \cancel{\text{cal}} \times \frac{1\ \text{kcal}}{1000\ \cancel{\text{cal}}}$$

$$= 65\ \text{kcal}, \quad \text{or } 65\ \text{Cal}$$

Since this is the heat absorbed by the water in the calorimeter, it must have come from the combustion of the bread. Thus we can say that the slice of bread also contained 65 Cal:

$$\text{Caloric value} = \frac{65\ \text{Cal}}{\text{slice of bread}}$$

Nutritional Food Values

The nutritional value of any food is determined by the presence of three major foodstuffs: carbohydrates, fats, and proteins. The caloric value of a foodstuff is the number of kilocalories per gram of food. Although each specific type of foodstuff has a slightly different caloric value, they are quite close. An average value for each major foodstuff is used for nutritional calculations. These values are given in Table 6.4.

Table 6.4 Average Values for Foodstuffs

Foodstuff	Caloric Value
Carbohydrate	$\frac{4\ \text{kcal}}{1\ \text{g carbohydrate}}$
Fat (lipid)	$\frac{9\ \text{kcal}}{1\ \text{g fat}}$
Protein	$\frac{4\ \text{kcal}}{1\ \text{g protein}}$

Example 6.15 A 5.0-g sample of glucose, a typical carbohydrate, is placed in a calorimeter. The mass of water in the container is 1000 g, and the initial temperature is 25°C. When the combustion of the glucose is complete, the water temperature has risen to 45°C. Calculate the caloric value of the glucose.

Solution: The amount of heat absorbed by the water is equal to the amount of heat lost by the glucose during combustion:

Heat loss (by glucose) = Heat gain (by water)

1. The heat gained by the water can be calculated as follows:

$$\underset{\text{Mass of water}}{1000\,\cancel{g}} \times \underset{\text{Temperature change for water}}{20\,\cancel{°C}} \times \underset{\text{Specific heat of water}}{\frac{1\text{ cal}}{\cancel{g}\,\cancel{°C}}} = 20{,}000\text{ cal}$$

$$\text{Kilocalories} = 20{,}000\,\cancel{\text{cal}} \times \frac{1\text{ kcal}}{1000\,\cancel{\text{cal}}} = 20\text{ kcal}$$

2. Since 20 kcal of heat were absorbed by the water, there must have been 20 kcal from the combustion of 5 g of glucose. The caloric value of glucose is calculated as

$$\text{Caloric value} = \frac{20\text{ kcal}}{5\text{ g glucose}} = \frac{4\text{ kcal}}{\text{g glucose}}$$

The caloric value of a food is the sum of the carbohydrate, fat, and protein. The general composition of some foods is given in Table 6.5.

Table 6.5 General Composition of Some Foods

Food	kcal	Protein (g)	Fat (g)	Carbohydrate (g)
Beans, red, $\frac{1}{2}$ cup cooked	125	8	—	25
Beef, lean, 3 oz	185	23	10	—
Steak, 3 oz	350	18	30	—
Cake, 1 piece	145	2	5	24
Chicken, no skin, 3 oz	115	20	3	—
Corn, cooked, $\frac{1}{2}$ cup	75	2	—	18
Egg, 1 large	80	6	6	—
Grains, enriched and whole 1 slice bread, $\frac{1}{2}$ cup cereal	65	2	1	16
Milk, 3.5% butterfat, 1 cup	160	9	9	12
Nonfat, 1 cup	90	9	—	13
Oils, cooking, 1 tbsp	125	0	14	0
Potato, 1 baked	80	2	—	18

Example 6.16 What is the energy value in kilojoules for one baked potato with a caloric value of 85 kcal?

Solution: The given caloric value is 85 kcal. The conversion factor of 4.184 J/cal can be used to convert kilocalories to kilojoules:

$$85\ \cancel{\text{kcal}} \times \frac{1000\ \cancel{\text{cal}}}{\cancel{\text{kcal}}} \times \frac{4.184\ \cancel{\text{J}}}{\cancel{\text{cal}}} \times \frac{1\ \text{kJ}}{1000\ \cancel{\text{J}}} = 356\ \text{kJ}$$

We can use the caloric values of food to convert from the grams of a foodstuff to number of kilocalories:

$$\text{Kilocalories} = \text{Grams of foodstuff} \times \text{Caloric value}$$

Example 6.17 A tablespoon of honey contains 17 g of sugar, which is a carbohydrate. How many kilocalories are in 1 tablespoon of honey?

Solution: The standard caloric value for a carbohydrate is 4 kcal/g:

$$\text{Kilocalories} = \underbrace{17\ \cancel{\text{g sugar}}}_{\text{Mass of honey}} \times \underbrace{\frac{4\ \text{kcal}}{\cancel{\text{g sugar}}}}_{\text{Caloric value of carbohydrate}} = 68\ \text{kcal}$$

Example 6.18 A sample of 10 french fries contains 20 g of carbohydrate, 7 g of fat, and 2 g of protein. Calculate the total number of kilocalories in 10 french fries.

Solution: Using the caloric values for the carbohydrate, fat, and protein portions of the food (Table 6.4), we can calculate the kilocalories in 10 french fries:

$$\text{Carbohydrate:} \quad 20\ \cancel{\text{g}} \times \frac{4\ \text{kcal}}{\cancel{\text{g}}} = 80\ \text{kcal}$$

$$\text{Fat:} \quad 7\ \cancel{\text{g}} \times \frac{9\ \text{kcal}}{\cancel{\text{g}}} = 63\ \text{kcal}$$

$$\text{Protein:} \quad 2\ \cancel{\text{g}} \times \frac{4\ \text{kcal}}{\cancel{\text{g}}} = 8\ \text{kcal}$$

$$\text{Total kilocalories in 10 french fries} = 151\ \text{kcal}$$

HEALTH NOTE

Loss and Gain of Weight

The number of kilocalories needed in your daily diet depends on your age and physical activity. Some general levels of energy needs are given in Table 6.6.

When food intake exceeds energy output, a person's body weight increases. Normally food intake is regulated by the hunger center in the hypothalamus located in the brain. The regulation of food intake is normally proportional to the nutrient stores in the body. If these nutrient stores get low, you feel hungry; if these stores are high, you do not feel like eating. You reduce your food intake to prevent overstorage.

Table 6.6 Energy Requirements in Daily Diet

Age (yr)	Weight (lb)	Kilocalories
Child		
0–1	16	820
2–3	29	1360
4–6	44	1830
7–9	62	2190
10–12	83	2480
Young Adult		
Female 13–19	115	2400
Male	130	2980
Adult		
Female	121	2200
Male	143	3000

The regulation of food intake does not operate in obese people. The causes of this failure may be psychological or physiological. One cause of obesity in adults may be childhood conditioning. In infancy, the rate of formation of new fat cells is rapid, as compared to that in later life. The number of fat cells produced in infancy determines the number of fat cells in the adult body. It is now believed that overfeeding children, especially infants, can lead to a lifelong struggle with obesity.

In stressful situations, some people overeat in an attempt to relieve tensions. Occasionally, tumors of the hypothalamus cause an excessive appetite and result in a weight increase.

Weight reduction occurs when food intake is less than energy output. Many diet products contain cellulose, which provides "bulk" and makes you feel "full" but contains no nutritive value. Some diet drugs include amphetamines, which depress the hunger center. These drugs must be used with great caution because they excite the nervous system and elevate blood pressure. Because muscular exercise is the most important means of expending energy, an increase in daily exercise greatly aids weight loss. (See Table 6.7.)

Table 6.7 Energy Expended (kcal/h) by a 70-kg (155-lb) Person

Activity	Energy Expended (kcal/h)
Sleeping	60
Sitting	100
Walking	200
Swimming	500
Running	550

The body uses carbohydrates for energy before using fats and proteins. However, your carbohydrate stores are quite small. There are only a few hundred grams of glycogen stored in the liver and muscles of the body. This glycogen provides enough energy for about a half a day. In severe dieting or fasting, this energy store is rapidly depleted. Fat stores then become the prime source of energy, and the amount of fat in the body decreases.

If the fat stores are depleted, as in starvation, the only energy store remaining is protein. Proteins can be converted to glucose by the liver to provide energy for the brain. However, because proteins are essential to cellular metabolism, their depletion eventually results in death.

Glossary

boiling The conversion of a liquid to gas that occurs when the molecules of gas form within the liquid.

boiling point The temperature at which boiling occurs.

calorie The amount of energy that will raise the temperature of 1 g of water 1°C.

Calorie The nutritional unit of energy, equal to 1000 cal, or 1 kcal.

calorimeter A device to measure heat absorbed or released by a sample that undergoes reaction.

change of state The change that occurs when a substance is transformed from one state to another, for example, liquid to solid, liquid to gas.

condensation The formation of liquid from a gas that occurs when a gas is cooled to the condensation point (same temperature as boiling).

cooling curve A graph that indicates the changes of state for a substance as the temperature is lowered.

crystallization The formation of a solid from a liquid caused by cooling; occurs at the same temperature as the melting point.

endothermic A reaction in which heat is absorbed.

energy The ability to do work.

evaporation The formation of gas (vapor) by the escape of molecules from the surface of a liquid.

exothermic A reaction in which heat is released.

gas A state of matter characterized by no definite shape or volume. Molecules in a gas move rapidly, with little or no attraction to each other.

heat The energy associated with the motion of particles in a substance.

heat of fusion The energy required to melt or freeze 1 g of a substance at its melting (freezing) point. For water, 80 cal are needed to melt 1 g of ice; 80 cal are released when 1 g of water freezes.

heat of vaporization The energy required to completely vaporize (condense) 1 g of substance. For water, 1 g of liquid requires 540 cal to completely vaporize; 1 g of steam will give off 540 cal when it condenses.

heating curve A graph that represents the transitions from one state to the next as heat is added to a substance.

joule An amount of heat equal to 4.184 cal.

kinetic energy A form of energy that is actively doing work. Also called *energy of motion.*

liquid A state of matter that takes the shape of its container but has its own volume. A liquid has moderate molecular motion and moderate attraction between molecules.

matter Anything that has mass and occupies space.

melting The conversion of a solid to a liquid.

melting point The temperature at which a solid becomes a liquid (melts).

potential energy An inactive, storage form of energy that is capable of doing work at a later time.

solid A state of matter that has its own shape and volume, with little molecular motion—only vibrations—and strong attractions between molecules.

sublimation The change of state in which a solid is transformed directly into a gas.

vapor The molecules of a gas above the liquid of the same substance.

Problems

Energy (Objective 6.1)

6.1 Indicate whether each statement describes potential or kinetic energy:

a. water at the top of a waterfall
b. a car speeding down the freeway
c. the energy in a lump of coal
d. a roller coaster at the top of a hill
e. the energy you will obtain from your food
f. a glacier on the side of a mountain
g. an earthquake
h. a ski jumper going down the ramp

Measuring Heat Energy (Objective 6.2)

6.2 Determine the amount of heat required to increase the temperatures of the substances in the following problems:
a. the number of calories to heat 40 g of water from 25°C to 50°C
b. the number of joules to heat 15 g of water from 10°C to 85°C
c. silver has a specific heat of 0.057 cal/g °C. How many calories are needed to raise the temperature of 10.0 g of silver from 15°C to 100°C?

6.3 An electric power plant released heat into a nearby stream. The temperature of 5.0 kg of stream water increased from 20°C to 28°C. How many kilocalories of heat were absorbed by the water?

6.4 Calculate the amount of heat that is lost by a substance in the following problems:
a. the number of calories lost when 200 g of water cools from 80°C to 20°C
b. the number of joules lost when 0.50 kg of water cools from 50°C to 5°C
c. the number of kilocalories lost when 250 g of beach sand cool from 45°C to 30°C (the specific heat of beach sand is 0.19 cal/g °C)

States of Matter (Objective 6.3)

6.5 Indicate whether each of the following statements describes a gas, a liquid, or a solid:
a. It has a definite volume, but no definite shape.
b. There are no attractions between the molecules.
c. The particles are held in a definite pattern.
d. The particles are very far apart.
e. It occupies the entire volume of the container.

Changes of State (Objective 6.4)

6.6 Draw a heating curve for ice (at −20°C) that is heated to 140°C. The melting point of ice is 0°C; the boiling point is 100°C. Indicate the portion of the curve that corresponds to each of the following:
a. solid
b. melting point
c. liquid
d. boiling point
e. gas

6.7 The melting point of benzene is 5°C and its boiling point is 80°C. Sketch a heating curve for benzene from 0°C to 100°C.
a. What is the state of benzene at 15°C?
b. What happens on the curve at 5°C?
c. What is the state of benzene at 60°C? At 90°C?
d. At what temperature can liquid and gas both be present?

6.8 What happens to a sample of steam, initially at 110°C, that is cooled to −10°C? Sketch the cooling curve.

6.9 Indicate whether each of the following statements describes melting, melting point, evaporation, boiling, or boiling point:
a. a temperature at which a solid is converted to a liquid
b. the formation of gaseous molecules at the surface of the liquid

c. a process in which energy is absorbed to break the bonds of a solid
d. the formation of bubbles of gas within the liquid

Energy in Changes of State (Objective 6.5)

6.10 Calculate the amount of heat needed in the following problems which involve the heat of fusion:

a. the number of calories needed to melt 15.0 g of ice
b. the number of joules needed to melt 75.0 g of ice
c. the number of calories needed to melt a 50.0-g ice cube at 0°C, and then to heat the water formed (0°C) to 65°C

6.11 A bag of ice was placed on a burn on a patient's hand. The ice bag contained 200 g of ice at 0°C. When the ice bag was removed from the burn, all of the ice inside had melted, and the water inside now had a temperature of 20°C. How many kilocalories were absorbed by the ice?

6.12 Calculate the heat change in the following problems which involve the heat of vaporization:

a. How many kilocalories are needed to vaporize (boil) 50.0 g of water at 100°C?
b. How many calories are released when 10.0 g of steam (gas at 100°C) condenses to liquid (at 100°C) and drops to 30°C?
c. How many kilocalories are released when 200 g of steam at 100°C condenses, cools, and finally freezes at 0°C?

6.13 An ice-cube tray holds 200 g of water. If tap water has a temperature of 25°C, how many calories must be removed in order for the water to cool and freeze at 0°C?

6.14 A 500-g sample of steam at 100°C escapes from a volcano. It condenses and cools and finally falls as snow at 0°C. How many kilocalories of heat are released?

6.15 Water is sprayed on the ground of an orchard to protect the fruit from freezing. How many kilocalories of heat are released if 5000 g of water at 20°C are sprayed on the ground and frozen at 0°C?

Calorimetry (Objective 6.6)

6.16 The combustion of the following foods has been determined by calorimetry. Use the data given below to calculate the kcal for each food:

Food	Water in Calorimeter (g)	Temperature (°C)	
		Initial	Final
a. Celery (1 stalk)	500	25	35
b. Waffle (7-in. diameter)	5000	20	62
c. Popcorn (1 cup, no oil)	1000	25	50

6.17 Use the known caloric values for carbohydrates, fats, and proteins to complete the following table:

Food	Serving	Carbohydrate (g)	Fat (g)	Protein (g)	Total (kcal)
a. Orange	1	16	0	1	______
b. Apple	1	______	0	0	72
c. Danish	1	30	15	5	______
d. Avocado	1	13	______	5	405

6.18 A diet consists of 220 g of carbohydrate, 100 g of lipid (fat), and 80 g of protein. How many total kilocalories per day does this diet provide?

6.19 A high-protein diet may contain the following:

70 g carbohydrate
150 g protein
5 g fat

How many total kilocalories does this diet provide daily?

6.20 A normal adequate diet in the United States provides 15 percent of its calories from protein, 45 percent from carbohydrates, and 40 percent from fats. Calculate the total grams of protein, carbohydrate, and fat to be included each day in diets with the following caloric requirements:

a. 1000 kcal
b. 1800 kcal
c. 2600 kcal

Gases

Objectives

7.1 Describe the characteristics of a gas according to the kinetic molecular theory.

7.2 Describe the units of measurement for pressure and change from one to another.

7.3 State the four properties on which gas behavior depends and give their units of measurement.

7.4 Use Boyle's law to solve for changes in pressure and volume.

7.5 Use Gay-Lussac's law to solve for changes in pressure and temperature.

7.6 Use Charles' law to solve for changes in volume and temperature.

7.7 Use the combined gas law to solve for changes in pressure, volume, and temperature.

7.8 Describe the relationship between the number of moles of a gas and its volume.

7.9 Use Dalton's law of partial pressures to calculate the partial or total pressure of a mixture of gases.

Scope

You live at the bottom of a sea of gases called the *atmosphere*. The most important of these gases is oxygen, which constitutes about 21 percent of all the gases in the atmosphere. The other gases include 78 percent nitrogen,

1 percent argon, and trace amounts of carbon dioxide (CO_2) and varying amounts of water vapor. Without oxygen, life on our planet would be impossible because oxygen is vital to the life processes of all plants and animals. Ozone (O_3), a gas that is formed in the upper atmosphere by the reaction of oxygen and light, absorbs harmful ultraviolet radiation from the sun before it can strike the earth's surface.

Carbon dioxide gas, produced by animals as a product of cellular metabolism, is needed by plants and photosynthesizing bacteria. They use the CO_2 in a process called *photosynthesis.*

Gases play many important roles externally in our environment and internally in our bodies. In this chapter you will learn more about the behavior of gases, some of the laws that govern gas behavior, and the exchange between the gases of the atmosphere, our lungs, and blood.

7.1 Kinetic Molecular Theory

OBJECTIVE Describe the characteristics of a gas according to the kinetic molecular theory.

Gases can be described in terms of the motion of their individual atoms or molecules. A model, called the *kinetic molecular theory*, helps us visualize and understand the way gases behave. The term *kinetic* is used because this model is based on the assumption that the particles of a gas are in motion.

Kinetic Molecular Theory

1. Gases are composed of very small particles (atoms or molecules) that are constantly moving in straight lines at high speeds in all directions.
2. The molecules of a gas are so far apart that there are no attractive or repulsive forces between them. Gases can be compressed or expanded easily.
3. Gas molecules collide with the walls of the container or other gas molecules without losing energy.
4. The kinetic energy of the gas molecules depends on the temperature of the gas. The molecules move faster at higher temperatures.

The kinetic molecular theory helps account for some of the characteristics we associate with gases. Because gases move rapidly, we can soon smell a perfume coming from the other side of a room. Since gases expand easily, we can heat the air in a hot air balloon to make it rise.

Example 7.1 State what the kinetic molecular theory says about the following characteristics of a gas:

1. attractive forces between gas molecules
2. speed of gas molecules
3. effect of temperature on gas molecules

Solution

1. Gas molecules are not attracted to each other because they are so far apart.
2. Gas molecules move very rapidly.
3. The motion of gas molecules increases as temperature rises.

7.2 Pressure of a Gas

OBJECTIVE Describe the units of measurement for pressure and change from one to another.

gas molecules create pressure when they bombard the walls of the container

When water boils in a pan covered with a lid, the collisions of the molecules of steam (gas) lift up the lid. The gas molecules in the pan hit the walls and lid with great force. The force of these collisions pushing against the lid of the pan is called *pressure*. Suppose we could observe a small area somewhere on the wall of a container as in Figure 7.1. We will say that if one molecule hits that area, the pressure is the force of one molecular collision per area. If four molecules hit that area, then we would say the pressure is the force of four collisions per area.

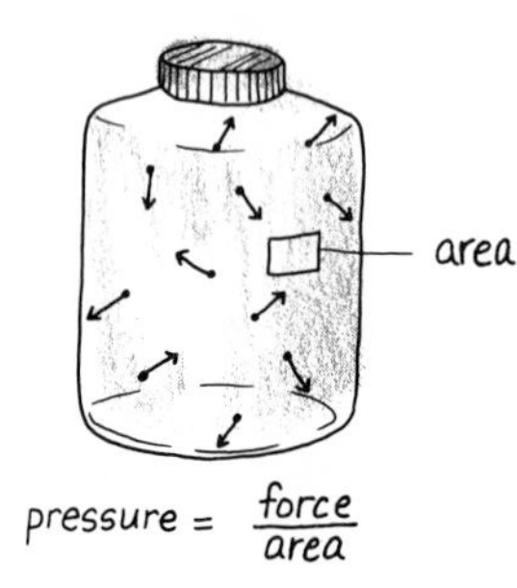

Figure 7.1 Collisions of gas molecules against the walls of the container create pressure.

Pressure is defined as the force per unit area:

$$\text{Pressure} = \frac{\text{Force}}{\text{Area}}$$

Example 7.2 Which phrase (or phrases) describes the pressure of a gas?

1. the force of the gas molecules
2. the force of gas molecules on a surface area of the container
3. the number of gas molecules in a container.

Solution: Phrase 2; pressure is defined as the force of the gas molecules per unit area.

Atmospheric Pressure

The air that covers the surface of the earth, which is referred to as the *atmosphere*, contains trillions of gas molecules. These gas molecules (mostly oxygen and nitrogen) have mass and exert pressure upon us and the earth. The pressure exerted by a column of air reaching from the top of the atmosphere is called *atmospheric pressure* and is measured by units called *atmospheres* (atm).

atmosphere

The atmospheric pressure is measured by a barometer, as seen in Figure 7.2. A long glass tube filled with mercury is inverted in an open dish of mercury. The mercury in the tube falls until its mass is supported by the atmosphere that is pressing down on the mercury surface in the dish. When the pressure exerted by the air supports a column of mercury 760 mm high, we say the atmospheric pressure is 1 atm. In the metric system, pressure is also expressed in units of *torr*, a pressure unit named in honor of Torricelli, the inventor of the barometer. One torr describes the same amount of pressure as 1 mmHg:

1 torr = 1 mmHg

$$1 \text{ mmHg} = 1 \text{ torr}$$

$$1 \text{ atm} = 760 \text{ mmHg} = 760 \text{ torr}$$

The U.S. Customary equivalent of 1 atm is 14.7 pounds per square inch (psi). This is what you read on a pressure gauge when you check the air pressure in the tires of a car or bicycle.

If you have a barometer in your home, it probably reads in inches of mercury. One atmosphere is the pressure of a column of mercury that is 29.9 in. high. Weather reports are given in inches of mercury. Low-pressure systems often indicate rain or snow, whereas high-pressure systems usually bring dry and sunny weather.

At sea level, the atmospheric pressure is approximately 1 atm. At higher altitudes, atmospheric pressure is lower because there is a decrease in the number of air molecules. Deep-sea divers must be concerned about the pressure conditions when they are below the surface of the ocean. Since water is much more dense than air, the pressure on a diver increases rapidly as the diver descends. At a depth of 33 ft below the surface of the ocean, there is an additional atmosphere of pressure. Thus, there is a total of 2 atm of pressure on a diver at a 33-ft depth. Divers must understand gas laws in order to properly

pressure under water

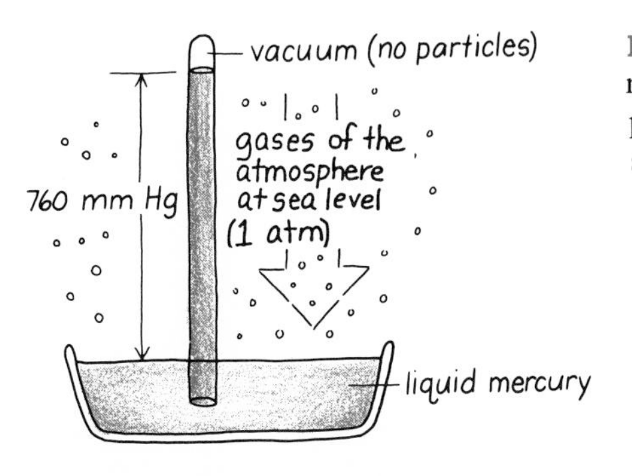

Figure 7.2 A barometer: A column of mercury 760 mm high is supported by the pressure exerted by atmospheric gases at sea level, $P = 1$ atm.

Table 7.1 Units for Measuring Pressure

Unit	Relationship to 1 atm of Pressure
atmosphere (atm)	1 atm
millimeters of mercury (mmHg)	760 mmHg
torr	760 torr
inches of mercury (in.Hg)	29.9 in.Hg
pounds per square inch (psi, lb/in.^{2})	14.7 psi

adjust the pressure of their air tanks to provide a proper pressure for the air they are breathing. Table 7.1 summarizes the various units used in the measurement of pressure.

Example 7.3 Change a pressure of 0.50 atm into the following pressure units:

1. mmHg
2. torr
3. inches of Hg

Solution

1. There are 760 mmHg in 1 atm. This can be written in the form of a conversion factor:

$$1 \text{ atm} = 760 \text{ mmHg}$$

$$\frac{760 \text{ mmHg}}{1 \text{ atm}} \quad \text{or} \quad \frac{1 \text{ atm}}{760 \text{ mmHg}}$$

Problem Setup

$$0.50 \cancel{\text{atm}} \times \frac{760 \text{ mm Hg}}{1 \cancel{\text{atm}}} = 380 \text{ mmHg}$$

2. One atmosphere is equal to 760 torr. This equality can be written as a conversion factor:

$$1 \text{ atm} = 760 \text{ torr}$$

$$\frac{760 \text{ torr}}{1 \text{ atm}} \quad \text{or} \quad \frac{1 \text{ atm}}{760 \text{ torr}}$$

Problem Setup

$$0.50 \cancel{\text{atm}} \times \frac{760 \text{ torr}}{1 \cancel{\text{atm}}} = 380 \text{ torr}$$

3. One atmosphere is the same as 29.9 in.Hg. We can write this as a conversion factor:

$$1 \text{ atm} = 29.9 \text{ in.Hg}$$

$$\frac{29.9 \text{ in.Hg}}{1 \text{ atm}} \quad \text{or} \quad \frac{1 \text{ atm}}{29.9 \text{ in.Hg}}$$

Problem Setup

$$0.50 \cancel{\text{atm}} \times \frac{29.9 \text{ in.Hg}}{1 \cancel{\text{atm}}} = 15 \text{ in.Hg}$$

Vapor Pressure

The pressure of the gas molecules above a liquid in a closed container is called *vapor pressure*. The vapor pressure of a pure liquid is constant at a given temperature. However, vapor pressure does change if the temperature changes. As temperature increases, more vapor forms, and the vapor pressure increases. Table 7.2 lists the vapor pressures of water at various temperatures.

Vapor Pressure and Boiling Point

A liquid begins boiling when the vapor pressure of the liquid becomes equal to the external pressure. Then bubbles of the gas form within the liquid and quickly rise to the surface. If the atmospheric pressure is 760 mmHg, then a vapor pressure of 760 mmHg must be reached before liquid starts to boil. For example, water boils at 100°C, the temperature at which the vapor pressure becomes equal to an atmospheric pressure of 760 mmHg.

External Pressure and Boiling Point

lower boiling points at higher altitudes

A change in the external pressure changes the boiling point of a liquid. At higher altitudes, the external pressure is lower. For example, the typical atmospheric pressure in Denver is 630 mmHg. Water has a vapor pressure of 630 mmHg at

Table 7.2 Vapor Pressure of Water at Various Temperatures

Temperature (°C)	Vapor Pressure (mmHg)
0	5.0
20	17.5
40	55.0
60	149
80	355
100	760

Table 7.3 Altitude, Atmospheric Pressure, and Boiling Point of Water

Location	Altitude (miles)	Altitude (km)	Atmospheric Pressure (torr)	Boiling Point (°C)
Mt. Everest	5.80	9.30	270	70
Mt. Whitney	2.50	4.00	467	87
Denver	1.00	1.60	630	95
Las Vegas	0.40	0.70	700	98
Los Angeles	0.05	0.09	752	99
Sea level	0	0	760	100

Table 7.4 Resulting Increases in Boiling Point of Water With Increases in Pressure

Pressure (torr)	Boiling Point (°C)
800	101.4
1075	110
1520 (2 atm)[a]	120
2026	130
7600 (10 atm)	180

[a] These pressures are attained by some autoclaves and pressure cookers.

95°C. That means that in Denver the temperature of water only has to reach 95°C for boiling to begin. Lower external pressures cause liquids to boil at lower temperatures. Some examples of the effect of altitude and reduced atmospheric pressure on boiling points are shown in Table 7.3. (See Figure 7.3.)

People who live in high altitudes often use pressure cookers to achieve higher cooking temperatures. If the pressure exceeds 1 atm, a temperature greater than 100°C is needed before water will boil. A pressure cooker and an autoclave (a device used in laboratories and hospitals to sterilize laboratory and surgical equipment) are two closed containers that develop pressures higher than atmospheric to create boiling points greater than 100°C. Table 7.4 shows how the boiling point of water change with increases in pressure.

Figure 7.3 Boiling of a liquid at different pressures. Formation of gas within a liquid (boiling) occurs when the vapor pressure of the liquid becomes equal to the atmospheric pressure.

Example 7.4 Identify the following phrases as descriptions of vapor pressure, boiling, or boiling point:

1. the pressure exerted by gas molecules above a liquid
2. the formation of bubbles of vapor within the liquid
3. the temperature at which vapor pressure equals the atmospheric pressure

Solution

1. vapor pressure
2. boiling
3. boiling point

7.3 Properties of Gases

OBJECTIVE State the four properties on which gas behavior depends and give their units of measurement.

To further describe gas behavior, we need to state the four properties of a gas: pressure (P), volume (V), temperature (T), and amount of gas (n).

Pressure

We stated in Section 7.2 that the pressure of a gas is the result of a force created by the molecules colliding with the walls of the container. When stating the pressure of a gas, the units you are most likely to use are atmosphere (atm), millimeters of mercury (mmHg), or torr.

Volume

A gas takes the shape and volume of its container. Therefore, the volume of a gas is the volume of the container. The volume of a gas in a 10-mL vial is 10 mL; the volume of a gas in a 40-L tank is 40 L. The units for the volume of a gas are usually milliliters (mL) or liters (L).

Temperature

When working with gases, all calculations must use absolute temperature values, or kelvins, to obtain correct results. On the kelvin (K) temperature scale, if a gas were to reach a temperature of absolute zero, the gas molecules would have neither energy nor motion. Then the volume of that gas would become zero. In Chapter 1 you learned that the relationship between degrees Celsius and kelvins is

$$T\text{ K} = T^\circ\text{C} + 273$$

HEALTH NOTE

Tidal Volume

A respiratory therapist might need to measure a patient's *tidal volume*, which is the volume of air that is moved into or out of the lungs while a person breathes. Tidal volume is measured with an instrument called a *spirometer*. Such a test is used to evaluate the functioning of the lungs, especially when there is an impairment of the lungs such as asthma, emphysema, and chronic bronchitis.

Tidal Volume of Air in One Breathing Cycle

Infant	75–125 mL
Child	200–250 mL
Adult	450–500 mL

For an adult, there are about 16–20 breathing cycles each minute, during which a total of 5–10 L of air moves into and out of the lungs.

Amount of Gas

In gas laws, the amount of gas is represented by the number of moles (n).

A summary of the four properties that describe a gas is found in Table 7.5.

Table 7.5 Properties of a Gas

Property	Units of Measurement	Feature
Pressure (P)	atm, mmHg, torr	The force exerted by gas molecules over a unit area
Volume (V)	L, mL	The space occupied by a gas (the size of the container)
Temperature (T)	K	The temperature of a gas sample
Amount of gas (n)	mol	The amount of gas in the sample

Example 7.5 Use the words *pressure*, *volume*, *temperature*, or *quantity* to indicate the property of a gas described in each of the following phrases or measurements:

1. 800 torr
2. 295 K
3. 4.0 L
4. the space occupied by a gas
5. the number of moles of gas in the sample

Solution

1. The measurement 800 torr describes the *pressure* of the gas in torrs.
2. The measurement 295 K gives the *temperature* of the gas in kelvins.
3. The measurement 4.0 L states the *volume* of the gas in liters.
4. The space occupied by a gas is its *volume*.
5. The number of moles of gas is the *quantity* of gas in the sample.

7.4 Pressure and Volume: Boyle's Law

OBJECTIVE Use Boyle's law to solve for changes in pressure and volume.

Imagine that you can see gas molecules hitting the walls of a bicycle pump. We can change the volume of the pump by pushing down on the handle. What happens to the pressure inside the pump when the volume of the gas is decreased? The volume of the container holding the air has been decreased. The total number of molecules has not changed; however, the molecules are crowded together and there are more collisions per unit area. The pressure of the gas increases because the force per unit area is greater. (See Figure 7.4.)

When one property (in this case, volume) changes another property (pressure), these properties are said to be dependent on each other. When changes occur in opposite directions (the pressure increased when volume decreased), we say that the two properties are *inversely dependent*. The relationship between the pressure and volume of a gas is given by Boyle's law. It states that the volume of a gas changes inversely with the pressure of the gas, when the temperature and number of moles remain constant. Accordingly, we can also state that the pressure of a gas decreases if the volume is increased. Gas laws such as Boyle's law work well for most gases. We will assume in this text that all gases are ideal and that they behave as such in accordance with these laws.

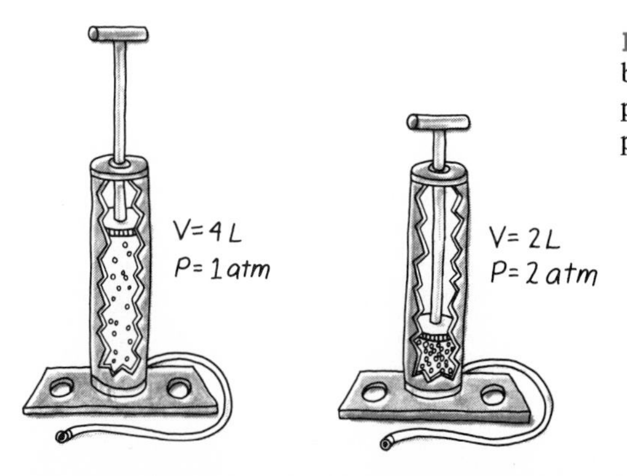

Figure 7.4 When the handle of the bicycle pump is pushed in, the pressure of the air inside the pump increases.

Example 7.6 Complete the information for the pressure and volume of a gas at constant temperature:

Pressure	*Volume*
1. increases	__________
2. __________	increases

Solution

1. Pressure and volume are inversely related. When the pressure of a gas increases, the volume decreases.
2. If the volume of a specific amount of gas increases, the pressure of the gas decreases.

Another way to look at Boyle's law is that the pressure multiplied by the volume of a gas sample will remain constant regardless of how the pressure or volume is changed:

$$P \times V = \text{constant}$$

In calculating changes in pressure or volume according to Boyle's law, we compare the pressure and volume of the sample at the initial conditions and at final conditions. Since PV for both conditions are equal, we can use the following equation as long as n and T remain unchanged (constant):

$$P_{\text{initial}}V_{\text{initial}} = P_{\text{final}}V_{\text{final}} \quad (n, T \text{ constant})$$

Example 7.7 A sample of gas has a volume of 100 mL and a pressure of 1 atm. What is the new pressure if the volume is decreased to 50 mL?

Solution: First, let us organize the given information. The initial conditions can be written as V_i and P_i and the final conditions as V_f and P_f. We are not including T and n, since they are unchanged.

Initial Conditions	*Final Conditions*
$P_i = 1$ atm	$P_f = ?$
$V_i = 100$ mL	$V_f = 50$ mL
	(Volume decreasing)

In this problem, we do not know the final pressure. Therefore, we begin our calculation with the initial pressure of 1 atm. This is multiplied by the effect of the volume changes:

New (final) pressure = Initial pressure × Fraction of volumes

$$P_f = 1 \text{ atm} \times \text{Volume effect}$$

In this problem, the volume is changing. There are two possible effects of volume. One effect is to decrease the pressure represented by a ratio of volumes that is less than 1 (<1). The other effect of the volume change would be to increase the pressure represented by a ratio of volume that has a value greater than 1 (>1):

$$\frac{100 \text{ mL}}{50 \text{ mL}} \quad \text{or} \quad \frac{50 \text{ mL}}{100 \text{ mL}}$$

[Greater than 1 (>1) will increase pressure] [Less than 1 (<1) will decrease pressure]

The choice of ratio depends on the change in the volume and its corresponding effect on the pressure. In this problem the volume has decreased. Boyle's law states that if the volume decreases, the pressure has to increase. Thus, the final pressure must be greater than the initial pressure. The initial pressure has to be multiplied by the ratio of volumes that will give a greater final pressure:

$$\text{New (final) pressure} = \text{Initial pressure} \times \text{Volume effect}$$

$$P_f = 1 \text{ atm} \times \frac{100 \cancel{\text{mL}}}{50 \cancel{\text{mL}}}$$

(Increases pressure)

$$= 2 \text{ atm}$$

Checking our answer, we see that the pressure doubled in value when the volume was decreased to half its initial value. (See Figure 7.5.) The product PV remains the same (constant) according to Boyle's law:

$$P_i V_i = P_f V_f$$

$$1 \text{ atm} \times 100 \text{ mL} = 2 \text{ atm} \times 50 \text{ mL}$$

$$100 \text{ atm} \cdot \text{mL} = 100 \text{ atm} \cdot \text{mL}$$

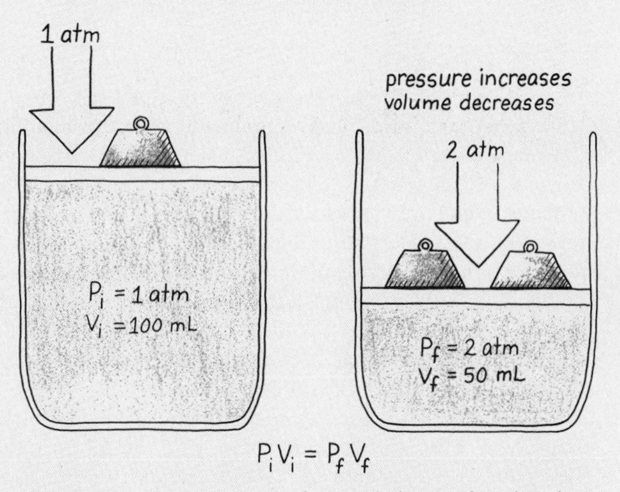

Figure 7.5 The pressure of a gas is inversely related to the volume: When pressure increases, volume decreases.

HEALTH NOTE

Boyle's Law and Ventilation

The importance of Boyle's law becomes more apparent when you consider the process of breathing. Your lungs are elastic structures which act like balloons that are open to the external atmosphere. They are held within an airtight chamber called the *thoracic cavity*. The diaphragm, a muscle, forms the flexible floor of the thoracic cavity. (See Figure 7.6.)

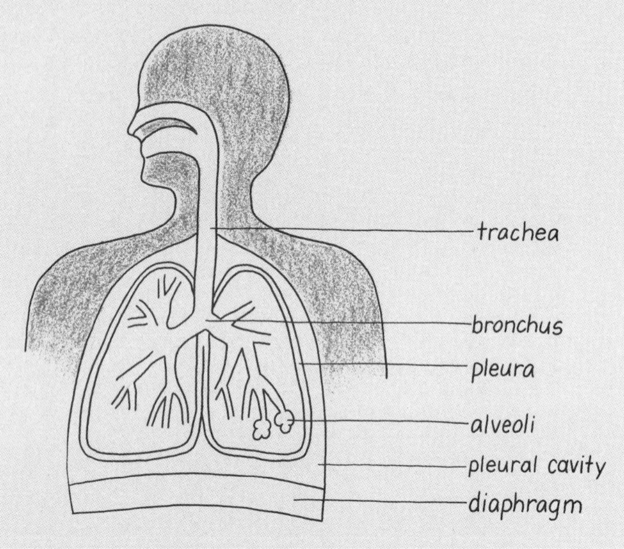

Figure 7.6 The airways of the respiratory tract.

When the diaphragm contracts, it flattens out, causing an increase in the volume of the thoracic cavity along with the expansion of the rib cage. The elasticity of the lungs allows them to expand when the thoracic cavity expands. According to Boyle's law, the pressure of a gas will decrease when the volume increases. When the volume of the lungs increases, the pressure inside the lungs decreases, falling below the pressure of the external (ambient) atmosphere. This difference in pressures creates a *pressure gradient* between the lungs and the external atmosphere. In a pressure gradient, molecules flow from the higher pressure (where there are more molecules) to the lower pressure. (See Figure 7.7.) Air flows into the lungs (*inspiration*) until the pressure within the lungs becomes equal to the pressure of the external atmosphere. You have just taken a breath of air.

pressure gradient

inspiration

expiration

The *expiration*, or exhalation phase of the ventilatory cycle, occurs when the diaphragm relaxes by arching back up in the thoracic cavity to its resting position. This reduces the volume of the thoracic cavity. The decrease in the volume in the

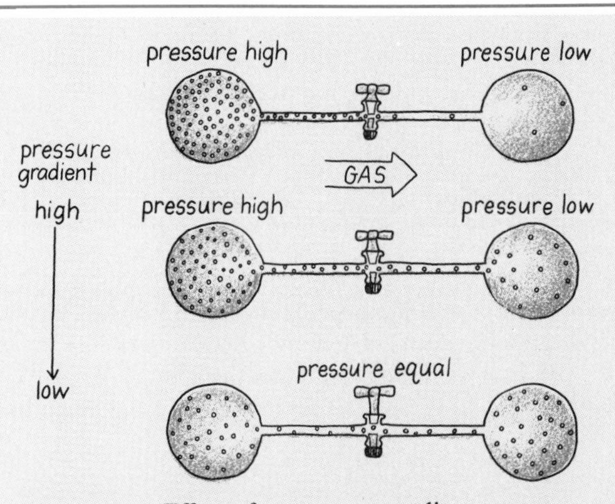

Figure 7.7 Effect of a pressure gradient on gas diffusion. Gas molecules move from a high pressure to a low pressure.

thoracic cavity squeezes the lungs so the volume of the lungs decreases. According to Boyle's law, a decrease in the volume of a gas will cause an increase in the pressure of the gas. Another pressure gradient is created. Now the pressure in the lungs is greater than the pressure of the external atmosphere, so air flows out of the lungs. Breathing is a process in which pressure gradients are continuously created between the lungs and the external environment as a result of the changes in the volume and pressure of the lungs. (See Figure 7.8.)

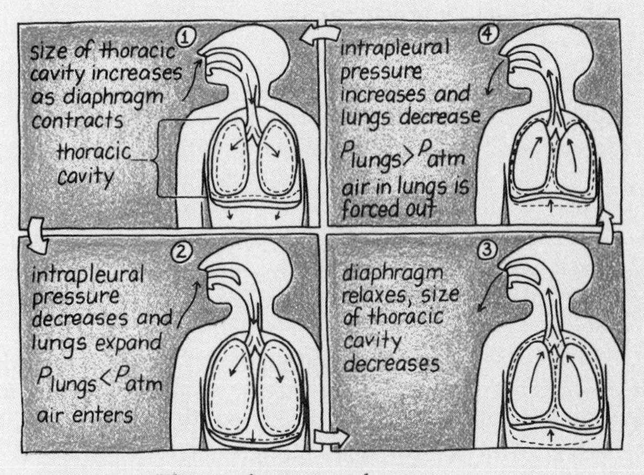

Figure 7.8 The respiratory cycle.

Example 7.8 A sample of oxygen has a volume of 12.0 L and a pressure of 200 torr. Find the new volume when the pressure becomes 800 torr (T and n constant).

Solution: The information can be organized in a table:

Initial Conditions	*Final Conditions*
$P_i = 200$ torr	$P_f = 800$ torr
$V_i = 12.0$ L	$V_f = ?$

We note that the pressure has increased. According to Boyle's law, an increase in the pressure will be accompanied by a decrease in the volume. The final volume must be smaller than the initial volume. We need to multiply the initial volume (V_i) by a ratio of the pressure values that decreases the volume. We will use a ratio of pressures that is less than 1:

New (final) volume = Initial volume × Fraction of pressures

$$V_f = 12.0\text{ L} \times \frac{200\text{ torr}}{800\text{ torr}}$$

(Decreases volume)

$$= 3.0\text{ L}$$

HEALTH NOTE

Mechanical Ventilators

Mechanical respirators use the principles of pressure and volume changes to develop pressure gradients between a patient's air supply and lungs. For example, the iron lung is one type of respirator that completely encloses a patient's body or covers the chest area. The iron lung operates like a cylinder with a movable piston at one end. When the piston is pushed in, the volume of air around the body is compressed. The increased pressure against the patient's chest decreases the volume in the lungs, causing the patient to exhale. When the piston is released, the volume of air around the patient increases and the pressure is reduced. The lungs expand and the patient inhales.

respirator

Other forms of respirators provide air or oxygen by the use of face masks or nasal tubes. One type of respirator produces pressure gradients by alternating the pressure of the air going to the patient. When pressure is higher in the respirator than in the patient's lungs, air flows into the lungs. When the pressure of the respirator drops, the high pressure in the lungs causes the patient to exhale. As the pressures are continually alternated, the patient inhales and exhales, allowing proper gas exchange between the lungs, blood, and tissues.

Example 7.9 Indicate if inspiration or expiration will occur in each of the following:

1. The diaphragm contracts (flattens out).
2. The pressure in the lungs is greater than the external atmosphere.

Solution

1. When the diaphragm contracts, the volume of the thoracic cavity and therefore the volume of the lungs increases. The pressure within the lungs will decrease to create a pressure gradient in which the external air has a higher pressure than the air in the lungs. *Inspiration* occurs.
2. When the pressure in the lungs is greater than the external atmosphere, a pressure gradient is created in which air flows out of the lungs. *Expiration* occurs.

7.5 Pressure and Temperature: Gay-Lussac's Law

OBJECTIVE Use Gay-Lussac's law to solve for changes in pressure and temperature.

Now we can consider another set of gas properties and their relationship to each other. We will compare the changes in temperature and pressure when we keep the volume and amount of gas constant. If we were able to see the molecules of a gas as the temperature rises, we would see them moving faster and hitting the sides of the container more often and with greater force. Since the container cannot change size (volume remains constant), the increase in the temperature will cause an increase in the pressure of the gas.

Gay-Lussac's law states that the pressure of a fixed amount of gas is directly related to its Kelvin temperature. This means that if the temperature of a gas increases, its pressure increases. It also means that a specific decrease in temperature will cause the pressure to also decrease by a definite amount.

When temperature is used in gas law calculations, it must always be in kelvins (absolute temperature scale). Then there is a direct relationship between temperature and pressure. A direct relationship means that both properties change in the same direction. If the Kelvin temperature doubles, the pressure will also double; if the Kelvin temperature drops to one-third of its initial value, the pressure will also drop to one-third of the initial value.

Example 7.10 Complete the following unknown changes for the temperature or pressure of a gas for a specific amount of gas when V is kept constant:

	T	P
1.	increases	________
2.	________	decreases

Solution

1. If the temperature of a specific amount of a gas increases, the pressure of the gas will also increase if V is held constant.
2. If the pressure of a gas decreases, it must be accompanied by a decrease in the temperature of the gas as long as volume is held constant.

Mathematically, Gay-Lussac's law is stated as

$$\frac{P_i}{T_i} = \frac{P_f}{T_f}$$

Note that temperature must be in kelvins. The units for pressure must be the same for the initial and final conditions.

Example 7.11 Aerosol cans can be dangerous if heated because the increase in pressure can cause an explosion. Suppose a can of hair spray has a pressure of 4.0 atm at a room temperature of 27°C. If the can were thrown into a fire, its temperature could reach 402°C. Can you determine the pressure of the gas inside the aerosol can? The aerosol can may explode if the pressure inside exceeds 8.0 atm. Would you expect the can to explode?

Solution: Since temperatures are given in degrees Celsius, they must be converted to kelvins:

$$T_i = 27 + 273 = 300 \text{ K} \qquad T_f = 402 + 273 = 675 \text{ K}$$

Now we can place the information for the temperature and pressure in a table:

Initial Conditions	*Final Conditions*
$P_i = 4.0$ atm	$P_f = ?$
$T_i = 300$ K	$T_f = 675$ K
	(Temperature increased)

Since the temperature of gas has increased, the final pressure also has to increase. In the calculation, we will use a ratio of the temperatures that is greater than 1.

$$P_f = 4.0 \text{ atm} \times \frac{675 \not{\text{K}}}{300 \not{\text{K}}}$$

(Increases pressure)

$$= 9.0 \text{ atm}$$

Since the new pressure exceeds 8.0 atm, we might expect the can to explode.

7.6 Volume and Temperature: Charles' Law

OBJECTIVE Use Charles' law to solve for changes in volume and temperature.

On a molecular level, we know that an increase in temperature increases the activity (kinetic energy) of the gas molecules. If we are going to keep the pressure constant, the volume of the container has to increase. Then the faster-moving molecules have to travel further before they collide with the container walls. This will keep the pressure constant. By contrast, if the temperature of a gas drops, the volume of the container must be reduced in order to maintain a constant pressure. (See Figure 7.9.)

Figure 7.9 The effect of different temperatures on the volume of air in a tire.

The relationship between temperature and volume is known as *Charles' law*. It states that the volume of a specific amount of gas changes directly with the Kelvin temperature when the pressure is kept constant. Mathematically, this can be expressed as

$$\frac{V_i}{T_i} = \frac{V_f}{T_f}$$

Example 7.12 Complete the following changes in volume and temperature for a specific amount of gas when P is kept constant:

	Volume	*Temperature*
1.	__________	decreases
2.	increases	__________

Solution

1. If the temperature of a gas decreases, its volume also decreases.
2. When the volume of a gas increases, the temperature also increases.

Example 7.13 A 20.0-L balloon is filled with helium at a temperature of 127°C. Find the new volume of the balloon when the temperature drops to −73°C while pressure remains constant.

Solution: First, we need to convert the Celsius temperatures into kelvins:

$$T_i = 127 + 273 = 400 \text{ K}$$

$$T_f = -73 + 273 = 200 \text{ K}$$

Now we can put the data in a table:

Initial Conditions	*Final Conditions*
$T_i = 400$ K	$T_f = 200$ K (Temperature decreases)
$V_i = 20.0$ L	$V_f = ?$

The temperature decreases from 400 K to 200 K. According to Charles' law, the volume also has to decrease. We need to multiply the initial volume by a ratio of the temperatures that is less than 1:

$$V_f = V_i \times \text{Ratio of temperatures (K)}$$

$$= 20.0 \text{ L} \times \frac{200 \text{ K}}{400 \text{ K}} \quad (T/T < 1)$$

$$= 10.0 \text{ L}$$

Example 7.14 A mountain climber inhales 500 mL of air at a temperature of −10°C. What volume will the air occupy in the lungs if his body temperature is 37°C? (Pressure is held constant.)

Solution: First, we need to convert the Celsius temperatures into kelvins:

$$T_i = -10 + 273 = 263\text{ K}$$

$$T_f = 37 + 273 = 310\text{ K}$$

The data can now be placed in a data table:

Initial Conditions	*Final Conditions*
$T_i = 263\text{ K}$	$T_f = 310\text{ K}$ (Temperature increases)
$V_i = 500\text{ mL}$	$V_f = ?$

Since temperature is increasing, we know from Charles' law that the volume will also increase. In the calculation, we multiply the initial volume by a ratio of temperature that is greater than 1:

$$V_f = 500\text{ mL} \times \frac{310\text{ K}}{260\text{ K}}$$

(Increases volume)

$$= 596\text{ mL}$$

7.7 Combined Gas Law

OBJECTIVE Use the combined gas law to solve for changes in pressure, volume, and temperature.

All of the gas laws we have studied regarding pressure, volume, and temperature may be combined into single relationship called *the combined gas law.* This expression is useful for studying the effect of changes in two of the three gas variables as long as the amount of gas remains constant:

$$P_i V_i = P_f V_f \quad \text{(Boyle's law)}$$

$$\frac{P_i}{T_i} = \frac{P_f}{T_f} \quad \text{(Gay-Lussac's law)}$$

$$\frac{V_i}{T_i} = \frac{V_f}{T_f} \quad \text{(Charles' law)}$$

$$\frac{P_i V_i}{T_i} = \frac{P_f V_f}{T_f} \quad \text{(Combined gas law)}$$

Example 7.15 A sample of argon gas has a pressure of 2.0 atm and a volume of 500 mL at a temperature of 200 K. Find the new (final) pressure when the volume is changed to 100 mL and the temperature is raised to 400 K.

Solution: Our data table will look like this:

Initial Conditions	*Final Conditions*
$P_i = 2.0$ atm	$P_f = ?$
$V_i = 500$ mL	$V_f = 100$ mL (Volume decreasing)
$T_i = 200$ K	$T_f = 400$ K (Temperature increasing)

To find the new pressure in this problem, we need to consider the effect of each change, one at a time. The effect of decreasing the volume would be to increase the pressure (Boyle's law):

$$\frac{500 \text{ mL}}{100 \text{ mL}} \quad \text{(Volume effect on pressure)}$$

The other property, temperature, also has an effect on the pressure in this system. The increase in temperature will cause an increase in pressure:

$$\frac{400 \text{ K}}{200 \text{ K}} \quad \text{(Temperature effect on pressure)}$$

Thus, the combined effects of a change in volume and a change in temperature can be calculated:

$$P_f = P_i \times \text{Volume effect} \times \text{Temperature effect}$$

$$= 2.0 \text{ atm} \times \frac{500 \cancel{\text{mL}}}{100 \cancel{\text{mL}}} \times \frac{400 \cancel{\text{K}}}{200 \cancel{\text{K}}} = 20 \text{ atm}$$

Example 7.16 A 50-mL bubble is released from a diver's air tank at a pressure of 3.0 atm and a temperature of 10°C. What is the volume of the bubble when it reaches the ocean surface, where the pressure is 1.0 atm and the temperature is 30°C?

Solution: Since the temperatures are given in degrees Celsius, they need to be changed to kelvins:

$$T_i = 10 + 273 = 283 \text{ K}$$

$$T_f = 30 + 273 = 303 \text{ K}$$

The pressures, volumes, and temperatures are placed in a data table:

Initial Conditions	*Final Conditions*
$P_i = 3.0$ atm	$P_f = 1.0$ atm (P decreases)
$V_i = 50$ mL	$V_f = ?$
$T_i = 283$ K	$T_f = 303$ K (T increases)

To set up the calculation, we need to consider the effects of the pressure change and the temperature change on the volume. The effect of the pressure decrease is an increase in volume (Boyle's law). The effect of the temperature increase is also an increase in volume:

$$V_f = 50\text{ mL} \times \underbrace{\frac{3.0\text{ atm}}{1.0\text{ atm}}}_{\text{Pressure effect}} \times \underbrace{\frac{303\text{ K}}{283\text{ K}}}_{\text{Temperature effect}} = 161\text{ mL}$$

7.8 Moles and Volume: Avogadro's Law

OBJECTIVE Describe the relationship between the number of moles of a gas and its volume.

In the previous sections, we studied the gas laws for a fixed quantity of gas. Now we will consider changes in the amount of gas. When you blow up a balloon, its volume increases because you add more air molecules. If some of the air leaks out of a balloon, its volume decreases. Thus, the volume of a gas is directly related to the number of moles of gas present at constant temperature and pressure. This is known as *Avogadro's law*:

$$\frac{V}{n} = \text{a constant} \qquad \text{and} \qquad \frac{V_i}{n_i} = \frac{V_f}{n_f}$$

Example 7.17 A balloon with a volume of 30.0 L is filled with 2.0 moles of helium. What is the new volume of the balloon when an additional 3.0 mol of helium is added to give a total of 5.0 mol? (P and T are held constant.)

Solution: A data table for volume and number of moles can be set up:

Initial Conditions	*Final Conditions*	
$V_i = 30.0$ L	$V_f = ?$	
$n_i = 2.0$ mol	$n_f = 5.0$ mol	(n increased)

Since the number of moles increased, the volume must increase according to Avogadro's law. The calculation is set up by multiplying the initial volume (30.0 L) by a ratio of the moles that is greater than 1:

$$V_f = 30.0\text{ L} \times \underbrace{\frac{5.0\text{ mol}}{2.0\text{ mol}}}_{\text{Quantity effect}} = 75\text{ L}$$

Molar Volume

Avogadro's law also indicates that two different gases having equal volumes at the same temperature and pressure will contain equal numbers of moles (or molecules). Suppose we have one container filled with oxygen gas (O_2) and

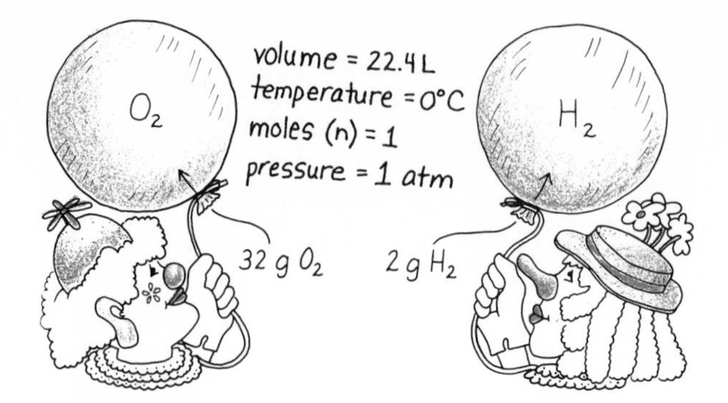

Figure 7.10 Avogadro's principle: equal gas volumes at the same temperature and pressure contain equal numbers of molecules (moles).

another filled with hydrogen gas (H_2). If each container has the same volume and they are at the same temperature and pressure, then each container has the same number of moles of gas. (See Figure 7.10.) This means that the number of molecules, not the mass of the molecules, is important in determining the behavior of a gas.

Standard Temperature and Pressure

Scientists have chosen a standard temperature and pressure (STP), namely, 0°C (273 K) and 1 atm (760 torr). At these standard conditions, 1 mol (6.02×10^{23} molecules) of any gas will occupy a volume of 22.4 L. This is called the *molar volume* of a gas. Thus, the molar volume of any gas is 22.4 L when 1 mol of gas is present at STP conditions:

$$1 \text{ mol of gas at STP} = 22.4 \text{ L}$$

This relationship between moles and volume can be used to convert between the number of moles of a gas and its volume as long as the gas has a temperature of 0°C (273 K) and a pressure of 1 atm.

Example 7.18 What is the volume occupied by 5.00 mol of oxygen at STP?

Solution: Since the gas is at STP, we can use the molar volume as a conversion factor:

$$\frac{1 \text{ mol}}{22.4 \text{ L}} \quad \text{or} \quad \frac{22.4 \text{ L}}{1 \text{ mol}}$$

The given is 5.0 mol of oxygen. We use the form of the conversion factor that will cancel moles and provide liters as the answer:

$$5.00\ \cancel{\text{mol}} \times \frac{22.4\ \text{L}}{1\ \cancel{\text{mol}}} = 112\ \text{L}\ O_2$$

Example 7.19 How many moles of neon gas are present in a volume of 5600 mL of gas at STP?

Solution: The given is 5600 mL. The solution pathway requires that we convert milliliters to liters and liters to moles using the molar volume factor:

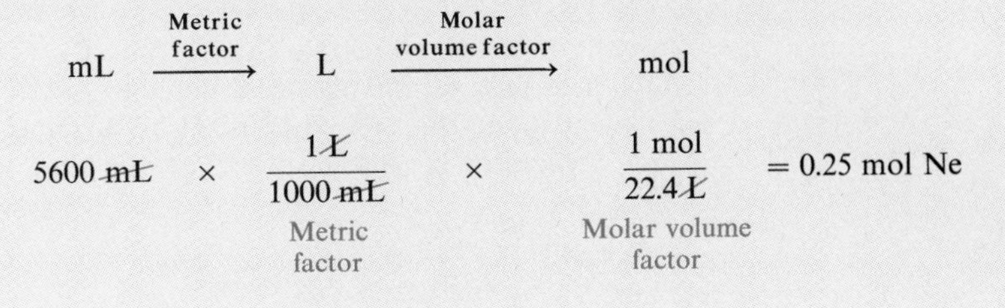

$$5600\ \cancel{\text{mL}} \times \underbrace{\frac{1\ \cancel{\text{L}}}{1000\ \cancel{\text{mL}}}}_{\text{Metric factor}} \times \underbrace{\frac{1\ \text{mol}}{22.4\ \cancel{\text{L}}}}_{\text{Molar volume factor}} = 0.25\ \text{mol Ne}$$

The Ideal Gas Law

All of the relationships concerning the four gas properties (P, V, T, and n) can also be combined into a single expression that applies to all gases under all conditions:

$$\frac{PV}{nT} = R = \text{Constant}$$

The gas constant, R, can be determined by using the molar volume values at standard conditions:

$$\frac{(1\ \text{atm})(22.4\ \text{L})}{(1\ \text{mol})(273\ \text{K})} = 0.0821\ \frac{\text{atm} \cdot \text{L}}{\text{mol} \cdot \text{K}}$$

The calculated gas constant, R, remains the same at all other conditions as well.

The ideal gas law states that the product of pressure (P) and volume (V) is equal to the product of moles of gas and kelvin temperature and the gas constant, R:

$$PV = nRT$$

This equation is useful when you know three of the four gas properties. It is necessary to use the same units as seen in the gas constant, R:

Property	*Unit*
P	atm
V	L
n	mol
T	K

Example 7.20 What is the pressure, in atmospheres, of 2.0 mol of Ne gas in a 10.0-L container that has a temperature of 400 K?

Solution: First, we can list the gas properties:

$$P = ?$$

$$V = 10.0 \text{ L}$$

$$n = 2.0 \text{ mol}$$

$$T = 400 \text{ K}$$

Since the unknown property is the pressure (P) of the gas, we can rearrange the ideal equation to solve for P. This means that the symbol P has to appear alone on one side of the equation. This is accomplished by dividing both sides of the equation by V. Since V cancels on the left side of the equation, the expression for P can now be solved:

$$\frac{P\cancel{V}}{\cancel{V}} = \frac{nRT}{V}$$

$$P = \frac{nRT}{V}$$

We can now insert the values for each of the known gas properties including the gas constant, R. All units cancel except for the pressure in atmospheres:

$$P = \frac{(2.0 \cancel{\text{mol}})\left(0.0821 \frac{\cancel{\text{L}} \cdot \text{atm}}{\cancel{\text{mol}} \cdot \cancel{\text{K}}}\right)(400 \cancel{\text{K}})}{10.0 \cancel{\text{L}}}$$

$$= 6.6 \text{ atm}$$

Example 7.21 How many moles of helium gas at a pressure of 1.50 atm and a temperature of 600 K are present in a 5.0-L container?

Solution: Our data list of the four properties will look like this:

$$P = 1.50 \text{ atm}$$

$$V = 5.0 \text{ L}$$

$$n = ?$$

$$T = 600 \text{ K}$$

The unkown is n, the number of moles. Now we need to rearrange the ideal gas equation to obtain n. This is accomplished by dividing both sides of the equation by RT:

$$PV = nRT$$

$$\frac{PV}{RT} = \frac{n\cancel{RT}}{\cancel{RT}}$$

$$\frac{PV}{RT} = n \quad \text{or} \quad n = \frac{PV}{RT}$$

Substituting the known gas properties gives the following calculation:

$$n = \frac{(1.50\ \cancel{\text{atm}})(5.0\ \cancel{\text{L}})}{(0.0821\ \cancel{\text{atm}} \cdot \cancel{\text{L}}/\text{mol} \cdot \cancel{\text{K}})(600\ \cancel{\text{K}})}$$

$$= 0.15 \text{ mol He}$$

7.9 Gas Mixtures

OBJECTIVE Use Dalton's law of partial pressures to calculate the partial or total pressure of a mixture of gases.

partial pressure

Dalton's law

When we use gas laws, we assume that molecules of all gases behave the same way. Therefore, we can mix different gases and calculate a pressure for the gas mixture. We then call the pressure of each gas in the mixture its *partial pressure*. The total pressure of the gas mixture is the sum of all the partial pressures of the gases in the mixture. This is known as *Dalton's law*:

$$P_{\text{total}} = P_A + P_B + P_C + \cdots$$

Example 7.22 A sample of air is composed of O_2 gas at a pressure of 160 torr and N_2 gas at a pressure of 600 torr. What is the total pressure of the mixture? (See Figure 7.11.)

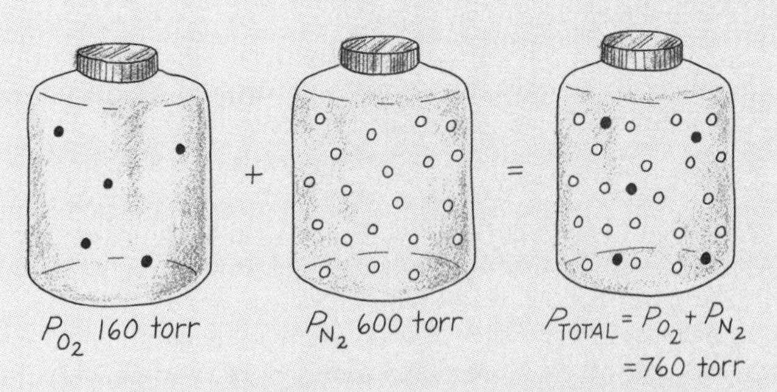

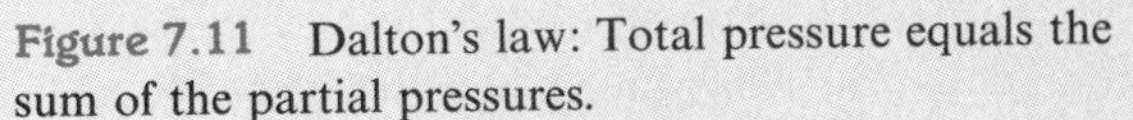

Figure 7.11 Dalton's law: Total pressure equals the sum of the partial pressures.

Solution: According to Dalton's law, the total pressure is the sum of the partial pressures of the gases in the mixture:

$$\begin{aligned} P_{\text{total}} &= P_{O_2} + P_{N_2} \\ &= 160 \text{ torr} + 600 \text{ torr} \\ &= 760 \text{ torr} \end{aligned}$$

Example 7.23 A gas mixture has a total pressure of 350 torr. In the mixture, there is O_2 gas at 80 torr, N_2 gas at 250 torr, and water vapor. What is the partial pressure of the water vapor?

Solution: We have the following information given in the problem:

$P_{total} = 350$ torr $\qquad P_{N_2} = 250$ torr

$P_{O_2} = 80$ torr $\qquad P_{H_2O} = ?$

We also know that the total pressure is equal to the sum of the partial pressures:

$$P_{total} = P_{O_2} + P_{N_2} + P_{H_2O}$$

To solve for the partial pressure of water, we need to rearrange the expression and solve for P_{H_2O}:

$$P_{total} - P_{O_2} - P_{N_2} = P_{H_2O}$$

$$350 \text{ torr} - 80 \text{ torr} - 250 \text{ torr} = 20 \text{ torr}$$

$$P_{H_2O} = 20 \text{ torr}$$

HEALTH NOTE

Hyperbaric Chambers

High pressures of oxygen can be achieved in a steel compartment called a *hyperbaric chamber*. Patients undergo treatment in a hyperbaric chamber to increase the oxygen tension in the blood and in the tissues. Such treatment is used for burns and infections, where increased tissue saturation of oxygen is desired. The oxygen aids in fighting infection caused by anaerobic bacteria. The hyperbaric chamber is also used during surgery and in the treatment of carbon monoxide poisoning and some cancers, where an increase in oxygen tension is beneficial.

If a chamber is filled with pure oxygen at a pressure of 3 atm, the oxygen exerts a pressure of 2280 torr on the patient and the alveoli. (An increase in external oxygen pressure increases the amount of oxygen dissolved in the patient's blood.) The blood is normally capable of dissolving up to 95 percent of the oxygen, which means that the oxygen tension of the blood can increase to 2160 torr. As a result, oxygen levels increase in the tissues, and oxygen replaces carbon monoxide in the hemoglobin in the case of carbon monoxide poisoning. (See Figure 7.12.)

It is critical that persons working inside a hyperbaric chamber take certain precautions. The patient must be observed for possible oxygen toxification. Oxygen supports combustion, so care must be taken to prevent sparks wherever there is a high concentration of oxygen. It is very important that a patient in the hyperbaric chamber undergo decompression to slowly reduce the concentration of dissolved oxygen in the blood. If decompression is too rapid, the high levels of oxygen dissolved in the blood will be reduced too quickly, and bubbles of gas may form within the capillary system.

Divers can suffer a similar condition, called the "bends," if they do not decompress slowly. When divers are under pressures below the surface of the ocean, they are breathing air at higher pressures. If pressure is decreased too rapidly, bubbles of nitrogen form in the blood and accumulate in the joints and tissues. The result is much pain as well as possible air emboli, which are dangerous. A diver suffering from

Partial Pressures of Gases in Air

The air you breathe is a mixture of gases. The atmospheric pressure at sea level is the sum of the partial pressures of oxygen, nitrogen, carbon dioxide, argon, and trace gases, which include neon, krypton, xenon, hydrogen, methane, and dinitrogen monoxide. (See Table 7.6.)

Table 7.6 Partial Pressure of Major Gases in the Atmosphere (Dry)

Gas	Percent	Partial Pressure (torr) at 760 torr
N_2	78.1	593.6
O_2	20.9	158.8
CO_2	0.03	0.2
Ar	0.93	7.1
Trace gases	0.04	0.3
	100.00	760.0

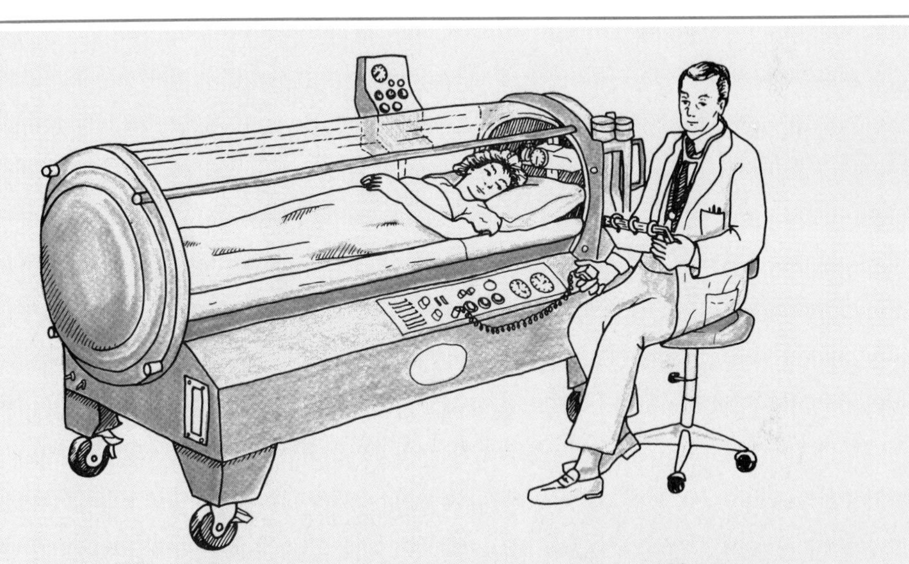

Figure 7.12 A hyperbaric chamber exposes the patient to 100 percent oxygen at two to three times the atmospheric pressure.

the bends is placed immediately in a decompression chamber, where pressures are increased and then slowly lowered again. The high levels of dissolved nitrogen gas can then be lost by way of the lungs until normal pressures are restored. Sometimes the nitrogen in an oxygen–nitrogen air mixture is replaced with helium because helium is much less soluble in the blood than nitrogen. Therefore, less gas dissolves, and the occurrence of the bends is not as likely.

HEALTH NOTE

Blood Gases

The same gases found in the atmosphere are found in the lungs. The pressure of water vapor increases within the lungs since the vapor pressure of water is 47 mmHg at body temperature (37°C).

alveoli

In the lungs, the gases from the air pass into the bloodstream through the membranes of the *alveoli*, the air sacs at the very end of the airways of the lungs. A capillary system surrounding each alveolus takes up oxygen from the lungs while releasing carbon dioxide. (See Figure 7.13.) During the process of gas exchange in the alveoli, the partial pressure of oxygen in the alveoli drops, while the partial pressure of carbon dioxide increases. In Table 7.7 we can see a comparison of partial pressure of these gases in the inspired air (air entering the lungs), alveolar air (air in the alveoli), and expired air (air that is exhaled into the atmosphere). The expired air shows higher partial pressures for oxygen and lower partial pressures for carbon dioxide because the alveolar air has mixed with the incoming air.

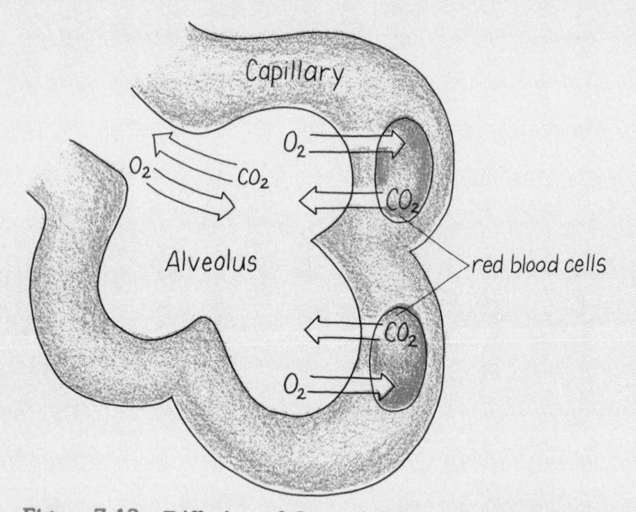

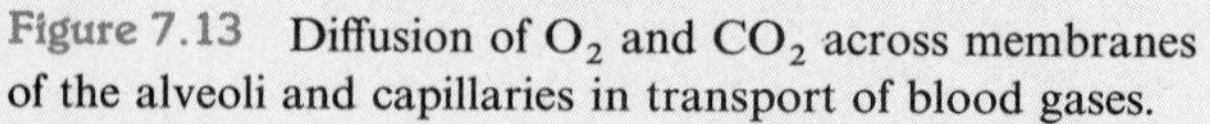

Figure 7.13 Diffusion of O_2 and CO_2 across membranes of the alveoli and capillaries in transport of blood gases.

Table 7.7 Partial Pressures of Major Gases in the Alveoli

	Inspired Air (torr)	Alveolar Air (torr)	Expired Air (torr)
N_2	564	573	569
O_2	149	100	116
H_2O	47	47	47
CO_2	0	40	28

blood gases
gas tension

Our cells continuously use oxygen and produce carbon dioxide. Both of these gases are carried in the bloodstream and hence are called *blood gases.* In discussions of dissolved gases, the term *gas tension* is frequently used. Gas tension is the partial pressure that is needed above a solution to give the concentration of the gas dissolved in the solution. The gas tensions for the blood gases oxygen and carbon dioxide are presented in Table 7.8.

Table 7.8 Gas Tensions (in torr) for Blood Gases

	Arterial	Venous	Tissue
P_{O_2}	100	40	<30
P_{CO_2}	40	46	>50

Oxygen normally has a partial pressure of 100 torr in the alveoli. The gas tension of oxygen in venous blood, however, is 40 torr. This difference creates a pressure gradient for oxygen, moving oxygen out of the alveoli into the bloodstream, and the blood becomes oxygenated. The circulatory system carries the oxygen, combined with its carrier hemoglobin, to the tissues of the body. Oxygen tension can be very low in the cells, usually less than 30 torr. In active cells, oxygen tension can be close to zero because oxygen is rapidly utilized in cellular processes. At the tissues, another pressure gradient moves oxygen out of the arterial bloodstream into the tissue cells.

Active tissues also produce carbon dioxide as an end product of metabolism, so carbon dioxide tension is higher in the tissues—as much as 50 torr or more. Carbon dioxide moves out of the tissues into the bloodstream, where its gas tension is lower. The capillaries leaving the tissues now carry venous blood, low in oxygen (40 torr) and higher in carbon dioxide (46 torr). At the lungs, carbon dioxide leaves the venous blood (46 torr) and goes into the alveoli (40 torr) to be exhaled. (See Figure 7.14.)

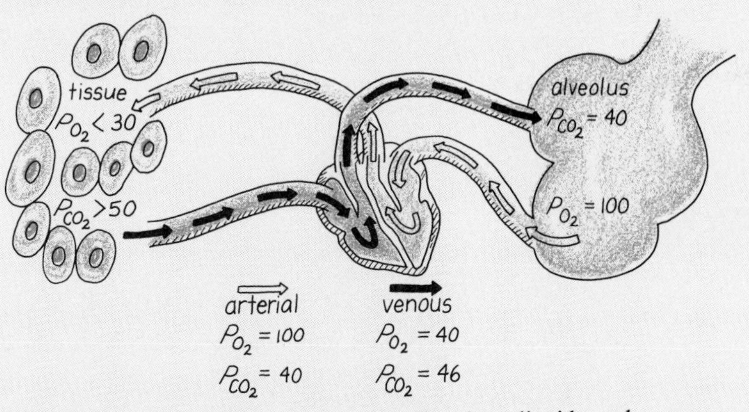

Figure 7.14 Transport and exchange of carbon dioxide and oxygen.

Glossary

alveolus The tiny air sac at the end of the airways in the lungs through which gas exchange of oxygen and carbon dioxide occurs.

atmosphere A unit of pressure exerted by a column of mercury 760 mm high.

atmospheric pressure The downward pressure exerted by the atmosphere at a particular altitude.

Avogadro's law A law stating that the volume of a gas depends on the number of moles of gas in the container when P and T are constant.

boiling The transition of liquid to vapor that occurs when the vapor pressure becomes equal to the atmospheric pressure.

boiling point The temperature at which boiling occurs. At higher altitudes, boiling points are lower than at sea level.

Boyle's law A law stating that the pressure of a gas varies inversely with the volume when T and n are held constant; that is, if volume decreases, pressures increases.

Charles' law A law stating that the volume of a gas changes directly with the changes in temperature (Kelvin) when P and n are constant.

Dalton's law The total pressure of a mixture of gases is the sum of the partial pressures of each of the gases in the mixture.

exhalation The part of ventilation in which air leaves the lungs and enters the atmosphere; this occurs when the pressure within the lungs exceeds the atmospheric pressure.

gas tension The measurement of dissolved gases in liquid relative to their partial pressures above the solution.

Gay-Lussac's law A gas law stating that the pressure of a gas changes directly with a change in temperature when n and V are constant.

hyperbaric chamber A steel chamber in which pressures of 3–4 atm can be attained.

ideal gas law A law that combines all gas laws by relating all four properties of gases: $PV = nRT$.

inhalation The part of ventilation in which air flows into the lungs from the atmosphere; this occurs when the atmospheric pressure is greater than the pressure within the lungs.

inverse relationship A relationship in which two properties change in opposite directions.

kinetic molecular theory A model that describes gases as small molecules that are far apart from each other and traveling at great speed.

partial pressure The pressure exerted by a single gas in a mixture of two or more gases.

pounds per square inch A U.S. Customary measure of pressure; 14.7 lb/in.^{2} is equal to 1 atm.

pressure The force exerted against the walls of a gas container by the bombardment of gas molecules.

pressure gradient A difference in pressures causing gas molecules to move from the area of high pressure to the area of low pressure.

tidal volume The volume of air moving into or out of the lungs while breathing.

torr A metric unit of pressure equal to 1 mmHg; 760 torr = 1 atm.

vapor pressure The pressure exerted by the vapor molecules above a liquid.

ventilation That part of the breathing process in which air is moved into and out of the lungs and which functions according to Boyle's law.

Problems

Kinetic Molecular Theory (Objective 7.1)

7.1 What does the kinetic molecular theory say about the following?

a. the types of particles in a gas
b. the path of gas particles
c. the distances between gas particles
d. the forces of attraction between gas particles
e. the speed of gas particles

Pressure of a Gas (Objective 7.2)

7.2 Which of the following phrases describe the pressure of a gas?
a. the force of the gas molecules on the walls of a container
b. the number of gas molecules in a container
c. the volume of the container
d. 3.00 atm
e. 700 torr

7.3 An oxygen tank contains oxygen at a pressure of 2.0 atm. Change this pressure into the following units:
a. torr
b. psi
c. mmHg

7.4 Change the following pressures into atmospheres:
a. 800 torr
b. 28.2 in.Hg
c. 740 mmHg

7.5 Indicate whether each of the following describes vapor pressure, atmospheric pressure, or boiling point:
a. the temperature at which bubbles of vapor appear within the liquid
b. pressure exerted by gaseous molecules above the surface of their liquid
c. the pressure exerted on the earth by the molecules in the air
d. the temperature reached when vapor pressure becomes equal to atmospheric pressure
e. a temperature that changes at higher altitudes

7.6 In which pair of pressures would boiling occur?

	Atmospheric Pressure	*Vapor Pressure*
a.	760 mmHg	700 mmHg
b.	640 torr	640 torr
c.	1.2 atm	912 mmHg
d.	800 mmHg	760 mmHg

7.7 Give an explanation for the following observations:
a. Water boils at 87°C atop Mt. Whitney.
b. Food cooks more quickly in a pressure cooker than in an open pan.
c. Water boils at 120°C at 2.0 atm in an autoclave used to sterilize surgical equipment.

Properties of Gases (Objective 7.3)

7.8 What are the four properties that describe a gas?

7.9 Use the words *pressure, volume, temperature,* or *quantity* to indicate the property of a gas that is described in each of the following phrases or measurements:
a. 350 K
b. 1200 torr
c. 4.00 mol
d. determines the kinetic energy of the gas molecules
e. space occupied by a gas
f. 10.0 L

Pressure and Volume: Boyle's Law (Objective 7.4)

7.10 Complete the following using Boyle's law:

	P	*V*
a.	____________	decreases
b.	decreases	____________
c.	____________	increases

7.11 Using Boyle's law, solve for the final pressure when the volume changes (*n* and *T* are constant):

a. A 10.0-L balloon contains helium gas at a pressure of 600 torr. What is the new pressure when the volume is increased to 20.0 L?
b. The volume of air inside an iron lung is 50.0 L at 725 torr. What is the new pressure when the piston compresses the volume of the chamber to 47.0 L?
c. A sample of helium has a volume of 425 mL at a pressure of 1.50 atm. If the volume is changed to 215 mL, what will the final pressure be?

7.12 Using Boyle's law, solve for the final volume when the pressure changes:

a. Cyclopropane, C_3H_6, is a general anesthetic that is inhaled. A 5.0-L sample of the anesthetic has a pressure of 1.0 atm. Determine the final volume of the gas when the pressure is changed to 4.0 atm.
b. The volume of air in a person's lungs is 600 mL at a pressure of 760 mmHg. During inhalation, the pressure in the lungs drops to 752 mmHg. To what new volume did the lungs expand?
c. An emergency tank of oxygen holds 20.0 L of O_2 gas at a pressure of 15 atm. What volume of O_2 can be delivered if the gas is released at a pressure of 1.0 atm?

7.13 Use the words *inspiration* or *expiration* to indicate the part of ventilation that occurs in each of the following statements:

a. The diaphragm relaxes, moving up into the thoracic cavity.
b. The pressure within the lungs drops.
c. The volume of the lungs decreases.
d. The diaphragm flattens out (contracts).
e. The pressure within the lungs is less than the atmospheric pressure.

Pressure and Temperature: Gay-Lussac's Law (Objective 7.5)

7.14 State the expected change in the following list when *n* and *V* are held constant:

	P	*T*
a.	____________	increases
b.	____________	decreases
c.	decreases	____________

7.15 Using Gay-Lussac's law, solve for the final pressure when the temperature changes (n and V constant):

a. A gas sample has a pressure of 1200 mmHg at a temperature of 150°C. What is the final pressure of the gas after the temperature has dropped to 0°C?

b. An aerosol can has a pressure of 1.40 atm at a room temperature of 22°C. What is the new pressure in the aerosol can when it is thrown into a fire that has a temperature of 600°C?

c. A fire extinguisher has a pressure of 150 psi at 25°C. What is the new pressure in atmospheres if the fire extinguisher is used at a temperature of 100°C? (*Hint*: 1 atm = 14.7 psi.)

7.16 Using Gay-Lussac's law, solve for the final temperature when pressure has changed:

a. A 10.0-L container of helium gas has a pressure of 250 torr at 0°C. To what temperature (°C) does the sample need to be heated to obtain a pressure of 1000 torr?

b. A sample of H_2 gas has a pressure of 2.00 atm at 80°C. To what temperature (°C) must the temperature drop to lower the pressure to 0.25 atm?

Volume and Temperature: Charles' Law (Objective 7.6)

7.17 State the change that will occur when P and n are held constant:

	V	T
a.	________	increases
b.	________	decreases
c.	decreases	________

7.18 A balloon contains 255 mL of helium gas at 150°C. What is the new volume when the temperature is changed to the following (n and P are constant)?

a. 50°C b. 1000 K c. −100°C

7.19 A gas has a volume of 4.00 L at 0°C. What final temperature is needed (°C) to cause the volume of the gas to change to the following volumes (n and P constant)?

a. 10.0 L b. 800 mL c. 2.50 L

Combined Gas Laws (Objective 7.7)

7.20 Solve the following problems for changes in P, V, and T (n is constant):

a. A 2500-mL sample of a gas at 300°C and 760 torr is compressed to a volume of 50 mL at a new temperature of 100°C. What is the new pressure of the gas?

b. A 1.35-L sample of carbon dioxide has a pressure of 0.85 atm and a temperature of 10°C. What is the final pressure at 5.00 L and 100°C?

c. A weather ballon has a volume of 500 L when filled with helium at 70°C to a pressure of 0.50 atm. What is the new volume of the balloon in an atmosphere where the pressure is 0.10 atm and the temperature is −50°C?

d. A 100-mL bubble of hot gases at 250°C and 2.5 atm has escaped from a volcano. What is the volume of the bubble when the temperature outside the volcano is 20°C and the pressure is 0.80 atm?

Moles and Volume: Avogadro's Law (Objective 7.8)

7.21 Solve the following problems at constant *P* and *T*:

a. A sample of 4.00 mol of argon has a volume of 10.0 L. A small leak causes one-half of the molecules to escape. What is new volume of the gas at the same pressure and temperature?

b. A balloon containing 1.00 mol of He has a volume of 440 mL. What is the volume when 2.00 mol of He are added to the balloon at the same pressure and temperature?

c. A 1500-mL sample of SO_2 contains 6.00 mol of SO_2. How many moles of SO_2 were lost when the volume decreased to 500 mL at the same pressure and temperature?

7.22 Use molar volume to solve the following problems at STP:

a. How many moles of O_2 are present in 44.8 L of O_2 gas at STP?

b. How many moles of CO_2 are present in 4.0 L of CO_2 gas at STP?

c. How many grams of Ne gas are present in 11.2 L of Ne gas at STP?

d. How many moles of H_2 gas are present in 1600 mL of the gas at STP?

7.23 Use molar volume to solve the following problems at STP:

a. What volume, in liters, is occupied by 2.5 mol of N_2 at STP?

b. What volume, in liters, is occupied by 0.40 mol of Cl_2 gas at STP?

c. What is the volume, in liters, of 32.0 g of O_2 at STP?

7.24 Use the ideal gas equation to solve the following problems:

a. What is the pressure, in atmospheres, of 2.0 mol of He gas in a 10.0-L container at 27°C?

b. What is the volume, in liters, of 5.0 mol of methane gas, CH_4, measured at a temperature of 0°C and a pressure of 2.0 atm?

c. How many moles of O_2 are present in a 20.0-L container when the pressure of the gas is 1.0 atm and the temperature is 227°C?

Gas Mixtures (Objective 7.9)

7.25 Use Dalton's law of partial pressure to solve the following problems for gas mixtures:

a. A steel cylinder contains a mixture of nitrogen gas at 400 torr, oxygen gas at 100 torr, and helium gas at 150 torr. What is the total pressure of the gas mixture?

b. What is the total pressure, in torr, of a gas mixture containing 0.25 atm of argon gas, 350 torr of helium gas, and 0.50 atm of nitrogen gas?

c. A gas mixture exerts a total pressure of 1250 torr. The mixture contains O_2 gas, He gas, and N_2 gas. If the partial pressure of the O_2 gas is 425 torr and the partial pressure of the helium gas is 320 torr, what is the partial pressure of the N_2 gas in the mixture?

d. A gas mixture contains 2.0 mol of He gas and 6.0 mol of O_2 gas. The total pressure of the mixture is 1200 torr. What is the partial pressure of each gas in the sample?

Solutions: Nature of Solute and Solvent

Objectives

8.1 Identify the solute and the solvent in a solution.

8.2 Given two substances and their polarity, predict if they will form a solution.

8.3 Describe solubility, the factors that affect solubility, and the rate of formation of a solution.

8.4 Describe the role of water in the process of dissolving a salt or a polar substance.

8.5 Write an ionic equation for a salt dissolving in water.

8.6 Given the heat of solution, state whether the process of dissolving a substance in water is exothermic or endothermic.

8.7 Write the ionic equation for a strong electrolyte, a weak electrolyte, and a nonelectrolyte dissolving in water.

Scope

Solutions are everywhere around us. Every solution consists of some substance dissolved in another. The air you breathe is a solution of oxygen and nitrogen gases. Carbon dioxide gas dissolved in water makes the soda in your carbonated drinks. Substances from coffee beans or tea leaves dissolve in hot water to make solutions of coffee or tea. The ocean is a solution of many minerals dissolved in water. In the hospital, the antiseptic tincture of iodine is a solution of iodine dissolved in alcohol.

Water serves many functions in the human body. It surrounds all living cells, acting as a solvent for many substances such as K^+, Na^+, Cl^-, HCO_3^-, glucose, and amino acids necessary for the survival of the living organism. These solutions in the body are called *body fluids.*

8.1 Solutions

OBJECTIVE Identify the solute and the solvent in a solution.

A solution is a uniform mixture of two or more substances. One of the substances is evenly distributed throughout the other substance so that samples taken from any part of the solution are identical. We have some special terms for the parts of a solution. The substance that dissolves is called the *solute.* The substance that does the dissolving is called the *solvent.* We usually think of the

solute

solvent

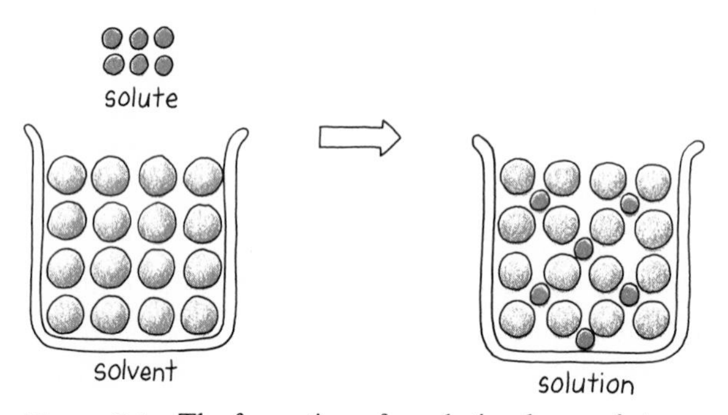

Figure 8.1 The formation of a solution by a solute and a solvent.

Table 8.1 Some Examples of Solutions

Solution	Example	Solute	Solvent
Gas/gas	Air	Oxygen	Nitrogen
Gas/liquid	Soda water	Carbon dioxide	Water
	Household ammonia	Ammonia (NH_3)	Water
Liquid/liquid	Vinegar	Acetic acid	Water
	Alcohol bath	Isopropanol (rubbing alcohol)	Water
Liquid/solid	Dental amalgam	Mercury	Silver
Solid/liquid	Saline	Salt	Water
	Tincture of iodine	Iodine	Alcohol
	Bleach	Sodium hypochlorite (NaOCl)	Water
Solid/solid	Steel (alloy)	Carbon	Iron
	Brass (alloy)	Zinc	Copper

solute as the substance that is present in smaller quantity. A solution forms when the particles of the solute become dispersed among the particles of the solvent. (See Figure 8.1.)

Any of the three states of matter—solid, liquid, or gas—can be either a solute or a solvent. When sodium chloride (solid) dissolves in a glass of water (liquid), a sodium chloride solution is formed. The sodium chloride is the solute, and the water is the solvent. Carbon dioxide gas dissolves in water to make a carbonated solution. The gas is the solute, and the water is the solvent. Table 8.1 lists some solutes and solvents that form solutions.

Example 8.1 Identify the solute and the solvent in each of the following solutions:

1. 1.0 g of sugar dissolved in 100 g of water
2. 50 mL of water mixed with 20 mL of isopropanol (rubbing alcohol)

Solution

1. Sugar is dissolving. Sugar is the solute; water is the solvent.
2. Since both are liquids, the one with the greater volume, water, is the solvent. Isopropanol is the solute.

8.2 Nature of the Solute and Solvent

OBJECTIVE Given two substances and their polarity, predict if they will form a solution.

The interaction between the solute and the solvent has much to do with whether or not a solution will form and how much of the solute will dissolve. When a solution forms, solute particles must break away from other solute particles and move into the solvent. At the same time, the solvent particles must allow solute particles to come between them.

There must be sufficient attractive forces between the solute and the solvent particles to allow a solution to form. These attractive forces occur when the polarities of the solute and the solvent are similar. A rule of thumb for predicting the formation of a solution is "like dissolves like." A polar solute is most likely to dissolve in a polar solvent. Both ionic salts and polar solutes are soluble in polar solvents.

"like dissolves like"

However, the ionic compound NaCl does not dissolve (insoluble) in a nonpolar solvent such as pentane, C_5H_{12}. (Compounds that consist of only carbon and hydrogen are nonpolar.) Oils and grease (nonpolar solutes) will not dissolve in water (polar), but will dissolve in nonpolar solvents such as pentane. Therefore we can also say that a nonpolar solute is only soluble in a nonpolar solvent. (See Figure 8.2.) Table 8.2 lists some soluble and insoluble combinations of solutes and solvents. Table 8.3 lists examples of solutes and solvents that will form solutions and those that will not.

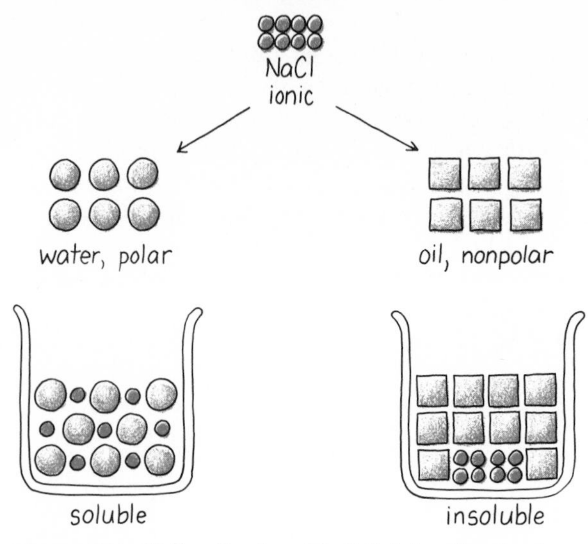

Figure 8.2 "Like dissolves like." Solutes dissolve in solvents of like or similar polarities.

Table 8.2 Soluble and Insoluble Combinations

Soluble (Will Form Solutions)		Insoluble (Will Not Form Solutions)	
Solute	Solvent	Solute	Solvent
Polar	Polar	Polar	Nonpolar
Nonpolar	Nonpolar	Nonpolar	Polar

Table 8.3 Formation of Solutions

	Solvent	
Solute	Water (Polar)	Pentane (Nonpolar)
NaCl (ionic)	Yes	No
Oil (nonpolar)	No	Yes
I_2 (nonpolar)	No	Yes
HCl (polar)	Yes	No
Sugar (polar)	Yes	No

Example 8.2 Predict whether a solution will form with the following solutes and solvents:

1. the salt KCl in water (polar)
2. the salt KCl in hexane (nonpolar)
3. oil (nonpolar) in water (polar)
4. oil (nonpolar) in hexane (nonpolar)

Solution

1. Yes. A salt such as KCl consists of ions of opposite charges. Compounds with electrical charges are most soluble in polar solvents.
2. No. An ionic compound such as KCl will not dissolve in a solvent that is not polar.
3. No. A nonpolar solute will not be soluble in a polar solvent such as water.
4. Yes. A nonpolar solute such as oil will dissolve in a nonpolar solvent such as hexane.

8.3 Solubility

OBJECTIVE Describe solubility, the factors that affect solubility, and the rate of formation of a solution.

$\frac{\text{g solute}}{\text{100 g solvent}}$

Most solutions reach a point where no more solute will dissolve. The maximum amount of solute that can dissolve in a solvent at a given temperature is called the *solubility* of that solute. Solubility is expressed as the maximum number of grams of solute that can dissolve in 100 g of solvent at a specific temperature. For example, the solubility of NaCl at 20°C is 36 g of NaCl in 100 g of water. Adding more NaCl to the solution will not result in a larger amount of dissolved solute. The additional NaCl will not dissolve but will remain in solid form (undissolved) at the bottom of the container.

Saturated Solutions

unsaturated

saturated

Any amount of dissolved solute less than the maximum solubility forms a solution that we call *unsaturated*. When more solute is added to the unsaturated solution, it dissolves. However, when we have a solution with the maximum amount of solute dissolved, we call it a *saturated solution*. If we try to add more solute to the saturated solution, the additional solute will cause some dissolved solute to return to the undissolved form. When the amount of solute particles dissolving becomes equal to the amount of solute particles returning to the undissolved form, a state of dynamic equilibrium has been reached (see Figure 8.3):

Saturated Solution (Solubility)

$$\text{Undissolved solute} \underset{\text{recrystallizes}}{\overset{\text{dissolves}}{\rightleftharpoons}} \text{Dissolved solute}$$

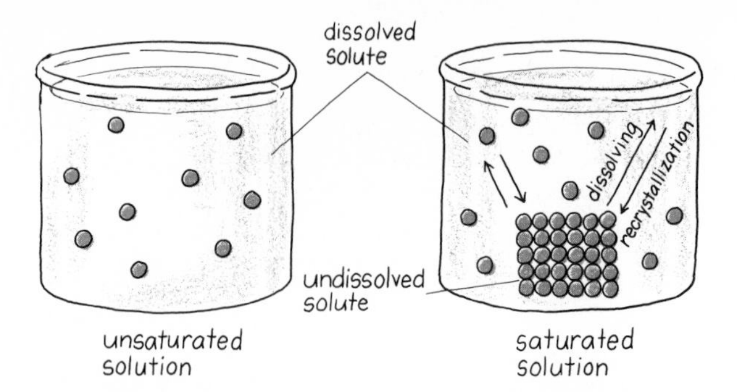

Figure 8.3 A saturated solution holds the maximum amount of solute dissolved at that temperature. Additional solute forms a solid whose rate of dissolving equals the rate of recrystallization. In unsaturated solutions, all solute dissolves because the solution does not contain its maximum amount of dissolved solute.

Supersaturated Solutions

Sometimes a solution in which the solute exceeds its maximum solubility level (supersaturated) can be prepared. A supersaturated solution can be made by slowly cooling a saturated solution. For example, a solution of sugar prepared at 80°C will dissolve 362 g of sugar in 100 g of water. This amount of sugar may remain dissolved as the solution slowly cools to 40°C. Now the solution contains more than 238 g, the solubility at 40°C; it is a supersaturated solution. However, a supersaturated solution is not stable, and disturbing the solution by shaking or adding a "seed" sugar crystal will cause the excess sugar to precipitate out of solution until just 238 g (saturated) remain in solution.

Example 8.3 The solubility of KCl at 70°C is 50 g per 100 g of water. Suppose a student adds 75 g of KCl to 100 g of water at 70°C:

1. How much of the KCl will dissolve?
2. What is the mass, in grams, of the undissolved KCl?

Solution

1. A total of 50 g of the KCl will dissolve in 100 g of water because that is its solubility and is therefore the maximum amount allowed in solution at 70°C.
2. The mass of the undissolved KCl will be 25 g.

Effect of Temperature on Solubility

most solids are more soluble at higher temperatures

Temperature often plays an important role in the solubility of a solute. Most solids become more soluble with increases in temperature. (See Table 8.4.)

You may have noticed the effect of temperature on the solubility of table sugar added to tea. More sugar can dissolve in hot tea than in iced tea. The hot tea tastes sweeter because more sugar has dissolved. Hot water is sometimes used to make iced tea because it helps dissolve the sugar. However, after ice is added to the tea, the solubility decreases as the temperature drops. A layer of undissolved sugar then forms at the bottom of the glass.

Table 8.4 Solubility at Various Tempertures (g/100 g Water)

Substance	20°C	40°C	60°C	80°C
Sugar (sucrose)	204	238	287	367
NH_4Cl	37	46	55	66
$AgNO_3$	222	376	525	669
K_2SO_4	11	15	18	21

Solubility of Gases

gases are less soluble at higher temperatures

The solubility of a gas in a liquid decreases with an increase in temperature. At higher temperatures, the dissolved gas molecules gain sufficient energy to escape from the surface of the solution, and fewer dissolved gas molecules remain. The decrease in solubility for carbon dioxide with an increase in temperature is listed in Table 8.5. Bottles of carbonated solutions may burst at higher temperatures as the number of gas molecules, and therefore the pressure inside the bottle, increases. Environmentalists are concerned with the effect of temperature on lakes and streams. Heat from industrial plants can cause an elevation in water temperature that reduces the solubility of oxygen in the water. The warmed water will support only a limited biological community. Figure 8.4 compares the solubility curves of solids and gases in water.

Table 8.5 Solubility of Carbon Dioxide at Various Temperatures

Temperature (°C)	Solubility (g/100 g Water)
0	0.34
25	0.15
50	0.08

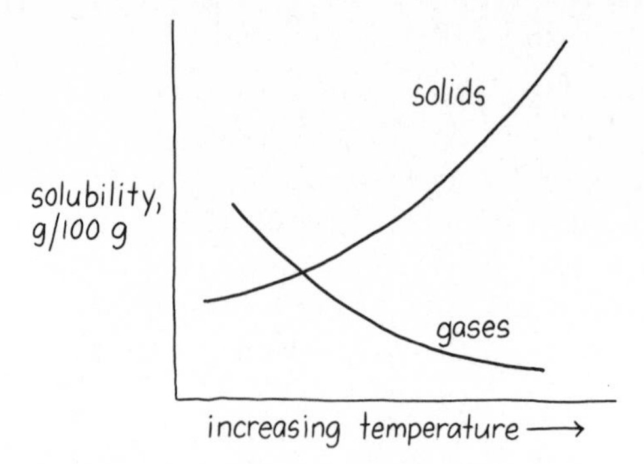

Figure 8.4 General solubility curves of solids and gases in water. The solubility of most solids increases as temperature rises; solubility of gases decreases at higher temperatures.

Figure 8.5 Henry's law: The amount of gas dissolved in solution depends on the pressure of the gas in contact with the solution.

Henry's Law: Solubility of Gases and Pressure

pressure increases the solubility of a gas in a liquid

Henry's law states that the amount of gas that will dissolve in a liquid at a given temperature depends on the pressure of that gas above the solution. You have probably noticed all the bubbling that occurs when you open a bottle of carbonated soda or champagne. When that beverage was prepared, the CO_2 gas pressure was high enough to dissolve some $CO_2(g)$ in the liquid. When you remove the cap, the CO_2 pressure above the solution is reduced. Now the carbon dioxide is less soluble, and many bubbles of gas escape from the beverage. Sometimes the amount of escaping gas is so great that liquid is actually sprayed out of the can or bottle. (See Figure 8.5.)

Example 8.4 Indicate whether the solubility of the solute will increase or decrease in each of the following situations:

1. increasing the temperature of river water (O_2 gas is the solute)
2. using 80°C water instead of 25°C water to dissolve sugar
3. removing the cork from a bottle of champagne [$CO_2(g)$ is the solute]

Solution

1. The solubility of the dissolved O_2 will decrease because gases are less soluble at higher temperatures.
2. Using warmer water to dissolve sugar will increase its solubility.
3. Removing the cork reduces the pressure of the $CO_2(g)$ above the liquid. According to Henry's law, this will decrease the solubility of the CO_2 in solution.

HEALTH NOTE

Gout: A Problem of Uric Acid Saturation

Gout is a disease that affects adults, primarily men over the age of 40. Attacks of gout occur when the uric acid concentration in the plasma exceeds its maximum solubility. Normal concentrations of uric acid are 2–6 mg in 100 mL of plasma. At body temperature (37°C), the solubility of uric acid is 7 mg in 100 mL of plasma.

When the solubility of uric acid is exceeded, insoluble crystals form. These crystals deposit in the joints, causing severe pain. The initial attacks of gout usually occur in one joint, often in the big toe. If left untreated, crystals of uric acid can appear in the cartilage, tendons, and soft tissues. When crystals deposit in the tissues of the kidneys, renal damage can occur. Uric acid crystals can even block the collecting tubules, or ureter, and prevent urine flow.

High levels of uric acid are found in diets rich in protein. Treatment for gout includes a reduction in protein intake and restriction of foods such as certain meats, sardines, mushrooms, and asparagus. Drugs may be used to increase the excretion of uric acid or to reduce uric acid production so that a lower uric acid concentration is maintained.

Rate of Solution

The time it takes for a solution to form depends on how fast the solute becomes dispersed evenly in the solvent. One way to change the rate at which a solute dissolves is to change the surface area of the solute. A more finely divided solute has more surface area, which allows more contact with the solvent molecules. Crushing the solute will enable more solute particles to enter solution, thereby increasing the rate at which the solution is formed. It is important to note here that we are not changing the amount of solute that dissolves but only affecting the time it takes for the solution to form. For example, a spoonful of powdered sugar will dissolve more rapidly in a glass of water than a sugar cube of equal amount. The sugar cube dissolves slowly because only a few molecules are on the surface to interact with the water. Eventually the final composition of the solution will be identical.

surface area

stirring

Another way to increase the rate at which a solute dissolves is to stir the solution. The dissolved solute particles that are close to the undissolved solute will be dispersed more quickly. Fresh solvent can then bombard the remaining solute. (See Figure 8.6.)

temperature

An increase in temperature also increases the rate of solution. Higher temperatures increase the motion of both the solute and solvent particles and aid in rapidly dispersing the solute in the solution.

Figure 8.6 Rate of solution may be increased by (a) crushing the solute or (b) stirring. As the surface area of the solute increases, the rate of solution increases.

Example 8.5 You are going to dissolve some sugar cubes in a glass of water. Indicate whether each of the following conditions will increase oı decrease the rate of solution for the sugar cubes:

1. using hot water
2. crushing some sugar cubes
3. placing the glass of water and sugar cubes in the refrigerator

Solution

1. An increase in the temperature of the water increases the rate of solution.
2. Increasing the surface area of the sugar that comes in contact with the water molecules will increase the rate of solution.
3. Cooling the solution will slow the movement of the solute into the solvent, causing a decrease in the rate of solution.

8.4 Water as a Solvent

OBJECTIVE Describe the role of water in the process of dissolving a salt or a polar substance.

Water is the most common solvent in nature and is sometimes referred to as the *universal solvent*. It is also the solvent found in biological systems. For this reason, we will take a closer look at the structure of water and its ability to dissolve a substance.

water is a polar solvent

As we saw in Chapter 3, water is a polar molecule. Therefore, in a solution, we call water a *polar solvent*. In the water molecule, the more electronegative oxygen atom has a stronger attraction for the shared electron pairs; this causes a charge separation in the molecule. The oxygen atom acquires a partial negative charge; the hydrogen part of the molecule becomes partially positive. (See Figure 8.7.)

Hydrogen Bonding

The electrical charges of polar water molecules cause water molecules to be attracted to each other. In the solid and liquid states, bonds form between water molecules. The oxygen of one water molecule is attracted to a hydrogen of another water molecule. Recall that opposite charges attract. This interaction,

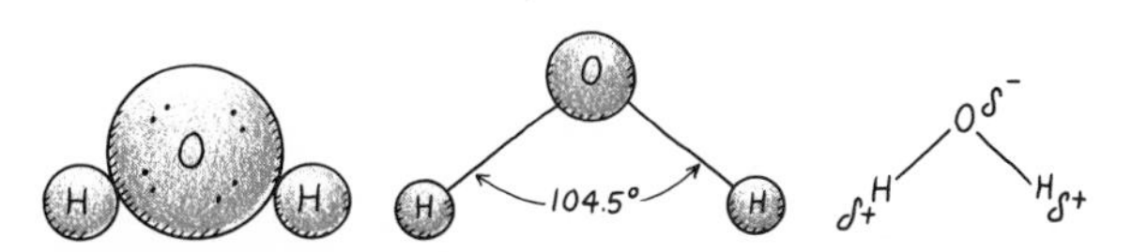

Figure 8.7 A polar water molecule.

hydrogen bonds are the attractive forces between water molecules

called *hydrogen bonding*, holds many water molecules together. Although hydrogen bonds are not as strong as either ionic or covalent bonds, they do play a major role in the chemical and physical properties of water. For example, water has an unusually high boiling point. More energy than expected is required to provide sufficient energy to break apart the hydrogen bonds that keep water in the liquid state. The hydrogen bonds of water are also important in maintaining the biologically active structure of the proteins and genetic substances in the body. (See Figure 8.8.)

Surface Tension

In the liquid state, hydrogen bonding accounts for some interesting features of water. Within the liquid, each water molecule is pulled equally in all directions by the surrounding water molecules. However, at the surface, the water molecules are pulled inward, drawing the surface molecules tighter together to form a "skin" on the top of the water. This physical property of water molecules

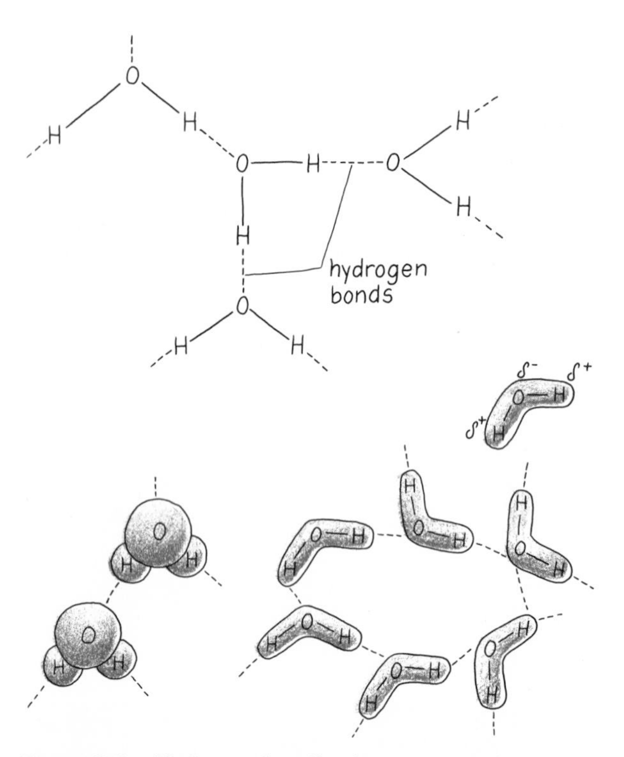

Figure 8.8 Hydrogen bonding in water. Weak bonds form between the hydrogen of one water molecule and the oxygen of a different water molecule.

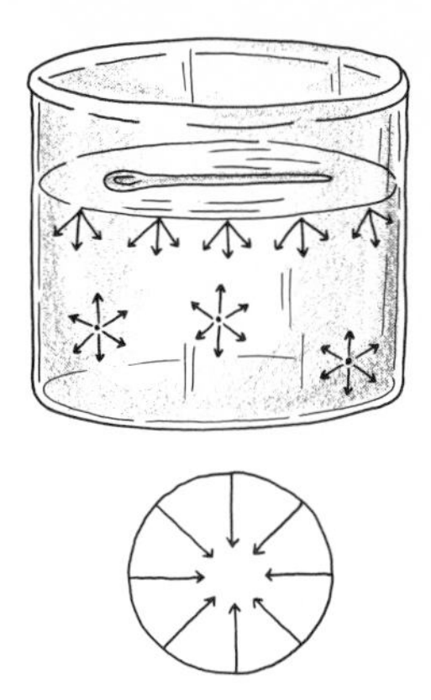

Figure 8.9 Surface tension in water permits a needle to float and makes raindrops spherical.

surface tension at the surface is called *surface tension*. This tight layer of water molecules makes it possible to float a needle on the top of the water. Once the needle is pushed through the surface layer, it sinks to the bottom of the container. It is surface tension that permits certain water bugs to travel across the surface of a pond or lake. The drops of water you see in rain or on a waxed surface are spherical in shape because of surface tension. (See Figure 8.9.)

The balance of intake and output of body fluids can be upset by high fevers, diarrhea, bleeding, burns, or ulcers. If the fluid output becomes excessive or if a patient cannot take fluids and food orally, fluids may be placed in the body by subcutaneous routes, intravenous feedings, or intramuscular injections. Fluids given in this way are called *parenteral* solutions and include any administration of fluid other than oral. When a patient has difficulty maintaining a balance between intake and output of fluids, the amount of fluids are recorded to ensure that fluid replacement is sufficient but not excessive. An intake–output record form is shown in Figure 8.12.

Water in the Solution Process

breaking bonds of solute

ionic compounds dissolve as ions

hydration

We can illustrate the way in which water dissolves a solute such as sodium chloride. During the solution process, the polar water molecules are constantly bombarding the surface of the NaCl crystal. The negative chloride ions are attracted to the partially positive hydrogen atoms of the water molecules. This weakens and overcomes the attraction of the chloride ion for the sodium ions in the salt, and the chloride ion is pulled into solution. Likewise, the sodium ions are attracted to the partially negative oxygen atoms of the water molecules. The ionic bonds holding the sodium ion in the salt crystal weaken, and the sodium ion becomes part of the solution. As the chloride ions and the sodium ions move into the solution, they are surrounded by more water molecules, a process called *hydration*. A sodium chloride solution has thus formed. (See Figure 8.13.)

HEALTH NOTE

Water in the Body

The average adult body contains about 60 percent water by weight, and the average infant contains about 75 percent. This water is distributed among the fluid-containing compartments of the body, where it acts as a solvent for salts, molecules, and gases.

About 60 percent of the body's water is found within the cells as intracellular fluids; the other 40 percent makes up the extracellular fluids. The extracellular fluid in the tissues is called *interstitial fluid*, and the extracellular fluid in the blood is the *plasma*. These external fluids aid in the exchange of nutrients and waste materials between the cells and the circulatory system. (See Figure 8.10.)

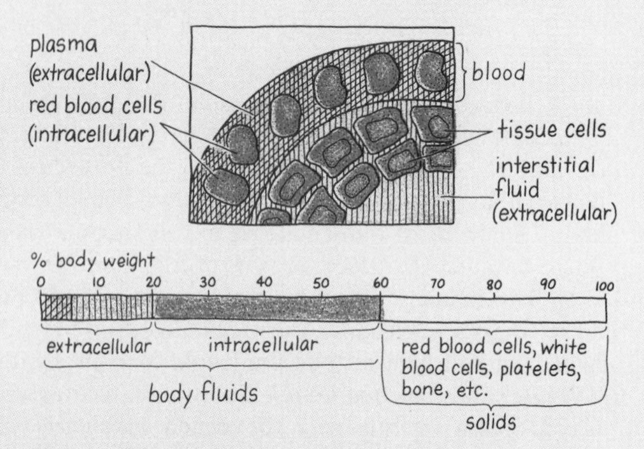

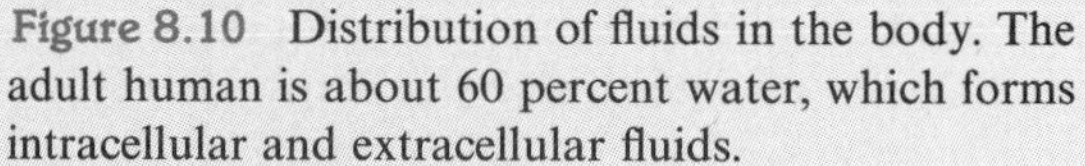

Figure 8.10 Distribution of fluids in the body. The adult human is about 60 percent water, which forms intracellular and extracellular fluids.

Every day you lose between 1500 and 3000 mL of water from your body. Water is lost in the formation of urine, by perspiration, by expiration of water vapor from the lungs, and in the formation of feces. If the water lost is not replaced, the body can be seriously dehydrated within a few days. Serious dehydration can occur when there is a 10 percent net loss in total body fluid. A 20-percent loss of fluid can be fatal for an adult, and an infant suffers severe dehydration when there is a 5- to 10-percent loss in body fluid.

To prevent dehydration, the body constantly replaces water loss. Water is replaced when you drink liquids and eat foods such as meats, fruits, and vegetables,

Table 8.6 Percentage of Water in Some Foods

Food	% Water	Food	% Water
Meats		**Fruits**	
Hamburger, broiled	60	Apple	85
Roast	40	Banana	76
Chicken, cooked	71	Grapefruit	89
		Orange	86
Vegetables		Strawberry	90
Carrots	88	Watermelon	93
Celery	94		
Cucumber	96	**Milk Products**	
Onion	89	Milk, whole	87
Potato	75	Cottage cheese	78
		Parmesan cheese	17
Grains			
French bread	31		
Cake	34		
Noodles, cooked	70		

which contain large amounts of water. An apple, for example, contains 85 percent water by weight. Table 8.6 lists the percentages of water contained in several foods. In addition, water is produced in the cells of the body as a product of the metabolism of food. (See Figure 8.11.)

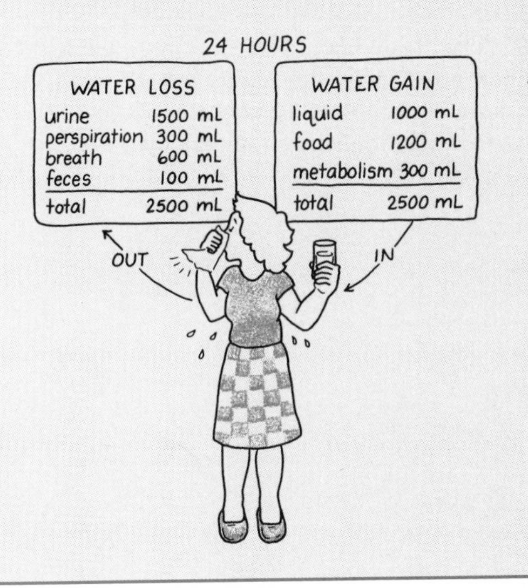

Figure 8.11 A balance in the loss and gain of water by the body.

HOSPITAL
Intake and Output Record

Name Patient ____ Ward ____ Hospital No. ____

Date 9/5 Intake 1800 mL Output 1800 mL Total ____

	Oral fluids, cc	Parental fluids, cc	Weight of patient	Urine, cc	Emesis, cc	Drainage (Wagenstein) etc., cc	Excess perspiration, diarrhea, or hemorrhage	Intake, cc	Output, cc
Date	9/1								
7-3	200	500	140	600	200		100	700	900
3-11	100	500	141	400		100		600	500
11-7	0	500	141	300			100	500	400
Total for 24 hrs	300	1500		1300	200	100	200	1800	1800

Figure 8.12 Intake and output record form.

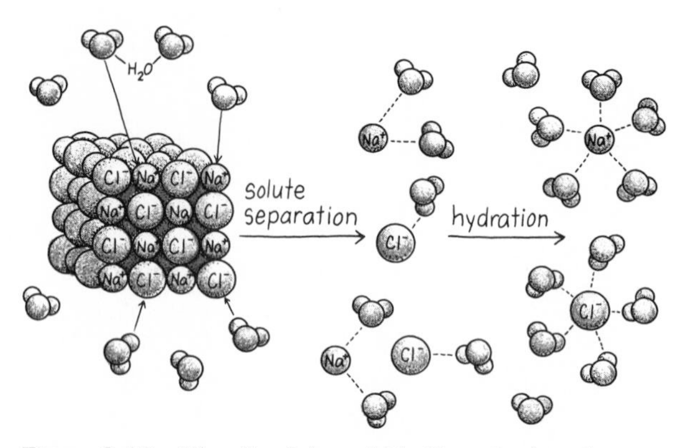

Figure 8.13 The dissolving of NaCl, an ionic solute, by water. Polar water molecules attract ions of NaCl, thereby breaking the bonds of the solute and allowing ions to move into solution where they are hydrated.

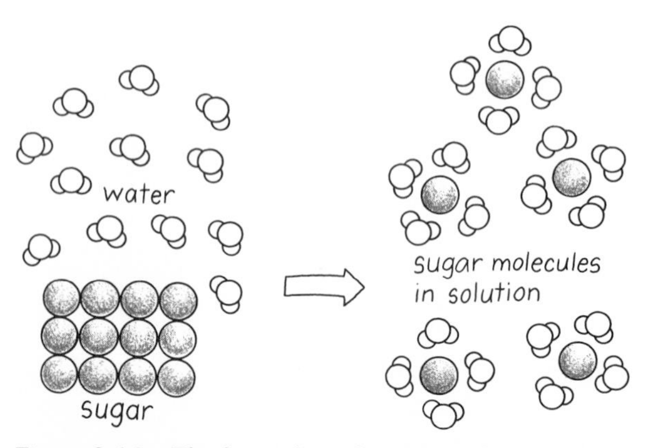

Figure 8.14 The formation of a sugar solution by polar covalent sugar molecules ($C_6H_{12}O_6$) and polar water molecules.

covalent compounds dissolve as molecules

Polar covalent solutes such as sugar will dissolve in water as individual molecules rather than as ions. The negative oxygen end of water attracts the positive portion of the sugar molecule, and the positive hydrogen part of water attracts the negative portion of the sugar molecule. In a similar manner, the sugar molecules break away from the other sugar molecules and enter the solvent to form a solution. (See Figure 8.14.)

Example 8.6 Fill in the following statements about the process of water dissolving the solute KCl:

The salt KCl consists of positive ________ and negative ________. The ________ end of the water molecule will be attracted to the K^+ ions, pulling the ions away from the salt crystal and into solution. The ________ end of the water molecule will be attracted to the Cl^- ions in the salt crystal. The dissolved potassium and chloride ions are surrounded by ________ molecules. This is called ________.

Solution: The salt KCl consists of positive *potassium ions* and negative *chloride ions.* The *oxygen* end of the water molecule will be attracted to the K^+ ions, pulling the ions away from the salt crystal and into solution. The *hydrogen* end of the water molecule will be attracted to the Cl^- ions in the salt crystal. The dissolved potassium and chloride ions are surrounded by *water* molecules. This is called *hydration.*

8.5 Equations for Solution Formation

OBJECTIVE Write an ionic equation for a salt dissolving in water.

A solution of NaCl appears to be just a clear liquid. As we now know, it actually consists of hydrated sodium ions and hydrated chloride ions. The process of forming the NaCl solution can be expressed in the form of an ionic equation. The solid form of the salt is written as NaCl(*s*). After dissolving in water (H_2O), the individual ions are shown to be hydrated as $Na^+(aq)$ and $Cl^-(aq)$ (see Figure 8.15):

$$NaCl(s) \xrightarrow{H_2O} Na^+(aq) + Cl^-(aq)$$

Other ionic compounds dissolve in a similar way to form aqueous solutions of their respective ions. When magnesium chloride ($MgCl_2$) dissolves in water, one magnesium ion is obtained from the salt along with two chloride ions. Recall that the formula for a salt represents a certain number of positive ions and negative ions. In a solution, these ions all separate from each other and become hydrated. In an ionic equation, the ions in solution maintain an electrical balance (see Figure 8.16):

$$MgCl_2(s) \xrightarrow{H_2O} Mg^{2+}(aq) + 2Cl^-(aq)$$

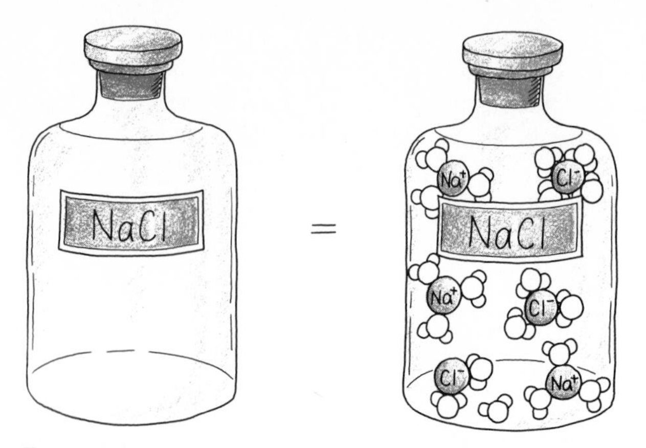

Figure 8.15 A solution of NaCl contains hydrated sodium and chloride ions.

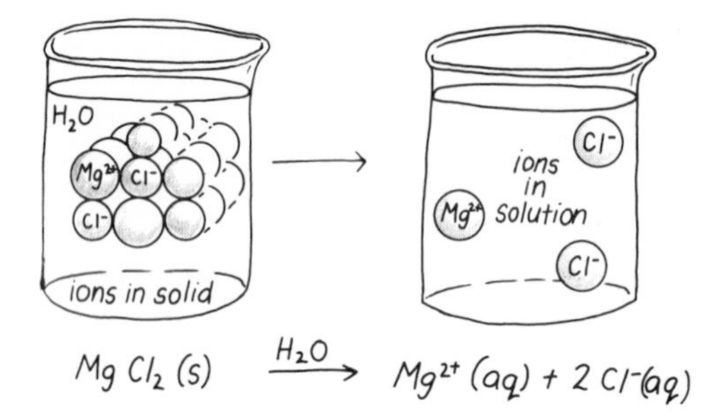

Figure 8.16 A diagram of the solution reaction of $MgCl_2$.

Equations for other ionic solutes that dissolve in water are given below:

$$KNO_3(s) \xrightarrow{H_2O} K^+(aq) + NO_3^-(aq)$$

$$BaCl_2(s) \xrightarrow{H_2O} Ba^{2+}(aq) + 2Cl^-(aq)$$

$$Na_3PO_4(s) \xrightarrow{H_2O} 3Na^+(aq) + PO_4^{3-}(aq)$$

Example 8.7 Write the ionic equation for the solution reaction for the following salts:

1. KCl
2. K_2SO_4

Solution

1. The formula of the salt KCl indicates one potassium (K^+) ion along with one chloride (Cl^-) ion. The ionic equation for its solution in water is

$$KCl(s) \xrightarrow{H_2O} K^+(aq) + Cl^-(aq)$$

2. The formula of the salt K_2SO_4 indicates two potassium K^+ ions along with one sulfate ion (SO_4^{2-}). The ionic equation for its solution in water is

$$K_2SO_4(s) \xrightarrow{H_2O} 2K^+(aq) + SO_4^{2-}(aq)$$

8.6 Heat of Solution

OBJECTIVE Given the heat of solution, state whether the process of dissolving a substance in water is exothermic or endothermic.

When a solute dissolves in water, there is often a change in temperature as the solute particles become dispersed among the solvent molecules. The energy involved in the temperature change is called the *heat of solution.* If the process of forming the solution removes energy (heat) from the surroundings, the temperature of the solution drops. The test tube or beaker holding the solution gets cold. We say that the process is *endothermic.* Then the ionic equation may include the heat of solution as a reactant in kilocalories per mole of salt:

endothermic

Endothermic Process (*Temperature Drops*)

$$KCl(s) + \underbrace{4.1\ \text{kcal/mol}}_{\text{Heat of solution}} \xrightarrow{H_2O} K^+(aq) + Cl^-(aq)$$

$$NaCl(s) + \underbrace{0.93\ \text{kcal/mol}}_{\text{Heat of solution}} \xrightarrow{H_2O} Na^+(aq) + Cl^-(aq)$$

If the formation of a solution releases heat to the surroundings, the temperature of the solution will rise. The test tube holding the solution will become warm or even hot. This is called an *exothermic* process, and the heat of solution will be shown as a product in the ionic equation:

exothermic

Exothermic Processes (*Temperature Rises*)

$$KOH(s) \xrightarrow{H_2O} K^+(aq) + OH^-(aq) + \underbrace{13.9\ \text{kcal/mol}}_{\text{Heat of solution}}$$

$$NaOH(s) \xrightarrow{H_2O} Na^+(aq) + OH^-(aq) + \underbrace{10.6\ \text{kcal/mol}}_{\text{Heat of solution}}$$

HEALTH NOTE

Hot Packs and Cold Packs

cold pack

In the hospital, at a first-aid station, or at an athletic event, a *cold pack* may be used to reduce swelling, remove heat from inflammation, or decrease capillary size to lessen the effect of hemorrhaging. In a cold pack, ammonium nitrate (NH_4NO_3) is placed in a vial inside a heavy cardboard or plastic pack filled with water. The cold pack is activated by hitting the pack to break the vial and release the salt. As the salt dissolves, the temperature drops, and the pack becomes cold. Each mole of NH_4NO_3 removes 6.3 kcal from the surrounding water:

$$NH_4NO_3(s) + \text{Heat (6.3 kcal/mol)} \xrightarrow{H_2O} NH_4NO_3(aq)$$

hot pack

In a hot pack, a vial containing the salt $CaCl_2$ is broken to permit the salt to dissolve in the water. Hot packs are used to relax muscles, lessen aches and cramps, and increase circulation by expanding capillary size. The reaction releases energy in the form of heat (18 kcal/mol), and the pack gets hot:

$$CaCl_2(s) \xrightarrow{H_2O} CaCl_2(aq) + \text{Heat (18 kcal/mol)}$$

Example 8.8 Sodium nitrate, $NaNO_3$, dissolves in water by absorbing 4.9 kcal/mol of salt. Write the ionic equation for the solution reaction including the heat of solution.

Solution: The salt, $NaNO_3$, consists of Na^+ and NO_3^-. Since heat is absorbed in the solution reaction, the reaction is endothermic and the heat of solution will be shown as a reactant:

$$NaNO_3(s) + 4.9 \text{ kcal/mol} \xrightarrow{H_2O} Na^+(aq) + NO_3^-(aq)$$

8.7 Electrolytes

OBJECTIVE Write the ionic equation for a strong electrolyte, a weak electrolyte, and a nonelectrolyte dissolving in water.

strong electrolytes
weak electrolytes
nonelectrolytes

Electrolytes are substances that produce ions when they dissolve in water. The ions move through the solution to carry electrical current. Substances that completely separate into ions in water are called *strong electrolytes.* If the dissolved substance produces only a few ions in water, it is a *weak electrolyte.* Other substances, called *nonelectrolytes,* dissolve only as molecules in water and cannot conduct electricity. In general, ionic compounds and many polar

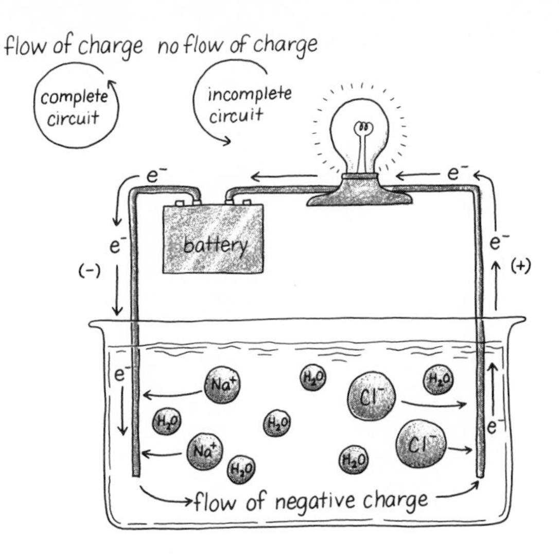

Figure 8.17 Electricity flowing through an electrolyte. The circuit is completed only when there is a flow of ions in the solution.

covalent compounds are electrolytes, whereas nonpolar covalent compounds are nonelectrolytes.

We can test for the presence of electrolytes using an apparatus consisting of two electrodes attached by wires to a light bulb. Current flows through the light bulb only when there are ions in the solution. The light bulb glows more brightly when there are greater numbers of ions in the solution. (See Figure 8.17.)

Writing Ionic Equations for Electrolytes

When a strong electrolyte forms a solution in water, the electrolyte is completely (100 percent) ionized. The ionic equation for a strong electrolyte is written with a single arrow to indicate that the reaction goes completely to the product side:

Examples of Strong Electrolytes

$$NaCl(s) \xrightarrow{H_2O} Na^+(aq) + Cl^-(aq)$$

$$K_2SO_4(s) \xrightarrow{H_2O} 2K^+(aq) + SO_4{}^{2-}(aq)$$

When a weak electrolyte dissolves in water, it produces only a few ions; most of the dissolved substance exists as molecules. A light bulb placed in a weak electrolyte will only give a dim glow because there are relatively few ions. In the equations for the ionization of weak electrolytes, a double or reverse arrow is used. This indicates that only a small number of molecules ionize at any given time. Any further ionization of a molecule will be matched by the recombining of ions to reform the molecular form. This is another example of a dynamic equilibrium:

Examples of Weak Electrolytes

$$\underset{\text{Acetic acid}}{HC_2H_3O_2} \xrightleftharpoons{H_2O} H^+(aq) + \underset{\text{Acetate ion}}{C_2H_3O_2^-(aq)}$$

$$HF \xrightleftharpoons{H_2O} H^+(aq) + F^-(aq)$$

The equation for the formation of a solution from a nonelectrolyte indicates the molecular form of the compound in both the initial state and in the aqueous solution. No ions are produced (see Figure 8.18):

Examples of Nonelectrolytes

$$\underset{\text{Glucose (solid)}}{C_6H_{12}O_6(s)} \xrightarrow{H_2O} \underset{\text{Glucose (in solution)}}{C_6H_{12}O_6(aq)}$$

$$\underset{\text{Carbon dioxide (gas)}}{CO_2(g)} \xrightarrow{H_2O} \underset{\text{Carbon dioxide (in solution)}}{CO_2(aq)}$$

HEALTH NOTE

Hard Water

Water in different parts of the country may contain variable amounts of minerals. Of these minerals, the ions Ca^{2+}, Fe^{3+}, and Mg^{2+} form insoluble substances when combined with soap, leaving a "scum" or "bathtub ring" in a container. Water containing these ions is called *hard water*. Greater amounts of soap are needed to produce a soapy solution because some of the soap is converted to the insoluble form (scum):

$$\underset{\text{Hard water}}{Ca^{2+}} + 2\,Soap^- \longrightarrow \underset{\text{Insoluble calcium soap "scum"}}{Ca(Soap)_2(s)}$$

Water that does not contain Ca^{2+}, Fe^{3+}, or Mg^{2+} is called soft water. With soft water, no scum forms with soap, and less soap is required to form a soapy solution. Phosphates were added to detergents to help "soften" water by tying up the Ca^{2+}, Fe^{3+}, and Mg^{2+} ions. They have been removed from several detergents now because of pollution problems. Phosphate aids the growth of algae and plants in rivers and streams, causing a decrease in available oxygen and endangering the survival of fish and other aquatic animals.

Hard water can be purified by removing the ions in several ways. Distillation is one way. In distillation, the hard-water sample is boiled; then the water vapor formed is condensed through a cooling apparatus, yielding soft water with no ions of calcium, iron, or magnesium. They remain in the distillation vessel as salts.

Several types of *ion exchangers* are on the market for commercial or home water softening. In ion exchangers, the offending ions are replaced by ions that do not affect water hardness. This is accomplished by passing the hard water through a column packed with a resin containing sodium ions. As the hard-water ions move through the

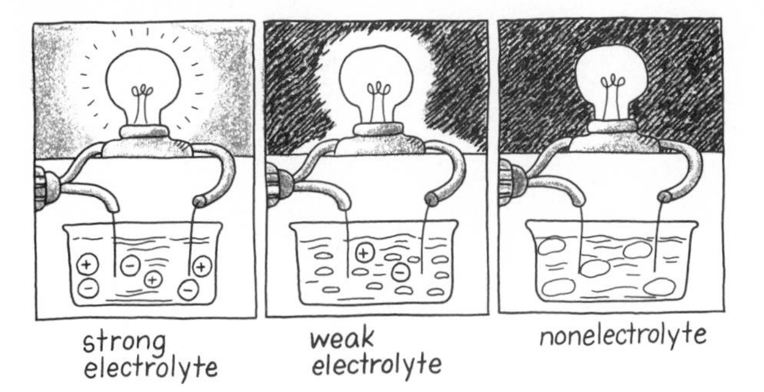

Figure 8.18 A strong electrolyte in solution produces ions and conducts an electrical current; a weak electrolyte produces only a few ions and conducts a weak electrical current; a nonelectrolyte produces no ions in solution, only molecules, and does not conduct electricity.

column, they are tied up by the resin and replaced by sodium ions; the sodium ions and the hard-water ions have been exchanged. The soft water leaving the ion exchanger now contains sodium ions instead of hard-water ions. This type of ion exchanger should be avoided by people who must maintain a low sodium diet. (See Figure 8.19.)

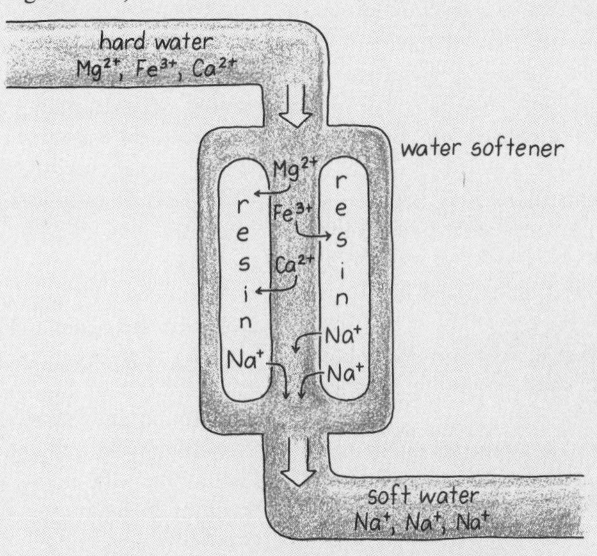

Figure 8.19 Softening water by means of an ion-exchange resin in which Mg^{2+}, Fe^{3+}, and Ca^{2+} ions are exchanged for Na^{+} ions.

Example 8.9 Write the ionic equation for the formation of a solution of each of the following compounds:

1. Na_2SO_4, a strong electrolyte
2. urea, $CH_4N_2O(s)$, a nonelectrolyte

Solution

1. A solution of a strong electrolyte such as Na_2SO_4 will contain only the ions of the dissolved salt, namely, $Na^+(aq)$ and $SO_4^{2-}(aq)$.

$$Na_2SO_4(s) \xrightarrow{H_2O} 2Na^+(aq) + SO_4^{2-}(aq)$$

2. A nonelectrolyte such as urea dissolves in water in the form of urea molecules only:

$$CH_4N_2O(s) \xrightarrow{H_2O} CH_4N_2O(aq)$$

Glossary

electrolyte A substance that produces ions when dissolved in water.

heat of solution The energy change as a solution forms.

Henry's law The amount of gas that dissolves in a liquid increases as the pressure of the gas increases above the solution at constant temperature.

hydration The surrounding of dissolved ions by water molecules.

hydrogen bonds Type of bond that forms between a partially positive hydrogen in a water molecule and the partially negative oxygen atom of other water molecules.

insoluble solute A solute that will not dissolve in a certain solvent.

nonelectrolyte A substance that dissolves in water but does not produce any ions; its soluton will not conduct electricity.

saturated solution A solution that attains its maximum solubility level at a given temperature. Addition of more solute will cause the formation of undissolved material in the container.

solubility The maximum amount of solute that will dissolve in a given quantity of solvent at a given temperature.

solute The component in a solution that changes state upon dissolving; if no change in state occurs, it is the component of smaller quantity.

solution A uniform mixture of two or more components.

solvent The dissolving material in a solution; the component whose state does not change.

strong electrolyte A substance that ionizes completely when it dissolves in water.

supersaturated solution An unstable solution in which the concentration of solute exceeds its solubility.

surface tension The property of water whereby surface molecules are pulled tightly together to form a "skin."

unsaturated solution A solution that contains less than its solubility level of solute. More solute can be added to the solution and dissolve.

weak electrolyte A substance that produces only a few ions along with many nonionized molecules when it dissolves in water. Its solution is a weak conductor of electricity.

Problems

Solutions (Objective 8.1)

8.1 Identify the solute and the solvent in the following solutions:
a. 10 g NaCl and 100 g H_2O
b. 50 mL ethanol and 10 mL H_2O
c. 2.0 L oxygen (O_2), 8.0 L nitrogen (N_2)
d. 100 g silver and 40 g mercury
e. 100 mL H_2O and 5.0 g sugar

Nature of Solute and Solvent (Objective 8.2)

8.2 State whether each of the following solutes will be more soluble in water (a polar solvent) or hexane (a nonpolar solvent):
a. KCl (a salt), ionic
b. sugar, polar
c. benzene, nonpolar
d. I_2, nonpolar
e. vegetable oil, nonpolar
f. potassium nitrate, a salt

Solubility (Objective 8.3)

8.3 State whether the following statements refer to saturated or unsaturated solutions:
a. A crystal added to a solution does not change in size.
b. A sugar cube dissolves when added to a cup of coffee.
c. The rate of dissolution of the solid is equal to the rate of formation of the undissolved solid.
d. The rate of crystal formation is lower than the rate of solution.
e. A layer of sugar forms at the bottom of a glass of iced tea.

8.4 Using the solubilities of KBr and KI given in the accompanying table, state whether the following solutions would be saturated or unsaturated:
a. 75 g KBr in 100 g H_2O at 40°C
b. 160 g KI in 100 g H_2O at 40°C
c. 100 g KBr in 100 g H_2O at 100°C
d. 100 g KBr in 200 g H_2O at 80°C
e. 50 g KI in 100 g H_2O at 20°C

Solubility (g Solute/100 g Water)

T (°C)	KBr	KI
20	65	145
40	80	160
60	90	175
80	100	190
100	110	210

8.5 A sample containing 100 g of KBr in 100 g H_2O at 100°C is cooled to 20°C. Using the solubility table for KBr and KI, answer the following:
a. Is the solution saturated or unsaturated at 100°C?
b. Is the solution saturated or unsaturated at 20°C?

8.6 a. Why does a bottle of carbonated beverage fizz when it is opened?
b. Tap water containing dissolved chlorine gas must sit at room temperature for at least a day or be boiled and cooled before it is used for a fish tank. Explain.

8.7 Indicate whether the following preparations will increase or decrease the rate of solution for KCl and water:
a. stirring the mixture
b. using large chunks of KCl
c. placing the beaker of KCl and water in ice
d. heating the beaker containing the KCl and water
e. crushing the KCl to a fine powder before mixing it with the water

Water as a Solvent (Objective 8.4)

8.8 Match the following phrases with the concepts of hydrogen bonding, surface tension, hydration, oxygen atom, or hydrogen atom:
a. the pulling together of water molecules at the surface to form a thick layer that will support the weight of small bugs
b. the portion of the water molecule attracted to the positive ions of a salt
c. attractive forces between water molecules
d. the surrounding of ions in a solution by several water molecules
e. the portion of the water molecule attracted to negative ions in a salt

Ionic Equations for Solution Formation (Objective 8.5)

8.9 Write an equation for the dissociation of the following salts in water:
a. KCl
b. Na_2SO_4
c. $CaCl_2$
d. $LiNO_3$
e. K_3PO_4
f. $Ba(NO_3)_2$

Heat of Solution (Objective 8.6)

8.10 State whether the following reactions are exothermic or endothermic:

a. $NaNO_3(s) + 4.9\text{ kcal/mol} \xrightarrow{H_2O} Na^+ + NO_3^-$

b. $KCl(s) + 4.1\text{ kcal/mol} \xrightarrow{H_2O} K^+ + Cl^-$

c. $LiI(s) \xrightarrow{H_2O} Li^+ + I^- + 15.1\text{ kcal/mol}$

d. $KNO_3(s) + 8.3\text{ kcal/mol} \xrightarrow{H_2O} K^+ + NO_3^-$

8.11 Write equations for the solution reactions of the following salts, including the heat of solution:
a. $AgNO_3(s)$, endothermic, 5.4 kcal/mol
b. $KBr(s)$, endothermic, 4.8 kcal/mol
c. $LiBr(s)$, exothermic, 11.7 kcal/mol
d. $CsF(s)$, exothermic, 8.8 kcal/mol

Electrolytes (Objective 8.7)

8.12 Indicate whether aqueous solutions of the following will contain (1) ions only, (2) molecules only, or (3) molecules and some ions:

a. KCl, a strong electrolyte
b. glucose, a nonelectrolyte
c. acetic acid (vinegar), a weak electrolyte
d. sodium bromide, a salt
e. urea, a nonelectrolyte

8.13 Indicate the type of electrolyte represented in the following equations:

a. $KI(s) \xrightarrow{H_2O} K^+(aq) + Cl^-(aq)$

b. $H_2O + NH_3(g) \xrightleftharpoons{H_2O} NH_4{}^+(aq) + OH^-(aq)$

c. $CH_3OH(l) \xrightarrow{H_2O} CH_3OH(aq)$

d. $C_6H_{12}O_6(s) \xrightarrow{H_2O} C_6H_{12}O_6(aq)$

e. $HF(g) \xrightleftharpoons{H_2O} H^+ + F^-$

9 Concentrations and Properties of Solutions

Objectives

9.1 Given the mass of the solute and the volume of the solution, calculate the percentage (weight/volume) concentration.

9.2 Given the volume and percent (w/v) concentration, calculate the mass of the solute in a solution; given the mass of solute and the percent (w/v) concentration, calculate the volume of a solution.

9.3 Calculate the final volume or the new concentration of a solution obtained by dilution.

9.4 Calculate the molarity of a solution; solve concentration problems using molarity as the conversion factor to find the moles of solute or the volume of the solution.

9.5 Given the mass of an electrolyte, calculate the number of equivalents or milliequivalents.

9.6 From the properties, identify a mixture as a true solution, a colloidal dispersion, or a suspension.

9.7 Given two solutions separated by a semipermeable membrane, indicate the side with the greater osmotic pressure, the direction in which water will flow, and the compartment that increases in volume.

9.8 Given the percent concentration for a solution, determine (1) whether that solution is isotonic, hypotonic, or hypertonic to a red blood cell and (2) whether a red blood cell in that solution would undergo hemolysis, crenation, or no change.

9.9 Given a substance in a dialysis bag, state whether it will dialyze through the semipermeable membrane.

Scope

The amount of solute that dissolves in a solvent is an important measurement for solutions. The concentration of a solution is the amount of solute dissolved in a certain amount of that solution. The concentrations of substances in the body fluids are normally quite constant for each type of body fluid. Because significant changes in concentrations can indicate illness or injury, the measurement of these concentrations is a valuable diagnostic tool.

In addition to solutions of small molecules or ions, there are mixtures of water and larger particles. Such mixtures are called *colloids*. The milk, mayonnaise, whipped cream, and Jell-O in the refrigerator are examples of colloids. In another type of mixture, called a *suspension*, the particles are so large that they eventually settle out. In the hospital, you may use suspensions such as liquid penicillin medication or calamine lotion. A suspension must be well shaken before its use.

Osmosis is a process of fluid transport in which water molecules flow through semipermeable membranes in plants and in the cells of the body. In the process of dialysis, small particles in solution also diffuse through semipermeable membranes. It is the principles of osmosis and dialysis that lead to an understanding to the way water, nutrients, and waste products move in and out of the cells of the body.

9.1 Percent Concentration

OBJECTIVE Given the mass of the solute and the volume of the solution, calculate the percentage (weight/volume) concentration.

The *concentration* of a solution indicates the amount of solute that is dissolved in a certain amount of that solution:

$$\text{Concentration} = \frac{\text{Amount of solute}}{\text{Amount of solution}}$$

There are several ways to express the concentration of a solution, depending on the units used to describe the amount of solute and the amount of solution. However, all of the expressions for concentration indicate a relationship that describes how much solute is dissolved in a certain amount of solution.

percent concentration (weight/volume)

One way to describe the concentration of a solution is by means of *percent concentration*. A percent is defined as "parts per hundred." Therefore, 2% means 2 parts out of 100 parts or 2/100. In the health area, the percent concentration most typically used is *weight/volume (w/v) percent*, in which the amount of solute is given in grams and the amount of solution is given in 100 mL. A 2% (w/v) solution means there are 2 g of solute in 100 mL of solution:

Percent Concentration	*Meaning in Numbers*	*Meaning in Words*
5% glucose solution	$\frac{5\text{ g glucose}}{100\text{ mL solution}}$	There are 5 g of glucose in 100 mL of solution
10% KCl solution	$\frac{10\text{ g KCl}}{100\text{ mL solution}}$	There are 10 g of KCl in 100 mL of solution
20% NaOH solution	$\frac{20\text{ g NaOH}}{100\text{ mL solution}}$	There are 20 g of NaOH in 100 mL of solution

The percent (w/v) concentration of a solution can be calculated by dividing the mass (grams) of solute by the given volume (milliliters) of the solution and multiplying by 100:

$$\text{Percent (w/v)} = \frac{\text{Grams of solute}}{\text{Milliliters of solution}} \times 100$$

Example 9.1 Calculate the percent (w/v) concentration of solutions prepared from 10 g of NaOH and water to give the following solution volumes (see Figure 9.1):

1. 100 mL of solution
2. 200 mL of solution
3. 40 mL of solution

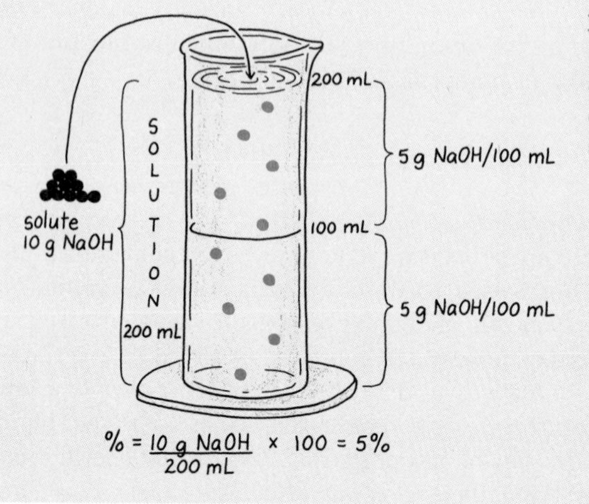

Figure 9.1 Calculation of percent (weight/volume) concentration of 10 g NaOH dissolved in 200 mL of solution.

Solution: The percent concentration is calculated using the ratio of the mass of the solute divided by the volume of the solution:

$$\%\,(\text{w/v}) = \frac{\text{Mass (g) solute}}{\text{Volume of solution}} \times 100$$

1. $$\%\,(\text{w/v}) = \frac{10\text{ g NaOH}}{100\text{ mL solution}} \times 100 = 10\%\,(\text{w/v})\text{ NaOH solution}$$

2. $$\%\,(\text{w/v}) = \frac{10\text{ g}}{200\text{ mL solution}} \times 100 = 5\%\,(\text{w/v})\text{ NaOH solution}$$

3. $$\%\,(\text{w/v}) = \frac{10\text{ g}}{40\text{ mL solution}} \times 100 = 25\%\,(\text{w/v})\text{ NaOH solution}$$

Example 9.2 You have obtained 20 mL of a KCl solution and evaporated it to dryness. The KCl salt has a mass of 0.40 g. What is the percent concentration of the KCl solution?

Solution: Since you know the quantity of the solute and the volume of the solution in milliliters, you can substitute the amounts into the expression for % concentration:

$$\%\,(\text{w/v}) = \frac{0.40\text{ g KCl}}{20\text{ mL solution}} \times 100 = 2\%\,(\text{w/v})\text{ KCl solution}$$

9.2 Calculations Using Percent Concentration

OBJECTIVE Given the volume and percent (w/v) concentration, calculate the mass of the solute in a solution; given the mass of solute and the percent (w/v) concentration, calculate the volume of a solution.

The percent (weight/volume) concentration of any solution gives a relationship between the grams of solute and a volume (100 mL) of solution. This relationship can be expressed in the form of a conversion factor. For example, we have seen that a 5% (w/v) glucose solution indicates there are 5 g of glucose in 100 mL of solution. This relationship can be written in the form of two percent conversion factors:

Percent Conversion Factors for a 5% Glucose Solution

$$\frac{5\text{ g glucose}}{100\text{ mL solution}} \quad\text{or}\quad \frac{100\text{ mL solution}}{5\text{ g glucose}}$$

Example 9.3 A bottle contains 400 mL of a 5% (w/v) glucose solution. How many grams of glucose are in the bottle? (See Figure 9.2.)

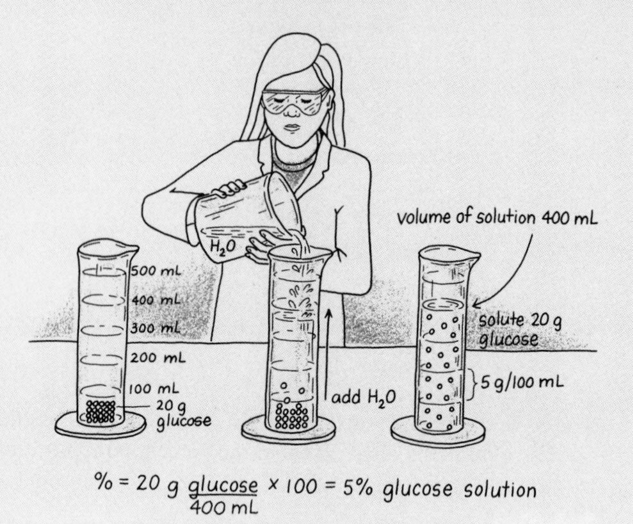

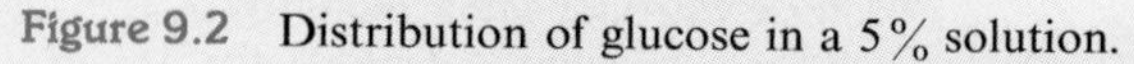

Figure 9.2 Distribution of glucose in a 5% solution.

Solution: Using the 5% (w/v) concentration as a percent conversion factor, we can calculate the mass in grams of glucose in the solution. The given is 400 mL of solution. The conversion factor is set up to cancel the unit mL.

$$400\ \cancel{\text{mL}} \times \underset{\substack{\text{\%(w/v) concentration used}\\ \text{as conversion factor}}}{\frac{5\text{ g glucose}}{100\ \cancel{\text{mL}}\text{ solution}}} = 20\text{ g glucose}$$

$$\text{Volume (mL) solution} \times \underset{\text{percent (w/v) concentration}}{\frac{\text{Mass (g) solute}}{100\text{ mL solution}}} = \text{Mass (g) of solute}$$

Example 9.4 How many grams of salt are needed to prepare 3.0 L of a 1% (w/v) NaCl solution?

Solution: The conversion factors for a 1% (w/v) NaCl concentration are

$$\frac{1\text{ g NaCl}}{100\text{ mL solution}} \quad \text{or} \quad \frac{100\text{ ml solution}}{1\text{ g NaCl}}$$

In the setup for the problem, the volume (3.0 L) must be changed to units (mL) that will cancel the volume units (mL) in the percent conversion factor:

$$\underset{\substack{\text{Volume}\\ \text{given}}}{3.0\ \cancel{\text{L}}} \times \underset{\substack{\text{Change}\\ \text{to mL}}}{\frac{1000\ \cancel{\text{mL}}}{1\text{ L}}} \times \underset{\substack{\text{Percent concentration}\\ \text{as conversion factor}}}{\frac{1\text{ g NaCl}}{100\ \cancel{\text{mL}}\text{ solution}}} = 30\text{ g NaCl}$$

Another type of solution calculation is that of finding the volume of solution that will provide a certain quantity of a solute.

Example 9.5 A patient requires 100 g of glucose. If the patient is receiving a 5% (w/v) glucose solution, how many liters (L) of the glucose solution should be given?

Solution: First, we can express the 5% (w/v) solution in the form of percent conversion factors:

$$5\%\ (\text{w/v})\ \text{solution} = \frac{5\ \text{g glucose}}{100\ \text{mL solution}} \quad \text{or} \quad \frac{100\ \text{mL solution}}{5\ \text{g glucose}}$$

In order to cancel out the units of the given, 100 g glucose, the percent conversion factor with g glucose in the denominator is used:

$$\underset{\text{Quantity of solute given}}{100\ \cancel{\text{g}}\ \text{glucose}} \times \underset{\text{Percent concentration (inverted)}}{\frac{100\ \text{mL glucose solution}}{5\ \cancel{\text{g}}\ \text{glucose}}} = \underset{\text{Volume of solution}}{2000\ \text{mL glucose solution}}$$

The volume in units of milliliters can now be changed to volume in liters:

$$2000\ \cancel{\text{mL}}\ \text{glucose} \times \frac{1\ \text{L}}{1000\ \cancel{\text{mL}}} = 2\ \text{L needed for patient}$$

9.3 Dilutions

OBJECTIVE Calculate the final volume or the new concentration of a solution obtained by dilution.

Sometimes you will need to prepare a solution by making a dilution. When you dilute a solution, you add more solvent to increase the volume. At the same time, you are lowering the concentration of the solute in that solution. For example, you can prepare orange juice for breakfast by making a dilution of the orange concentrate. (See Figure 9.3.)

Figure 9.3 Dilution of orange juice at breakfast time.

Suppose we have 100 mL of a 10% (w/v) sucrose solution. By adding more water to this solution, we are lowering the concentration of the solute. If we add 400 mL of water to make a new volume of 500 mL, we have diluted our solution from a 10% (w/v) to a 2% (w/v) sucrose solution. (See Figure 9.4.) Note that there has been no change in the mass of solute. We started with 10 g of sucrose, and there is still 10 g of sucrose in the diluted solution. However, we have increased the volume by 5 times and diluted the solution to 1/5 of the original concentration:

	Initial Solution	*Diluted Solution*
Mass of solute:	10 g sucrose	10 g sucrose
Volume of solution:	100 mL solution	500 mL solution
Water added:	+400 mL	
Percent (w/v) concentration:	10% (w/v)	2% (w/v)

Dilution Factor

The increase in volume from 100 mL to 500 mL is called a *1:5 dilution*. This ratio of the volumes is called the *dilution factor*:

$$\frac{100\text{ mL}}{500\text{ mL}} = \frac{1}{5} \quad \text{or} \quad \underbrace{1:5}_{\text{Dilution factor}}$$

Suppose you need to make a 1:2 dilution of 250 mL of a 10% (w/v) NaCl solution. This means that the new volume will be 500 mL, or two times larger than the original volume:

$$\text{New, diluted volume} = \underbrace{250\text{ mL}}_{\text{Old volume}} \times \underbrace{\frac{2}{1}}_{\text{Dilution factor}} = \underbrace{500\text{ mL}}_{\text{New volume}}$$

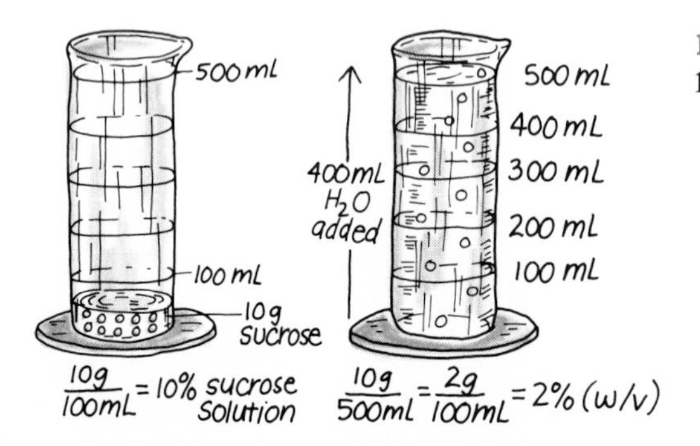

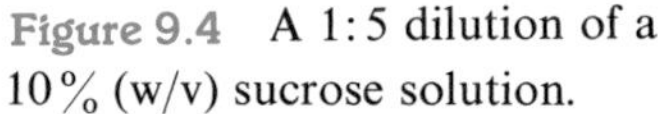

Figure 9.4 A 1:5 dilution of a 10% (w/v) sucrose solution.

Since we already have 250 mL in our initial sample, we need to add another 250 mL of water to bring the new diluted volume up to 500 mL. The percent concentration of the diluted solution is one-half of the initial concentration, or 5% after dilution:

$$\underset{\text{Initial concentration}}{10\% \text{ (w/v)}} \times \underset{\text{Dilution factor}}{\tfrac{1}{2}} = \underset{\text{New concentration}}{5\% \text{ (w/v) after dilution}}$$

Example 9.6 Calculate the final volume and the new concentration of the diluted solution when a student carries out a 1:3 dilution of 500 mL of a 12% KCl solution.

Solution: The final volume of the diluted solution will be three times greater than the original volume of 500 mL:

$$\underset{\text{Old Volume}}{500 \text{ mL}} \times \underset{\text{Dilution factor}}{\tfrac{3}{1}} = \underset{\text{Volume of diluted solution}}{1500 \text{ mL, final volume}}$$

The dilution of a solution means that its concentration is lowered:

$$\underset{\text{Original concentration}}{12\% \text{ KCl solution}} \times \underset{\text{Dilution factor}}{\tfrac{1}{3}} = \underset{\text{Concentration of diluted solution}}{4\% \text{ KCl solution}}$$

9.4 Molarity

OBJECTIVE Calculate the molarity of a solution; solve concentration problems using molarity as the conversion factor to find the number of moles of solute or the volume of the solution.

The molarity of a solution is an important concentration for chemists. It describes the moles of solute in a liter of solution that are available for a chemical reaction. The molarity (M) of a solution is defined as the moles of solute in 1 L of solution:

$$\text{Molarity (M)} = \frac{\text{moles of solute}}{\text{liter of solution}}$$

Molarity	*Meaning*
5-M glucose	$\frac{5 \text{ mol glucose}}{1 \text{ L}}$
0.20-M KI	$\frac{0.20 \text{ mol KI}}{1 \text{ L}}$

The molarity of a solution can be calculated when the number of moles of solute and the volume of solution are known:

Moles of Solute	*Liters of Solution*	*Calculation*	*Molarity*
1 mol KCl	1 L	$\frac{1 \text{ mol KCl}}{1 \text{ L}}$	1-*M* KCl
6 mol NaOH	2 L	$\frac{6 \text{ mol NaOH}}{2 \text{ L}}$	3-*M* NaOH
3 mol KI	0.5 L	$\frac{3 \text{ mol KI}}{0.5 \text{ L}}$	6-*M* KI

Example 9.7 What is the molarity of each of the following solutions?

1. 2.0 mol $CaCl_2$ in 400 mL of solution
2. 60.0 g NaOH in 3.0 L of solution

Solution

1. The expression for molarity calculations calls for solute in units of moles and volume in liters:

$$\text{Molarity } (M) = \frac{\text{Moles of solute}}{\text{Liters of solution}}$$

The volume of 400 mL must be changed to liters:

$$400 \cancel{\text{mL}} \times \frac{1 \text{ L}}{1000 \cancel{\text{mL}}} = 0.4 \text{ L}$$

Now we can substitute the amounts of solute and volume into the molarity expression:

$$\text{Molarity} = \frac{2 \text{ mol } CaCl_2}{0.4 \text{ L solution}} = 5\text{-}M \text{ } CaCl_2 \text{ solution}$$

2. Since the molarity requires moles of solute, we must convert the grams of solute to moles. Recall that the molar mass of a compound gives us the relationship of grams per mole. The molar mass of NaOH is 40.0 g/mol:

$$60.0 \cancel{\text{g}} \text{ NaOH} \times \frac{1 \text{ mol NaOH}}{40.0 \cancel{\text{g}} \text{ NaOH}} = 1.5 \text{ mol NaOH}$$

Now we can substitute the values for solute and volume into the molarity expression:

$$\text{Molarity} = \frac{1.5 \text{ mol NaOH}}{3 \text{ L solution}} = 0.5\text{-}M \text{ NaOH solution}$$

Calculations Using Molarity

The molarity of a solution is useful in the form of a conversion factor that relates moles of solute and volume of solution. For example, a 4.0-*M* KCl solution means that there are 4.0 mol of KCl in 1 L of solution. It can be written in the form of the following molarity conversion factors:

Molarity Conversion Factors for 4.0-*M KCl*

$$\frac{4.0 \text{ mol KCl}}{1 \text{ L solution}} \quad \text{or} \quad \frac{1 \text{ L solution}}{4.0 \text{ mol KCl}}$$

When you know the volume of a solution and its molarity, you can calculate the amount (moles or grams) of solute contained in that volume of solution.

Example 9.8 How many moles of NaCl are present in 5 L of a 2-*M* NaCl solution?

Solution: The setup for this problem starts with the given volume of solution (5 L) multiplied by the conversion factor derived from the molarity of the solution

$$\underset{\text{Volume given}}{5 \cancel{\text{L}}} \times \underset{\text{Molarity factor}}{\frac{2 \text{ mol NaCl}}{1 \cancel{\text{L}} \text{ solution}}} = \underset{\text{Number of moles of solute}}{10 \text{ mol NaCl}}$$

Example 9.9 How many grams of KCl are needed to prepare 250 mL of a 2.0-*M* KCl solution?

Solution: We can first find the moles of KCl by converting the volume to liters and multiplying by the molarity factor:

$$250 \cancel{\text{mL}} \text{ solution} \times \frac{1 \cancel{\text{L}}}{1000 \cancel{\text{L}}} \times \frac{2.0 \text{ mol KCl}}{1 \cancel{\text{L}}} = 0.50 \text{ mol KCl}$$

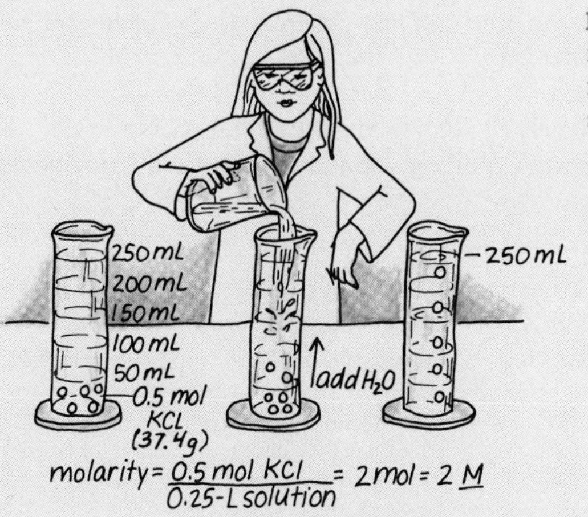

Figure 9.5 A 2-*M* KCl solution is obtained when 37.4 g KCl (0.50 mol) dissolve in 250 mL of solution.

When the moles of solute have been calculated, the number of grams can be determined by using the molar mass of the solute. For KCl, the molar mass is 74.7 g/mol:

$$0.50\ \cancel{\text{mol KCl}} \times \frac{74.7\ \text{g KCl}}{1\ \cancel{\text{mol KCl}}} = 37.4\ \text{g KCl needed}$$

If you were to prepare this solution, you would place 37.4 g of KCl in a 250-mL graduated cylinder or volumetric flask and add water to bring the solution up to the 250-mL volume mark, as shown in Figure 9.5.

When the number of moles of solute is given along with the molarity of the solution, the volume of the solution needed can be calculated.

Example 9.10 What volume, in milliliters of a 6.0-*M* HCl solution will contain 3.0 mol of HCl?

Solution: Using the inverted form of the molarity factor, we can set up the following calculation:

$$\underset{\text{Given moles}}{3.0\ \cancel{\text{mol HCl}}} \times \underset{\substack{\text{Molarity factor} \\ \text{(inverted)}}}{\frac{1\ \text{L solution}}{6.0\ \cancel{\text{mol HCl}}}} \times \underset{\text{Metric factor}}{\frac{1000\ \text{mL}}{1\ \cancel{\text{L}}}} = 500\ \text{mL HCl}$$

9.5 Equivalents

OBJECTIVE Given the mass of an electrolyte, calculate the number of equivalents or milli-equivalents.

The levels of the electrolytes such as Na^+, Cl^-, K^+, and HCO_3^- in the body fluids are usually given in terms of their ionic charges. One mole of ionic charge

Table 9.1 Equivalents of Electrolytes

Moles of Ion	Moles of Charge	Number of Equivalents
1 mol Na^+	1 mol positive charge	1
1 mol Cl^-	1 mol negative charge	1
1 mol Ca^{2+}	2 mol positive charge	2
1 mol Fe^{3+}	3 mol positive charge	3

one equivalent contains 1 mol of ionic charge

is called one *equivalent* (eq). For example, 1 mol of Na^+ ion carries 1 mol of positive charge. We say that we have 1 eq of Na^+ ion. One mole of sulfide ion, S^{2-}, carries 2 mol of negative charge. We say that 1 mol of sulfide ion contains 2 eq of ionic charge:

$$1 \text{ eq} = 1 \text{ mol positive charge or } 1 \text{ mol negative charge}$$

Table 9.1 lists the equivalent amounts of some important ions in the body.

We can use the number of equivalents in a mole of positive or negative ions as a conversion factor. For example, 1 mol of Mg^{2+} contains 2 eq of charge:

$$\frac{1 \text{ mol } Mg^{2+}}{2 \text{ eq}} \quad \text{or} \quad \frac{2 \text{ eq } Mg^{2+}}{1 \text{ mol}}$$

Example 9.11 Calculate the number of equivalents in 1.5 g of Zn^{2+}.

Solution: We can convert 1.5 g of Zn^{2+} into moles by using the molar mass of zinc. For Zn^{2+}, there are 2 eq per mole. Written as a conversion factor, this becomes

$$\frac{2 \text{ eq}}{1 \text{ mol}} \quad \text{or} \quad \frac{1 \text{ mol}}{2 \text{ eq}}$$

The number of grams can now be converted to equivalents as follows:

$$1.5 \text{ g } Zn^{2+} \times \underbrace{\frac{1 \text{ mol } Zn^{2+}}{65.4 \text{ g}}}_{\text{Molar mass}} \times \underbrace{\frac{2 \text{ eq}}{1 \text{ mol}}}_{\text{Ionic charge}} = 0.046 \text{ eq } Zn^{2+}$$

The concentrations of electrolytes present in body fluids and in parenteral solutions are often given in milliequivalents per liter (meq/L) of solution. Table 9.2 gives the concentrations of some electrolytes typical for blood plasma.

Table 9.2 Normal Concentrations of Some Electrolytes in Blood Plasma

Electrolyte	Concentration
Sodium (Na^+)	138–146 meq/L
Potassium (K^+)	4.1–5.4 meq/L
Chloride (Cl^-)	98–108 meq/L
Bicarbonate (HCO_3^-)	21–27 meq/L

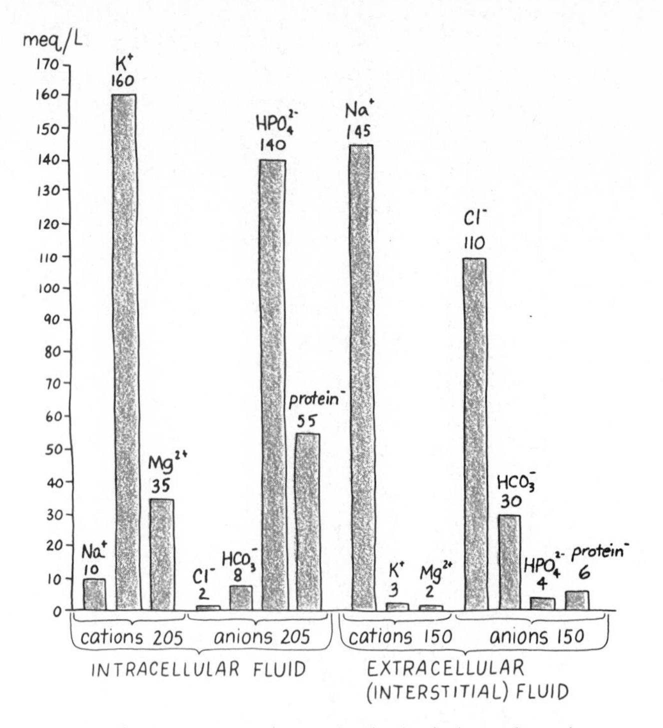

Figure 9.6 Concentrations of principal electrolytes in body fluids. The number of equivalents of positive ions (cations) is equal to the number of equivalents of negative ions (anions).

Figure 9.6 compares the electrolyte concentrations of intracellular and extracellular fluids.

Example 9.12 A typical concentration for potassium ion (K^+) is 5 meq/L of body fluid. How many grams of potassium ion are there in 1.0 L of body fluid?

Solution: Using the volume and the concentration (in milliequivalents per liter), we can find the number of milliequivalents in the 1.0 L of body fluid (1 eq = 1000 meq):

$$1.0\,\cancel{L} \times \frac{5\,\cancel{meq}}{1\,\cancel{L}} \times \frac{1\text{ eq}}{1000\,\cancel{meq}} = 0.005\text{ eq K}^+$$

The number of grams of K^+ is calculated by changing equivalents to moles, then moles to grams using the molar mass:

$$0.005\,\cancel{eq}\text{ K}^+ \times \underbrace{\frac{1\,\cancel{mol}\text{ K}^+}{1\,\cancel{eq}}}_{\text{Ionic charge}} \times \underbrace{\frac{39.1\text{ g K}^+}{1\,\cancel{mol}}}_{\text{Molar mass}} = 0.20\text{ g K}^+$$

HEALTH NOTE

Electrolytes in Parenteral Solutions

The use of a specific kind of parenteral solution depends on the nutritional, electrolyte, and fluid needs of the individual patient. Examples of various types of parenteral solutions are given in Table 9.3.

Conditions Affecting Electrolyte Levels

Certain diseases or conditions can cause electrolyte levels to become abnormal. Table 9.4 lists some of these conditions and the changes in electrolyte balance.

Table 9.3 Electrolyte Concentrations and Uses of Parenteral Solutions

Solution	Electrolytes (meq/L)	Cations (meq/L)	Anions (meq/L)	Use
Sodium chloride (0.9%)	Na^+ 154, Cl^- 154	154	154	Replaces fluid losses
Potassium chloride (5% dextrose)	K^+ 40, Cl^- 40	40	40	Treatment of malnutrition with low potassium levels
Ringer's solution	Na^+ 147, K^+ 4, Ca^{2+} 4, Cl^- 155	155	155	Replaces fluids and electrolytes lost through dehydration
Maintenance solution (5% dextrose)	Na^+ 40, K^+ 35, Cl^- 40, lactate$^-$ 20, HPO_4^{2-} 15	75	75	Maintains fluid and electrolyte levels
Replacement solution (extracellular)	Na^+ 140, K^+ 10, Ca^{2+} 5, Mg^{2+} 3, Cl^- 103, acetate$^-$ 47, citrate^{3-} 8	158	158	Replaces electrolytes of extracellular fluids

Table 9.4 Causes of Fluctuation in Levels of Electrolytes and Body Fluids[a]

Cause	Electrolytes					Body Fluids
	Na^+	K^+	Cl^-	H^+	HCO^{3-}	
Vomiting, gastric suction	↓	↓	↓	↓	↓	↓
Diuretic	↓	↓	↓	↓		↓
Diarrhea	↓	↓		↑	↓	↓
Diabetes	↓	↓		↑	↓	
Renal disease	↓	↓		↑		
Edema	↓					↑
Dehydration	↓	↓	↓			↓
Cerebrovascular injury	↑		↑			
Heart failure, congestive	↑		↑			↑
Digitalis medication	↓	↓	↓			↓
Sweating	↓		↓			↓
Burns	↓	↓	↓			↓
Surgery	↓	↓				↓

[a] ↑ increases; ↓ decreases.

9.6 Solutions, Colloids, and Suspensions

OBJECTIVE From the properties, identify a mixture as a true solution, a colloidal dispersion, or a suspension.

true solution
small particles
homogeneous
transparent

The size of the solute particles in a mixture plays an important role in determining the properties of that mixture. We have seen that in a *true solution* the solute dissolves to give very small particles—atoms, molecules, or ions. The particles of a true solution are dispersed uniformly throughout the solvent and do not settle out when the solution is stored. We say that a true solution is a *homogeneous* solution. When you observe a true solution, such as salt water, you cannot distinguish the solute from the solvent. The solution appears transparent even when a light shines through it.

Each particle of a true solution is so small that it will pass through filters and membranes such as cell walls. The diameter of a true solution particle is usually less than one nanometer (1 nm). Examples of true solutions include salt water and sugar water.

Colloids

colloids
large particles
usually homogeneous
stopped by membranes
Tyndall effect

The particles of *colloids* are larger than true solution particles, with diameters between 1 and 100 nm. They may be large molecules, such as proteins, or they may be aggregates of molecules, atoms, or ions.

Colloidal particles are evenly dispersed in a dispersing medium and usually do not settle out. They are homogeneous mixtures. Colloidal particles are still small enough to pass through filters, but too large to pass through membranes such as cell walls.

When a beam of light shines through a colloid, the particles scatter the light, and the beam of light is visible. The reflection of light by colloidal particles is called the *Tyndall effect*, shown in Figure 9.7. The path of a light beam can be seen in fog because of the Tyndall effect of the dust and water droplets in the air.

Types of Colloids

aerosols
foams
emulsions

There are several types of colloids which differ in colloidal particles and dispersing medium. *Aerosols* consist of liquid or solid particles dispersed in gas. Fog, clouds, and mist are aerosols containing liquid colloidal particles, whereas smoke is an aerosol consisting of fine, airborne solid particles.

Foams are mixtures of gases dispersed in liquid or solid. Shaving creams and whipped creams are foams in which a gas is dispersed in a liquid. Styrofoam is a foam consisting of a gas dispersed in a solid.

Emulsions are mixtures of a liquid dispersed in another liquid. Mayonnaise is an emulsion consisting of eggs dispersed in oil. When milk is not homogenized, the suspended particles of cream separate. The process of homogenization

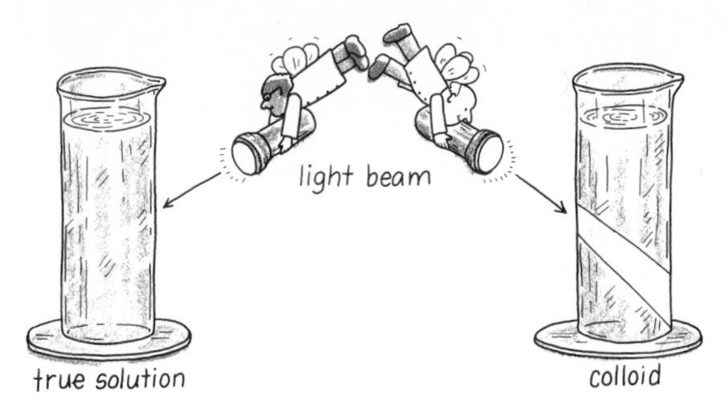

Figure 9.7 The Tyndall effect: Light is reflected by colloidal particles, making a light bream visible; true solution particles are too small and do not produce the Tyndall effect.

breaks up the cream particles to form colloids that remain dispersed within the solution. Table 9.5 lists several examples of colloids.

Table 9.5 Examples of Some Colloids

Example	Type	Colloidal Particles	Dispersion Medium
Mayonnaise, butter, homogenized milk, hand lotions	Emulsion	Liquid	Liquid
Cheese, gelatin	Gel	Liquid	Solid
Fog, clouds, sprays	Aerosol	Liquid	Gas
Dust, smoke	Aerosol	Solid	Gas
Shaving cream, whipped cream	Foam	Gas	Liquid
Styrofoam	Foam	Gas	Solid
Blood plasma, paints (latex)	Sol	Solid	Liquid
Cement, gemstones	Sol	Solid	Solid

Suspensions

suspension A *suspension* is a heterogeneous mixture that is very different from a true solution or a colloid. By comparison, the particles of a suspension are very large, with diameters greater than 100 nm. The suspended particles are visible and can be seen with the naked eye. A suspension particle is large enough to be trapped in a filter, so the components of a suspension can be separated by both filters and membranes.

The mass of a suspension particle causes it to settle out soon after mixing. If you stir muddy water, it quickly separates, and the liquid portion becomes clear as the suspension particles settle to the bottom. Waste treatment plants make use of the properties of suspensions in water purification. The addition of chemicals such as aluminum sulfate or ferric sulfate causes the formation of large particles that settle rapidly to the bottom of the treatment tanks. When the

Table 9.6 Characteristics of True Solutions, Colloids, and Suspensions

True Solutions	Colloids	Suspensions
	Type of particles	
Single atoms, ions, small molecules	Aggregates of atoms, molecules, or ions; macromolecules	Very large particles
	Size of particles	
0–1 nm	1–100 nm	Greater than 100 nm
	Examples	
Salt water	Mayonnaise, fog	Muddy water
	Effect of a beam of light	
Transparent	Beam is visible; opaque; Tyndall effect	Beam is visible; opaque; Tyndall effect
	Settling	
None	Very little	Settle rapidly
	Separation	
Cannot be separated by filters or membranes	Separated by membranes but not by filters	Can be separated by filters as well as by membranes

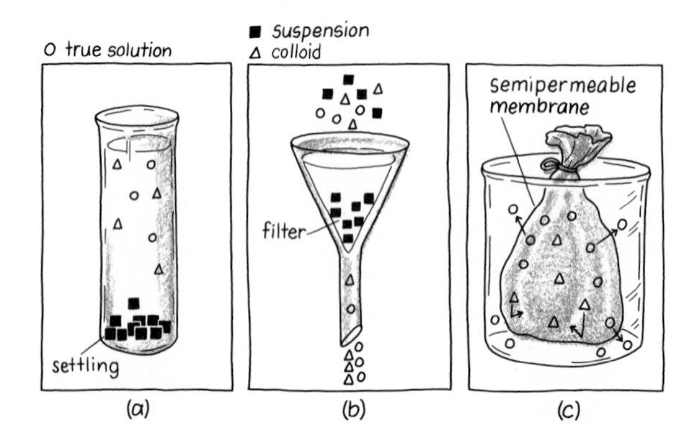

Figure 9.8 Comparison of properties of solutions: (a) Suspended particles settle out: (b) only suspended particles are removed by filters; (c) only true solution particles move through a semipermeable membrane and become separated from colloidal and suspended particles.

settling-out process is complete, the water layer is drawn off. You can also find suspensions among the medications in a hospital or in your medicine cabinet. These include Kaopectate, calamine lotion, antacid mixtures, and liquid penicillin. It is important to "shake well before using" in order to suspend the particles before taking a dose.

Table 9.6 summarizes the similarities and differences of true solutions, colloids, and suspensions. (See also Figure 9.8.)

Example 9.13 Classify the solution in the following descriptions as true solutions, colloids, or suspensions:

1. a liquid sample in which a beam of light becomes visible, but does not settle out
2. a sample that settles upon standing
3. a liquid sample whose particles pass through filters and membranes

Solution

1. Particles of colloids are evenly dispersed and do not settle out. A beam of light is visible when it is directed through the colloid.
2. The very large particles in a suspension will eventually settle out, forming a layer on the bottom and leaving a clear portion on top.
3. Particles of a true solution are small enough to pass through filters and membranes.

HEALTH NOTE

Separation of Colloidal and True Solutions in the Body

Membranes of the body separate colloids from true solutions. For example, the intestinal lining, a membrane, allows true solution particles to pass into the blood and lymph circulatory systems. However, the colloids from foods are too large to pass through the membrane and are retained in the intestinal tract. The process of digestion breaks down large colloidal food particles, such as starches and proteins, into true solution particles, such as glucose and amino acids, which enter the circulatory system for transport to the body cells for utilization. Certain foods, such as bran, a fiber, cannot be broken down by human digestive processes and remain in the intestinal tract until they are eliminated, thereby providing bulk for feces.

The membranes of the cells in the tissues and organs of the body also separate true solution particles from colloids. The nutrients for the cell, such as oxygen, water, amino acids, electrolytes, glucose, and minerals, all form true solutions, which can pass through the cellular membrane into the cell. The waste products from the cells, such as urea and carbon dioxide, are also true solutions and can leave the cell to be excreted.

9.7 Diffusion and Osmosis

Given two solutions separated by a semipermeable membrane, indicate the side with the greater osmotic pressure, the direction in which water will flow, and the compartment that increases in volume.

The movement of materials in a solution, up the stem of a plant and in and out of cells, depends on the process of *diffusion*. Diffusion takes place wherever there is a difference in concentration—a *concentration gradient*. Particles of solute and solvent are constantly moving from areas where their concentrations are high to areas of lower concentration. Eventually, the particles become evenly distributed, and there is no longer any concentration gradient. Figure 9.9 illustrates particles of a solute and a solvent undergoing the diffusion process in the formation of a solution.

Semipermeable Membrane

In the body, the various body fluids are separated by membranes that regulate the flow of water, nutrients, and waste products in and out of the cells. These membranes are called *semipermeable membranes*. A semipermeable membrane permits the passage of water molecules and small, dissolved solutes but blocks the passage of larger molecules.

Osmosis

Osmosis is a diffusion process that occurs in the cells of plants and animals. It occurs when water flows across a semipermeable membrane from a more dilute solution to a more concentrated solution. Figure 9.10 shows what happens in a system in which two solutions of different concentration are separated by a semipermeable membrane.

Compartment A contains a sucrose solution, and compartment B contains an equal volume of water. Osmosis occurs as water diffuses from compartment

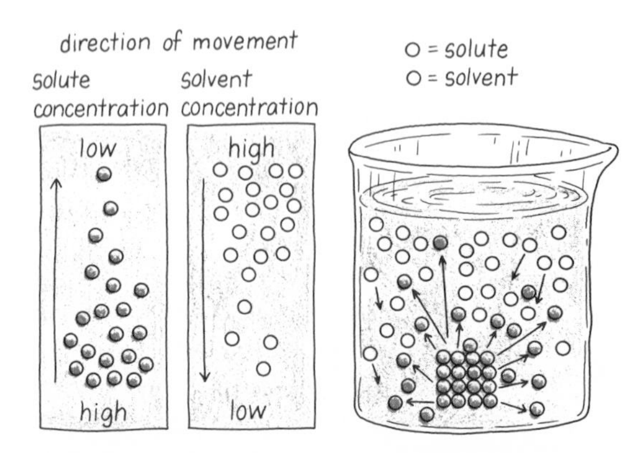

Figure 9.9 Effect of a concentration gradient on the process of solution. Particles diffuse from areas of high concentration to areas of low concentration.

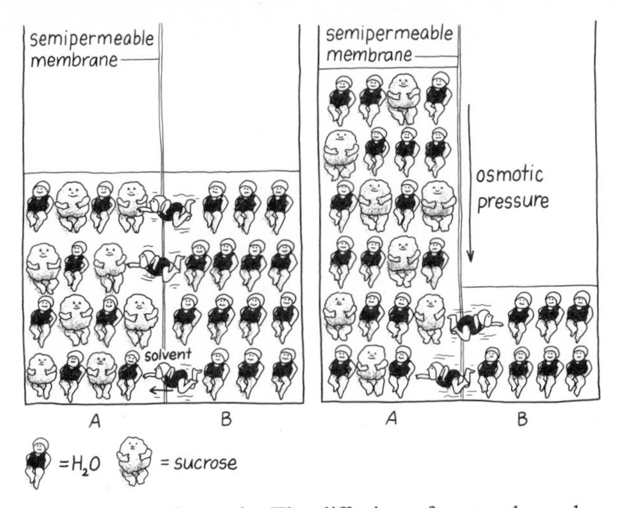

Figure 9.10 Osmosis: The diffusion of water through a semipermeable membrane toward the compartment with the higher solute concentration.

B into compartment A, since water moves through the membrane from an area of higher water concentration (more dilute) to an area where the concentration of water is lower. (The sucrose, a colloid, cannot pass through the membrane, even though a concentration gradient exists.)

The effect of the diffusion of water into compartment A is to dilute the sucrose solution. At the same time, the level of the fluid in compartment A rises as the water level in compartment B drops. The column of water that forms in compartment A causes a pressure that prevents any further increase in volume in compartment A. This pressure, called *osmotic pressure*, has the effect of squeezing or pushing water molecules out of compartment A and back into compartment B. At this point, the number of water molecules moving in and out of compartment A is equal, and a state of dynamic equilibrium exists. We can also define osmotic pressure as the pressure we would have to apply to prevent water from moving across the semipermeable membrane when there is a concentration difference.

osmotic pressure

Let us look at one more example of osmosis as shown in Figure 9.11. A semipermeable membrane separates a 4% starch mixture from a 10% starch mixture. Water diffuses by osmosis from the side with the 4% starch into the side with the 10% starch. Eventually, the concentrations of starch in both compartments will become equal at 7%. Osmosis is often defined as the movement of water in the direction of the greater solute concentration and therefore greater osmotic pressure—in this case, the 10% starch.

Osmotic pressure depends on the number of particles in the solution. Water, with no dissolved particles, has a zero osmotic pressure. If a solute is dissolved in water, the solution has an osmotic pressure. The greater the number of particles dissolved, the higher the osmotic pressure.

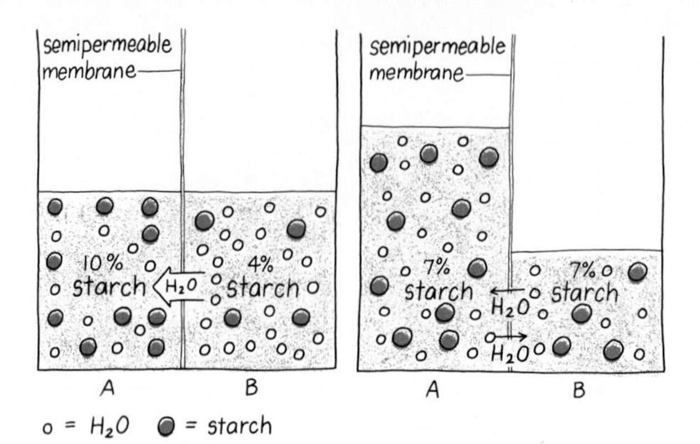

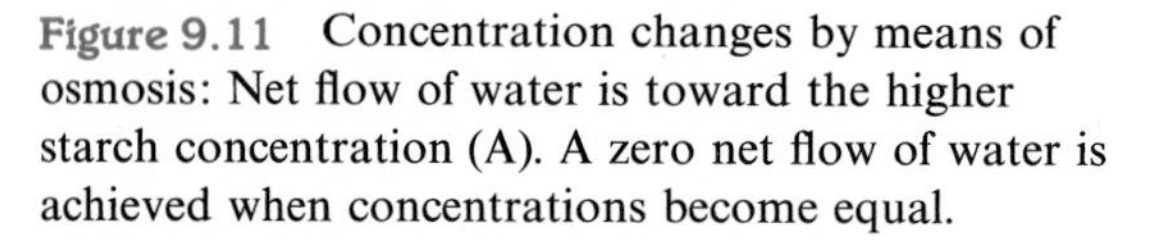

Figure 9.11 Concentration changes by means of osmosis: Net flow of water is toward the higher starch concentration (A). A zero net flow of water is achieved when concentrations become equal.

Example 9.14 Two aqueous solutions, a 2% sucrose solution and an 8% sucrose solution, are separated by a semipermeable membrane.

1. Which sucrose solution exerts the greater osmotic pressure?
2. What is the net movement of water during osmosis?
3. Which side will increase in solution volume?

Solution

1. The 8% sucrose solution has the higher concentration and therefore has more solute particles. It will exert the greater osmotic pressure.
2. The net flow of water, the solvent, in osmosis, will be away from the 2% solution and into the 8% solution.
3. The 8% solution will increase its solvent and therefore its volume as its solute concentration is diluted by osmosis.

Osmotic Pressure of Physiological Solutions

physiological solutions have osmotic pressures equal to body fluids

In the body, the osmotic pressure of the blood, tissue fluids, lymph, and plasma depends on the dissolved particles in each fluid. Solutions that are identical in osmotic pressure to those of body fluids are called *physiological solutions.* Physiological saline solution (0.9% NaCl) and a 5% glucose solution have the same osmotic pressure as body fluids. Both are used to replace fluids in the body or to carry other components into the body. Since the saline and glucose solutions have the same osmotic pressures as body fluids, they do not upset the diffusion of water between the various fluid compartments within the body.

Calculating the Osmotic Pressure of Physiological Solutions

osmole

Every mole of particles dissolved in a solution creates a unit of osmotic pressure called an *osmole* (abbreviated osmol). One mole of NaCl dissolved in 1 L of solution produces 1 mol of sodium ions and 1 mol of chloride ions to give a total of 2 mol of particles:

$$\begin{aligned} 1 \text{ mol NaCl} \xrightarrow{H_2O} & \ 1 \text{ mol Na}^+(aq) + 1 \text{ mol Cl}^-(aq) \\ & = 2 \text{ mol particles (ions)} \\ & = 2 \text{ osmol} \end{aligned}$$

osmolarity

The concentration of particles present in a solution is expressed in terms of *osmolarity*, which is the total number of moles of all the particles in a liter of solution. When 1 mol of NaCl dissolves in 1 L of solution, the osmolarity is 2 osmolar (2 osM):

$$\text{Osmolarity} = \frac{2 \text{ osmol}}{1 \text{ L}} = 2 \text{ osM}$$

osmolarity of a 0.9% saline solution

We can now calculate the osmolarity of the physiological solution of saline. First, we need to calculate the number of moles of NaCl in 1 L of 0.9% NaCl:

Grams of NaCl in 1 L

$$1 \cancel{\text{L}} \times \frac{100 \cancel{\text{mL}}}{1 \cancel{\text{L}}} \times \frac{0.9 \text{ g NaCl}}{100 \cancel{\text{mL}}} = 9 \text{ g NaCl}$$

Moles of NaCl

$$9 \cancel{\text{g}} \times \frac{1 \text{ mol NaCl}}{58.5 \cancel{\text{g}}} = 0.15 \text{ mol NaCl}$$

Osmoles of NaCl

$$0.15 \cancel{\text{mol NaCl}} \times \frac{2 \text{ osmol}}{1 \cancel{\text{mol NaCl}}} = 0.30 \text{ osmol}$$

Osmolarity

$$\frac{0.30 \text{ osmol NaCl}}{1 \text{ L}} = 0.30 \text{ osM}$$

Thus, the osmolarity of the 0.9% NaCl solution is 0.30 osM. The osmolarity of a 5% glucose solution can be calculated in the same way to give a similar osmolarity of 0.28 osM. The two physiological solutions are very close in osmolarity and exert about the same osmotic pressure as compared to that of the fluids in the body, which also have an osmolarity of 0.3 osM. Remember, we are not necessarily giving intravenous solutions with the same kinds of particles as the solutions within our cells, but we are using solutions with the same osmolarity or osmotic pressure as that existing within the cells of the body. (See Figure 9.12.)

HEALTH NOTE

Transport of Body Fluids

The capillary systems in the body are the most important portions of the circulatory system. At the capillaries, nutrients and waste products are exchanged with the tissues. The direction in which fluids and small particles flow depends on the pressures inside and outside the capillaries. These pressures change as the fluids in the blood and plasma move through the capillaries. The blood enters the arterial end of the capillaries as arterial blood and leaves the venular end as venous blood.

blood pressure is hydrostatic pressure
filtration

Pressures in the capillaries and tissues are of two types: (1) hydrostatic and (2) osmotic. *Hydrostatic pressure*, or blood pressure, is the force placed upon the blood by the pumping of the heart. This has the effect of pushing fluids and small molecules out of the capillaries, a process called *filtration*. The greater the force on a fluid, the greater its hydrostatic pressure. This is analogous to a coffee filter. When the filter is full of water, the liquid runs out rapidly. As the level of water in the filter lowers, the coffee drips more slowly.

osmotic pressure

The other type of pressure, *osmotic* or *oncotic pressure*, is a result of the colloidal particles in the blood plasma. These particles, such as red blood cells and proteins, are unable to permeate the capillary walls but affect the fluid transport by creating osmotic pressure within the capillary vessels.

The forces of hydrostatic pressure and osmotic pressure oppose each other. Hydrostatic forces "push" fluids out, and osmotic forces "pull" fluids in. The relationship of these pressures along the capillaries determines the net flow of fluid and solutes between the capillaries and the tissues. At the arterial end of the capillary, the hydrostatic pressure of the blood, 35 torr, is greater than the osmotic pressure, 25 torr, so fluid and nutrients are pushed into the tissues. At the venular end, however, the blood pressure is much lower, so the osmotic pressure, 25 torr, is now greater than the hydrostatic pressure, which has dropped to 15 torr. Now, the osmotic pressure controls the flow, and fluids with their waste products move back into the capillary. (See Figure 9.13.)

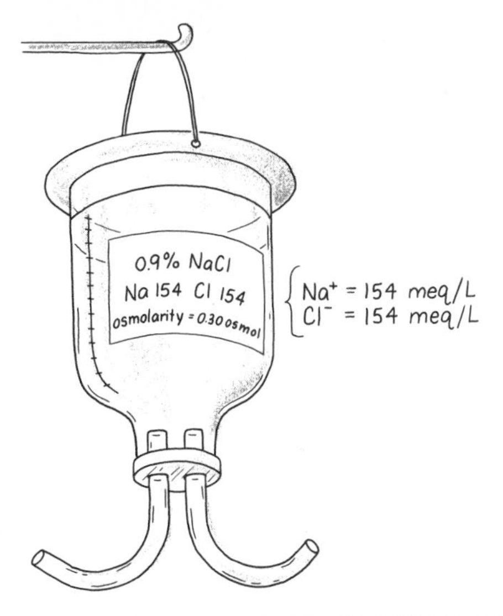

Figure 9.12 The label on an IV bottle gives percent concentration of solutes, concentration of electrolytes in meq/L, and the osmotic pressure of all dissolved particles in osmolarity.

In kidney disease, shock associated with injury, serious burns, or surgery, proteins may enter the tissues. The abnormal level of protein in the tissues decreases the osmotic pressure in the capillary and increases the osmotic pressure in the tissues. As a result, less fluid is returned to the capillary. The higher osmotic pressure of the tissues causes more fluid to flow into the tissues. When fluid collects in tissues, they swell, thereby creating a condition called *edema.*

Edema can also result from high blood pressure, when the hydrostatic pressure pushes additional fluids into the tissues. If the blood pressure remains high at the venular end, the osmotic pressure of the blood is not great enough to draw sufficient fluid back into the capillary. The net effect is the excess accumulation of fluid in the tissues—edema.

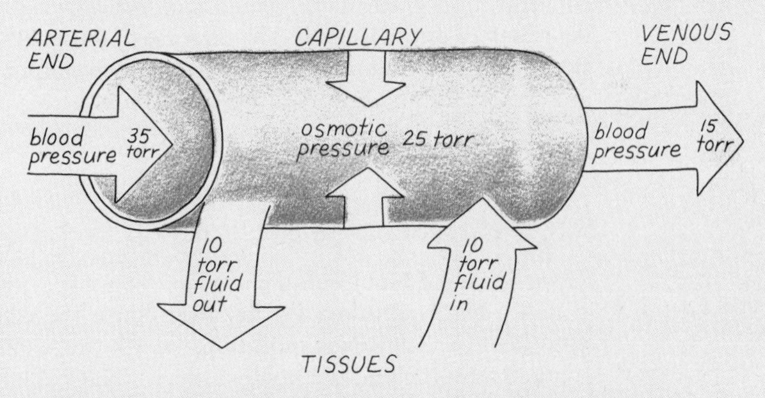

Figure 9.13 At the arterial end of a capillary, blood pressure (hydrostatic) exceeds osmotic pressure, and fluid is pushed out of the capillary into tissues. At the venous end, osmotic pressure exceeds blood pressure, and fluids from tissue return to the capillary.

9.8 Isotonic Solutions

OBJECTIVE Given the percent concentration for a solution, determine (1) whether that solution is isotonic, hypotonic, or hypertonic to a red blood cell and (2) whether a red blood cell in that solution would undergo hemolysis, crenation, or no change.

isotonic

When a cell is placed in a solution having an osmotic pressure equal to its own, the cell maintains its normal volume. We call such a solution an *isotonic solution* (*iso* means "equal to," and *tonic* refers to the biological osmotic pressure). Because an isotonic solution has the same osmotic pressure as the cells contained in it, there is no concentration gradient. The flow of water into the cell is equal to the flow of water out of the cell, and so the net flow of water is zero.

Generally, injections and other parenteral solutions are prepared from isotonic solutions such as physiological saline (0.9% NaCl) or a 5% glucose solution so that osmotic pressures and cellular volumes are not disturbed.

If a cell is placed in a solution that is not isotonic, concentration gradients are created. Water moves into and out of the cell at unequal rates, thereby changing osmotic pressures and cellular volumes.

Effect of Hypotonic Solutions on Red Blood Cells

hypotonic

hemolysis

Consider a red blood cell that has been placed in pure water. The water has an osmotic pressure lower than that of the cell, and so we say that the water (or any solution with an osmotic pressure lower than that of a cell) is *hypotonic* to the cell. *Hypo* means "lower than," and *tonic* again refers to the biological osmotic pressure of the cell. In this case, the solute concentration is greater within the cell; the cell has a higher osmotic pressure than the surrounding fluid. A concentration gradient is created, and osmosis occurs: More water flows into the cell than out of the cell. The increase of fluid within the cell causes the cell to swell in volume and possibly to burst. The swelling and bursting of a cell placed in a hypotonic solution is called *hemolysis*.

Effect of Hypertonic Solutions on Red Blood Cells

hypertonic

crenation

Suppose the red blood cell is placed in a solution with a higher osmotic pressure, in a 4% NaCl solution, for instance. Since the cell has an osmotic pressure equivalent to a 0.9% NaCl solution, the 4% NaCl solution has a greater osmotic pressure than the cell. We say that the 4% NaCl solution is *hypertonic* to the cell (*hyper* meaning "greater than"). In this case, the concentration gradient causes water to flow out of the cell into the surrounding hypertonic solution. The cell shrinks as fluid is lost, a process called *crenation*. Figure 9.14 shows how isotonic, hypotonic, and hypertonic solutions affect red blood cells.

The process of making pickles is another example illustrating the effect of hypertonic solutions. (See Figure 9.15.)

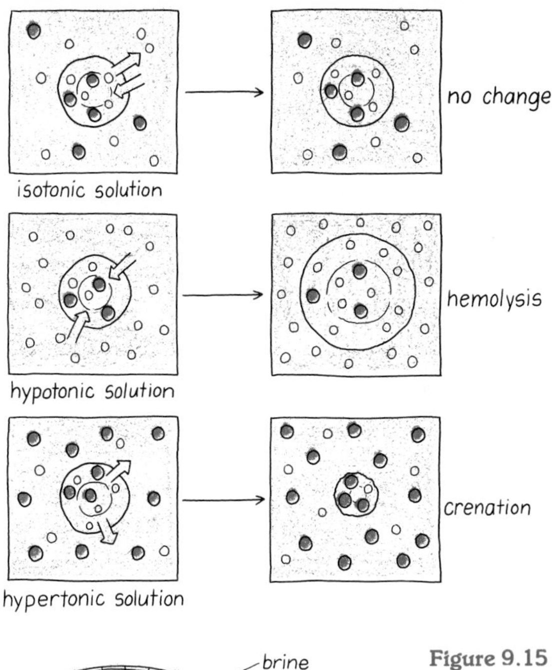

Figure 9.14 Effects of isotonic, hypotonic, and hypertonic solutions on red blood cells. In hypotonic solutions, red blood cells swell (hemolysis) because more water flows into the cell; in hypertonic solutions, more water leaves the red blood cell, causing it to shrink (crenation).

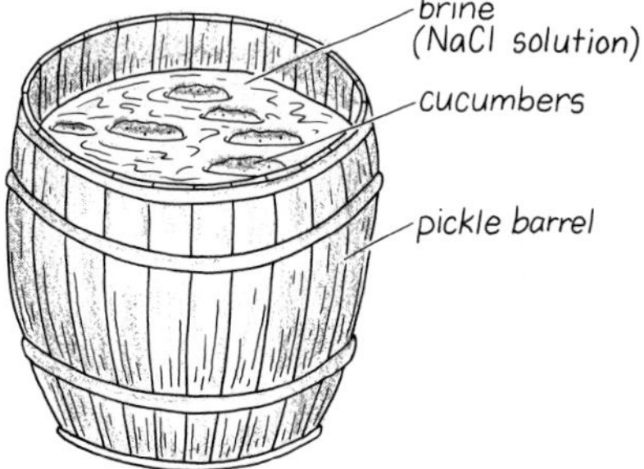

Figure 9.15 Why does a cucumber shrivel up and become a pickle in a strong brine solution?

Example 9.15 Indicate whether each of the following solutions is isotonic, hypotonic, or hypertonic to red blood cells, and state whether a red blood cell placed in the solution will undergo hemolysis, crenation, or no change:

1. a 5% glucose solution
2. a 0.2% NaCl solution
3. a 10% glucose solution

Solution

1. A 5% glucose solution is an isotonic solution because it has the same osmotic pressure as body fluids. A red blood cell will not undergo any change in an isotonic solution.
2. A 0.2% NaCl solution is a hypotonic solution and will cause a red blood cell to absorb water and undergo hemolysis.
3. A 10% glucose solution is hypertonic compared to body fluids. In this solution, a red blood will lose water and undergo crenation.

9.9 Dialysis

OBJECTIVE Given a substance in a dialysis bag, state whether it will dialyze through the semipermeable membrane.

Dialysis is the term used to describe the transport process whereby small molecules, electrolytes, and water molecules diffuse through a semipermeable membrane or *dialyzing membrane.* Colloidal particles are retained within the membrane.

Consider a cellophane bag (a dialyzing membrane) filled with a solution of NaCl, glucose, starch, and protein. The NaCl (in the form of Na^+ and Cl^- ions) and the glucose are small, true solution particles, whereas starch and protein are large, colloidal particles. When the cellophane bag is placed in a beaker of pure water, only the Na^+, Cl^-, and the molecules of glucose dialyze through the bag into the water. The cellophane bag is acting as a dialyzing membrane. (See Figure 9.16.) Since the concentration of NaCl and glucose is higher within the bag than outside it, a concentration gradient is created. The ions and small molecules move across the membrane, increasing the concentration outside the bag and decreasing the concentration inside. Eventually, the concentrations of NaCl and glucose inside and outside the bag become equal, and no further change occurs. The only way to continue to remove more NaCl or glucose from inside the cellophane container is to place the sample in a fresh quantity of pure water and reestablish a concentration gradient.

In addition to the NaCl and glucose undergoing dialysis, water is moving by osmosis into the cellophane bag. The colloidal particles of starch and protein remain within the cellophane bag. Dialysis is a way in which true solution particles may be separated from colloidal particles.

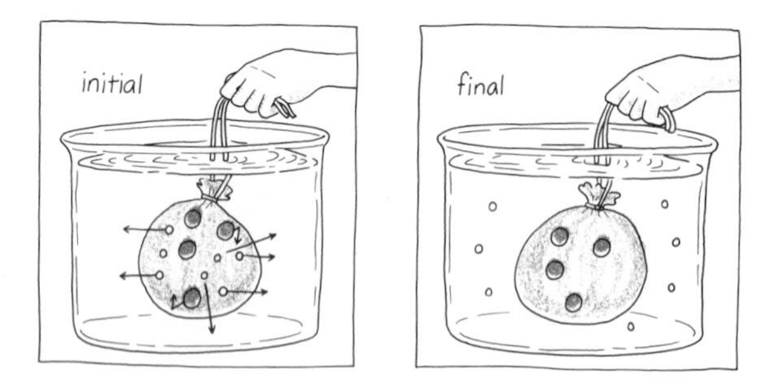

o true solution particle such as Na^+, Cl^-, glucose, urea
● colloidal particle such as protein, red blood cell

Figure 9.16 Dialysis: The separation of true solution particles from colloidal particles in water.

Example 9.16 KCl, starch, and a protein are placed in an aqueous solution in a dialysis bag. How will the concentration of each change when the dialysis bag is placed in distilled water?

Solution: KCl, as K^+ and Cl^-, dialyzes through the bag into the water. The concentration of KCl within the bag decreases. Eventually, the KCl concentration becomes equal inside and outside the bag. Starch and protein remain inside the dialysis bag as colloidal particles. Water flowing into the bag will dilute its contents, thereby lowering the concentration of the starch and protein.

Dialysis by the Kidneys

The kidneys play a major role in maintaining the concentrations of molecules and electrolytes and the volume of water in the blood. The working unit of the kidney is called the *nephron.* (See Figure 9.17.) The average adult with two normally functioning kidneys is estimated to have about 2 million nephron units. Each nephron is like a funnel. At the top of the funnel, there is a network of arterial capillaries called the *glomerulus.*

nephron

glomerulus

Blood flows into the glomerulus, where the hydrostastic pressure of the blood pushes water, small molecules, and ions—urea, amino acids, glucose, and electrolytes—through the capillary walls. The resulting solution, called a *filtrate,* enters Bowman's capsule, a double-walled membrane that surrounds the glomerulus. The filtrate moves through a long, convoluted tubule where several of the substances in the filtrate still of value to the body are reabsorbed by a capillary system surrounding the tubule. The amino acids, most of the water,

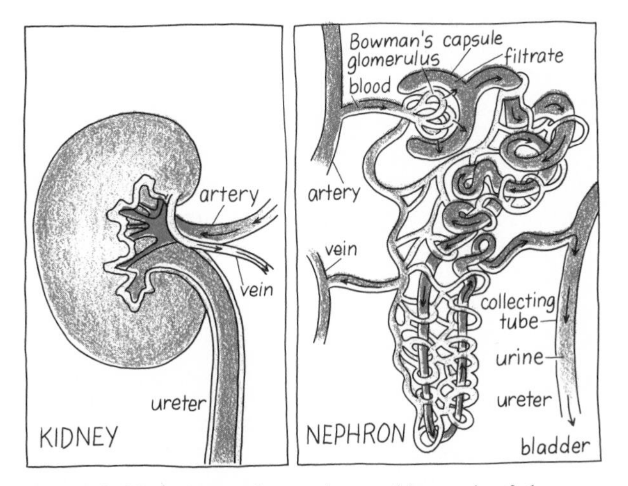

Figure 9.17 The nephron, the working unit of the kidney. Water and true solution particles diffuse out of the blood across the membrane of the glomerulus to form a filtrate. Most of the water, electrolytes, and usuable solutes are resorbed from filtrate, while urea and other waste products are retained and excreted in urine.

HEALTH NOTE

Hemodialysis

If the kidneys should fail to filter toxic waste products from the blood, these substances accumulate to a fatal level in a relatively short period of time. A patient with kidney failure may become a candidate for a kidney transplant or may be put on a dialysis unit called an *artificial kidney*. The cleansing of the blood by the artificial kidney is called *hemodialysis*. (See Figure 9.18.)

A typical artificial kidney machine contains a large tank filled with about 100 L of distilled water to which are added substances such as electrolytes, usually at physiological concentrations. In the center of this dialyzing bath, or *dialysate*, is a dialyzing coil. The dialyzing coil consists of cellulose tubing, which is the dialyzing membrane. The arterial blood from the arm or leg of the patient flows through the dialyzing coil, where it is dialyzed and then returned to the patient through a vein.

As the blood passes through the coil, the highly concentrated waste products move out of the blood. No blood is lost because the membrane is not permeable to large particles such as red blood cells. Electrolyte levels are measured as dialysis begins, and the dialysate may be adjusted to establish normal electrolyte levels. For example, if a patient is retaining high levels of potassium in the blood, the dialysate is prepared without any potassium ion. The concentration gradient between the potassium in the blood and the dialysate causes potassium to dialyze out of the blood. On the other hand, a patient may be losing too much potassium. In this case, potassium is added to the dialysate, so potassium dialyzes into the patient's blood. Adjustments in the concentrations of the dialysate may be made throughout dialysis. The purpose of dialysis is to establish normal solute concentrations in the blood and to remove toxic waste products.

Dialysis patients do not usually produce very much urine. As a result, they retain large amounts of water between dialysis treatments. In fact, the intake of fluids

normally all the glucose, and certain electrolytes are reabsorbed and returned to the tissues. The primary waste product, urea, is retained within the filtrate, which is collected and stored as urine in the bladder.

Control of Osmotic Pressure by the Kidneys

It is estimated that about 99 percent of the water in the filtrate is reabsorbed and the remaining 1 percent is retained to form urine. This reabsorption of water is sensitive to the osmotic pressure of the blood, which in turn regulates the secretion of an antidiuretic hormone (called ADH, or vasopressin), which regulates the absorption of water by the kidneys.

If the osmotic pressure of the blood increases, the secretion of ADH also increases, and reabsorption of water by the kidneys increases. As the increased absorption of water occurs, the solute concentration of the blood decreases by dilution, and the osmotic pressure drops. When the osmotic pressure of the

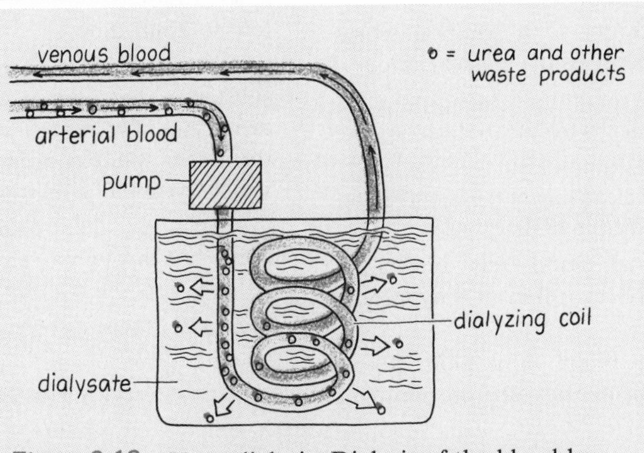

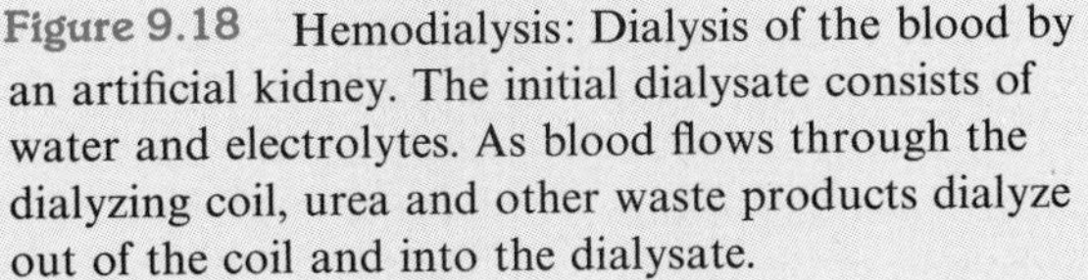

Figure 9.18 Hemodialysis: Dialysis of the blood by an artificial kidney. The initial dialysate consists of water and electrolytes. As blood flows through the dialyzing coil, urea and other waste products dialyze out of the coil and into the dialysate.

by a dialysis patient may be restricted to as little as a few teaspoons of water a day. By increasing the pressure of the blood as it circulates through the dialyzing coil, the hydrostatic pressure is increased, and more water can be squeezed out of the blood. For some patients, 2–10 L of water may be removed from the blood during one dialysis treatment. Dialysis patients have from two to three treatments a week, each treatment requiring about 5–7 h. Some of the newer treatments require less time.

blood drops, there is a decrease in the secretion of ADH, and less water is reabsorbed by the kidneys. The solute concentration of the blood increases, and so does the osmotic pressure. At the same time, there is a greater amount of water in the urine, and urine output increases.

Glossary

colloidal particles Solute particles that are very large in solution. They will pass through filters but will not pass through semipermeable membranes.

colloids Solutions containing colloidal particles.

concentration A measure of the quantity of solute dissolved in a specified amount of solution.

concentration gradient A gradient created by a difference in the concentrations of solutes; solute particles diffuse from high solute concentration to low solute concentration. If the solute is blocked, water will flow into the compartment of greater solute concentration.

crenation The shriveling of a cell due to water flowing out when the cell is placed in a hypertonic solution.

dialysis A process in which water and true solution particles move through a semipermeable membrane from solutions of high concentration to low concentration.

diffusion Movement of a substance caused by a difference in concentration.

dilution The addition of water to a solution which increases the volume and decreases the concentration of solute.

equivalent The amount of an electrolyte that carries one mole of electrical charge.

hemodialysis A mechanical cleansing of the blood by an artifical kidney using the principle of dialysis.

hemolysis A bursting of red blood cells due to an increase in water flowing into the cell from a hypotonic solution.

hydrostatic pressure The pressure of a fluid. Blood pressure is a hydrostatic pressure caused by the pumping action of the heart.

hypertonic A solution that has a higher osmotic pressure (greater solute concentration) than the cells of the body.

hypotonic A solution that has a lower osmotic pressure (lower solute concentration) than the cells of the body.

isotonic A solution that has the same osmotic pressure (equal solute concentration) as the cells of the body.

molarity The moles of solute in 1 L of solution.

osmosis The movement of water through a semipermeable membrane from its area of high concentration to low concentration.

osmotic pressure The pressure created by a dissolved substance that cannot pass through a semipermeable membrane. The flow of water into the compartment with the higher solute concentration builds a column of fluid that prevents a further increase in volume.

parenteral solution Fluids that are not administered orally but through a vein or muscle.

percent concentration (weight per volume) The mass of solute, in grams, dissolved in each 100 mL of solution.

semipermeable membrane A membrane that allows the passage of certain substances while retaining others.

suspension A mixture in which the particles are so large that they settle out. Suspensions will not pass through filters or semipermeable membranes.

true solutions Solutions in which the solute is in the form of small particles. True solutions will pass through filters and semipermeable membranes.

Tyndall effect The cloudiness that appears when a beam of light shines through a colloidal dispersion.

Problems

Percent Concentration (Objective 9.1)

9.1 Calculate the percent concentration (w/v) of the following solutions:
a. 2.0 g of sucrose in 100 mL of solution
b. 20.0 g of KCl in 400 mL of solution
c. 75.0 g of Na_2SO_4 in 0.50 L of solution
d. 0.300 kg of glucose in 5.00 L of solution
e. 4.0 g of $CaCl_2$ in 20 mL of solution

Calculations Using Percent (w/v) Concentration (Objective 9.2)

9.2 Calculate the number of grams of solute needed to prepare the following solutions:
a. 50 mL of a 5% (w/v) KCl solution
b. 400 mL of a 1% (w/v) NaCl solution
c. 5.0 L of a 4.0% (w/v) NH_4Cl solution
d. 250 mL of a 10% (w/v) glucose solution

9.3 How many grams of glucose are obtained by a patient when the patient has received 2.0 L of a 5% (w/v) glucose solution?

9.4 A patient receives 100 mL of 20% (w/v) mannitol solution every hour.
a. How many grams of mannitol are given in 1 h?
b. How many grams of mannitol does the patient receive in 1 day?

9.5 Calculate the volume (in milliliters) of solution needed to provide the given mass of solute from the following solutions:
a. 150 g of glucose from a 5% (w/v) glucose solution
b. 400 g of NaOH from a 10% (w/v) NaOH solution
c. 50 g of NaCl from a 2.0% (w/v) NaCl solution
d. 2.0 g of KBr from a 8.0% (w/v) solution

9.6 A patient needs 100 g of glucose in the next 12 h. How many liters of a 5% (w/v) glucose solution should be given?

Dilutions (Objective 9.3)

9.7 Calculate the final volume for the following dilutions:
a. a 1:10 dilution of 10 mL of a 5% (w/v) KCl solution
b. a 1:5 dilution of 200 mL of a 20% (w/v) mannitol solution
c. a 1:2 dilution of 400 mL of a 10% (w/v) NaCl solution

9.8 Calculate the new concentrations of the diluted solutions in Problem 9.7.

Molarity (Objective 9.4)

9.9 Calculate the molarity (M) of the following solutions:
a. 4.0 mol of KOH in 2.0 L of solution
b. 2.0 mol of glucose in 4.0 L of solution
c. 5.0 mol of NaOH in 500 mL of solution
d. 0.10 mol of NaCl in 40 mL of solution

9.10 Calculate the molarity (M) of the following solutions:
a. 36.5 g of HCl in 1.0 L of solution
b. 8.0 g of NaOH in 200 mL of solution
c. 320 g of glucose, $C_6H_{12}O_6$, in 500 mL of solution

9.11 Calculate the number of moles provided by the following volumes of solution:
a. 1.0 L of a 3.0-M NaCl solution
b. 5.0 L of a 2.0-M $CaCl_2$ solution
c. 200 mL of a 4.0-M glucose solution
d. 50 mL of a 10-M sucrose solution

9.12 Calculate the grams of solution in each of the following solutions:
a. 1.0 L of a 1.0-M NaOH solution
b. 4.0 L of a 2.0-M KCl solution
c. 500 mL of a 1.0-M NaCl solution
d. 200 mL of a 6-M NaOH solution
e. 750 mL of a 1.0-M glucose ($C_6H_{12}O_6$) solution

9.13 What volume (in liters) will contain the following amounts of solute?
a. 2.0 mol of NaOH from a 2.0-M NaOH solution
b. 10 mol of NaCl from a 1-M NaCl solution
c. 80.0 g of NaOH from a 1-M NaOH solution
d. 1.0 mol of glucose from a 2.0-M glucose solution

Equivalents (Objective 9.5)

9.14 A physiological saline solution contains 154 meq/L Na^+ and 154 meq/L Cl^-. Calculate the grams of Na^+ and the grams of Cl^- in 1 L of the solution.

9.15 A Ringer's solution contains the following cations (in milliequivalents per liter): Na^+ 147, K^+ 4, and Ca^{2+} 4. If Cl^- is the only anion in the solution, what is its concentration in milliequivalents per liter?

Types of Solution (Objective 9.6)

9.16 Identify the following as characteristic of a true solution, colloidal solution, or suspension:
a. The solution is clear and cannot be separated by filters or membranes.
b. The solution appears cloudy when a beam of light is passed through it.
c. Particles of this solution remain inside a semipermeable membrane.
d. The solution settles out upon standing.
e. The solute in this solution can be separated by filtering.
f. A beam of light does not appear opaque in this solution.
g. The particles of solute in this solution are very large and are visible in the solvent.

9.17 Two solutions, a 5% (w/v) starch solution and a 1% (w/v) starch solution, are separated by a semipermeable membrane.
a. Which solution has the greater osmotic pressure?
b. In which direction will water flow initially?
c. Which compartment will increase in volume?

9.18 Two solutions, a 0.1% (w/v) albumin solution and a 2% (w/v) albumin solution, are separated by a semipermeable membrane. (Albumins are proteins that are colloids.)
a. Which solution has the greater osmotic pressure?
b. In which direction will water flow initially?
c. Which compartment will increase in volume?

9.19 How many osmoles are contained in 1 mol of each of the following?
a. glucose, a nonelectrolyte
b. NaCl
c. $CaCl_2$
d. urea, a nonelectrolyte

Isotonic Solutions (Objective 9.8)

9.20 Consider the following solutions: water, 1% (w/v) glucose, 4% (w/v) NaCl, 5% (w/v) glucose, 0.05% (w/v) NaCl, 0.9% (w/v) NaCl, and 10% (w/v) glucose. Compared to red blood cells, which of these solutions would
a. be hypotonic?
b. be hypertonic?
c. be isotonic?
d. cause crenation?
e. cause hemolysis?
f. cause no change?

9.21 If the red blood cells and body fluids have an osmolarity of 0.30 osmol/L, state whether the following solutions would be hypotonic, isotonic, or hypertonic:
a. 5% (w/v) glucose
b. 0.50 osmol/L NaCl
c. 1.00 osmol/L KCl
d. 0.10 osmol/L glucose
e. 0.30 osmol/L NaCl

Dialysis (Objective 9.9)

9.22 Each of the following solutions is placed inside a dialyzing bag, and the bag is immersed in distilled water. State which components will dialyze through the bag into the distilled water.

a. starch and NaCl
b. albumin (colloidal protein), KCl (salt), and glucose
c. urea (true solution) and NaCl

9.23 A patient on dialysis has a high level of urea, a high level of sodium, and a low level of potassium in the blood. Why is the dialysate prepared with a high level of potassium but no sodium and urea?

Acids and Bases

Objectives

10.1 Write an equation for the ionization of an acid in water.

10.2 Write an equation for the dissociation of a base in water.

10.3 Write the concentrations of H^+ and OH^- in pure water; calculate the value of the $[H^+]$ or $[OH^-]$ in an acidic or basic solution.

10.4 Identify a pH value as acidic, neutral, or basic.

10.5 Determine the pH, hydrogen ion concentration, or hydroxide ion concentration of a solution.

10.6 Complete and balance an equation for the neutralization of an acid and a base.

10.7 Calculate the molarity of an acidic solution, given the volume and molarity of the basic solution used in the titration.

10.8 Identify the components of a buffer and the role of each component in maintaining the pH of a solution.

Scope

A lemon tastes sour, and too much vinegar on a salad is unpleasant. Both the lemon and vinegar contain *acids*, which taste sour. Antacids, including milk of magnesia, taste bitter and metallic because they contain compounds called *bases*. Solutions that do not have the properties of acids or bases are neutral. Water is a neutral liquid.

When a solution has acidic properties, hydrogen ions predominate in the solution. A basic solution has a prevalence of hydroxide ions. Body fluids, including blood and urine, have very specific levels of hydrogen ions. The measurement of the concentration of hydrogen ions is called the pH of the solution. The pH of body fluids, including blood and urine, is regulated primarily by the lungs and the kidneys. Major changes in the pH of the body fluids can severely affect biological activities within the cells.

10.1 Acids

OBJECTIVE Write an equation for the ionization of an acid in water.

Acids are substances that release hydrogen ions (H^+) when they dissolve in water. For example, the compound HCl ionizes in water to form hydrogen ions (H^+) and chloride ions (Cl^-):

$$\underset{\text{Hydrogen chloride}}{HCl(g)} \xrightarrow{H_2O} \underset{\text{Hydrogen ion}}{H^+(aq)} + \underset{\text{Chloride ion}}{Cl^-(aq)}$$

hydronium ions

Actually, the hydrogen ion does not exist by itself in water; it combines with water molecules (hydration). The positively charged ions that form are called *hydronium ions*. They are represented by the formula H_3O^+. You will often see a hydrogen ion (H^+) written for convenience, but it is the hydronium ion that is present in acidic solutions (see Figure 10.1):

$$HCl(g) + H_2O(l) \longrightarrow H_3O^+(aq) + Cl^-(aq)$$

Properties of Acids

There are several characteristics that help identify acids:

1. Acids supply hydrogen ions in water.
2. Acids have a sour taste.
3. Acids turn blue litmus (a vegetable dye) red.

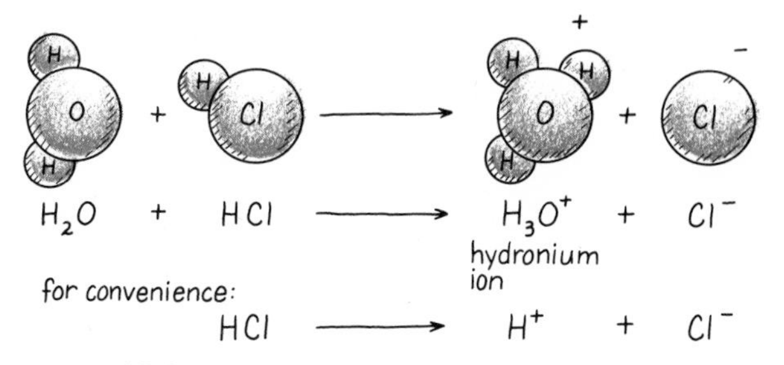

Figure 10.1 Hydrochloric acid ionizes in water to produce a hydrogen ion (H^+) and a chloride ion (Cl^-). The hydrogen ion is hydrated to give a hydronium ion (H_3O^+).

4. Acids behave as electrolytes in solution.
5. Acids neutralize solutions containing hydroxide ions (OH^-).
6. Acids react with several metals, releasing H_2 gas.
7. Acids react with carbonates, releasing CO_2 gas.

Strong Acids

A *strong acid* is a strong electrolyte and ionizes almost completely in water to produce a relatively large number of hydrogen ions. There are not very many strong acids. The three important strong acids you will need to remember are HCl, HNO_3, and H_2SO_4:

Hydrochloric acid: $HCl \xrightarrow{H_2O} H^+(aq) + Cl^-(aq)$

Nitric acid: $HNO_3 \xrightarrow{H_2O} H^+(aq) + NO_3^-(aq)$

Sulfuric acid: $H_2SO_4 \xrightarrow{H_2O} H^+(aq) + HSO_4^-(aq)$

It is important to be aware of the concentrations of strong acids you are using in the laboratory or hospital. At high concentrations, there is a sufficient concentration of hydrogen ions to severely burn the skin and damage the eyes. If you spill an acid, be sure to immediately dilute it with water. This lowers the concentration of hydrogen ions and lessens the damage it causes. Low concentrations of strong acids are safer to work with than the high concentrations.

Example 10.1 Write an equation for the ionization of the strong acid HNO_3 in water.

Solution: A strong acid ionizes completely to form hydrogen ions (H^+) and an anion:

$$\underset{}{HNO_3} \xrightarrow{H_2O} \underset{\text{Hydrogen ion}}{H^+(aq)} + \underset{\text{Nitrate ion}}{NO_3^-(aq)}$$

Weak Acids

A weak acid is a weak electrolyte and dissolves in water mainly as molecules. Only a few, about 1–2 percent, of the dissolved molecules ionize to produce hydrogen ions and anions. Most acids in nature are weak acids. Even at high concentrations, weak acids produce relatively few hydrogen ions. They are not nearly as damaging to the skin and eyes. In fact, you use many weak acids, such as lemon juice and vinegar (a 5 percent acetic acid solution), in the home. Carbonic acid, H_2CO_3, is a weak acid used in the preparation of carbonated soft drinks. The sour taste in some fruits and vegetables is due to the presence of weak acids such as citric acid.

The equations for the ionization of some important weak acids use a double arrow to indicate that both the molecular and ionic forms of the acid are present in solution. Acetic acid is often represented by the abbreviation HAc:

Abbreviation	*Formula*	*Name*
HAc	$HC_2H_3O_2$	Acetic acid
Ac^-	$C_2H_3O_2^-$	Acetate ion (anion)

Some important weak acids are listed below:

Acetic acid: $$HAc \xrightleftharpoons{H_2O} H^+(aq) + Ac^-(aq)$$

Carbonic acid: $$H_2CO_3 \xrightleftharpoons{H_2O} H^+(aq) + HCO_3^-(aq)$$

Phosphoric acid: $$H_3PO_4 \xrightleftharpoons{H_2O} H^+(aq) + H_2PO_4^-(aq)$$

10.2 Bases

OBJECTIVE Write an equation for the dissociation of a base in water.

A base or alkali is a substance that causes an increase in hydroxide ions (OH^-) when dissolved in water. This means that a base is also an electrolyte because ions, one of which is a hydroxide ion, are produced in aqueous solution. For example, sodium hydroxide (NaOH) is a base because it releases sodium ions (Na^+), which are metallic cations, along with hydroxide ions (OH^-):

$$\underset{\text{Sodium hydroxide}}{NaOH(s)} \xrightarrow{H_2O} \underset{\text{Sodium ion}}{Na^+(aq)} + \underset{\text{Hydroxide ion}}{OH^-(aq)}$$

Properties of Bases

There are several properties that we associate with bases:

1. Bases release, or cause an increase in, hydroxide ions (OH^-) in aqueous solutions.
2. Bases taste bitter.
3. Bases have a slippery, soapy feeling.
4. Bases turn red litmus (a vegetable dye) blue.
5. Bases are electrolytes in aqueous solution.
6. Bases neutralize solutions containing hydrogen ions.

Example 10.2 Identify each of the following solutions as characteristic of acids, bases, or both:

1. contain hydroxide ions
2. taste sour
3. turn blue litmus red
4. feel slippery or soapy

Solution

1. Bases contain hydroxide ions.
2. Acids taste sour.
3. Acids turn blue litmus red.
4. Bases feel slippery or soapy.

Strong Bases

A strong base is a strong electrolyte and is completely ionized into metallic ions and hydroxide ions in aqueous solutions. Two strong bases, NaOH and KOH, are very soluble and produce solutions with many hydroxide ions. At high concentrations, hydroxide ions are very damaging to the skin and eyes. Strong bases such as NaOH (also known as "lye") are used in household products to dissolve grease in ovens and to clean drains. The use of such products in the home should be carefully supervised.

Some important strong bases are

Sodium hydroxide: $NaOH(s) \xrightarrow{H_2O} Na^+(aq) + OH^-(aq)$

Potassium hydroxide: $KOH(s) \xrightarrow{H_2O} K^+(aq) + OH^-(aq)$

Barium hydroxide: $Ba(OH)_2 \xrightarrow{H_2O} Ba^{2+}(aq) + 2OH^-(aq)$

The hydroxides of Group IIA are not very soluble. Therefore, the concentration of hydroxide ions produced in solution by these bases is also quite small. For example, $Mg(OH)_2$ is used in the antacid milk of magnesia, taken to counteract the effects of too much acid in the stomach.

Example 10.3 Write an equation for the dissociation of the strong base KOH in water.

Solution: A strong base dissociates in water, providing metallic cations (K^+) and hydroxide ions (OH^-):

$$KOH(s) \xrightarrow{H_2O} K^+(aq) + OH^-(aq)$$

Weak Bases

Weak bases, which are weak electrolytes, do not typically contain hydroxide ions but react with water to produce a small number of hydroxide ions. Ammonia (NH_3) is a base according to the Brønsted–Lowry theory because its lone pair of electrons can accept a proton from water. The hydroxide ion that is left from the water molecule makes the solution basic. The shorter arrow pointing toward the products' side of the equation indicates that only a few of the dissolved ammonia molecules react to give ammonium and hydroxide ions:

$$\underset{\text{Ammonia}}{NH_3(g)} + \underset{\text{Water}}{H_2O(l)} \rightleftharpoons \underset{\text{Ammonium ion}}{NH_4^+(aq)} + \underset{\text{Hydroxide ion}}{OH^-(aq)}$$

10.3 Ionization of Water

OBJECTIVE Write the concentrations of H^+ and OH^- in pure water; calculate the value of the $[H^+]$ or $[OH^-]$ in an acidic or basic solution.

We usually consider pure water as a nonelectrolyte that contains only molecules of water. However, if we were to measure very carefully, we would find that there are a few water molecules that do ionize. One water molecule in 10 million ionizes to produce a hydrogen ion (H^+) and a hydroxide ion (OH^-). When water molecules ionize, a proton is transferred from one water molecule to another (see Figure 10.2):

$$H_2O + H_2O \longrightarrow H_3O^+ + OH^-$$

or

$$H_2O \longrightarrow H^+ + OH^-$$

Ion Product for Water

In pure water at 25°C, the concentration of H^+ is 1×10^{-7} M, which is equal to the concentration of the hydroxide ion, OH^-. When the H^+ and the OH^-

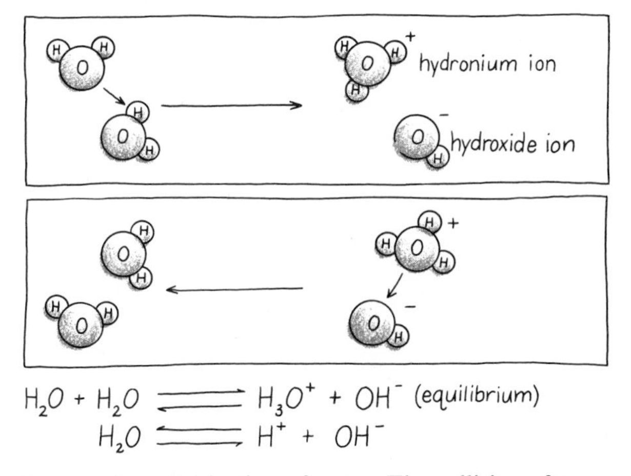

Figure 10.2 Ionization of water. The collision of two water molecules transfers a proton to form a hydronium ion (H_3O^+) and a hydroxide ion (OH^-). The reverse reaction produces two water molecules.

concentrations are the same (equal) and the solution shows no acidic or basic properties, we say that the water is *neutral*. The molar concentrations of these ions are indicated by using square brackets:

for pure water, $[H^+] = [OH^-]$

$$[H^+] = 1 \times 10^{-7}\ M$$
$$[OH^-] = 1 \times 10^{-7}\ M$$

The product of the H^+ and OH^- concentrations gives a value called the *ion product* for water, which is 1×10^{-14}:

ion product for water

$$\text{Ion product} = [H^+][OH^-]$$
$$= [1 \times 10^{-7}][1 \times 10^{-7}]$$
$$= 1 \times 10^{-14}$$

The ion product is constant for pure water as well as for any other kind of aqueous solution, acidic or basic. However, in acids or bases, the concentrations of H^+ and OH^- are no longer equal, even though their ion product $[H^+][OH^-]$ is still 1×10^{-14}.

For example, if an acid is added to water, the hydrogen ion concentration will increase and the hydroxide ion will decrease. In acids, the hydrogen ion concentration will be greater than the hydroxide ion concentration. For example, if the $[H^+]$ is $1 \times 10^{-4}\ M$, the $[OH^-]$ is $1 \times 10^{-10}\ M$, so that the ion product, $[H^+][OH^-]$, still equals $1 \times 10^{-14}\ M$.

for acids, $[H^+] > [OH^-]$

When a base is added to water, the hydroxide ion becomes greater than the hydrogen ion concentration. A basic solution with an $[OH^-]$ of $1 \times 10^{-2}\ M$ has a $[H^+]$ of $1 \times 10^{-12}\ M$. Table 10.1 gives some examples of ion product in acidic, neutral, and basic solutions. You can calculate the concentration of one of the ions in an acid or a base if you know the concentration of the other by using the ion product of water, 1×10^{-14}.

bases $[H^+] < [OH^-]$

Table 10.1 Examples of Ion Product in Acidic, Neutral, and Basic Solutions

	$[H^+]$	$\times\ [OH^-]$	= Ion Product
Acidic solutions, $[H^+] > [OH^-]$	$1 \times 10^{-2}M$ $1 \times 10^{-5}M$	$\times\ 1 \times 10^{-12}M$ $\times\ 1 \times 10^{-9}M$	$= 1 \times 10^{-14}$ $= 1 \times 10^{-14}$
Neutral solutions, $[H^+] = [OH^-]$	$1 \times 10^{-7}M$	$\times\ 1 \times 10^{-7}M$	$= 1 \times 10^{-14}$
Basic solutions, $[H^+] < [OH^-]$	$1 \times 10^{-8}M$ $1 \times 10^{-11}M$	$\times\ 1 \times 10^{-6}M$ $\times\ 1 \times 10^{-3}M$	$= 1 \times 10^{-14}$ $= 1 \times 10^{-14}$

Example 10.4 What is the $[H^+]$ in a basic solution where the $[OH^-]$ is 1×10^{-4} *M*?

Solution: Using the ion product of water and the given $[OH^-]$, we can solve for the $[H^+]$:

$$[H^+][OH^-] = 1 \times 10^{-14}$$

Rearranging the expression for $[H^+]$, we obtain

$$[H^+] = \frac{1 \times 10^{-14}}{[OH^-]}$$

$$= \frac{1 \times 10^{-14}}{1 \times 10^{-4}}$$

$$= 1 \times 10^{-10}\ M$$

The division of exponential numbers is done by subtracting the denominator exponent from the numerator exponent:

$$\frac{1 \times 10^{-14}}{1 \times 10^{-4}} = 1 \times 10^{(-14)-(-4)} = 1 \times 10^{(-14+4)} = 1 \times 10^{-10}$$

See Appendix B.

Example 10.5 What is the $[OH^-]$ in an acidic solution when the $[H^+]$ is 1×10^{-1}?

Solution: The ion product for water is

$$[H^+][OH^-] = 1 \times 10^{-14}$$

Rearranging the expression for $[OH^-]$ gives

$$[OH^-] = \frac{1 \times 10^{-14}}{[H^+]}$$

Substituting in the given value of $[H^+]$ completes the calculation:

$$[OH^-] = \frac{1 \times 10^{-14}}{1 \times 10^{-1}}$$

$$= 1 \times 10^{-13}\ M$$

10.4 The pH Scale

OBJECTIVE Identify a pH value as acidic, neutral, or basic.

The acidity or basicity of a solution is indicated by its pH. The pH scale is a way of describing the $[H^+]$; this scale ranges from 0 to 14. Low pH values, 0–7, correspond to acidic solutions. Water, a neutral solution, has a pH of exactly 7.00, the midpoint of the pH scale. The pH values 7–14 correspond to basic solutions. (See Figure 10.3.) Table 10.2 lists the pH values for some common solutions.

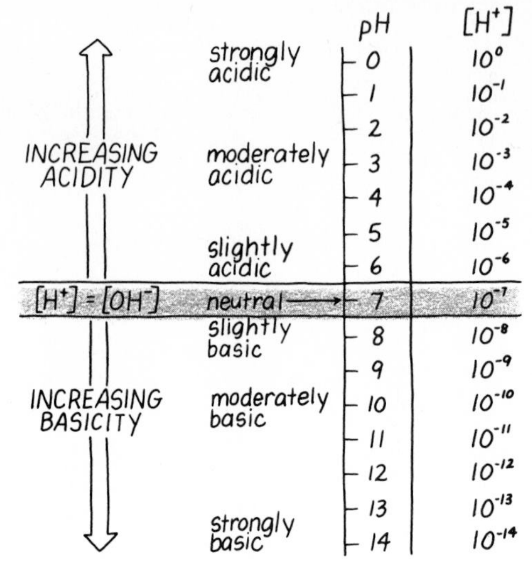

Figure 10.3 The pH scale ranges from values of 0 to 14. Values less than 7 are acidic, a value of 7 is neutral, and values above 7 are basic.

Table 10.2 Some Typical pH Values

Solution	pH
1.0-*M* HCl	0
Gastric juice	1.6
Lemon juice	2.2
Vinegar	2.8
Carbonated drinks	3.0
Coffee	5.0
Urine	6.0
Water, pure	7.0
Blood	7.4
Bile	8.0
Detergents	8.0–9.0
Milk of magnesia	10.5
Ammonia	11.0
Bleach	12.0
1.0-*M* NaOH (lye)	14.0

Example 10.6 Classify the following solutions as acidic, basic, or neutral:

1. rainwater, pH 4.5
2. saliva, pH 6.6
3. antacid, pH 8.5

Solution

1. The rainwater is acidic because a pH of 4.5 is between 0 and 7 on the pH scale, which is in the acidic range.
2. The saliva is acidic because a pH value of 6.6 is below pH 7.
3. The antacid is basic. A pH of 8.5 is between 7 and 14, which is the basic range of the pH scale.

HEALTH NOTE

Stomach Acid, HCl

The gastric glands of the stomach secrete an HCl solution, which is isotonic with body fluids. The pH of the solution is 0.8, which means that it is highly concentrated. The sight, smell, thoughts, or taste of food begins the stimulation of gastric secretions. Once food enters the stomach, there are additional secretions of gastric juices which continue while the food remains in the stomach. In one day, a person may secrete as much as 2000 mL of gastric juice.

The presence of HCl in the stomach activates the enzyme pepsin for the digestion of proteins in the stomach. The secretion of acid in the stomach continues until a pH of 2 is reached. Further secretions of gastric juice are blocked. This inhibitory effect keeps the pH at the optimum levels for activation of the digestion enzymes and aids in preventing ulceration of the stomach lining. Normally, the stomach is protected from acid and enzyme damage by the secretion of large quantities of a viscous mucus that serves as a protective barrier for the stomach lining during the digestion processes.

10.5 Calculation of pH

OBJECTIVE Determine the pH, hydrogen ion concentration, or hydroxide ion concentration of a solution.

The pH is related to the hydrogen ion concentration, $[H^+]$, of a solution. In mathematical terms, pH is defined as "the negative logarithm of the hydrogen ion concentration":

$$pH = -\log [H^+]$$

Our calculations will only concern concentrations that have 1 as a coefficient, such as 1×10^{-2} *M* or 1×10^{-8} *M*, although it is common to have concentrations such as 3.2×10^{-4} *M* or 8.4×10^{-10} *M*. Then the negative logarithm, or pH, is simply the exponent without the negative sign:

$[H^+]$	pH
1×10^{-②} →	2
1×10^{-8}	8

Table 10.3 Relationship Between $[H^+]$, $[OH^-]$, and pH Value

pH	$[H^+]$	$[OH^-]$
0	1×10^{0} *M*	1×10^{-14} *M*
2	1×10^{-2} *M*	1×10^{-12} *M*
4	1×10^{-4} *M*	1×10^{-10} *M*
7	1×10^{-7} *M*	1×10^{-7} *M*
8	1×10^{-8} *M*	1×10^{-6} *M*
11	1×10^{-11} *M*	1×10^{-3} *M*
14	1×10^{-14} *M*	1×10^{0} *M*

The hydrogen ion concentration always determines the pH of a solution. This is true even in basic solutions, where the hydrogen ion concentration is lower than the hydroxide ion concentration. This means that a very basic solution has very few hydrogen ions. Table 10.3 gives the relationship between some $[H^+]$, $[OH^-]$, and pH values.

Example 10.7 Calculate the pH of the following solutions:

1. $[H^+] = 1 \times 10^{-5}$ *M*
2. $[OH^-] = 1 \times 10^{-2}$ *M*

Solution

1. Since the $[H^+]$ is given, we can calculate the pH of the solution directly:

$$\text{pH of } 1 \times 10^{-5} = 5$$

2. Since hydroxide ion concentration $[OH^-] = 1 \times 10^{-2}$ *M* is given, we have to first calculate the corresponding $[H^+]$. We use the relationship of $[H^+]$ and $[OH^-]$ in the ion product of water:

$$[H^+][OH^-] = 1 \times 10^{-14}$$

$$[H^+] = \frac{1 \times 10^{-14}}{[OH^-]}$$

$$= \frac{1 \times 10^{-14}}{1 \times 10^{-2}}$$

$$= 1 \times 10^{-12}$$

Now we can calculate the pH from the above value for the $[H^+]$ in the solution:

$$\text{pH of } 1 \times 10^{-12} = 12$$

Example 10.8 Determine the $[H^+]$ for solutions with the following pH values:

1. 3
2. 10

Solution: The pH of a solution is related to the exponent of the hydrogen ion concentration. Placing a negative sign in front of the pH value gives the correct exponent for the hydrogen ion concentration:

$$[H^+] = 1 \times 10^{-pH}$$

1. $[H^+] = 1 \times 10^{-3}\ M$ pH
2. $[H^+] = 1 \times 10^{-10}\ M$

10.6 Neutralization Reactions

OBJECTIVE Complete and balance an equation for the neutralization of an acid and a base.

Neutralization is a reaction of an acid and a base to form a salt and water. In the neutralization reaction, the H^+ from the acid combines with an OH^- from the base to form a molecule of water. We say that the H^+ and the OH^- have neutralized each other:

$$H^+ + OH^- \longrightarrow H_2O$$

Let us mix equal amounts of hydrochloric acid and sodium hydroxide. The H^+ from the HCl molecules and the OH^- from the NaOH molecules neutralize each other, leaving the ions Na^+ and Cl^-, a salt:

$$\underset{\text{Acid}}{HCl} + \underset{\text{Base}}{NaOH} \longrightarrow \underset{\text{Water}}{H_2O} + \underset{\text{Salt}}{NaCl}$$

The ionic equation can be written

$$H^+ + \cancel{Cl^-} + \cancel{Na^+} + OH^- \longrightarrow H_2O + \cancel{Na^+} + \cancel{Cl^-}$$

Since the Na^+ and Cl^- are unchanged by the reaction, they can be omitted from the *net ionic equation*:

$$H^+ + OH^- \longrightarrow H_2O$$

Balancing an Acid–Base Neutralization Equation

Consider the neutralization of sulfuric acid, H_2SO_4, and sodium hydroxide, NaOH. Since the products of neutralization of an acid and a base are a salt and water, we can write the following unbalanced equation:

$$\underset{\text{Acid}}{H_2SO_4} + \underset{\text{Base}}{NaOH} \longrightarrow \underset{\text{Salt}}{\text{Salt}} + \underset{\text{Water}}{H_2O}$$

The H_2O is the result of the combination of the H^+ from H_2SO_4 and the OH^- from NaOH. Since the H_2SO_4 produces two H^+, we place a coefficient 2 in front of the NaOH to provide an equal number of OH^- from the base. The reaction of $2H^+$ and $2OH^-$ will result in the formation of $2H_2O$:

$$H_2SO_4 + 2NaOH \longrightarrow \text{Salt} + 2H_2O$$

HEALTH NOTE

Antacids

Antacids are basic substances used to neutralize the effects of excess stomach acid. Several antacids such as Di-Gel, Maalox, Mylanta, and milk of magnesia use mixtures of aluminum hydroxide and magnesium hydroxide:

$$\underset{\text{Antacid}}{Al(OH)_3} + \underset{\text{Stomach acid}}{3HCl} \longrightarrow AlCl_3 + 3H_2O$$

$$\underset{\text{Antacid}}{Mg(OH)_2} + \underset{\text{Stomach acid}}{2HCl} \longrightarrow MgCl_2 + 2H_2O$$

Use of aluminum hydroxide has the side effect of causing constipation. It also binds phosphate in the intestinal tract, leading to weakness and anorexia in the patient. Magnesium hydroxide has the effect of causing diarrhea. These side effects are less likely when a combination of the antacids is used.

Some antacids such as Pepto-Bismol and Tums use calcium carbonate:

$$\underset{\text{Antacid}}{CaCO_3} + \underset{\text{Stomach acid}}{2HCl} \longrightarrow CaCl_2 + H_2O + CO_2(g)$$

The calcium is absorbed into the bloodstream, where it elevates the levels of serum calcium. The calcium ion can also stimulate the secretion of stomach acid and is not recommended for patients with peptic ulcers.

Still other antacids such as Alka-Seltzer and Bromo Seltzer include the acid-absorbing compound sodium bicarbonate:

$$NaHCO_3 + HCl \longrightarrow NaCl + CO_2 + H_2O$$

This antacid has a tendency to increase blood pH and elevate sodium levels in the body fluids. It is also not recommended in the treatment of peptic ulcers.

The salt results from the remaining ions, $2Na^+$ and SO_4^{2-}. The correct formula for the salt of sodium and sulfate would be Na_2SO_4. Since the salt actually exists as ions in solution, we could obtain solid Na_2SO_4 by evaporation of the water. Placing the formula for the salt in our equation, we can write the balanced equation:

$$H_2SO_4 + 2NaOH \longrightarrow Na_2SO_4 + 2H_2O$$

Table 10.4 lists some typical acid–base reactions.

Table 10.4 Neutralization Reactions of Acids and Bases

Acid	+ Base	$\longrightarrow$ Water	+ Salt
HCl	+ KOH	$\longrightarrow$ H_2O	+ KCl
$2HNO_3$	+ $Ca(OH)_2$	$\longrightarrow$ $2H_2O$	+ $Ca(NO_3)_2$
H_3PO_4	+ $3NaOH$	$\longrightarrow$ $3H_2O$	+ Na_3PO_4

Example 10.9 Complete and balance the following equation for the neutralization of HCl and $Ba(OH)_2$:

$$\_\_ HCl + \_\_ Ba(OH)_2 \longrightarrow \_\_\_\_\_ + \_\_\_\_\_$$

Solution: In a neutralization reaction an acid (HCl) and a base [$Ba(OH)_2$] react to form a salt and water:

$$HCl + Ba(OH)_2 \longrightarrow \text{Salt} + H_2O$$

Since the base, $Ba(OH)_2$, produces $2OH^-$, we need to place a coefficient 2 in front of the HCl to provide the $2H^+$ for neutralization. The reaction will also produce $2H_2O$:

$$2HCl + Ba(OH)_2 \longrightarrow \text{Salt} + 2H_2O$$

The salt will be composed of the Ba^{2+} from the base and $2Cl^-$ from the acid. The correct formula would be $BaCl_2$:

$$2HCl + Ba(OH)_2 \longrightarrow BaCl_2 + 2H_2O$$

10.7 Titration

OBJECTIVE Calculate the molarity of an acidic solution, given the volume and molarity of the basic solution used in the titration.

Titration is a laboratory procedure in which a base is added to an acid solution until neutralization is complete. The molarity of the acid solution can be calculated from the volume and molarity of the base that was added in the titration.

Suppose we wish to find the molarity of a 10.0-mL sample of HCl. We would place the sample in a flask and add several drops of an indicator, usually phenolphthalein. In the acidic solution, this indicator is colorless. We would use a buret to slowly add NaOH (base) of known molarity, for example, 1.00 *M*. We are going to neutralize the acid with the base:

$$HCl + NaOH \longrightarrow NaCl + H_2O$$

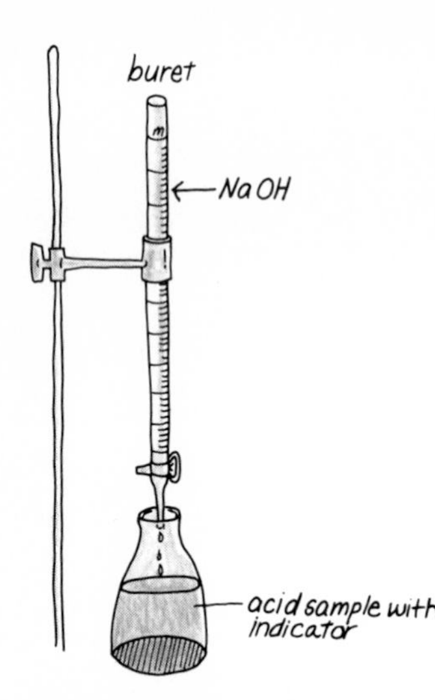

Figure 10.4 The titration of an acid. The acid sample in the flask is titrated by the base in the buret.

To completely neutralize the acid with the base, we need to add an equal number of moles of NaOH. The way we know that neutralization is complete is by a color change in the indicator. The indicator in our example will turn a faint pink upon adding a drop of the NaOH solution. At this point, a slight excess of NaOH causes the solution to be slightly basic, and the phenolphthalein is pink in basic solution. Now we read the buret to see what volume of NaOH was added. In this analysis, a total of 20.0 mL of NaOH was used. We can calculate the number of moles of NaOH that were used in the titration (see Figure 10.4):

$$20.0\ \cancel{\text{mL}} \times \frac{1\ \cancel{\text{L}}}{1000\ \cancel{\text{mL}}} \times \frac{1.0\ \text{mol}}{1\ \cancel{\text{L}}} = 0.020\ \text{mol NaOH}$$

From the equation for the neutralization, we can determine that 1 mol of NaOH will neutralize 1 mol of HCl. This can be expressed as a conversion factor:

$$\frac{1\ \text{mol NaOH}}{1\ \text{mol HCl}} \quad \text{or} \quad \frac{1\ \text{mol HCl}}{1\ \text{mol NaOH}}$$

Using the known amount of NaOH (0.20 mol) and the mole–mole conversion factor, we can set up the calculation:

$$0.020\ \cancel{\text{mol NaOH}} \times \frac{1\ \text{mol HCl}}{1\ \cancel{\text{mol NaOH}}} = 0.020\ \text{mol HCl}$$

This means that the sample in the flask contains 0.020 mol of HCl. The molarity of the acid sample can now be calculated using the expression for molarity. The volume of 10.0 mL of HCl would be 0.010 L:

$$\text{Molarity} = \frac{\text{moles}}{\text{liter}}$$

$$= \frac{0.020\ \text{mol HCl}}{0.010\ \text{L}}$$

$$= 2.0\text{-}M\ \text{HCl}$$

Therefore, the molarity of the acid sample in the flask is 2.0 M.

Example 10.10 A 10.0-mL sample of H_2SO_4 is placed in a flask. A total of 40.0 mL of 0.10 M NaOH was used to titrate the acid. What is the molarity of the acid? Note the reacting quantities of base and acid in the balanced equation:

$$H_2SO_4 + 2NaOH \longrightarrow Na_2SO_4 + 2H_2O$$

Solution: First, let us place our information in tabular form:

	H_2SO_4	NaOH
Volume	10.0 mL	40.0 mL
Molarity	?	0.10 M

The number of moles of NaOH can be calculated from the volume and molarity of the NaOH:

$$\text{Moles of NaOH} = 40.0\ \text{mL} \times \frac{1\ \text{L}}{1000\ \text{mL}} \times \frac{0.10\ \text{mol}}{1\ \text{L NaOH}}$$

$$= 0.0040\ \text{mol NaOH}$$

From the equation, we find that it takes 2 mol of NaOH to titrate 1 mol of H_2SO_4. This can be used as a conversion factor to calculate the number of moles of H_2SO_4 in the sample:

$$\text{Moles of } H_2SO_4 = 0.0040\ \text{mol NaOH} \times \frac{1\ \text{mol } H_2SO_4}{2\ \text{mol NaOH}}$$

$$= 0.0020\ \text{mol } H_2SO_4$$

Finally, the molarity (concentration) of the acid can be calculated from the volume of the sample and the number of moles of H_2SO_4:

$$\text{Molarity} = \frac{\text{moles}}{\text{liter}}$$

$$= \frac{0.0020\ \text{mol}}{0.010\ \text{L}}$$

$$= 0.20\text{-}M\ H_2SO_4$$

10.8 Buffers

OBJECTIVE Identify the components of a buffer and the role of each component in maintaining the pH of a solution.

A buffer solution contains components that allow the solution to keep a specific pH value. The pH of water changes greatly when an acid or base is added. However, adding an acid or base to a buffer solution does not appreciably change its pH. The function of a buffer is to keep the pH fairly constant by preventing changes in the hydrogen ion concentration, $[H^+]$. A buffer system is operating continuously to maintain a pH of 7.4 (7.35–7.45) in the blood. If the blood pH varies even slightly, it can cause such drastic effects on respiration and metabolism that death may occur. Even though we are constantly eating foods and juices that are acidic or basic, the blood buffer system acts to absorb the acid or base and leaves the $[H^+]$ unchanged.

In order to be a buffer, a solution must contain substances that can neutralize an acid or base added to the solution. These substances that make up a buffer can be a weak acid and its salt or a weak base and its salt.

For example, a mixture of acetic acid (HAc) and sodium acetate (NaAc) would act as a buffer. Since acetic acid is a weak acid, only a small amount of it ionizes:

$$\text{HAc} \rightleftharpoons H^+ + Ac^-$$

However, the buffer system needs more acetate ions than the weak acid provides by itself, so we help the system out by adding more acetate ions, Ac^-, in the form of a salt, NaAc. The salt NaAc dissociates into ions, Na^+ and Ac^-, when it dissolves. Now the system has sufficient amounts of the Ac^- ion for buffering action.

Action of a Buffer with Hydroxide Ions (OH^-)

Suppose some hydroxide ions from a base enter the buffer system. The weak acid (HAc) part of the buffer will react with the OH^-:

$$HAc + OH^- \longrightarrow H_2O + Ac^- \text{ (neutralizes } OH^-\text{)}$$

The $[H^+]$ is unchanged and the pH of the solution remains the same.

Action of a Buffer with Hydrogen Ions (H^+)

The other part of the buffer, the acetate ion (Ac^-), is capable of reacting with hydrogen ions (H^+). If extra hydrogen ions enter the buffer system, the Ac^-

HEALTH NOTE

Carbonic Acid–Bicarbonate, a Blood Buffer System

In the blood plasma, a pH close to 7.4 is maintained primarily by a carbonic acid–bicarbonate buffer system. This blood buffer consists of carbonic acid (H_2CO_3), a weak acid, and bicarbonate ions (HCO_3^-), the salt of the weak acid.

Blood Buffer System

$$\underset{\text{(weak acid)}}{\underset{\text{Carbonic acid}}{H_2CO_3}} \rightleftharpoons \underset{\text{Hydrogen ion}}{H^+} + \underset{\text{(salt of weak acid)}}{\underset{\text{Bicarbonate ion}}{HCO_3^-}}$$

buffer neutralizes OH^-

Should the blood become too basic and the pH rise above 7.4, the carbonic acid molecules will neutralize the excess hydroxide ions (OH^-):

$$\underset{\text{Carbonic acid}}{H_2CO_3} + \underset{\text{Excess hydroxide ion}}{OH^-} \longrightarrow \underset{\text{Water}}{H_2O} + \underset{\text{Bicarbonate ion}}{HCO_3^-}$$

In other cases, such as certain metabolic diseases, the kidneys fail to remove excess hydrogen ions from the blood; thus the pH of the blood falls below 7.4. Then the bicarbonate ion of the blood buffer combines with the overabundant hydrogen ions (see Figure 10.5):

combines with the H^+ to form HAc molecules. The H^+ concentration is preserved, and the pH of the solution is not altered:

$$H^+ + Ac^- \longrightarrow HAc \qquad \text{(removes excess } H^+\text{)}$$

In summary, when OH^- enters a buffer system, it is neutralized by the weak acid part of the buffer. On the other hand, if H^+ enters the buffer solution, the Ac^- from the salt part of the buffer ties up the H^+ to form HAc molecules. Both parts of a buffer act to preserve the H^+ concentration of the solution. Therefore, the pH of the solution is not altered and remains constant:

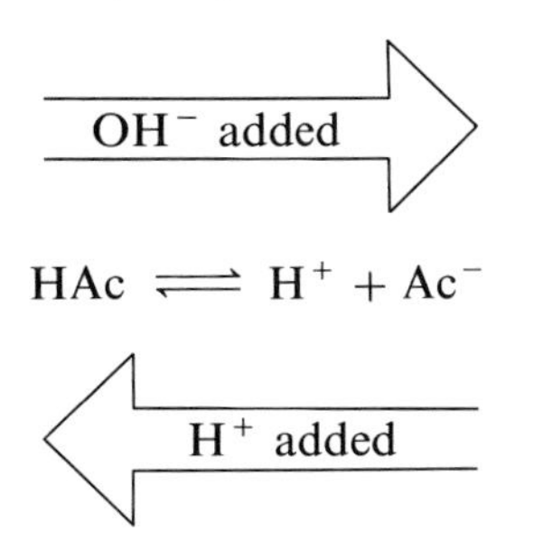

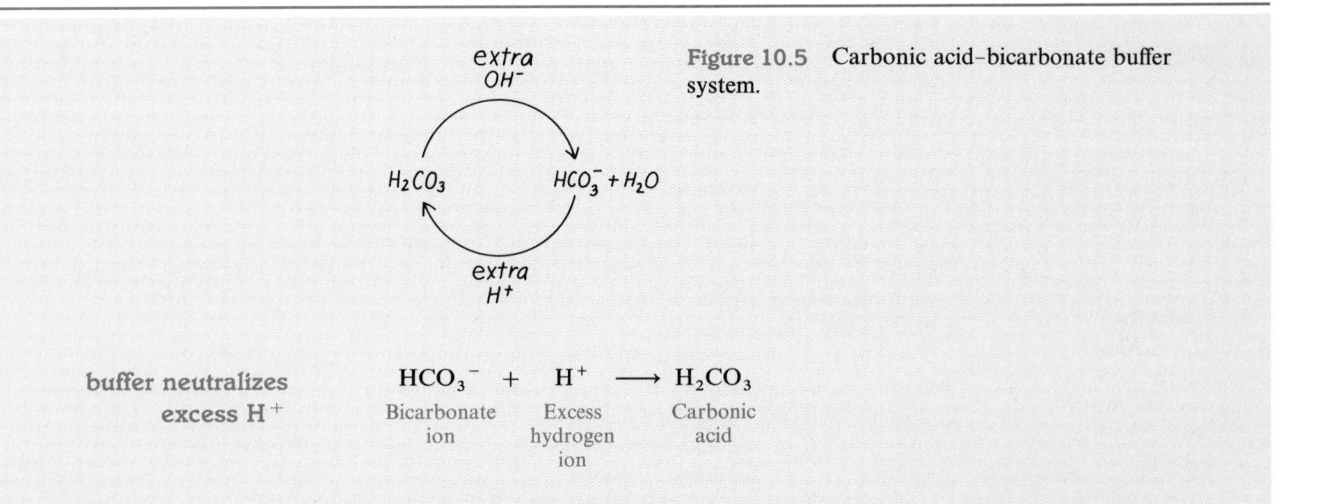

Figure 10.5 Carbonic acid–bicarbonate buffer system.

buffer neutralizes excess H^+

$$\underset{\text{Bicarbonate ion}}{HCO_3^-} + \underset{\text{Excess hydrogen ion}}{H^+} \longrightarrow \underset{\text{Carbonic acid}}{H_2CO_3}$$

Control of pH in the Blood

In the cells of your body, carbon dioxide (CO_2) is continuously produced as an end product of cellular metabolism. Some CO_2 is carried to the lungs for elimination, and the rest dissolves in the fluids of the body. The dissolved CO_2 combines with water to form carbonic acid, H_2CO_3, the weak acid of the blood buffer. The gas tension of CO_2, P_{CO_2}, regulates the level of carbonic acid, which in turn affects the pH of the blood:

$$\underset{\text{Carbon dioxide}}{CO_2} + \underset{\text{Water}}{H_2O} \rightleftharpoons \underset{\text{Carbonic acid}}{H_2CO_3} \rightleftharpoons \underset{\text{Hydrogen ion}}{H^+} + \underset{\text{Bicarbonate ion}}{HCO_3^-}$$

Table 10.5 lists the normal values for the components of the blood buffer in arterial blood.

Table 10.5 Values for Blood Buffer in Arterial Blood

P_{CO_2}	40 torr
H_2CO_3	1.2 meq/L of plasma
HCO_3^-	24 meq/L of plasma
pH	7.4

When the CO_2 level deviates from normal, the pH of the blood also changes. If the body retains CO_2, then the P_{CO_2} also increases. The body retains CO_2 when the gas cannot properly diffuse across the lung membrane, as happens in emphysema, or when the activity of the respiratory center in the medulla of the brain is affected by an accident or by depressive drugs. Poor ventilation or below-normal ventilation (called *hypoventilation*) may lead to a retention of CO_2 in the blood:

$$CO_2 + H_2O \rightleftharpoons H_2CO_3 \rightleftharpoons H^+ + HCO_3^-$$

Increases ⇒ Increases ⇒ Increases

The increase in CO_2 in the blood causes more carbonic acid to form, which leads to an increase in the H^+ concentration. This causes the blood pH to drop below 7.4, a condition called *acidosis*. When the origin of acidosis is respiratory, the condition is called *respiratory acidosis*.

acidosis

The P_{CO_2} can be lowered by *hyperventilation* (rapid breathing), which may be brought on by excitement, trauma, or high temperatures. To compensate for the lowered CO_2 level, some carbonic acid is converted to CO_2 and H_2O. This change causes a drop in the hydrogen ion concentration, and the pH goes above 7.4, a condition called *respiratory alkalosis*:

alkalosis

$$CO_2 + H_2O \rightleftharpoons H_2CO_3 \rightleftharpoons H^+ + HCO_3^-$$

Decreases ⇐ Decreases ⇐ Decreases

pH control by lungs and kidneys

The blood pH is regulated by the lungs and the kidneys. The level of CO_2 in the blood depends on the respiratory processes. The kidneys also control H^+ concentration by excreting or retaining H^+ in the urine. The pH of urine varies considerably and can range from 4.5 to 8.2. In severe conditions, even these mechanisms fail, and critical changes in blood pH occur. Table 10.6 lists some of the conditions that can lead to change in the blood pH and their possible treatment.

Table 10.6 Acidosis and Alkalosis: Symptoms, Causes, and Treatments

Respiratory Acidosis: CO_2 ⇧ pH ⇩	
Symptoms:	Failure to ventilate; suppression of breathing. Disorientation, weakness, coma.
Causes:	Lung disease blocking gas diffusion, e.g., emphysema, pneumonia, bronchitis, and asthma. Depression of respiratory center by drugs, cardiopulmonary arrest, stroke, poliomyelitis, or nervous system disorders.
Treatment:	Correction of disorder. Infusion of bicarbonate.
Respiratory Alkalosis: CO_2 ⇩ pH ⇧	
Symptoms:	Increased rate and depth of breathing. Numbness, light-headedness, tetany.
Causes:	Hyperventilation due to anxiety, hysteria, fever, exercise. Reaction to drugs such as salicylate, quinine, and antihistamines. Conditions causing hypoxia, e.g., pneumonia, pulmonary edema, and heart disease.
Treatment:	Elimination of anxiety-producing state; rebreathing into a paper bag.
Metabolic Acidosis: H^+ ⇧ pH ⇩	
Symptoms:	Increased ventilation. Fatigue and confusion.
Causes:	Renal disease, including hepatitis and cirrhosis. Increased acid production in diabetic mellitus, hyperthyroidism, alcoholism, and starvation. Loss of alkali in diarrhea. Acid retention in renal failure.
Treatment:	Sodium bicarbonate may be given orally. Dialysis for renal failure. Insulin treatment for diabetic ketosis.
Metabolic Alkalosis: H^+ ⇩ pH ⇧	
Symptoms:	Depressed breathing. Apathy, confusion.
Causes:	Vomiting. Diseases of the adrenal glands. Ingestion of excess alkali.
Treatment:	Infusion of saline solutions. Treatment of underlying diseases.

Example 10.11 Indicate whether each of the following would be a buffer:

1. $HCl + NaCl$
2. NH_4OH
3. $H_2CO_3 + NaHCO_3$

Solution

1. No. This is a solution of a strong acid and its salt. A buffer must contain a weak acid and its salt.
2. No. A weak base is not sufficient for a buffer. The mixture must include the salt of the weak base.
3. Yes. This mixture consists of the necessary components for a buffer: a weak acid and its salt.

Glossary

acid A substance that releases H^+ in aqueous solution; a hydrogen donor.

acidosis A physiological condition in which the blood pH is lower than the normal pH 7.4.

alkalosis A physiological condition in which the blood pH is higher than the normal pH 7.4.

base A substance that releases or causes an increase in OH^- in an aqueous solution; a hydrogen acceptor.

buffer A mixture of a weak acid (or weak base) and its salt; a buffer maintains the pH of a solution.

hydronium ion The combination of a proton and a water molecule, H_3O^+.

indicator A substance added to a sample during titration that changes color when the pH of the sample changes.

neutralization A reaction between an acid and a base to form a salt and water.

pH A measure of the $[H^+]$ in a solution.

strong acid An acid that is completely ionized in solution.

strong base A base that is completely ionized in solution.

titration The addition of a measured amount of base to an acid to determine the concentration of the acid.

weak acid An acid that only slightly ionizes in solution.

weak base A base that only slightly ionizes in solution.

Problems

Acids (Objective 10.1)

10.1 Write an equation for the dissociation of the following strong acids in water:
a. HCl b. HNO_3

10.2 Write an equation for the dissociation of the following weak acids:
a. H_3BO_3 (boric acid) b. H_2CO_3 (carbonic acid)

10.3 Identify the following as strong or weak acids:
a. HBr
b. H_2SO_4
c. H_2S
d. $H_2C_2O_4$ (oxalic acid)

Bases (Objective 10.2)

10.4 Write an equation for the dissociation of the following bases in water:
a. LiOH b. $Mg(OH)_2$ c. KOH

10.5 Write an equation for the reaction of the weak base ammonia (NH_3) in water.

10.6 Indicate whether each of the following descriptions is characteristic of an acid or a base:
a. has a sour taste
b. turns red litmus blue
c. neutralizes acids
d. has a bitter taste
e. releases hydrogen ions in aqueous solution

Ionization of Water (Objective 10.3)

10.7 a. Write the $[H^+]$ and $[OH^-]$ in pure water.
b. Write the ion product expression and its value for water.

10.8 Find the $[H^+]$ of a solution with the following $[OH^-]$:
a. $[OH^-] = 1 \times 10^{-11}\ M$
b. $[OH^-] = 1 \times 10^{-5}\ M$
c. $[OH^-] = 1 \times 10^{-1}\ M$
d. $[OH^-] = 1 \times 10^{-12}\ M$

10.9 Calculate the $[OH^-]$ of a solution when the $[H^+]$ has the following values:
a. $[H^+] = 1 \times 10^{-11}\ M$
b. $[H^+] = 1 \times 10^{-2}\ M$
c. $[H^+] = 1 \times 10^{-4}\ M$
d. $[H^+] = 1 \times 10^{-8}\ M$

pH Scale (Objective 10.4)

10.10 State whether the following solutions are acidic, neutral, or basic:
a. blood, pH = 7.4
b. coffee, pH = 5.5
c. pancreatic juice, pH = 8.2
d. vinegar, pH = 2.8
e. soda, pH = 3.2
f. milk, pH = 7.0
g. pH-balanced shampoo, pH = 6.0
h. hot tub water, pH = 7.8
i. drain cleaner, pH = 11.2
j. laundry detergent, pH = 9.5

10.11 Arrange the pH values of the following groups in order, from the most acidic to the least acidic (most basic):
a. 4.5, 13.0, 0.4, 6.8
b. 1.6, 11.7, 7.1, 2.3, 8.5
c. 14.0, 9.8, 3.3, 4.4, 2.9
d. 8.8, 9.7, 11.4, 13.4, 7.4

Calculation of pH (Objective 10.5)

10.12 State whether the following solutions are acidic, neutral, or basic:
a. $[H^+] = 1.0 \times 10^{-4}\ M$
b. $[H^+] = 1.0 \times 10^{-7}\ M$
c. $[OH^-] = 1.0 \times 10^{-3}\ M$
d. $[OH^-] = 1.0 \times 10^{-10}\ M$
e. $[H^+] = 1.0 \times 10^{-2}\ M$
f. $[H^+] = 0.01\ M$
g. $[OH^-] = 1.0 \times 10^{-4}\ M$
h. $[OH^-] = 1.0 \times 10^{-7}\ M$
i. $[H^+] = 0.0001\ M$
j. $[OH^-] = 0.001\ M$

10.13 Calculate the pH of each of the solutions in Problem 10.12.

10.14 Complete the following table:

$[H^+]$	$[OH^-]$	pH	Acidic, Basic, or Neutral?
________	1×10^{-6}	________	________
________	________	2	________
1×10^{-5}	________	________	________
________	________	10	________
________	________	________	Neutral

Neutralization Reactions (Objective 10.6)

10.15 Balance the following acid–base reactions:

a. $HCl + KOH \longrightarrow KCl + H_2O$
b. $HNO_3 + Ca(OH)_2 \longrightarrow Ca(NO_3)_2 + H_2O$
c. $H_3PO_4 + LiOH \longrightarrow Li_3PO_4 + H_2O$
d. $HBr + Al(OH)_3 \longrightarrow AlBr_3 + H_2O$

10.16 Complete and balance the following acid–base equations:

a. $NaOH + H_2SO_4 \longrightarrow$ ________ + ________
b. $KOH + HCl \longrightarrow$ ________ + ________
c. ________ + ________ $\longrightarrow CaSO_4 + 2H_2O$
d. ________ + ________ $\longrightarrow NaBr + H_2O$
e. $H_3PO_4 + NaOH \longrightarrow$ ________ + ________
f. $Al(OH)_3 + H_2SO_4 \longrightarrow$ ________ + ________

10.17 Write an equation that would produce the following salts using an acid–base reaction:

a. K_2SO_4 b. $LiNO_3$ c. $Ca_3(PO_4)_2$

Titration (Objective 10.7)

10.18 Calculate the molarity of the acid in each of the following titrations:

a. a 5.0-mL sample of HCl that is titrated with 22.0 mL of 0.50-*M* NaOH:

$$HCl + NaOH \longrightarrow NaCl + H_2O$$

b. a 25.0-mL sample of H_2SO_4 that is titrated with 38.0 mL of 1.0-*M* NaOH:

$$H_2SO_4 + 2NaOH \longrightarrow Na_2SO_4 + 2H_2O$$

c. a 10.0 mL sample of H_3PO_4 that is titrated with 16.0 mL of 1.0-*M* NaOH.

$$H_3PO_4 + 3NaOH \longrightarrow Na_3PO_4 + H_2O$$

Buffers (Objective 10.8)

10.19 Which of the following represent a buffer system? Why?

a. $HCl + NaCl$
b. HAc
c. HAc + NaAc
d. $KCl + NaCl$
e. $H_2CO_3 + NaHO_3$

10.20 Consider the buffer system of the weak acid HAc and its salt NaAc:

$$HAc \rightleftharpoons H^+ + Ac^-$$

a. What is the purpose of the buffer system?
b. What is the function of the salt (NaAc)?
c. Which component, HAc or Ac^-, reacts with excess H^+ in the system?
d. Which component, HAc or Ac^-, reacts with excess OH^- in the system?

Alkanes: An Introduction to a Study of Organic Compounds

Objectives

11.1 Given a list of properties, indicate whether each is more typical of organic or inorganic compounds.

11.2 Write a structural formula for a hydrocarbon, given the number of carbon atoms.

11.3 Write the correct condensed formula from the structural formula of a hydrocarbon.

11.4 Write the structure and name of the first 10 straight-chain alkanes.

11.5 Write the structure and IUPAC name of an alkane with one or more side groups.

11.6 Given the molecular formula of an alkane, write the condensed formulas of its isomers.

11.7 Write the name and structure of a cycloalkane.

11.8 Write a chemical equation for the combustion and halogenation of an alkane.

11.9 Write the name and structure of a haloalkane.

Scope

Compounds that contain the element carbon, C, are called *organic compounds*, and the chemistry of carbon compounds is called *organic chemistry*. At one time, scientists believed that organic compounds had to be derived from living sources; hence the name "organic." It was thought that only living matter possessed the essential quality, or "vital force," needed to produce organic material. Nonliving substances were considered to be lacking in this vital force and were classified as "inorganic" compounds. Eventually, an experiment showed that the inorganic ammonium cyanate could be converted into urea, an organic compound without a "vital force":

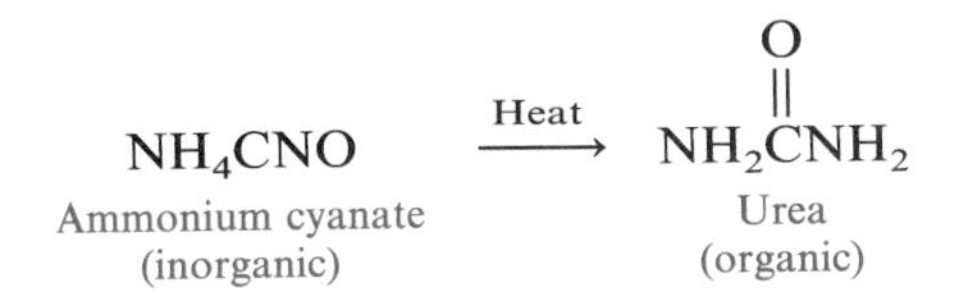

Carbon compounds make up many of the compounds of our environment as well as our bodies. There are carbon compounds in plastic bags and wrappers, in fabrics, in vitamins and drugs, and in the food you eat. All proteins, fats, and carbohydrates are organic compounds. The genetic material that determined your physical makeup and will determine that of your children is also composed of organic compounds.

Some carbon compounds were classified as inorganic compounds before organic chemistry came to be defined as the chemistry of carbon. They include the oxides of carbon such as CO, CO_2, and CO_3^{2-}, as well as H_2CO_3, HCO_3^-, and CNO^-.

11.1 Properties of Organic Compounds

OBJECTIVE Given a list of properties, indicate whether each is more typical of organic or inorganic compounds.

For the most part, the compounds we have studied in the past 10 chapters have been inorganic compounds, which are held together by ionic or polar covalent bonds. In general, inorganic compounds have rather high melting and boiling points. They are often electrolytes, which dissolve in polar solvents, usually water, to form ions in solution.

carbon

Organic compounds are compounds that contain carbon. They generally have low melting and boiling points. Many organic compounds burn easily; most inorganic compounds do not. Most organic compounds are nonpolar and are soluble in nonpolar solvents. Those that are polar enough to dissolve in water dissolve as nonelectrolytes. There are many more organic compounds than inorganic compounds, perhaps 10 times as many. See Table 11.1 for a summary of the general properties of organic and inorganic compounds.

Table 11.1 General Properties of Inorganic and Organic Compounds

Properties	Inorganic	Organic
Electrolytes	Yes	No
Bonding	Ionic	Covalent
Melting point	High	Low
Boiling point	High	Low
Solubility	In water	In nonpolar solvents, some in water
Flammable	No	Yes
Quantity	200,000	2,000,000 or more
Elements	All	C, H, O, N

Example 11.1 State whether each of the following properties is most likely to describe an organic or an inorganic compound:

1. soluble in water
2. low boiling point
3. very flammable

Solution

1. Most *inorganic compounds* are soluble in water.
2. Most *organic compounds* have low boiling points.
3. *Organic compounds* are often very flammable.

11.2 Structural Formulas of Organic Compounds

OBJECTIVE Write a structural formula for a hydrocarbon, given the number of carbon atoms.

One of the reasons for the large number of carbon compounds is the unique ability of carbon to form covalent bonds with many other carbon atoms. As a result, stable chains of carbon atoms are produced. Carbon can also form covalent bonds with other elements—particularly hydrogen, nitrogen, oxygen, sulfur, and the halogens.

To understand the bonding of carbon in compounds, we need to look at its electron arrangement. Carbon has atomic number 6 and therefore has six electrons, two in the first energy level and four in the second. Thus, carbon has four valence electrons, and its outer energy level is half-filled:

	Energy Level	
	1	2
Carbon:	$2e^-$	$4e^-$

four bonds to carbon

A carbon atom needs to share four electrons to bring the total number of electrons in the outer level to eight. By sharing four electrons with other atoms, carbon will form four covalent bonds. In hydrocarbons, carbon atoms combine with hydrogen atoms. A hydrogen atom has one valence electron. A single carbon atom will combine with four hydrogen atoms to form the hydrocarbon *methane*:

$$\cdot\dot{\underset{\cdot}{\text{C}}}\cdot + 4\text{H}\cdot \longrightarrow \text{H}:\overset{\text{H}}{\underset{\text{H}}{\overset{\cdot\cdot}{\underset{\cdot\cdot}{\text{C}}}}}:\text{H} = \text{H}-\overset{\text{H}}{\underset{\text{H}}{\overset{|}{\underset{|}{\text{C}}}}}-\text{H}$$

tetrahedron

A shared pair of electrons is represented by a single line drawn between two atoms. However, you need to know that carbon compounds such as methane, CH_4, are not really flat as some drawings suggest. Methane actually has a three-dimensional shape, as indicated in Figure 11.1. The three-dimensional shape, a *tetrahedron* with the carbon atom at the center, is represented by a ball-and-stick model such as one you might build in the lab. Another model, called a *space-filling model*, shows the spaces occupied by the carbon and hydrogen atoms.

alkanes

When there are two or more carbon atoms in a hydrocarbon, electrons are also shared between the carbon atoms. When single bonds form between the carbon atoms, the hydrocarbon is called an *alkane*. The remaining valence electrons of each carbon atom combine with hydrogen atoms. All alkanes have single bonds between the carbon atoms, with the remaining valence electrons sharing electrons with hydrogen atoms.

structural formula

There are several ways to represent an organic compound. When we draw a diagram that shows every atom and where that atom is attached in the compound, we are drawing a *structural formula*. In this type of formula, a line is

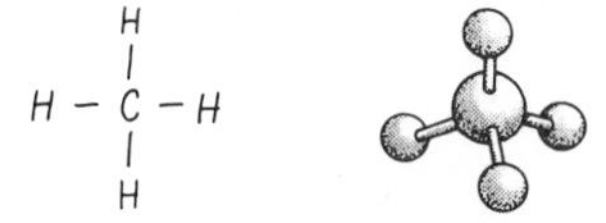

Figure 11.1 Structural representations of methane, CH_4.

tetrahedron

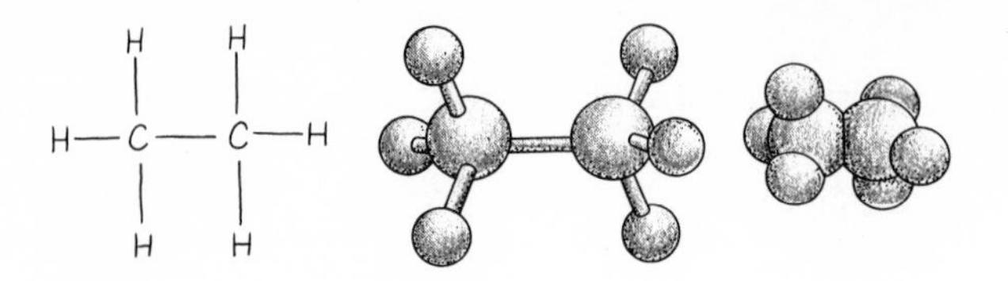

Figure 11.2 Structural representations of ethane, C_2H_6.

drawn for each bond. In the hydrocarbon *ethane*, there are two carbon atoms bonded to six hydrogen atoms:

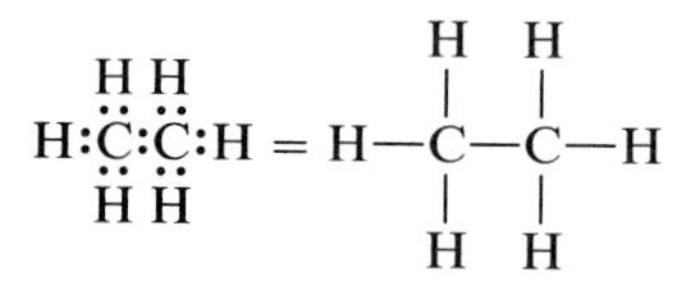

Figure 11.2 illustrates the structural formula of ethane, C_2H_6, along with the ball-and-stick model and the space-filling model.

If we increase the chain to three carbon atoms, we need a total of eight hydrogen atoms to complete the four bonds for every carbon atom. This hydrocarbon is called *propane*:

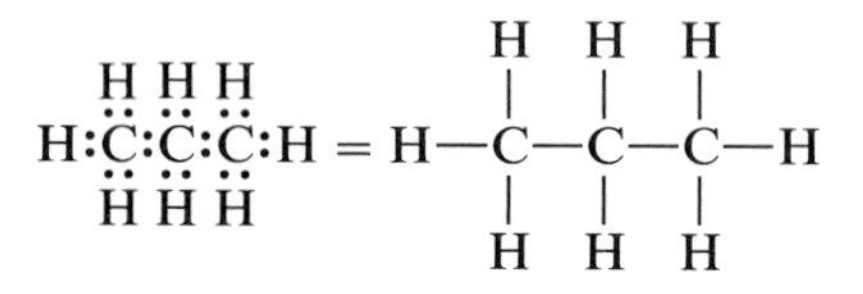

Figure 11.3 illustrates the structural representation of propane along with the ball-and-stick model and its space-filling model.

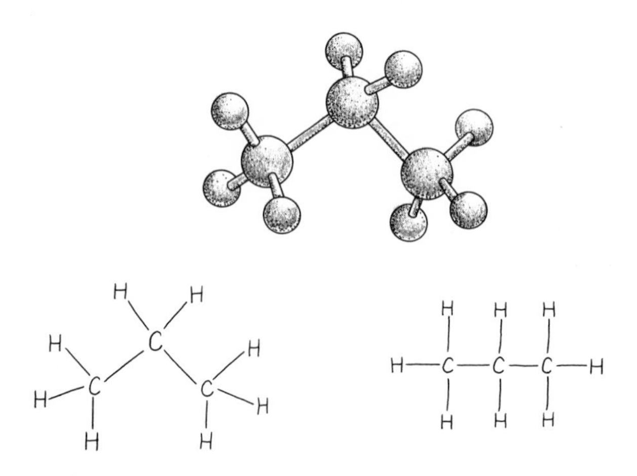

Figure 11.3 Structural representations of propane, C_3H_8.

Example 11.2 Assuming that the only element missing in the following compound is hydrogen, complete the structural formula using a single line to represent a single bond:

C—C—C—C—C

Solution: Hydrogen atoms are added to the chain of five carbon atoms to give each carbon atom in the chain four bonds:

```
   H  H  H  H  H
   |  |  |  |  |
H—C—C—C—C—C—H
   |  |  |  |  |
   H  H  H  H  H
```

The alkanes methane, ethane, and propane contain carbon atoms attached in a continuous chain. Such compounds are referred to as *straight-chain compounds*. As the number of carbon atoms increases, it is likely that there will be groups of carbon atoms called *side chains* or *branches* attached to a longer chain of carbon atoms. Such compounds are called *branched-chain compounds*. Let us compare a straight-chain compound with a branched-chain one:

a straight-chain compound has no branches

a branched-chain compound has side groups attached

Straight-Chain Compound	*Branched-Chain Compound*
All five carbons are in a continuous chain:	Four carbons make up a continuous chain with a one-carbon branch:

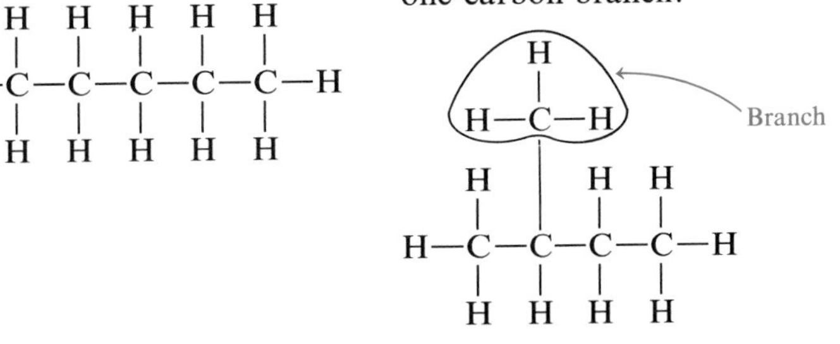

Example 11.3 Write out a structural formula for the following alkanes:

1. a straight-chain alkane with four carbon atoms
2. a branched-chain alkane with four carbon atoms

Solution

1. A straight-chain compound consists of four carbon atoms attached in a continuous chain. The hydrogen atoms are added to give four bonds to each carbon atom:

```
   H  H  H  H
   |  |  |  |
H—C—C—C—C—H
   |  |  |  |
   H  H  H  H
```

2. In a branched-chain compound, at least one carbon atom must be attached to a longer chain of carbon atoms:

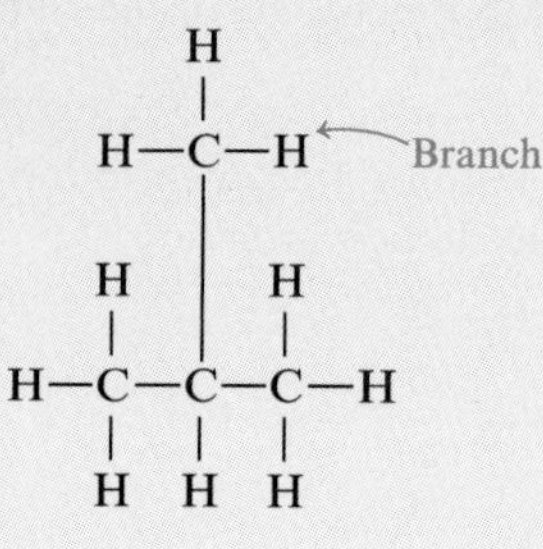

11.3 Condensed Formulas

OBJECTIVE Write the correct condensed formula from the structural formula of a hydrocarbon.

Structural formulas are very helpful in showing the way that the individual atoms are bonded in an organic compound. However, they become cumbersome to write as the number of carbon atoms increases.

condensed formula

A *condensed formula* is often used because it simplifies the writing of the structure. In a condensed formula, the order of the carbon atoms is still shown, but the number of hydrogen atoms attached to each carbon atom is indicated by a subscript. The condensed formula for ethane, CH_3CH_3, indicates that a carbon atom with three hydrogen atoms is bonded to another carbon atom with three hydrogen atoms. Lines or dashes may be used to separate the carbon atoms. However, they are optional and are usually omitted. (See Table 11.2.)

Table 11.2 Types of Formulas

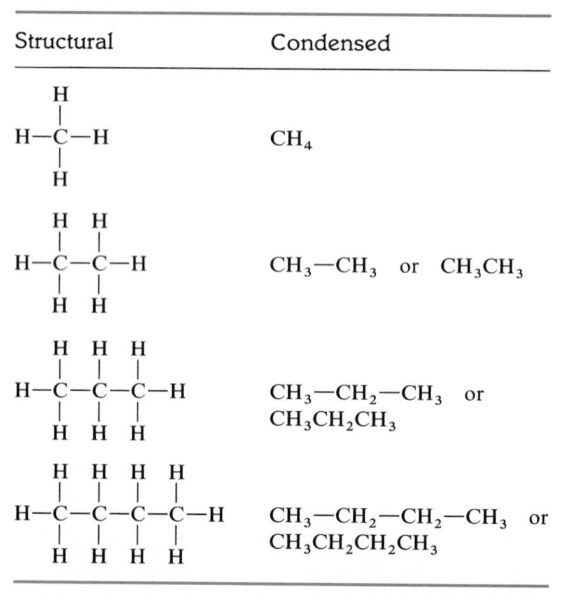

Structural	Condensed
H | H—C—H | H	CH_4
H H | | H—C—C—H | | H H	$CH_3—CH_3$ or CH_3CH_3
H H H | | | H—C—C—C—H | | | H H H	$CH_3—CH_2—CH_3$ or $CH_3CH_2CH_3$
H H H H | | | | H—C—C—C—C—H | | | | H H H H	$CH_3—CH_2—CH_2—CH_3$ or $CH_3CH_2CH_2CH_3$

Writing a Condensed Formula

In order to write the condensed formula, you need to look at one carbon at a time and count the number of hydrogen atoms attached to it. In the following compound, the structure is divided up to illustrate the individual carbon atoms and their respective hydrogen atoms. The first carbon has three hydrogen atoms, the second carbon atom has two hydrogen atoms, the third carbon has two hydrogen atoms, and the last carbon atom is attached to three hydrogen atoms:

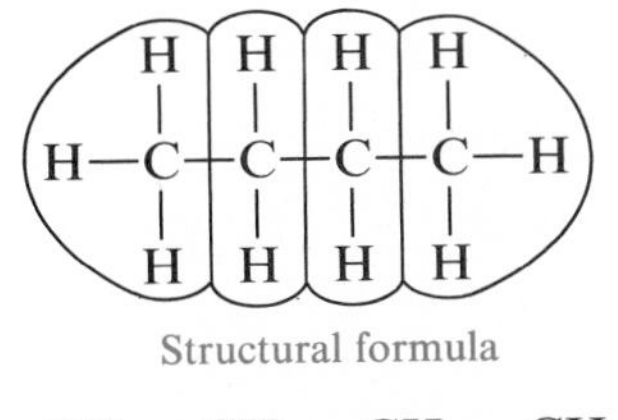

Structural formula

$$CH_3—CH_2—CH_2—CH_3$$

Condensed formula

or

$$CH_3CH_2CH_2CH_3$$

The condensed formula is also used for branched-chain compounds. In this text, we will use condensed formulas that display the side group above or below the carbon chain. Figure 11.4 compares some structural formulas with their corresponding condensed formulas.

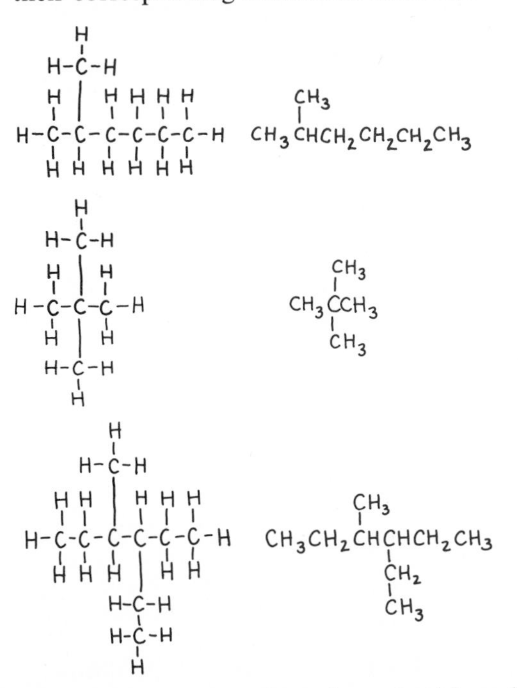

Figure 11.4 A comparison of structural formulas and their corresponding condensed formulas.

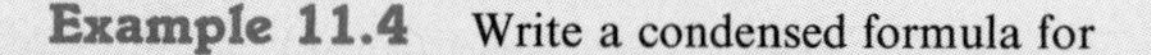

Example 11.4 Write a condensed formula for

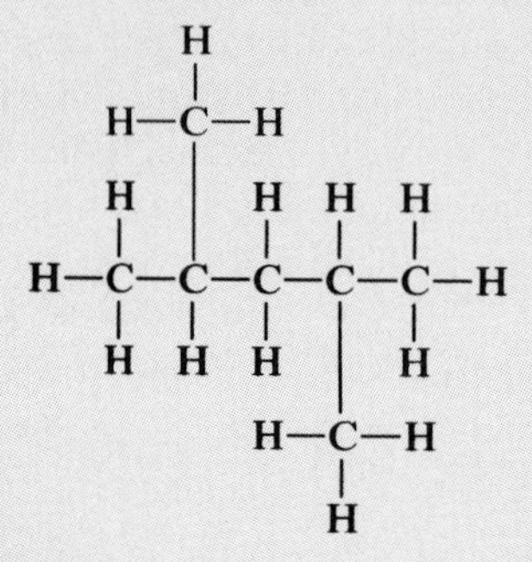

Solution

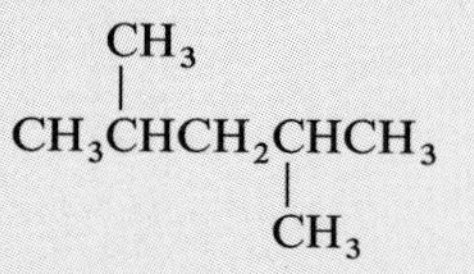

Example 11.5 Write a structural formula for

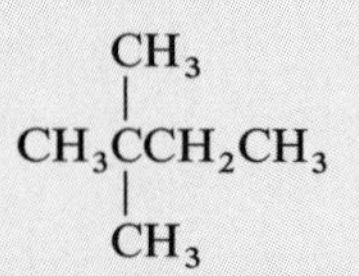

Solution: In a structural formula, a bond is drawn from the carbon atoms to each of the attached hydrogen atoms:

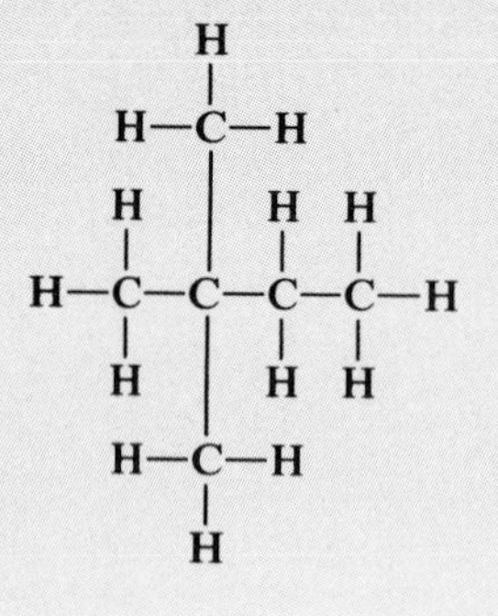

11.4 Straight-Chain Alkanes

OBJECTIVE Write the structure and name of the first 10 straight-chain alkanes.

At one time, organic compounds were named in a random fashion. A system of naming became necessary because of the increasing number of organic compounds that were being discovered. Rules for a systematic method of naming organic compounds were officially formulated by the International Union of Pure and Applied Chemistry (IUPAC). Our emphasis in this text is on the IUPAC system of naming, but we cannot completely ignore, as yet, the common names. For certain compounds, common names have persisted; these names are included in this text.

IUPAC

The naming of organic compounds is based on the names given to the alkanes, which, as stated earlier, are those hydrocarbons that have only single bonds between the carbon atoms. We also call alkanes *saturated hydrocarbons.* The IUPAC names use a *root* and a *suffix*. The root indicates the number of carbon atoms in the longest chain in the compound. The suffix, in this case, *-ane*, classifies the compound as part of the alkane family. The first four alkanes were in use before the IUPAC system was established. Thereafter the roots are the Greek names of the number of carbon atoms in the longest chain. Table 11.3 lists the names, condensed structures, and molecular formulas of the first 10 straight-chain alkanes.

saturated hydrocarbons

Table 11.3 The First Ten Alkanes

Name (C_nH_{2n+2})	Number of Carbon Atoms	Condensed Structural Formula	Molecular Formula
Methane	1	CH_4	CH_4
Ethane	2	CH_3CH_3	C_2H_6
Propane	3	$CH_3CH_2CH_3$	C_3H_8
Butane	4	$CH_3CH_2CH_2CH_3$	C_4H_{10}
Pentane	5	$CH_3CH_2CH_2CH_2CH_3$	C_5H_{12}
Hexane	6	$CH_3CH_2CH_2CH_2CH_2CH_3$	C_6H_{14}
Heptane	7	$CH_3CH_2CH_2CH_2CH_2CH_2CH_3$	C_7H_{16}
Octane	8	$CH_3CH_2CH_2CH_2CH_2CH_2CH_2CH_3$	C_8H_{18}
Nonane	9	$CH_3CH_2CH_2CH_2CH_2CH_2CH_2CH_2CH_3$	C_9H_{20}
Decane	10	$CH_3CH_2CH_2CH_2CH_2CH_2CH_2CH_2CH_2CH_3$	$C_{10}H_{22}$

Example 11.6 Write the names of the following alkanes:

1. $CH_3CH_2CH_2CH_3$
2. $CH_3CH_2CH_2CH_2CH_2CH_2CH_3$

Solution

1. The alkane that contains four carbon atoms in a continuous chain is butane.
2. The alkane that contains seven carbon atoms in a continuous chain is heptane.

We will stop here and look at some other ways of writing formulas. So far, we have represented the carbon atoms in a linear carbon chain. We have done so primarily for convenience. However, any pattern of five carbon atoms in a continuous chain is acceptable as a representation of pentane. For instance, all of the patterns in Figure 11.5 would be correct for five carbon atoms in a row.

$CH_3 - CH_2 - CH_2 - CH_2 - CH_3$ (zigzag) $\quad$ CH_2, CH_2, CH_2, CH_3, CH_3 (bent) $\quad$ CH_3 over $CH_2-CH_2-CH_2$ over CH_3

not a side chain → CH_3 over $CH_2-CH_2-CH_2-CH_3$ $\quad$ $CH_3CH_2CH_2CH_2CH_3$

Figure 11.5 All condensed formulas in this figure represent the same straight-chain isomer of C_5H_{12}.

Properties of Alkanes

The first four alkanes—methane, ethane, propane, and butane—are widely used as fuels. These compounds are gases at room temperature and pressure and act as anesthetics. Those alkanes with 5–18 carbon atoms exist as liquids at room temperatures, while the alkanes with 20 or more carbon atoms are solids. Mineral oil, a mixture of hydrocarbon compounds, is used as a cathartic and a lubricant. Petroleum jelly is also a mixture of hydrocarbon compounds and is used as a lubricant and a solvent.

All alkanes are nonpolar and insoluble in water. Most alkanes such as gasoline and oil are less dense than water and thus float on water.

11.5 Naming the Branched-Chain Alkanes

OBJECTIVE Write the structure and IUPAC name of an alkane with one or more side groups.

branches with carbon atoms are named alkyl groups

Most organic compounds consist of a chain of carbon atoms to which a side group is attached. A carbon side group is named an *alkyl group*. An alkyl group cannot exist by itself because there would be one hydrogen missing. The root of its name is derived from the alkane with the same number of carbon atoms. However, the suffix changes from *-ane* to *-yl*. The names of some commonly used side groups are listed in Table 11.4. Note that there are two alkyl groups derived from propane. A *n*-propyl group is attached at the end carbon, whereas the isopropyl group is attached at the center carbon.

Table 11.4 Some Alkanes and Their Alkyl Groups

Hydrocarbon	Name of Hydrocarbon	Alkyl Group	Name of Group
CH_4	Methane	CH_3-	Methyl
CH_3CH_3	Ethane	CH_3CH_2-	Ethyl
$CH_3CH_2CH_3$	Propane	$CH_3CH_2CH_2-$	*n*-Propyl
		CH_3CHCH_3 (bond down from central C)	Isopropyl

Step-by-Step Naming of Branched-Chain Alkanes

Suppose you wanted to name the following structure:

$$\begin{array}{l} \quad\;\; CH_3 \\ \quad\;\;\; | \\ CH_3CHCH_2CH_2CH_3 \end{array}$$

The procedure for naming this compound occurs in the following steps:

1. Find the longest continuous chain of carbon atoms (parent chain):

$$\begin{array}{l} \quad\;\; CH_3 \\ \quad\;\;\; | \\ CH_3CHCH_2CH_2CH_3 \longleftarrow \text{Parent chain (longest chain)} \end{array}$$

The longest chain in the compound contains five carbon atoms. The name associated with five carbons is pentane. This is the base, or *parent name*, for the compound:

pentane

Pentane

2. Identify the alkyl side group attached to the pentane chain and name it. In this structure, the side group is —CH_3, the methyl group. The name *methyl* is placed in front of the name of the parent alkane. No space is left between the name of the branch and the parent chain:

methylpentane

Methylpentane

3. Number the parent chain from the end nearest the branch. Then the particular carbon atom to which the branch is attached can be determined. Indicate the location of the side group on carbon atom 2 by writing the number 2 followed by a dash in front of the name of the side group:

Correct Name

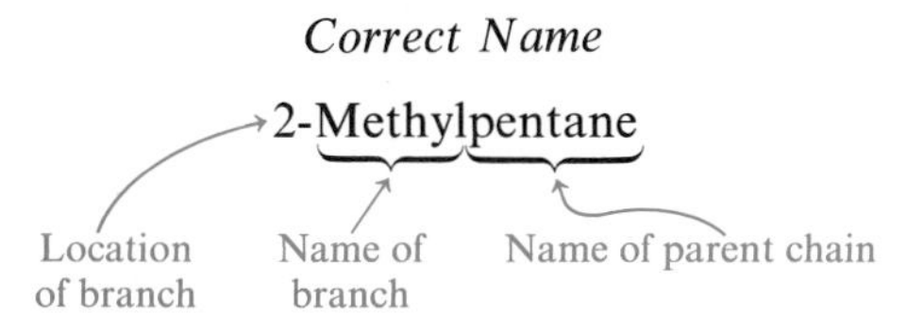

Note that the numbering of the carbon chain is done from the end of the parent chain that gives the side group the lowest possible number. If this were done incorrectly for our compound, we would obtain the incorrect name *4-methylpentane*:

4-Methylpentane (Incorrect)	*2-Methylpentane (Correct)*
$\begin{array}{l} \quad\;\; CH_3 \\ \quad\;\;\; \vert \\ CH_3CHCH_2CH_2CH_3 \\ 5 \quad 4 \quad 3 \quad 2 \quad 1 \end{array}$	$\begin{array}{l} \quad\;\; CH_3 \\ \quad\;\;\; \vert \\ CH_3CHCH_2CH_2CH_3 \\ 1 \quad 2 \quad 3 \quad 4 \quad 5 \end{array}$
Incorrect numbering (This puts the branch on the fourth carbon of the parent chain.)	Correct numbering (This puts the branch on the second carbon of the parent chain.)

The numbering of a side group becomes necessary when we need to differentiate the compound from another methylpentane having the methyl group on the third carbon atom:

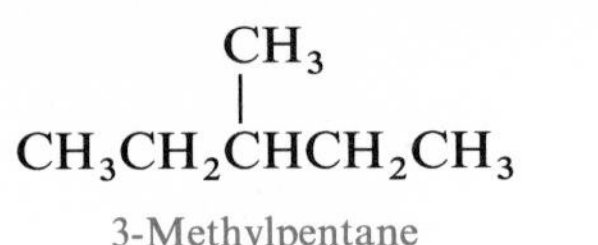

3-Methylpentane

4. If there are two or more side groups, each one must be designated by a number to show its location on the parent chain. All of the names of the alkyl groups, each with their number and a hyphen, are written in front of the parent name. In the following compound, the side groups consist of a methyl group and an ethyl group. They are listed in alphabetical order in front of the name of the parent chain:

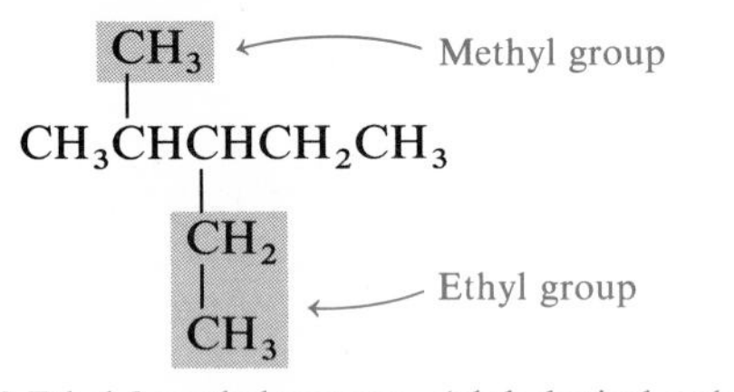

3-Ethyl-2-methylpentane (alphabetical order)

If two or more side groups are the same, the name of that alkyl group is used with a prefix that indicates the number of times that alkyl group appears. The prefixes used are the following:

Number of Alkyl Groups	*Prefix*
2	di-
3	tri-
4	tetra-
5	penta-

There must be a separate number for each side group to indicate its location on the parent chain, even if the side groups are attached to the same carbon atom. The numbers are separated by commas:

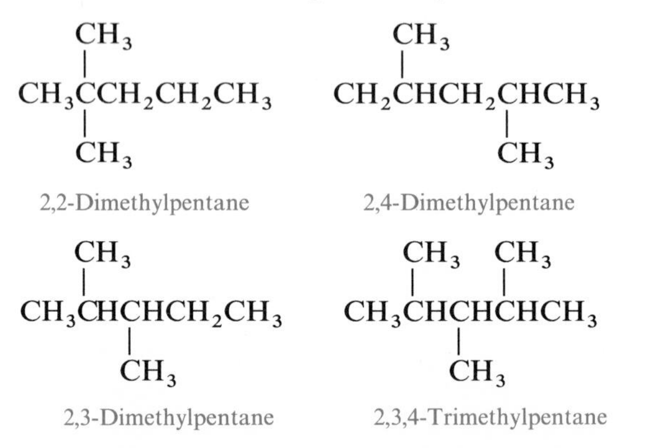

2,2-Dimethylpentane   2,4-Dimethylpentane

2,3-Dimethylpentane   2,3,4-Trimethylpentane

Be sure that you always find the longest continuous carbon chain. Watch out: The longest chain need not be the most obvious horizontal one. Look at the following structure:

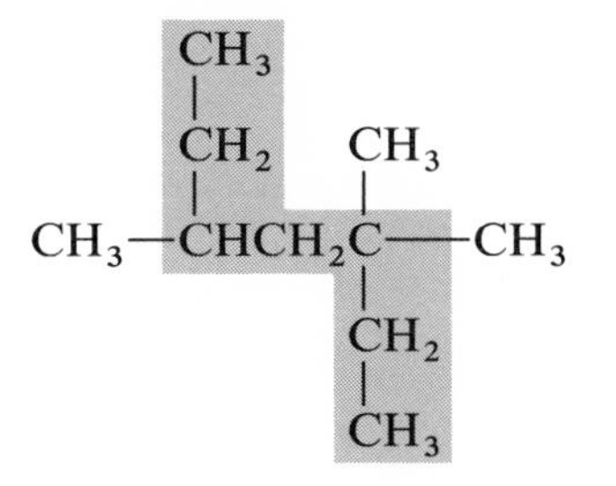

Longest carbon chain = Parent chain

How many carbon atoms do you see in the longest continuous chain? If you said five, look again. You should be able to find a continuous chain of seven carbon atoms. Numbering the chain will help you find the longest chain. We have rewritten the structure to display the longest chain horizontally. Note that the numbering always begins at the end of the parent chain that gives the lowest numbers to the branches. Had you numbered from the other end of the structure, you would have named the compound 3,5,5-trimethylheptane. The sum of the location numbers is higher than in 3,3,5-trimethylheptane, the correct name:

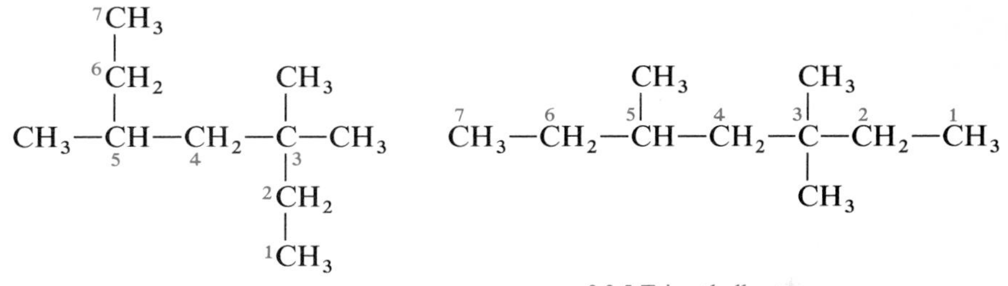

3,3,5-Trimethylheptane

Let us look at another example of selecting the correct order of numbering with many branches. Once you have decided in which direction you will number the chain, you must continue in that direction:

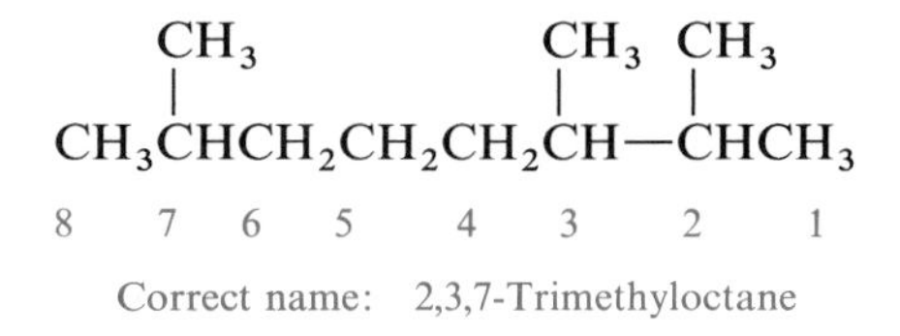

Correct name: 2,3,7-Trimethyloctane

Summary of IUPAC Naming Rules

1. Find the parent or longest chain and assign it a name.
2. Name each side group.

3. Number the parent chain from the end nearest a branch to give the lowest set of numbers to the branch(es).
4. Assign each branch a number for its location on the parent chain. Use prefixes di-, tri-, and so on to indicate two or more of the same branch. (These prefixes do not determine alphabetical order.)

Example 11.7 Write the IUPAC name for the following alkanes:

1. $CH_3CH_2CH_2CH_3$

2.
$$\begin{array}{c} CH_3 \\ | \\ CH_3CCH_3 \\ | \\ CH_3 \end{array}$$

3.
$$\begin{array}{l} \quad\;\; CH_3 \;\; CH_3 \\ \quad\;\; | \qquad\;\; | \\ CH_3CHCHCHCH_2CH_3 \\ \qquad\quad\; | \\ \qquad\quad CH_3 \end{array}$$

Solution

1. This is a straight-chain compound with four carbon atoms:

   Butane

2. This is a branched-chain compound. The parent chain is propane. The one-carbon side groups are methyl groups. Since there are two of these methyl groups, we will identify them as dimethyl. The locations of the methyl groups are on the second carbon atom of the parent chain:

   2,2-Dimethylpropane

3. This is also a branched-chain compound in which the parent chain is composed of six carbon atoms. There are three methyl groups located on the second, third, and fourth carbon atoms:

   2,3,4-Trimethylhexane

Writing the Structural Formula from the Name of the Compound

You should now be able to write the structural formula from the name of an organic compound. Suppose you are given the name 3-ethyl-2-methylhexane. How would you write the structure for that compound? You could proceed as follows:

1. Start with the alkane name—in this case, *hexane*. Write the carbon skeleton of the parent chain. Hexane has six carbon atoms, so the chain would look like

$$-C-C-C-C-C-C-$$

2. Number the chain you have written. Identify the number of carbon atoms in each branch and attach the alkyl group(s) to the parent chain. In this example, there is a methyl group, CH_3—, on the second carbon atom in the chain, and there is an ethyl group, CH_3CH_2—, on the third carbon atom:

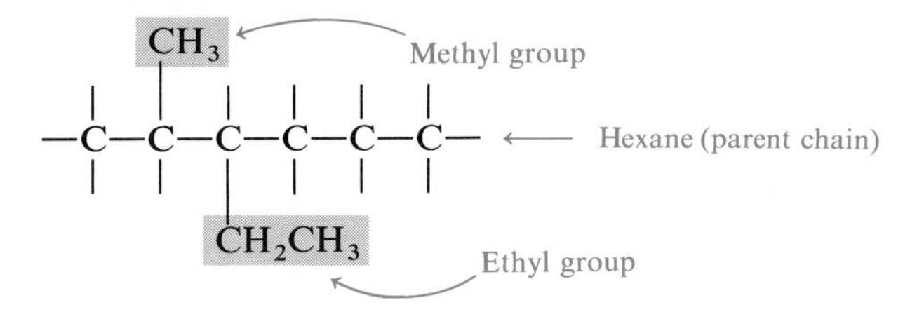

3. Add hydrogen atoms to complete the structural formula. Write the condensed formula:

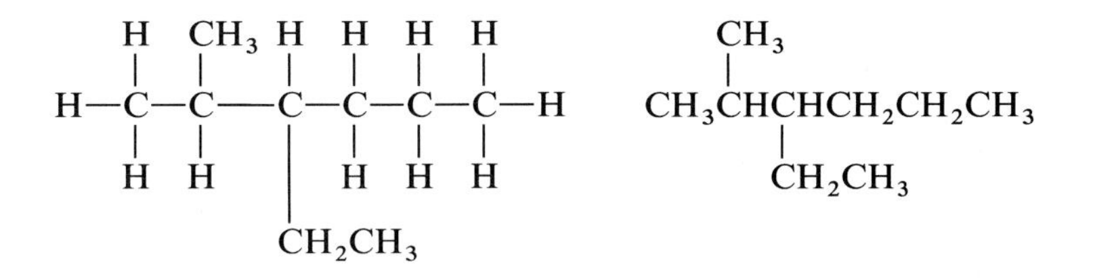

4. In order to check your answer, name the structure that you have written and compare it with the name given in the original problem:

   3-Ethyl-2-methylhexane

Example 11.8 Write the condensed formula for the following alkane:

2,2,4-Trimethylpentane

Solution: The parent chain is pentane, a five-carbon chain. There are three methyl groups, CH_3—, attached to the parent chain. Two methyl groups are attached to carbon atom 2 and the other methyl group is attached to carbon atom 4:

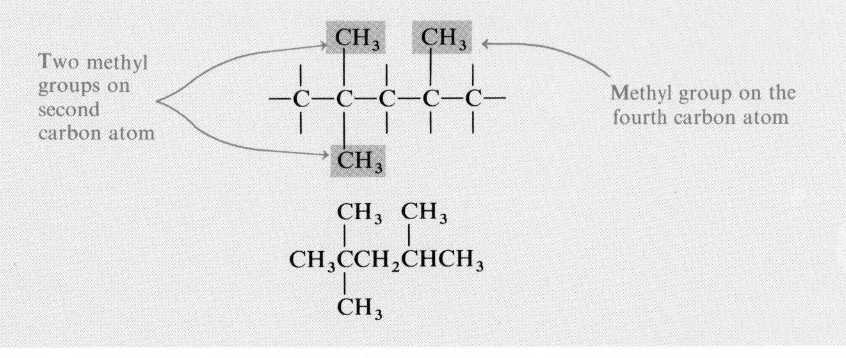

11.6 Isomers

OBJECTIVE Given the molecular formula of an alkane, write the condensed formulas of its isomers.

A *molecular formula*, such as Fe_2O_3 or C_2H_6 or $C_6H_{12}O_6$, lists only the number and types of atoms in a compound and not the order of attachment of the atoms. The structural and condensed formulas we have used for the alkanes do indicate the sequence of the carbon atoms. In organic chemistry, the formulas that show structure are preferred because it is often possible to write more than one structure for the same molecular formula by changing the arrangement of the carbon atoms.

isomers have different arrangements of atoms

Compounds with identical molecular formulas but different arrangement of atoms are called *isomers*. The structural or condensed formulas of isomers can look quite different, and the compounds themselves can vary in physical and chemical characteristics such as boiling point, solubility, and reactivity. Yet compounds that are isomers have exactly the same number of atoms of each element and therefore the same molecular formula and molecular weight.

For example, C_4H_{10} has two isomers. Both compounds have a total of 4 carbon atoms and 10 hydrogen atoms. However, one compound is a straight-chain isomer and the other is a branched-chain isomer. The isomers of C_4H_{10} and some of their properties are shown in Table 11.5.

Table 11.5 Isomers of C_4H_{10}

	$CH_3CH_2CH_2CH_3$	$CH_3CH(CH_3)CH_3$
Name:	Butane	Methylpropane
mol wt:	58.1	58.1
mp:	−138°C	−145°C
bp:	−0.5°C	−12°C

Writing the Structural and Condensed Formulas of Isomers

Alkanes with five carbons have the molecular formula C_5H_{12}. We can show the five carbon atoms in a chain by writing the carbon atoms in a row. Then we add the hydrogen atoms to complete the rest of the four bonds to each carbon atom:

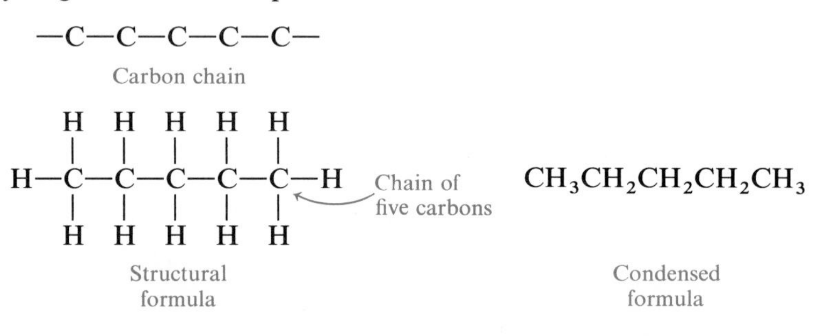

To find the formula of another isomer, write a continuous carbon chain that has four carbon atoms, one less than in the preceding straight chain. Use the remaining carbon atom as a side chain or branch. Add the appropriate number of hydrogen atoms to complete the structural formula and condensed formula.

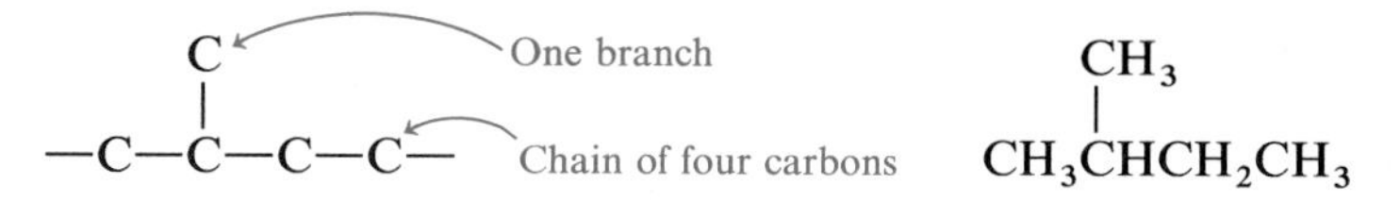

Reducing the continuous carbon chain to three carbons leaves two carbon atoms that can be attached as side chains. A different structure is obtained, which means that it is another isomer of C_5H_{12}:

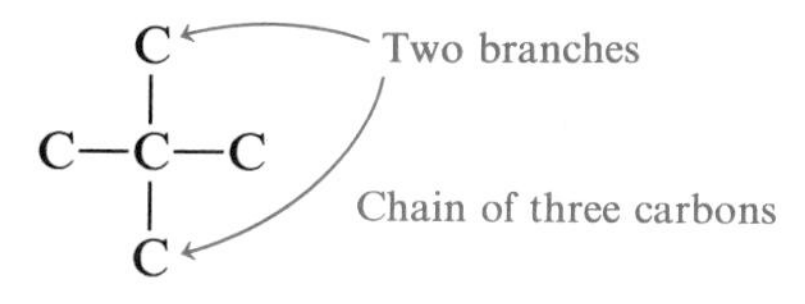

Since we were able to write three different structures with the same molecular formula, we can say that C_5H_{12} has three isomers. (See Figure 11.6.)

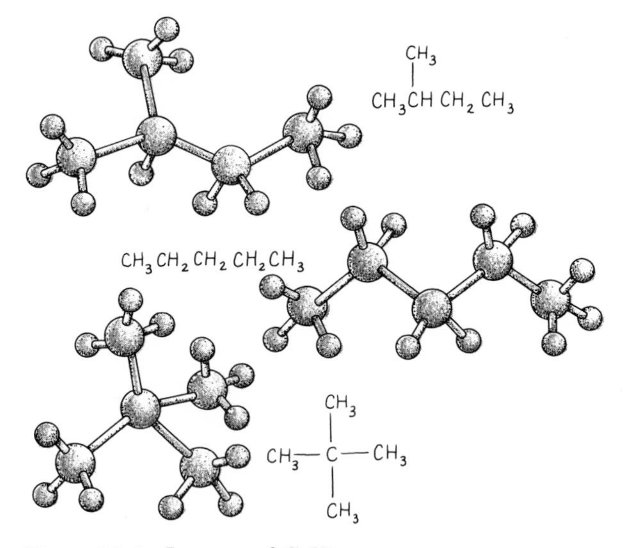

Figure 11.6 Isomers of C_5H_{12}.

You also need to become aware of the location of a side group relative to the end of a carbon chain. The following condensed formulas represent the same compound. They are identical and are not isomers. Both show a single side group on a carbon atom next to one end of the chain:

$$\begin{array}{c} CH_3 \\ | \\ CH_3CHCH_2CH_3 \end{array} \quad \text{is the same as} \quad \begin{array}{c} \qquad\quad CH_3 \\ \qquad\quad | \\ CH_3CH_2CHCH_3 \end{array}$$

Naming Isomers

Since isomers have different structural formulas, they can be differentiated by naming them. For example, we will write the isomers and names of C_5H_{12}:

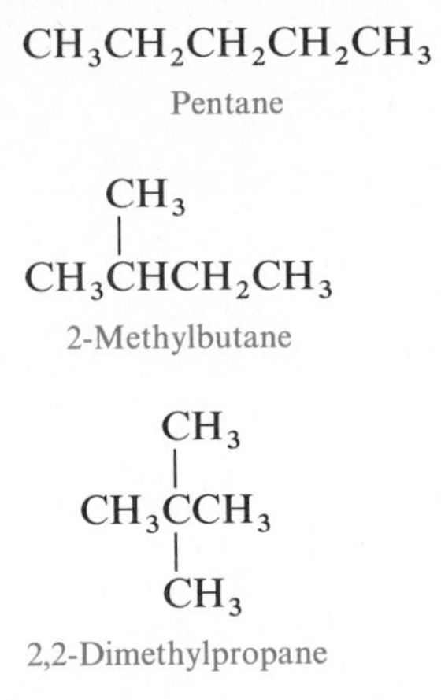

Pentane

2-Methylbutane

2,2-Dimethylpropane

Example 11.9 There are five isomers with the molecular formula C_6H_{14}. Write out the condensed formulas and names of the isomers.

Solution: One of the isomers is the straight-chain structure with six carbon atoms:

$$CH_3CH_2CH_2CH_2CH_2CH_3$$

Hexane

Two isomers can be written with one-carbon branches attached to a chain of five carbon atoms. These are different structurally, because the CH_3— branch is on a different carbon atom in the longest continuous carbon chain:

$$\begin{array}{c} CH_3 \\ | \\ CH_3CHCH_2CH_2CH_3 \end{array} \qquad \begin{array}{c} CH_3 \\ | \\ CH_3CH_2CHCH_2CH_3 \end{array}$$

2-Methylpentane 3-Methylpentane

Two more isomers can be written with two methyl branches attached to a chain of four carbon atoms:

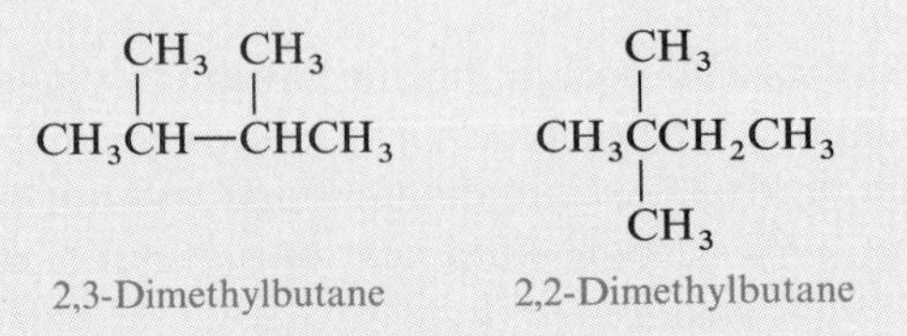

2,3-Dimethylbutane 2,2-Dimethylbutane

Any other combinations of atoms will be identical to one of the above structures.

11.7 Cycloalkanes

OBJECTIVE Write the name and structure of a cycloalkane.

Sometimes the ends of an alkane are joined, forming a ring structure. An alkane having a ring structure is called a *cycloalkane*. These cyclic compounds are often represented on paper by geometrical shapes. Each corner of the figure represents one carbon atom and the appropriate number of hydrogen atoms. The single lines indicate that the compound is a saturated hydrocarbon. The cycloalkanes have two fewer hydrogen atoms because of the cyclic structure. Their names are derived from the longest carbon chain, with the prefix cyclo- placed in front of the alkane. Some examples of cycloalkanes are listed below:

Alkane	*Structural Formula*	*Cycloalkane*	*Structural Formula*	*Geometric Formula*
Propane	$CH_3CH_2CH_3$	Cyclopropane	CH_2 bonded to $CH_2—CH_2$ (three-membered ring)	△
Butane	$CH_3CH_2CH_2CH_3$	Cyclobutane	$CH_2—CH_2$ / $CH_2—CH_2$ (four-membered ring)	□
Pentane	$CH_3CH_2CH_2CH_2CH_3$	Cyclopentane	CH_2 / CH_2 CH_2 / $CH_2—CH_2$ (five-membered ring)	⬠

Naming Cycloalkanes with Side Groups

Side groups on a cycloalkane are named by their alkyl names. With just one side group, no numbers are needed because every carbon atom in the ring is equal. By rotating the ring, the branch can appear on each corner:

Name	*Meaning*	*Formula*
Methylcyclopropane	A $CH_3—$ attached to a ring of three carbons	CH_3 on a triangle
Ethylcyclopentane	A $CH_3CH_2—$ attached to a ring of five carbons	CH_2CH_3 on a pentagon

However, if there are two or more branches on a cyclic structure, they can have more than one possible arrangement, that is, there can be more than one isomer. Therefore, the carbon atoms in the ring are numbered. The carbon attached to one of the side groups is designated as carbon atom 1. The numbers are assigned to the rest of the ring by numbering from carbon 1 to give the lowest possible numerical order as you go around the ring:

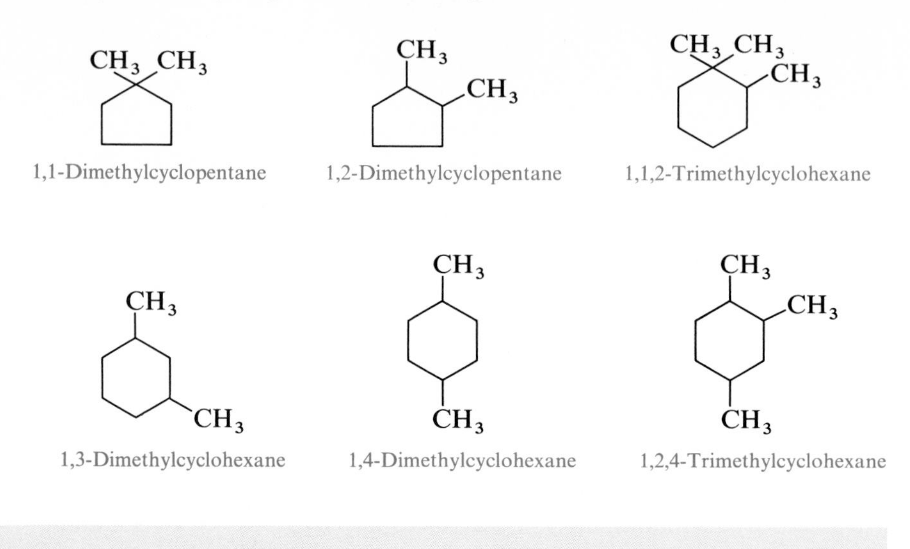

1,1-Dimethylcyclopentane 1,2-Dimethylcyclopentane 1,1,2-Trimethylcyclohexane

1,3-Dimethylcyclohexane 1,4-Dimethylcyclohexane 1,2,4-Trimethylcyclohexane

Example 11.10 Name the following cycloalkanes:

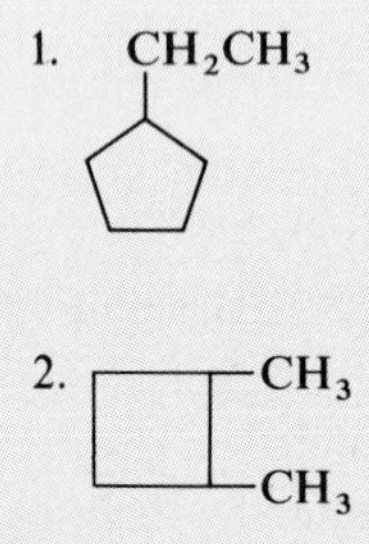

Solution

1. The geometric figure, a pentagon, indicates the five-carbon cycloalkane cyclopentane. The two-carbon branch is an ethyl group. No numbering is needed with a single branch on a cyclic alkane. Thus, the name of the cycloalkane is

   Ethylcyclopentane

2. The geometric figure is a square, which represents cyclobutane. The two branches are methyl groups, which should be numbered to indicate their relationship to each other. The carbon holding one of the methyl groups is designated as carbon 1. The cyclobutane ring is numbered from carbon 1 toward the next branch. Thus, the compound is called

   1,2-Dimethylcyclobutane

Example 11.11 Write the structure for 1,3-dimethylcyclohexane.

Solution: The parent chain of this compound is a ring structure with six carbons, which is cyclohexane. The rest of the name tells us that there are two methyl groups attached to the cyclohexane. We attach one methyl group to a carbon, which we designate as carbon 1. Counting around the figure brings us to carbon 3, on which we place another methyl group:

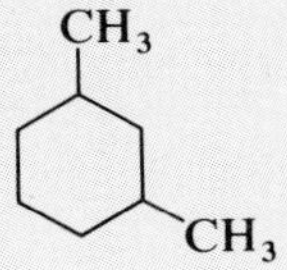

CH_3

CH_3

11.8 Chemical Properties of Alkanes

OBJECTIVE Write a chemical equation for the combustion or halogenation of an alkane.

The alkanes are relatively inert to chemical reaction; however, they are reactive with oxygen and the halogens fluorine, chlorine, and bromine.

In the presence of oxygen, alkanes burn at high temperatures to produce carbon dioxide, water, and energy in the form of heat. Most of the fuels we use for heating, cooking, running a car, or producing electricity are fossil fuels and include many alkanes. Methane is the major component of natural gas, the fuel used in gas appliances and in many laboratories. Liquefied propane gas (LPG) is used as a fuel for camping equipment and in homes without direct gas lines. Butane, another alkane, is the fuel in butane lighters. The burning of an alkane is called *combustion* and can be expressed by the following equations:

combustion

Combustion

$$\text{Alkane} + O_2 \longrightarrow CO_2 + H_2O + \text{Heat}$$

Examples of Combustion

$$CH_4 + 2O_2 \longrightarrow CO_2 + 2H_2O + \text{Heat}$$

$$C_3H_8 + 5O_2 \longrightarrow 3CO_2 + 4H_2O + \text{Heat}$$

Example 11.12 One of the components of gasoline is pentane, C_5H_{12}. Write and balance the equation for the complete combustion of pentane as it undergoes reaction in the engine of a car.

Solution: The reactants of this combustion reaction are pentane and oxygen. The products of complete combustion are carbon dioxide, water, and energy:

Unbalanced Equation

$$C_5H_{12} + O_2 \longrightarrow CO_2 + H_2O + \text{Energy}$$

The equation is balanced by placing coefficients in front of the appropriate formulas:

Balanced Equation

$$C_5H_{12} + 8O_2 \longrightarrow 5CO_2 + 6H_2O + \text{Energy}$$

Halogenation of Alkanes

In the *halogenation* reaction, a halogen atom (fluorine, chlorine, or bromine) is substituted for a hydrogen atom in the alkane. The halogenation of an alkane requires the presence of heat or light; it will not occur in the dark. The products

HEALTH NOTE

Combustion Reactions

As we discuss the various reactions of organic compounds, we will look at similar reactions in the body. For example, the body uses the fuel glucose along with oxygen to provide heat and energy for the body tissues. The metabolic combustion of glucose occurs in many small steps, but the end products are carbon dioxide and water—the same as in the combustion reactions we have just studied:

$$\underset{\text{Glucose (from foods)}}{C_6H_{12}O_6} + \underset{\text{Oxygen (from air)}}{6O_2} \longrightarrow \underset{\text{Released from lungs and kidneys}}{6CO_2 + 6H_2O} + \underset{\text{Utilized by cells}}{\text{Energy}}$$

You probably know that it is dangerous to use gas appliances in a closed room or to run a car in a closed garage, where ventilation is not adequate. This limited supply of oxygen causes an incomplete combustion of the fuel, resulting in the production of carbon monoxide and water:

$$\underset{\text{Methane}}{2CH_4} + \underset{\text{Oxygen (limited)}}{3O_2} \longrightarrow \underset{\text{Carbon monoxide}}{2CO} + 4H_2O + \text{Energy}$$

The danger of carbon monoxide is caused by its strong affinity for the hemoglobin in the blood—an affinity 200 times greater than that of carbon dioxide for hemoglobin. Carbon dioxide (CO_2) dissociates from the hemoglobin and diffuses out of the lungs; carbon monoxide (CO) binds so tightly that the sites for oxygen on the hemoglobin are not available to pick up the oxygen needed for the survival of the cells. Oxygen starvation and death may result if carbon monoxide inhalation is not treated immediately.

of the reaction are a haloalkane and a hydrogen halide. The use of heat or light to produce the reaction is indicated at the reaction arrow:

Halogenation of an Alkane

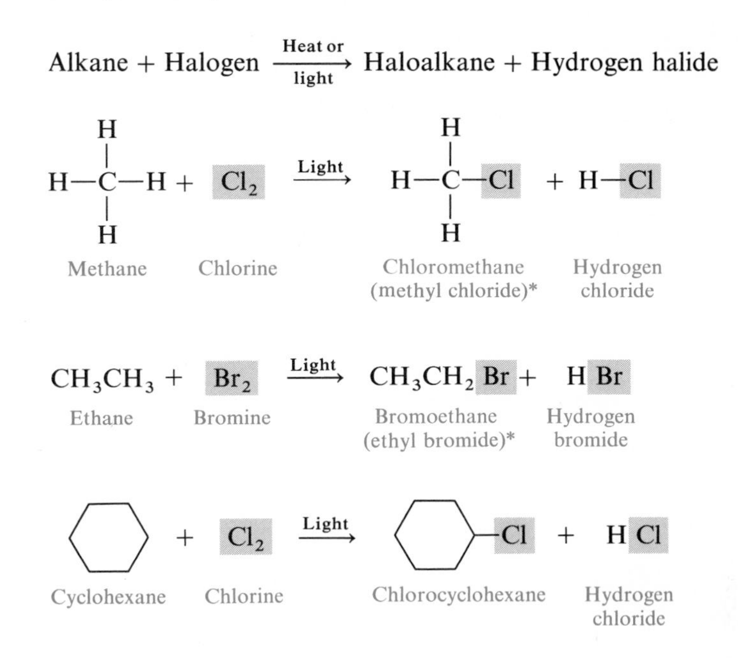

The halogenation reaction does not need to stop with the substitution of just one hydrogen atom of the alkane. The halogenated products may continue to react with more halogen atoms and form a mixture of halogenated compounds:

$$CH_4 + Cl_2 \xrightarrow{\text{Light}} \underset{\text{Chloromethane}}{CH_3Cl} + HCl$$

$$CH_3Cl + Cl_2 \xrightarrow{\text{Light}} \underset{\text{Dichloromethane}}{CH_2Cl_2} + HCl$$

$$CH_2Cl_2 + Cl_2 \xrightarrow{\text{Light}} \underset{\text{Trichloromethane (chloroform)*}}{CHCl_3} + HCl$$

$$CHCl_3 + Cl_2 \xrightarrow{\text{Light}} \underset{\text{Tetrachloromethane}}{CCl_4} + HCl$$

*The names in parentheses are the common names.

Example 11.13 Complete the following equations for the halogenation of alkanes. Give formulas of monohalogenated products only:

1. $CH_4 + Br_2 \xrightarrow{\text{Light}}$

2. 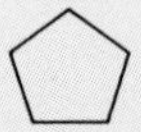 (cyclopentane) $+ Cl_2 \xrightarrow{\text{Light}}$

Solution

1. The reaction of methane with bromine in the presence of light will cause the substitution of a hydrogen atom on the alkane by a bromine atom:

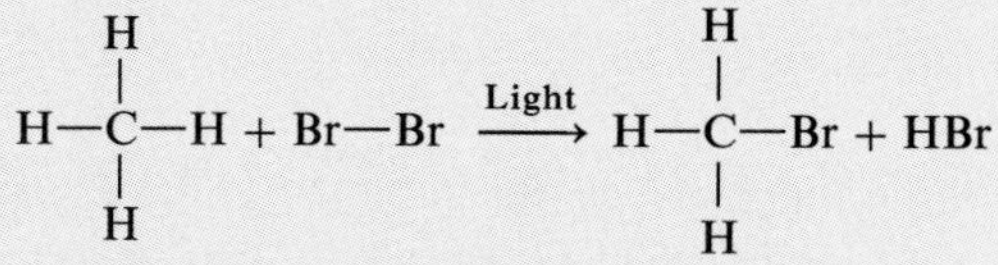

$$\text{H—}\overset{\text{H}}{\underset{\text{H}}{\text{C}}}\text{—H} + \text{Br—Br} \xrightarrow{\text{Light}} \text{H—}\overset{\text{H}}{\underset{\text{H}}{\text{C}}}\text{—Br} + \text{HBr}$$

$$CH_4 + Br_2 \xrightarrow{\text{Light}} CH_3Br + HBr$$

2. A cycloalkane will react with chlorine in the presence of light to produce a halogenated cycloalkane. Since all of the carbon atoms in the ring structure are equal, any of the five carbon atoms in the structure can undergo halogenation:

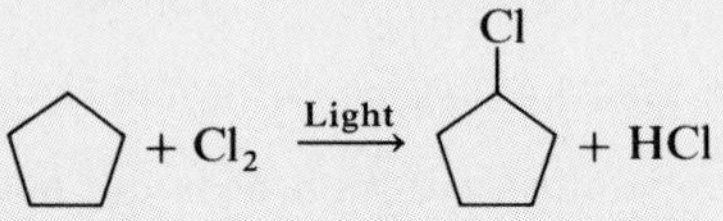

(cyclopentane) $+ Cl_2 \xrightarrow{\text{Light}}$ (chlorocyclopentane, Cl) $+ HCl$

11.9 Haloalkanes (Alkyl Halides)

OBJECTIVE Write the name and structure of a haloalkane.

The halogenation reaction of an alkane produces derivatives of alkanes called *haloalkanes* or *alkyl halides*. In the IUPAC system, the halogen is named as a substituent on the longest carbon chain:

F = *fluoro-*
Cl = *chloro-*
Br = *bromo-*
I = *iodo-*

The carbon atom to which the halogen atom is bonded is numbered. There are some common names that still persist in which the compound is named as an alkyl halide. The carbon group is named as an alkyl group in front of the name of the halogen atom. Common names are listed in parentheses:

Haloalkane (Alkyl Halide)

Group	X (X = F, Cl, Br, I)
IUPAC Name	*halo*alkane
Common Name	*alkyl* halide
CH_3Cl	Chloromethane (methyl chloride)
$CH_3CH_2CH_2Br$	1-Bromopropane (*n*-propyl bromide)
$CH_3CH(F)CH_2CH_2F$	1,3-Difluorobutane
$CH_3CH(CH_3)CH_2CH(Cl)CH_2CH_3$	4-Chloro-2-methylhexane (alphabetical order)

Example 11.14 Name the following haloalkanes:

1. $CH_3CH_2CH_2Br$

2. 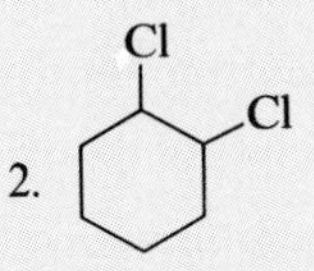

Solution

1. The IUPAC name is obtained from the name of the parent chain, *propane.* The bromine atom is indicated by the prefix *bromo-*. The location of the bromine atom is on the first carbon atom (carbon 1) of the carbon chain. (Remember that the carbon chain is numbered from the carbon atom nearest the bromine atom.) Thus we name the haloalkane

   IUPAC name: 1-Bromopropane

   The common name of this compound is derived from the alkyl name of a three-carbon group attached at the end carbon atom, *n*-propyl. The bromine atom is named bromide. Thus we have

   Common name: *n*-Propyl bromide

2. The parent chain is the cyclic chain with six carbon atoms, which is cyclohexane. The ring must be numbered because there are two chlorine atoms bonded to it. Thus we obtain the following:

   IUPAC name: 1,2-Dichlorocyclohexane

   There is no common name for this compound.

HEALTH NOTE

Chlorinated Hydrocarbons in Medicine, Industry, and Agriculture

Some haloalkanes have been used as anesthetics. *Trichloromethane*, or *chloroform*, ($CHCl_3$), was used for many years as an inhaled anesthetic, but its use was discontinued because of its toxicity and possible carcinogenic (cancer-causing) effects. Another haloalkane, called *halothane*, is in wide use today as an inhaled anesthetic:

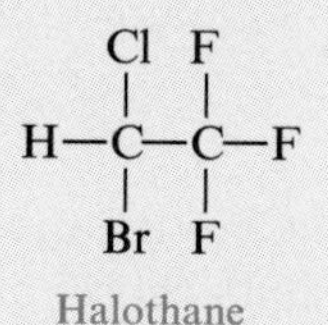

Halothane

Chloromethane (methyl chloride, CH_3Cl), and chloroethane (ethyl chloride, CH_3CH_2Cl) are used as topical anesthetics. When sprayed on the skin, they produce local anesthesia by lowering the temperature in the treated area below 0°C. This freezes the nerve endings and makes local surgery near the surface of the skin painless.

Other haloalkanes are used as refrigerants. These include the freon gases, which are a group of fluorinated and chlorinated methane compounds:

$CFCl_3$ Freon 11
CF_2Cl_2 Freon 12

Some of the major components of insecticide sprays and dusts are chlorinated hydrocarbons such as DDT, Aldrin, Chlordane, Dieldrin, and Lindane. When they are absorbed by the body, they act as poisons. Symptoms of pesticide poisoning in humans include dizziness, weakness, and nervous system excitability. Paralysis and coma may follow.

In the 1940s, DDT was effective in controlling malaria, typhus, and sleeping sickness and was hailed as a "miracle." By the 1950s, many insects had developed a

The following tables summarize the naming and the reactions of the alkanes:

Summary of Naming the Alkanes

Class	Example	Structure
Alk*ane*	Prop*ane*	$CH_3CH_2CH_3$
*Cyclo*alk*ane*	*Cyclo*prop*ane*	△
*Halo*alk*ane*	2-*Bromo*prop*ane*	$CH_3CH(Br)CH_3$

resistance to the pesticide. In addition, DDT was found to be concentrated in the food chain. The structures of DDT and Lindane are given below:

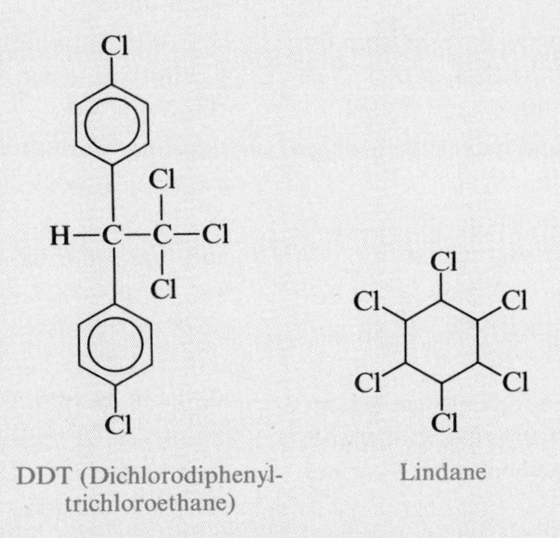

DDT (Dichlorodiphenyl-trichloroethane)   Lindane

DDT, a nonpolar compound, is soluble in nonpolar substances. In lakes and ponds where spraying of DDT occurred, small amounts of the pesticide were stored in the lipids of microorganisms. Because these microorganisms served as food for larger organisms and small fish, the DDT became more concentrated in the body fat of larger animals. Predatory birds and animals at the top of the food chain had the greatest concentrations of DDT. These elevated DDT levels were thought to contribute to the decline of some species of hawks, eagles, falcons, and brown pelicans. Recently, many of the chlorinated hydrocarbons, including DDT, have been banned from use in widespread spraying

DDT is also stored in human adipose tissue (body fat) and has been found in human breast milk.

Summary of Reactions of the Alkanes

Combustion

$$\text{Alkane} + O_2 \longrightarrow CO_2 + H_2O + \text{Heat}$$
$$CH_4 + 2O_2 \longrightarrow CO_2 + 2H_2O + \text{Heat}$$

Halogenation (substitution)

$$\text{Alkane} + \text{Halogen} \xrightarrow{\text{Light}} \text{Haloalkane} + \text{Hydrogen halide}$$

$$CH_4 + Cl_2 \xrightarrow{\text{Light}} CH_3Cl + HCl$$

Glossary

alkane An organic compound that consists only of carbon and hydrogen atoms and has only single carbon-to-carbon bonds.

alkyl groups The name of a side group derived from the alkane having the same number of carbon atoms. The alkane name is changed to -yl to indicate the branch.

branch Portions of organic compounds that are attached to a longer continuous chain.

branched-chain compound A hydrocarbon containing at least one side group attached to a parent chain.

combustion A chemical reaction in which a hydrocarbon reacts with oxygen to produce CO_2, H_2O, and heat.

condensed formula A formula that illustrates the arrangement of carbon atoms in an organic compound. The number of hydrogen atoms attached to each carbon atom is expressed as a subscript.

cycloalkane An alkane consisting of a ring structure.

haloalkane A derivative of an alkane in which a hydrogen atom has been replaced by a halogen atom.

halogenation A reaction in which a halogen atom replaces a hydrogen atom in an alkane. Light or heat is required for the reaction.

hydrocarbon An organic compound consisting of only carbon and hydrogen atoms.

isomers Two or more organic compounds that have the same molecular formula but different arrangements of the atoms.

IUPAC International Union of Pure and Applied Chemistry, which devised a system of naming organic compounds.

molecular formula A formula that indicates the types of atoms in a molecule, but not their arrangement.

organic compound Compounds that contain carbon and that are usually composed of covalent bonds. Most are nonpolar, insoluble in water, soluble in nonpolar solvents, and have low melting and boiling points.

parent chain The longest continuous chain of carbon atoms in an organic compound.

saturated hydrocarbon A compound of carbon and hydrogen atoms with only single bonds between the carbon atoms; an alkane.

straight chain A hydrocarbon consisting of carbon atoms in a continuous chain.

structural formula A formula in which there is a line to represent each bond that exists in a compound, thereby showing the arrangement of atoms in the compound.

tetrahedron The three-dimensional arrangement of atoms attached to a central carbon atom.

Problems

Properties of Organic Compounds (Objective 11.1)

11.1 Indicate whether each of the following statements describes an inorganic or organic compound:

a. A compound that melts at 800°C and conducts electricity when it dissolves in water.
b. A liquid that dissolves in cyclohexane.
c. A compound that vaporizes at 35°C and burns with oxygen.
d. Ampicillin, a compound whose formula is $C_{16}H_{19}N_3O_4S$.

Structural Formulas of Organic Compounds (Objective 11.2)

11.2 Write the structural formulas for the following organic structures by adding the correct number of hydrogen atoms:

a. C—C—C—C—C

b.
```
     C
     |
C—C—C—C
```

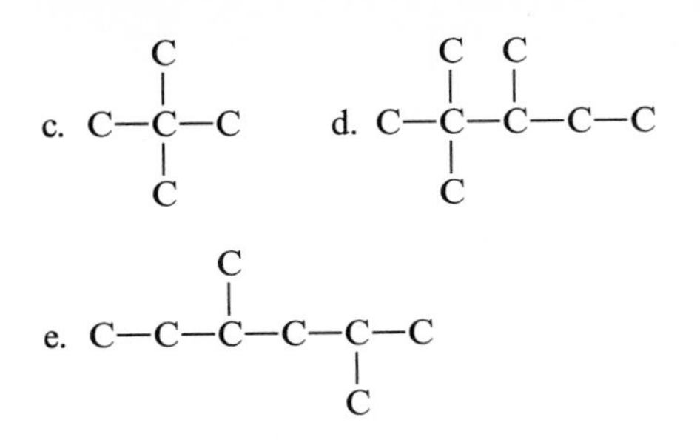

Condensed Formulas (Objective 11.3)

11.3 Write the condensed formula for each of the following structural formulas:

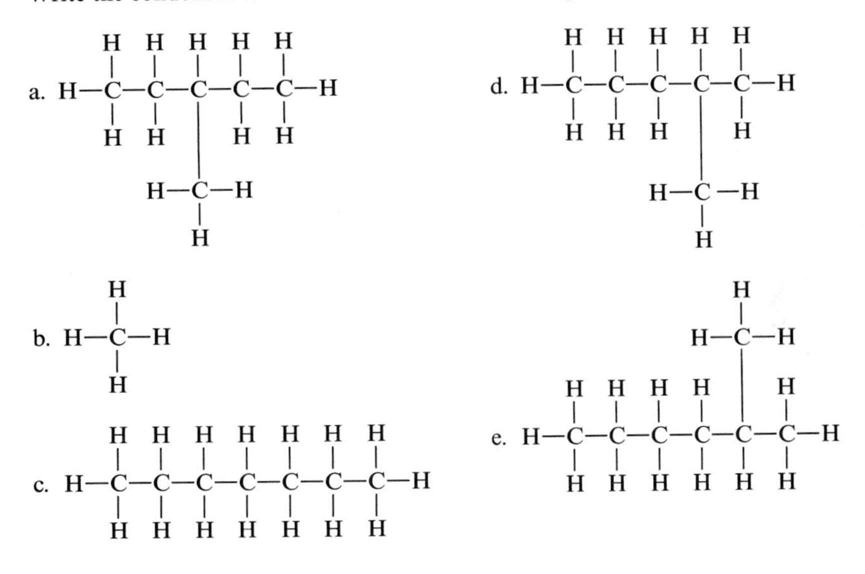

11.4 Write the structural formula for each of the following:

a. CH_3CH_3

b. $CH_3CH_2CH_2CH_2CH_3$

c.
$$\begin{array}{l} \quad CH_3 \\ \quad | \\ CH_3CHCHCH_3 \\ \qquad\ | \\ \qquad CH_3 \end{array}$$

d.
$$\begin{array}{l} \quad CH_3 \\ \quad | \\ CH_3CHCH_2CHCH_2CH_2CH_3 \\ \qquad\qquad\ | \\ \qquad\qquad CH_2 \\ \qquad\qquad\ | \\ \qquad\qquad CH_3 \end{array}$$

Straight-Chain Alkanes (Objective 11.4)

11.5 Write the IUPAC name for the following alkanes:

a. $CH_3CH_2CH_2CH_3$ c. $CH_3CH_2CH_2CH_2CH_2CH_3$
b. CH_3CH_3 d. CH_4

11.6 Write the condensed formula of each of the following:

a. propane c. pentane
b. heptane d. octane

Naming the Branched-Chain Alkanes (Objective 11.5)

11.7 Write the IUPAC name for the following alkanes:

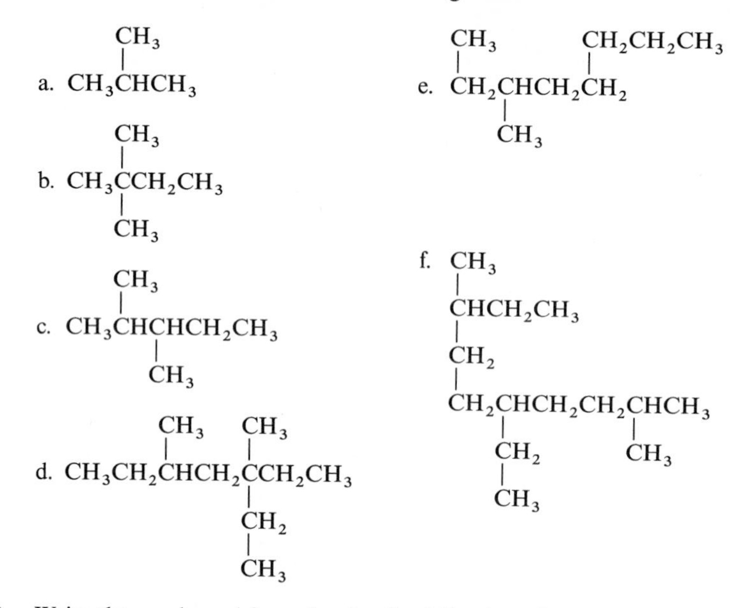

11.8 Write the condensed formulas for the following alkanes:

a. 2-methylpentane d. 3-ethyl-2-methylhexane
b. 3,3-dimethylhexane e. 2-methyl-4-isopropylheptane
c. 2,3-dimethylbutane f. 2,2,4,4-tetramethylhexane

Isomers (Objective 11.6)

11.9 Write all the isomers of C_5H_{12}.

11.10 Write the structures of all the isomers of heptane that have a five-carbon parent chain with two methyl side groups.

11.11 Indicate whether the following structures are isomers or identical:

a. $CH_3CH_2CH_2CH_3$ $\quad$ $CH_3\overset{\overset{\large CH_3}{|}}{C}HCH_3$

b. $CH_3CH_2CH_2CH_2CH_2CH_3$ $\quad$ $CH_3\underset{\underset{\large CH_3}{|}}{\overset{\overset{\large CH_3}{|}}{C}}CH_2CH_3$

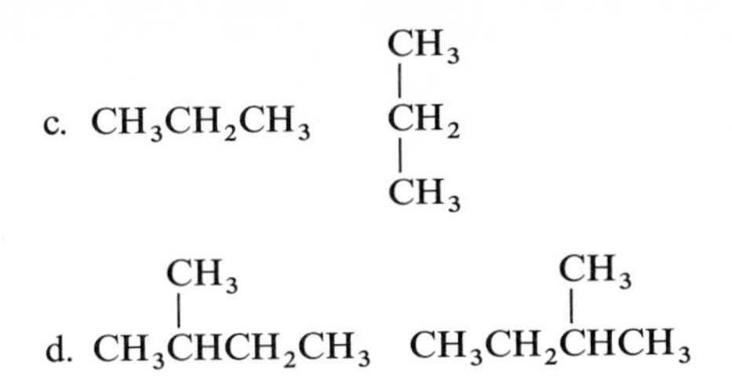

Cycloalkanes (Objective 11.7)

11.12 Write the IUPAC name for the following cycloalkanes:

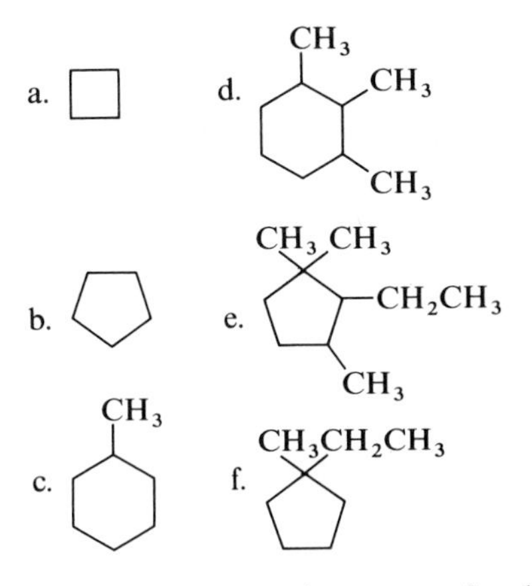

11.13 Draw the condensed structures for the following cycloalkanes:

a. cyclopentane
b. methylcyclohexane
c. ethylcyclobutane
d. 1,1-dimethylcyclobutane
e. 1,3,5-trimethylcyclohexane

11.14 Draw and name all the isomers of dimethylcyclopentane.

11.15 Draw and name all the cyclic isomers of C_5H_{10}.

Chemical Properties of Alkanes (Objective 11.8)

11.16 Write balanced equations for the complete combustion of the following:

a. methane, CH_4
b. propane, C_3H_8
c. butane, C_4H_{10}
d. cyclobutane, C_4H_8
e. octane C_8H_{18}

11.17 Write the formulas of the monohalogenated products of the following reactions:

a. $CH_3CH_3 + Cl_2 \xrightarrow{\text{Light}}$

b. ⬡ $+ Br_2 \xrightarrow{\text{Light}}$

c. $CH_3CH_2CH_3 + Cl_2 \xrightarrow{\text{Light}}$

d. □ $+ Br_2 \xrightarrow{\text{Light}}$

e. $CH_4 + Cl_2 \xrightarrow{\text{Dark}}$

Haloalkanes (Objective 11.9)

11.18 Give the IUPAC name for the following compounds:

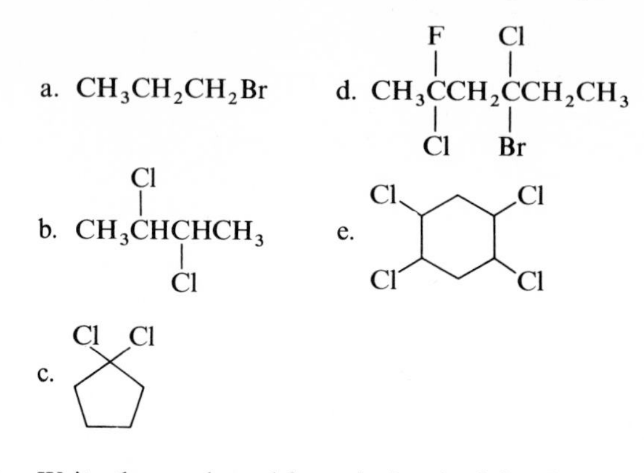

11.19 Write the condensed formula for the following:

a. ethyl chloride
b. 1-bromo-2-chloropropane
c. 1,1-dichlorocyclopropane
d. 2-chloro-3-methylhexane
e. 3,3-dichloro-2-methylpentane
f. 1,1,2,4-tetrachlorocyclopentane

12 Alkenes, Alkynes, Aromatics, Alcohols, and Ethers

Objectives

12.1 Write the name and structure of an alkene or alkyne.

12.2 Given the name and/or structure of an alkene, write the equation for the addition of hydrogen, halogens, or hydrogen halide.

12.3 Write the name and structure of an aromatic hydrocarbon.

12.4 Write the name and structure of an alcohol, phenol, or thiol.

12.5 Classify an alcohol as primary, secondary, or tertiary.

12.6 Write an equation for the hydration of an alkene to form an alcohol.

12.7 Write the equation for the dehydration of an alcohol.

12.8 Write the name and structure of an ether.

12.9 Write an equation for the formation of an ether.

Scope

The presence of a double bond in an alkene or a triple bond in an alkyne provides an organic compound with a reactive site. In alcohols and ethers, the reactive site contains oxygen. This reactive area is the portion of the molecule where various reactants can attach and cause changes in the compound. Such a reactive area is called a *functional group*.

functional group

The types of reactions that alkenes and alcohols undergo are determined by their functional groups. A functional group is an atom or a group of atoms that gives a particular and consistent set of characteristics to that compound.

Compounds containing the same kind of functional group enter into the same kinds of reactions. The presence of functional groups allows us to organize organic compounds into classes or families. We will begin to study these families, their functional groups, and chemical reactions.

12.1 Alkenes and Alkynes: Unsaturated Hydrocarbons

OBJECTIVE Write the name and structure of an alkene or alkyne.

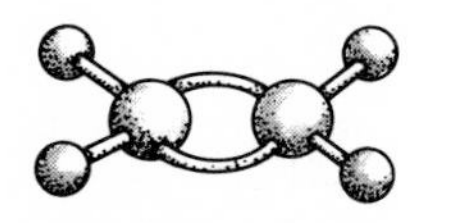

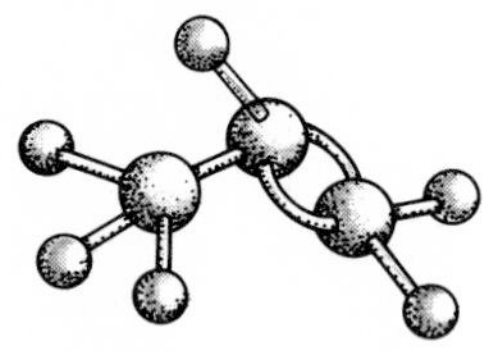

Figure 12.1 Ball-and-stick models of alkenes, ethene and propene.

A hydrocarbon containing a double bond is called an *alkene*. This double bond constitutes a reactive site or functional group, which gives the alkene greater chemical reactivity than the corresponding alkane. The double bond occurs when two pairs of electrons are shared between two carbon atoms (see Figure 12.1):

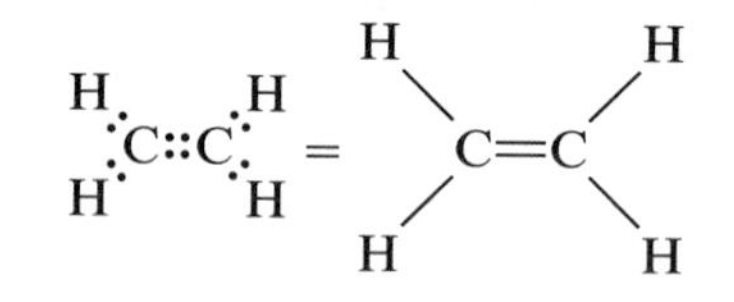

Naming Alkenes

Alkene

Functional group:

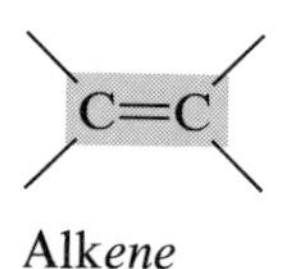

IUPAC name: Alk*ene*
Common name: Alkyl*ene*

The name of an alkene is derived from the alkane with the same number of carbon atoms.

Determine the name of an alkene as follows:

1. Find the longest carbon chain that includes the double bond.
2. Use the root of the corresponding alkane, adding the suffix *-ene* for the alkene name. Some of the simple alkenes also have a common name consisting of the alkyl prefix followed by the *-ene* ending:

$H_2C{=}CH_2$
Ethene
(ethylene)

$CH_3\overset{CH_3}{\overset{|}{C}}{=}CH_2$
2-Methylpropene
(isobutylene)

$CH_3CH{=}CH_2$
Propene
(propylene)

△
Cyclopropene
(cyclopropylene)

3. For alkenes with a carbon chain of four or more carbon atoms, there is more than one possible location for the double bond. Number the chain to give the first carbon atom of the double bond the lowest possible number.

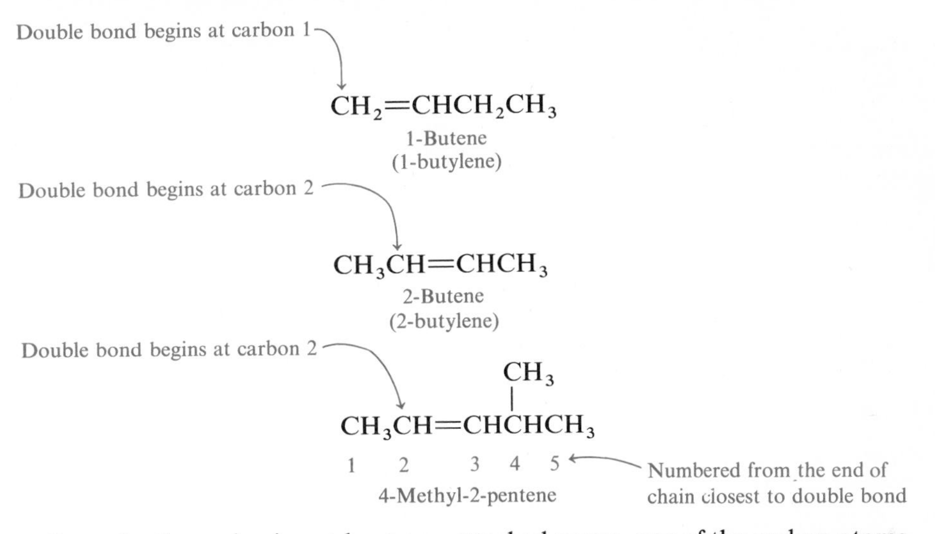

In cycloalkenes having at least one attached group, one of the carbon atoms in the double bond is designated as carbon 1; the other carbon atom becomes carbon 2. The order of numbering proceeds to give the lowest numbers to the attached group(s):

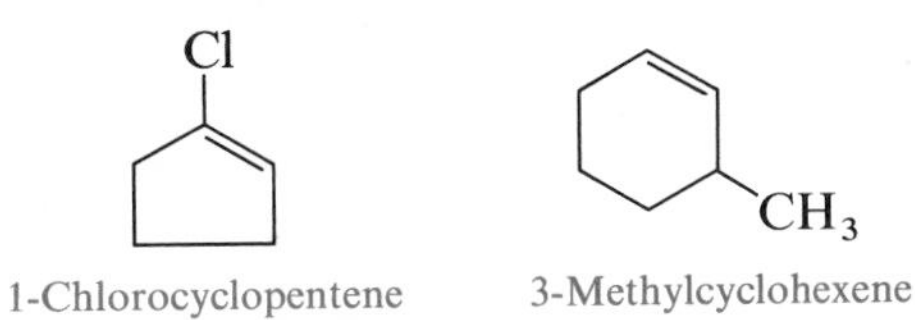

Example 12.1 Write the IUPAC or common names of the following alkenes:

1. $CH_3CH_2CH{=}CH_2$

2. $CH_3CH{=}C(CH_3)CH_2CH_3$ (with CH_3 on the third carbon)

3. Cl (chlorine attached to a cyclohexene ring)

Solution

1. The parent chain is an alkene with four carbon atoms, which is butene. Since the double bond begins at the first carbon, the number 1 is used with the alkene name:

   IUPAC name: 1-Butene

2. This compound is a five-carbon chain with a double bond, which is pentene. When the chain is numbered from the end closest to the double bond, the double bond begins at carbon 2, and the branch, a methyl group, is on carbon 3:

   3-Methyl-2-pentene

3. This compound is a cyclic compound with six carbon atoms and a double bond. The cyclic portion is cyclohexene. The chlorine atom is located on the fourth carbon when the numbering of the carbon atoms begins at the double bond and goes in the direction to give the lowest number to the chlorine:

   4-Chlorocyclohexene

Example 12.2 Write all the isomers of monochlorocyclopentene and name each.

Solution: The name indicates a cyclic parent chain with five carbons and a double bond. A single chlorine atom can be placed on the ring in three different positions to give three different isomers:

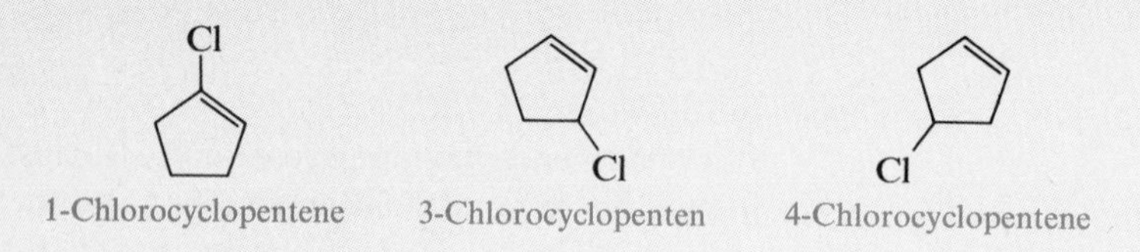

Geometrical Isomers

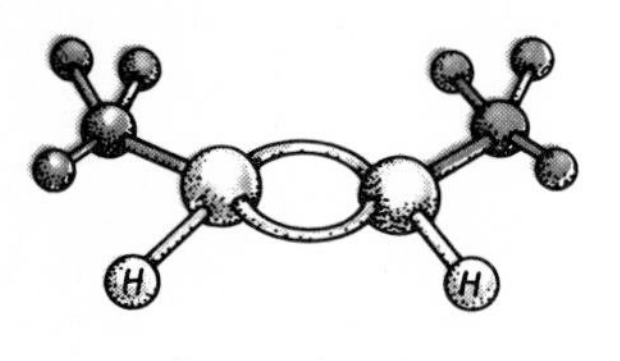

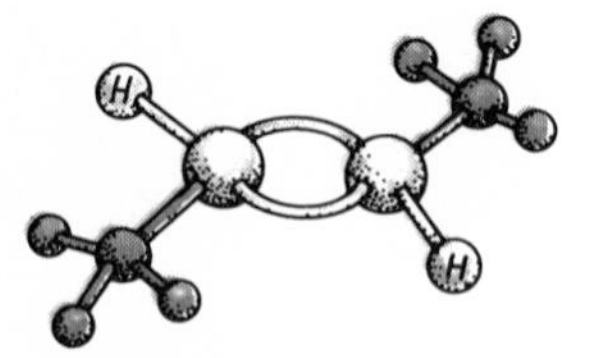

Figure 12.2 Geometric isomers: Ball-and-stick models and figures of cis and trans isomers of ClCH=CHCl.

Whereas there is free rotation about a single bond, the atoms on both sides of the double bond in an alkene are limited in their ability to move freely (rotate) about the double bond. As a result, the attached groups remain on one side of the double bond or the other. When there are two different groups on both carbon atoms in a double bond, we can describe two possible isomers, called *geometrical isomers*. The attachment of the atoms is the same, but they may be attached on the same side of the double bond or on opposite sides. Let us look at the geometrical isomers of 2-butene:

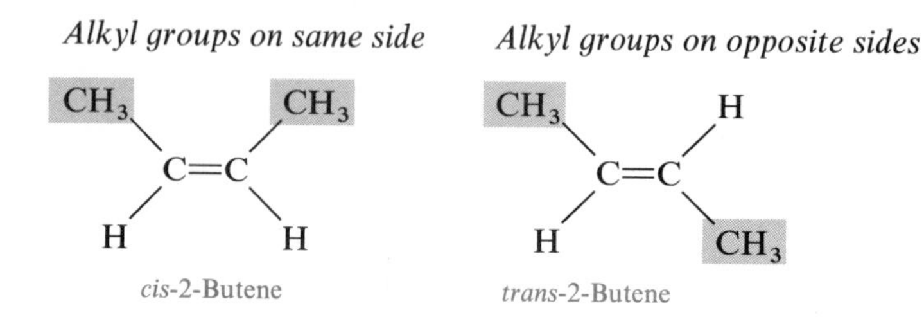

cis and trans

When the methyl groups appear on the same side of the double bond, they are said to be *cis* to each other. If they appear on opposite sides of the double bond, they are *trans* to each other. (See Figure 12.2.)

Example 12.3 Name the following structure and indicate whether it is the cis or trans isomer:

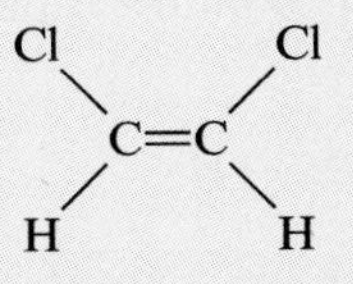

Solution: The carbon chain consists of two carbon atoms that form the double bond. The parent chain is ethene. The two chlorine atoms attached to the double bond are located on the same side. This is the cis isomer of 1,2-dichloroethene:

cis-1,2-Dichloroethene

NATURE NOTE

Insect Attractants

pheromones

In the animal kingdom, organic compounds called *pheromones* are emitted in very low quantities into the air to be carried to other members of the species. Such pheromones may indicate danger, mark a trail, or be sex attractants. Of the many pheromones that have been identified for various species of insects, several contain cis or trans double bonds:

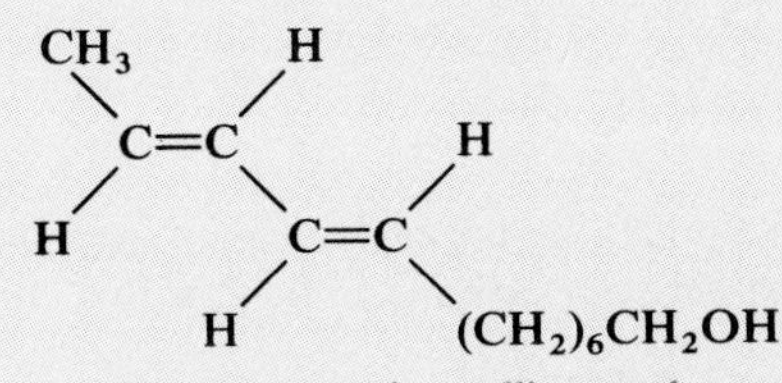

Sex attractant for codling moth

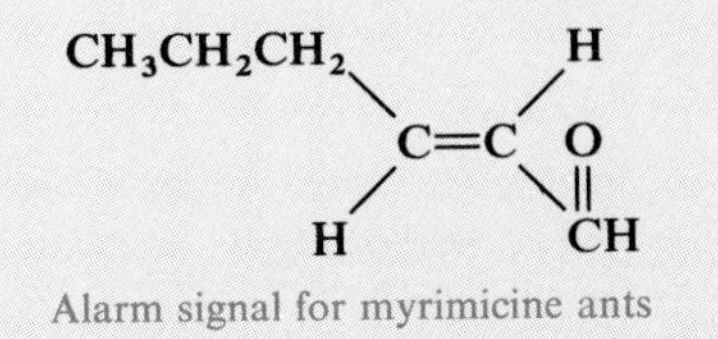

Alarm signal for myrimicine ants

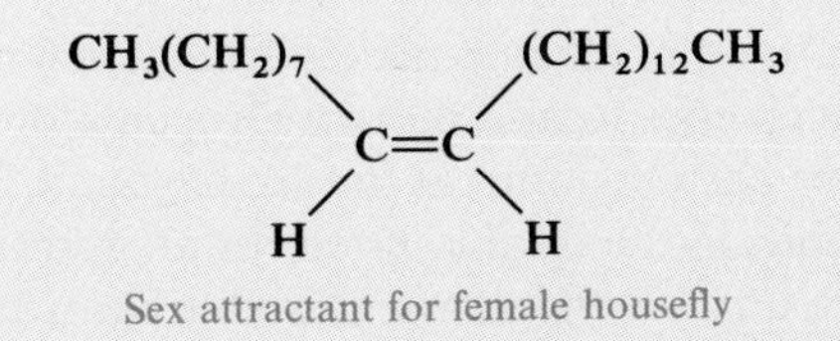

Sex attractant for female housefly

Alkynes

Alkynes are hydrocarbons that contain a triple bond in which two carbon atoms share three pairs of electrons:

$$H{:}C{:::}C{:}H \qquad H{-}C{\equiv}C{-}H$$

In the IUPAC system, alkynes are named by changing the suffix *-ane* of the parent alkane name to *-yne*. In chains of four or more carbon atoms, the alkyne group must be numbered to indicate its location. There are no geometrical isomers for alkynes.

$$CH{\equiv}CH$$
Ethyne (acetylene)

$$CH{\equiv}CCH_3$$
Propyne

$$\begin{array}{c} \qquad\quad CH_3 \\ \qquad\quad | \\ CH_3C{\equiv}CCHCH_3 \end{array}$$
4-Methyl-2-pentyne

Example 12.4 Name the following alkyne:

$$\begin{array}{c} \quad Cl \\ \quad | \\ CH{\equiv}CCHCH_3 \end{array}$$

Solution: Numbering the carbon chain from the end nearest the triple bond puts the triple bond between carbons 1 and 2, with a chlorine atom attached to carbon 3:

3-Chloro-1-butyne

12.2 Reactions of Alkenes

OBJECTIVE Given the name and/or structure of an alkene, write the equation for the addition of hydrogen, halogens, or hydrogen halide.

addition reaction

The double bond of alkenes makes them much more reactive than alkanes. It is possible for alkenes to add more hydrogen atoms, halogen atoms, or hydrogen halides to the carbon atoms at the double bond. The addition of these reagents is going to *saturate* the double bond. In an addition reaction, one of the double bonds is broken and the newly available electrons pair with the electrons of the newly added atoms to form new single bonds. In general, we can picture two atoms (X and Y) bonding to the carbon atoms in the double bond—an addition reaction:

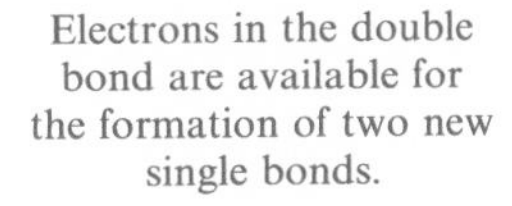

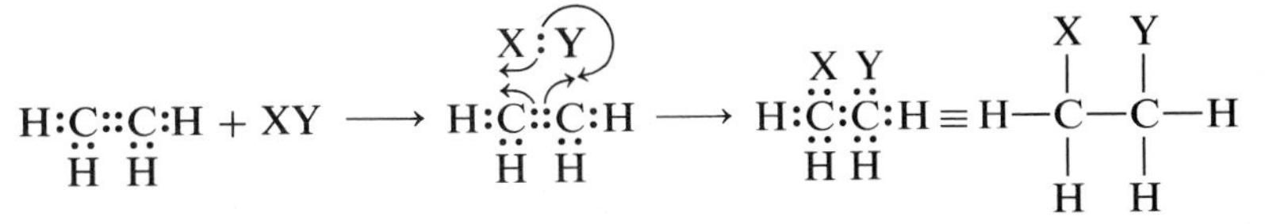

Hydrogenation

addition of hydrogen In the hydrogenation of an alkene, two atoms of hydrogen are added to the double bond. A platinum catalyst is used to facilitate the reaction, but it is not consumed in the overall process. The presence of a catalyst such as platinum is indicated over the arrow for the equation. The effect of hydrogenation is to saturate the double-bond site. In this reaction, an alkene is converted to an alkane:

Hydrogenation

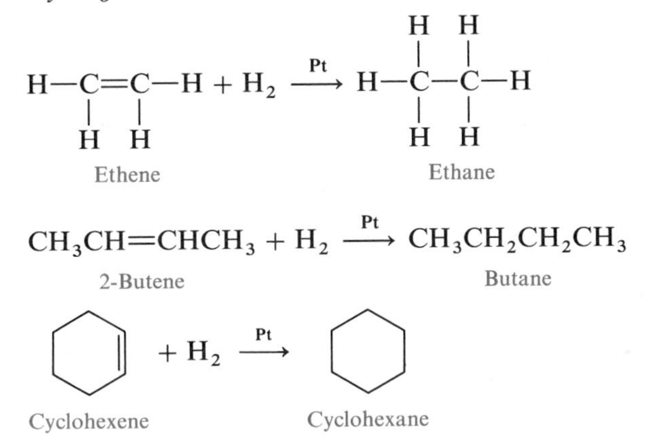

Example 12.5 Write the structure of the product of the following hydrogenation reaction:

$$CH_2{=}CHCH_2CH_2CH_3 + H_2 \xrightarrow{Pt}$$

Solution: The reactant, which is an alkene, will undergo an addition of hydrogen atoms, one at each carbon of the double bond. The product is an alkane:

$$CH_2{=}CHCH_2CH_2CH_3 + H_2 \xrightarrow{Pt} CH_3CH_2CH_2CH_2CH_3$$

(Hydrogen added)

HEALTH NOTE

Hydrogenation of Unsaturated Fats

The process of hydrogenation is used commercially to convert low-melting-point vegetable oils (unsaturated fats), which contain double bonds, into fats with fewer unsaturated sites. The saturated fats have higher melting points and will be solid at room temperatures. Hydrogenation of liquid vegetable oils, such as corn oil or safflower oil, results in the production of more solid fats, such as soft margarine, margarine, and shortening, which are used for various cooking and dietary purposes.

Halogenation

addition of a halogen

When the addition compound is a halogen (chlorine or bromine), both atoms add to the double bond, one halogen atom for each carbon atom. The halogenation of an alkene occurs readily, without any catalysts. The product is a dihaloalkane:

Halogenation of Alkenes

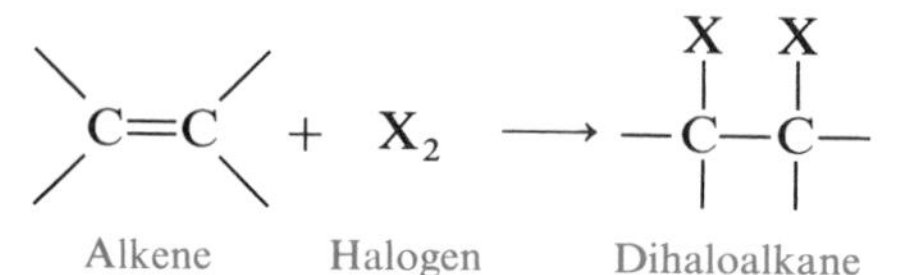

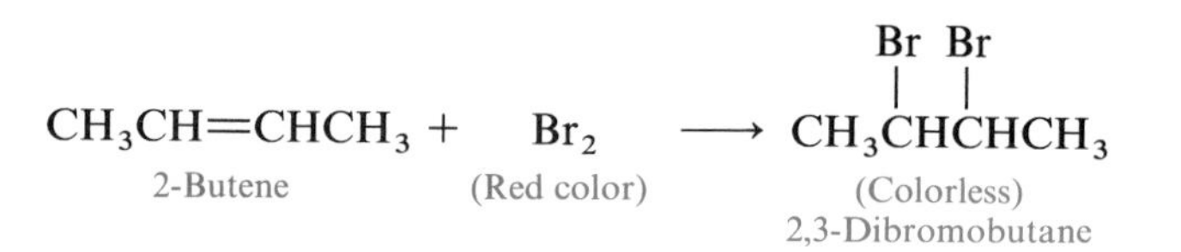

The addition of bromine can be used as a test for the presence of double bonds in organic compounds. If a red bromine solution (Br_2 dissolved in CCl_4) reacts with a compound containing double bonds, the red color quickly disappears, and the solution appears colorless.

Example 12.6 Write the structure of the product in the following reaction:

$$CH_3CH{=}CH_2 + Br_2 \longrightarrow$$

Solution: The addition of bromine to an alkene places a bromine atom on each carbon of the double bond. The bond between the carbons is now a single bond:

$$\underset{\text{Propene}}{CH_3CH{=}CH_2} + Br_2 \longrightarrow \underset{\text{1,2-Dibromopropane}}{CH_3\overset{Br}{\overset{|}{C}}H{-}\overset{Br}{\overset{|}{C}}H_2}$$

Hydrohalogenation

addition of a hydrogen halide

The hydrogen halides (HCl, HBr, and HI) also add easily to the unsaturated (double-bond) site of the alkenes. The hydrogen atom bonds to one of the carbon atoms, and the halogen adds to the other. The addition of the hydrogen halide produces a haloalkane:

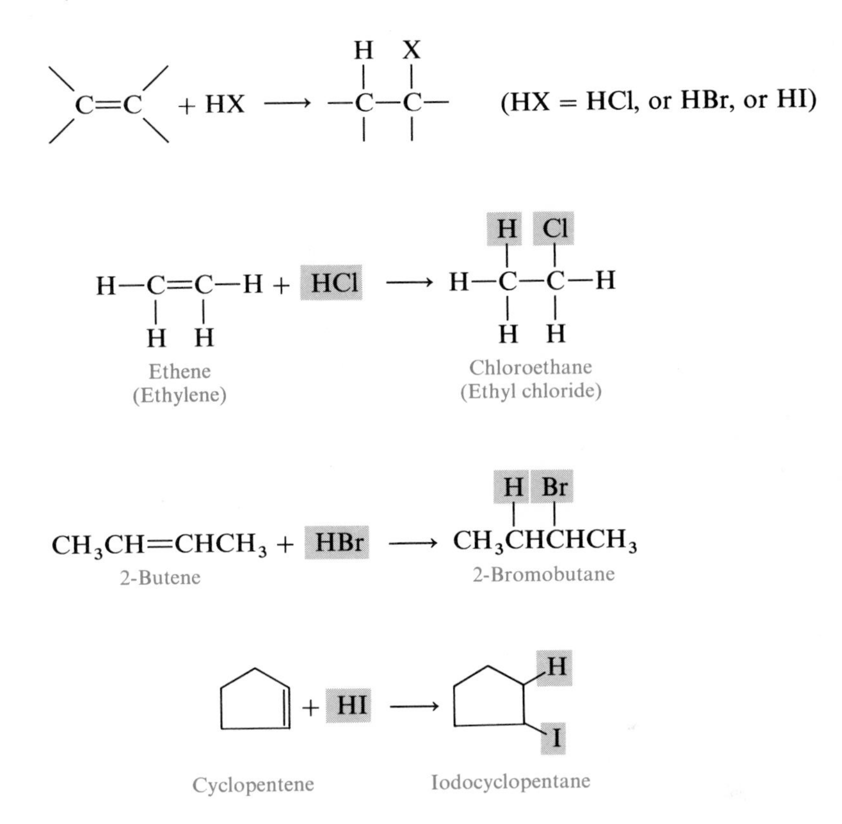

Markovnikov's Rule

Because the two atoms in a hydrogen halide are not identical, some problems can arise in determining which carbon atom of the double bond is to receive the hydrogen atom, as well as which is to receive the halogen atom. We can use a rule formulated by a Russian chemist Vladimir Markovnikov.

Markovnikov's Rule *When adding a compound HX to a double bond, place the hydrogen atom from HX on the carbon atom of the double bond with the most hydrogen atoms.*

HEALTH NOTE

Polymers Made From Alkenes

monomers

Polymers are long-chain molecules made up of repeating smaller units called *monomers.* Small alkene molecules have been used to produce various kinds of polymers. Molecules of the simplest alkene, ethene (ethylene), will react among themselves under conditions of pressure and heat. The symbol *n* preceding the formula indicates a large number (1000–1,000,000) of ethylene molecules taking part in the polymerization reaction (see Figure 12.3):

$$\underset{\text{Ethylene}}{nCH_2{=}CH_2} \xrightarrow{\text{Polymerization}} \underset{\text{Polyethylene}}{-(CH_2{-}CH_2)_n-}$$

polyethylene

Polyethylene is a polymer that is used in the manufacturing of plastic bottles, films, and plastic dinnerware.

polyvinyl chloride

Another monomer unit is used in the production of polyvinyl chloride, a polymer used in the production of plastics, raincoats, plastic pipes and tubing, body replacement parts, phonograph records, and vinyl material. The monomer unit, vinyl chloride, has been determined to be a carcinogen. Those who work with vinyl chloride follow a strict set of safety rules to limit their exposure to the compound.

$$\underset{\text{Vinyl chloride}}{nCH_2{=}\underset{\displaystyle Cl}{\underset{|}{CH}}} \xrightarrow{\text{Polymerization}} \underset{\text{Polyvinyl chloride (PVC)}}{-(CH_2{-}\underset{\displaystyle Cl}{\underset{|}{CH}})_n-}$$

If you were to look at a beginning section of this polymer, it would have the following structure:

$$CH_3{-}\underset{\displaystyle Cl}{\underset{|}{CH}}{-}CH_2{-}\underset{\displaystyle Cl}{\underset{|}{CH}}{-}CH_2{-}\underset{\displaystyle Cl}{\underset{|}{CH}}{-}CH_2{-}\underset{\displaystyle Cl}{\underset{|}{CH}}{-}CH_2{-}\underset{\displaystyle Cl}{\underset{|}{CH}}{-}CH_2{-}\underset{\displaystyle Cl}{\underset{|}{CH}}{-}\text{etc.}$$

Other kinds of alkenes have been used to make polymers used as food wrapping such as Saran Wrap; they have also been used in the production of synthetic fabrics such as Orlon and polyester. Another polymer you might have in your home is Teflon, which is used to make nonstick pans and cooking utensils. The monomer unit for Teflon is tetrafluoroethylene:

$$\underset{\text{Tetrafluoroethylene}}{nCF_2{=}CF_2} \xrightarrow{\text{Polymerization}} \underset{\text{Teflon}}{-(CF_2{-}CF_2)_n-}$$

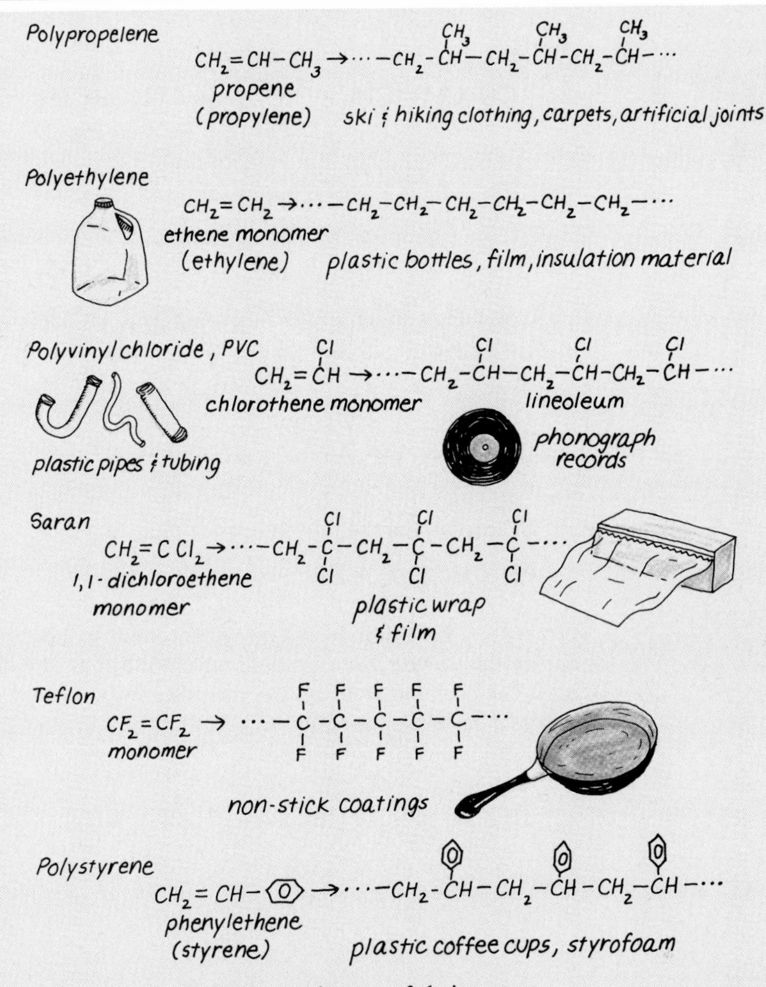

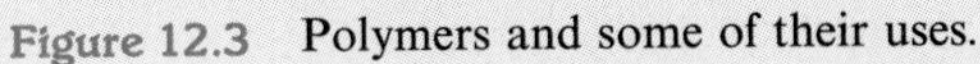

Figure 12.3 Polymers and some of their uses.

Consider the addition of the compound HCl to the following alkenes, which have double bonds with different numbers of hydrogens:

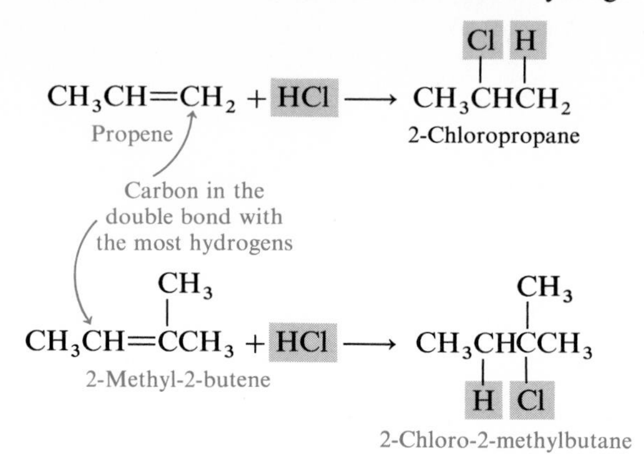

Example 12.7 Write the product of the following reaction:

$$CH_3CH{=}CH_2 + HBr \longrightarrow$$

Solution: The addition of hydrogen bromide to a double bond places a hydrogen atom on the carbon atom in the double bond that already has the most hydrogen atoms. The bromine atom goes to the other carbon in the double bond, which is the center carbon in this compound:

$$CH_3CH{=}CH_2 + HBr \longrightarrow CH_3\underset{\,}{\overset{Br}{\overset{|}{C}}}H{-}\overset{H}{\overset{|}{C}}H_2$$

Carbon in the double bond with the most hydrogens

12.3 Aromatic Compounds

OBJECTIVE Write the name and structure of an aromatic hydrocarbon.

benzene The aromatic hydrocarbons are organic compounds that contain a cyclic structure called *benzene*. This family was originally named for the fragrant odors of its compounds. Benzene has the formula C_6H_6. At first, scientists wrote a six-sided figure for benzene with three double bonds:

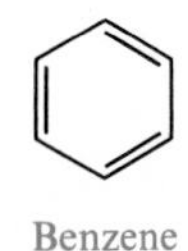

Benzene

However, an alternating pattern of single and double bonds suggests that a molecule with this structure would be very reactive. Yet experiments showed that benzene was very stable and did not undergo the typical addition reactions of the unsaturated hydrocarbons. Therefore, the electrons in benzene must be in an especially stable bonding pattern. In 1865, August Kekulé proposed that the structure of benzene was a combination of two structures:

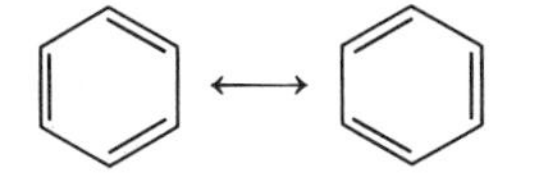

Today, scientists draw a structure for benzene that represents an average of the Kekulé structures. The electrons are distributed equally over the carbon atoms. This helps bind the electrons more strongly, thus accounting for the stability of the benzene ring. The symbol for the benzene ring is usually drawn using a hexagon with a circle inside. This will be the preferred structure for benzene in this text:

Benzene

Naming Aromatic Compounds

The parent compound of the aromatics is benzene. In the IUPAC system, a group or atom bonded to the benzene ring is identified, and its name is placed in front of the word benzene:

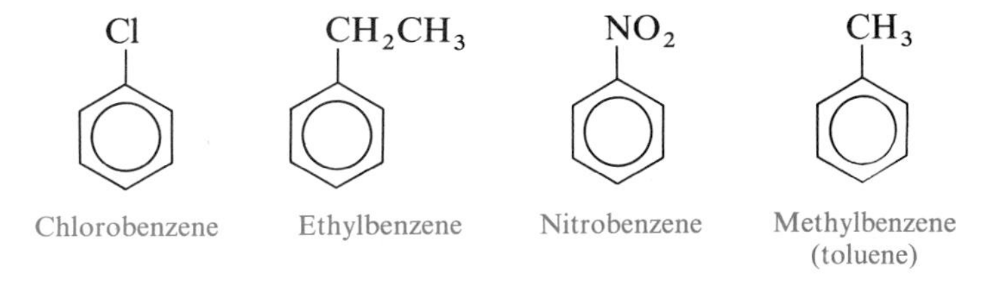

Chlorobenzene   Ethylbenzene   Nitrobenzene   Methylbenzene (toluene)

In naming the aromatic compounds, many common names are also used. For example, the common name toluene is used rather than methylbenzene.

HEALTH NOTE

Effects of Benzene and Toluene on Health

Benzene and toluene are substances found in paint removers, in rubber, and plastic cements. Inhalation or ingestion of these compounds can cause poisoning that affects the central nervous system. Initial symptoms include euphoria followed by convulsions, coma, and respiratory failure. Long-term exposure to benzene or toluene can eventually produce bone marrow depression followed by aplastic anemia.

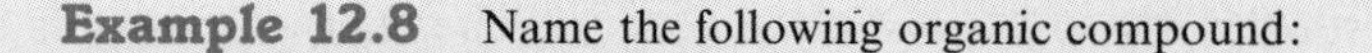

Example 12.8 Name the following organic compound:

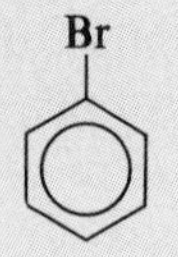

Solution: The hexagon structure with the circle is the symbol for the compound benzene. The bromine is named as a substituent (prefix bromo-). Thus the compound is called

Bromobenzene

Naming Two Groups on the Benzene Ring

ortho, meta, para

When there are two substituents on a benzene ring, we must number the ring in the IUPAC system to indicate the relative positions of the side groups just as we did for the cycloalkanes. Another, more common, naming system is possible for indicating the position of two substituents. A 1,2-pattern is *ortho* (*o*-), a 1,3-pattern is *meta* (*m*-), and a 1,4-pattern is *para* (*p*-):

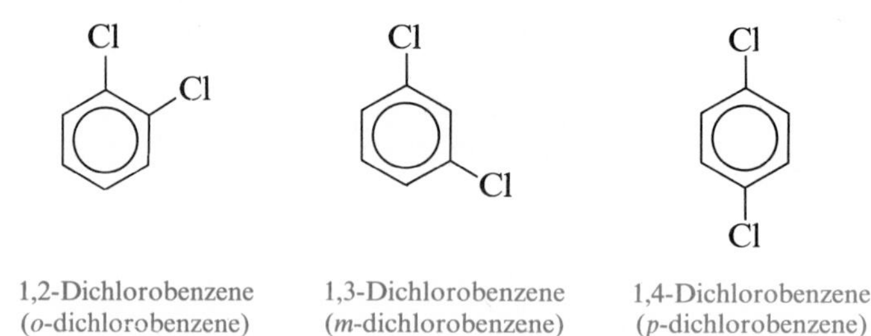

1,2-Dichlorobenzene (*o*-dichlorobenzene) 1,3-Dichlorobenzene (*m*-dichlorobenzene) 1,4-Dichlorobenzene (*p*-dichlorobenzene)

A benzene ring with a methyl side group uses the common name toluene; benzene with two methyl groups is named xylene:

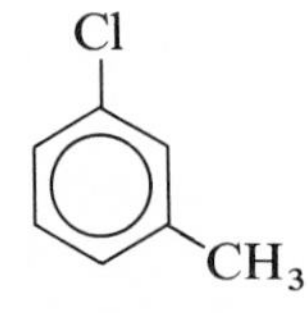

1-Chloro-3-methylbenzene (*m*-chlorotoluene)

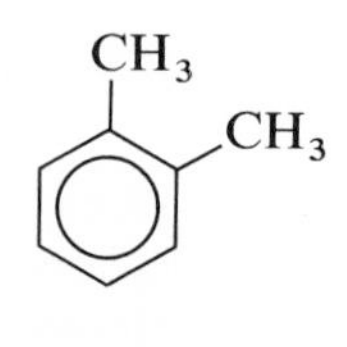

1,2-Dimethylbenzene (*o*-xylene)

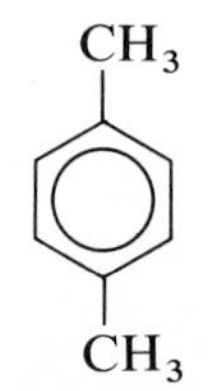

1,4-Dimethylbenzene (*p*-xylene)

Example 12.9 Name the following aromatic compounds:

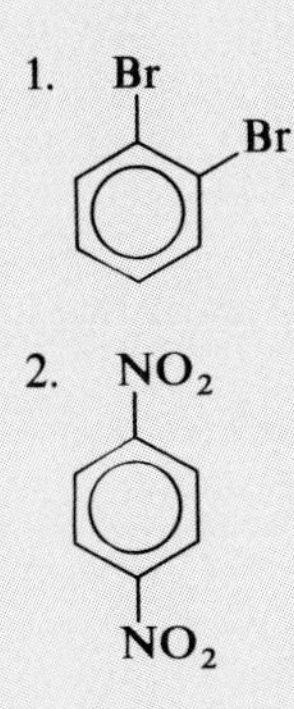

Solution

1. The parent structure is the benzene ring. The two bromine atoms are numbered in the IUPAC naming system. Thus we have

   IUPAC name: 1,2-Dibromobenzene

   The common name uses the relative location of the two bromine atoms. When they occur on adjacent carbons, it can be indicated as ortho (*o*- in the name):

   o-Dibromobenzene

2. This is a benzene ring with two nitro groups attached. In the IUPAC naming system, each nitro group is numbered.

   IUPAC name: 1,4-Dinitrobenzene

   A benzene ring with two nitro groups can also use the para (*p*-) prefix to indicate the 1,4 position of the nitro groups:

   p-Dinitrobenzene

Naming Three Groups on a Benzene Ring

When there are three or more side groups, the benzene ring must be numbered to show the location of each. The names of the side groups are listed in alphabetical order:

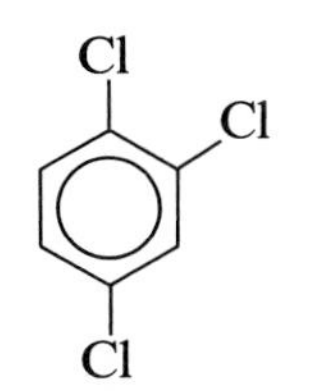

1,2,4-Trichlorobenzene

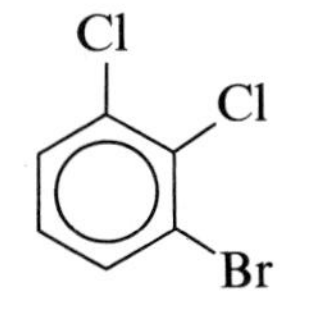

3-Bromo-1,2-dichlorobenzene

If the compound includes the toluene form or the xylene form, the carbon attached to a methyl group is designated as carbon 1 and the rest of the side groups are numbered to give the lowest possible numerical sequence:

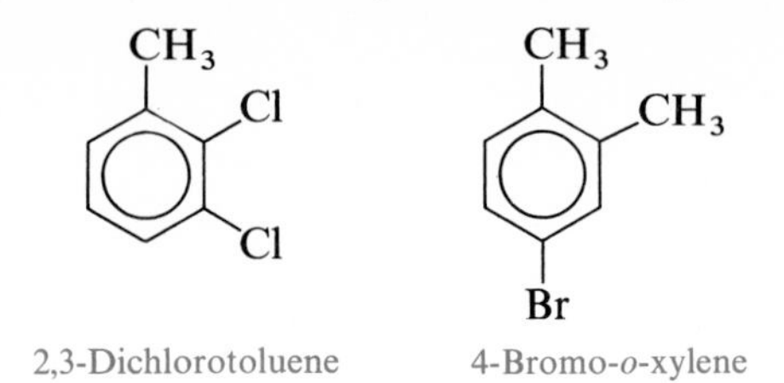

2,3-Dichlorotoluene 4-Bromo-*o*-xylene

HEALTH NOTE

Fused Benzene Rings in Nature and Cancer

Some of the larger aromatic systems contain *fused* benzene rings. Common names are generally used for these. The crystalline form of naphthalene is often found in mothballs. Anthracene is used in the manufacture of dyes.

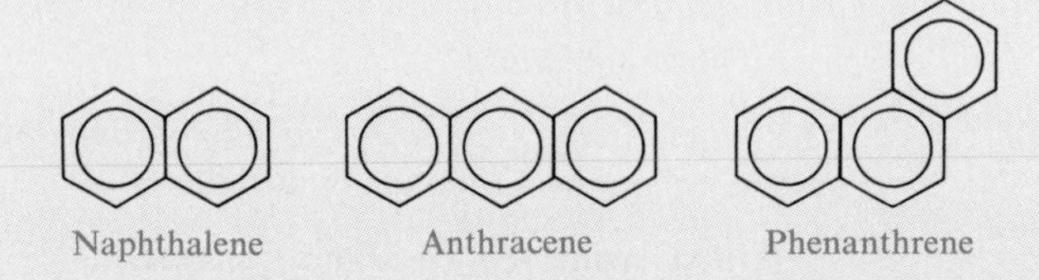

Naphthalene Anthracene Phenanthrene

carcinogens

Some of the large polycyclic benzene compounds include the phenanthrene portion and are known to cause cancer; they are *carcinogens*. Several carcinogens containing fused benzene rings have been identified in soot, coal tar, dyes, and cigarette smoke:

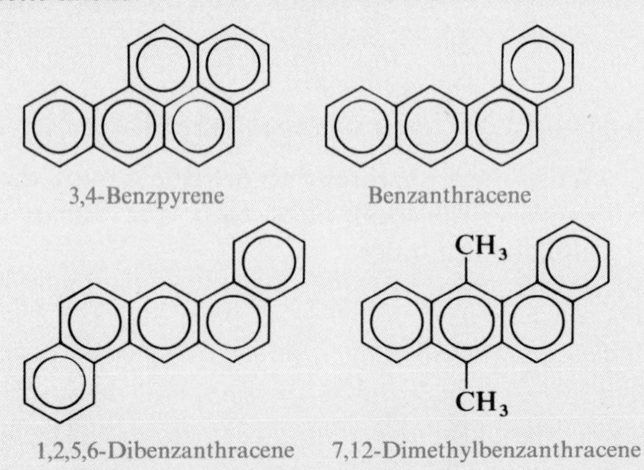

3,4-Benzpyrene Benzanthracene

1,2,5,6-Dibenzanthracene 7,12-Dimethylbenzanthracene

Some scientists believe that a chemical carcinogen must interact in some way with molecules in a cell to alter the growth of that cell. It is thought that the proteins and nucleic acids in the cells may serve as targets for several chemical carcinogens.

12.4 Alcohols

OBJECTIVE Write the name and structure of an alcohol, phenol, or thiol.

The alcohols are a family of organic compounds that contain a hydroxyl (—OH) or alcohol group. We can think of the alcohol group as replacing a hydrogen atom of an alkane:

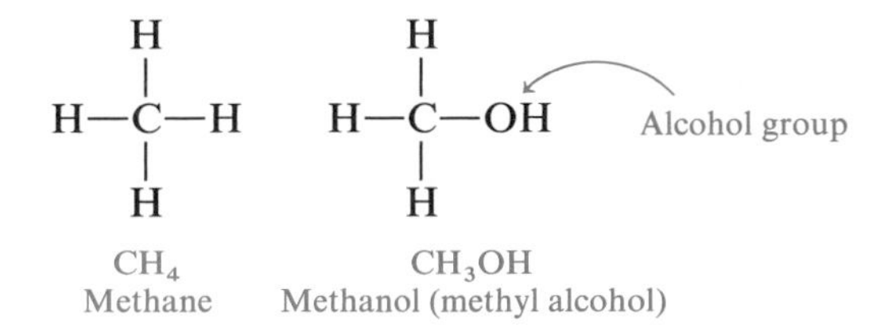

CH_4
Methane

CH_3OH
Methanol (methyl alcohol)

Names of Alcohols

To name an alcohol by the IUPAC system, select the longest carbon chain to which the —OH is attached. Change the *-e* ending of that alkane name to *-ol*. For example, the alcohol derived from methane is methan*ol*. The common name for an alcohol is obtained by naming the group of carbon atoms attached to the —OH as an alkyl group followed by the word *alcohol*. The common name of methanol is methyl alcohol. The method of naming an alcohol is summarized below:

Alcohol

Functional group: Hydroxyl, —OH
IUPAC name: Alkan*ol*
Common name: Alkyl alcohol

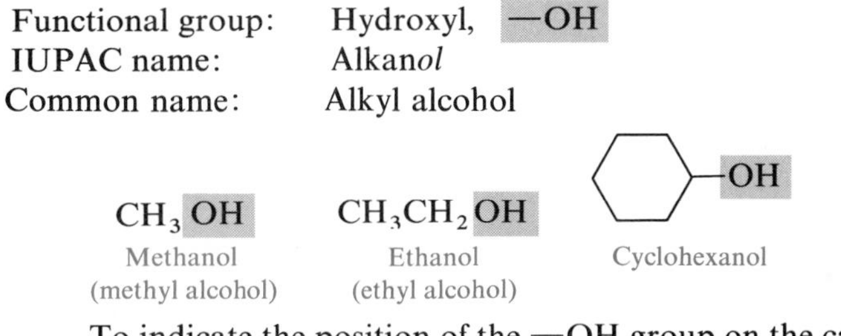

Methanol (methyl alcohol) — Ethanol (ethyl alcohol) — Cyclohexanol

To indicate the position of the —OH group on the carbon chain, number the chain to give the lowest possible number to the carbon atom bonded to the —OH group. If only one structural isomer is possible, the number for the alcohol group may be omitted:

$CH_3CH_2CH_2OH$
1-Propanol
(*n*-propyl alcohol)

2-Propanol
(isopropyl alcohol)

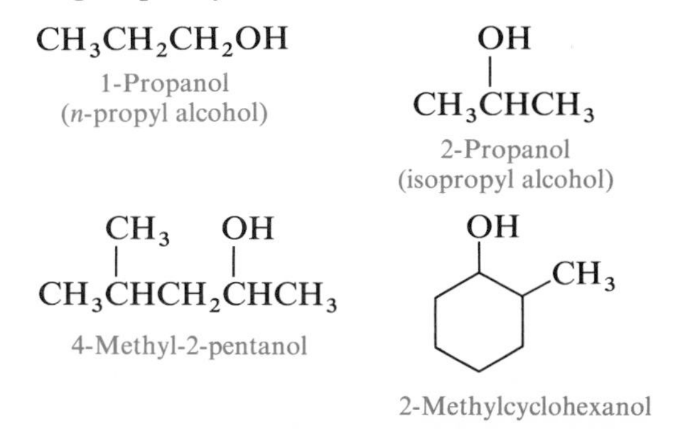

4-Methyl-2-pentanol

2-Methylcyclohexanol

Example 12.10 Write the name for each of the following alcohols:

1. CH_3OH

2.
$$\begin{array}{c} OH \\ | \\ CH_3CH_2CHCH_2CH_3 \end{array}$$

Solution

1. The hydroxyl group (—OH) is attached to a carbon portion derived from methane (CH_4). The IUPAC name requires the use of the *-ol* suffix:

   IUPAC name: Methanol

   In the common name, the carbon portion is named as an alkyl group, methyl.

   Common name: Methyl alcohol

2. The parent chain is a five-carbon chain named pentane. The hydroxyl group (—OH) means that the compound is in the alcohol family. We change the ending of the alkane name to pentanol. The location of the hydroxyl group is on the third carbon (carbon 3):

   3-Pentanol

Solubility of Alcohols in Water

The alcohols contain a functional group (—OH) that is polar because of the presence of the oxygen atom. This polar part of the alcohol has an effect on solubility in water. The low-molecular-weight alcohols (methanol, ethanol, and propanol) are very soluble in water. They have sufficient polarity to form hydrogen bonds with water molecules. As the number of carbon atoms in the carbon chain of the alcohol increases, the polar —OH group loses its effect. Alcohols with four or more carbon atoms behave more like alkanes and are considerably less soluble in water than the smaller alcohols. (See Figure 12.4.)

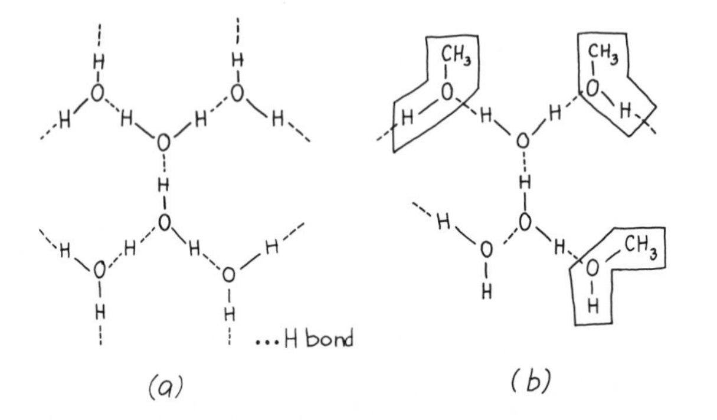

Figure 12.4 Solubility of alcohols in water: (a) Hydrogen bonding in water. (b) Low-molecular-weight molecules of methanol take part in the hydrogen-bonding patterns of water which make them soluble.

HEALTH NOTE

Alcohols in Health and Medicine

Methanol, CH_3OH, is used as a solvent and is found in duplicator fluids. It is very toxic to the human body. The toxicity is caused by the oxidation of methanol to formaldehyde within the tissues of the body. When even very small amounts of methanol are ingested, headaches, fatigue, blindness, coma, or death may occur.

A solution that is 70% by volume ethanol (CH_3CH_2OH) and 30% water is widely used as an antiseptic. It destroys bacteria by coagulating the protein in the bacteria. Ethanol is also used as a solvent for some medicine in tincture form. In the hospital, ethanol may be given to treat a patient with methanol poisoning. The ethanol takes the place of the methanol on the enzymes, thus allowing the methanol to be eliminated without forming its dangerous oxidation product, formaldehyde. In small amounts, ethanol produces a feeling of euphoria in the body. However, if large amounts of ethanol are ingested, it acts as a depressant, which interferes with mental and physical coordination. If the ethanol blood concentration exceeds 0.5%, death may occur.

Isopropyl alcohol (rubbing alcohol) is used in astringents because it evaporates rapidly and cools the skin, reducing the size of blood vessels near the surface and decreasing pore size.

Glycerol (glycerine) is a trihydroxy alcohol, a natural component of fats and oils. It is used in skin lotions and some soaps as a skin softener:

CH_2OH
|
CHOH
|
CH_2OH

Glycerol

Menthol is used in throat lozenges and sprays. It causes the mucous membranes to increase their secretions and thus soothes the respiratory tract:

CH_3
(cyclohexane ring)
OH
CH_3CH
|
CH_3

Menthol

Phenol

The alcohol of benzene, an aromatic alcohol, is called *phenol*:

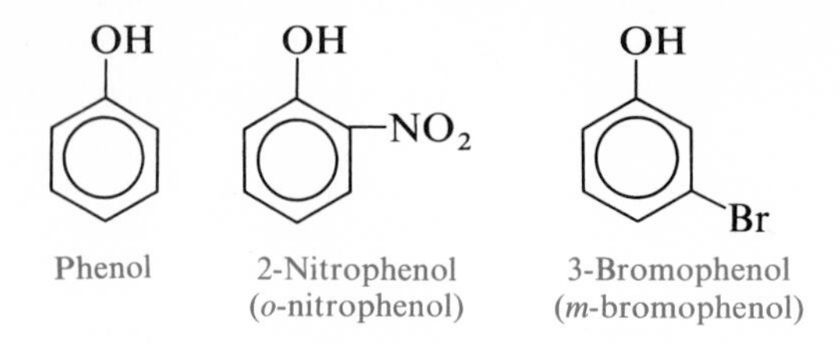

Phenol 2-Nitrophenol (*o*-nitrophenol) 3-Bromophenol (*m*-bromophenol)

The behavior of an alcohol group on an aromatic compound (phenol) differs considerably from the alcohol of an alkane. It is more soluble in water than the corresponding alkyl alcohol with the same number of carbon atoms. A phenol dissociates slightly in water to give an acidic solution. Sometimes phenol is called *carbolic acid*:

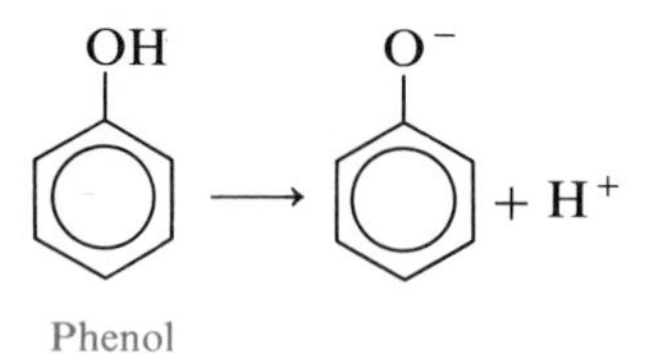

Phenol

A dilute solution of phenol is used in hospitals as an antiseptic solution. Small amounts of phenol are used in throat lozenges. However, concentrated samples of phenol are very corrosive and highly irritating to the skin. Solutions containing phenol should be used with care (use gloves where possible).

Several relatives of phenol are used as disinfectants. Resorcinol and thymol are used in bactericides and fungicides:

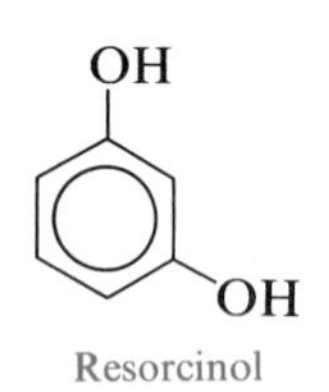

Resorcinol

Thymol (from thyme) has a pleasant, minty taste and is also used in mouthwashes. It is used by dentists to disinfect a cavity in preparation for filling compound.

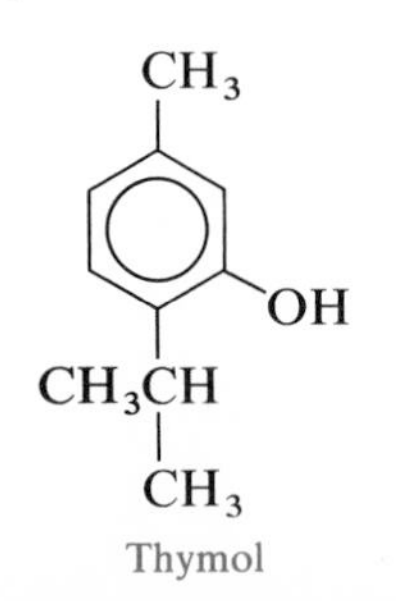

Thymol

One of the most typical ways to preserve foods such as cereals and oils is to add an antioxidant such as BHT, a derivative of phenol:

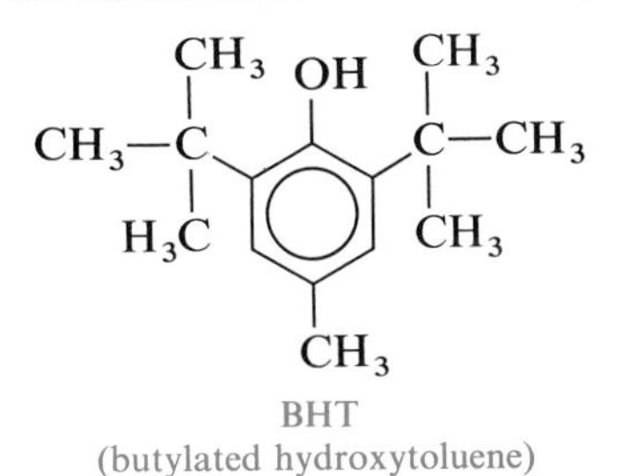

BHT
(butylated hydroxytoluene)

The BHT undergoes reaction with the oxygen in the air instead of the food products themselves.

Example 12.11 Name the following compound:

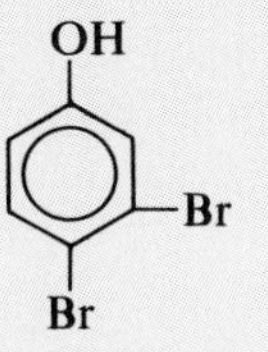

Solution: The —OH group on a benzene ring classifies this compound as a phenol. The phenol carbon is counted as carbon 1, with the bromine atoms on carbon 3 and 4. Thus, the compound is called

3,4-Dibromophenol

Thiols

The thiols are similar to alcohols, except that in thiols there is an —SH group rather than an —OH group. Many of the common thiols have strong, often disagreeable, odors. The fragrance that a skunk uses as a protective device is a thiol; the odor of garlic is from another thiol.

Thiols are named by adding the suffix *-thiol* to the parent name. The common names use an older term, *mercaptan*, for the sulfur compound.

Thiol

Functional group:	—SH
IUPAC name:	Alkanethiol
Common name:	Alkyl mercaptan
CH_3SH	Methanethiol (methyl mercaptan)
CH_3CH_2SH	Ethanethiol (ethyl mercaptan)
$CH_3CH_2CH_2SH$	1-Propanethiol (*n*-propyl mercaptan) ("onions")
$CH_3\underset{}{\overset{CH_3}{\overset{|}{C}}}HCH_2CH_2SH$	3-Methyl-1-butanethiol ("Skunk odor")

Example 12.12 Give an IUPAC name for the following thiols:

1. CH_3SH
2. $CH_3CH(SH)CH_3$

Solution

1. The parent chain in this thiol is derived from methane. To name the thiol, the suffix thiol is added to the parent name:

   IUPAC name: Methanethiol
   Common name: Methyl mercaptan

2. The parent chain in this thiol has three carbons, propane. To indicate the —SH group, add the suffix *-thiol* to the parent name. The location of the —SH group is on the second carbon (carbon 2) of the parent chain:

   IUPAC name: 2-Propanethiol
   Common name: Isopropyl mercaptan

12.5 Classes of Alcohols

OBJECTIVE Classify an alcohol as primary, secondary, or tertiary.

The alkyl alcohols may be classified as primary, secondary, or tertiary, according to the number of alkyl groups bonded to the carbon attached to the —OH group. We shall refer to this carbon atom as the "alcohol carbon." If the alcohol carbon is attached to *one* alkyl group and therefore has two hydrogen atoms attached, the alcohol is classified as a *primary (1°) alcohol*:

primary alcohol

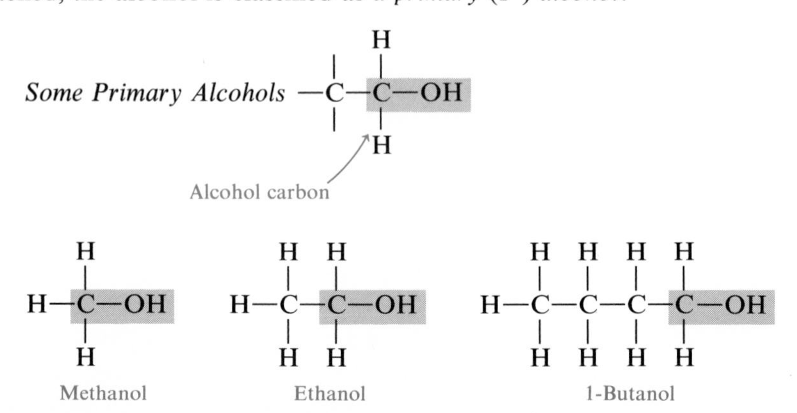

(Methanol has no alkyl groups attached to the alcohol carbon. Since it reacts in the same way as primary alcohols, it is included in the primary alcohol category.)

secondary alcohol

When the alcohol carbon is attached to *two* alkyl groups and has only 1 hydrogen atom attached, the alcohol is a *secondary* (2°) *alcohol*:

Some Secondary Alcohols

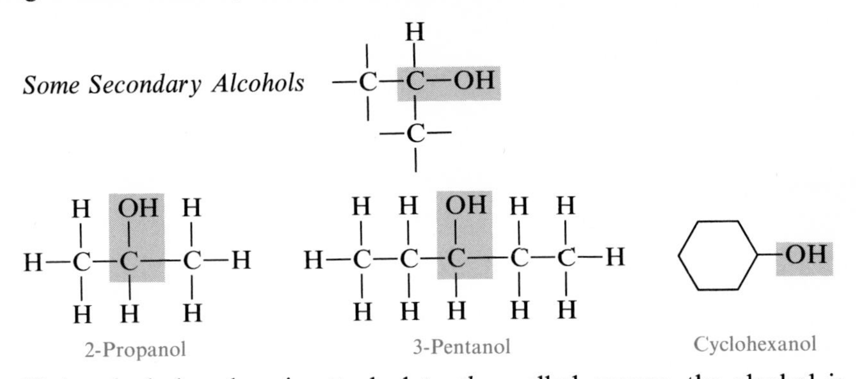

2-Propanol 3-Pentanol Cyclohexanol

tertiary alcohol

If the alcohol carbon is attached to *three* alkyl groups, the alcohol is classified as a *tertiary* (3°) *alcohol.* Note that in a tertiary alcohol there are no hydrogen atoms left on the alcohol carbon. The prefix *t*- (or *tert*-) is sometimes used to indicate a tertiary alcohol group.

Some Tertiary Alcohols

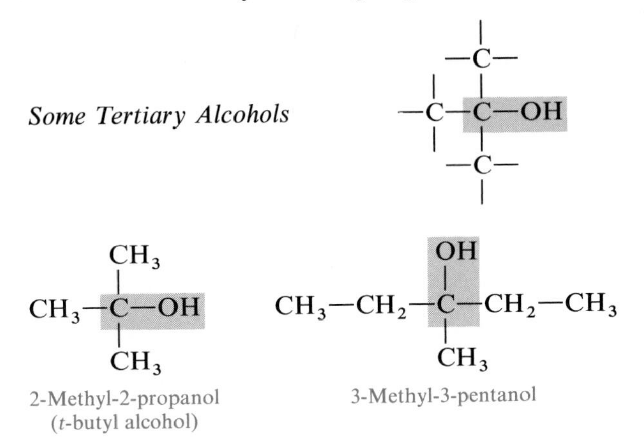

2-Methyl-2-propanol (*t*-butyl alcohol) 3-Methyl-3-pentanol

Example 12.13 Identify the following alcohols as primary (1°), secondary (2°), or tertiary (3°) alcohols:

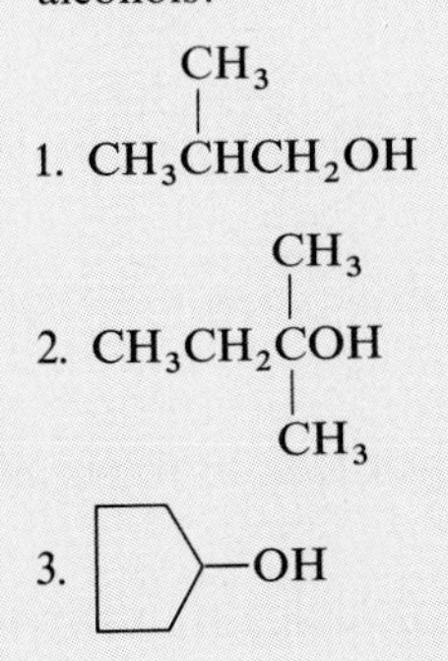

Solution

1. The hydroxyl group is bonded to a carbon that is attached to just one alkyl group. This is a primary (1°) alcohol.
2. The alcohol carbon is bonded to three alkyl groups. This is a tertiary (3°) alcohol.
3. The alcohol carbon is bonded to two other carbon atoms in the cyclic structure. This is a secondary (2°) alcohol.

12.6 Formation of Alcohols

OBJECTIVE Write an equation for the hydration of an alkene to form an alcohol.

An alcohol can be produced by adding water, HOH, to an alkene. This reaction is called *hydration* and requires an acid (H^+) catalyst. The hydration of an alkene is similar to the other addition reactions for alkenes we saw in Section 12.2:

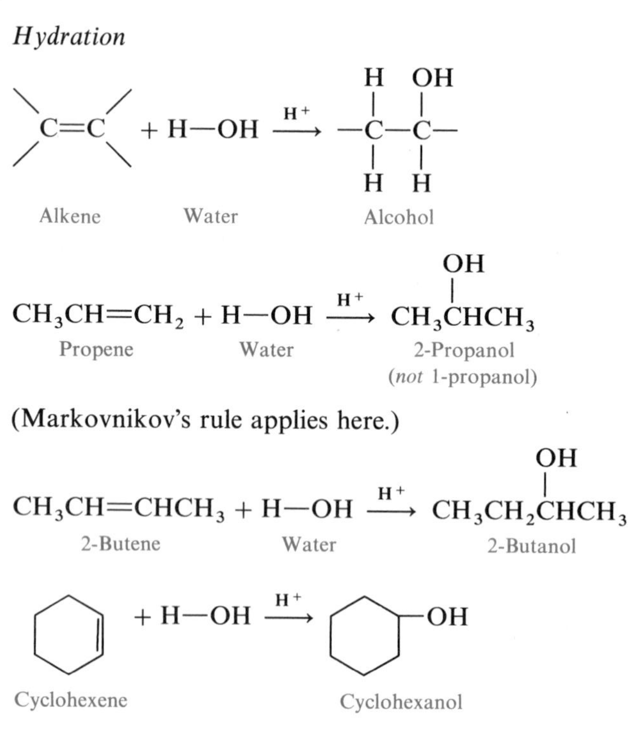

Hydration

$$\text{C=C (Alkene)} + \text{H—OH (Water)} \xrightarrow{H^+} \text{—C(H)(H)—C(OH)(H)— (Alcohol)}$$

$$CH_3CH{=}CH_2 + H{-}OH \xrightarrow{H^+} CH_3CH(OH)CH_3$$

Propene, Water, 2-Propanol (*not* 1-propanol)

(Markovnikov's rule applies here.)

$$CH_3CH{=}CHCH_3 + H{-}OH \xrightarrow{H^+} CH_3CH_2CH(OH)CH_3$$

2-Butene, Water, 2-Butanol

$$\text{Cyclohexene} + H{-}OH \xrightarrow{H^+} \text{Cyclohexanol}$$

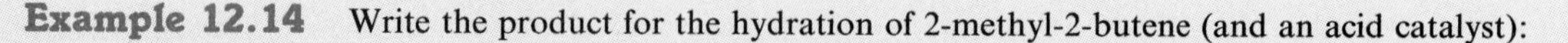

Example 12.14 Write the product for the hydration of 2-methyl-2-butene (and an acid catalyst):

$$CH_3C(CH_3){=}CHCH_3 + H_2O \xrightarrow{H^+}$$

Solution: We can think of the water molecule as being composed of two parts, H and OH. These two parts must attach to the two carbons in the double bond. The H will attach to the carbon atoms in the double bond that already has a hydrogen atom. The OH portion will attach to the other carbon atom of the double bond:

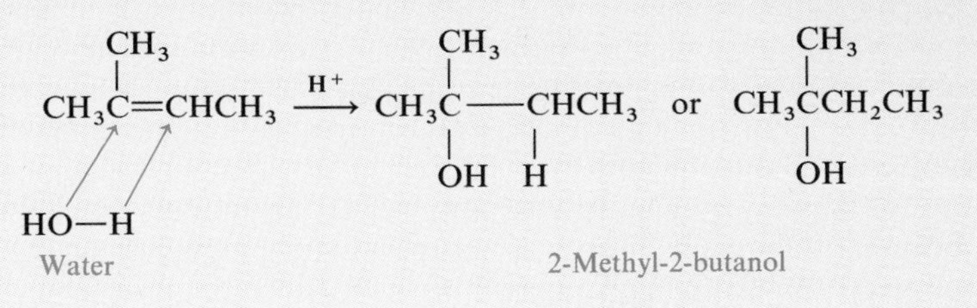

12.7 Reactions of Alcohols

OBJECTIVE Write the equation for the dehydration of an alcohol.

dehydration of alcohols

In a *dehydration* reaction, a molecule of water is removed from an alcohol. When the components of water (H and OH) are taken from just one alcohol molecule, the products are an alkene and water.

Dehydration of an Alcohol to Form an Alkene

When an alcohol is subjected to heat (175°C) and an acid catalyst such as H_2SO_4, the components of the water (H and OH) are removed from two carbon atoms that are next to each other. The adjacent carbon atoms that were bonded to the H and OH form a double bond, and an alkene is produced. The dehydration reaction is the reverse reaction to hydration discussed in Section 12.6:

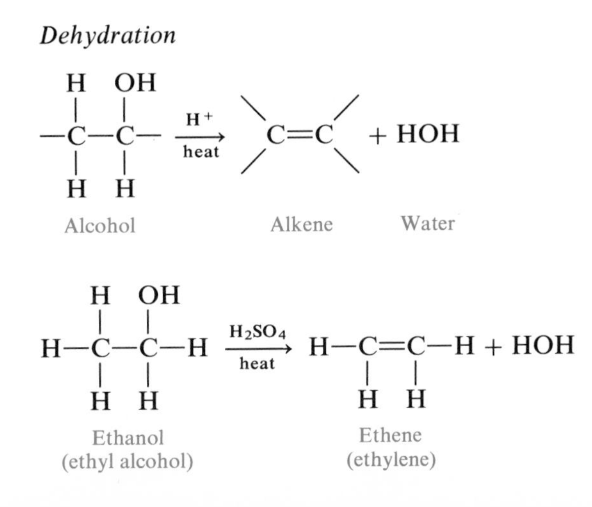

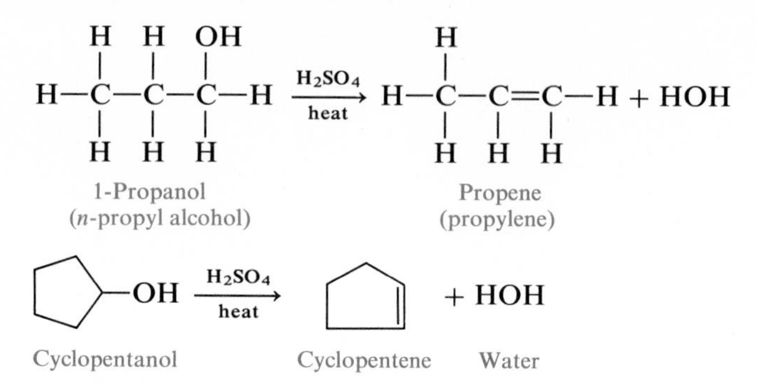

Sometimes the hydrogen part of the water molecule can be taken from carbon atoms on both sides of the alcohol carbon. Then there is the possibility of different locations of the double bond. The major product is the one in which the hydrogen atom for the water molecule is removed from the carbon that has the lowest number of hydrogen atoms:

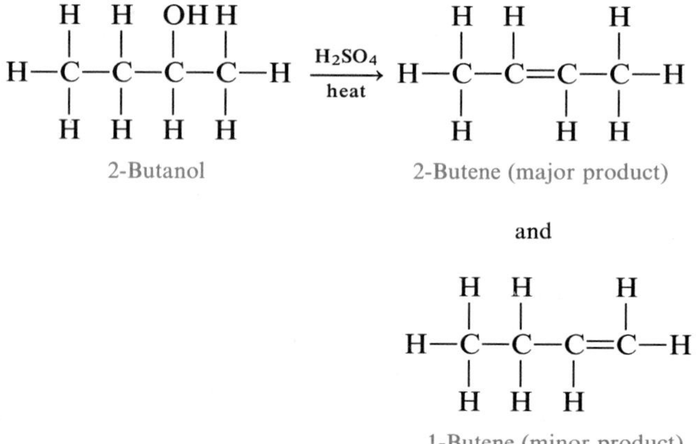

Example 12.15 Write the products for the dehydration of the following alcohols:

1. $CH_3CH_2CH_2CH_2OH \xrightarrow[\text{heat}]{H_2SO_4}$

2. OH $\xrightarrow[\text{heat}]{H_2SO_4}$

Solution

1. Dehydration of an alcohol occurs in the presence of an acid. The components of water (H and OH) must be removed from adjacent carbon atoms. Since the OH is on the first carbon in the chain, the H must be taken from the next or second

carbon in the chain. The two carbon atoms that have lost the parts of the water molecule form a second bond which makes the double bond:

$$\underset{\text{1-Butanol}}{CH_3CH_2\overset{\overset{\displaystyle H}{|}}{C}H{-}\overset{\overset{\displaystyle OH}{|}}{C}H_2} \xrightarrow[\text{heat}]{H_2SO_4} \underset{\text{1-Butene}}{CH_3CH_2CH{=}CH_2} + \underset{\text{Water}}{HOH}$$

2. The cyclic alcohol will dehydrate by removing the OH from its carbon atom and a H from one of the adjacent carbon atoms in the cyclic structure. The double bond forms between the same two carbon atoms:

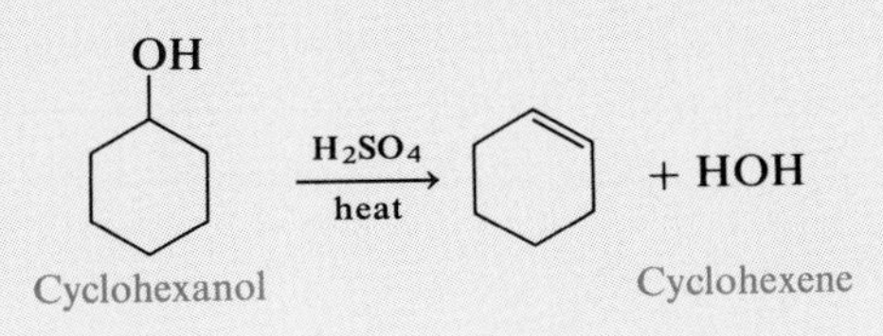

12.8 Ethers

OBJECTIVE Write the name and structure of an ether.

Ethers are a family of organic compounds that contain two alkyl or aromatic groups attached to a single oxygen atom. In the common names of ethers, the names of the two hydrocarbon groups are placed in front of the word *ether*. If the groups are identical, the prefix *di-* precedes the name of the group. The common names are most often used, especially with the smaller ethers. There is an IUPAC name for ethers, but it is complex and will not be used in this text:

Ether	—C—O—C—	
Common name:	Alkyl alkyl ether or dialkyl ether	
CH_3OCH_3 Dimethyl ether	$CH_3CH_2OCH_3$ Ethyl methyl ether	$CH_3CH_2OCH_2CH_2CH_3$ Ethyl *n*-propyl ether

Ethers cannot undergo hydrogen bonding, which causes their boiling points to be low. They are also rather insoluble in water. However, ethers are very volatile at room temperatures and there is always a danger of their combustion. When stored for long periods of time, ethers may form peroxides, compounds with an —O—O— group. These peroxides are quite explosive and dangerous in the laboratory. Caution must be exercised when working with ethers.

HEALTH NOTE

Uses of Ethers

Ethers, diethyl ether in particular, are commonly used as solvents because they are less reactive than many other organic materials and thus do not react with the substances they dissolve. Diethyl ether (or, simply, "ether") has been in use as an anesthetic for many years. It has remained a popular anesthetic because it has minimal side effects and has anesthetic properties over a wide concentration range. A disadvantage of ether is its flammability. Great care must be exercised to avoid any flames or even sparks wherever ethers are being used.

$CH_3CH_2OCH_2CH_3$
Diethyl ether
("ether")

Example 12.16 Write the common names of the following ethers:

1. $CH_3OCH_2CH_3$
2. 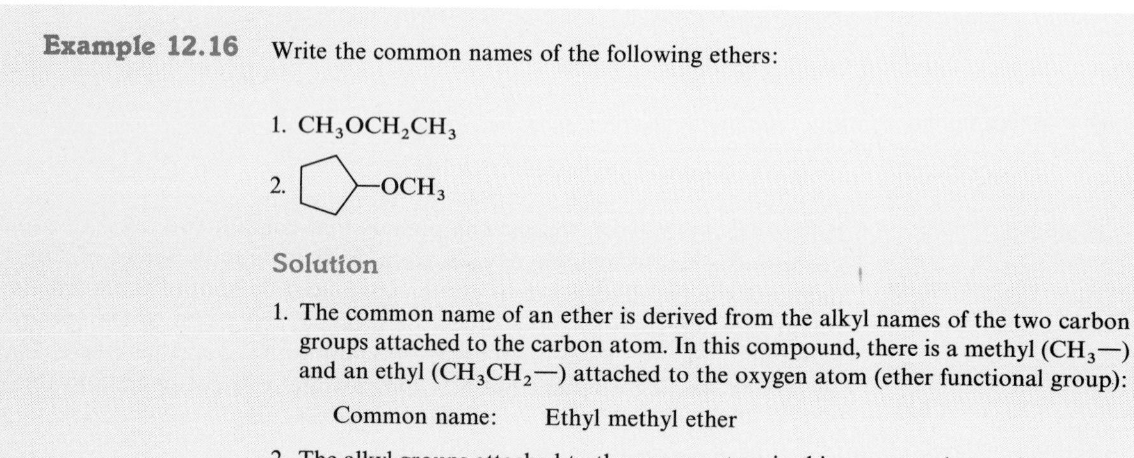

Solution

1. The common name of an ether is derived from the alkyl names of the two carbon groups attached to the carbon atom. In this compound, there is a methyl (CH_3—) and an ethyl (CH_3CH_2—) attached to the oxygen atom (ether functional group):

   Common name: Ethyl methyl ether

2. The alkyl groups attached to the oxygen atom in this compound are a five-carbon cyclic group (cyclopentyl) and a methyl (CH_3—) group:

   Common name: Cyclopentyl methyl ether

12.9 Dehydration of an Alcohol to Form an Ether

OBJECTIVE Write an equation for the formation of an ether.

When dehydration occurs at lower temperatures (140°C), the components of water are removed from two alcohol molecules. The OH group of one alcohol molecule and the H of the other alcohol group are used to produce the water molecule. The remaining portions of the alcohol molecules join together to form an ether. When we wish to show the formation of an ether, the equation will start with two molecules of alcohol:

Ether Formation (Dehydration)

$$\text{Alcohol} + \text{Alcohol} \xrightarrow[140°C]{H_2SO_4} \text{Ether} + \text{HOH}$$

$$\underset{\text{Methyl alcohol}}{CH_3\ OH} + \underset{\text{Methyl alcohol}}{\boxed{H}\ OCH_3} \xrightarrow{H^+} \underset{\text{Dimethyl ether}}{CH_3OCH_3} + \boxed{HOH}$$

This may also be written as

$$2CH_3OH \xrightarrow{H^+} CH_3OCH_3 + HOH$$

Ethyl ether, the anesthetic, is formed from ethyl alcohol:

$$\underset{\text{Ethyl alcohol}}{2CH_3CH_2OH} \xrightarrow{H^+} \underset{\text{Diethyl ether}}{CH_3CH_2OCH_2CH_3} + HOH$$

Example 12.17 Complete the following reactions for ether formation:

1. $2CH_3OH \xrightarrow[140°C]{H_2SO_4}$

2. $\_\_\_\_\_\_ \xrightarrow[140°C]{H_2SO_4} CH_3CH_2OCH_2CH_3$

Solution

1. This equation suggests the dehydration of two alcohol molecules to form an ether. The structure of the ether product is written by attaching the alkyl portions of the reacting alcohols to an oxygen atom. The removal of the components of water is shown as a water molecule in the products:

$$\underset{\text{Methanol}}{CH_3OH + HOCH_3} \xrightarrow[140°C]{H_2SO_4} \underset{\text{Dimethyl ether}}{CH_3OCH_3} + HOH$$

2. This equation requires the structures of the reacting alcohols. The alkyl portions of the ether are derived from the corresponding alcohols. Since the alkyl groups are both ethyl groups, the alcohol molecules that reacted must have both been ethyl alcohol:

$$\underset{\text{Ethanol}}{2CH_3CH_2OH} \xrightarrow[140°C]{H_2SO_4} \underset{\text{Diethyl ether}}{CH_3CH_2OCH_2CH_3} + HOH$$

The following table summarizes the naming of alkenes, alkynes, aromatics, alcohols, thiols, and ethers:

Summary of Naming

Class	Example	Structure
Alkene	Ethene (ethylene)	$CH_2{=}CH_2$
Alkyne	Ethyne (acetylene)	$CH{\equiv}CH$
Aromatic	Benzene	
Alcohol	Ethanol (ethyl alcohol)	CH_3CH_2OH
Thiol	Methane thiol (methyl mercaptan)	CH_3SH
Ether	Dimethyl ether	CH_3OCH_3

The table below gives a summary of the reactions of alkenes and alcohols:

Summary of Reactions

Hydrogenation of Alkenes

$$\text{Alkene} + \text{Hydrogen} \xrightarrow{\text{Pt}} \text{Alkane}$$

$$CH_2{=}CH_2 + H_2 \xrightarrow{\text{Pt}} CH_3CH_3$$

Halogenation of Alkenes

$$\text{Alkene} + \text{Halogen} \longrightarrow \text{Dihaloalkane}$$

$$CH_2{=}CH_2 + Br_2 \longrightarrow \overset{Br\quad Br}{\overset{|\qquad |}{CH_2CH_2}}$$

Hydrohalogenation of Alkenes

$$\text{Alkene} + \text{Hydrogen halide} \longrightarrow \text{Haloalkane}$$

$$CH_2{=}CH_2 + HCl \longrightarrow CH_3CH_2Cl$$

Hydration of Alkenes

$$\text{Alkene} + \text{HOH} \xrightarrow{H^+} \text{Alcohol}$$

$$CH_2{=}CH_2 + HOH \xrightarrow{H^+} CH_3CH_2OH$$

Dehydration of Alcohols

$$\text{Alcohol} \xrightarrow{H^+} \text{Alkene} + \text{HOH}$$

$$\overset{H\quad OH}{\overset{|\qquad |}{CH_2CH_2}} \xrightarrow{H^+} CH_2{=}CH_2 + HOH$$

Ether Formation

$$\text{Two alcohols} \xrightarrow{H^+} \text{Ether} + \text{HOH}$$

$$CH_3OH + HOCH_3 \xrightarrow{H^+} CH_3OCH_3 + HOH$$

Glossary

addition The adding of new atoms to the groups of atoms already present in a compound.

alcohol A compound in which a hydrogen is replaced by the —OH group.

alkene An unsaturated hydrocarbon containing a double bond.

alkyne An unsaturated hydrocarbon containing a triple bond.

aromatic A hydrocarbon that contains a benzene ring in its structure.

benzene A six-carbon ring with six hydrogen atoms, C_6H_6, which is very stable.

dehydration A reaction in which the components of water (H—OH) are removed from an alcohol in the presence of an acid to form alkenes or ethers.

ether A compound in which an oxygen atom is bonded to two hydrocarbon groups.

functional group An atom or group of atoms that has a strong influence on the overall chemical behavior of a compound.

geometric isomers The cis and trans arrangements of two different groups on the carbon atoms of a double bond. In a cis arrangement, similar groups are on the same side of the double bond; in the trans arrangement, they appear on opposite sides.

halogenation The name of an addition reaction for an alkene in which two halogen atoms are added to a double bond, one to each of the carbon atoms in the double bond.

hydration An addition reaction in which the components of water attach themselves to an alkene to form an alcohol.

hydrogenation An addition reaction in which hydrogen atoms are added to an alkene.

hydrohalogenation The addition of a hydrogen halide to the double bond of an alkene.

primary alcohol An alcohol that has one alkyl group (or none) attached to the alcohol carbon atom.

secondary alcohol An alcohol that has two alkyl groups attached to the alcohol carbon atom.

tertiary alcohol An alcohol that has three alkyl groups attached to the alcohol carbon atom.

Problems

Alkenes and Alkynes: Unsaturated Hydrocarbons (Objective 12.1)

12.1 Compare the compounds cyclohexane and cyclohexene.

12.2 Write the IUPAC name or common name for each of the following alkenes:

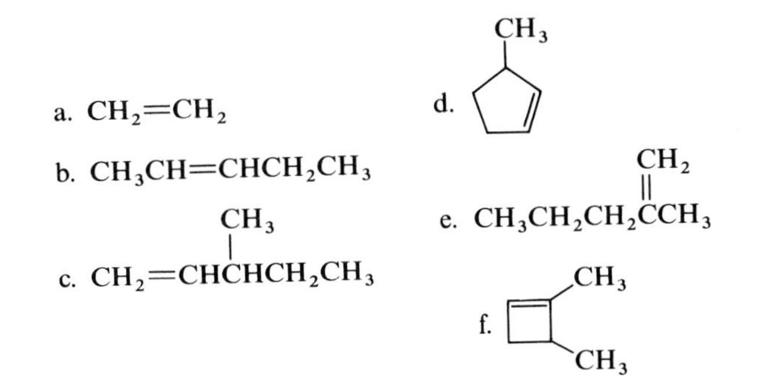

12.3 Write the condensed structural formula for each of the following alkenes:

a. propene
b. 2-methylpropene
c. cyclopentene
d. 1-chlorocyclopentene
e. 3,4-dimethylcyclohexane
f. 2,3-dimethyl-2-butene
g. 2,4-dichloro-2-pentene

12.4 Write the cis and trans isomers of the following:

a. 2-butene
b. 1,2-dibromoethene
c. 2-pentene

12.5 Write the IUPAC name or common name for the following alkynes:

a. $CH\equiv CH$

b. $CH_3CH_2CH_2C\equiv CCH_3$

c. 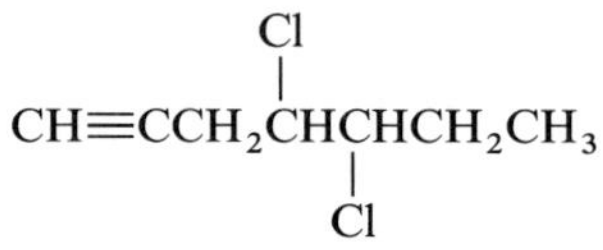
$$\begin{array}{c} \text{Cl} \\ | \\ CH\equiv CCH_2CHCHCH_2CH_3 \\ | \\ \text{Cl} \end{array}$$

d. 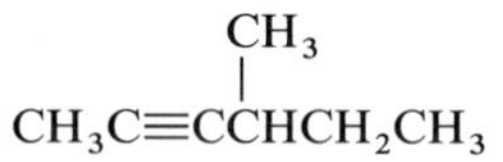
$$\begin{array}{c} CH_3 \\ | \\ CH_3C\equiv CCHCH_2CH_3 \end{array}$$

12.6 Draw a condensed structure for each of the following alkynes:

a. 2-butyne
b. acetylene
c. 4,4-dichloro-2-pentyne

Reactions of Alkenes (Objective 12.2)

12.7 Write the structural formulas of the major product of the following reactions of alkenes:

a. $CH_2{=}CH_2 + HBr \longrightarrow$

b. 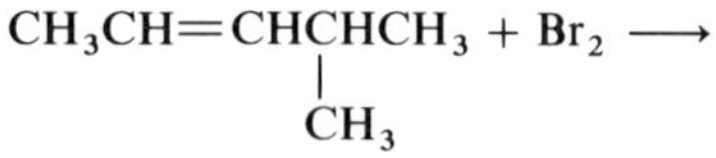 $CH_3CH{=}CHCHCH_3 + Br_2 \longrightarrow$ (with CH_3 on the fourth carbon)
$$\begin{array}{c} CH_3CH{=}CHCHCH_3 \\ \phantom{CH_3CH{=}CH}| \\ \phantom{CH_3CH{=}CH}CH_3 \end{array} + Br_2 \longrightarrow$$

c. $CH_3CH{=}CHCH_3 + H_2 \xrightarrow{Pt}$

d. $\triangle + HCl \longrightarrow$

e. 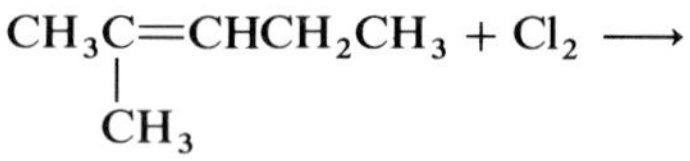
$$\begin{array}{c} CH_3C{=}CHCH_2CH_3 \\ |\phantom{{=}CHCH_2CH_3} \\ CH_3\phantom{{=}CHCH_2CH_3} \end{array} + Cl_2 \longrightarrow$$

f. 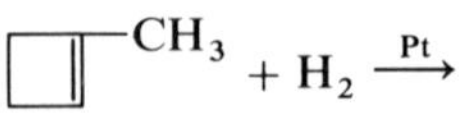 $\square{-}CH_3 + H_2 \xrightarrow{Pt}$

g. 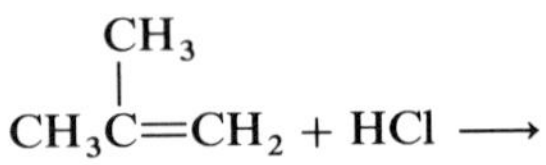
$$\begin{array}{c} CH_3\phantom{{=}CH_2} \\ |\phantom{{=}CH_2} \\ CH_3C{=}CH_2 \end{array} + HCl \longrightarrow$$

12.8 Write an equation using condensed formulas and catalysts (if needed) for the following reactions:

a. hydrogenation of 2-methylpropene
b. addition of chlorine to cyclopentene
c. addition of hydrogen bromide to 2-methyl-2-butene

Aromatic Compounds (Objective 12.3)

12.9 How does benzene differ from cyclohexane and cyclohexene?

12.10 Write the IUPAC name or common name for each of the following aromatic compounds:

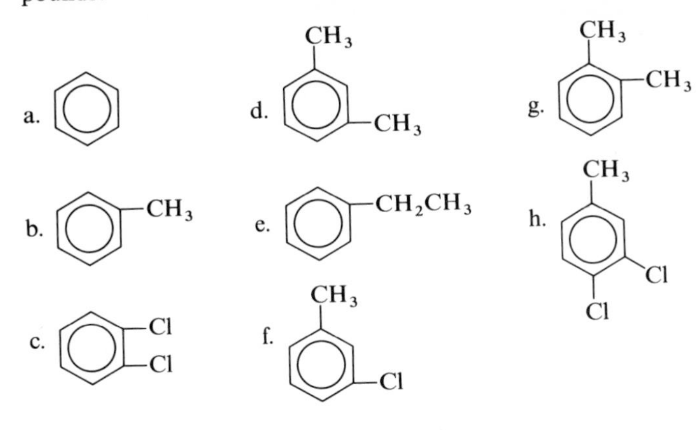

12.11 Write the structure of the following organic compounds:

a. benzene
b. toluene
c. bromobenzene
d. *o*-bromotoluene
e. *p*-dichlorobenzene
f. 1,3,5-trichlorobenzene

Alcohols (Objective 12.4)

12.12 Write the name of the following alcohols:

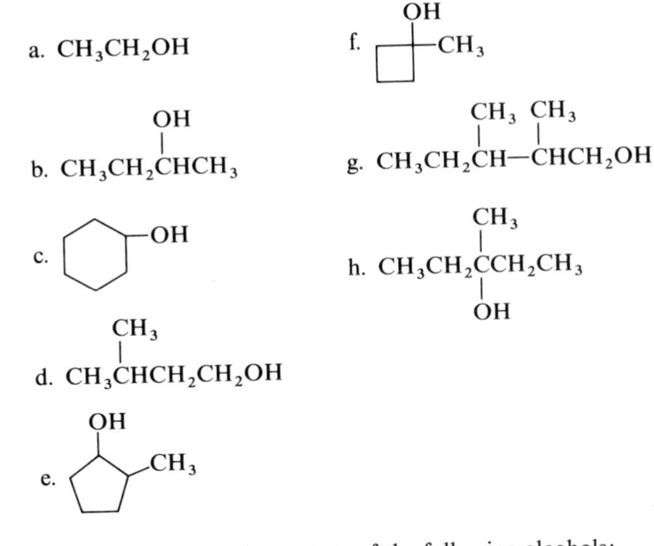

12.13 Write the condensed structure of the following alcohols:

a. 1-propanol
b. phenol
c. 2,4-dimethyl-2-pentanol
d. 2,4-dinitrophenol
e. 3-ethylcyclopentanol
f. 2,4-dichlorocyclohexanol
g. isopropyl alcohol

12.14 Write the names of the following thiols:

a. CH_3SH

b. CH_3CH_2SH

c. $CH_3CH_2\overset{SH}{\overset{|}{C}}HCH_3$

d. $CH_3\overset{CH_3}{\overset{|}{C}}HCH_2SH$

Class of Alcohols (Objective 12.5)

12.15 Classify the following alcohols as primary (1°), secondary (2°), or tertiary (3°):

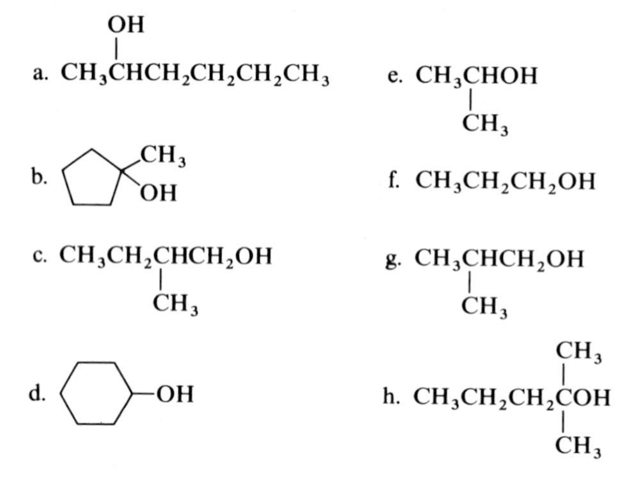

a. $CH_3\overset{OH}{\overset{|}{C}}HCH_2CH_2CH_2CH_3$

b. (cyclopentane ring bearing CH_3 and OH)

c. $CH_3CH_2\underset{CH_3}{\underset{|}{C}}HCH_2OH$

d. (cyclohexane ring)$-OH$

e. $CH_3\underset{CH_3}{\underset{|}{C}}HOH$

f. $CH_3CH_2CH_2OH$

g. $CH_3\underset{CH_3}{\underset{|}{C}}HCH_2OH$

h. $CH_3CH_2CH_2\overset{CH_3}{\underset{CH_3}{\overset{|}{\underset{|}{C}}}}OH$

Formation of Alcohols (Objective 12.6)

12.16 Complete the following hydration reactions:

a. $CH_3CH{=}CHCH_3 + HOH \xrightarrow{H_2SO_4}$

b. (cyclohexene) $+ HOH \xrightarrow{H^+}$

c. $CH_3CH{=}CH_2 + HOH \xrightarrow{H_3PO_4}$

d. (benzene ring bearing $CH{=}CH_2$) $+ HOH \xrightarrow{H_2SO_4}$

e. ______ $\xrightarrow{H_2SO_4}$ (cyclohexane ring bearing CH_3 and OH)

Reactions of Alcohols (Objective 12.7)

12.17 Complete the following dehydration reactions:

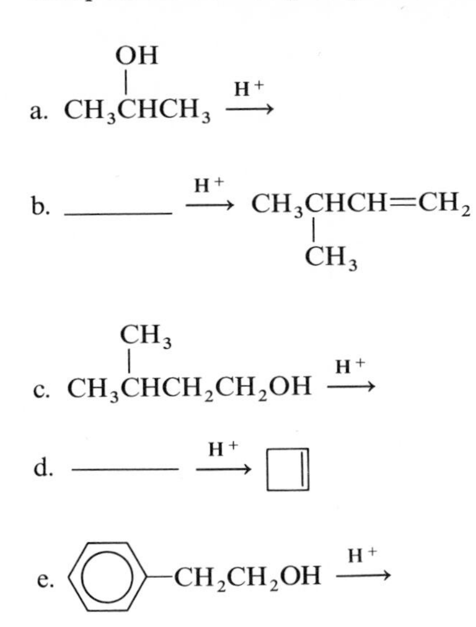

12.18 Write the equation for the following reactions:
a. hydration of cyclopentene
b. hydration of 2-methyl-1-butene
c. dehydration of cyclobutanol
d. dehydration of 4-methyl-2-pentanol

Ethers (Objective 12.8)

12.19 Write the names of the following ethers:

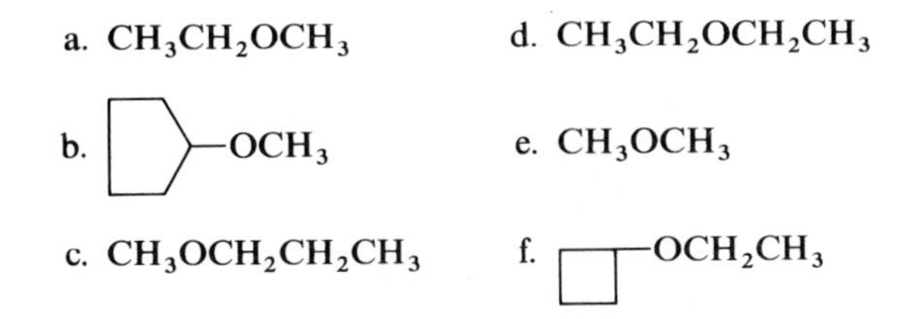

12.20 Write the condensed structure of the following ethers:
a. methyl propyl ether
b. cyclopropyl ethyl ether
c. dimethyl ether
d. diethyl ether

Formation of Ethers (Objective 12.9)

12.21 Complete the following reactions for ether formation:

a. $2CH_3OH \xrightarrow{H^+}$

b. 2 —OH $\xrightarrow{H^+}$

c. $2CH_3CH_2OH \xrightarrow{H^+}$

12.22 Identify the following as an alkane, haloalkane, alkene, alkyne, aromatic, alcohol, ether, or thiol. Then write an IUPAC name or common name:

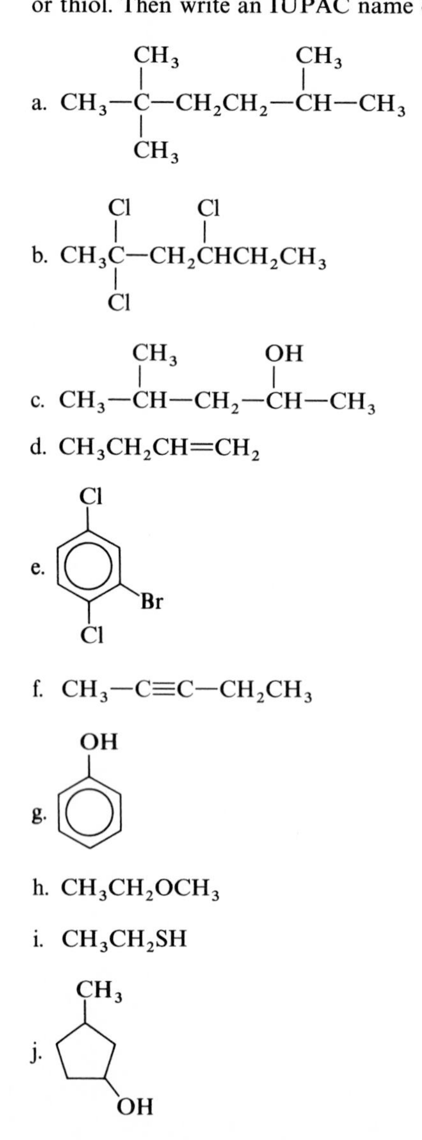

12.23 Write the structure of the following:

a. 1,2-dimethylcyclopentane
b. *p*-dichlorobenzene
c. propylbromide
d. *o*-chlorotoluene
e. 2,5-dichloro-2-hexene
f. *p*-chlorophenol
g. diethyl ether
h. *n*-propanethiol
i. 3-methyl-2-butanol

12.24 Write the equations for the following reactions:

a. complete combustion of propane
b. hydrogenation (Pt) of 1-methylcyclohexene
c. addition of Cl_2 to 2-methyl-2-butene
d. addition of HCl to 2-methyl-2-pentene
e. addition of H_2O to 1-methylcyclopentene

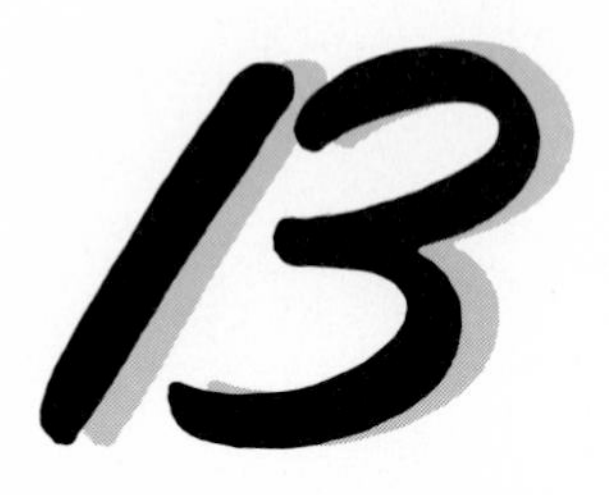

Aldehydes, Ketones, Carboxylic Acids, and Esters

Objectives

13.1 Write the name and structure of an aldehyde and ketone.

13.2 Given the structure of an alcohol, write the expected oxidation products.

13.3 Write the reduction products of aldehydes and ketones.

13.4 Write the products of acetal and ketal formation.

13.5 Write the name and structure of a carboxylic acid.

13.6 Write the equation for the formation of a carboxylic acid by the oxidation of an aldehyde.

13.7 Write the equation for the ionization and neutralization of a carboxylic acid.

13.8 Write the equation for the formation of an ester.

13.9 Write the name and structure of an ester.

13.10 Write the products of ester hydrolysis and saponification.

Scope

Of the aldehydes, the simplest, formaldehyde, is used for preserving tissue specimens. The presence of ketones in body fluids is a danger signal: Ketone bodies are produced when the body fails to utilize available carbohydrates, when diabetes is present, or when a condition of starvation exists.

When you think of an acid, you think of something that tastes sour, produces hydronium ions, and turns blue litmus red. These are the properties of organic as well as inorganic acids. Organic acids are typically weak and dissociate only slightly in aqueous solutions. The vinegar you use in your salad is acetic acid; the tartness of citrus fruits is caused by the presence of citric acid. Proteins are formed from amino acids, which are derivatives of carboxylic acids.

When an acid combines with an alcohol, a compound called an ester is produced. Aspirin contains an ester group. The pleasant aromas of fruits and flowers are characteristic of compounds containing ester functional groups. Fats in both animals and vegetables are esters of the alcohol glycerol combined with one to three acid molecules.

13.1 Aldehydes and Ketones

OBJECTIVE Write the name and structure of an aldehyde and ketone.

carbonyl group Both aldehydes and ketones contain a functional group called the *carbonyl group*:

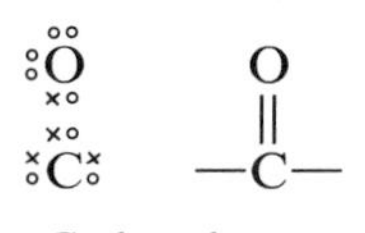

Carbonyl group

aldehyde In aldehydes, the carbonyl group is attached to at least one hydrogen atom. The carbonyl group is also attached to one hydrocarbon group, and sometimes another hydrogen atom:

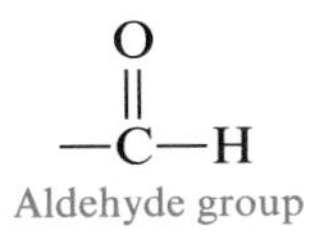

Aldehyde group

In the IUPAC system, the aldehyde is named by replacing the *-e* in the corresponding alkane name with the suffix *-al*. An aldehyde group always appears at the end of the carbon chain. Since it is the functional group, its carbon is counted as carbon 1 although no number is needed. Any side groups on the carbon chain are located by numbering the carbon chain from the carbon atom in the aldehyde group:

Aldehyde

Functional group: 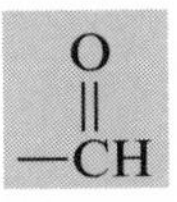

IUPAC name: Alkan*al*

Common name: The word *aldehyde* preceded by the common name prefix: 1 C, *form-*; 2 C, *acet-*; 3 C, *propion-*; 4 C, *butyr-*

For the simple aldehydes with one to four carbon atoms, the common names are generally used. The prefixes for the common names of the first four aldehydes are *form-*, *acet-*, *propion-*, and *butyr-*. The prefixes are followed by the word *aldehyde* to indicate the aldehyde functional group. When there is an aldehyde group on benzene, the compound is called *benzaldehyde*. The position of a side group in a common name is indicated by the symbols α (alpha), β (beta), and γ (gamma), which designate the carbon atoms *adjacent* to the aldehyde group:

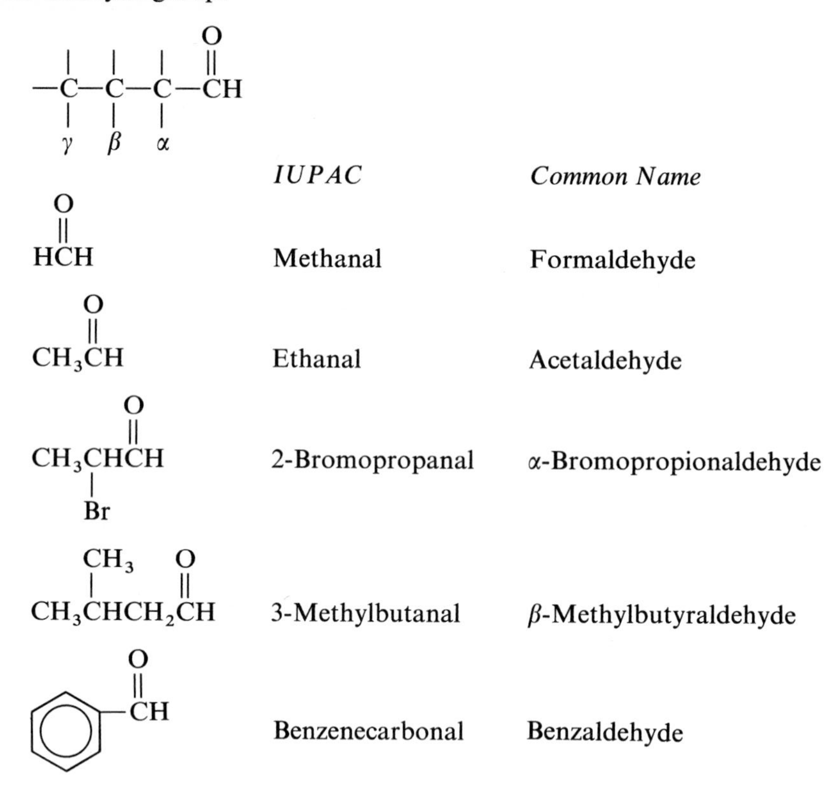

	IUPAC	*Common Name*
HCHO	Methanal	Formaldehyde
CH_3CHO	Ethanal	Acetaldehyde
$CH_3CHBrCHO$	2-Bromopropanal	α-Bromopropionaldehyde
$(CH_3)_2CHCH_2CHO$	3-Methylbutanal	β-Methylbutyraldehyde
C_6H_5CHO	Benzenecarbonal	Benzaldehyde

Example 13.1 Write a name for the following aldehyde:

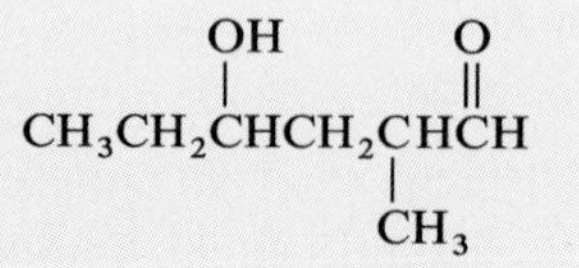

Solution: This is a long-chain aldehyde. Since the parent chain consists of six carbon atoms, it is named *hexanal*. When the carbon chain is numbered from the carbon atom of the aldehyde group, there is a methyl group on carbon 2 and a hydroxy group on carbon 4. There is no common name:

IUPAC name: 4-Hydroxy-2-methylhexanal

Example 13.2 Write the structure for β-chloropropionaldehyde.

Solution: This is a common name for an aldehyde with three carbon atoms. The chlorine atom is on the beta carbon of the chain:

Beta carbon → $ClCH_2CH_2\overset{O}{\overset{\|}{C}}H$

Ketones

Ketones are organic compounds in which the carbonyl group is attached to two hydrocarbon groups. To name a ketone by the IUPAC system, find the longest carbon chain containing the carbonyl group and change the last letter in the corresponding alkane name to *-one*. The location of the ketone group is indicated when there are five or more carbon atoms in the compound:

naming ketones

Ketone

Functional group:	$-\overset{O}{\overset{\|}{C}}-$
IUPAC name:	Alkan*one*
Common name:	Alkyl (two) ketone

$CH_3\overset{O}{\overset{\|}{C}}CH_3$

Propanone (dimethyl ketone, "acetone")

$CH_3\overset{O}{\overset{\|}{C}}\underset{CH_3}{\underset{|}{C}}HCH_3$

3-Methylbutanone (isopropyl methyl ketone)

$CH_3\underset{Br}{\underset{|}{C}}H\overset{O}{\overset{\|}{C}}CH_2CH_3$

2-Bromo-3-pentanone

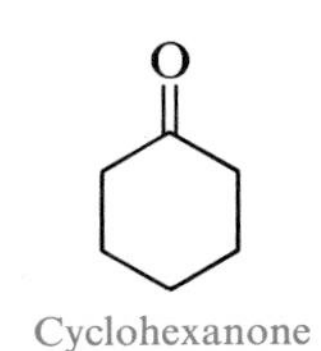

Cyclohexanone

HEALTH NOTE

Aldehydes and Ketones in Health

The simplest aldehyde is formaldehyde,

$$\overset{\displaystyle O}{\overset{\|}{HCH}}$$

Aqueous solutions of formaldehyde (formalin) are used to preserve tissues and as germicides. It is used in the manufacturing of paper, insulation materials, and cosmetics such as shampoos. There is concern now that formaldehyde is a potential carcinogen (cancer-causing compound) and should be eliminated from household substances.

Propanone (acetone or dimethyl ketone),

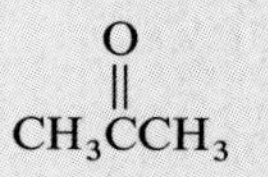

is an excellent solvent for many organic materials. It is found in paint removers and fingernail polish remover. In the body, acetone can be produced when fats are metabolized for energy production. This happens in uncontrolled diabetes and fasting. The fruity odor of acetone can be detected on the person's breath.

The monosaccharides, part of the carbohydrate family that makes up starches and cellulose, are compounds with either an aldehyde or ketone group and several alcohol groups:

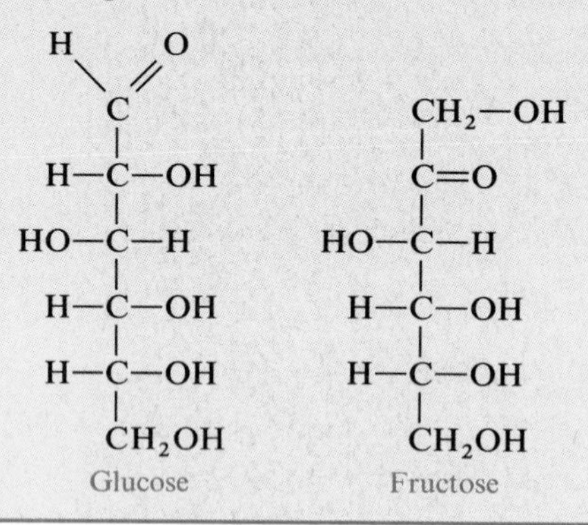

Glucose Fructose

Example 13.3 Write the name for the following ketones:

1. $CH_3CH_2\overset{O}{\overset{\|}{C}}CH_2CH_3$

2. (cyclopentane ring)$=O$

Solution

1. In the IUPAC name, the ketone group is on the third carbon atom in a chain of five. Since the ketone group could be at different locations, it is numbered:

   IUPAC name: 3-Pentanone

   The common name for this ketone indicates each of the hydrocarbon groups attached to the ketone group:

   Common name: Diethyl ketone

2. This is a ketone of the cyclic alkane cyclopentane. We can name it as a ketone by changing the *-e* to *-one*:

   IUPAC name: Cyclopentanone

Example 13.4 Write the structure of 4-methyl-2-pentanone.

Solution: The parent name indicates that there is a chain of five carbons and that the second carbon has a ketone group. A methyl side chain is attached at the fourth carbon. The carbonyl group has the lowest possible number:

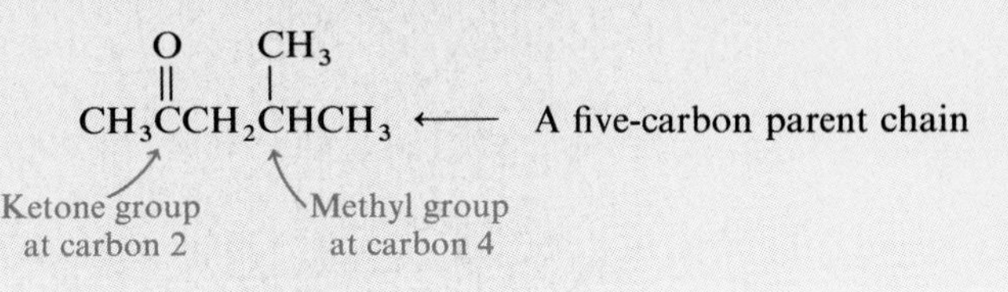

13.2 Oxidation of Alcohols to Form Aldehydes and Ketones

OBJECTIVE Given the structure of an alcohol, write the expected oxidation products.

The oxidation of an alcohol requires the removal of two hydrogen atoms from the alcohol. An oxidation reaction requires the presence of an *oxidizing agent* such as $KMnO_4$ or $K_2Cr_2O_7$. When an alcohol is oxidized, one hydrogen atom is taken from the —OH group and the second hydrogen atom is removed from the alcohol carbon. In equations for the oxidation of alcohols in this text, the presence of an oxidizing agent will be indicated by the symbol [O]:

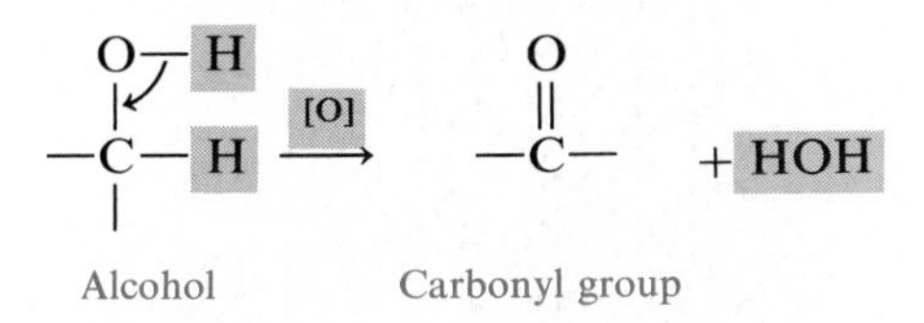

When a primary alcohol is oxidized, the product is an aldehyde.

Oxidation of Primary Alcohols

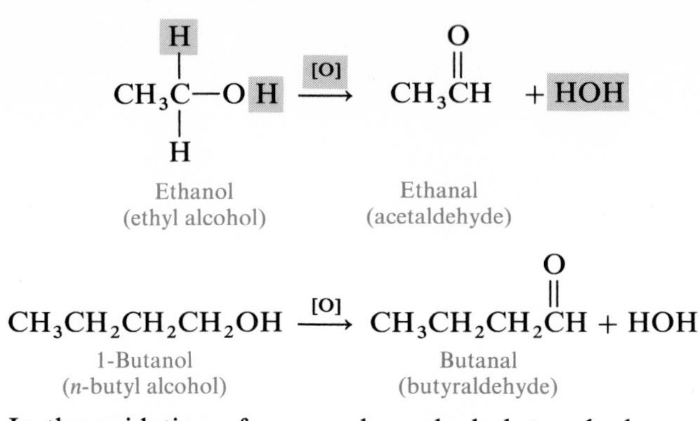

In the oxidation of a secondary alcohol, two hydrogen atoms are again removed by the oxidizing agent [O]. The resulting carbonyl group now lies between two hydrocarbon groups, thus a ketone is produced.

Oxidation of Secondary Alcohols

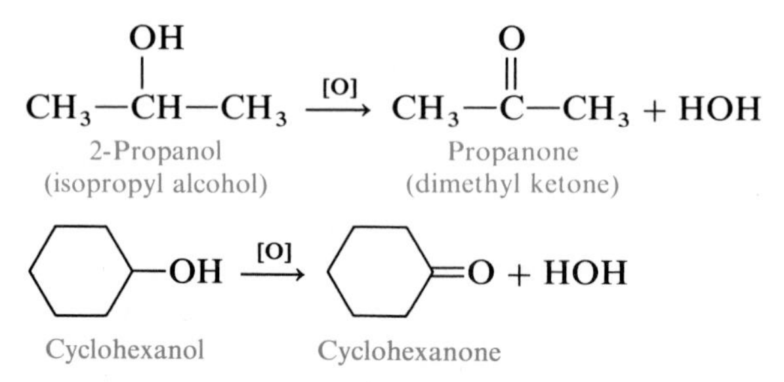

Tertiary alcohols do not oxidize. There is no hydrogen atom to remove from the alcohol carbon; therefore, no oxidation can occur. We say that tertiary alcohols are stable with regard to oxidation:

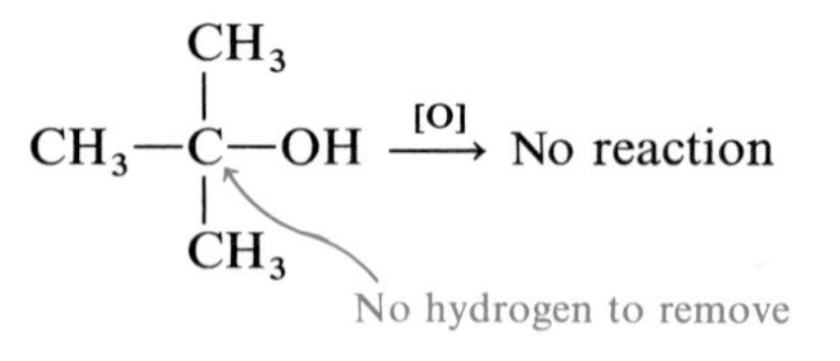

Example 13.5 Write the structure of the oxidation products of the following alcohols:

1. $CH_3CH_2\overset{\overset{\displaystyle OH}{|}}{C}HCH_3 \xrightarrow{[O]}$

2. $CH_3\underset{\underset{\displaystyle CH_3}{|}}{C}HCH_2OH \xrightarrow{[O]}$

Solution

1. This alcohol is a secondary alcohol. Its oxidation produces a ketone which appears on the same carbon that held the hydroxyl group:

$$\underset{\text{2-Butanol}}{CH_3CH_2\overset{\overset{\displaystyle OH}{|}}{C}HCH_3} \xrightarrow{[O]} \underset{\text{Butanone}}{CH_3CH_2\overset{\overset{\displaystyle O}{\|}}{C}CH_3} + HOH$$

2. This alcohol is a primary alcohol. It will oxidize to produce an aldehyde:

$$\underset{\text{2-Methyl-1-propanol}}{CH_3\underset{\underset{\displaystyle CH_3}{|}}{C}HCH_2OH} \xrightarrow{[O]} \underset{\text{2-Methyl propanal}}{CH_3\underset{\underset{\displaystyle CH_3}{|}}{C}H\overset{\overset{\displaystyle O}{\|}}{C}H} + H_2O$$

13.3 Reduction of Aldehydes and Ketones

OBJECTIVE Write the reduction products of aldehydes and ketones.

Aldehydes and ketones can be reduced to produce their corresponding alcohols. A typical reduction uses molecular hydrogen in the presence of a platinum catalyst. In the reduction reaction, the hydrogen atoms are added to the carbon and oxygen atoms of the carbonyl group. The carbonyl group is reduced to an alcohol:

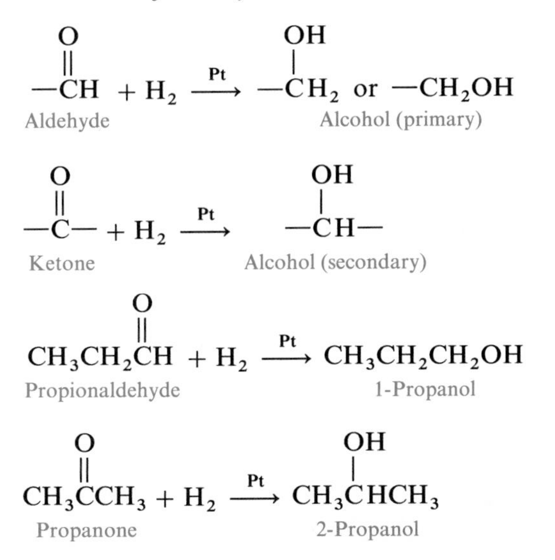

Example 13.6 Write the product for the reduction of cyclopentanone.

Solution: The reacting molecule is a ketone of the cyclic five-carbon ring. The ketone will reduce to give the corresponding alcohol:

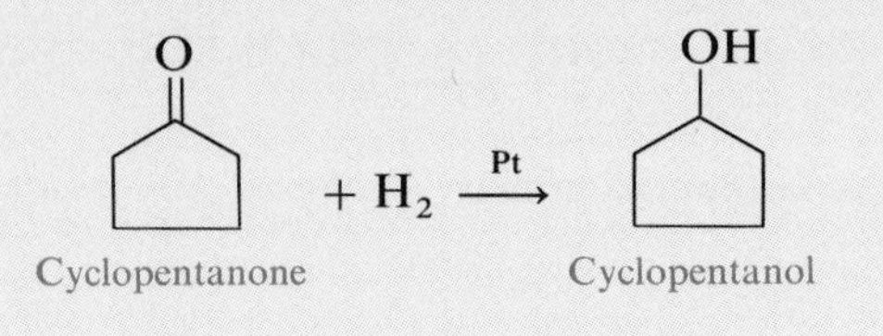

13.4 Acetals and Ketals

OBJECTIVE Write the products of acetal and ketal formation.

Since the carbonyl group in aldehydes and ketone has a double bond, it can undergo addition reactions. When alcohols are added to the carbonyl groups in the presence of acid, compounds called *hemiacetals* and *hemiketals* form. Such compounds are important in carbohydrate chemistry.

This reaction is similar to the addition reactions for alkenes. In this case, an alcohol adds across the carbonyl group. The hydrogen atom from the alcohol group attaches itself to the oxygen atom of the carbonyl group; the rest of the alcohol attaches to the carbon atom. When the reaction involves an aldehyde, the product is called a hemiacetal; when it is a ketone, the product is a hemiketal:

hemiacetals and hemiketals

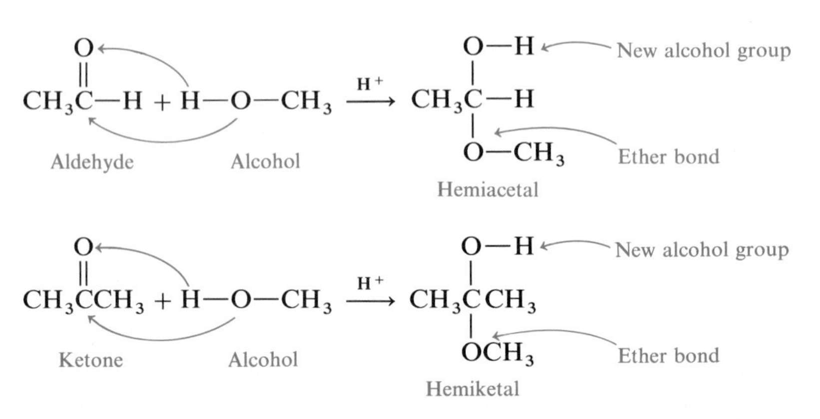

acetals and ketals

When a hemiacetal or a hemiketal adds another molecule of alcohol, an acetal or a ketal is formed. In this reaction, the alcohol groups (—OH) react to form an ether bond. In the final products, the carbon atom from the carbonyl group is attached by two ether bonds to hydrocarbon groups:

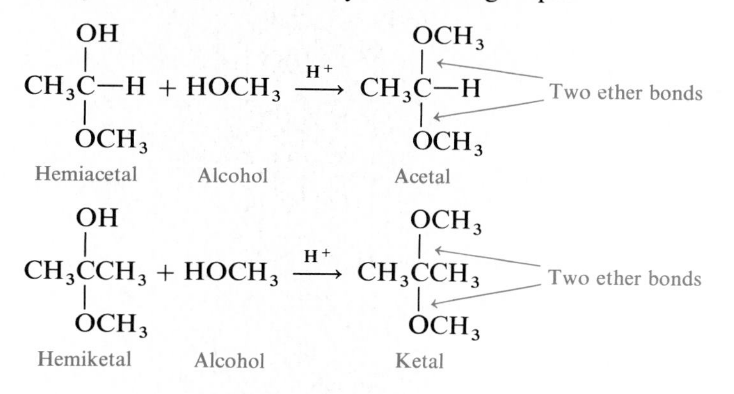

Example 13.7 Write the hemiacetal product for the following reaction:

$$CH_3CH_2\overset{\overset{\displaystyle O}{\|}}{C}H + CH_3OH \xrightarrow{H^+}$$

Solution: The oxygen of the alcohol adds to the carbon atom of the carbonyl group. The hydrogen left from the alcohol forms a new alcohol group by bonding with the O atom from the old carbonyl group:

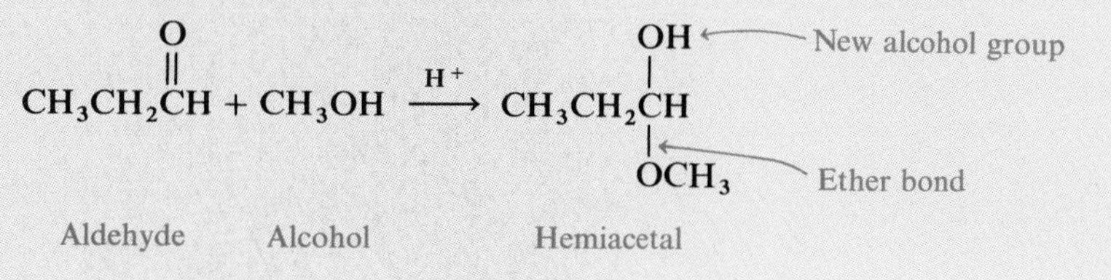

13.5 Carboxylic Acids

OBJECTIVE Write the name and structure of a carboxylic acid.

carboxyl group

Carboxylic acids are organic compounds that contain a functional group called a *carboxyl group*:

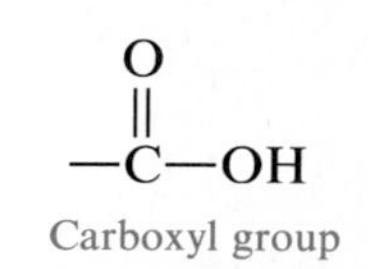

Carboxyl group

This group consists of an alcohol group attached to a carbonyl group. The carboxylic group always appears at one end of the compound.

naming carboxylic acids

To name a carboxylic acid using the IUPAC system, the suffix *-oic acid* replaces the *-e* in the corresponding alkane name. Common names for the first four carboxylic acids are often used. The prefixes used for the common names of aldehydes are the same for carboxylic acids:

Carboxylic Acid

Functional group: $-\overset{\text{O}}{\overset{\|}{\text{C}}}\text{OH}$

IUPAC name: Alkan*oic acid*

Common name: Common prefixes of *form-* (1 C), *acet-* (2 C), *propion-* (3 C), and *butyr-* (4 C), with ending *-ic acid*

$$\text{H}\overset{\text{O}}{\overset{\|}{\text{C}}}\text{OH}$$

Methanoic acid (formic acid)
("ant stings")

$$\text{CH}_3\overset{\text{O}}{\overset{\|}{\text{C}}}\text{OH}$$

Ethanoic acid (acetic acid)
("vinegar")

$$\text{CH}_3\text{CH}_2\overset{\text{O}}{\overset{\|}{\text{C}}}\text{OH}$$

Propanoic acid (propionic acid)

$$\text{CH}_3\text{CH}_2\text{CH}_2\overset{\text{O}}{\overset{\|}{\text{C}}}\text{OH}$$

Butanoic acid (butyric acid)
("rancid butter")

In the IUPAC system, the carbon atom in the carboxyl group is counted as carbon 1 when there are side groups to name and number. When you want to use the common name of a carboxylic acid, the locations of side groups are indicated by the symbols α, β, γ for carbon atoms adjacent to the carboxyl group. We used this system for the aldehydes also:

$$\text{CH}_3\overset{\text{O}}{\overset{\|}{\text{C}}}\text{CH}_2\overset{\text{O}}{\overset{\|}{\text{C}}}\text{OH}$$

3-Ketobutanoic acid (β-ketobutyric acid)

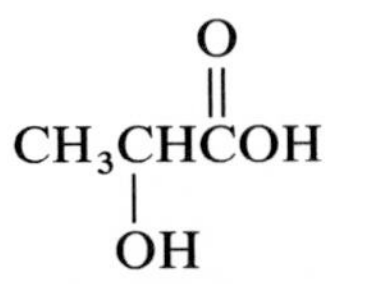

2-Hydroxypropanoic acid
(α-hydroxypropionic acid) ("lactic acid")

$$CH_3CH_2\overset{CH_3}{\overset{|}{C}}HCH_2\overset{O}{\overset{\|}{C}}OH$$

3-Methylpentanoic acid

Sometimes carboxylic acid groups can appear on both ends of the carbon chain. Then the name of the acid ends in *-dioic acid*:

$$HO\overset{O}{\overset{\|}{C}}\overset{O}{\overset{\|}{C}}OH$$

Ethanedioic acid (oxalic acid)

$$HO\overset{O}{\overset{\|}{C}}CH_2CH_2\overset{O}{\overset{\|}{C}}OH$$

Butanedioic acid (succinic acid)

HEALTH NOTE

Carboxylic Acids in Nature

Formic acid is the toxic substance released under the skin from a bee sting, ant sting, and other insect bites. Acetic acid forms as the oxidation product of ethanol in wines and apple cider. Vinegar is a 5% solution of acetic acid.

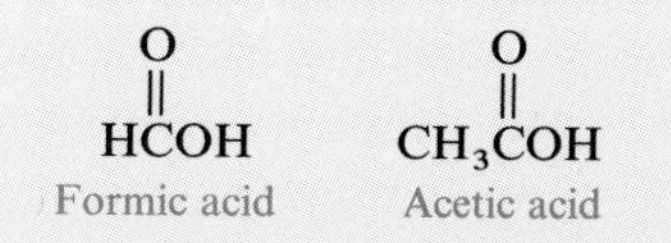

Formic acid   Acetic acid

Lactic acid is found in sour milk. It is also produced during the contraction of muscles when energy is liberated in a series of chemical reactions called the *lactic acid cycle*. Benzoic acid and its salts are used as food preservatives. Citric acid is a widely occurring natural product. It is found in fruits and is largely responsible for their tart tastes.

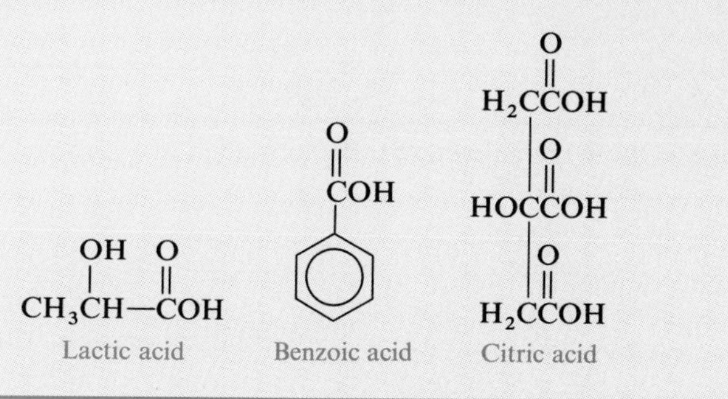

Lactic acid   Benzoic acid   Citric acid

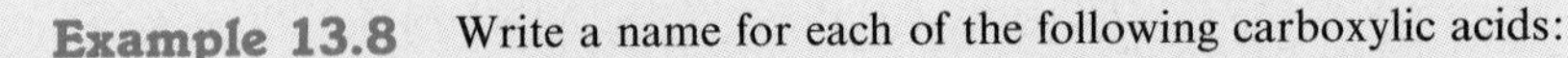

Example 13.8 Write a name for each of the following carboxylic acids:

1. $\mathrm{CH_3\overset{\overset{O}{\|}}{C}OH}$

2. $\mathrm{CH_3\overset{\overset{OH}{|}}{C}HCH_2\overset{\overset{O}{\|}}{C}OH}$

3. $\mathrm{CH_3\overset{\overset{O}{\|}}{C}\overset{\overset{O}{\|}}{C}OH}$

Solution

1. This is the carboxylic acid of the corresponding two-carbon compound ethane. In the IUPAC system, the *-e* is replaced by *-oic acid*:

   IUPAC name: Ethanoic acid

   The common name of this carboxylic acid has the prefix *acet-*, which is used for two-carbon compounds with carbonyl groups on the end. The common name of an acid ends in *-ic acid*:

   Common name: Acetic acid

2. This compound will be named as an acid with a hydroxy group on the third carbon. The carbon of the acid group is always counted as carbon 1. When the IUPAC name is used for the acid, the carbon chain is numbered. When the common name is given, the Greek letters α, β, and γ are used for the first, second, and third carbon atoms adjacent to the carbonyl group:

   IUPAC name: 3-Hydroxybutanoic acid

   Common name: β-Hydroxybutyric acid

3. This is a carboxylic acid with three carbons atoms in the chain. The carbon next to the acid group is another functional group, a ketone. When a ketone is named as a substituent, it is called *keto-*:

   IUPAC name: 2-Ketopropanoic acid

   Common name: α-Ketopropionic acid

13.6 Formation of Carboxylic Acids

OBJECTIVE Write the equation for the formation of a carboxylic acid by the oxidation of an aldehyde.

We learned in the last chapter that the oxidation of alcohols occurs with the loss of two hydrogen atoms. It is also possible for aldehydes formed from primary alcohols to continue the oxidation process, this time by the addition of oxygen to the aldehyde. In the oxidation of an aldehyde, an oxygen atom converts the aldehyde group to an acid group. A ketone does not oxidize any further:

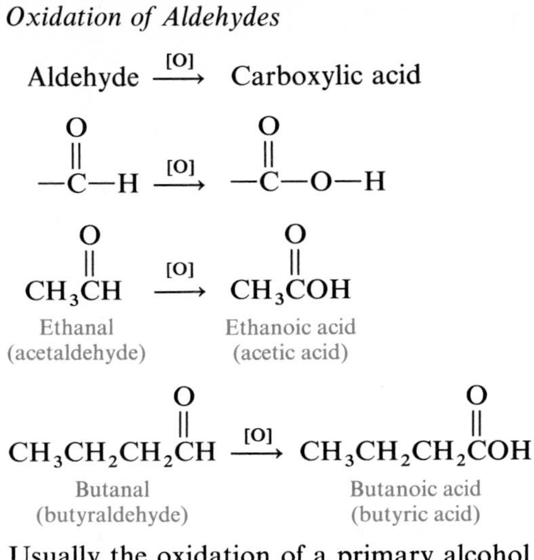

Usually the oxidation of a primary alcohol will proceed to the aldehyde and on to the carboxylic acid:

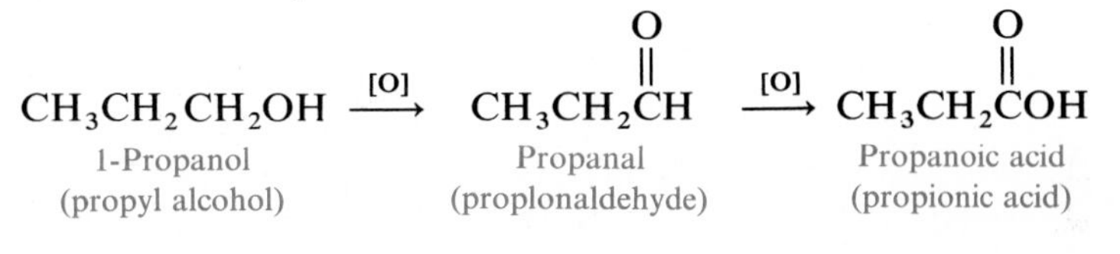

Example 13.9 Write the structure of the oxidation products for the following:

1. $HCH \xrightarrow{[O]}$ (with C=O)

2. $CH_3CH_2CH_2OH \xrightarrow{[O]}$

Solution

1. The oxidation of formaldehyde will produce a carboxylic acid. The structure can be written by changing the hydrogen atom attached to the end carbon to an —OH:

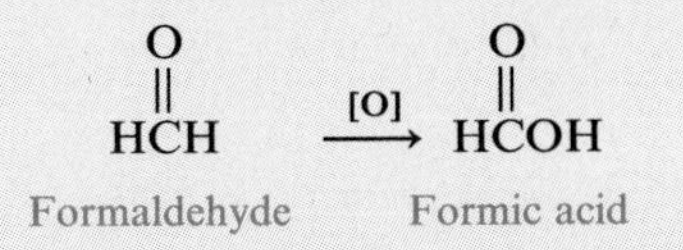

2. The primary alcohol group of this compound will be oxidized to an aldehyde, which can be oxidized further to give a carboxylic acid:

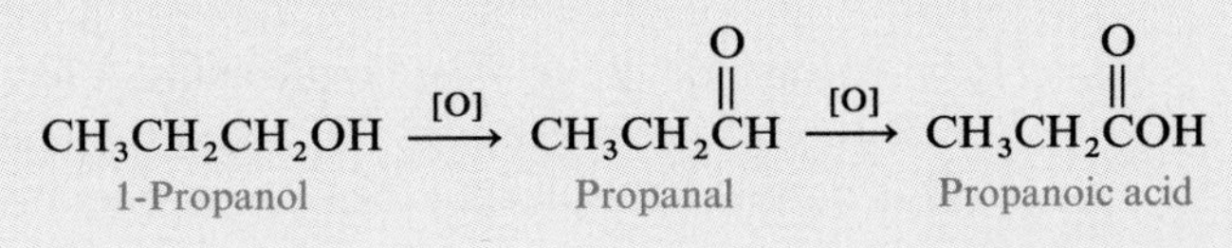

HEALTH NOTE

Oxidation of Alcohol in the Body

Ethanol, the alcohol contained in alcoholic beverages, undergoes oxidation in the liver by an enzyme (catalyst) called *alcohol dehydrogenase*. The product is acetaldehyde, which is further oxidized to acetic acid. Acetic acid takes part in chemical reactions in the cell and eventually is converted to carbon dioxide and water:

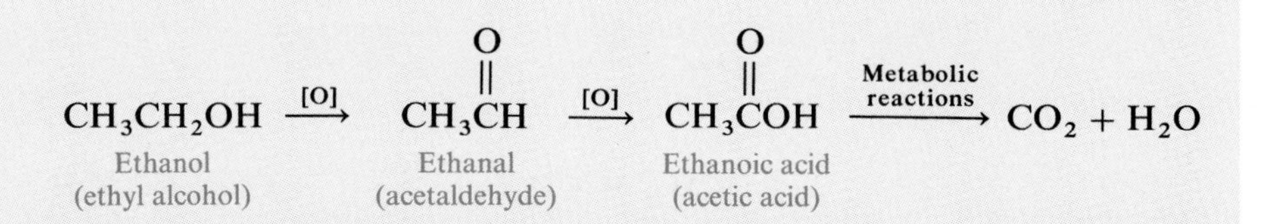

The drug Antabuse is used to treat alcoholism by increasing the rate of oxidation of the ethanol to acetaldehyde. High concentrations of acetaldehyde in the tissues cause nausea and ill effects that are so unpleasant for the patient that the use of alcohol is curtailed.

Methyl alcohol, also known as "wood alcohol," is used as a solvent, as antifreeze, and as mimeographic fluid. A small amount (15–20 mL) can cause blindness and possible death. Methanol is also oxidized in the liver by the dehydrogenase enzyme. The oxidation products, formic acid and particularly formaldehyde, are extremely toxic, causing damage to the retina of the eye and optic nerve in addition to severe acidosis:

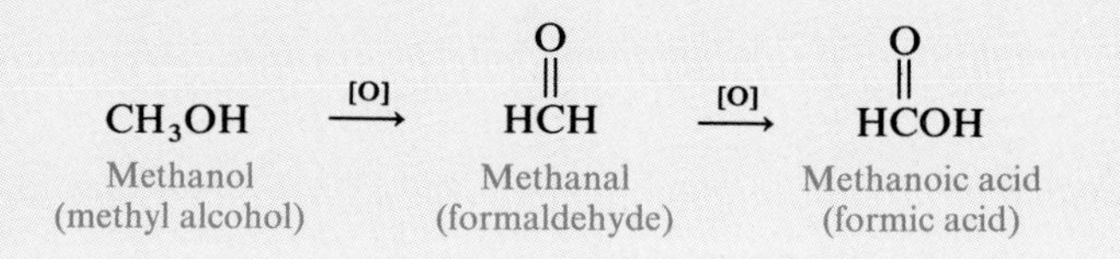

The body does not have a metabolic route for dissipating the formic acid as it does for the acetic acid from the oxidation of ethanol.

13.7 Ionization and Neutralization of Carboxylic Acids

OBJECTIVE Write the equation for the ionization and neutralization of a carboxylic acid.

naming carboxylic acids

In water, carboxylic acids are weak acids. A few molecules of the carboxylic acid ionize to give hydrogen ions (H^+) and the anion of the acid. The anion of the acid is named by replacing the *-ic acid* part of the name to *-ate*:

Ionization of Carboxylic Acids

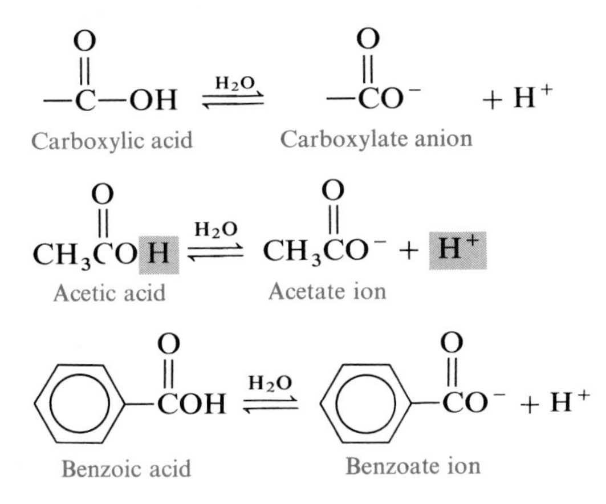

Example 13.10 Write the equation for the ionization of propionic acid in water.

Solution: The ionization of propionic acid produces a hydrogen ion and an anion:

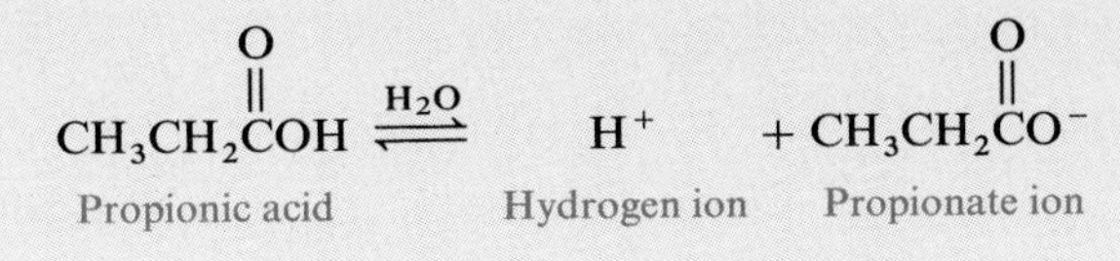

Neutralization of a Carboxylic Acid

We saw in Chapter 10 that an acid can neutralize a base to produce a salt and water. Carboxylic acids also react with bases to give the salt of the acid and water:

Neutralization Reactions of Carboxylic Acids

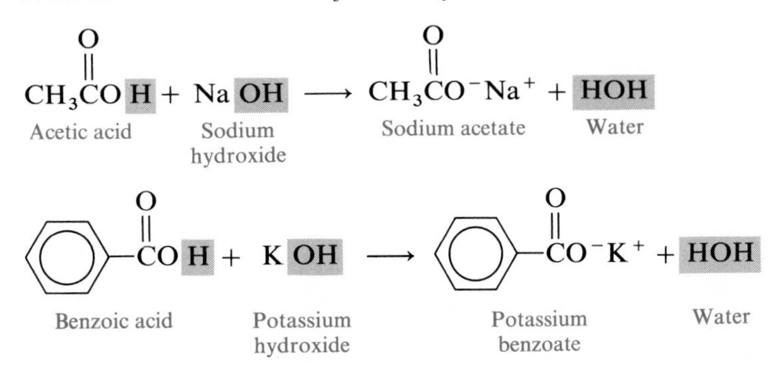

Example 13.11 Write the equation for the neutralization reaction of propionic acid with potassium hydroxide.

Solution: The neutralization of an acid with a base produces the salt of the acid and water. In this reaction, the salt produced would be potassium propionate:

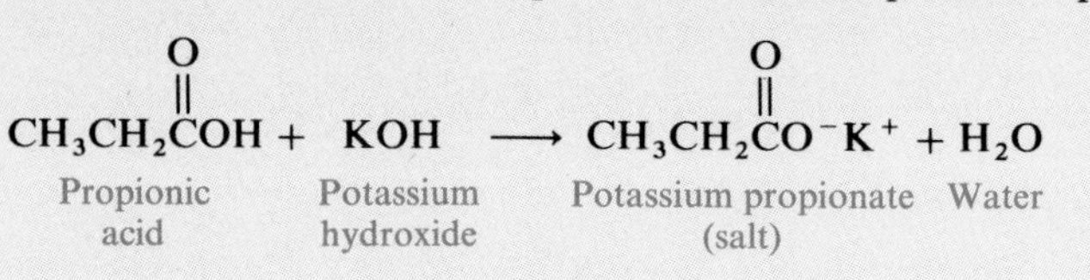

HEALTH NOTE

Some Esters in Medicine

aspirin

Aspirin (acetylsalicyclic acid), is a compound widely used as an analgesic (pain reliever), antipyretic (fever reducer), and anti-inflammatory agent. We can write its formation from salicylic acid and acetic acid as follows:

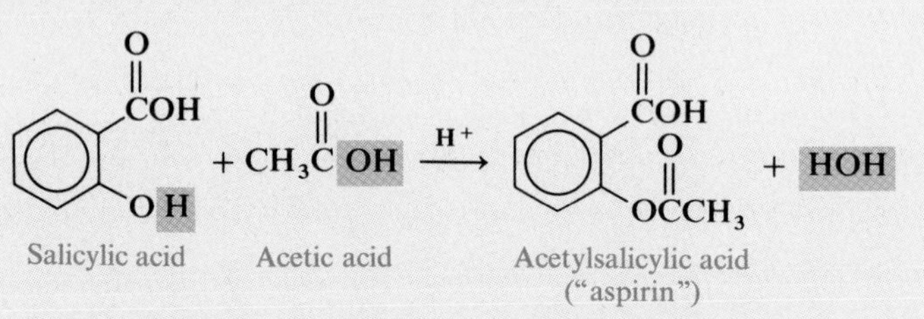

oil of wintergreen

Oil of wintergreen, or methyl salicylate, has a spearmint odor. It is used in skin ointments as a counterirritant that produces heat and soothes muscle pain:

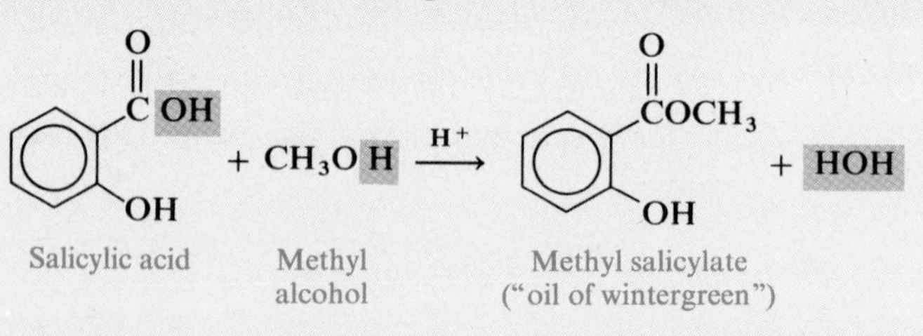

Nitroglycerine, or glyceryl nitrate, is an ester of glycerol and nitric acid. It reduces high blood pressure by causing dilation of the small blood vessels:

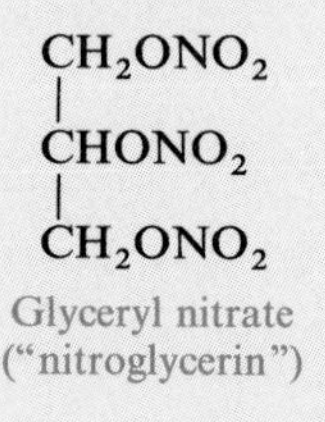

Glyceryl nitrate ("nitroglycerin")

13.8 Formation of an Ester

OBJECTIVE Write the equation for the formation of an ester.

Esters are organic compounds that are produced from a reaction of a carboxylic acid and an alcohol. In the reaction, the —OH group of the carboxylic acid reacts with the —H from the alcohol group to produce a molecule of water; the fragments combine to produce an ester, which has a hydrocarbon group replacing the —H of the carboxylic acid.

Vitamin C, or ascorbic acid, is an ester. In this case, the reacting acid group and alcohol group are on the same molecule, and an ester (also called an inner ester, or lactone) forms within the molecule, making a ring. Vitamin C is water soluble and acidic. It is found in fresh fruits and vegetables:

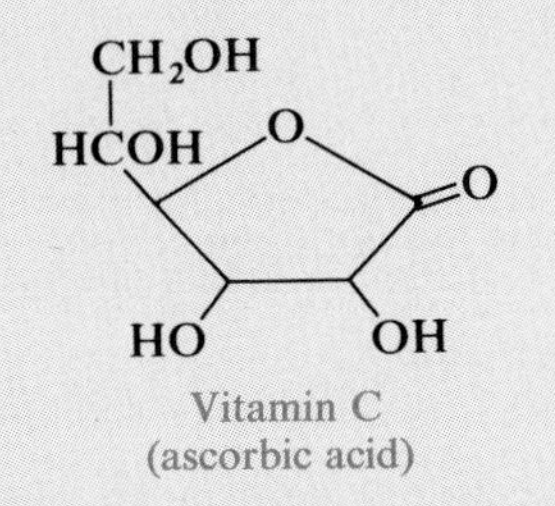

Vitamin C
(ascorbic acid)

Ester groups are also found in fats. Fats are made from an alcohol called *glycerol*, which has three hydroxyl groups and long-chain carboxylic acids (often 16–18 carbons long) called fatty acids:

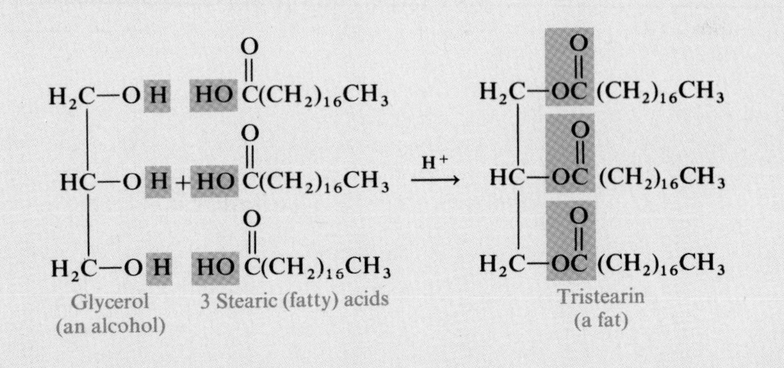

Glycerol
(an alcohol)

3 Stearic (fatty) acids

Tristearin
(a fat)

When a carboxylic acid reacts with an alcohol, the reaction is called *esterification.* A catalyst such as H_2SO_4 is required:

Esterification

$$\text{Carboxylic acid} + \text{Alcohol} \xrightarrow{H^+} \text{Ester} + \text{HOH}$$

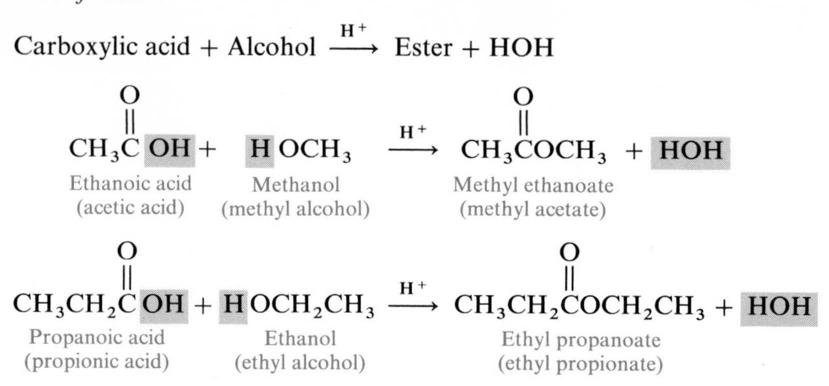

Example 13.12 Write the products of the following esterification reactions:

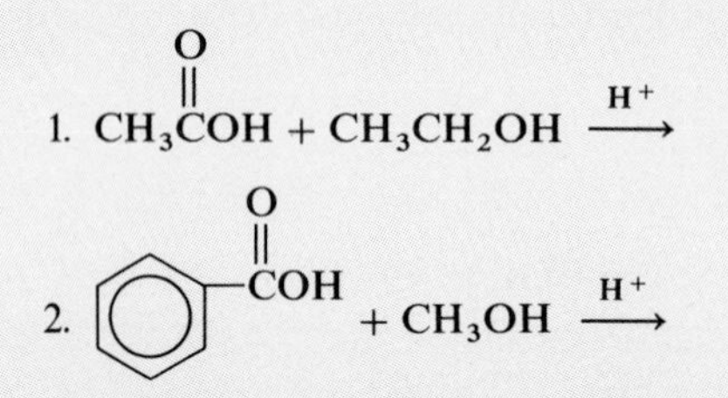

Solution

1. A carboxylic acid and an alcohol will form an ester in the presence of a strong acid. The structure of the ester can be written by replacing the hydrogen in the carboxylic acid by the ethyl portion of the alcohol:

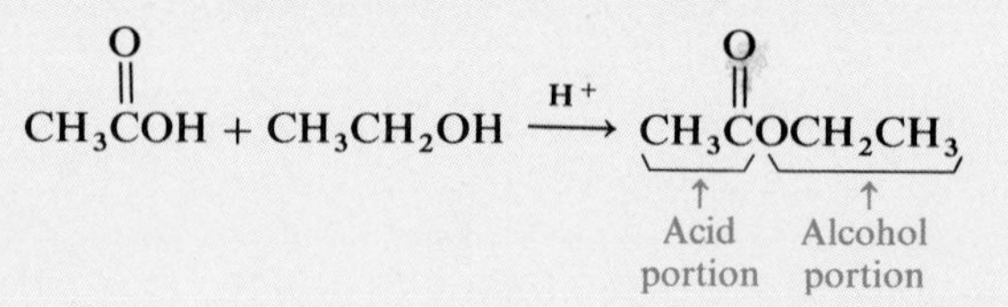

2. The structure of the ester that forms in this esterification can be written by replacing the —H of the benzoic acid by the methyl portion of the alcohol:

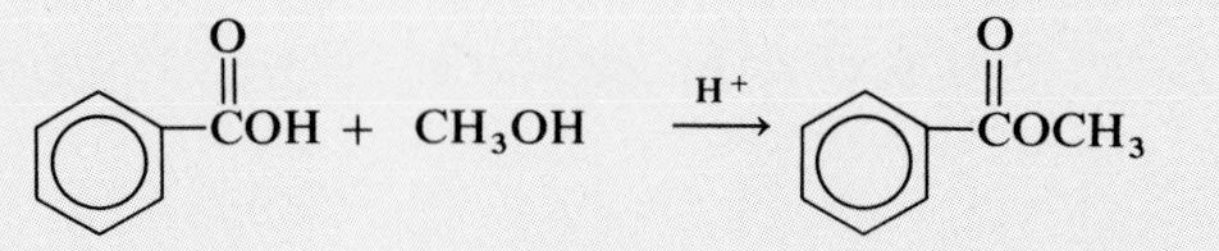

13.9 Naming Esters

OBJECTIVE Write the name and structure of an ester.

Esters are named by specifying the alkyl group derived from the alcohol, followed by the parent name derived from the carboxylic acid. The suffix of the acid name is changed to *-ate*. The common names are used for acids with one to four carbon atoms:

Ester

Functional group: 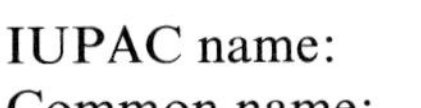

IUPAC name: Alkyl alkanoate
Common name: Alkyl alkanate (common name for esters with 1–4 carbons)

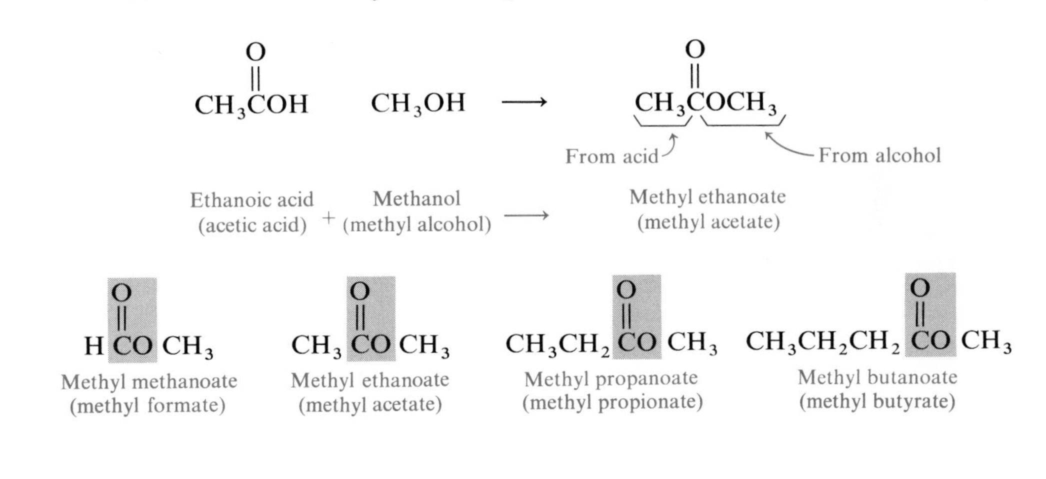

Example 13.13 Write the names of the following esters:

1. $CH_3CH_2\overset{O}{\overset{\|}{C}}OCH_2CH_3$

2. 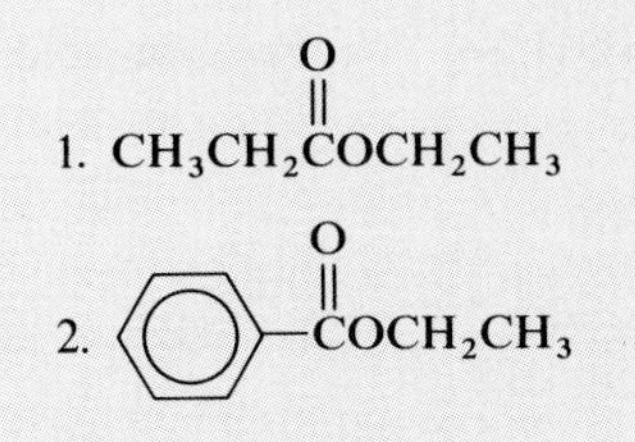

Solution

1. The acid portion of the ester comes from propanoic acid (propionic acid). The alcohol portion is from ethyl alcohol. The ester is named by giving the alkyl name

of the alcohol first. The second part of the name comes from the acid portion. The acid ending is changed to *-ate*:

IUPAC name: Ethyl propanoate
Common name: Ethyl propionate

2. The alkyl portion from the alcohol is ethyl. The acid portion is benzoic acid. To name the ester, the *-ic* acid is replaced by *-ate*:

IUPAC name: Ethyl benzoate

Example 13.14 Write the condensed structure of the following ester:

Methyl butyrate

Solution: The name of the ester indicates that the alkyl portion is a methyl group ($—CH_3$). The acid portion has four carbon atoms (butyric acid):

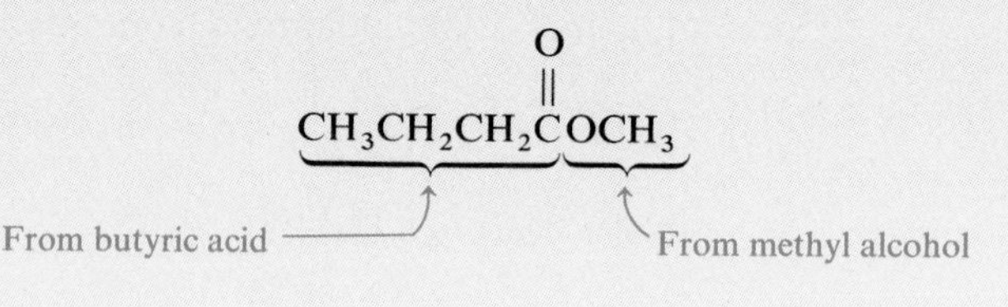

13.10 Reactions of Esters

OBJECTIVE Write the products of ester hydrolysis and saponification.

The process of breaking apart a compound by adding the components of water is called *hydrolysis*. In the body, enzymes (biological catalysts) direct the process of digestion, a series of hydrolysis reactions in which large, complex foodstuffs are broken down into small molecules. In the laboratory, an ester can be hydrolyzed by a strong acid or base catalyst. The products of ester hydrolysis are a carboxylic acid and an alcohol:

hydrolysis

Hydrolysis of an Ester

$$\text{Ester} + \text{HOH} \xrightarrow{\text{Acid}} \text{Carboxylic acid} + \text{Alcohol}$$

$$CH_3\overset{\overset{O}{\|}}{C}OCH_3 + HO{-}H + HOH \xrightarrow{\text{Acid}} CH_3\overset{\overset{O}{\|}}{C}OH + HOCH_3$$

Methyl ethanoate (methyl acetate) — Ethanoic acid (acetic acid) — Methanol (methyl alcohol)

$$C_6H_5{-}\overset{\overset{O}{\|}}{C}OCH_2CH_3 + HOH \xrightarrow{\text{Acid}} C_6H_5{-}\overset{\overset{O}{\|}}{C}OH + HOCH_2CH_3$$

Ethyl benzoate — Benzoic acid — Ethanol (ethyl alcohol)

HEALTH NOTE

Flavors and Odors of Esters

Esters are major components of the chemicals that give a characteristic flavor and odor to fruits and flowers. Some examples are shown in Table 13.1.

Table 13.1 Esters of Essential Oils in Fruits and Flowers

Essence	Ester	Common Name
Rum	$HC(=O)OCH_2CH_3$	Ethyl formate
Raspberry	$HC(=O)OCH_2CH(CH_3)CH_3$	Isobutyl formate
Banana	$CH_3C(=O)O(CH_2)_4CH_3$	Amyl acetate
Orange	$CH_3C(=O)O(CH_2)_7CH_3$	Octyl acetate
Pear	$CH_3C(=O)OCH_2CH_2CH(CH_3)CH_3$	Isoamyl acetate
Pineapple	$CH_3CH_2CH_2C(=O)OCH_2CH_3$	Ethyl butyrate
Apricot, strawberry	$CH_3CH_2CH_2C(=O)O(CH_2)_4CH_3$	Amyl butyrate
Grape, jasmine	$C_6H_4(NH_2)C(=O)OCH_3$ (benzene ring with $-COCH_3$ (C=O) and $-NH_2$ on adjacent positions)	Methyl anthranilate

Example 13.15 Write the products of hydrolysis for the following ester:

$$CH_3CH_2CH_2\overset{\overset{\large O}{\|}}{C}OCH_3$$

Solution: To write the hydrolysis products, we need to determine the carboxylic acid and the alcohol that formed the ester. The alkyl group attached to the oxygen atom of the ester bond came from the alcohol. The rest of the molecule including the carbonyl part of the bond came from the carboxylic acid:

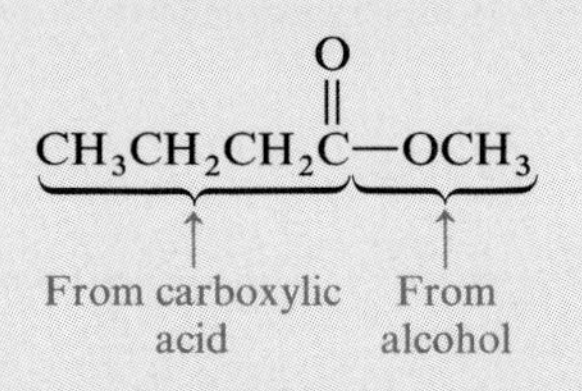

We can complete the hydrolysis products by adding the —OH from water to the acid part and the —H from water to the alcohol part:

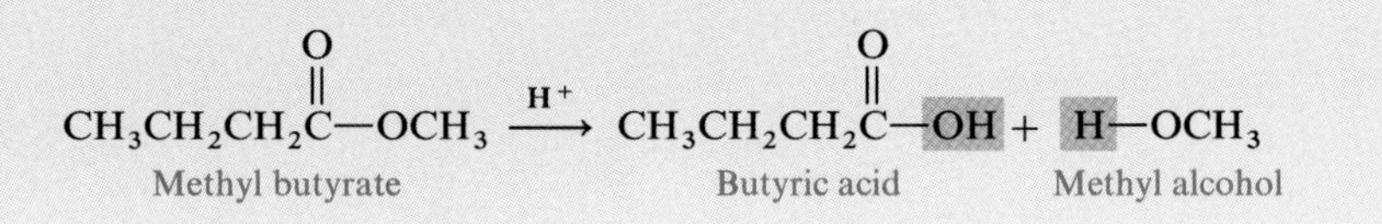

HEALTH NOTE

Hydrolysis in the Digestion of Fats in the Body

lipases

In the body, fats, which are esters, are hydrolyzed to fatty acids and glycerol, an alcohol. The hydrolysis of fats during the digestion process occurs through the action of enzymes called *lipases*:

$$\text{Fat} + 3H_2O \xrightarrow{\text{Lipase}} 3\ \text{Fatty acid molecules} + \text{Glycerol}$$

$$\begin{array}{l} C_{17}H_{35}\overset{\overset{\displaystyle O}{\|}}{C}OCH_2 \\ C_{17}H_{35}\overset{\overset{\displaystyle O}{\|}}{C}OCH \\ C_{17}H_{35}\overset{\overset{\displaystyle O}{\|}}{C}OCH_2 \end{array} + 3H_2O \xrightarrow{\text{Lipase}} \begin{array}{l} C_{17}H_{35}\overset{\overset{\displaystyle O}{\|}}{C}OH \\ C_{17}H_{35}\overset{\overset{\displaystyle O}{\|}}{C}OH \\ C_{17}H_{35}\overset{\overset{\displaystyle O}{\|}}{C}OH \end{array} + \begin{array}{l} HOCH_2 \\ HOCH \\ HOCH_2 \end{array}$$

Tristearin — Stearic acid (3 molecules) — Glycerol (alcohol)

Saponification

When an ester is heated with a strong base such as NaOH or KOH, the hydrolysis reaction is called *saponification.* The products of saponification are the salt of the carboxylic acid and the alcohol:

Saponification of an Ester

Ester + NaOH $\longrightarrow$ Salt of carboxylic acid + Alcohol

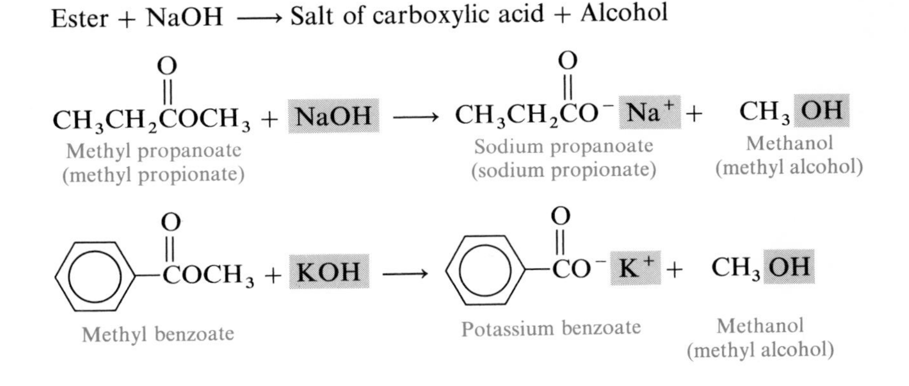

Soaps

When fats are heated with a strong base, the saponification products are salts of the acid and an alcohol. The salts of the fatty acids released during hydrolysis are commonly known as *soaps*, and the saponification of fats is the process used in the preparation of soap. The alcohol produced in saponification of a fat is glycerol:

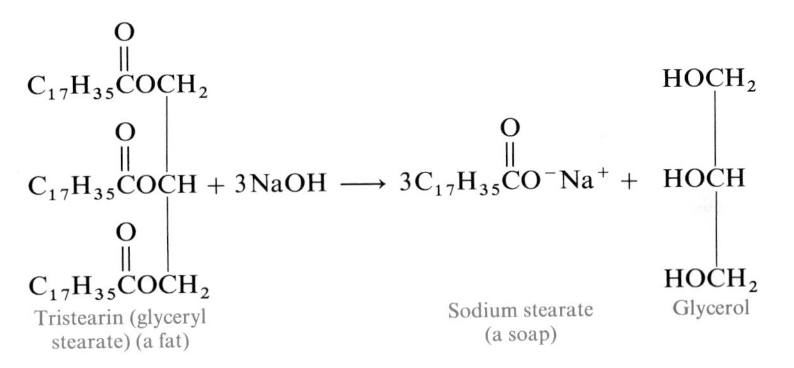

Example 13.16 Write the products of the following reaction:

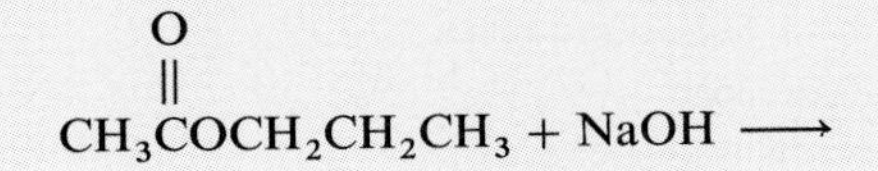

Solution: An ester reacts with a base (NaOH) to form the salt of the acid and an alcohol. In the salt, the metal ion from the base has replaced the hydrogen atom of the carboxylic acid. The other product is the alcohol. The formula of the alcohol is derived from the alkyl portion of the ester:

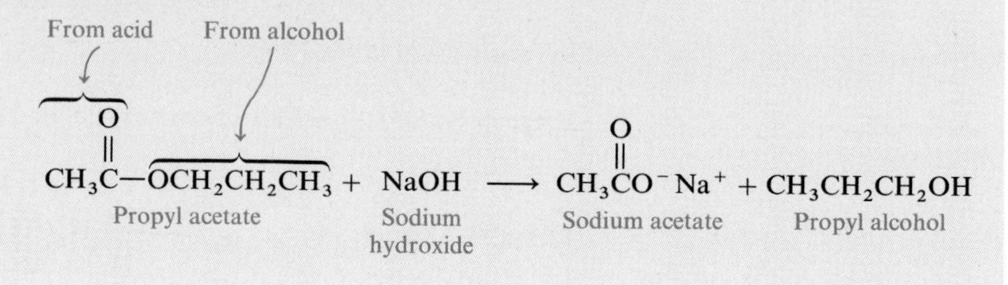

The following table summarizes the naming of aldehydes, ketones, carboxylic acids, acid salts and esters:

Summary of Naming

Class	IUPAC Example	Common Name	Structure
Aldehyde	Propanal	Propionaldehyde	$CH_3CH_2\overset{O}{\overset{\|\|}{C}}H$
Ketone	Propanone	Dimethyl ketone (acetone)	$CH_3\overset{O}{\overset{\|\|}{C}}CH_3$
Carboxylic acid	Ethanoic acid	Acetic acid	$CH_3\overset{O}{\overset{\|\|}{C}}OH$
Acid salt	Sodium ethanoate	Sodium acetate	$CH_3\overset{O}{\overset{\|\|}{C}}O^-\ Na^+$
Ester	Methyl ethanoate	Methyl acetate	$CH_3\overset{O}{\overset{\|\|}{C}}OCH_3$

The reactions of alcohols, aldehydes, and ketones are given in the table below:

Summary of Reactions

Oxidation of Alcohols

Alcohol (1°) $\xrightarrow{[O]}$ Aldehyde + H_2O

$$CH_3CH_2OH \xrightarrow{[O]} CH_3\overset{O}{\overset{\|}{C}}H + H_2O$$

$$\text{Alcohol } (2^\circ) \xrightarrow{[O]} \text{Ketone} + H_2O$$

$$CH_3\overset{OH}{\overset{|}{C}}HCH_3 \xrightarrow{[O]} CH_3\overset{O}{\overset{\|}{C}}CH_3 + H_2O$$

Oxidation of Aldehydes

$$\text{Aldehyde} \xrightarrow{[O]} \text{Carboxylic acid}$$

$$CH_3\overset{O}{\overset{\|}{C}}H \xrightarrow{[O]} CH_3\overset{O}{\overset{\|}{C}}OH$$

Reduction of Aldehydes

$$\text{Aldehyde} \xrightarrow{H_2/Pt} \text{Alcohol } (1^\circ)$$

$$CH_3\overset{O}{\overset{\|}{C}}H \xrightarrow{H_2/Pt} CH_3CH_2OH$$

Hemiacetal Formation

$$\text{Aldehyde} + \text{Alcohol} \xrightarrow{H^+} \text{Hemiacetal}$$

$$CH_3\overset{O}{\overset{\|}{C}}H + HOCH_3 \xrightarrow{H^+} CH_3\underset{OCH_3}{\underset{|}{\overset{OH}{\overset{|}{C}}}}H$$

Reduction of Ketones

$$\text{Ketone} \xrightarrow{H_2/Pt} \text{Alcohol } (2^\circ)$$

$$CH_3\overset{O}{\overset{\|}{C}}CH_3 \xrightarrow{H_2/Pt} CH_3\overset{OH}{\overset{|}{C}}HCH_3$$

Hemiketal Formation

$$\text{Ketone} + \text{Alcohol} \xrightarrow{H^+} \text{Hemiketal}$$

$$CH_3\overset{O}{\overset{\|}{C}}CH_3 + HOCH_3 \xrightarrow{H^+} CH_3\underset{OCH_3}{\underset{|}{\overset{OH}{\overset{|}{C}}}}CH_3$$

Ionization of Carboxylic Acids

$$\text{Carboxylic Acid} \xrightarrow{H_2O} \text{Anion}^- + H^+$$

$$CH_3\overset{O}{\overset{\|}{C}}OH \xrightarrow{H_2O} CH_3\overset{O}{\overset{\|}{C}}O^- + H^+$$

Neutralization of Carboxylic Acids

$$\text{Carboxylic acid} + \text{Base} \longrightarrow \text{Salt} + HOH$$

$$CH_3\overset{O}{\overset{\|}{C}}OH + NaOH \longrightarrow CH_3\overset{O}{\overset{\|}{C}}O^-\ Na^+ + HOH$$

Esterification of Carboxylic Acids

Carboxylic acid + Alcohol ⟶ Ester + HOH

$$CH_3\overset{\overset{O}{\|}}{C}OH + CH_3OH \longrightarrow CH_3\overset{\overset{O}{\|}}{C}OCH_3 + HOH$$

Hydrolysis of Esters

Ester + HOH $\xrightarrow{H^+}$ Carboxylic acid + Alcohol

$$CH_3\overset{\overset{O}{\|}}{C}OCH_3 + HOH \longrightarrow CH_3\overset{\overset{O}{\|}}{C}OH + CH_3OH$$

Saponification of Esters

Ester + Base ⟶ Salt + Alcohol

$$CH_3\overset{\overset{O}{\|}}{C}OCH_3 + NaOH \longrightarrow CH_3\overset{\overset{O}{\|}}{C}O^- Na^+ + CH_3OH$$

Glossary

acetal The product of the addition of two alcohol molecules to an aldehyde.

aldehyde An organic compound containing the $-\overset{\overset{O}{\|}}{C}H$ group.

carboxylic acid An organic compound containing the carboxyl group:

$$-\overset{\overset{O}{\|}}{C}OH$$

ester An organic compound containing the carboxyl group between two hydrocarbon groups:

$$-\overset{\overset{O}{\|}}{C}-O-$$

esterification The formation of an ester by the reaction of a carboxylic acid and an alcohol during the process of which a molecule of water is removed. This reaction requires an acid catalyst.

hemiacetal The product of an alcohol added to an aldehyde.

hemiketal The product formed by adding an alcohol to a ketone.

hydrolysis The separation of a molecule into two smaller molecules by a reaction with water. Hydrolysis of an ester produces a carboxylic acid and an alcohol.

ionization The dissociation of a molecule in water to produce ions in solution.

ketal The product from the addition of two alcohol molecules to a ketone.

ketone Organic compound containing the carbonyl group

$$-\overset{\overset{O}{\|}}{C}-$$

between two hydrocarbon groups.

neutralization The reaction of an acid and a base to produce the salt of the acid and water.

oxidation The loss of two hydrogen atoms, as in the oxidation of alcohols to aldehydes or ketones, or the addition of an oxygen atom, as in the oxidation of aldehydes to carboxylic acids.

reduction Addition of hydrogen atoms to a carbonyl bond. Aldehydes reduce to yield primary alcohols; ketones yield secondary alcohols.

saponification The hydrolysis of an ester with a strong base to produce the salt of the acid ("soaps" when the ester is a fat) and an alcohol.

Problems

Aldehydes and Ketones (Objective 13.1)

13.1 Write the name of the following aldehydes:

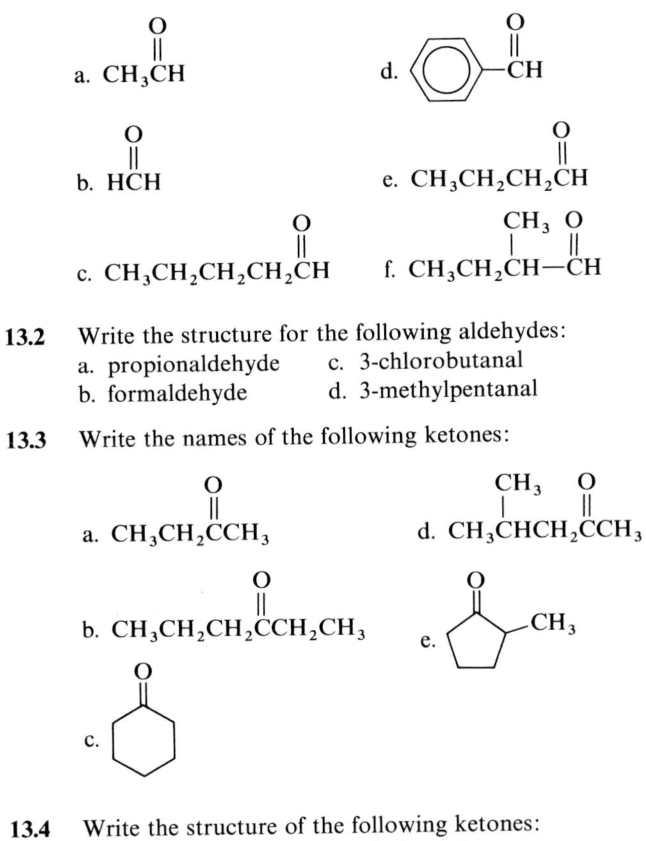

a. $CH_3\overset{O}{\overset{\|}{C}}H$

b. $H\overset{O}{\overset{\|}{C}}H$

c. $CH_3CH_2CH_2CH_2\overset{O}{\overset{\|}{C}}H$

d. $C_6H_5-\overset{O}{\overset{\|}{C}}H$

e. $CH_3CH_2CH_2\overset{O}{\overset{\|}{C}}H$

f. $CH_3CH_2\overset{CH_3}{\overset{|}{C}}H-\overset{O}{\overset{\|}{C}}H$

13.2 Write the structure for the following aldehydes:

a. propionaldehyde
b. formaldehyde
c. 3-chlorobutanal
d. 3-methylpentanal

13.3 Write the names of the following ketones:

a. $CH_3CH_2\overset{O}{\overset{\|}{C}}CH_3$

b. $CH_3CH_2CH_2\overset{O}{\overset{\|}{C}}CH_2CH_3$

c. (cyclohexanone ring, C=O)

d. $CH_3\overset{CH_3}{\overset{|}{C}}HCH_2\overset{O}{\overset{\|}{C}}CH_3$

e. (cyclopentanone ring with CH_3)

13.4 Write the structure of the following ketones:

a. diethyl ketone
b. acetone
c. 3-ethylcyclohexanone
d. methyl isopropyl ketone

13.5 Identify each of the following compounds as an alcohol, an aldehyde, or a ketone. Give a name for each.

a. $CH_3CH_2\overset{O}{\overset{\|}{C}}H$

b. $CH_3\overset{OH}{\overset{|}{C}}HCH_3$

c. $CH_3CH_2CH_2OH$

d. $CH_3\overset{O}{\overset{\|}{C}}CH_3$

Oxidation of Alcohols to Aldehydes and Ketones (Objective 13.2)

13.6 Write the oxidation products of the following:

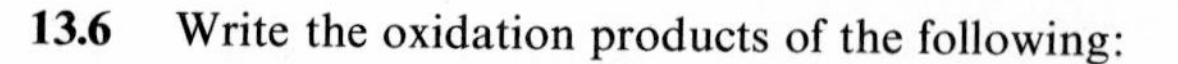

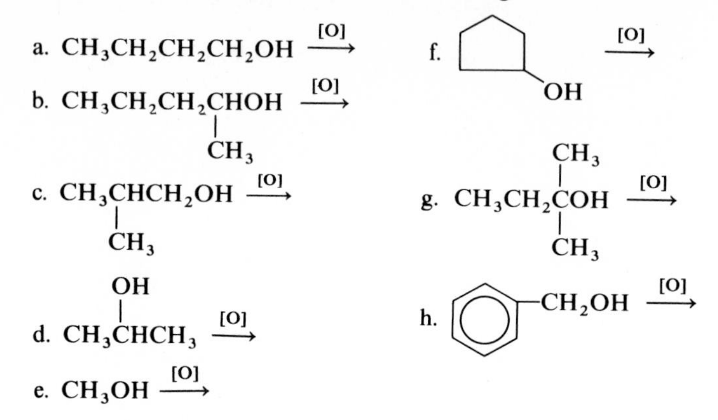

13.7 Write the products of the following reactions:
a. oxidation of cyclopentanol
b. dehydration of cyclopentanol
c. oxidation of 1-propanol
d. dehydration of 1-propanol

13.8 Give the product of the reduction of:

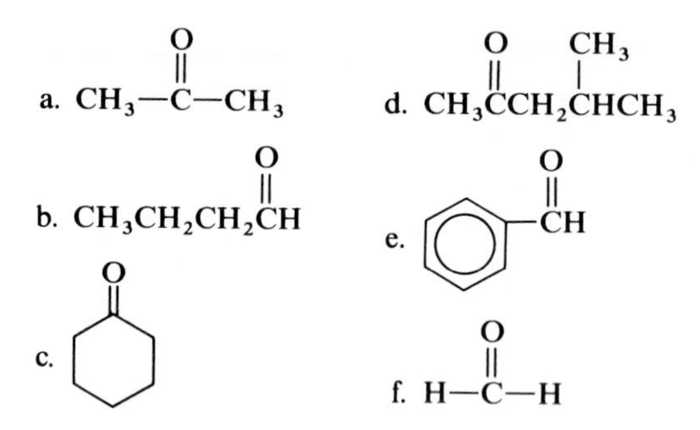

13.9 Write the hemiacetal or hemiketal products of the following:

a. $CH_3CH_2\overset{O}{\overset{\|}{C}}H + HOCH_3 \xrightarrow{H^+}$

b. $CH_3\overset{O}{\overset{\|}{C}}CH_3 + HOCH_2CH_3 \xrightarrow{H^+}$

c. $CH_3CH_2CH_2\overset{O}{\overset{\|}{C}}H + CH_3OH \xrightarrow{H^+}$

Carboxylic Acids (Objective 13.5)

13.10 Write the IUPAC name or common name for the following carboxylic acids:

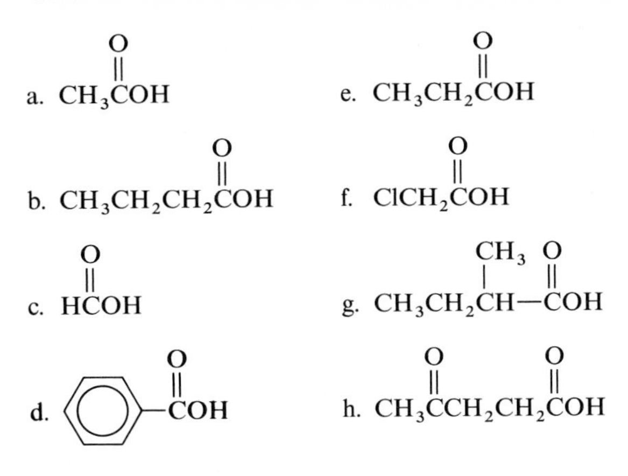

13.11 Write condensed structures for the following names of carboxylic acids:
a. β-methylbutyric acid
b. 3-chloropentanoic acid
c. 4-bromobutanoic acid
d. α-hydroxypropionic acid
e. 3-bromobenzoic acid
f. *p*-aminobenzoic acid

Formation of Carboxylic Acids (Objective 13.6)

13.12 Write the structure of the expected oxidation products of the following:

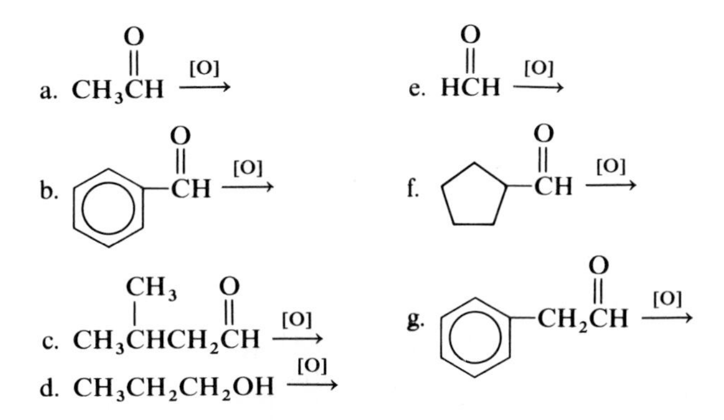

Ionization and Neutralization of Carboxylic Acids (Objective 13.7)

13.13 Identify each reaction as an ionization or neutralization of a carboxylic acid. Write the structure of the expected products.

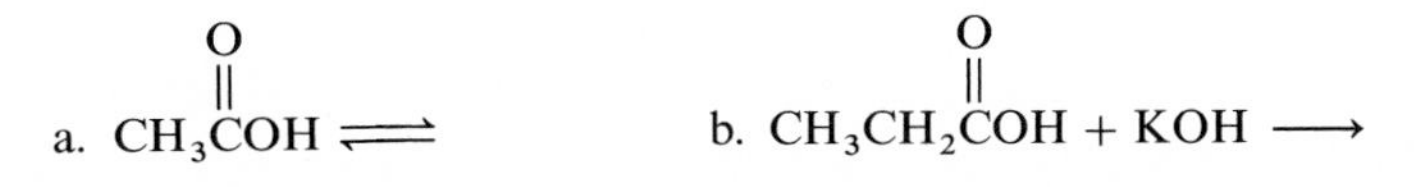

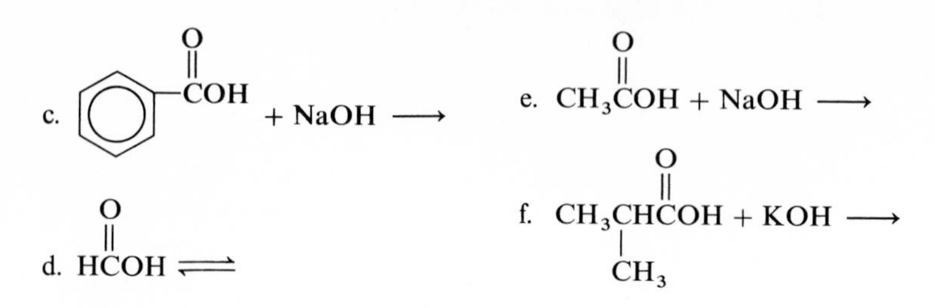

Formation of An Ester (Objective 13.8)

13.14 Write the structure of the esters formed by the following reactions of carboxylic acids and alcohols:

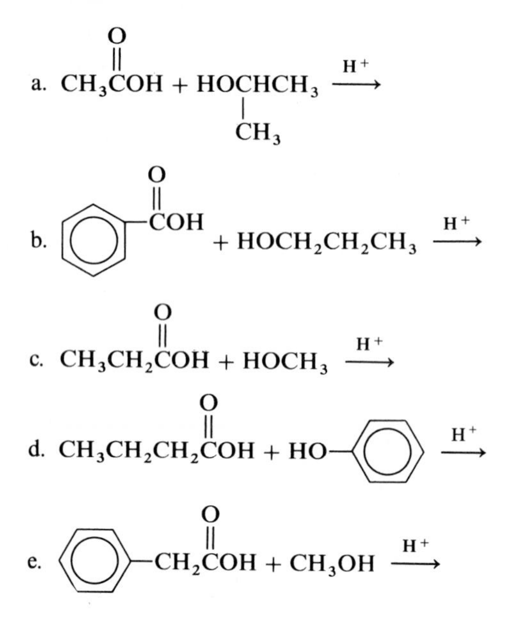

Naming Esters (Objective 13.9)

13.15 Write the IUPAC name or common name of each of the following esters:

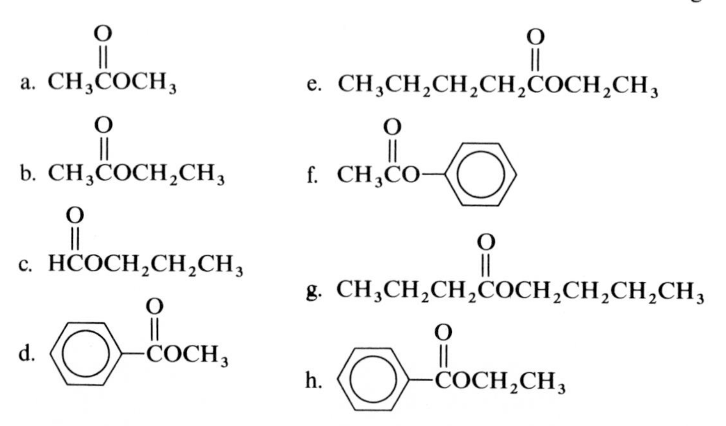

13.16 Many esters are the source of flavors and odors of fruits. Write the structure for some of the following esters used as flavoring agents:

a. ethyl butyrate (pineapple)
b. ethyl formate (rum)
c. pentyl butyrate (apricot)
d. octyl acetate (orange)

Reactions of Esters (Objective 13.10)

13.17 Identify each of the following reactions as hydrolysis or saponification of an ester. Write the expected products of the reaction:

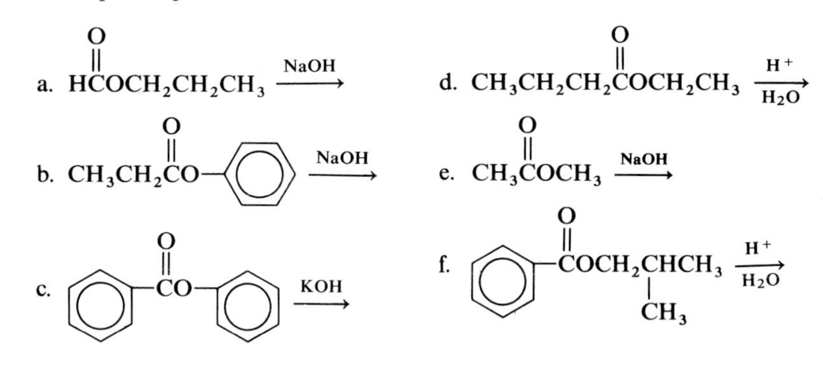

14 Nitrogen-Containing Organic Compounds: Amines and Amides

Objectives

14.1 Classify an amine as a primary, secondary, or tertiary amine.

14.2 Write the name and structure of an amine.

14.3 Identify a heterocyclic ring in a heterocyclic amine.

14.4 Write an equation for the ionization of an amine in water; write an equation for the reaction of an amine with an acid.

14.5 Write the name and structure of an amide.

14.6 Write the structure for the amide product of a reaction between an amine and a carboxylic acid.

14.7 Write the equation for the hydrolysis of an amide.

Scope

Amines are a group of organic compounds that contain nitrogen. They are responsible for the odor you associate with fish. The high nitrogen content of the amines in fish is the reason for their wide use in plant fertilizers. The amides are nitrogen-containing compounds with carbonyl groups. The amide bond holds together all the amino acids in a protein.

Some amines and amides are physiologically active. The adrenalin you produce as a response to fear is an amine. The amphetamines are a group of amine compounds that act as stimulants and antidepressants; they were once used for manufacturing diet pills. Other nitrogen-containing compounds are depressants. Drugs such as phenolbarbital are used as sedatives and anti-convulsant drugs.

14.1 Classification of Amines

OBJECTIVE Classify an amine as a primary, secondary, or tertiary amine.

Amines are organic compounds of ammonia, NH_3, in which one, two, or three of the hydrogen atoms in ammonia are replaced by an alkyl group or an aromatic group:

$$\begin{array}{c} \mathrm{H{-}\ddot{N}{-}H} \\ | \\ \mathrm{H} \end{array}$$

primary, secondary, and tertiary amines

An amine is a *primary* amine if just one of the hydrogen atoms of ammonia is replaced by a hydrocarbon group. A *secondary* amine has two hydrocarbon groups attached to the nitrogen, and a *tertiary* amine has three organic groups replacing all three hydrogen atoms of ammonia. Figure 14.1 shows ball-and-stick models of primary, secondary, and tertiary amines. Table 14.1 lists examples of the classes of amines.

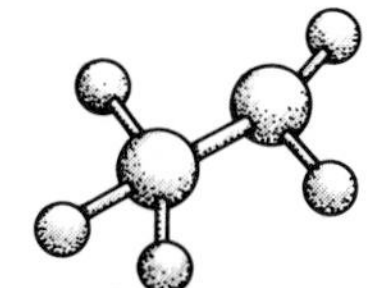

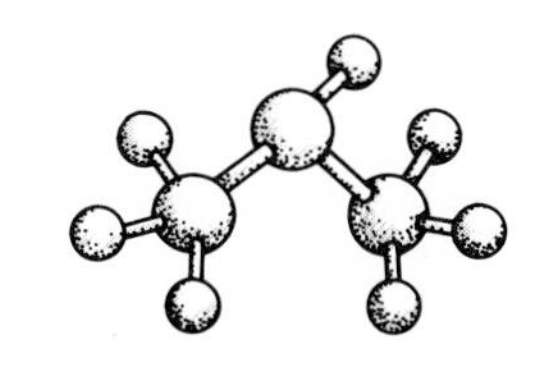

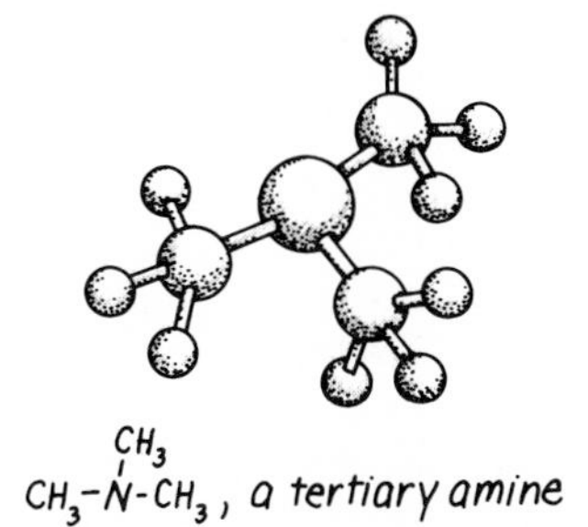

Figure 14.1 Ball-and-stick models of primary, secondary, and tertiary amines.

Table 14.1 Classification of Amines

Ammonia	Primary (1°)	Secondary (2°)	Tertiary (3°)
H | H—N—H	H | H—N—CH_3 (Nitrogen atom; Hydrocarbon group)	H | CH_3NCH_3	CH_3 | CH_3NCH_3
	$CH_3CH_2CH_2NH_2$	(benzene ring)—$NHCH_3$	CH_3—N (ring)

Example 14.1 Classify the following amines as primary, secondary, or tertiary:

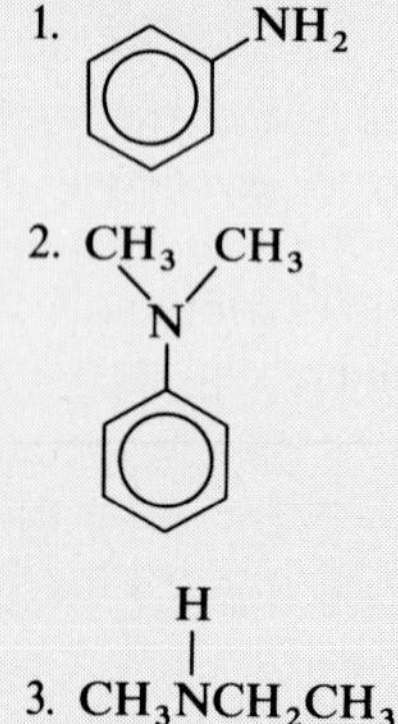

Solution

1. This compound is a primary amine because there is one hydrocarbon group attached to the nitrogen atom.
2. In this compound, all three hydrogen atoms of the NH_3 molecules have been replaced by hydrocarbon groups. Thus it is a tertiary amine.
3. In this compound, there are two hydrocarbon groups attached to the nitrogen, which makes it a secondary amine.

14.2 Naming Amines

OBJECTIVE Write the name and structure of an amine.

Amines are generally named by their common names. The names of simple amines consist of the names of each hydrocarbon group attached to the nitrogen atom followed by the ending *-amine*:

Amine

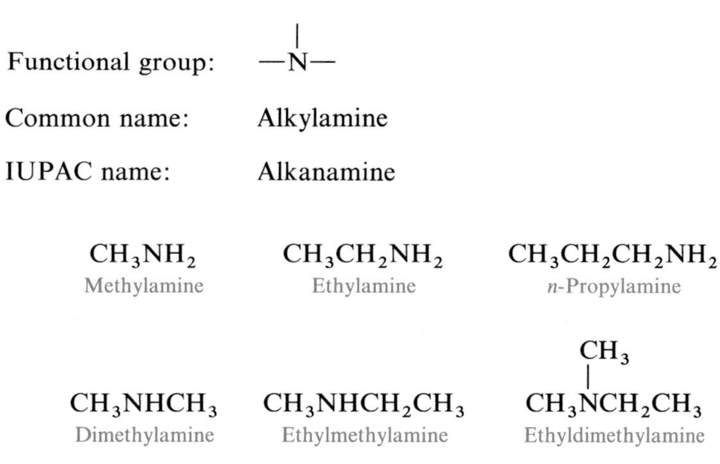

In the IUPAC system, an amine is named by finding the longest carbon chain attached to the nitrogen atom. The parent chain is named by using its corresponding alkane name and replacing the *-e* by *-amine*. As in naming alcohols, a number is used to locate the amine group:

$CH_3CH_2CH_2NH_2$

1-Propanamine

NH_2 (on C-2)

$CH_3CH_2CHCH_3$

2-Butanamine

For secondary and tertiary amines, the smaller hydrocarbon groups attached to the nitrogen atom are listed as substituents of the nitrogen atom (*N*-) in front of the parent name:

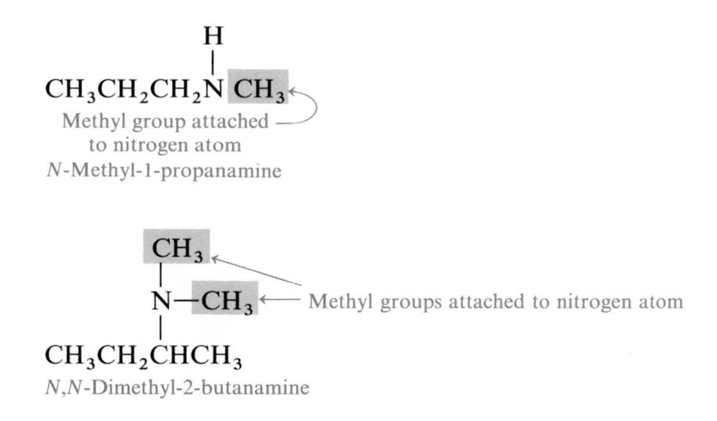

HEALTH NOTE

Amines in Health and Medicine

ephedrine
norepinephrine

Ephedrine and norepinephrine are part of a family of compounds used in remedies for colds, hay fever, and asthma. They contract the capillaries in the mucous membranes of the respiratory passages and elevate blood pressure:

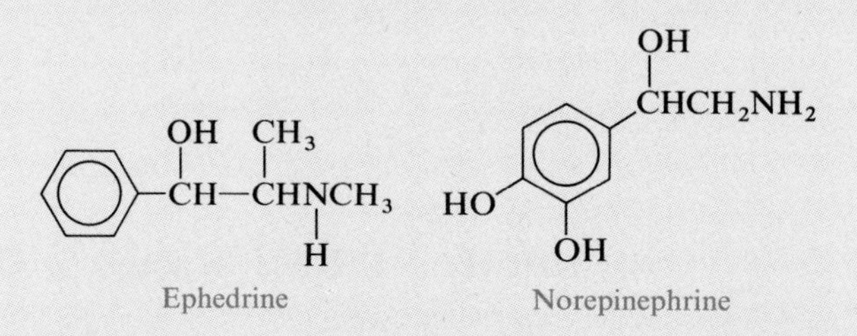

Ephedrine Norepinephrine

Benzedrine

Benzedrine™, or amphetamine, is used in medications that are inhaled to reduce respiratory congestion from colds, hay fever, and asthma. Sometimes Benzedrine is taken internally to combat the desire to sleep, but it has side effects and can be habit-forming:

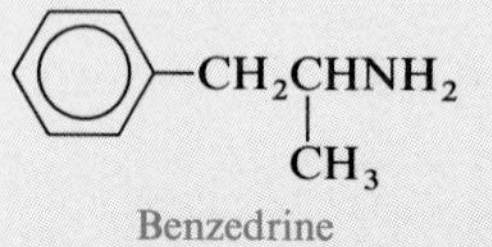

Benzedrine (amphetamine)

nicotine

Nicotine is a cyclic amine found in tobacco leaves. It affects the central nervous system and causes changes in blood pressure:

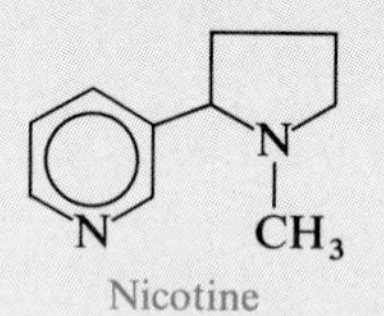

Nicotine

histamine
antihistamine

Histamine may be found in tissues, usually in an inactive form. The active form of histamine may be responsible for certain allergic reactions. An antihistamine such as diphenylhydramine helps block the allergic effects of the histamines:

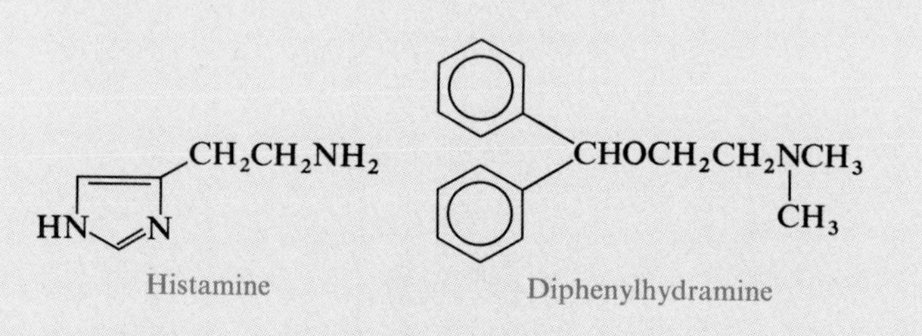

Histamine Diphenylhydramine

aniline The aromatic amines are named as derivatives of *aniline*, the amine of benzene. Additional alkyl group on the nitrogen atom are indicated by using the letter *N*- preceding the alkyl names.

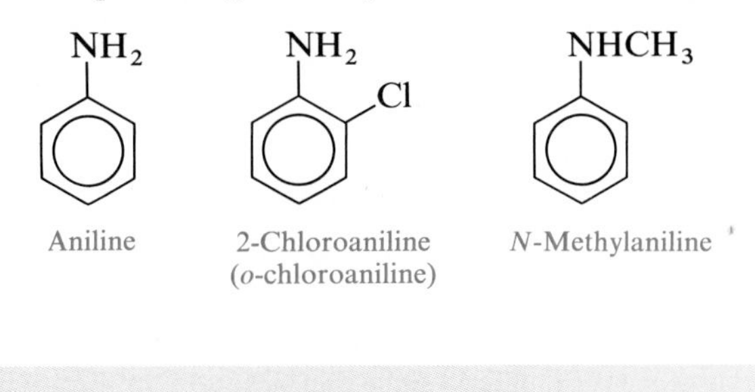

Example 14.2 Write a name for each of the following amines:

1. $CH_3CH_2CH_2\overset{\displaystyle H}{\overset{|}{N}}CH_2CH_3$

2. 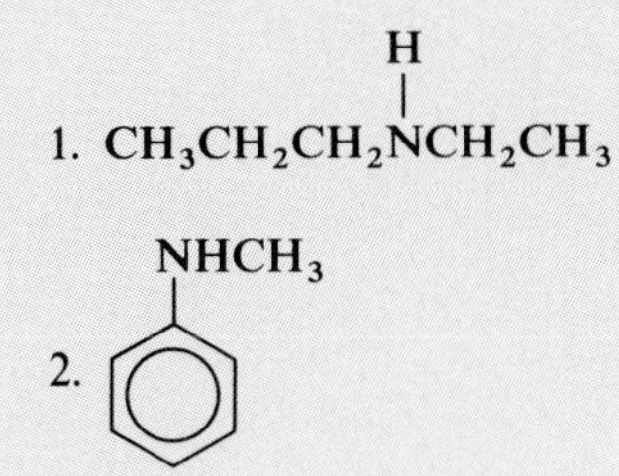

Solution

1. The two hydrocarbon groups attached to the nitrogen atom are an ethyl group and an *n*-propyl group. The common name of the amine is

   Ethyl *n*-propylamine

   In the IUPAC system, the parent chain is propanamine with an ethyl group attached to the nitrogen atom:

   N-Ethyl-1-propanamine

2. The parent compound is aniline, the amine of benzene. The methyl group on the nitrogen atom is named *N*-methyl to indicate its location:

   N-Methylaniline

14.3 Heterocyclic Amines

OBJECTIVE Identify a heterocyclic ring in a heterocyclic amine.

When at least one or more nitrogen atom appears in a ring structure, the compound is classified as a heterocyclic amine. The ring is typically composed of five or six atoms and is often unsaturated.

Pyrrolidine is a ring structure containing one nitrogen atom and four carbon atoms. Pyrrole contains the same number and type of atoms, but its ring

HEALTH NOTE

Alkaloids

Many heterocyclic amines occur naturally in plants and have significant physiological effects. Some of these nitrogen-containing compounds are called *alkaloids* because of their basic (alkalilike) characteristics. Although their function in plants is not known, we do know something about their effects on humans as anesthetics and hallucinogens, with many having habituating effects.

quinine

One of the earliest alkaloids used in medicine was probably quinine, which is obtained from the Cinchona tree in South America. Quinine has been used in the treatment of malaria since the 1600s:

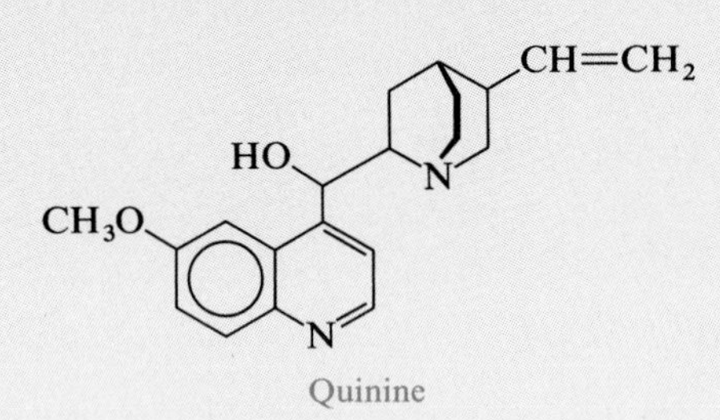

Quinine

Nicotine (from the leaves of the tobacco plant) and *coniine* (from hemlock) are alkaloids that are extremely toxic:

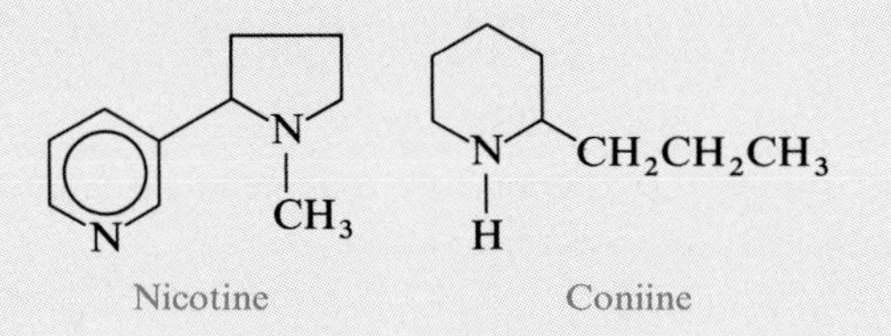

Nicotine Coniine

atropine
cocaine
procaine

Atropine (belladonna) and cocaine (coca plant) are used in low concentrations as anesthetics for the eyes and sinuses. However, in greater doses, they produce a level of euphoria followed by depression that necessitates the need for additional quantities of the drug. Chemists used the structures of atropine and cocaine to develop a synthetic alkaloid, procaine, that retained the anesthetic qualities without the addictive side effects:

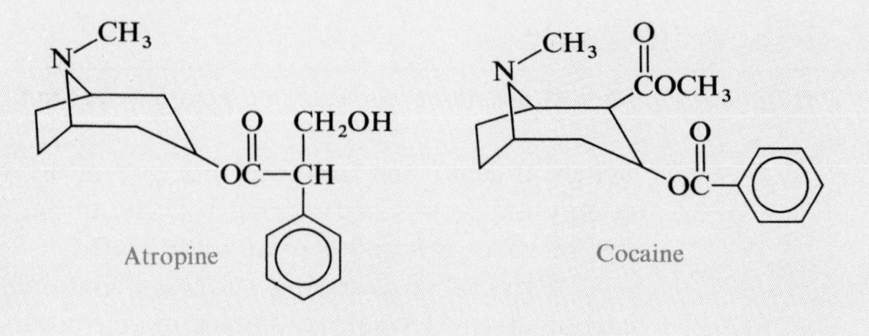

Atropine Cocaine

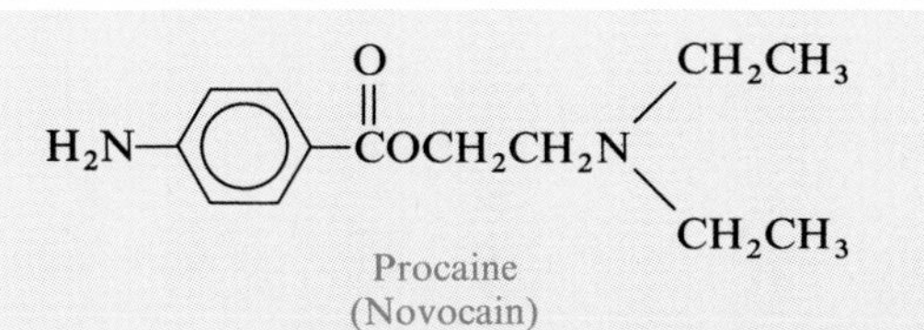

Procaine
(Novocain)

painkillers: morphine, Demerol

For many centuries, morphine, an alkaloid from the white poppy, has been used as a painkiller. It is still considered to have excellent analgesic effects but has strong hallucinogenic and addictive side effects. A synthetic alkaloid, meperidine, or Demerol™, which possesses some of the chemical structure of morphine, was developed in an attempt to reduce the side effects associated with morphine. Although the side effects have been reduced, they have not been eliminated. One of the areas of research in pharmacology is the search for a morphinelike compound that has no side effects and can be used safely as a painkiller.

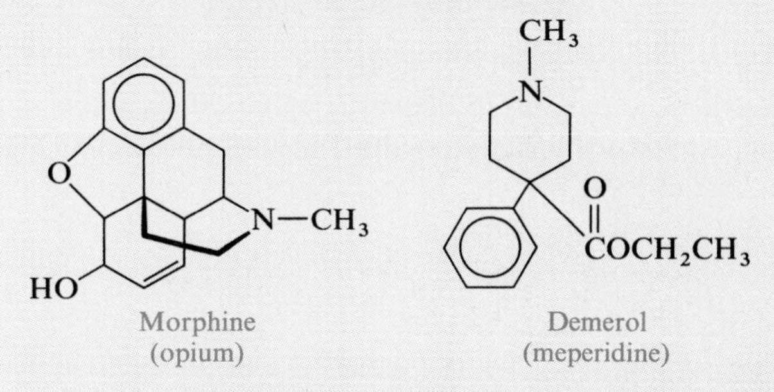

Morphine
(opium)

Demerol
(meperidine)

hallucinogens: mescaline, LSD

Alkaloids are prevalent among the compounds known as hallucinogens. Examples of hallucinogens are mescaline, from the peyote cactus, and LSD (lysergic acid diethylamide), which is prepared from the lysergic acid produced by a fungus that grows on rye. Lysergic acid itself has some hallucinogenic effects.

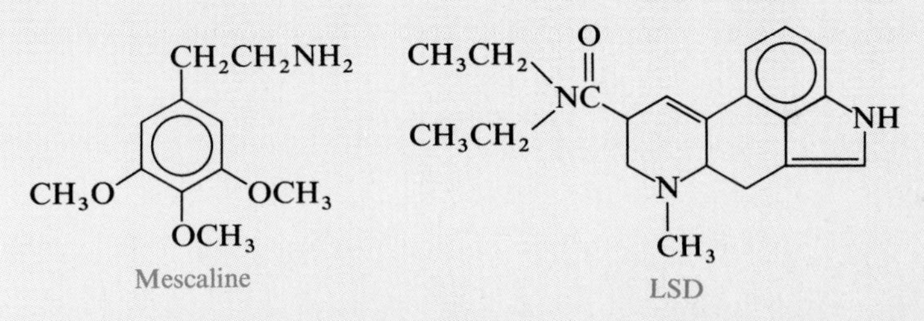

Mescaline

LSD

serotonin

Several amines that are tranquilizers act on the brain to reduce transmission of nerve impulses or to increase the activity of the brain. Low levels of serotonin in the brain appear to be associated with depressive states:

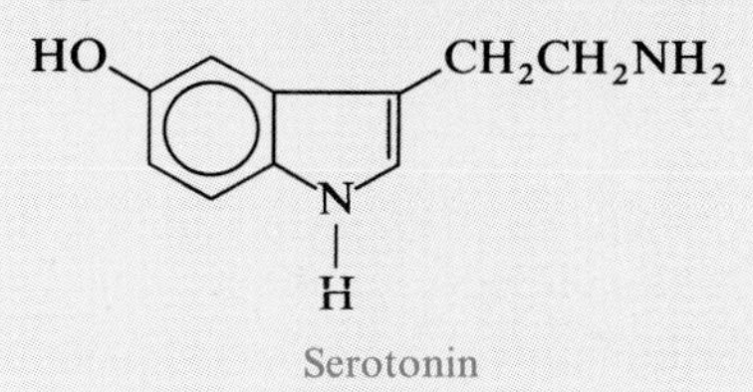

Serotonin

structure is unsaturated. Imidazole is also a five-atom ring, but it contains two nitrogen atoms:

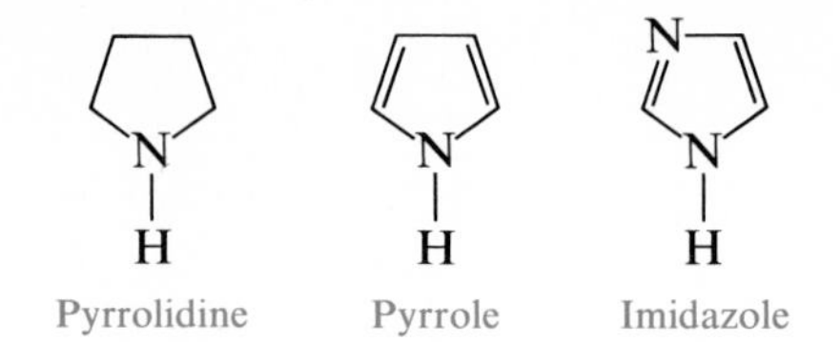

Pyrrolidine Pyrrole Imidazole

Two other heterocyclic amines contain six atoms in the ring. Pyrridine has a benzenelike structure with one nitrogen atom in the ring. Pyrimidine contains two nitrogen atoms in its unsaturated ring:

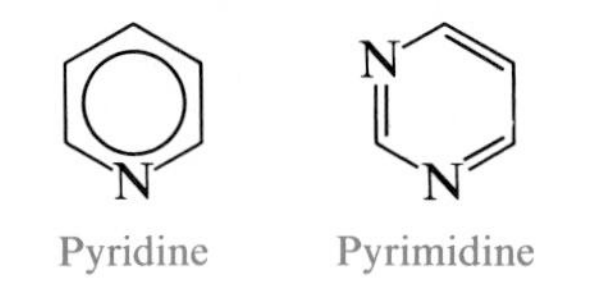

Pyridine Pyrimidine

The heterocyclic amine purine contains two rings and four nitrogen atoms in its structure:

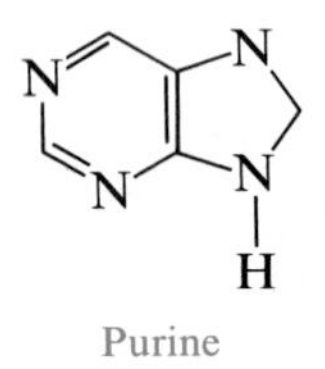

Purine

Example 14.3 Indicate whether each of the following is an amine or a heterocyclic amine:

1. 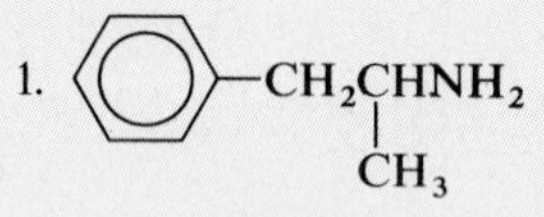

Benzedrine

2. 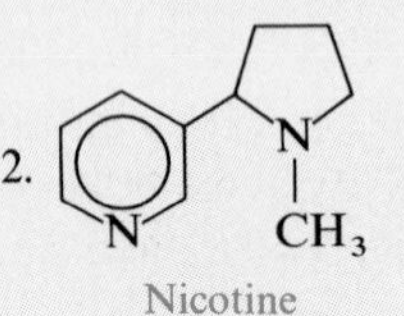

Nicotine

Solution

1. Benzedrine contains an NH_2 group.Thus it is an amine.
2. Nicotine contains nitrogen atoms in ring structures. Therefore nicotine is a heterocyclic amine.

14.4 Reactions of Amines

OBJECTIVE Write an equation for the ionization of an amine in water; write an equation for the reaction of an amine with an acid.

In water, amines act as weak bases attracting the hydrogen from water molecules. They form alkylammonium ions and hydroxide ions:

Ionization of Amines

$$\text{Amine} + \text{HOH} \longrightarrow \text{Alkylammonium ion}^+ + \text{OH}^-$$

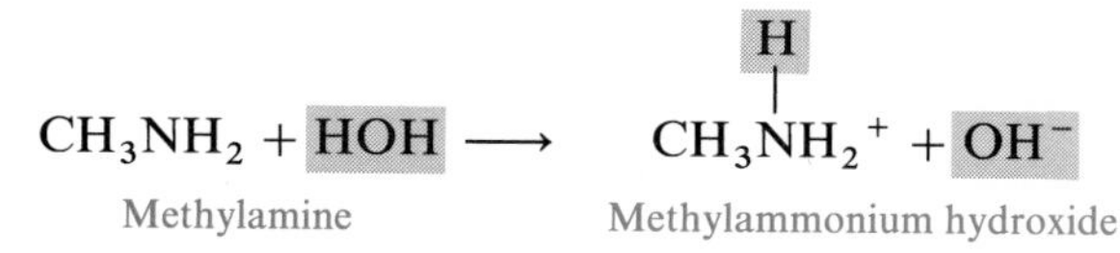

Example 14.4 Write the equation for the ionization of propylamine in water.

Solution: In water, the nitrogen atom will attract a hydrogen from water to give an alkylammonium ion and a hydroxide ion:

$$\underset{\text{Propylamine}}{CH_3CH_2CH_2NH_2} + H_2O \longrightarrow \underset{\text{Propylammonium hydroxide}}{CH_3CH_2CH_2NH_3^+ + OH^-}$$

Reaction with Acids

When an amine reacts with an acid, an alkylammonium salt is formed. When you use lemon juice on fish, you counteract the fishy odor of the amine by forming the ammonium salt, which has no fishy odor.

Reaction of Amines with Acids

$$\text{Amine} + \text{Acid} \longrightarrow \text{Alkylammonium salt}$$

$$\underset{\text{Ethylamine}}{CH_3CH_2NH_2} + \underset{\text{Hydrochloric acid}}{HCl} \longrightarrow \underset{\text{Ethylammonium chloride}}{CH_3CH_2NH_3^+\,Cl^-}$$

$$\underset{\text{Aniline}}{C_6H_5{-}NH_2} + \underset{\text{Hydrochloric acid}}{HCl} \longrightarrow \underset{\text{Anilinium chloride}}{C_6H_5{-}NH_3^+\,Cl^-}$$

Example 14.5 Write the equation for the reaction of dimethylamine and HBr.

Solution: In the reaction, the base (dimethylamine) accepts a proton from the acid (HBr). A salt is formed, which consists of the alkylammonium ion and the bromide anion:

$$\underset{\text{Dimethylamine}}{CH_3\underset{\underset{H}{|}}{N}CH_3} + HBr \longrightarrow \underset{\text{Dimethylammonium bromide}}{CH_3\overset{\overset{H}{|}}{\underset{\underset{H}{|}}{N^+}}CH_3} + Br^-$$

14.5 Amides

OBJECTIVE Write the name and structure of an amide.

Amides are nitrogen derivatives of carboxylic acids. In an amide, the —OH portion of the carboxylic acid is replaced by an amino (—NH_2) group. They are named as derivatives of the carboxylic acid. The common name of an amide is derived from the common name of the carboxylic acid. The *-ic acid* ending is replaced with *-amide*. The IUPAC name of the amide is derived from the IUPAC name of the carboxylic acid. The *-oic acid* ending is replaced with *-amide*:

Naming Amides

Functional group: $-\overset{\overset{O}{\|}}{C}-NH_2$

IUPAC name: Alkanamide (replace *-oic acid* of IUPAC acid name)

Common name: Alkanamide (replace *-ic acid* of common acid name)

Carboxylic Acid *Amide*

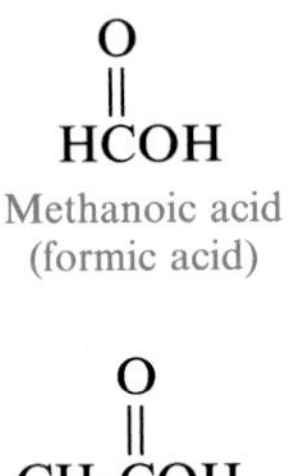

Methanoic acid (formic acid)

Methanamide (formamide)

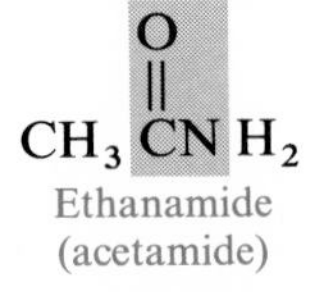

Ethanoic acid (acetic acid)

Ethanamide (acetamide)

The hydrogen atoms on the nitrogen atom may also be substituted in an amide. The group(s) attached to the nitrogen atom is (are) preceded by the letter *N*-:

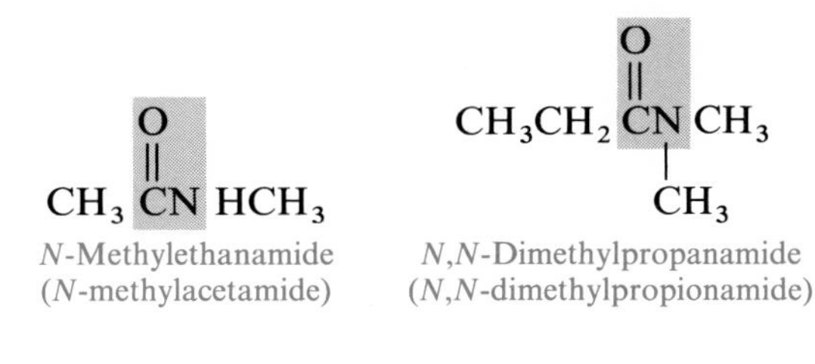

Example 14.6 Write the name of the following amides:

1. $CH_3CH_2\overset{O}{\overset{\|}{C}}NH_2$

2. $CH_3\overset{O}{\overset{\|}{C}}NHCH_2CH_3$

Solution

1. This amide is derived from the carboxylic acid with three carbon atoms, which is propionic acid or propanoic acid. The amide is named by replacing -(*o*)*ic acid* with the functional group name -*amide*:

IUPAC name:	Propanamide
Common name:	Propionamide

2. The parent chain in this amide is the carbon portion with the carbonyl group,

$$-\overset{O}{\overset{\|}{C}}-$$

that is, ethanamide or acetamide. The two-carbon organic group (ethyl) attached to the nitrogen atom is indicated by the letter *N*- preceding its name, that is, *N*-ethyl:

IUPAC name:	*N*-Ethylethanamide
Common name:	*N*-Ethylacetamide

Example 14.7 Write the structure of *N*-methylbenzamide.

Solution: The amide portion would be derived from benzoic acid. The *N*-methyl part of the name indicates there is a methyl (CH_3—) group attached to the nitrogen atom:

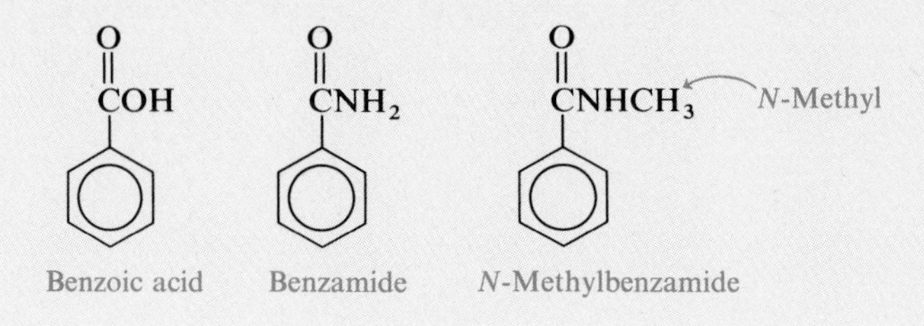

Benzoic acid Benzamide *N*-Methylbenzamide

14.6 Formation of Amides

OBJECTIVE Write the structure for the amide product of a reaction between an amine and a carboxylic acid.

An amide may be prepared by heating a mixture of a carboxylic acid and ammonia or an amine. A molecule of water is eliminated, and the fragments of the two molecules combine to form the amide. (It is much like ester formation, in which a carboxylic acid and an alcohol combine with the elimination of a molecule of water.) The reaction between a carboxylic acid and ammonia or an amine is called *amidation* and results in the formation of an *amide bond*. In proteins, amide bonds are called *peptide bonds* because they join together many amino acids (peptides). Some examples of amidation are shown below:

amidation, amide bond

Amidation

$$\text{Carboxylic acid} + \text{Ammonia (or amine)} \xrightarrow{\text{Heat}} \text{Amide} + \text{HOH}$$

$$\underset{\text{Acid}}{R{-}\overset{\overset{\displaystyle O}{\|}}{C}OH} + \underset{\text{Amine}}{H{-}NH_2} \xrightarrow{\Delta} \underset{\text{Amide}}{R{-}\overset{\overset{\displaystyle O}{\|}}{C}NH_2} + \underset{\text{Water}}{HOH}$$

$$R{-}\overset{\overset{\displaystyle O}{\|}}{C}OH + \underset{\underset{\displaystyle H}{|}}{NH}{-}R' \xrightarrow{\Delta} R{-}\overset{\overset{\displaystyle O}{\|}}{C}NHR' + HOH$$

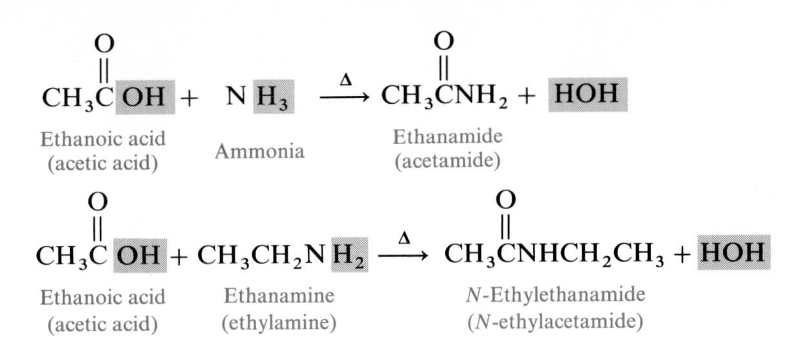

Example 14.8 Write the structure of the products for the reaction of benzoic acid and methylamine:

$$C_6H_5\text{—}\overset{O}{\overset{\|}{C}}OH + CH_3\text{—}NH_2 \xrightarrow{\text{Heat}}$$

Solution: A carboxylic acid and an amine react to form an amide. The structure can be written by attaching the carbonyl group of the carboxylic acid to the nitrogen atom of the amine. The —OH of the acid and an —H from the amine are removed and appear as H_2O in the products:

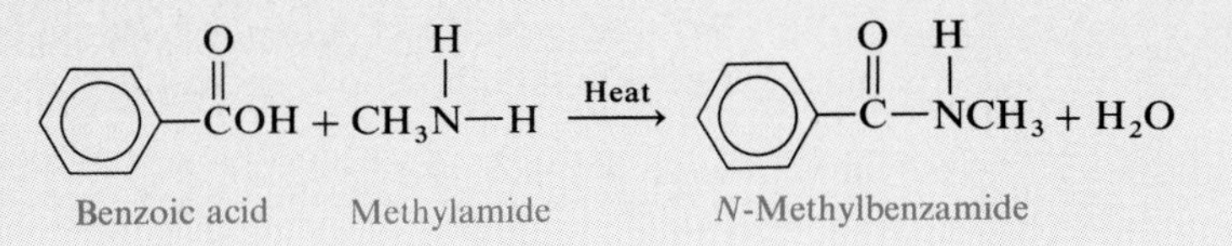

14.7 Hydrolysis of Amides

OBJECTIVE Write the equation for the hydrolysis of an amide.

The splitting of an amide by water (hydrolysis) must occur in the presence of a strong acid or strong base catalyst. The products of amide hydrolysis are the initial carboxylic acid and amine; if a strong base is used, the salt of the carboxylic acid is obtained.

Hydrolysis of an Amide

Amide + HOH ⟶ Carboxylic acid + Amine

$$R\text{—}\overset{O}{\overset{\|}{C}}\text{—}NH_2 + HOH \longrightarrow R\text{—}\overset{O}{\overset{\|}{C}}\text{—}OH + NH_3$$

Examples of Hydrolysis

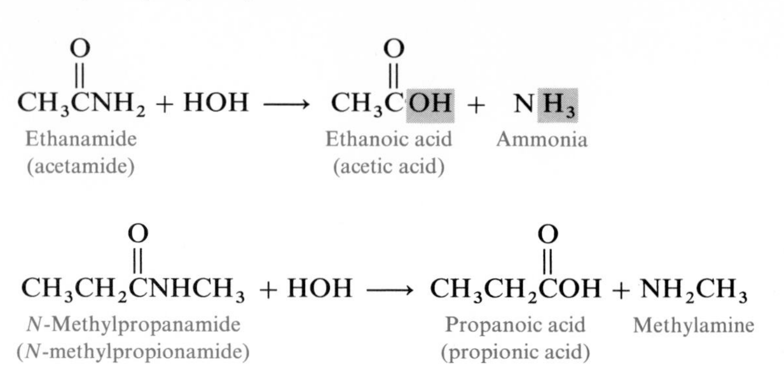

Example 14.9 Write the structure of the products of the following hydrolysis reaction:

$$\mathrm{CH_3\overset{O}{\overset{\|}{C}}NHCH_3 + HOH \longrightarrow}$$

Solution: Water will split the amide molecule between the carbonyl group and the nitrogen atom to give the following pieces:

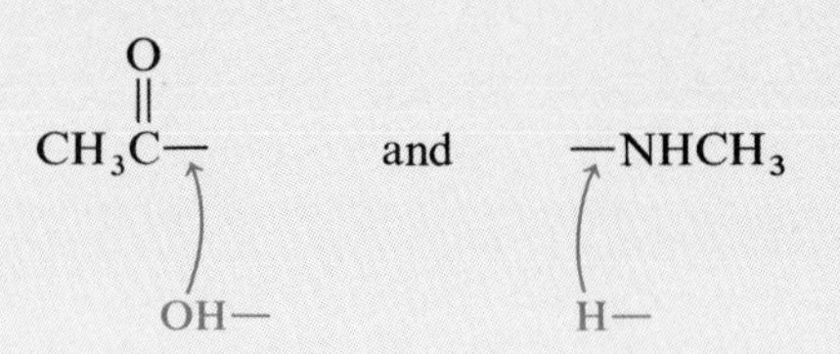

The components of water are now added. The —OH portion is added to the carbonyl group to yield a carboxylic acid. The H— is added to the nitrogen atom:

$$\mathrm{CH_3\overset{O}{\overset{\|}{C}}{-}OH + H{-}NHCH_3}$$

or

$$\mathrm{CH_3\overset{O}{\overset{\|}{C}}OH + NH_2CH_3}$$

Acetic acid Methylamine

HEALTH NOTE

Amides in Health and Medicine

Saccharin is a very powerful sweetener and is often used as a sugar substitute. Phenacetin is an analgesic and antipyretic and may be used as an aspirin substitute:

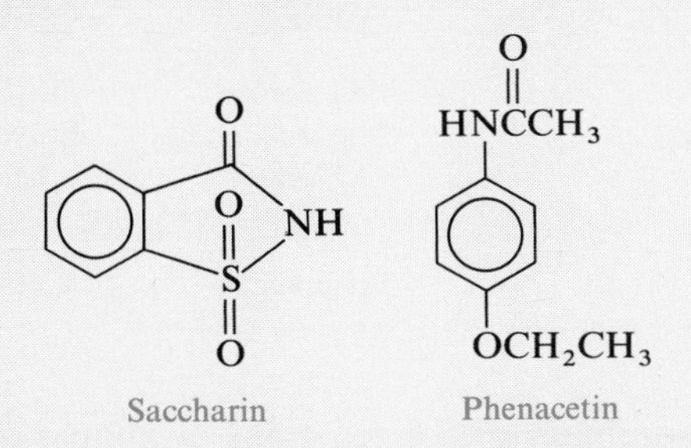

Saccharin   Phenacetin

Urea is the end product of protein metabolism in the body:

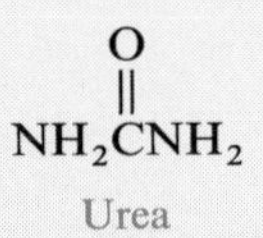

Urea

The kidneys remove urea from the blood and provide for its excretion in urine. In malfunctioning kidneys, the urea is not removed from the blood and it builds to levels that are toxic—a condition called *uremic poisoning*.

Many of the barbiturates, which act as depressants on the central nervous system, are amides in a cyclic form:

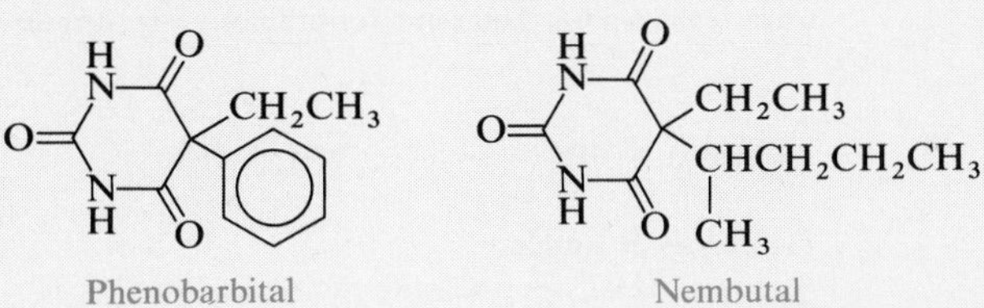

Phenobarbital   Nembutal (pentobarbital)

H N O, O=, N H O, $CH_2CH{=}CH_2$, $CHCH_2CH_2CH_3$, CH_3

Seconal
(secobarbital)

The following table summarizes the naming of amines, ammonium salts, and amides:

Summary of Naming

Class	IUPAC Example	Common Name	Structure
Amine	Ethanamine	Ethylamine	$CH_3CH_2NH_2$
Ammonium salt	Ethylammonium chloride	Ethylamine hydrochloride	$CH_3CH_2NH_3^+\ Cl^-$
Amide	Ethanamide	Acetamide	$CH_3\overset{\overset{\displaystyle O}{\|}}{C}{-}NH_2$

A summary of the reactions of amines and amides is given in the table below:

Summary of Reactions

Ionization of Amines

$$\text{Amine} + \text{HOH} \longrightarrow \text{Ammonium cation} + OH^-$$

$$CH_3NH_2 + HOH \longrightarrow CH_3NH_3^+ + OH^-$$

Neutralization of Amines

$$\text{Amine} + \text{Acid} \longrightarrow \text{Ammonium salt}$$

$$CH_3NH_2 + HX \longrightarrow CH_3{-}NH_3^+\ X^-$$

Amidation of Amines

$$\text{Carboxylic acid} + \text{Ammonia (or amine)} \xrightarrow{\Delta} \text{Amide} + H_2O$$

$$CH_3\overset{\overset{\displaystyle O}{\|}}{C}OH + NH_3 \xrightarrow{\Delta} CH_3\overset{\overset{\displaystyle O}{\|}}{C}NH_2 + H_2O$$

Hydrolysis of Amides

$$\text{Amide} + \text{HOH} \longrightarrow \text{Carboxylic acid} + \text{Amine}$$

$$CH_3\overset{\overset{\displaystyle O}{\|}}{C}NHCH_3 + HOH \longrightarrow CH_3\overset{\overset{\displaystyle O}{\|}}{C}OH + NH_2CH_3$$

Glossary

alkaloid Plant extract that contains heterocyclic amines known to cause physiological effects.

amidation A reaction by which an amide is formed from a carboxylic acid and ammonia (or an amine).

amide An organic compound containing the carbonyl group attached to an amino group or to a substituted nitrogen atom:

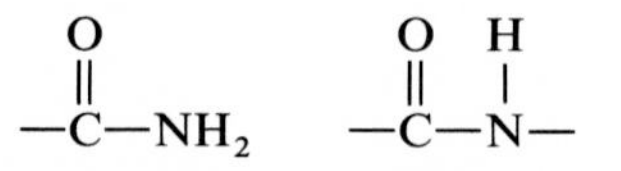

amine An organic compound in which the hydrogen atoms of ammonia, NH_3, are replaced with one, two, or three hydrocarbon groups.

heterocyclic amine A ring structure containing at least one nitrogen atom.

hydrolysis A splitting apart of one molecule into two smaller molecules involving the reaction with water. Hydrolysis of an amide produces a carboxylic acid and an amine.

primary amine An amine in which one hydrogen atom of ammonia has been replaced by a hydrocarbon group.

secondary amine An amine in which two hydrogen atoms of ammonia have been replaced by hydrocarbon groups.

tertiary amine An amine in which three hydrogen atoms of ammonia have been replaced by hydrocarbon groups.

Problems

Classification of Amines (Objective 14.1)

14.1 Classify the following amines as primary, secondary, or tertiary:

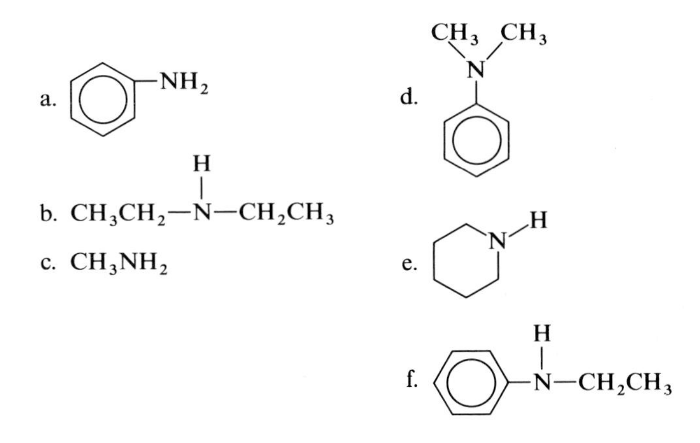

Naming Amines (Objective 14.2)

14.2 Write the common name for each of the following amines:

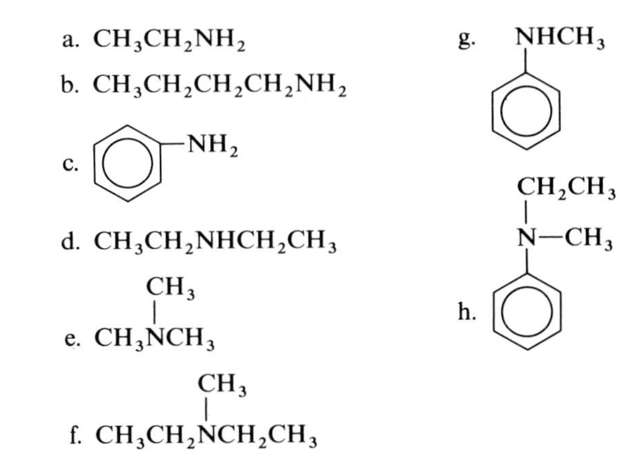

14.3 Write the condensed structure of the following amines:

a. ethylamine
b. aminocyclohexane
c. *p*-aminobenzoic acid
d. dimethylamine
e. methylisopropylamine
f. 4-aminobutanoic acid

Heterocyclic Amines (Objective 14.3)

14.4 Using the text as a reference, state the type(s) of heterocyclic rings in each of the following compounds:

a. methioprim (tumor antagonist)

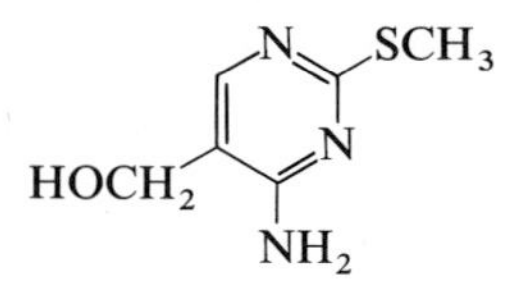

b. pyrrolnitrin (antifungicide)

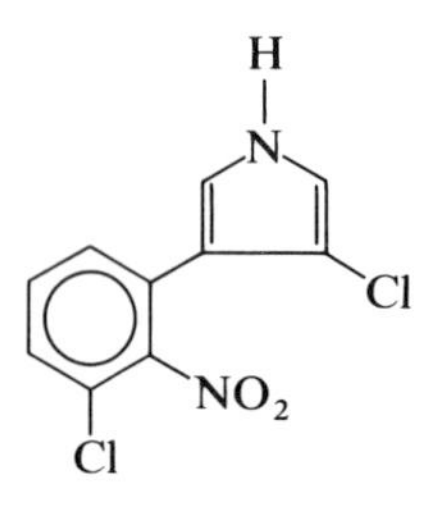

c. pyrimethamine (antimalarial)

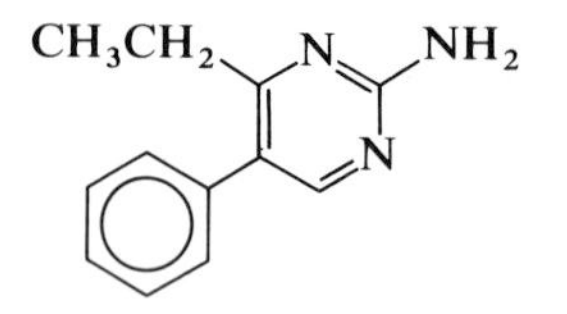

d. adrenoglomerulotropin (hormone)

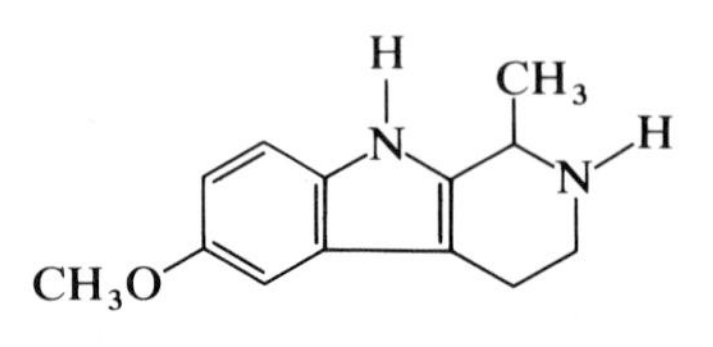

e. nicotine

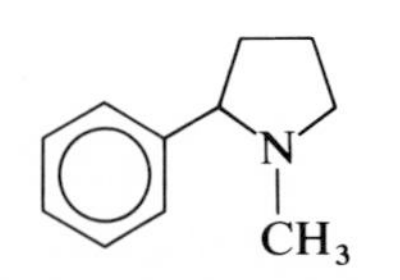

Reactions of Amines (Objective 14.4)

14.5 Indicate whether each of the following reactions is an ionization or neutralization of an amine. Write the structure(s) of the expected products:

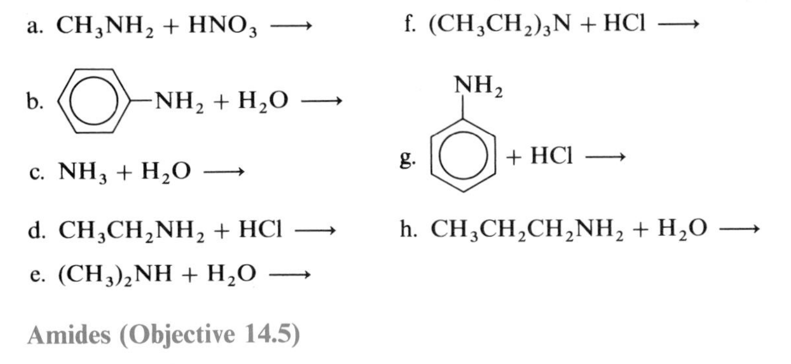

a. $CH_3NH_2 + HNO_3 \longrightarrow$

b. C_6H_5—$NH_2 + H_2O \longrightarrow$

c. $NH_3 + H_2O \longrightarrow$

d. $CH_3CH_2NH_2 + HCl \longrightarrow$

e. $(CH_3)_2NH + H_2O \longrightarrow$

f. $(CH_3CH_2)_3N + HCl \longrightarrow$

g. NH_2 (on benzene ring) $+ HCl \longrightarrow$

h. $CH_3CH_2CH_2NH_2 + H_2O \longrightarrow$

Amides (Objective 14.5)

14.6 Write the name of each amide:

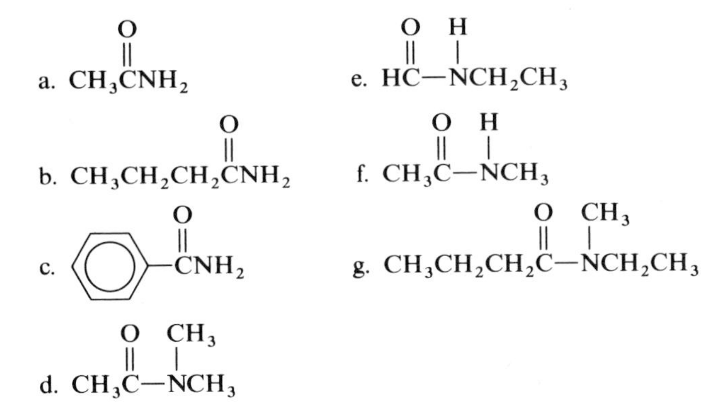

a. $CH_3\overset{O}{\overset{\|}{C}}NH_2$

b. $CH_3CH_2CH_2\overset{O}{\overset{\|}{C}}NH_2$

c. C_6H_5—$\overset{O}{\overset{\|}{C}}NH_2$

d. $CH_3\overset{O}{\overset{\|}{C}}—\overset{CH_3}{\overset{|}{N}}CH_3$

e. $H\overset{O}{\overset{\|}{C}}—\overset{H}{\overset{|}{N}}CH_2CH_3$

f. $CH_3\overset{O}{\overset{\|}{C}}—\overset{H}{\overset{|}{N}}CH_3$

g. $CH_3CH_2CH_2\overset{O}{\overset{\|}{C}}—\overset{CH_3}{\overset{|}{N}}CH_2CH_3$

14.7 Write the structures of the following amides:

a. acetamide

b. pentanamide

c. 3-methylbutanamide

d. *N*-methylbutanamide

e. *N*,*N*-dimethylethanamide

f. 3-chloropropanamide

14.8 Identify the numbered functional groups in the following compounds as carboxylic acid, ester, amine, or amide:

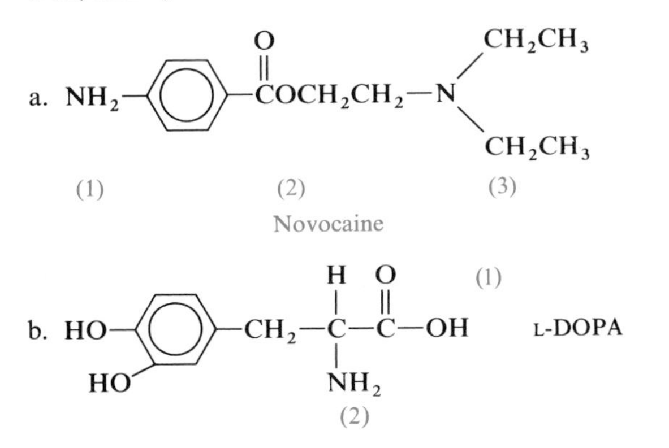

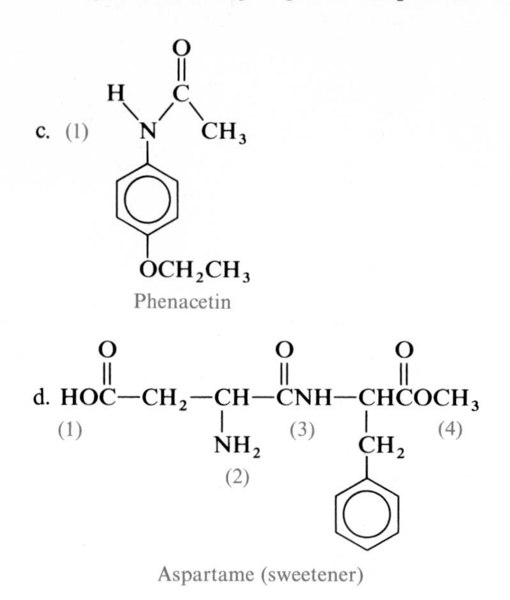

Phenacetin

Aspartame (sweetener)

Formation of Amides (Objective 14.6)

14.9 Write the structure of the amide formed in the following amidation reactions:

a. $CH_3\overset{O}{\overset{\|}{C}}OH + NH_3 \longrightarrow$

b. $C_6H_5\text{—}\overset{O}{\overset{\|}{C}}OH + NH_3 \longrightarrow$

c. $H\overset{O}{\overset{\|}{C}}OH + H_2NCH_2CH_2CH_3 \longrightarrow$

d. $CH_3CH_2\overset{O}{\overset{\|}{C}}OH + H_2NCH_2CH_3 \longrightarrow$

e. $CH_3\overset{O}{\overset{\|}{C}}OH + H\underset{}{\overset{CH_3}{\overset{|}{N}}}CH_3 \longrightarrow$

Hydrolysis of Amides (Objective 14.7)

14.10 Write the products of hydrolysis for the following amides:

a. $CH_3\overset{O}{\overset{\|}{C}}NHCH_3 + HOH \longrightarrow$

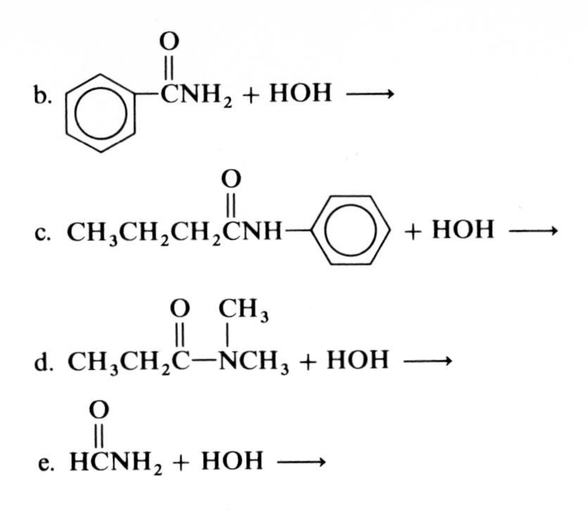
b. CNH2 + HOH
c. CH3CH2CH2CNH + HOH
d. CH3CH2C—NCH3 + HOH
e. HCNH2 + HOH

Carbohydrates

Objectives

15.1 Given the structure of a monosaccharide, classify it as an aldo- or keto-triose, tetrose, pentose, or hexose.

15.2 Identify a linear structural formula for a monosaccharide as its D or L isomer and write its mirror image.

15.3 Write the Haworth structure for a monosaccharide, given its Fischer projection. Write the open- and closed-ring structure for the dietary monosaccharides D-glucose, D-galactose, and D-fructose and state the food sources of each.

15.4 For the dietary disaccharides maltose, lactose, and sucrose, state the monosaccharide units, describe the structure, and identify the food sources of each.

15.5 Given the name or structure of a polysaccharide, describe its monosaccharide units, type(s) of glycosidic linkages, food sources, and biological functions.

15.6 Predict the results a carbohydrate will give in the Benedict's, fermentation, and iodine tests.

Scope

The carbohydrates make up an important part of our daily dietary requirements. Obtained from plants and digested into a usable form called *glucose*, carbohydrates serve as our major source of energy. Carbohydrates are also called *saccharides*, which comes from the Latin word *saccharum*, meaning "sugar." Cellulose is a carbohydrate used by plants to build rigid cell walls. Although cellulose is not digestible by humans, it does play an important role in providing fiber in the diet. Other carbohydrates in our diet include sucrose, which we use to sweeten our food, and lactose, a sugar found in milk.

The recommended dietary allowances (RDA) established by the Food and Nutrition Board of the National Academy of Sciences suggest that about 45 percent of the total caloric intake of a healthy individual should be in the form of complex carbohydrates such as the starches. Complex carbohydrates are digested more slowly than the smaller carbohydrates such as sucrose. This slow process helps to maintain a constant level of glucose in the blood and tissues. Foods high in complex carbohydrates include potatoes and grains such as rice and wheat.

15.1 Classification of Carbohydrates

OBJECTIVE Given the structure of a monosaccharide, classify it as an aldo- or keto- triose, tetrose, pentose, or hexose.

Carbohydrates are produced in plants and bacteria through the process of photosynthesis. The green pigment *chlorophyll* contained in the leaves of plants traps energy from sunlight for the conversion of carbon dioxide and water into carbohydrate molecules. The general equation for the photosynthesis of glucose can be written as

$$6CO_2 + 6H_2O \xrightarrow{\text{Sunlight}} C_6H_{12}O_6 + 6O_2$$

The energy from sunlight is converted to the potential energy in the bonds of the glucose molecules. In the body, glucose enters the bloodstream and body tissues. In the tissues, glucose is oxidized in several steps to carbon dioxide (CO_2) and water while the chemical energy of its bonds is released and made available for the work in the cells. (See Figure 15.1.)

A carbohydrate can be defined as a *polyhydroxy* (many —OH groups) *aldehyde* or *ketone*. The monosaccharides (simple sugars) are a family of carbohydrates that cannot be hydrolyzed into any smaller carbohydrate molecules. They can be classified according to the carbonyl group (aldo- or keto-) and the number of carbon atoms contained in the molecule. The names of most carbohydrates end in the suffix *-ose*.

The ratio of oxygen to hydrogen in many carbohydrates is the same as it is in water. It was thought that these compounds were hydrates of carbon in which a molecule of water was attached to each carbon atom. We now know that there are no molecules of water in these compounds; however, the name *carbohydrate* is still used. The elements of carbon, hydrogen, and oxygen in a monosaccharide have the general formula $(CH_2O)_n$, where the value of n is usually 3, 4, 5, or 6.

Figure 15.1 Carbon cycle in nature depicts the interdependence of photosynthesis and respiration.

Number of Carbon Atoms in Monosaccharides

3	Triose
4	Tetrose
5	Pentose
6	Hexose

aldose, ketose

In all sugar molecules, one carbon atom is bonded to an oxygen atom to form a carbonyl group; the rest of the carbon atoms are bonded to hydroxyl (—OH) groups. If the carbonyl group is an aldehyde, the sugar is called *aldose*; *ketose* is the name given to monosaccharides with ketone groups. Most of the naturally occurring carbohydrates are aldoses, with the most prevalent being aldopentoses and aldohexoses. There are a few ketoses that are biologically important, such as ribulose (a ketopentose) and fructose (a ketohexose). Examples of some important monosaccharides are listed in Table 15.1.

Table 15.1 Some Monosaccharide Units

	General Formula	Aldoses	Ketoses
Triose	$C_3H_6O_3$	CHO[a] | H—C—OH | CH_2—OH Glyceraldehyde	CH_2—OH | C=O | CH_2—OH Dihydroxyacetone
Tetrose	$C_4H_8O_4$	CHO | H—C—OH | H—C—OH | CH_2—OH Erythrose	CH_2—OH | C=O | H—C—OH | CH_2—OH Erythulose
Pentose	$C_5H_{10}O_5$	CHO | H—C—OH | H—C—OH | H—C—OH | CH_2—OH Ribose	CH_2—OH | C=O | H—C—OH | H—C—OH | CH_2—OH Ribulose
Hexose	$C_6H_{12}O_6$	CHO | H—C—OH | HO—C—H | H—C—OH | H—C—OH | CH_2—OH Glucose	CH_2—OH | C=O | HO—C—H | H—C—OH | H—C—OH | CH_2—OH Fructose

[a] CHO = —C(=O)—H

Example 15.1 Identify the following monosaccharides as aldo- or keto- triose, tetrose, pentose, or hexose:

1.
CH_2OH
|
C=O
|
HCOH
|
HOCH
|
CH_2OH

2.
CHO
|
HCOH
|
CH_2OH

3.
CHO
|
HOCH
|
HCOH
|
HCOH
|
HCOH
|
CH_2OH

Solution

1. This carbohydrate is a pentose because it contains a backbone of five carbon atoms. The carbonyl group on the second carbon from the top makes the compound a ketone. Thus it is classified as a ketopentose.
2. This three-carbon carbohydrate is a triose. Since the first carbon is an aldehyde group (—CHO is another way of writing the aldehyde group), it is classified as an aldotriose.
3. This compound is a polyhydroxy compound containing six carbon atoms and an aldehyde group. Thus it is classified as an aldohexose.

15.2 Optical Isomers

OBJECTIVE Identify a linear structural formula for a monosaccharide as its D or L isomer and write its mirror image.

Optical isomerism is related to the symmetry of a molecule. We can explain the idea of symmetry by first looking at some common objects. When a plane is drawn through the center of a symmetrical object, it divides that object into two equal sections. For example, a drinking glass, a ball, a plate, and a tennis racket are symmetrical objects. Both halves are identical. (See Figure 15.2.) Other objects, such as baseball gloves, golf clubs, scissors, shoes, and your hands are asymmetrical; they cannot be divided into identical halves. They do not have a plane of symmetry. They are usually in left- or right-handed forms.

We need to stop here and explain the way in which we are writing these structures. Let us consider the simplest carbohydrate, glyceraldehyde. Recall that the three-dimensional shape of carbon bonded to four other atoms is a tetrahedron. By convention, we look at the glyceraldehyde molecule with the aldehyde group (—CHO) at the top and the $—CH_2OH$ group at the bottom. Two of the attached groups (—CHO and CH_2OH) would be behind the page,

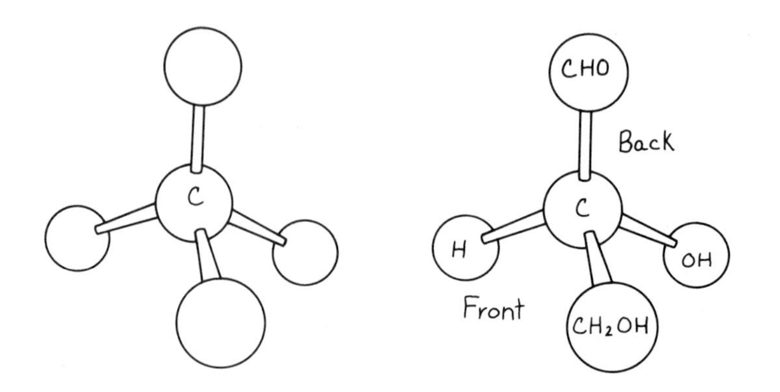

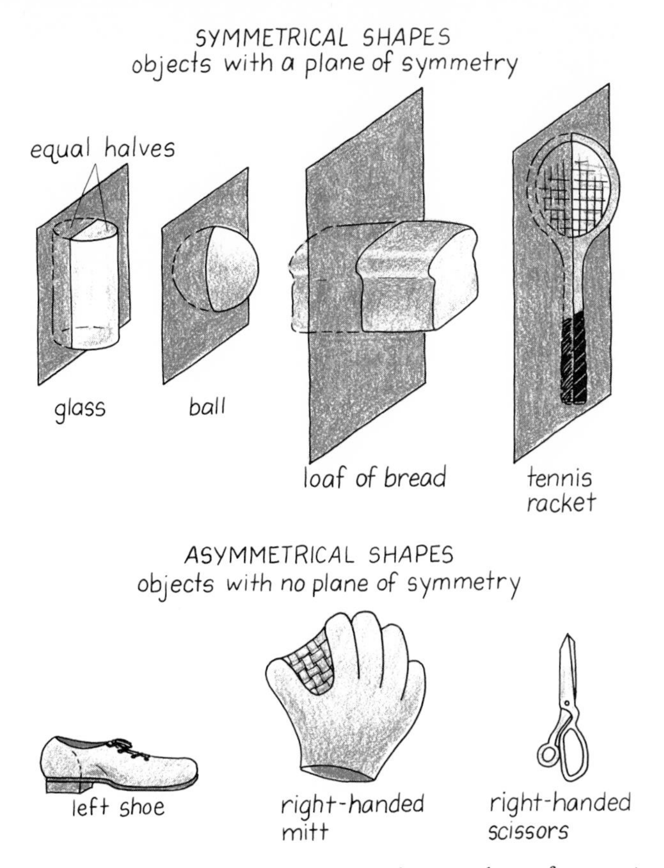

Figure 15.2 Symmetrical shapes have a plane of symmetry, whereas asymmetrical shapes do not.

while the other two groups (—H and —OH) would project in front of the page, toward the observer. On paper, the atoms behind the page are drawn on vertical lines, whereas the groups to the front are drawn on horizontal lines. This method of illustrating a carbohydrate structure is called a *Fischer projection*:

Fischer projection

Glyceraldehyde

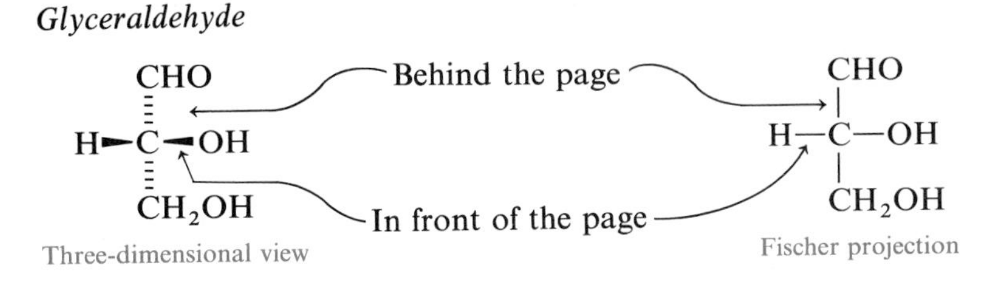

In molecules such as glyceraldehyde, if all four groups attached to a carbon atom are different, the molecule is *chiral*; it has no plane of symmetry.

chiral

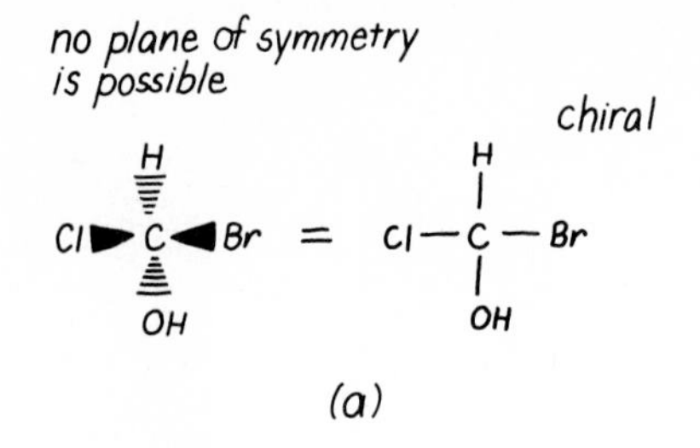

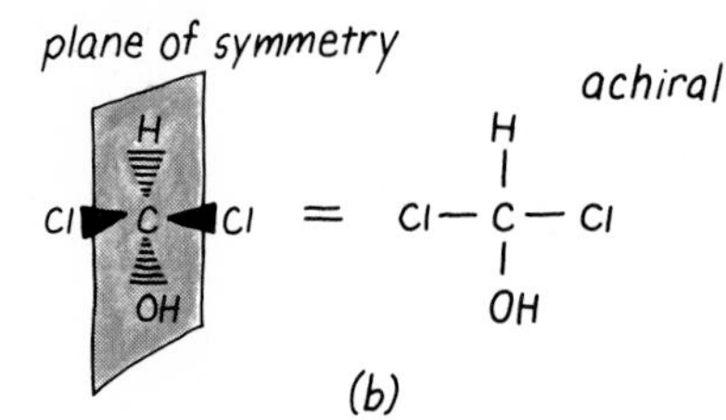

Figure 15.3 Symmetrical and asymmetrical organic compounds: (a) A chiral compound has a carbon atom attached to four unlike groups or atoms, with no plane of symmetry. (b) An achiral compound has at least two identical groups or atoms attached to a central carbon atom, with a plane of symmetry dividing the compound into two equal halves.

The carbon atom attached to the four different groups is a *chiral carbon atom.* When two or more identical groups are attached to a carbon atom, the resulting molecule has a plane of symmetry. It can be divided in half by a plane to give two identical halves. Such a molecule is an *achiral* molecule. (See Figure 15.3.)

Example 15.2 Indicate whether each of the following is a chiral or achiral molecule:

1. H—C—CH_3 (with H above and H below the central C)
2. HO—C—CH_3 (with H above and NH_2 below the central C)
3. HO—C—COOH (with H above and CH_3 below the central C)

Solution

1. This molecule has three identical atoms attached to the central atom. Therefore, a plane of symmetry could be drawn through the horizontal bonds to give two identical halves. It is achiral.
2. This molecule has four different groups of atoms attached to the central carbon. There is no plane of symmetry. It is chiral.
3. This molecule has four different groups attached to the central atom. It is chiral.

Mirror Images

If you place the palms of your hands together, the thumbs and the fingers match. If you turn one hand over so your thumbs are going in the same direction, one hand has its palm facing you. We say that the hands are not superimposable:

there is no way to make the thumbs, fingers, palms and backs of the hand all line up in the same order.

However, if you looked at one hand in a mirror, its mirror image would match your other hand. Therefore, we can say that our hands are *mirror images* of each other. The same thing happens with chiral molecules. Chiral molecules have mirror images that cannot be superimposed. (See Figure 15.4.)

Optical Isomers

We can use the glyceraldehyde molecule to illustrate mirror images of carbon compounds. If we sketch a three-dimensional figure for glyceraldehyde, we find that there are two distinct possibilities for arranging the four different groups attached to the central atom. When the —CHO groups are matched at the top and the —CH_2OH groups are matched at the bottom, the —OH groups and the —H atoms can be placed on opposite sides. The two arrangements are mirror images; each is the mirror reflection of the other. (If you have a model of glyceraldehyde, look at its reflection in a mirror.) It is impossible to superimpose

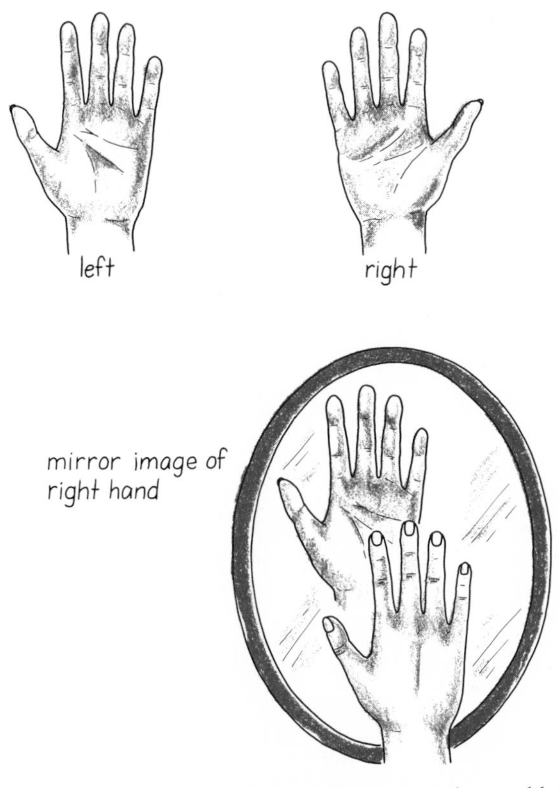

Figure 15.4 Why are the hands not superimposable?

the two mirror images. When a pair of mirror images are not superimposable, they are called *optical isomers.*

Example 15.3 Write the mirror images of the following compound:

$$\begin{array}{c} CHO \\ | \\ Cl-C-H \\ | \\ CH_2OH \end{array}$$

Solution: The mirror image of this compound can be drawn by placing the same groups at the top and the bottom and switching the —Cl and —H groups to opposite sides of the middle carbon atoms:

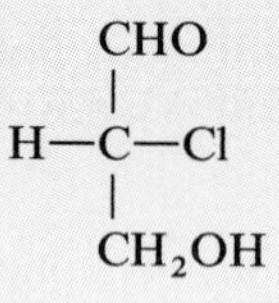

D and L Optical Isomers

The mirror images of a chiral molecule can be distinguished by designating them as D or L isomers. The glyceraldehyde molecule written with the aldehyde group at the top is our model for determination of the D and L isomers of monosac-

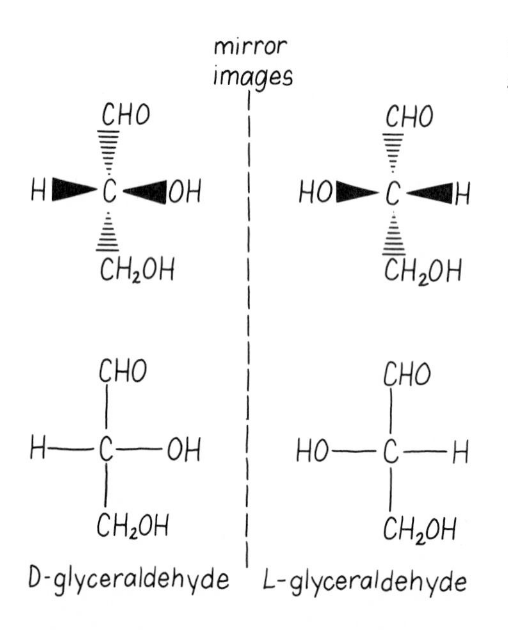

Figure 15.5 The D and L isomers of glyceraldehyde.

charides. When the —OH group on the central carbon is written on the left, the optical isomer is designated the L isomer, L-glyceraldehyde. With the —OH written on the right, the optical isomer is the D isomer, D-glyceraldehyde. This designation distinguishes between the mirror images of the molecule. (See Figure 15.5.)

For carbohydrates with more than three carbon atoms such as glucose, it is the hydroxyl group on the next to the last carbon atom that determines the D or L status of the isomer. For a hexose, this would be the hydroxyl group on the fifth carbon. The D isomer has the hydroxyl group (on the fifth carbon) written on the right, whereas the L isomer has the hydroxyl group (fifth carbon) written on the left. For the most part, only the D isomers of the monosaccharides occur in nature. Few of the L isomers are used in biological systems.

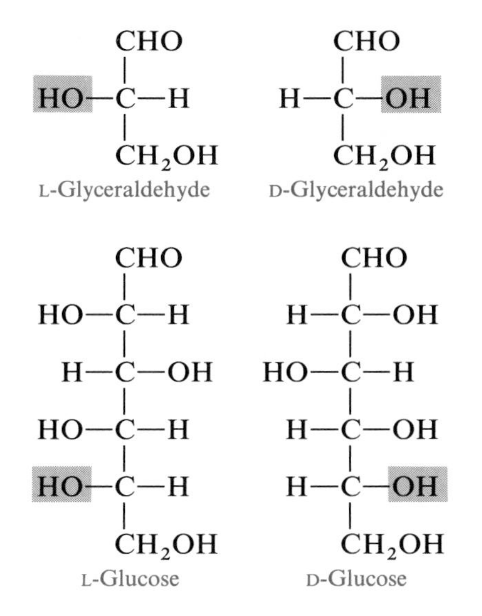

HEALTH NOTE

Optical Isomers in Nature

Many compounds in biological systems consist of only one of the possible optical isomers. Rarely do both the D and L forms of a compound exist in nature. Most biologically active monosaccharides are D isomers; the L forms cannot be utilized. The isomer known as blood glucose is the D-glucose isomer. Among the amino acids, only the L forms are active. The amino acid lysine is active as L-lysine; the body cannot use D-lysine. The compound dopamine is used in the treatment of Parkinson's disease. The L isomer of dopamine is used in treatment, whereas the D form has no effect. One form of LSD strongly affects the production of serotonin in the brain and can cause hallucinations. The other isomer causes little change in the levels of serotonin.

Example 15.4 Indicate whether each of the following carbohydrates is a D or L isomer:

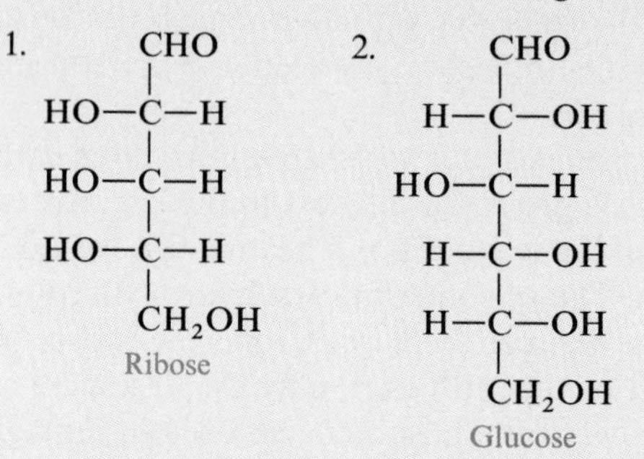

Ribose
Glucose

Solution

1. Our guide to determining a D or L isomer is the position of the hydroxyl group (—OH) on the next to the last carbon atom. In this Fischer projection, the hydroxyl group on the next to the bottom carbon atom is on the left. This structure is the L isomer of ribose, L-ribose.
2. In this structure, the hydroxyl (—OH) on the next to the bottom carbon (away from the —CHO group) is on the right side of the carbon chain. This is the D isomer of glucose, D-glucose.

Optical Activity

Ordinary light consists of electromagnetic vibrations in many planes. When it passes through a polarizing filter, the emerging light is polarized—the electromagnetic vibrations occur in only one plane. If this plane-polarized light travels through a substance that consists of one of a pair of mirror image isomers, the plane of polarized light is bent and emerges at an angle to the original plane. If the light is rotated to the right (clockwise), we say that the chiral isomer is dextrorotatory and it is assigned a (+) sign. If the light rotates to the left (counterclockwise), the chiral isomer is levorotatory and designated as (−).

polarized light

The ability of a substance to bend the plane of polarized light is called *optical activity*; the substance is *optically active*. Only chiral molecules are optically active; the two nonsuperimposable mirror images rotate light in opposite directions.

15.3 Monosaccharides

OBJECTIVE Write the Haworth structure of a carbohydrate, given its Fischer projection. Write the open- and closed-ring structures for the dietary monosaccharides D-glucose, D-galactose, and D-fructose and state the food sources of each.

The three most important monosaccharides, also known as *dietary monosaccharides*, are glucose, galactose, and fructose:

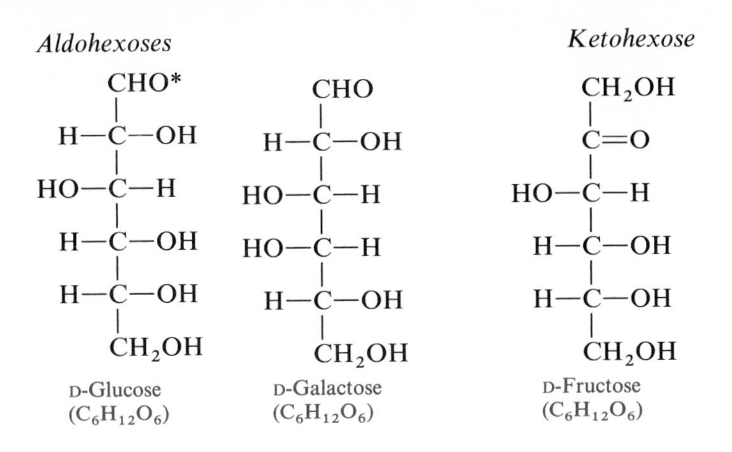

D-Glucose ($C_6H_{12}O_6$) D-Galactose ($C_6H_{12}O_6$) D-Fructose ($C_6H_{12}O_6$)

Glucose

The most abundant hexose in our diet is glucose. Also known as dextrose, grape sugar, and blood sugar, glucose is found in fruits, vegetables, corn syrup, and honey. It is the building block of the more complex carbohydrates such as starch and cellulose.

In the blood and tissues of animals, glucose is found at a normal concentration of 70–90 mg/100 mL. If the glucose concentration exceeds 160 mg/100 mL, glucose is excreted in the urine. The maximum glucose concentration that the blood can carry without glucose appearing in the urine is called the *renal threshold.* The presence of glucose in the urine indicates abnormally high levels of glucose in the blood.

The amount of glucose in the blood depends on the time that has passed since the last meal was eaten. In the first hour after a meal, the level of glucose rises, reaching a peak of about 130 mg/100 mL. The concentration of glucose then decreases over the next 2–3 h until it returns to the previous level. The glucose is utilized in the production of energy in the tissues, in metabolic reactions, and in the synthesis of biological compounds. Some glucose is also converted to energy storage as glycogen in the liver and muscle.

The Haworth Structure (Closed-Ring) for Glucose

The preferred geometry of a hexose molecule such as glucose is not truly linear, as the Fischer projections suggest. Most of the time, the hexose exists as a cyclic, closed-ring structure called a *Haworth structure.*

Haworth structures represent the stable, cyclic forms of carbohydrates

Recall that carbohydrates consist of several alcohol groups and a carbonyl group (aldehyde or ketone). As we saw in Chapter 13, an alcohol group can add to the carbonyl portion of the molecule, thus forming a hemiacetal or hemiketal.

A hemiacetal or hemiketal can also form when the hydroxyl group and the carbonyl group are in the same molecule such as a monosaccharide. This gives a

* CHO = —C(=O)—H

HEALTH NOTE

Hyperglycemia and Hypoglycemia

If a patient shows abnormally high or low blood glucose, a doctor may order a *glucose tolerance test*. In a glucose tolerance test, the patient fasts for 12 h and then drinks 100 g of glucose. Blood samples are drawn over a 4-h period to monitor the blood glucose levels. If the level of glucose in the blood goes above 130 mg/100 mL and maintains a relatively high level, *hyperglycemia*, a sign of possible diabetes mellitus, may be indicated. In diabetes mellitus, insufficient quantities of insulin are produced by the pancreas, so that the tissues cannot make use of the glucose in the blood. Typical symptoms of juvenile diabetes (in people under the age of 40) include thirst, excessive urination, increased appetite, and weight loss with the possibility of ketoacidosis. In maturity-onset diabetes, the patient is most likely to be overweight.

hyperglycemia

Epinephrine is also known to block insulin secretion and to activate the breakdown of glycogen (stored form of glucose) in the liver. High levels of blood glucose result, impairing the storage capacity of the liver. Anxiety during the testing procedure can release sufficient epinephrine to give false positive results for blood glucose.

If the blood glucose levels fall to abnormally low levels, another condition, *hypoglycemia*, may exist. After glucose is ingested, its level in the blood rises but then decreases at an accelerated rate to levels as low as 40 mg/100 mL. In some cases the pancreas appears to overproduce insulin, which uses up glucose too quickly. Low blood glucose can cause dizziness, general weakness, and muscle tremors. A diet consisting of several small meals high in protein and low in carbohydrate is sometimes prescribed to prevent an overstimulation of insulin production by the pancreas. Some hypoglycemic patients are recently finding success with more balanced diets that include complex carbohydrates rather than simple sugars. (See Figure 15.6.)

hypoglycemia

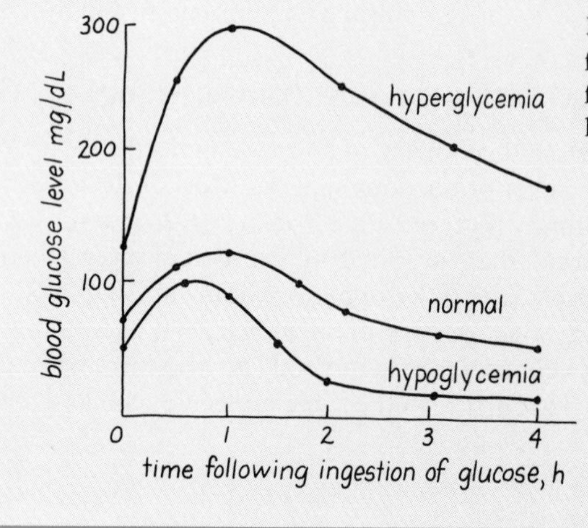

Figure 15.6 Blood glucose levels following ingestion of 100 g glucose, for normal, hyperglycemic, and hypoglycemic conditions.

product that has a cyclic, closed-ring structure. In an aldohexose, a cyclic, closed-ring structure is formed when the hydroxyl group on the fifth carbon reacts with the carbonyl group. The six-member ring (five carbon atoms and an oxygen atom) has been found to be more stable than the open-chain form:

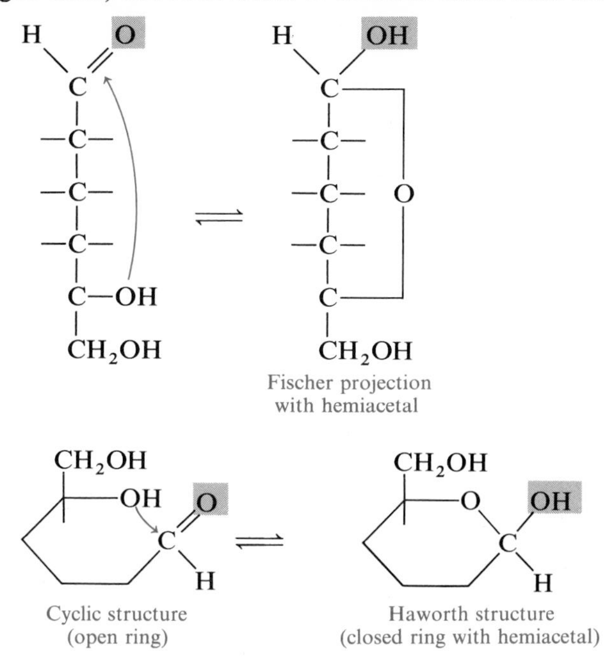

If we think of the closed ring as lying in the same plane as this page, the alcohol groups will appear above or below the page. In the formation of the closed-ring structure, a new alcohol group is formed on the first carbon (formerly the carbonyl group). This new alcohol group can appear above or below the plane of the closed-ring structure. When the alcohol group is below the plane, the glucose is referred to as the α isomer; if the alcohol group appears above, the glucose is the β isomer:

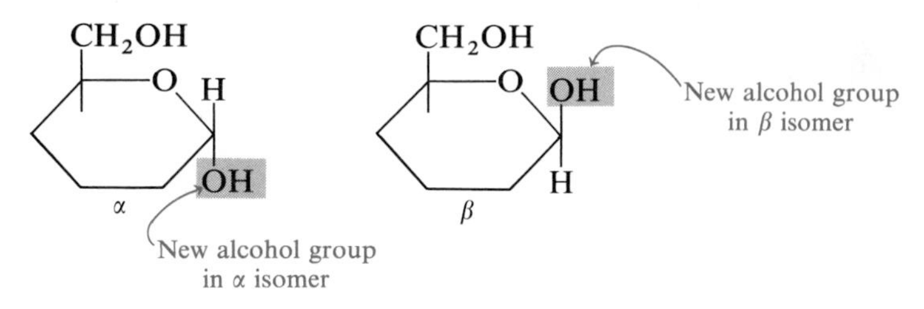

We can write the closed-ring structures for glucose by using the following rules:

Rules for Writing a Haworth (Closed-Ring) Structure for Aldohexoses

1. Number the open-chain compound (Fischer projection) with carbon 1 at the top.

2. Draw a hexagon containing an oxygen atom in the upper right corner; number the carbons in the ring.
3. Write the OH groups on the left of the Fischer projection (open) upward in the Haworth structure, and write the OH groups to the right downward.
4. Write the hydrogen of carbon 5 below the plane and the carbon 6 group above the plane. This variation occurs during the closing of the ring as the hemiacetal bond forms.
5. Write the new alcohol group on carbon 1 for the α or β isomeric form.

A simpified structure indicates only the alcohol groups for clarity:

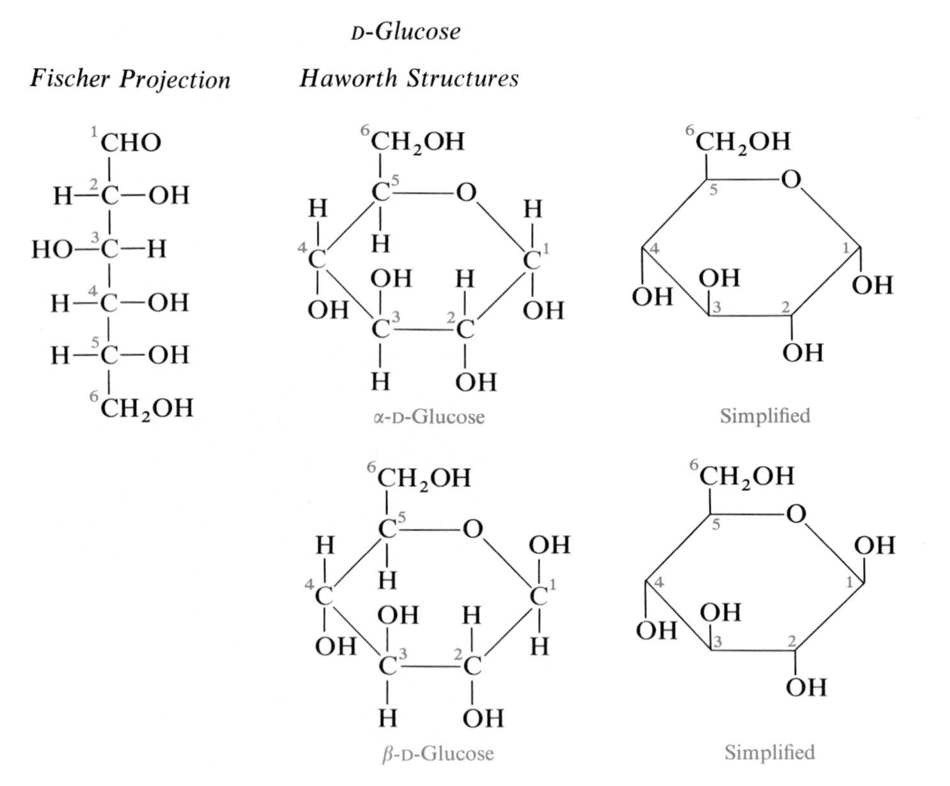

α- and β-glucose

A sample of glucose contains both the α and β isomers. An equilibrium occurs with the α form converting to the β form and back again. This occurs because the ring opens and closes again to reform the hemiacetal bond. In doing so, the —OH group shifts from the α position to the β position, a process called *mutarotation*. The open ring is necessary for mutarotation, but there is only a trace amount present at any given time. Glucose exists primarily as α and β isomers, with the β isomer more prevalent:

mutarotation of glucose

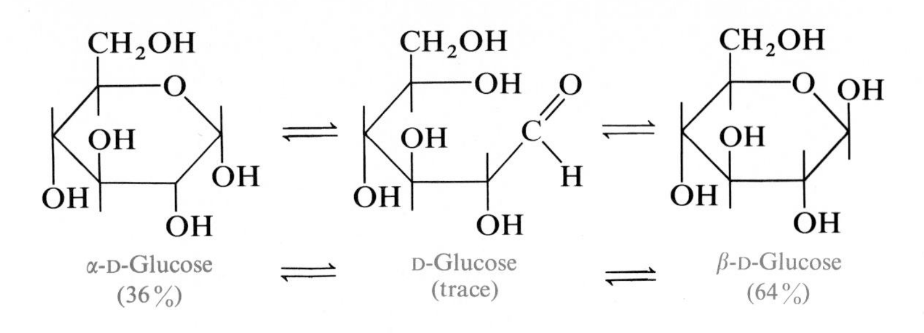

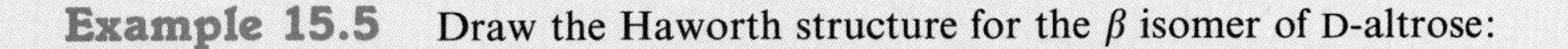

Example 15.5 Draw the Haworth structure for the *β* isomer of D-altrose:

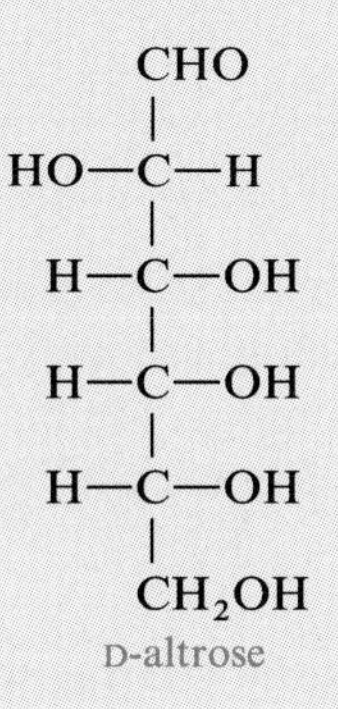

Solution: First, we can number the carbon atoms in the linear structure. We can also draw and number the cyclic, closed-ring structure of an aldohexose that contains five carbon atoms and one oxygen atom:

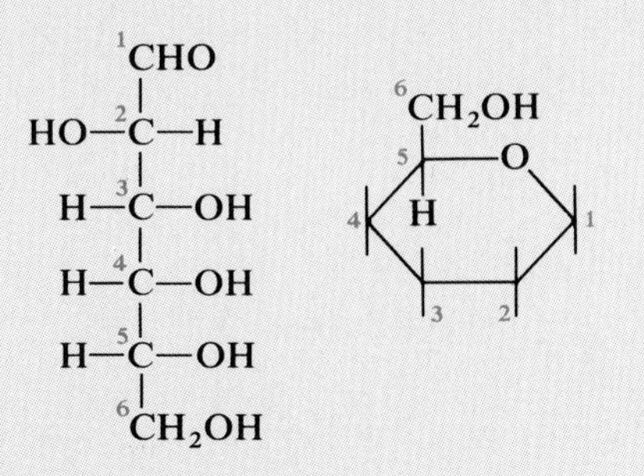

The alcohol groups can now be placed on the cyclic structure. The —OH groups on the right of the Fischer projection represent those facing "down"; that is, they are below the cyclic structure. The —OH groups on the left represent those facing "up";

that is, they are above the cyclic structure. At carbon 1, the newly produced alcohol group is written up or above the cyclic structure, thus giving the β isomer:

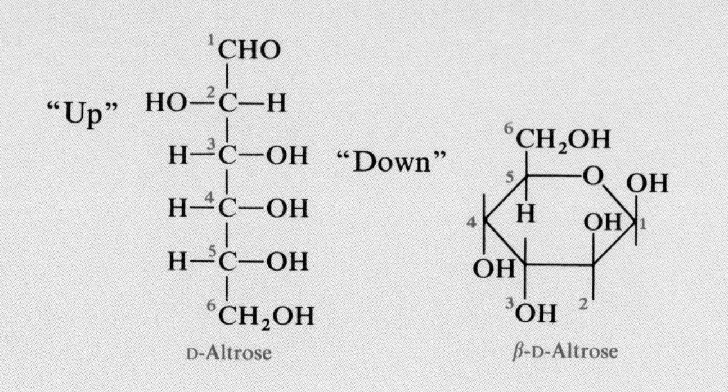

Galactose

Galactose does not occur in the free state in nature; it is obtained as a hydrolysis product of the disaccharide lactose, which is found in milk and milk products. Galactose is the prevalent monosaccharide in the cellular membranes of the brain and nervous system. The Fischer projection and Haworth structures for D-galactose are given below:

Fischer Projection *Haworth Structures*

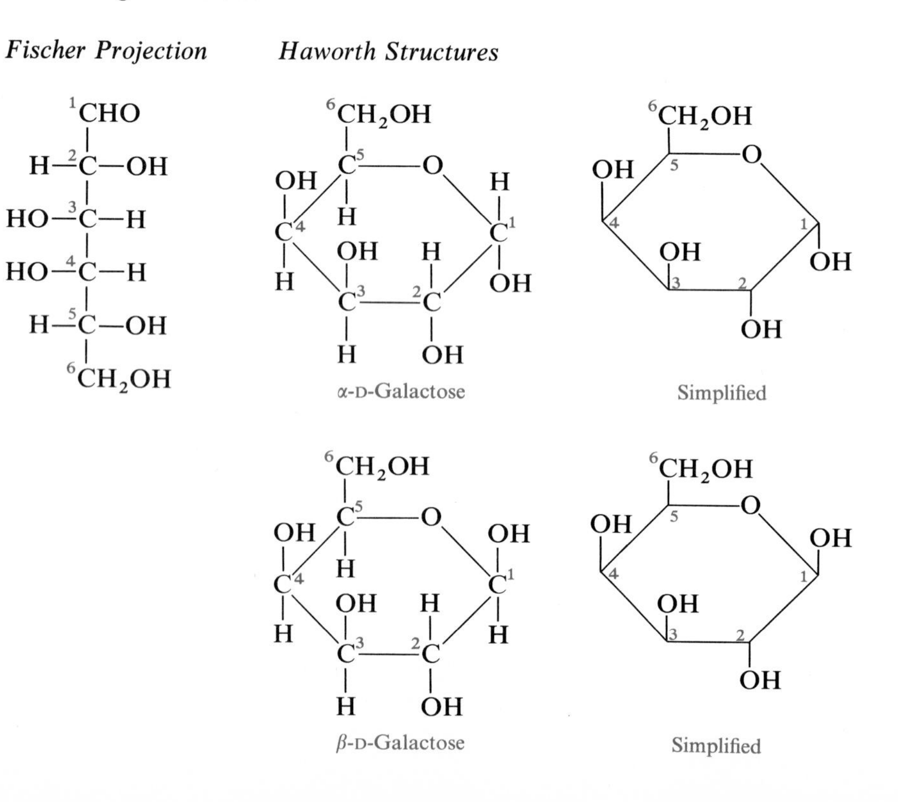

In a condition called *galactosemia*, an infant cannot metabolize galactose properly. As galactose accumulates in the blood and tissues, the child suffers cataract formation, mental retardation, and cirrhosis. The treatment for galactosemia is the removal of all galactose-containing foods, mainly milk and milk products, from the diet. Milk substitutes are used to give an infant a galactose-free diet. If this is done immediately after birth, no ill effects occur.

Example 15.6

1. Draw the open- and closed-ring structure for α- and β-D-galactose.
2. Give the food sources for D-galactose.

Solution

1. The Fischer projection and the Haworth structures for the α and β isomers are written below:

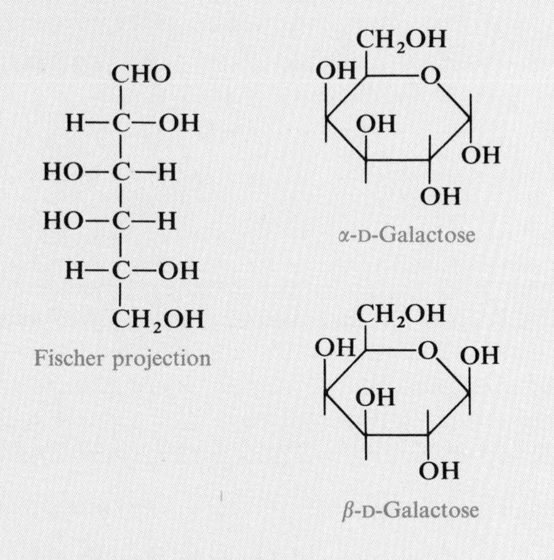

Fischer projection

α-D-Galactose

β-D-Galactose

2. D-Galactose is obtained by the hydrolysis of lactose, which is found in milk and milk products.

Fructose

Fructose is a very sweet hexose found in fruit juices and in honey; it is also called *levulose* and *fruit sugar*. Fructose is also obtained as a hydrolysis product of sucrose, the disaccharide that is also known as *table sugar*.

Fructose is the sweetest carbohydrate, twice as sweet as the most common sweetener, sucrose. This makes fructose popular with dieters since less fructose and, therefore, fewer calories are needed to provide a pleasant taste. After fructose is absorbed in the bloodstream, it is converted to its isomer glucose for use as energy by the tissues. The Fischer projection and Haworth structures for D-fructose are given below:

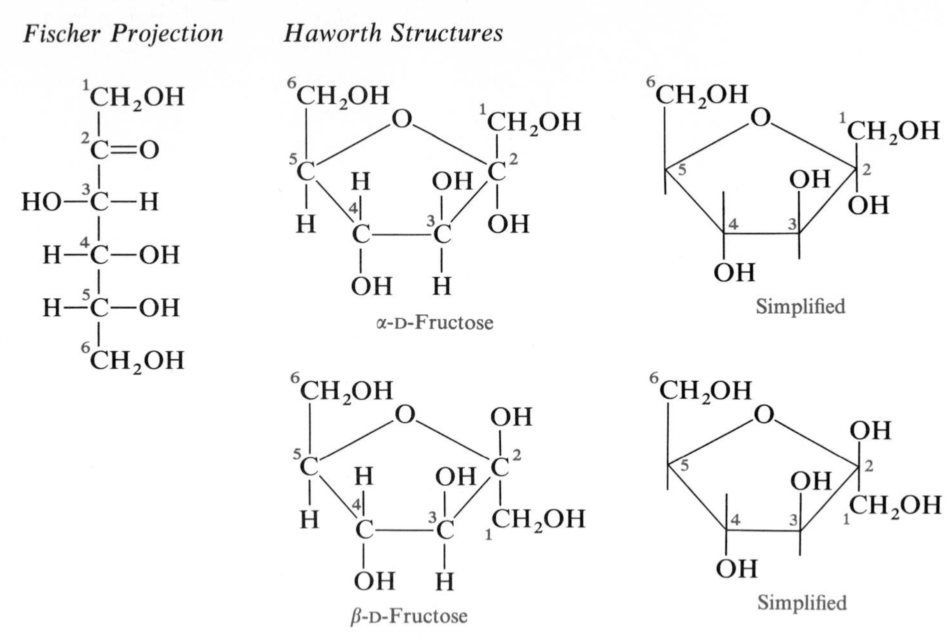

15.4 Disaccharides

OBJECTIVE For the dietary disaccharides maltose, lactose, and sucrose, state the monosaccharide units, describe the structure, and identify the food sources of each.

The *disaccharides* are sugars composed of two monosaccharide units. The dietary disaccharides maltose, lactose, and sucrose can each be hydrolyzed to give a glucose unit and another monosaccharide:

$$\text{Disaccharide} + H_2O \longrightarrow \text{Two monosaccharides}$$

$$\text{Maltose} + H_2O \longrightarrow \text{Glucose} + \text{Glucose}$$

$$\text{Lactose} + H_2O \longrightarrow \text{Glucose} + \text{Galactose}$$

$$\text{Sucrose} + H_2O \longrightarrow \text{Glucose} + \text{Fructose}$$

Glycosidic Bonds

The two monosaccharides in a disaccharide combine to form an acetal. Recall that a monosaccharide exists primarily in the closed-ring hemiacetal form, which is capable of reacting with the alcohol group of another monosaccharide to form an acetal. The acetal bond that forms between two carbohydrate molecules is called a *glycosidic bond.* For an aldohexose, the glycosidic bond forms when the —OH on carbon 1 of the hemiacetal reacts with an —OH group on carbon 4 of the second monosaccharide, thus producing the acetal and a molecule of water.

Glycosidic bonds are designated α or β, as determined by the oxygen atom in the glycosidic bond. In the α-glycosidic bond, the oxygen atom is below the plane. In the β-glycosidic bond, the oxygen atom appears above the plane:

Formation of a Disaccharide

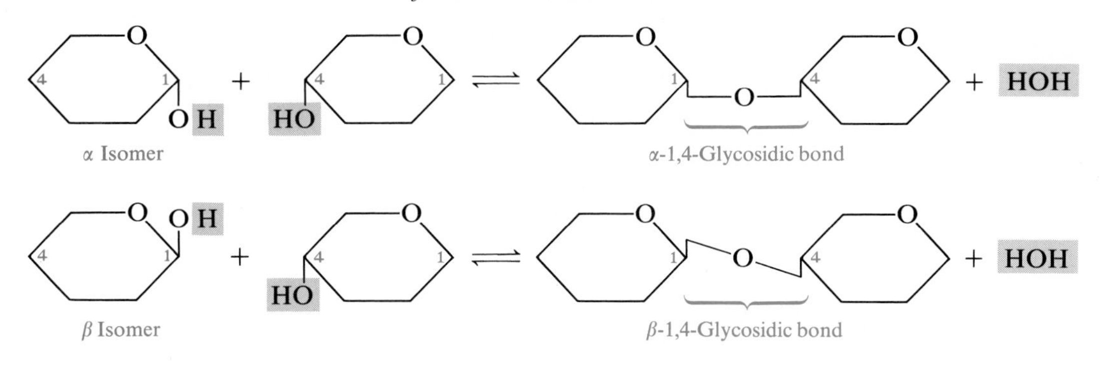

Maltose

The disaccharide maltose is not usually found free in nature but is formed by the hydrolysis of starch molecules. It is produced in germinating grains and finds commericial use in brewing. Maltose is hydrolyzed by yeast enzymes to form glucose molecules, which undergo fermentation to produce ethanol and carbon dioxide. Maltose is also used in malts, cereals, and some candies.

In maltose, there is an α-1,4-glycosidic bond between two glucose units. There are two isomers of maltose because there is a free —OH group on carbon 1 at the end of the molecule. Both α-maltose and β-maltose exist, but β-maltose is more prevalent:

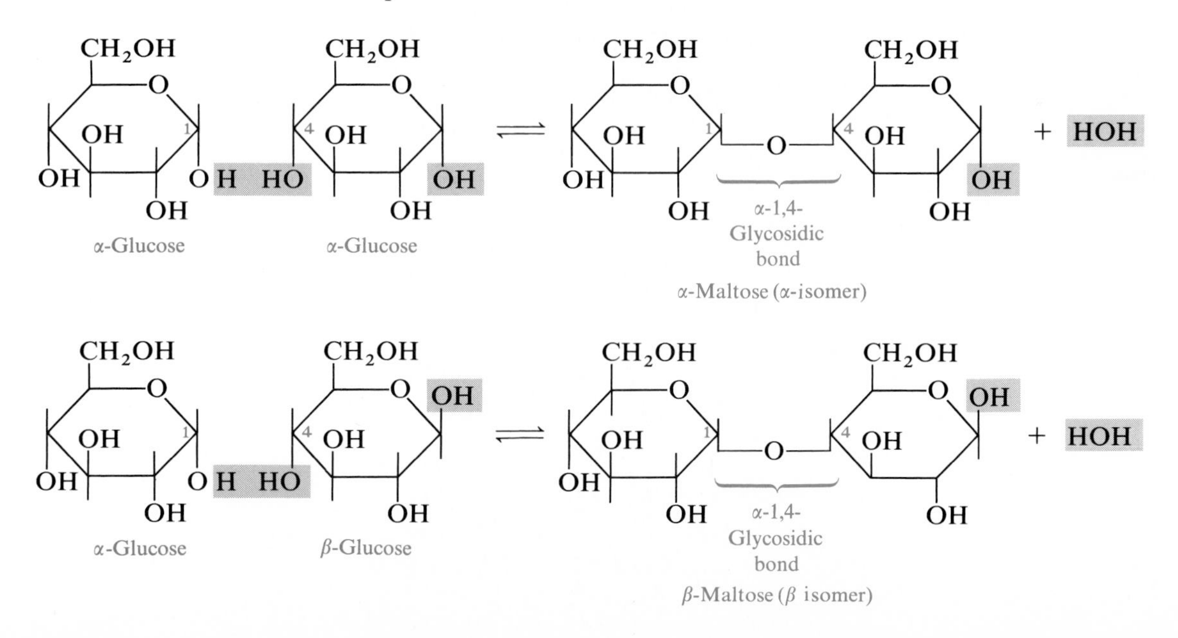

Lactose

The only sources of lactose are milk and milk products. It makes up 6 to 8 percent of human milk and about 4 to 5 percent of cow's milk. Lactose is the disaccharide used in products that attempt to duplicate mother's milk. Many people are unable to digest lactose. This lactose intolerance occurs in adulthood when the body no longer produces sufficient quantities of the lactase enzyme needed to hydrolyze lactose. The sugar remains undigested in the stomach and intestinal tract, causing abdominal cramps and diarrhea.

The hydrolysis of lactose produces glucose and galactose. Lactose has a β-1,4-glycosidic bond which connects carbon 1 of β-galactose with carbon 4 of glucose. The α-lactose and β-lactose isomers both exist:

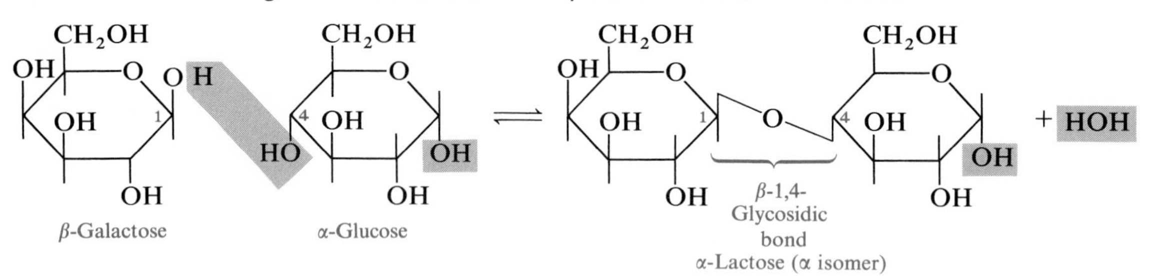

Sucrose

Sucrose is a found in sugar cane and sugar beets. It is produced in greater quantities than any other organic compound. As table sugar, it is consumed in vast amounts. When hydrolyzed, sucrose yields glucose and fructose.

In sucrose, the glycosidic bond forms between carbon 1 of glucose and the —OH group on carbon 2 of the fructose molecule. Sucrose has no α or β isomers, since there is no free —OH group on the isomeric carbon:

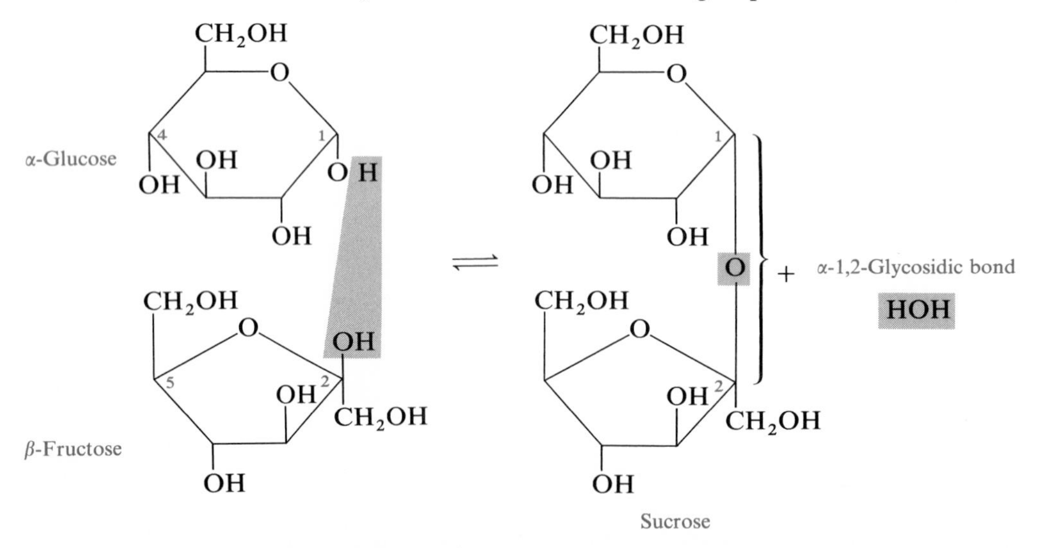

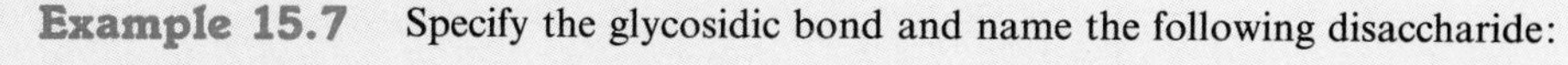

Example 15.7 Specify the glycosidic bond and name the following disaccharide:

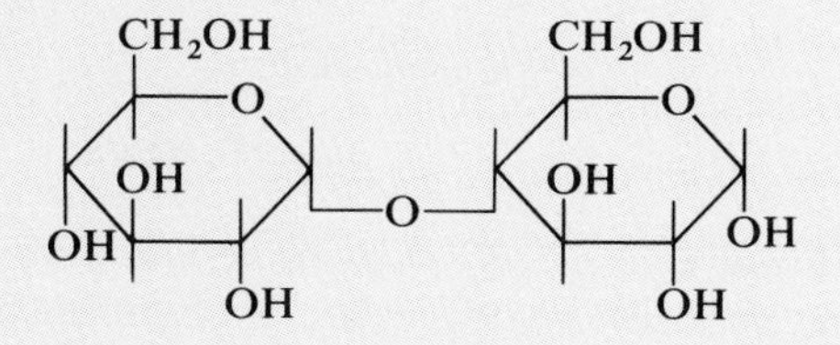

Solution: The glycosidic bond is formed between the —OH group (α isomer) on carbon 1 of the first molecule and the —OH group on carbon 4 of the second molecule. Therefore, we say that the glycosidic bond is an α-1,4-glycosidic bond. Hydrolysis of the disaccharide produces two glucose molecules. Two glucose molecules with an α-1,4-glycosidic bond form the disaccharide maltose. The free —OH group on the second molecule indicates that this is the α-isomer of the disaccharide:

α-Maltose

15.5 Polysaccharides

OBJECTIVE Given the name or structure of a polysaccharide, describe its monosaccharide units, type(s) of glycosidic linkages, food sources, and biological functions.

The polysaccharides starch, cellulose, and glycogen make up the largest group of carbohydrates. All of these polysaccharides consist of long-chain polymers of glucose. The differences occur with regard to the type of glycosidic bonds and in the amount of branching in the molecules.

Starches

The starches are a storage form of glucose in plants. They are found in rice, wheat, root vegetables, grains, and cereals. Starches consist of two kinds of molecules. One type of starch, called *amylose*, is an unbranched polymer made up of glucose molecules all connected by α-1,4-glycosidic bonds. The other type of starch, called *amylopectin*, is the more prevalent form, making up 80 percent of the starches. Amylopectin is a branched form of starch. Glucose units are bonded by α-1,4-glycosidic bonds to form long-chain portions; branches attach to these chains by α-1,6-glycosidic bonds. (See Figure 15.7.)

When the starches undergo partial hydrolysis, smaller sections of glucose units called *pectins* are obtained. Pectins provide the toasty brown color of breads and cookies.

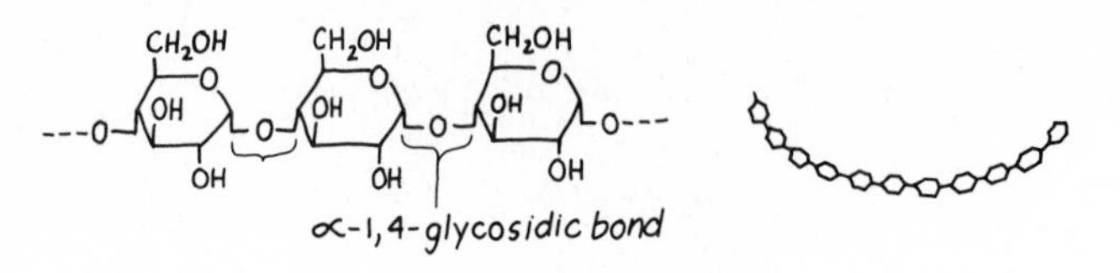

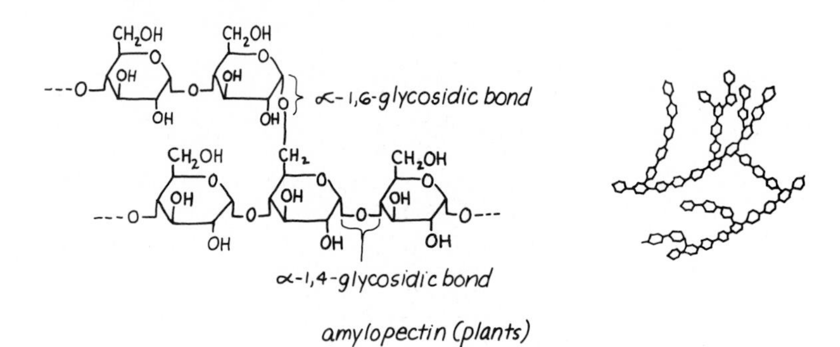

Figure 15.7 Diagrams of the polysaccharides amylose and amylopectin.

Glycogen

Glycogen, found in the liver and muscles of animals, is the storage form of glucose in animals. Glycogen in the liver is hydrolyzed to produce glucose; this maintains the blood level of glucose and provides energy between meals. The structure of glycogen is very similar to that of amylopectin, except glycogen has fewer glucose units and is more highly branched.

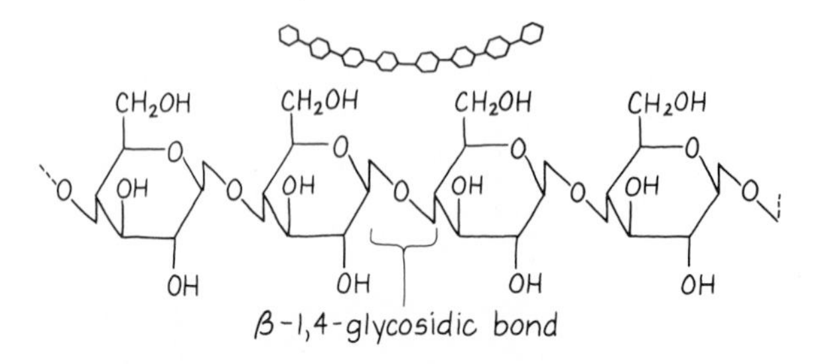

Figure 15.8 Cellulose: A polysaccharide found in the cell wall structure of plants.

Cellulose

Cellulose is used by plants to build the structural components of the cell. Cotton is almost pure cellulose. In cellulose, the glucose molecules form an unbranched chain similar to that of amylose. However, the glucose units in cellulose are bonded by β-1,4-glycosidic bonds. The saliva and pancreatic juices contain enzymes that can hydrolyze the α-1,4-glycosidic bonds in starches, but there are

Table 15.2 Summary of Carbohydrates

Carbohydrate Family	Compound	Food Sources	Monosaccharides
Monosaccharides	Glucose	Fruit juices, honey, vegetables, corn syrup Starch hydrolysis	
	Galactose	Milk and milk products Lactose hydrolysis	
	Fructose	Fruit juices, honey, Sucrose hydrolysis	
Disaccharides	Maltose	Germinating grains Starch hydrolysis	Glucose + glucose
	Lactose	Milk and milk products	Glucose + galactose
	Sucrose	Sugar cane, sugar beets	Glucose + fructose
Polysaccharides	Amylose	Rice, wheat, grains, cereals	Many glucose molecules; straight-chain α-1,4 bonds
	Amylopectin	Rice, wheat, grains, cerals	Many glucose molecules; branched-chain α-1,4 and α-1,6 bonds
	Glycogen	Liver, muscles	Many glucose molecules; branched-chain α-1,4 and α-1,6 bonds
	Cellulose	Plant fiber	Many glucose molecules in a straight chain of β-1,4 bonds; not digestible by humans

no enzymes in humans that can break the β-1,4 bonds in cellulose. Therefore, we cannot digest cellulose. Some animals such as horses, goats, and cows, as well as certain insects such as termites, are able to break down cellulose. Their intestinal tracts contain microorganisms that have enzymes that can hydrolyze β-1,4-glycosidic bonds. (See Figure 15.8.) Table 15.2 summarizes the various carbohydrates discussed in this chapter.

HEALTH NOTE

Fiber in the Diet

The term *dietary fiber* includes all plant materials that are not digestible by humans. One type of dietary fiber is cellulose, which is used by plants to build cells. Food sources include whole grains, bran, fruits, and vegetables. Recently, there has been an increased awareness of the importance of fiber with regard to the health of an individual.

Fiber aids in the formation of bulk in the intestinal tract, which in turn increases the absorption of water. This increases the rate at which digestive wastes move through the intestinal tract and is believed to reduce the time of contact of any carcinogens with the membranes of the intestine and colon. The uptake of water has a laxative effect by producing softer stools. Some forms of diverticulitis (inflammation of the colon) have been relieved by increasing the quantity of fiber in the diet.

The absorptive effects of fiber may also be beneficial in weight maintenance. Fiber increases the bulk in the stomach and intestines without contributing to caloric intake. Fiber may also absorb some of the carbohydrates and cholesterol from the diet, thus decreasing the quantity that diffuses through the intestinal walls.

Example 15.8 Describe the monosaccharide units, the structure, and the food sources of amylose.

Solution: Amylose is a straight-chain polymer of glucose molecules connected by α-1,4-glycosidic bonds. Amylose is found in plants such as rice, wheat, and other grains.

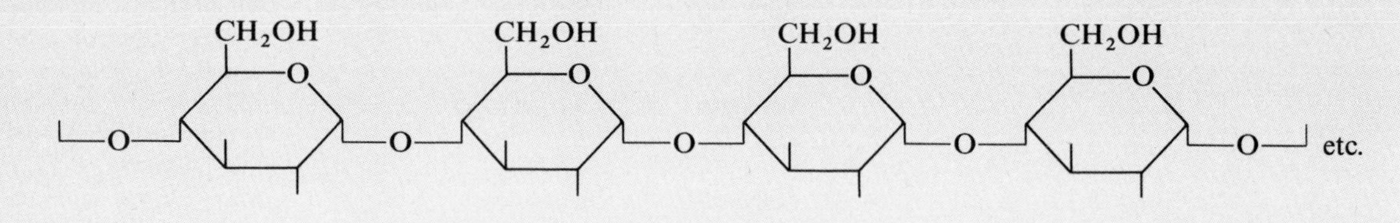

15.6 Tests for Carbohydrates

OBJECTIVE Predict the results a carbohydrate will give in the Benedict's, fermentation, and iodine tests.

Benedict's test

Several sugars, such as glucose, exist in equilibrium with a free aldehyde group. The free aldehyde group is capable of undergoing oxidation. In the *Benedict's test*, cupric hydroxide, $Cu(OH)_2$, is added to the sugar solution, and the mixture is placed in a boiling water bath for 5 min. The aldehyde group is oxidized and the cupric ion reduced to cuprous ion. The blue-colored solution changes to green or orange as the reddish-orange precipitate of cuprous oxide forms.

Carbohydrates that give a positive Benedict's test are called *reducing sugars* because they reduce the copper ion from cupric (2+) to cuprous (1+):

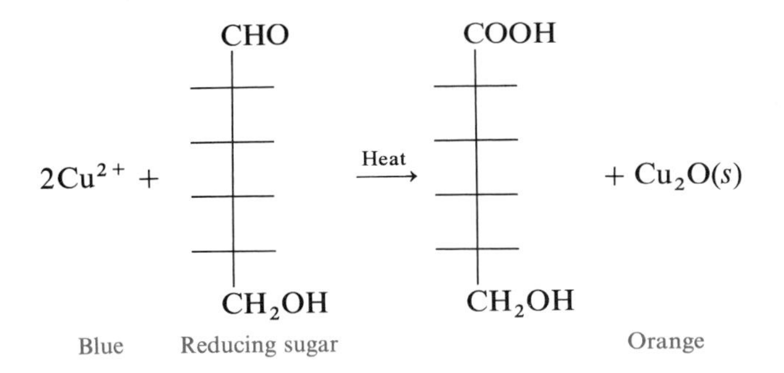

All monosaccharides and disaccharides (except sucrose) are reducing sugars. Although fructose is a ketohexose, its ketone group influences the adjacent alcohol group so that it is also oxidized. Sucrose is not a reducing sugar and gives a negative test with Benedict's reagent. Polysaccharides give negative results with Benedict's reagent because only a few of the glucose molecules in the polysaccharide structure can provide a free aldehyde group.

fermentation test

When treated with yeast, the hexoses glucose and fructose, but not galactose, will undergo fermentation:

$$\underset{\text{Hexose}}{C_6H_{12}O_6} \xrightarrow{\text{Yeast}} \underset{\text{Ethanol}}{2C_2H_5OH} + 2CO_2$$

Lactose will not ferment because the enzyme lactase it needs for its hydrolysis is not present in yeast. However, the enzymes for the hydrolysis of maltose and sucrose are present, so that samples containing these sugars produce glucose and fructose to give positive fermentation tests. Fermentation does not occur with polysaccharides.

iodine test

The polysaccharide amylose in starch reacts strongly with iodine to form a characteristic deep blue-black complex. Cellulose, glycogen, and amylopectins produce reddish-purple and brown colors. Such color does not develop with mono- or disaccharides.

Example 15.9 Indicate the results that fructose, sucrose, and starch give in the iodine, fermentation, and Benedict's tests.

Solution: Iodine will not produce a color change with fructose or sucrose, whereas starch will give a blue-black color. The fermentation test will produce CO_2 with fructose and glucose, but not with starch. Benedict's reagent with fructose will give a positive color change to blue-green or orange. Benedict's test will be negative (no color change) with starch and sucrose. These results are summarized in the following table:

Carbohydrate	Iodine Test	Fermentation Test	Benedict's Test
Fructose	Negative	Positive	Positive
Sucrose	Negative	Positive	Negative
Starch	Positive	Negative	Negative

HEALTH NOTE

Testing for Glucose in Urine

Benedict's test is used to determine the presence and quantity of glucose in urine samples. Glucose does not normally appear in the urine. However, if the level of blood glucose exceeds about 160 mg/100 mL (renal threshold for glucose), there is more glucose present than can be reabsorbed; glucose then appears in the urine, indicating a condition called *glucosuria.* The amount of cuprous oxide formed is proportional to the amount of reducing sugar (glucose) in the sample. If the glucose level is high, the specimen turns bright orange. Lower levels of glucose turn the solution green or yellow.

In the hospital, Benedict's reagent is often used in pellet form. The pellet is added to 5 drops of urine and 10 drops of water. A plastic test strip coated with Benedict's reagent can also be used. This test strip is placed in the urine sample and can be read in 10–30 s. The level of glucose present in the urine is found by matching the color produced to the standard color chart found on the container. (See Table 15.3.)

Table 15.3 Glucose Test Strip Results

	Glucose Present	
Color	%	mg/dL
Blue	<0.1	<100
Blue-green	0.25	250
Green	0.50	500
Yellow	1.00	1000
Orange	2.00	2000

Glossary

aldose Monosaccharides that contain an aldehyde group.

amylopectin A branched-chain polysaccharide (starch) that is composed of glucose units and is produced in plants.

amylose A straight-chain polysaccharide (starch) that is composed of glucose units and is produced in plants.

Benedict's test A chemical test in which cupric hydroxide is reduced to reddish-orange cuprous oxide by all monosaccharides and disaccharides except sucrose.

carbohydrate A polyhydroxyl compound that contains an aldehyde or ketone group.

cellulose A polysaccharide composed of glucose units in a straight-chain polymer that is undigestible by humans.

disaccharide A compound formed by the combination of two monosaccharides.

fermentation test A test for glucose or fructose and any disaccharide that breaks down to give glucose or fructose. A positive test occurs when carbon dioxide is formed upon the addition of yeast to the sample.

fructose A monosaccharide found in honey and fruit juices; it is combined with glucose in sucrose. Also called *levulose* and *grape sugar*.

galactose A monosaccharide that occurs combined with glucose in lactose.

glucose The most prevalent of the monosaccharides in the diet. Found in fruits, vegetables, corn syrup, and honey. Also known as *blood glucose* and *dextrose*. Combines in polymers to form most of the polysaccharides.

glycogen A polysaccharide formed in the liver and muscles for the storage of glucose as an energy reserve. It is composed of glucose in a highly branched polymer.

glycosidic bond The acetal bond that holds monosaccharide units together to form disaccharides and polysaccharides.

hemiacetal A carbon bond attached to both an alcohol group and an ether group.

iodine test A test for starch (amylose). A positive test is the formation of a blue-black color after iodine is added to the sample.

ketose A monosaccharide that contains a ketone group.

lactose A disaccharide consisting of glucose and galactose; it is found in milk.

maltose A disaccharide consisting of two glucose units; it is obtained from the hydrolysis of starch and in germinating grains.

monosaccharide A simple carbohydrate most commonly consisting of three to six carbon atoms.

optical isomerism The type of isomerism characteristic of chiral compounds that have a carbon bonded to four different groups.

photosynthesis The formation of carbohydrates by plants from water and carbon dioxide in the presence of light and chlorophyll.

polysaccharide A carbohydrate formed by the combination of many monosaccharide units, usually glucose.

reducing sugar A sugar with a free aldehyde or ketone group capable of reducing cupric hydroxide to give a positive test with Benedict's reagent.

saccharide A term from the Latin word *saccharum*, meaning "sugar"; it is used to describe the carbohydrate family.

sucrose A disaccharide composed of glucose and fructose; a nonreducing sugar.

Problems

Classes of Carbohydrates (Objective, 15.1)

15.1 Classify each structure as an aldo- or keto- triose, tetrose, pentose, or hexose:

a. CH_2OH / C=O / HOCH / HCOH / HCOH / CH_2OH
Fructose

b. CHO / HCOH / HCOH / HCOH / CH_2OH
Ribose

c. CH_2OH / C=O / CH_2OH
Dihydroxyacetone

d. CHO / HCOH / HCOH / HOCH / CH_2OH
Xylose

e. CHO / HCOH / HOCH / HOCH / HCOH / CH_2OH
Galactose

Optical Isomers (Objective 15.2)

15.2 Indicate whether each of the following molecules contains a chiral carbon:

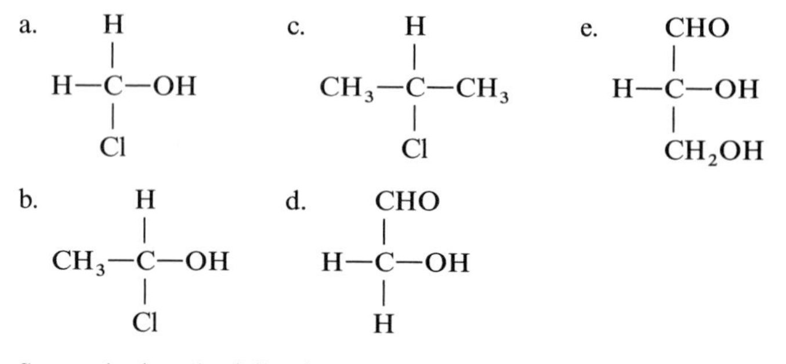

15.3 State whether the following are D or L isomers of carbohydrate molecules:

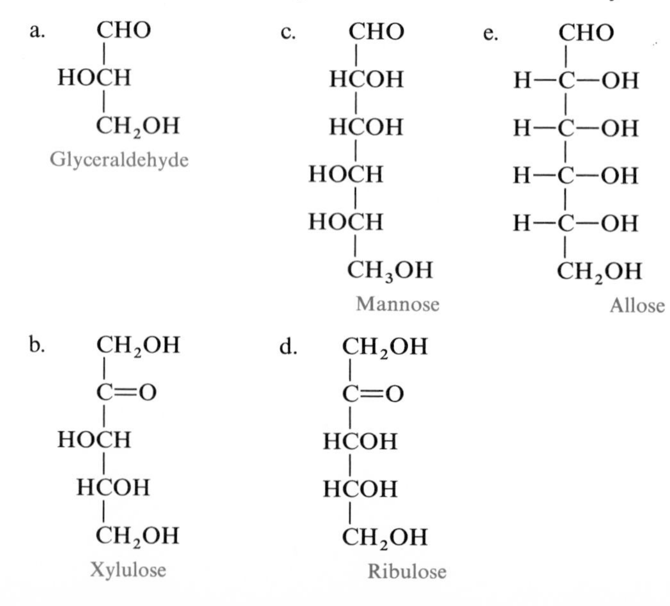

15.4 Write the mirror image of each of the following isomers:

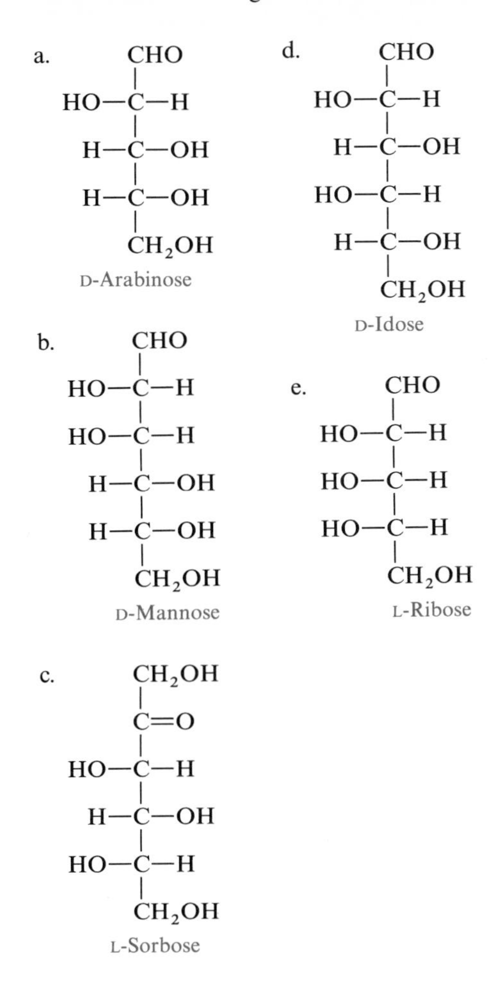

Monosaccharides (Objective 15.3)

15.5 Give the structure of the cyclic hemiacetal for the following hexoses:

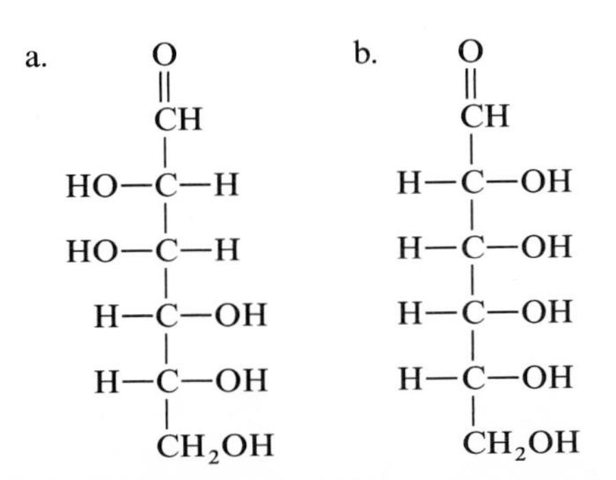

15.6 a. Number the carbon atoms in the open-chain structure of D-mannose.
b. Write the Haworth (closed-ring) structure for α and β isomers of D-mannose.

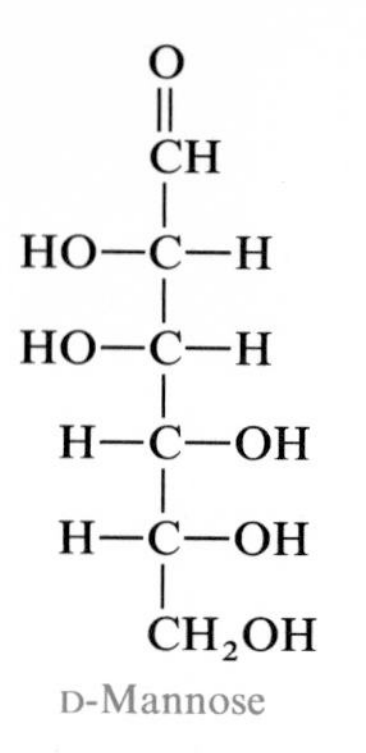

D-Mannose

15.7 The following are the Haworth structures of some D-sugars. Name each monosaccharide including the α or β isomer.

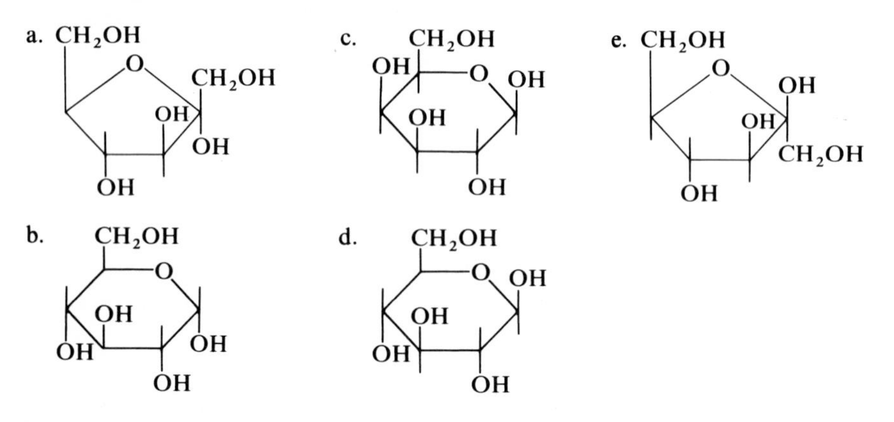

15.8 Draw the Haworth structure of
a. β-D-galactose b. α-D-fructose

15.9 Using Haworth structures, write the equation for the mutarotation of α-D-galactose to β-D-galactose.

15.10 Draw the Haworth structures (α and β isomers) of the hexose, D-talose:

O
||
CH
|
HO—C—H
|
HO—C—H
|
HO—C—H
|
H—C—OH
|
CH_2OH

D-Talose

15.11 Name the dietary food sources for
a. glucose b. fructose

Disaccharides (Objective 15.4)

15.12 For each of the following disaccharides state (1) the monosaccharide units (including α or β form), (2) the type of glycosidic bond, and (3) the name of the disaccharide:

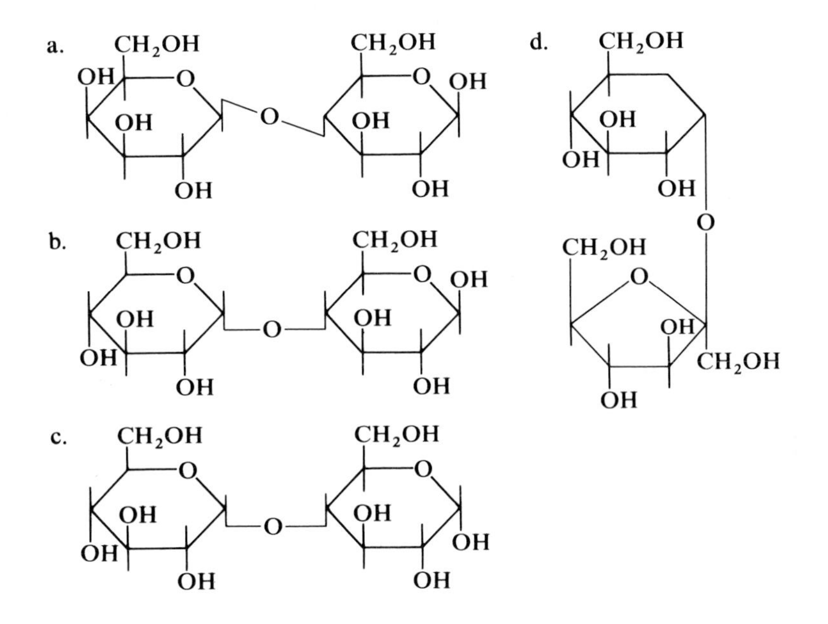

Polysaccharides (Objective 15.5)

15.13 For each of the polysaccharides listed in the table below, indicate the following structural features: (1) the subunit and its isomeric form, (2) straight-chain or branched chain, (3) the type(s) of glycosidic bond(s) in each.

Polysaccharide	Subunit α- or β-Glucose	Straight or Branched Chain	Glycosidic Bond(s) α-1,4, β-1,4, or α-1,6
Amylose			
Amylopectin			
Glycogen			
Cellulose			

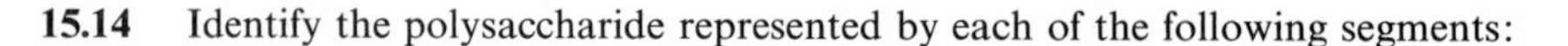

15.14 Identify the polysaccharide represented by each of the following segments:

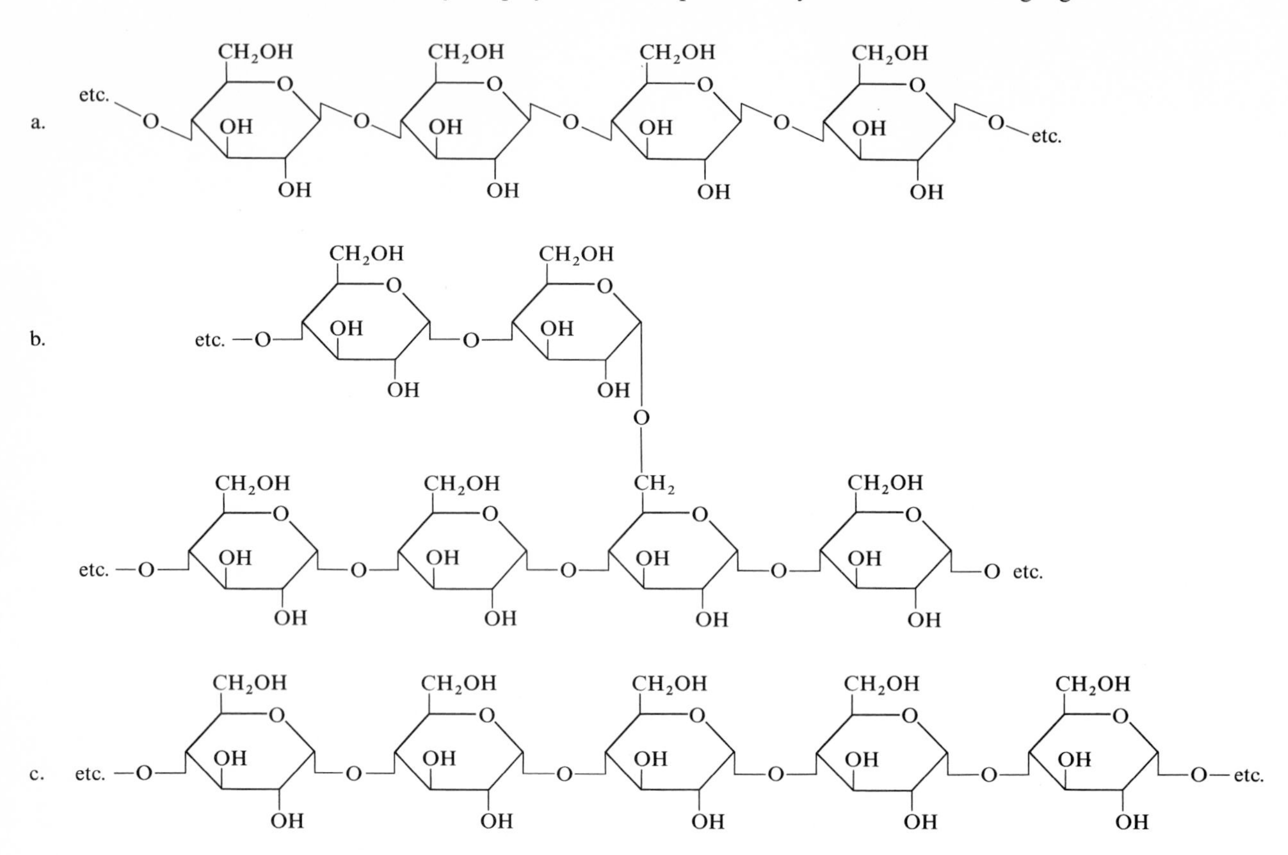

Tests for Carbohydrates (Objective 15.6)

15.15 Describe the test reagent, the sugars that give a positive test, and the results of a positive test for the following carbohydrate tests:
a. Benedict's test
b. fermentation test
c. iodine test

15.16 Indicate the results of the following tests for the carbohydrates indicated:

Carbohydrates	Iodine	Fermentation	Benedict's Test
Galactose			
Maltose			
Sucrose			
Starch			

15.17 You are given two disaccharides to identify. Sugar A will not ferment and gives a positive Benedict's test. Sugar B is fermented and gives no reaction (negative) with Benedict's test. Identify sugars A and B.

15.18 Give the name of a carbohydrate for each of the following descriptions:

a. a disaccharide that is not a reducing sugar
b. a disaccharide that occurs as a breakdown product of starch
c. a disaccharide composed of glucose and fructose
d. a carbohydrate that is produced as a storage form of energy in plants
e. the storage form of carbohydrates in animals
f. a carbohydrate that is used to build cell walls in plants
g. a monosaccharide that combines with glucose to form lactose
h. a monosaccharide that does not ferment
i. a disaccharide composed of two glucose units
j. a carbohydrate that is not digestible by humans

16 Lipids

Objectives

16.1 Identify a fatty acid as saturated or unsaturated. Predict whether that fatty acid is liquid or solid at room temperature and of animal or vegetable origin.

16.2 Write the structure of a wax or a triacylglycerol that is produced when a particular fatty acid reacts with a long-chain alcohol or with glycerol; name the triacylglycerol.

16.3 Given the structure or name of a triacylglycerol, write the products of (1) hydrogenation, (2) acid or enzyme hydrolysis, (3) oxidation, and (4) saponification.

16.4 Describe the iodine number for unsaturation in simple lipids; describe the acrolein test for the presence of glycerol.

16.5 Describe the components of phosphoglycerides, sphingolipids, and glycolipids and state their biological function.

16.6 Given the structure or name of a simple lipid, determine if it is related to the steroid or terpene family of lipids.

Scope

The animal fats and vegetable oils you obtain from your diet are lipids. In the body, lipids are found as cholesterol, fat-soluble vitamins, hormones, and that ever-present adipose tissue commonly known as "body fat." The compounds that comprise the lipid family are vastly different and have seemingly unrelated structures. However, there is one similarity among them all—they are not soluble in water.

Today there is much study of lipids such as the triacylglycerides and cholesterol. Many researchers believe that a high level of cholesterol in the blood is related to arteriosclerosis, a condition in which deposits of lipid materials accumulate in the coronary blood vessels. These deposits restrict the flow of blood to the tissue, causing *necrosis* (death) of the tissue. In the heart, this could result in a *myocardial infarction* (heart attack). If some of the deposited material breaks away, it can produce an *embolism*, which may block off smaller vessels of the lungs or brain.

Fats in the body serve as a rich store of energy, insulate the body, and protect the inner organs.

16.1 Fatty Acids

OBJECTIVE Identify a fatty acid as saturated or unsaturated. Predict whether that fatty acid is liquid or solid at room temperature and of animal or vegetable origin.

lipids are not very soluble in water

Lipids vary considerably in composition, more than any other family of biological compounds. The only feature that is common to all lipids is their poor solubility in water. As waxes, lipids are found as coatings on leaves and the feathers of birds. In the body, lipids can be used for energy, for insulation, and protection of inner organs. Lipids are found in many cell membranes, including the brain and nerves. The fat-soluble vitamins A and D, certain hormones, and cholesterol are all lipids.

Classes of Lipids

Lipids can be divided into several subgroups: Some contain fatty acids in their structures; others do not. Those containing fatty acids are complex lipids that can be hydrolyzed to yield fatty acids and some type of alcohol. (See Table 16.1.) Lipids without fatty acids are classified as *simple lipids.*

Fatty Acids

The fatty acids occurring in vegetables and animals are usually long-chain carboxylic acids with an even number (12–24) of carbon atoms. Fatty acids containing 16–18 carbon atoms are the most prevalent in biological systems.

saturated

unsaturated

Fatty acids may or may not have double bonds in the carbon chain. If there are no double bonds, we say that the fatty acid is *saturated.* In *unsaturated* fatty acids, there are one to four double bonds in the chain. Figure 16.1 illustrates the structures of saturated and unsaturated fatty acids. (See Table 16.2 for a list of some important fatty acids.)

Table 16.1 Classes of Lipids

Complex Lipids	Products of Hydrolysis
Waxes	Fatty acids, long-chain alcohol
Triacylglycerols	Fatty acids, glycerol
Phosphoglycerides	Fatty acids, glycerol, phosphate, amino alcohol
Sphingolipids	Fatty acids, sphingosine (an alcohol), phosphate, amino alcohol
Glycolipids	Fatty acids, sphingosine, monosaccharide
Simple Lipids	**Structure**
Steroids	Multicyclic ring structure
Terpenes	Isoprenoid units

Properties of Fatty Acids

Fats and oils obtained from vegetable and animal sources contain both saturated and unsaturated fatty acids. In general, the fatty acids in vegetables are more unsaturated than the fatty acids in animals.

As a rule, the melting point of a fatty acid increases with an increase in saturation. Since the structure of a saturated fat is linear, the molecules are able to fit together in a tighter arrangement that allows greater attractions among the molecules. More energy is required to melt the saturated fatty acids; this accounts for a higher melting point. Generally, animal fats contain more saturated fatty acids, have a higher melting point, and are solid at room temperature.

The unsaturated fatty acid molecules have a bent shape at the cis double bond; this does not allow the molecules to fit so close together. Since the molecules are further apart, there is less attraction among them. Less energy is needed to melt unsaturated fatty acids. This means that unsaturated fatty acids have lower melting points. Oils, usually derived from vegetable sources, have

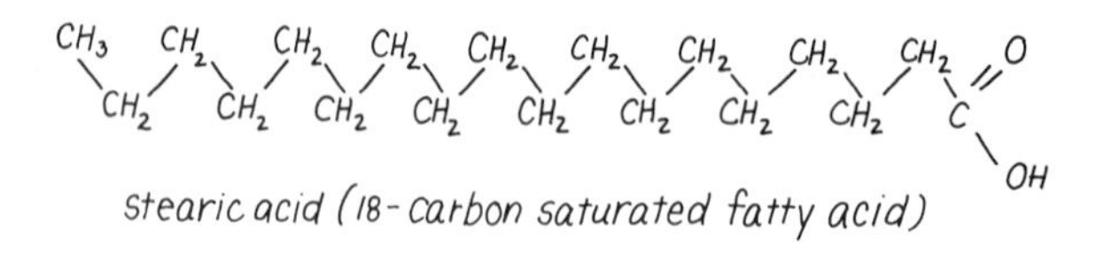

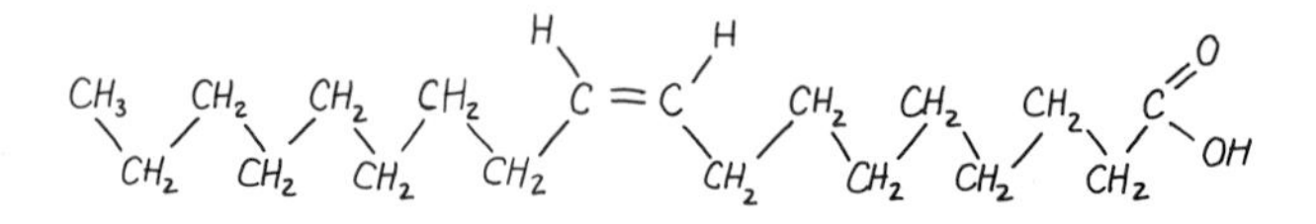

Figure 16.1 The structure of stearic acid, an 18-carbon saturated acid, and oleic acid, an 18-carbon unsaturated fatty acid with a cis double bond.

Table 16.2 Important Fatty Acids

Name	Number of Carbon Atoms	Structure	Melting Point (°C)	Source
Saturated				
Butyric	4	$CH_3CH_2CH_2\overset{O}{\overset{\|\|}{C}}OH$	−8	Butter
Lauric	12	$CH_3(CH_2)_{10}\overset{O}{\overset{\|\|}{C}}OH$	44	Coconut oil
Myristic	14	$CH_3(CH_2)_{12}\overset{O}{\overset{\|\|}{C}}OH$	58	Nutmeg
Palmitic	16	$CH_3(CH_2)_{14}\overset{O}{\overset{\|\|}{C}}OH$	64	Fats and oils
Stearic	18	$CH_3(CH_2)_{16}\overset{O}{\overset{\|\|}{C}}OH$	69	Fats and oils
Unsaturated				
Oleic	18	$CH_3(CH_2)_7CH{=}CH(CH_2)_7\overset{O}{\overset{\|\|}{C}}OH$	14	Olive oil
Linoleic	18	$CH_3(CH_2)_4CH{=}CHCH_2CH{=}CH(CH_2)_7\overset{O}{\overset{\|\|}{C}}OH$	−5	Vegetable oils
Linolenic	18	$CH_3CH_2CH{=}CHCH_2CH{=}CHCH_2CH{=}CH(CH_2)_7\overset{O}{\overset{\|\|}{C}}OH$	−11	Vegetable oils
Arachidonic	20	$CH_3(CH_2)_4CH{=}CHCH_2CH{=}CHCH_2CH{=}CHCH_2CH{=}CH(CH_2)_3\overset{O}{\overset{\|\|}{C}}OH$	−50	Animal organs

more unsaturated fatty acids and are liquid at room temperature. Figure 16.2 illustrates the fatty acid composition of some fats and oils.

Essential Fatty Acids

The body is capable of synthesizing all but a few fatty acids from carbohydrates. Those that cannot be synthesized are called *essential fatty acids*, that is, they must be included in the diet. One such fatty acid, linoleic acid, has been shown to be important in the prevention of dry skin dermatitis in infants. Its role in adult nutrition is not well understood. Two other fatty acids considered important and often included in the group of essential fatty acids are linolenic and arachidonic acids.

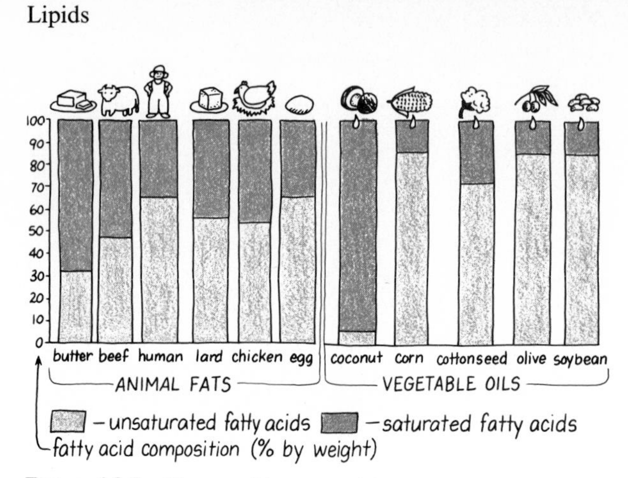

Figure 16.2 Fatty acid composition of some common fats and oils (solid and liquid triacyclglycerols).

HEALTH NOTE

Prostaglandins

Prostaglandins were first discovered in the prostate gland and are now known to be present in almost all tissues of the body. Prostaglandins are 20-carbon unsaturated fatty acids synthesized from essential fatty acids such as arachidonic acid. Their general structure comes from prostanoic acid:

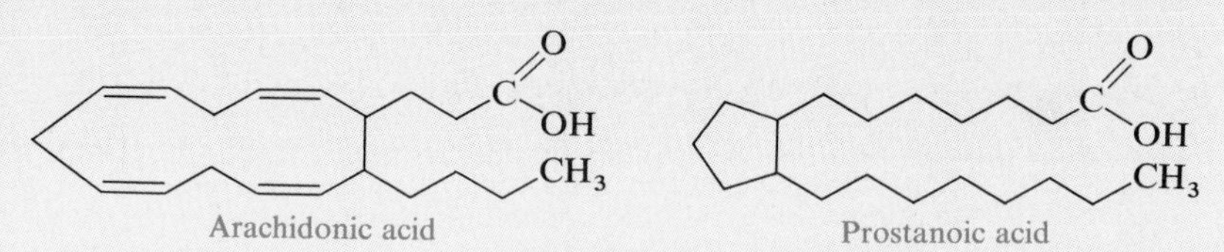

Prostaglandins acting as tissue hormones in low concentrations are released from the tissues into nearby circulating blood. Some increase blood pressure (*vasoconstrictors*); others lower blood pressure (*vasodilators*). Some prostaglandins stimulate contraction of smooth muscle and play a role in reproduction and in blood-clotting processes. Many of the effects of the prostaglandins are inhibited by aspirin, which acts to prevent their synthesis. The structures of prostaglandins E_1 and E_2 (PGE_1 and PGE_2) are

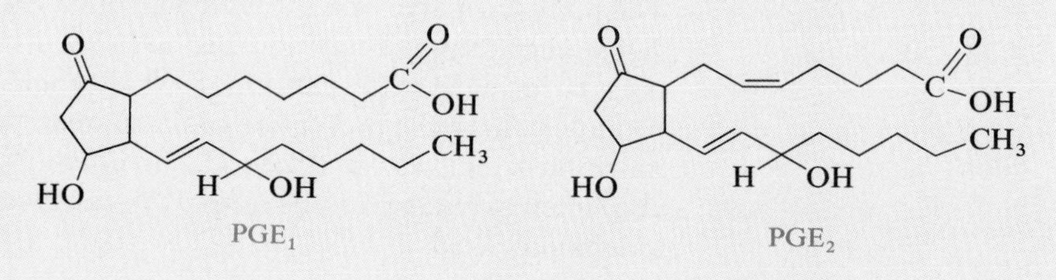

Example 16.1 Write the formula of oleic acid. Answer the following questions:

1. How many carbons are in oleic acid?
2. Is it a saturated or unsaturated fatty acid?
3. Is it most likely to be a solid or a liquid at room temperature?
4. Is it most likely from an animal or vegetable source?

Solution: The formula of oleic acid is as follows:

$$CH_3(CH_2)_7CH{=}CH(CH_2)_7\underbrace{\overset{\overset{\displaystyle O}{\|}}{C}OH}_{\text{Carboxylic acid group}}$$

1. Oleic acid contains 18 carbon atoms.
2. It is an unsaturated fatty acid.
3. Since it is an unsaturated fatty acid, it is most likely to be a liquid at room temperature.
4. It is most likely to come from vegetable sources.

16.2 Lipids Containing Fatty Acids

OBJECTIVE Write the structure of a wax or a triacylglycerol that is produced when a particular fatty acid reacts with a long-chain alcohol or with glycerol; name the triacylglycerol.

Waxes

The *waxes* are one kind of lipid in which an ester is formed between a long-chain alcohol and a saturated fatty acid. (These waxes are different from the "paraffin" waxes obtained from petroleum.)

$$\text{Alcohol}-O\overset{\overset{\displaystyle O}{\|}}{C}-\text{Fatty acid}$$

Wax

Wax coatings on skin, fur, feathers, fruits, and leaves prevent excessive loss of water. Lanolin, a wax from wool, is used in creams and lotions to aid retention of water, which helps soften the skin. Beeswax and carnauba wax are used in furniture, car, and floor polishes and waxes. The oil of the sperm whale contains a wax called *spermaceti*, which is used in candles and cosmetic products. A typical wax has a 15- to 30-carbon fatty acid and a 16- to 30-carbon alcohol molecule. Table 16.3 provides additional information on some waxes.

Triacylglycerols (Triglycerides)

Fats and oils are the most prevalent form of lipid storage material in the adipose tissue. They are commonly known as *triglycerides* and are esters of glycerol and three fatty acids. A simple triglyceride is composed of three identical fatty acids.

Table 16.3 Some Typical Waxes

Wax	Structure	Source	Uses
Beeswax	$CH_3(CH_2)_{29}O\overset{O}{\overset{\|\|}{C}}(CH_2)_{14}CH_3$	Honeycomb	Candles, polishes
Carnauba	$CH_3(CH_2)_{29}O\overset{O}{\overset{\|\|}{C}}(CH_2)_{24}CH_3$	Carnauba palm	Furniture, car, floor waxes
Spermaceti	$CH_3(CH_2)_{15}O\overset{O}{\overset{\|\|}{C}}(CH_2)_{14}CH_3$	Sperm whale oil	Candles, cosmetics

However, it is common to have triglycerides with two or three different fatty acids; these are called *mixed triglycerides*:

General Structure of Triacylglycerol (Triglyceride)

a triacylglycerol is an ester of glycerol and three fatty acids

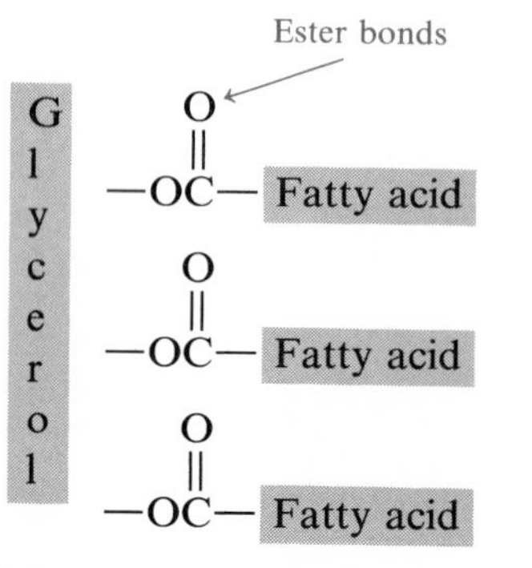

A triglyceride can be formed using stearic acid as the specific fatty acid, as shown in the following equation. It is commonly named *tristearin*; its IUPAC name is *glyceryl tristearate*:

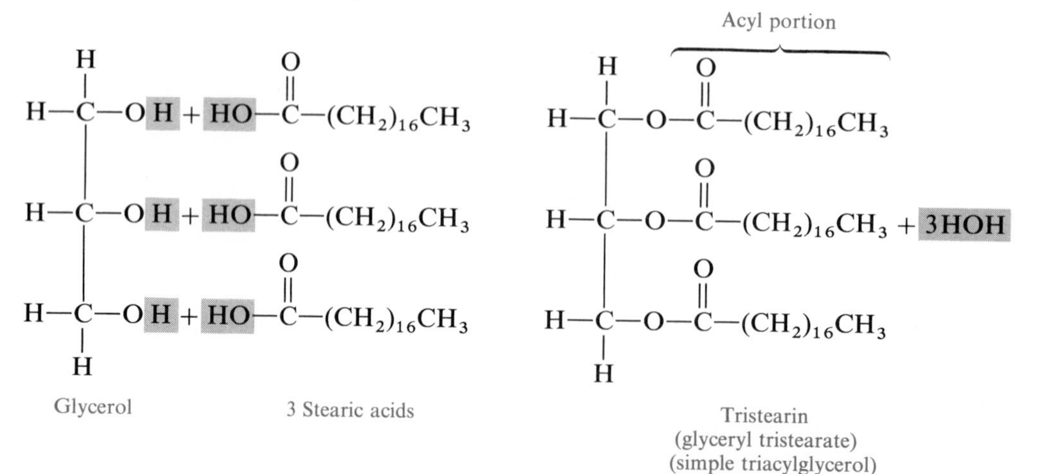

A mixed triglyceride might contain one molecule of palmitic acid and two molecules of myristic acid. Some possibilities for the arrangement of these different fatty acids are shown below:

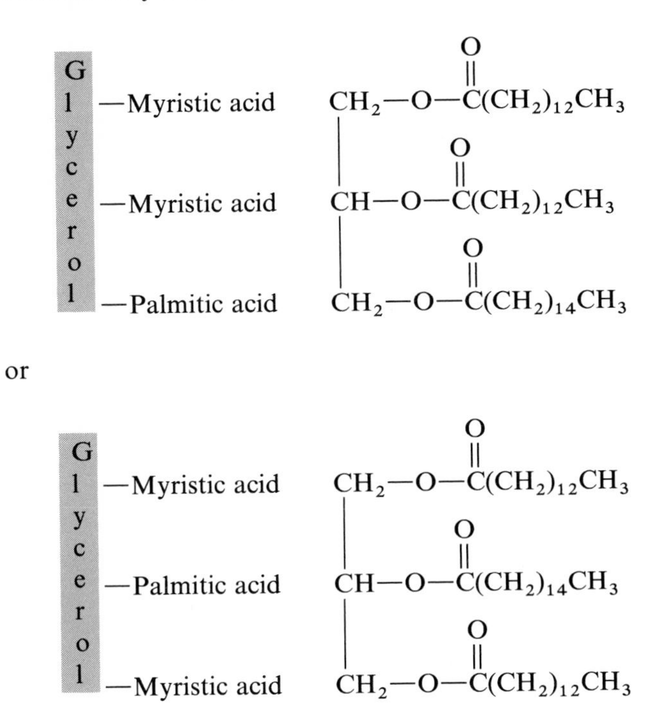

Example 16.2 Write the structure and name of the triacylglycerol that uses three oleic acid molecules.

Solution: A triacylglycerol is the ester of three hydroxyl groups of glycerol and three oleic acid molecules. Its name would be triolein or glyceryl trioleate:

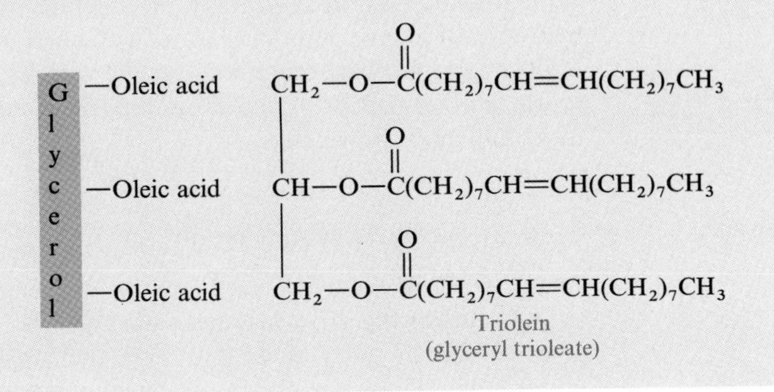

Triolein
(glyceryl trioleate)

16.3 Reactions of Triacylglycerols

OBJECTIVE Given the structure or name of a triacylglycerol, write the products of (1) hydrogenation, (2) acid or enzyme hydrolysis, (3) oxidation, and (4) saponification.

Hydrogenation

hydrogenation converts liquid oils to solid fats

The process of *hydrogenation*, which we discussed in Chapter 12, involves the addition of molecular hydrogen, H_2, to a double bond, an unsaturated site. Platinum (Pt) or nickel (Ni) is required to catalyze the reaction:

$$-CH{=}CH- + H_2 \xrightarrow{Pt} -CH_2-CH_2-$$

Hydrogenation of a triacylglycerol is the same process: The double bonds of an unsaturated fatty acid are converted to single, saturated bonds. Usually in the hydrogenation of vegetable oils, the introduction of hydrogen is stopped before all the double bonds are saturated:

Hydrogenation of an Unsaturated Oil

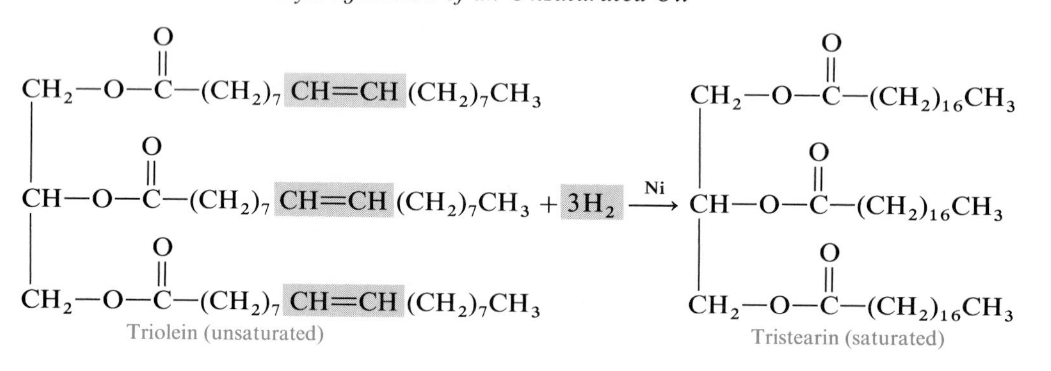

Triolein (unsaturated) Tristearin (saturated)

Complete hydrogenation gives a very brittle product. However, by partial hydrogenation, a liquid vegetable oil can be changed into a soft or semisolid fat. With this increase in saturation, the melting point, too, increases. Continued hydrogenation gives a more solid fat with an even higher melting point. Control of the degree of hydrogenation gives the various types of vegetable oil products on the market today—liquid shortenings; soft margarines; solid, cube margarines; and solid shortenings.

Acid or Enzyme Hydrolysis

lipases

Triacylglycerols are hydrolyzed (split by reacting with water) in the presence of strong acids or digestive enzymes called *lipases*. The products of hydrolysis are glycerol and the fatty acids that were tied up in the ester bonds. The polar glycerol will dissolve in the water, but the nonpolar fatty acids will not.

Hydrolysis of a Triacylglycerol with Strong Acid or Lipase Action

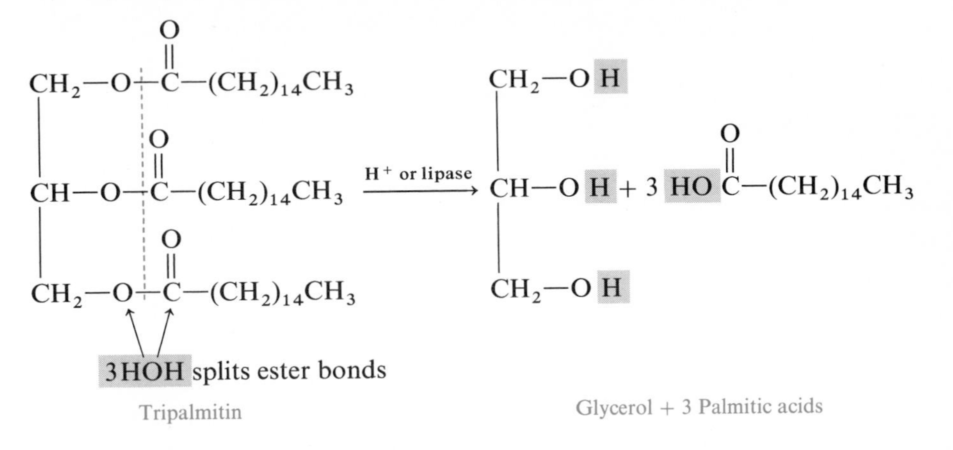

Saponification

Saponification is the hydrolysis of a triacylglycerol in the presence of a strong base. The reaction produces glycerol and the salts of the fatty acids, which are soaps. When NaOH is used in the saponification, the soap produced is a solid that can be molded into a desired shape; KOH produces softer or liquid soaps. The softness of the soap is also related to the degree of unsaturation. Oils having a higher degree of unsaturation tend to produce softer soaps. Names like "coconut" or "avocado shampoo" tell you the source of the oil used in making the soap.

Triacylglycerol + Strong base ⟶ Salts of fatty acids (soaps) + Glycerol

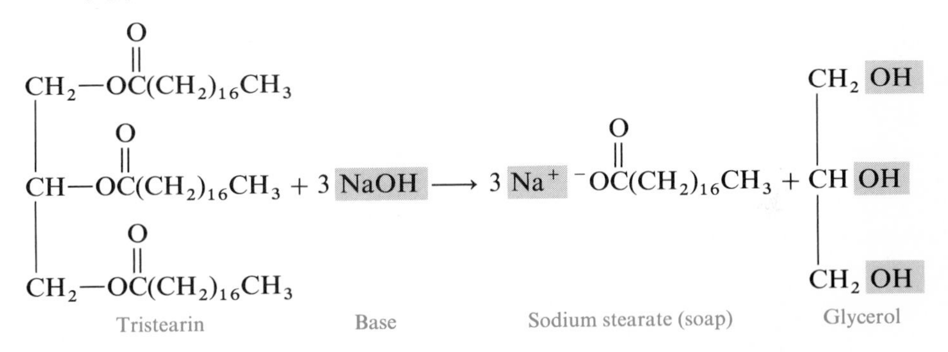

Cleaning Action of Soap

Since soaps are the salts of long-chain fatty acids, the soap molecules have two very different ends. The nonpolar carbon chain of the fatty acid is *hydrophobic* (water-fearing) and is not soluble in water. The ionic carboxylate end is *hydrophilic* (water-attracting) and does dissolve in water. In water, soap molecules form clusters called *micelles* with the hydrophobic carbon chains on the inside and the hydrophilic carboxylate groups on the outside.

micelles

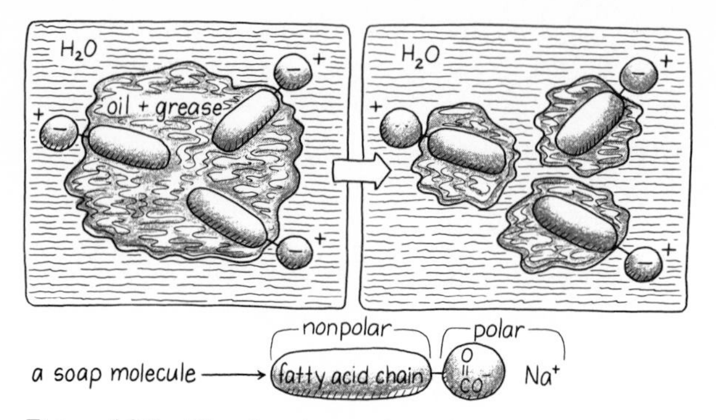

Figure 16.3 The cleaning action of soap. The nonpolar portion of the soap molecule dissolves in the grease and oil accompanying dirt on clothing; the polar portion of the molecule is attracted to the water molecules and pulls the grease and oil into the water.

During the cleaning action of soap, the nonpolar greases and oils that accompany dirt dissolve in the hydrocarbon portion of the micelle. The oil and grease, along with the dirt, are broken up into small globules that break free and become suspended in water. They are then rinsed away. (See Figure 16.3.)

Example 16.3 Write the equation for the reaction of lipase with tripalmitin (glyceryl tripalmitate).

Solution: The fat in this reaction is the ester of glycerol and three molecules of palmitic acid which has 16 carbon atoms. Lipase is a hydrolytic enzyme that splits the molecule into a molecule of glycerol and three molecules of palmitic acid:

$$\begin{array}{l} CH_2{-}O{-}\overset{\overset{\displaystyle O}{\|}}{C}(CH_2)_{14}CH_3 \\ | \\ CH{-}O{-}\overset{\overset{\displaystyle O}{\|}}{C}(CH_2)_{14}CH_3 \\ | \\ CH_2{-}O{-}\overset{\overset{\displaystyle O}{\|}}{C}(CH_2)_{14}CH_3 \end{array} + 3H_2O \xrightarrow{\text{Lipase}} \begin{array}{l} CH_2{-}OH \\ | \\ CH{-}OH \\ | \\ CH_2{-}OH \end{array} + 3HO{-}\overset{\overset{\displaystyle O}{\|}}{C}(CH_2)_{14}CH_3$$

Tripalmitin (glyceryl tripalmitate) — Glycerol — 3 Palmitic acid molecule

Oxidation

Oxidation takes place in the presence of oxygen and microorganisms when a fat or oil is exposed to the air. Unsaturated sites are particularly susceptible to oxidation. Vegetable oils are more susceptible because of their greater percentage of unsaturated sites. The products of oxidation of triacylglycerols are short-chain fatty acids and aldehydes with unpleasant odors:

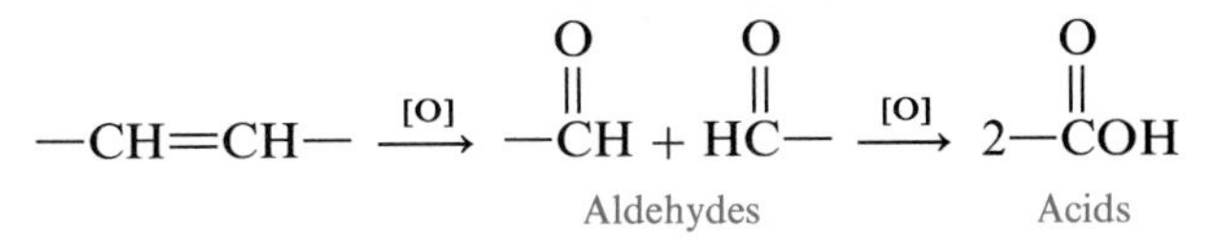

If vegetable oils have no antioxidant preservatives, such as vitamin E, BHA, or BHT, they should be tightly covered and refrigerated to prevent oxidation.

Oxidation also occurs in the oils that accumulate on the surface of the skin. The high temperature of the body stimulates rapid oxidation of these oils in the presence of oxygen and water. The resulting compounds account for the odor associated with perspiration.

16.4 Tests for Triacylglycerols

OBJECTIVE Describe the iodine number for unsaturation in simple lipids; describe the acrolein test for the presence of glycerol.

iodine number The degree of unsaturation in a fat or oil can be determined by the *iodine number*. The iodine number is the number of grams of iodine (I_2) that will react with 100 g of the fat or oil. Actually a combination of halogens, such as ICl, which is more reactive than I_2, is used in the reaction, and the iodine number is calculated as though I_2 were used:

$$\underset{\text{Unsaturated fat}}{—CH{=}CH—} + ICl \longrightarrow \underset{\text{Halogenated fat}}{—\underset{|}{\overset{I}{CH}}—\overset{Cl}{CH}—}$$

Most animal fats have lower iodine numbers because their fatty acids contain fewer double bonds. Vegetable oils generally have higher iodine numbers because their fatty acids are usually more unsaturated. Here are some typical iodine number values:

		Fatty Acid Percent Composition					
		Saturated				Unsaturated	
Fat or Oil	Iodine Number	Lauric Acid	Myristic Acid	Palmitic Acid	Stearic Acid	Oleic Acid	Linoleic Acid
Animal							
Butterfat	30	5	14	30	10	35	6
Lard	60		1	30	12	50	7
Vegetable							
Olive oil	90			8	2	85	5
Cottonseed oil	100		1	23	2	24	50
Corn oil	120		1	10	4	50	35
Soybean oil	140			12	3	30	55

Example 16.4 From the following iodine numbers, state which is the most saturated and which is the most unsaturated fat or oil:

Fat or Oil	*Iodine Number*
Coconut oil	10
Safflower oil	150
Chicken fat	75

Solution: The most saturated fat or oil is the one with the least amount of unsaturation and therefore the lowest iodine number: coconut oil. The most unsaturated fat or oil is the one with the highest degree of unsaturation and therefore the highest iodine number: safflower oil.

Acrolein Test

The acrolein test is a qualitative test for the presence of glycerol. It is used as a general test for fats and oils because glycerol is present in both. Glycerol dehydrates at high temperatures in the presence of a dehydrating agent such as $KHSO_4$ (potassium bisulfate), giving a product called *acrolein*:

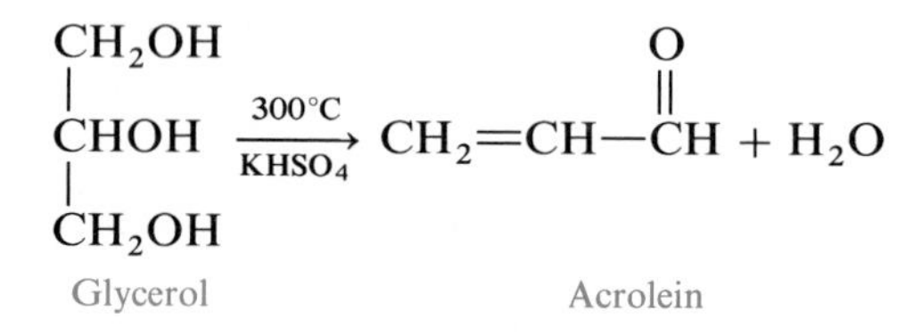

The detection of acrolein's strong, irritating odor is considered positive identification of the presence of glycerol and, therefore, of fats and oils.

16.5 Phospholipids

OBJECTIVE Describe the components of phosphoglycerides, sphingolipids, and glycolipids and state their biological functions.

A phospholipid is a lipid containing a phosphate group. In this phospholipid family, there are the phosphoglycerides, (which contain glycerol) and the sphingolipids (which use another alcohol called *sphingosine*).

Phosphoglycerides

Phosphoglycerides are important constituents of cellular membranes. They play a role in enzyme systems and help transport lipids in the body. The phosphoglycerides contain glycerol that has formed ester bonds with two fatty acids (just like triglycerides) and a phosphate group attached to an amino alcohol. The ionization of the phosphate portion and the amino group provides a polar section in the phosphoglyceride, whereas the fatty acid portions are nonpolar:

Phosphoglyceride

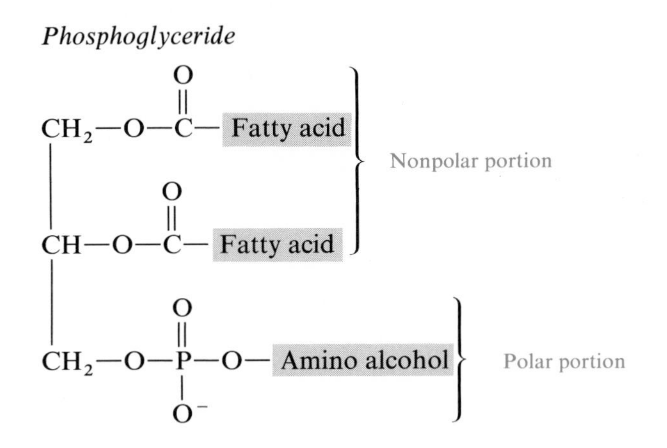

The polarity of the phosphoglycerides gives them more solubility in water than most lipids. The amino alcohol determines the particular type of phosphoglyceride:

Amino Alcohols Present in Phosphoglycerides

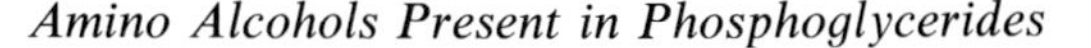

$$HO{-}CH_2CH_2N^+(CH_3)_3$$

Choline

$$HO{-}CH_2CH(NH_3^+){-}COO^-$$

Serine

$$HO{-}CH_2CH_2{-}NH_3^+$$

Ethanolamine

lecithin *Lecithin* is a phosphoglyceride that uses choline as its amino alcohol. It appears in all plant and animal cells and is particularly abundant in brain and nerve tissues. It occurs in high concentrations in egg yolks. The chemical name for lecithin is phosphatidyl choline:

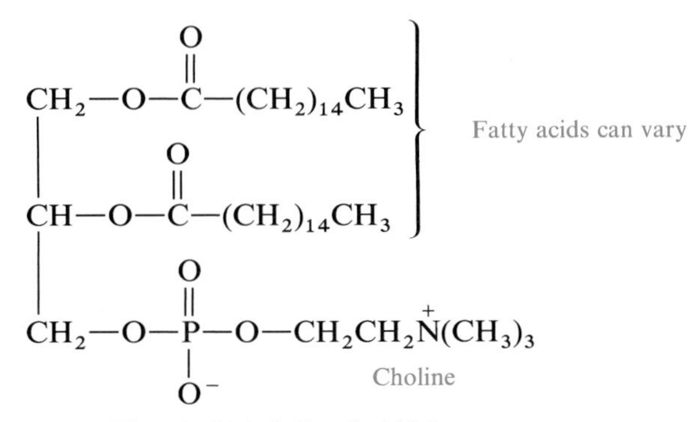

Phosphatidyl choline (lecithin)

cephalin The cephalins are another major group of phosphoglycerides. In the cephalins, the amino group is ethanolamine or serine. The cephalins are found in greatest concentrations in brain tissues:

Cephalins

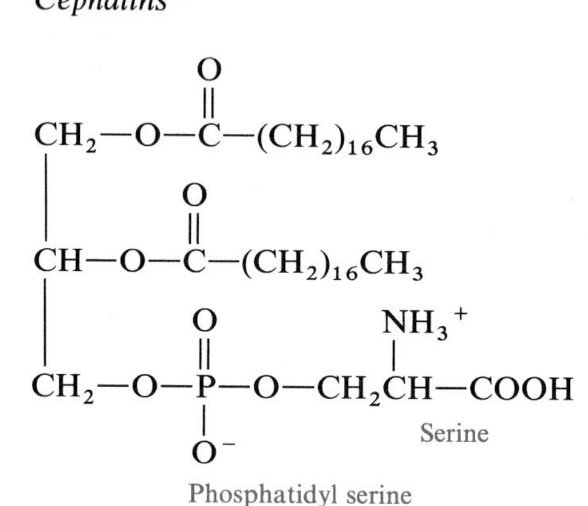

Phosphatidyl serine

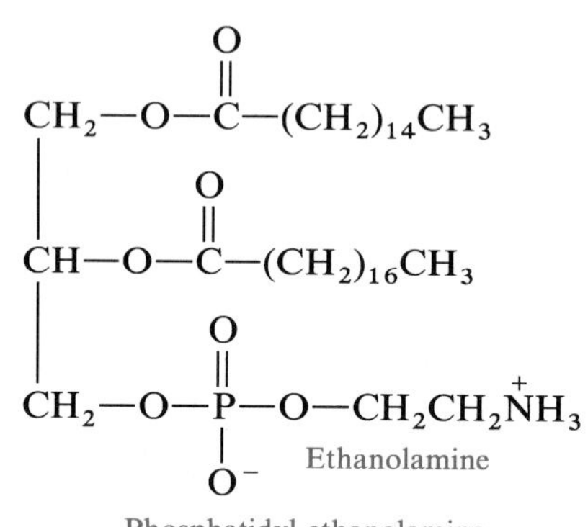

Phosphatidyl ethanolamine

Sphingolipids

The sphingolipids are a group of phospholipids that contain the alcohol *sphingosine* rather than glycerol. The spingolipids are found in brain and nerve tissues and in most cellular membranes. In a sphingolipid, a fatty acid is bonded to the amino group of sphingosine by an amide linkage:

$$
\begin{array}{l}
CH_3(CH_2)_{12}-CH=CH-CH-OH \\
\qquad\qquad\qquad\qquad\qquad\quad | \\
\qquad\qquad\qquad\qquad\qquad\quad CH-NH_2 \\
\qquad\qquad\qquad\qquad\qquad\quad | \\
\qquad\qquad\qquad\qquad\qquad\quad CH_2-OH
\end{array}
$$

Sphingosine

One of the typical sphingolipids is *sphingomyelin*:

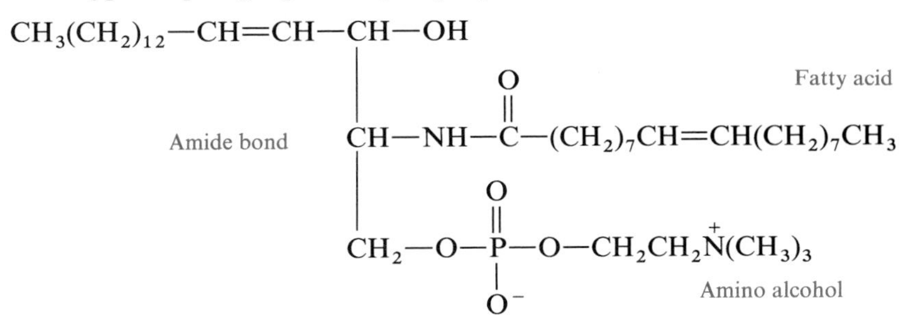

Sphingomyelin (a sphingolipid)

Glycolipids

Glycolipids, also called *cerebrosides*, are another kind of sphingolipid that is abundant in the brain and in the myelin sheaths of nerves. They are similar to the sphingolipids except that a monosaccharide replaces the phosphate and amino alcohol. The carbohydrate portion is usually D-galactose and sometimes D-glucose:

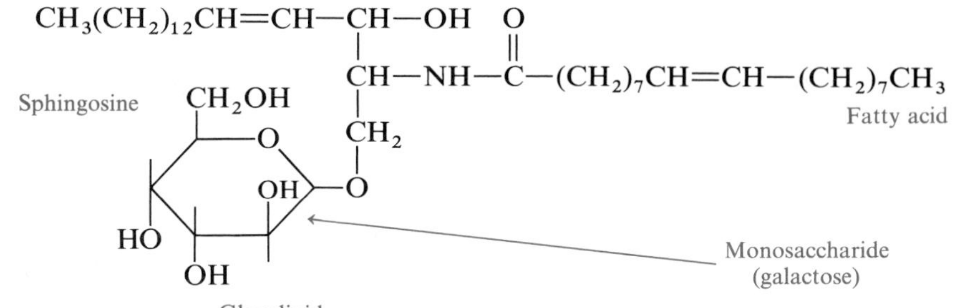

Glycolipid

HEALTH NOTE

Lipid Diseases (Lipidoses)

Genetic diseases that involve the excessive accumulation and storage of lipids are called *lipidoses*. Some lipidoses result from a deficiency or absence of one of the enzymes necessary for the complete degradation of a particular compound lipid. In Gaucher's disease, the enzyme responsible for the splitting of glucose from a glycolipid is deficient or absent. The glycolipid glucosylceramide accumulates in the spleen, liver, and kidneys, causing an enlargement of those organs. The same lipid accumulates in the bone marrow, causing an enlargement of the bone-marrow cells—an effect that is used as a diagnostic tool.

In Nieman-Pick disease, an enzyme is missing that is needed for the breakdown of sphingomyelins. The accumulation of sphingomyelins in the brain leads to mental retardation or death.

In Tay-Sachs disease, a glycolipid that contains galactose accumulates because the enzyme β-galactosidase is deficient. The accumulation of this lipid material in the brain causes death within the first few years of life.

HEALTH NOTE

Importance of Phospholipids in Cell Membranes

The phospholipids play an important role in the formation of cellular membranes that separate one fluid compartment from another. Because the phospholipids have both a polar and nonpolar portion, they can be both attracted to and repelled by water. A current model suggests that a cell membrane consists of a double row of phospholipid molecules called the *bilipid layer*. In this bilipid layer, the phospholipid molecules are lined up so that the nonpolar, hydrophobic (water-fearing) tails meet at the center of the layer away from the water. The polar, hydrophilic (water-attracting) heads extend out into the water. This bilipid layer allows water and nonpolar molecules to pass through it, but it is impermeable to ions and polar molecules. Proteins that are embedded in the bilipid layer form tunnels that allow ions and polar molecules to move across a membrane. (See Figure 16.4.)

bilipid layer

Cells in the body differ greatly in the amount of lipid contained in their membranes. Some cells have large quantities of phospholipids (80 percent). As a result, few polar molecules move through the membranes. Other cells have low quantities of lipids (25 percent), permitting the passage of many more polar molecules and ions in and out of the cell.

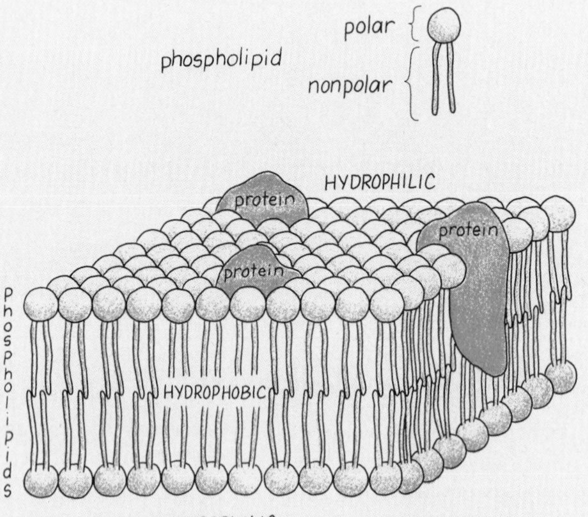

Figure 16.4 Bilipid model of a cell membrane. A double row of phospholipids forms a barrier with polar ends in contact with fluids and nonpolar ends at the center away from the water.

Example 16.5 Draw the structure of a cephalin that contains ethanolamine as its amino alcohol and stearic acid for the fatty acids. Describe each of the components of the cephalin.

Solution: Cephalin is a phospholipid. In general, phospholipids are composed of a glycerol molecule in which the first two carbon atoms are attached to fatty acids. The third carbon atom is attached to a phosphate group that is bonded to an amino alcohol. When the amino alcohol is ethanolamine or serine, the phospholipid is a cephalin, a lipid important in brain tissues:

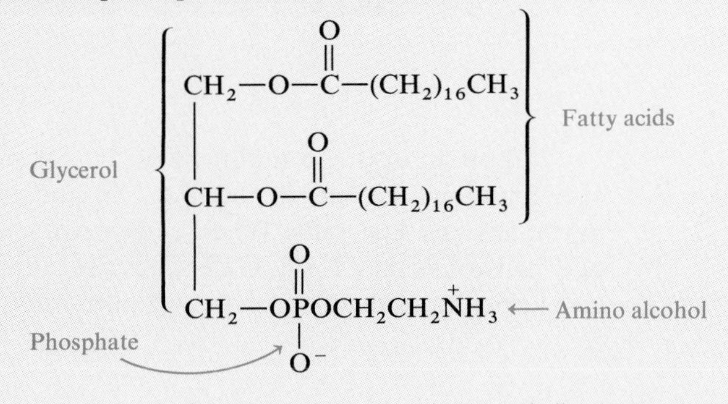

Phosphatidyl ethanolamine (a cephalin)

16.6 Steroids and Terpenes

OBJECTIVE Given the structure or name of a simple lipid, determine if it is related to the steroid or terpene family of lipids.

Steroids and terpenes are two important families of simple lipids, the lipids without ester bonds to fatty acids. They are included in the lipid family because they are biological compounds that are not soluble in water.

Steroids

Steroids contain a fused-ring structure called the *steroid nucleus*:

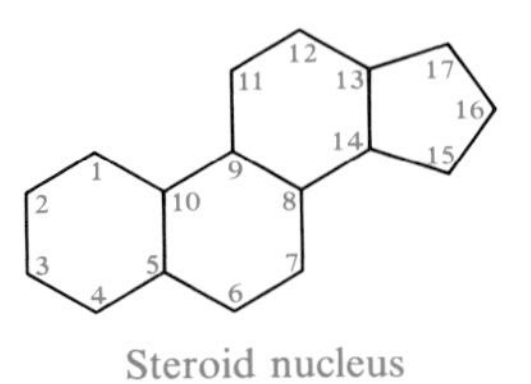

Steroid nucleus

Certain substitutents can be added to the steroid nucleus to form the various steroid compounds. One of the most important steroid compounds in the body is cholesterol. There is an alcohol group attached to carbon atom 3. Cholesterol is also called a *sterol* because it contains an alcohol group. There are also methyl

sterol

groups and a branched chain of carbon atoms. Small changes in these side groups can lead to great changes in the behavior of the compound in the body.

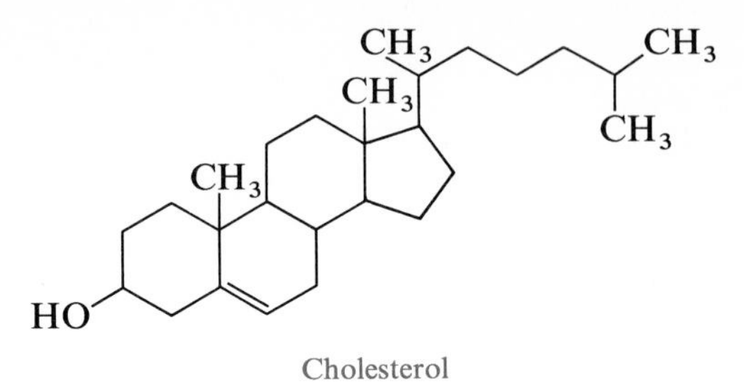

Cholesterol

Cholesterol is a component of cellular membranes and nerve tissue. It is also found in the blood and the bile and is necessary for the synthesis of hormones and bile salts. When cholesterol exceeds its saturation levels in the bile, gallstones may form. Gallstones are composed largely of cholesterol with some calcium salts, fatty acids, and phospholipids.

Example 16.6 Consider the structure of cholesterol shown above:

1. Indicate the part of the molecule that is the steroid nucleus.
2. What additional features have been added to the steroid nucleus to form the cholesterol molecule?
3. Indicate the part of the molecule that classifies cholesterol as a sterol.
4. Why is cholesterol classified as a lipid?

Solution

1. The four-ring system represents the steroid nucleus.
2. The structure for cholesterol contains several attached side groups, an alcohol group (—OH), two methyl groups (—CH_3), and a branched carbon side chain along with an unsaturated site in one of the cyclic rings.
3. The alcohol group is responsible for the sterol classification.
4. Cholesterol is not soluble in water; it is classified with the lipid family.

Steroid Hormones

The sex hormones and the adrenocortical hormones are closely related in structure to cholesterol and depend on cholesterol for their synthesis. The *estrogens* such as estradiol are produced in the ovaries to regulate the menstrual cycle and effect changes during pregnancy. They are also responsible for female secondary sex characteristics including breast and uterus enlargement. The male hormones are called *androgens*. The most important male hormone is testosterone, which is responsible for male secondary characteristics such as muscle

estrogens

androgens

development and facial hair, as well as the development of the reproductive organs and the formation of sperm.

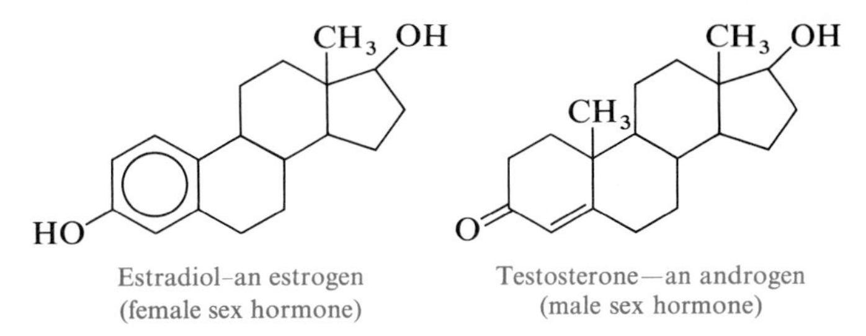

Estradiol-an estrogen (female sex hormone)

Testosterone—an androgen (male sex hormone)

Corticosteroids

aldosterone

glucocorticoids

cortisol

Adrenocortical hormones called *corticosteroids* are produced by the cortex of the adrenal gland. The *mineralcorticoids* are responsible for electrolyte and water balance in the body. The major mineralcorticoid in humans is *aldosterone*, which regulates the absorption and excretion of sodium by the kidneys. The *glucocorticoids* regulate carbohydrate and protein metabolism in the body. A major glucocorticoid is *cortisol*, which decreases the amount of protein synthesized by muscle, thereby increasing the concentration of compounds that can be converted to glucose:

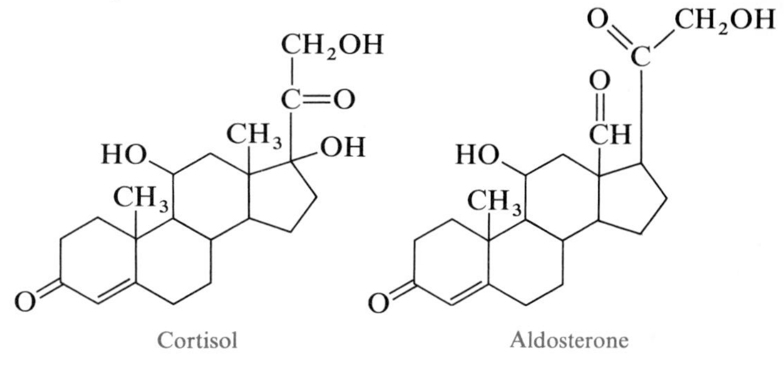

Cortisol

Aldosterone

Bile Acids

Bile acids such as cholic acid are also synthesized from cholesterol: In the small intestine, bile acids aid in the emulsification of fats to form smaller globules that are more accessible to enzymes.

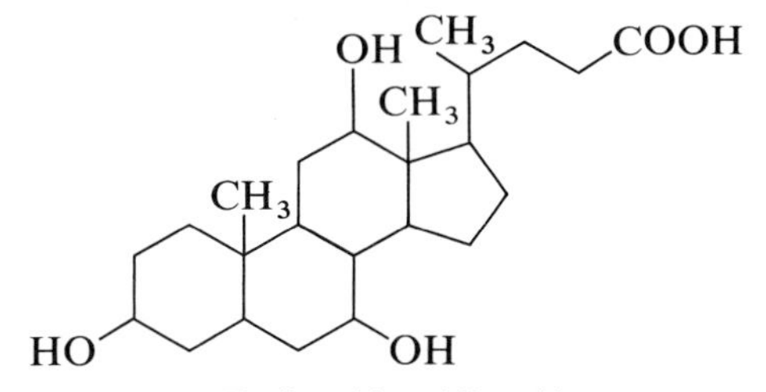

Cholic acid—a bile acid

HEALTH NOTE

Serum Cholesterol

About 40 percent of the cholesterol in the body is obtained from the diet from animal sources. Plants do not generally produce cholesterol. The remainder is synthesized by the body from fats, carbohydrates, and proteins.

According to some theories, high levels of serum cholesterol are associated with the accumulation of lipid deposits that line the coronary arteries, thus making them narrower and less efficient. Clinically, cholesterol levels are considered elevated if the total plasma cholesterol level exceeds 200 mg/100 mL in an individual below the age of 20 and 140 mg/100 mL in an individual above the age of 20.

Suggestions for reducing a high serum cholesterol level include decreasing the intake of foods containing cholesterol and changing the type of fat ingested. However, some research indicates that dietary intake of cholesterol does not greatly affect the blood levels of cholesterol. A diet that contains food low in saturated fats appears to be more helpful in reducing the serum cholesterol level.

Terpenes

isoprene units

Terpenes are lipid compounds found in plants and flowers. They are responsible for the characteristic odors and color of plants. For example, the smell of a geranium and of mint, lemon, or pine oil, as well as the color of carrots, are due to terpenes. Terpenes are made from *isoprene* units that consist of a five-carbon group of atoms that is unsaturated:

$$CH_2{=}\overset{\displaystyle CH_3}{\overset{|}{C}}{-}CH{=}CH_2$$

Isoprene

To form a terpene, these isoprene units are joined together, head-to-tail. Dashed lines are used to set apart the individual isoprene units:

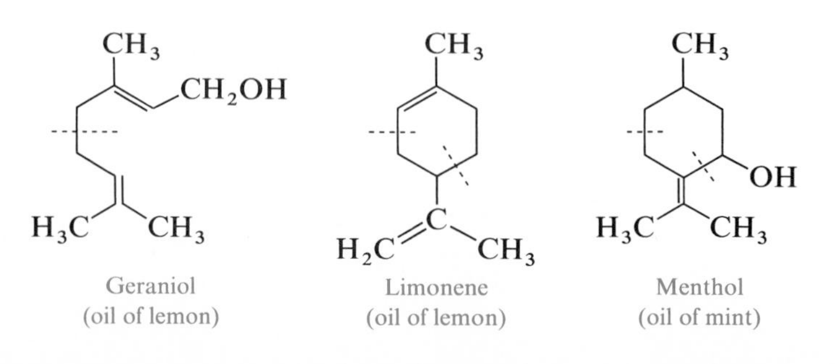

Geraniol (oil of lemon) | Limonene (oil of lemon) | Menthol (oil of mint)

Fat-Soluble Vitamins: A, D, E, and K

vitamin D

Vitamin D is a fat-soluble vitamin and a steroid derivative. Vitamin D is necessary for the absorption of calcium from the small intestine and for the utilization of calcium and phosphate in bone formation. A deficiency of vitamin D in children can result in a change in the bone and teeth structure—a condition called *rickets*. Vitamin D is prevalent in fish liver oils, herring, tuna, egg yolks, and liver:

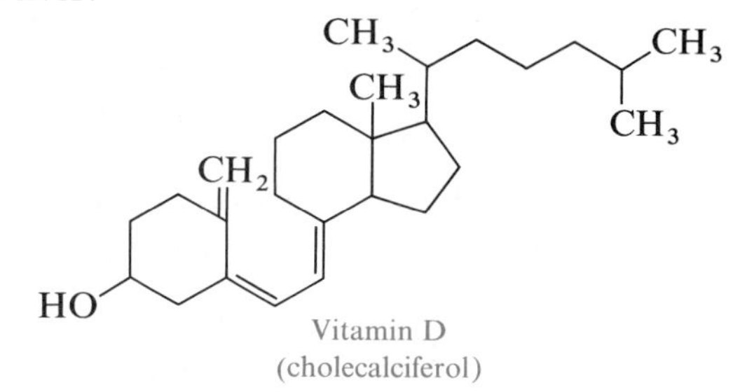

Vitamin D
(cholecalciferol)

vitamin A

Vitamin A is most closely related to the terpenes. It is required for proper growth and reproductive function and the maintenance of the tissues of the retina. An early sign of vitamin A deficiency is night blindness, in which the eyes cannot distinguish objects in a dim light. Sources of vitamin A include milk, cheese, eggs, liver, green vegetables, and tomatoes. Because it is a fat-soluble vitamin, vitamin A can be stored in the liver. The structure of vitamin A is

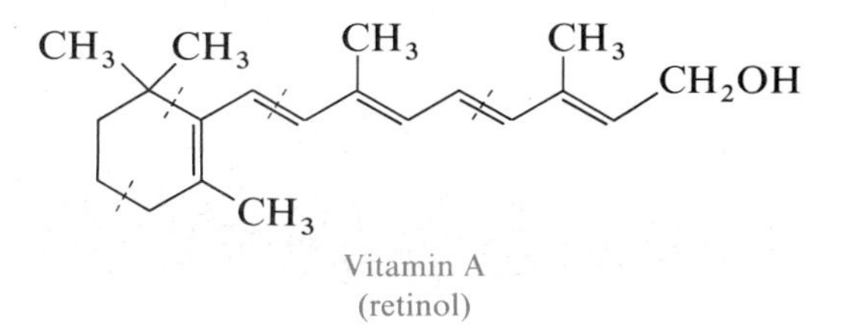

Vitamin A
(retinol)

tocopherols
vitamin E

Vitamins E and K are compounds that belong to a subgroup of the terpenes called *tocopherols*. These compounds possess an aromatic ring structure with a side chain of isoprenoid units. Vitamin E is abundant in wheat germ oil, corn oil, and cottonseed oil as well as in egg yolks, meat, and green vegetables.

The function of vitamin E in the body is not yet known, although it appears to prevent the oxidation of unsaturated fatty acids in the tissue lipids. Deficiency symptoms have not yet been observed in humans. The structure of vitamin E is

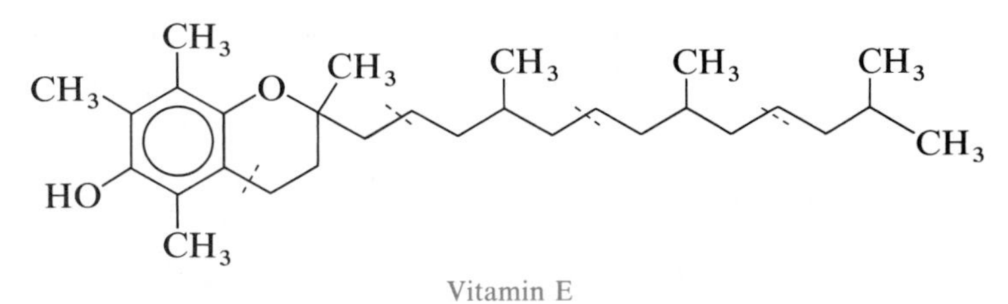

Vitamin E

HEALTH NOTE

Lipoproteins

Since trigylcerides and cholesterol are usually nonpolar, they are not transported easily in the blood. To improve the solubility of nonpolar lipids in the blood, they attach to proteins and phospholipids to form a globular complex called a *lipoprotein*, which is polar. (See Figure 16.5.)

High- and Low-Density Lipoproteins

VLDL

There are different kinds of lipoproteins that vary in density, size, the type of lipid carried, and function. The VLDL (very low density lipoprotein) are 5–10 times bigger than the other lipoproteins and have densities less than 1.006 g/mL. Their major

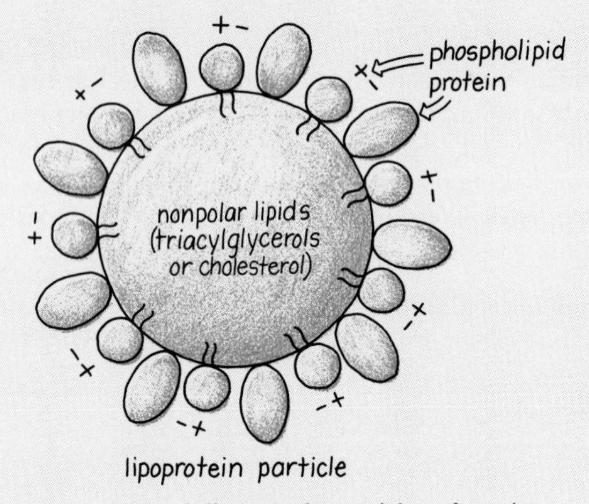

Figure 16.5 A lipoprotein particle, a form in which nonpolar lipids are carried through the bloodstream.

function is to carry triglycerides (triacylglycerols) from the liver into the bloodstream to the capillaries where the triglycerides are hydrolyzed. The resulting fatty acids are used for the synthesis of fats (triglycerides) within the adipose tissues. The LDL (low density lipoproteins) are smaller, with densities between 1.02 and 1.06 g/mL. They are composed primarily of cholesterol, which they carry to cells for use in the synthesis of cell membranes and steroid hormones. About one-fifth of the serum cholesterol is carried in the form of HDL (high density lipoproteins). These are the smallest of the lipoproteins but are the most dense, with densities between 1.06 and 1.20 g/mL. The HDL particles function in the transport of cholesterol to the liver, where it is mixed with the bile to be excreted.

LDL

HDL

Since cholesterol has been associated with the onset of atherosclerosis and heart disease, the measurement of the LDL and HDL blood serum levels is of much interest to medical researchers. Free cholesterol can deposit in the blood vessels, causing restriction of blood flow. When the LDL levels are increased, there is an increase in the triglycerides and serum cholesterol; this indicates an increased risk of myocardial infarction. However, high levels of HDL have been associated with decreased risk of myocardial infarction because more cholesterol is being removed from the cells and excreted. This leaves less cholesterol available for accumulation on the walls of blood vessels. HDL levels appear to increase in persons who exercise regularly, and they are higher in women whose estrogen levels are adequate. Some properties of plasma lipoproteins are listed in the following table:

Some Properties of Plasma Lipoproteins

Lipoprotein	Diameter (nm)	% Protein	Density (g/mL)	Major Lipids
VLDL	30–80	5–10	<1.006	Dietary triglycerides
LDL	15–30	20–25	1.02–1.06	Cholesterol (to cells for use in synthesis; accumulation)
HDL	5–15	35–50	1.060–1.21	Cholesterol (to liver for removal)

vitamin K Vitamin K plays an important role in blood coagulation, where it is required in the synthesis of prothrombin. If there is a deficiency of vitamin K, the blood takes longer to clot. People with vitamin K deficiency bruise easily and tend to hemorrhage under the skin. Sources of vitamin K include spinach, cabbage, cauliflower, and tomatoes. The structure of vitamin K is

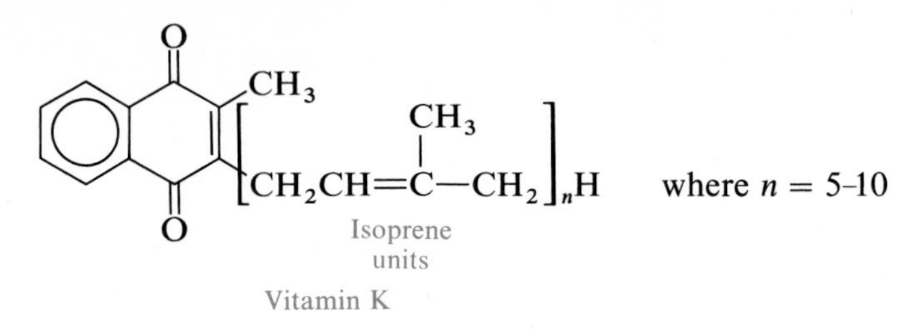

Vitamin K

Example 16.7 Identify the following lipid structures as belonging to the steroid or terpene family:

1.

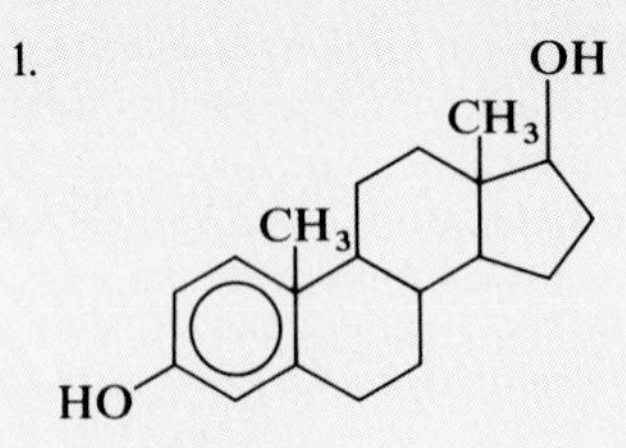

2. aldosterone

3. $CH_2{=}C(CH_3){-}CH_2{-}CH_2{-}CH_2{-}CH(CH_3){-}CH_2{-}CH_2OH$

Solution

1. This structure contains a steroid nucleus, which makes it a part of the steroid family.
2. Aldosterone is one of the sex hormones whose structure is derived from the steroid nucleus. It is a part of the steroid family.
3. This structure consists of two isoprene units, which makes it a part of the terpene family.

Glossary

acrolein test A test in which glycerol is heated to produce acrolein, whose strong, irritating odor is a positive test.

cephalin A phospholipid, found in brain and nerve tissues, in which the amino alcohol is serine or ethanolamine.

cholesterol The most prevalent of the steroid compounds known as sterols; it is found in cellular membranes and is necessary for the synthesis of vitamin D, hormones, and bile acids.

fat Another term for solid triacylglycerols.

fatty acid A long-chain carboxylic acid usually having an even number of carbon atoms. Found in fats.

glycolipid A phospholipid that contains a monosaccharide in place of the phosphate and amino alcohol.

hydrogenation The addition of hydrogen to unsaturated fats.

hydrolysis The splitting of bonds by water, usually in the presence of acid or enzymes.

lecithin A phospholipid containing choline as the amino alcohol.

lipase An enzyme that hydrolyzes fats in the digestive tract.

lipid A family of compounds that vary in structure but are nonpolar in nature and soluble in nonpolar solvents. Includes fats, waxes, phospholipids, steroids, and terpenes.

lipidosis A genetic disease that involves a deficiency in an enzyme necessary for the hydrolysis of a compound lipid.

lipoproteins Nonpolar lipids combine with proteins to form a more polar component for the transport of lipids through body fluids.

mixed triacylglycerol A fat or oil that contains two or more different fatty acids.

oil Another term for liquid triacylglycerol.

phosphoglyceride A polar lipid in which two hydroxyl groups of glycerol are bonded as esters to fatty acids and the third hydroxyl group is bonded to a phosphate group attached to an amino group (choline, serine, or ethanolamine).

phospholipid A lipid that contains phosphate.

prostaglandins Compounds that may be derived from the essential fatty acid arachidonic acid and appear to regulate several physiological processes, including muscle contraction and blood clotting.

saponification The reaction of a fat with a strong base to form the salt of the acid, which is used as soap.

saturated Saturated lipids are those composed of saturated fatty acids (having no double bonds); they have higher melting points than do unsaturated lipids and are solid at room temperatures.

sphingolipid A phospholipid in which glycerol is replaced by another alcohol, sphingosine.

steroid A type of lipid composed of a multicyclic ring system.

terpenes A type of lipid derived from a five-carbon unsaturated unit called an *isoprene unit.*

triacylglycerols A family of lipids that are composed of three fatty acids bonded through ester bonds to an alcohol called *glycerol.*

unsaturated A lipid containing at least one fatty acid that is unsaturated (having double bonds).

wax The ester of a long-chain monohydroxylic alcohol and a long-chain saturated fatty acid.

Problems

Fatty Acids (Objective 16.1)

16.1 Describe some similarities and differences with regard to the structure of a saturated fatty acid and an unsaturated fatty acid.

16.2 Why do unsaturated fatty acids have lower melting points than saturated fatty acids?

16.3 Write the structure of the following saturated fatty acids:
a. myristic acid (14 carbon atoms)
b. palmitic acid (16 carbon atoms)
c. stearic acid (18 carbon atoms)

16.4 For the following fatty acids, state:
a. whether they are saturated or unsaturated
b. their source (animal or vegetable)

c. their state at room temperature (liquid or solid)

$$CH_3(CH_2)_4CH{=}CHCH_2CH{=}CH(CH_2)_7\overset{\overset{\displaystyle O}{\|}}{C}OH$$

Linoleic acid

$$CH_3(CH_2)_{14}\overset{\overset{\displaystyle O}{\|}}{C}OH$$

Palmitic acid

Lipids Containing Fatty Acids (Objective 16.2)

16.5 Beeswax is formed from myricyl alcohol, $CH_3(CH_2)_{29}OH$, and palmitic acid. Write the structure of beeswax.

16.6 a. Draw the structure of glycerol.
b. A simple triacylglycerol forms stearic acid and glycerol upon hydrolysis. Write the structure of this triacylglycerol.

16.7 What fatty acids are present in the following triacylglycerol?

$$\begin{array}{l} CH_2{-}O{-}\overset{\overset{\displaystyle O}{\|}}{C}(CH_2)_{14}CH_3 \\ | \\ CH{-}O{-}\overset{\overset{\displaystyle O}{\|}}{C}(CH_2)_7CH{=}CHCH_2CH{=}CH(CH_2)_4CH_3 \\ | \\ CH_2{-}O{-}\overset{\overset{\displaystyle O}{\|}}{C}(CH_2)_7CH{=}CH(CH_2)_7CH_3 \end{array}$$

16.8 A mixed triacylglycerol contains two palmitic acid molecules to every one oleic acid molecule. Write a possible structure for the compound. What other structures are possible?

16.9 Why does triolein have a lower melting point than tristearin?

Reactions of Triacylglycerols (Objective 16.3)

16.10 Write the equation for the acid hydrolysis of
a. trimyristin b. triolein

16.11 Write the equation for the NaOH saponification of
a. tripalmitin b. triolein

16.12 Write the equation for the hydrogenation of the following fat:

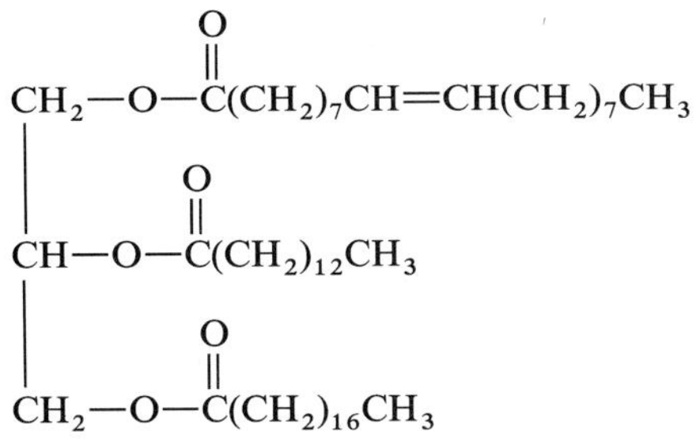

$$\begin{array}{l} \quad\quad\quad\quad\ \ \text{O} \\ \quad\quad\quad\quad\ \ \| \\ \text{CH}_2\text{—O—C(CH}_2)_7\text{CH=CH(CH}_2)_7\text{CH}_3 \\ | \quad\quad\quad\quad \text{O} \\ | \quad\quad\quad\quad \| \\ \text{CH—O—C(CH}_2)_{12}\text{CH}_3 \\ | \quad\quad\quad\quad \text{O} \\ | \quad\quad\quad\quad \| \\ \text{CH}_2\text{—O—C(CH}_2)_{16}\text{CH}_3 \end{array}$$

Tests for Triacylglycerols (Objective 16.4)

16.13 A fat has a low iodine number. What does that mean? When does a fat have a high iodine number?

16.14 The iodine number is given below for several fats and oils. Arrange them in order of increasing saturation (most unsaturated first).

Fat or Oil	*Iodine Number*
Soybean oil	140
Coconut oil	10
Peanut oil	90
Butter	40

16.15 Determine which fat or oil in each pair will have a higher iodine number:
a. triolein or tristearin
b. a triacylglycerol with two molecules of oleic acid and one molecule of linoleic acid or a fat formed from two molecules of linoleic acid and one molecule of oleic acid

16.16 Which of the following will give a positive acrolein test:
a. oleic acid
b. tristearin
c. glycerol
d. beeswax

Phospholipids (Objective 16.5)

16.17 How are phosphoglycerides different from triacylglycerols?

16.18 What is the polar section of a phosphotriglyceride? What is the nonpolar section?

16.19 A phosphoglyceride contains two molecules of palmitic acid and the amino alcohol ethanolamine. Write the structure of the compound.

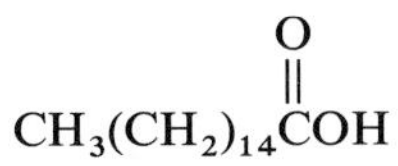

$$\begin{array}{c} \quad\quad\quad\quad\quad \text{O} \\ \quad\quad\quad\quad\quad \| \\ \text{CH}_3(\text{CH}_2)_{14}\text{COH} \end{array}$$

Palmitic acid

$$\text{HOCH}_2\text{CH}_2\overset{+}{\text{N}}\text{H}_3$$

Ethanolamine

16.20 How does lecithin differ from a cephalin?

16.21 How do sphingolipids differ from phosphoglycerides?

16.22 How does the polarity of the phospholipids contribute to their function in cell membranes?

Steroids and Terpenes (Objective 16.6)

16.23 a. What is the characteristic structural feature of a steroid?
b. What is the characteristic structural feature of a terpene?

16.24 Identify the following lipids as steroids or terpenes:

a.

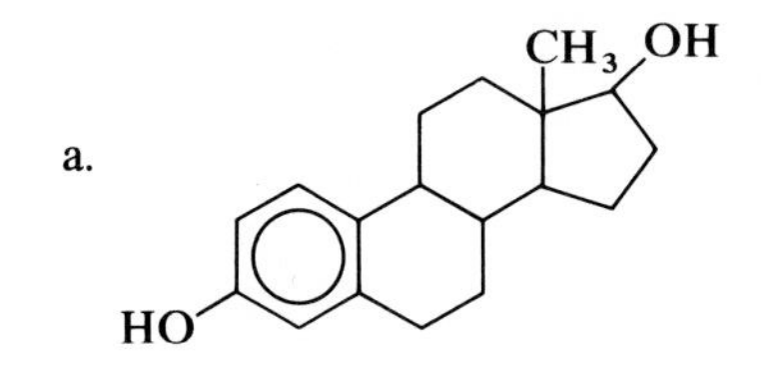

b. vitamin D
c. cholic acid

d.

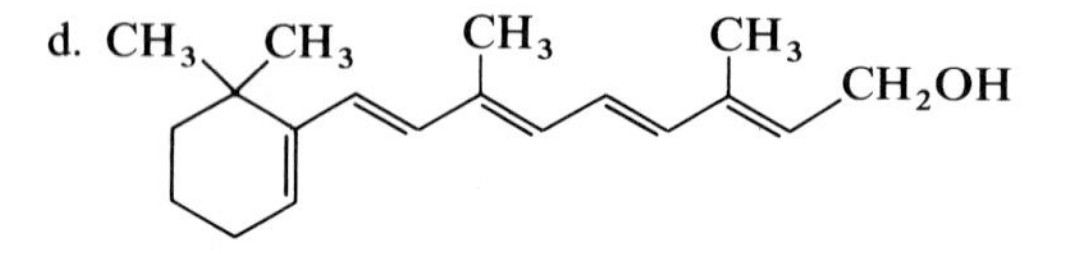

e. vitamin K
f. cortisol

16.25 The following synthetic estrogen is used in some oral contraceptive pills:

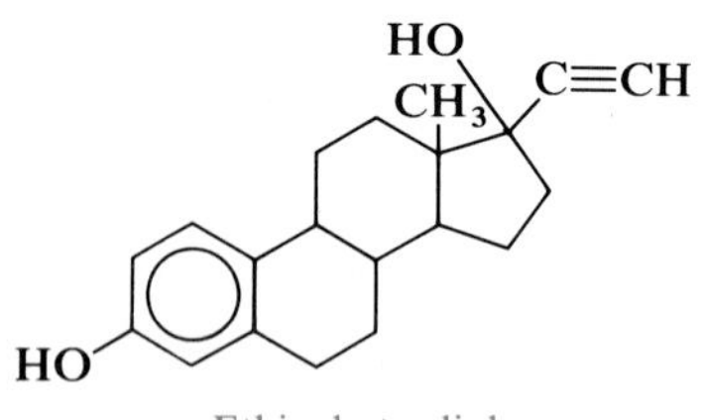

Ethinylestradiol

a. Why is it classified as a steroid?
b. Why would its physiological effects be similar to those of the natural hormone, estradiol?

16.26 Match one of the following classes of lipids with each of the structures a through f: (1) triacylglycerol, (2) wax, (3) phospholipid, (4) sphingolipid, (5) terpene, (6) steroid.

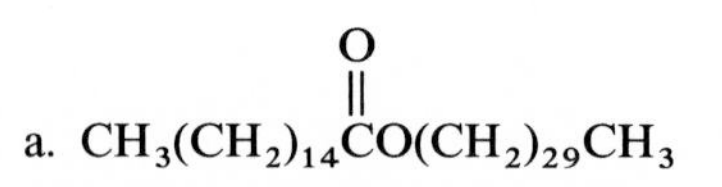

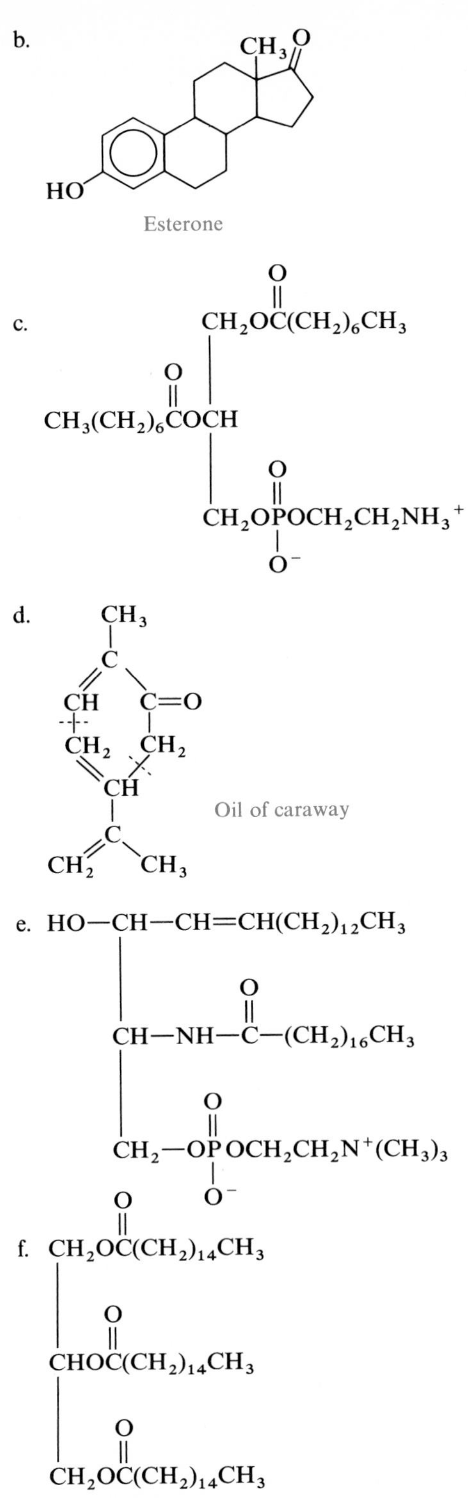

Esterone

Oil of caraway

16.27 Match one of the structural features with each of the lipids a through i: (1) fatty acids, (2) salts of fatty acids, (3) glycerol, (4) phosphate, (5) long-chain alcohol, (6) amino alcohol, (7) sphingosine, (8) monosaccharide, (9) steroid nucleus, (10) isoprene.

a. beeswax
b. cholesterol
c. lecithin
d. tripalmitin
e. essential oil of lemon
f. soap
g. glycolipid
h. testosterone
i. vitamin E

17 Proteins

Objectives

17.1 State the functions of protein in the body.

17.2 Given the R group for a specific amino acid, write the structure of the amino acid.

17.3 Distinguish between complete and incomplete proteins.

17.4 Write the ionic form of an amino acid as the zwitterion; write the structure of an amino acid in acidic and basic solutions.

17.5 Write the structure of a di- or tripeptide, given the individual amino acids.

17.6 Describe the primary, secondary, tertiary, and quaternary structural levels of proteins.

17.7 Discuss denaturation and the factors that cause denaturation of a protein.

17.8 Describe some chemical tests for amino acids and proteins.

Scope

Thousands of different compounds are necessary for the normal functioning of a living system. The absence of even one of these compounds can jeopardize the system's well-being. Among the essential compounds are the proteins, whose name is derived from the Greek word *proteios*, meaning "first."

Made from amino acids, proteins do many things in the body: They build cartilage and connective tissue, and they transport oxygen in blood and muscle. In the form of enzymes, proteins catalyze biological reactions; as antibodies, they defend the body against infection; as hormones, they control metabolic processes.

Compared to many of the compounds we have studied, the proteins are gigantic in size. One protein, insulin, has a molecular weight of 5700; another, hemoglobin, has a molecular weight of about 64,000. Still larger are some of the virus proteins, which have molecular weights of over 40 million. But despite the size and complexity of proteins, the same chemical concepts that hold true for simpler compounds also hold true for them.

Within these huge protein molecules there is simplicity of structure. A group of about 20 amino acids forms the building blocks for all proteins. In your body alone there are thousands of different proteins, each made up of specific quantities and sequences of amino acids. By looking at the amino acid building blocks and how they join together, we can understand some characteristics of protein structure and function.

17.1 Functions of Proteins

OBJECTIVE State the functions of protein in the body.

Proteins perform many functions in the body. Some, called *enzymes*, regulate biological reactions such as digestion and cellular metabolism. Other proteins make up structural components such as cartilage, hair, and nails. Still other proteins, hemoglobin and myoglobin, carry oxygen in the blood and muscle. Proteins known as *antibodies* provide immunity to disease. Table 17.1 lists some of the various functions of proteins.

Example 17.1 Match a function of proteins with the following proteins:

1. catalyzes metabolic reactions of lipids
2. carries oxygen in the bloodstream
3. stores amino acids in milk

a. casein
b. lipase
c. hemoglobin

Solution

1. The protein lipase (choice b), an enzyme, catalyzes metabolic reactions of lipids.
2. The transport protein hemoglobin (choice c) carries oxygen through the bloodstream.
3. The storage protein casein (choice a) stores amino acids in milk.

Table 17.1 Biological Functions of Proteins

Type	Example	Functions
Enzyme—Proteins that catalyze chemical reactions in the cells	Sucrase	Breaks down the sugar sucrose
	Lipase	Hydrolyzes lipids
	Trypsin	Hydrolyzes proteins
Structural—Proteins that provide support for the body	Keratin	Forms hair, skin, nails, horns, feathers, and wool
	Collagen	Builds connective tissue and cartilage (makes up over 50% of body protein)
	Fibroin	Forms cocoons and spider webs
Transport—Proteins that carry small molecules through the circulatory system	Hemoglobin	Carries oxygen in the blood
	Myoglobin	Carries oxygen in muscles
	Lipoprotein	Carries lipids in body fluids
Protective—A type of protein that protects the body	Prothrombin	Causes blood clotting
	Fibrinogen	Causes blood clotting
	Antibodies	React with antigens (foreign proteins)
Storage—Proteins formed within the cells as a store for amino acids	Casein	Protein in milk
	Egg albumin	Protein in egg white
	Zein	Protein in corn seed
Hormonal—Protein messengers that regulate metabolic processes	Insulin	Regulates the metabolism of glucose
	Growth hormone	Regulates growth of bones

17.2 Amino Acids

OBJECTIVE Given the R group for a specific amino acid, write the structure of the amino acid.

Amino acids are the building blocks of proteins. An amino acid contains two functional groups: an α-amino group ($-NH_2$) and a carboxyl group ($-C(=O)OH$ or $-COOH$). These two functional groups are bonded to the alpha carbon atom in the amino acid along with a hydrogen atom (—H) and a fourth group designated as R. The general formula for an amino acid can be written as follows:

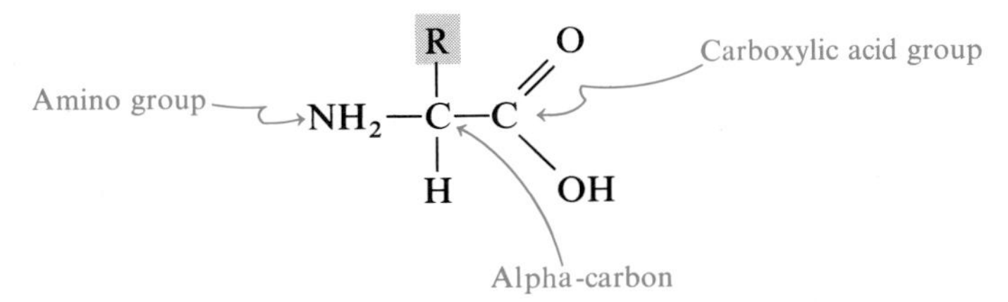

R group

The R group is different in each amino acid and is responsible for the unique characteristics of the molecule. We can classify amino acids according to the R group. The R groups in alanine, valine, leucine, isoleucine, and proline are hydrocarbon groups. The R groups in phenylalanine and tryptophan contain aromatic groups. All of the hydrocarbon and aromatic R groups are nonpolar, which means that the R groups in these amino acids are hydrophobic (water-fearing).

hydrophobic

hydrophilic

Other R groups contain substituents that make them polar and hydrophilic (water-attracting). The R groups in serine, threonine, and tyrosine contain alcohol groups. The sulfur-containing R groups are found in cysteine and methionine. The amino acids aspartic acid and glutamic acid contain acidic R groups with a carboxyl group. Basic R groups containing an amino group are found in lysine, arginine, and histidine. There are also two amide R groups found in asparagine and glutamine. Table 17.2 presents the structures of the R groups for 20 amino acids, as well as their names and their three-letter abbreviations.

Example 17.2 Write the structures for the following amino acids:

1. alanine (R = $-CH_3$)
2. serine (R = $-CH_2OH$)
3. aspartic acid (R = $-CH_2\overset{\overset{\large O}{\|}}{C}OH$)

Solution: The structures of the amino acids are written by using the given R group with the general structure of an amino acid:

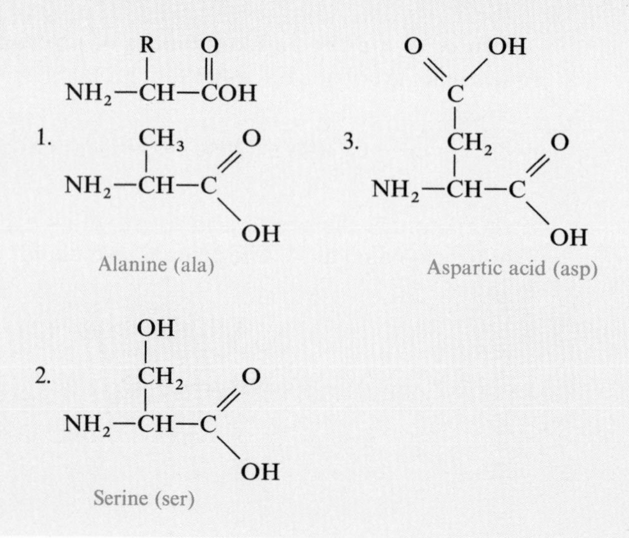

Alanine (ala)

Aspartic acid (asp)

Serine (ser)

Table 17.2 Classification of Amino Acids by R Groups

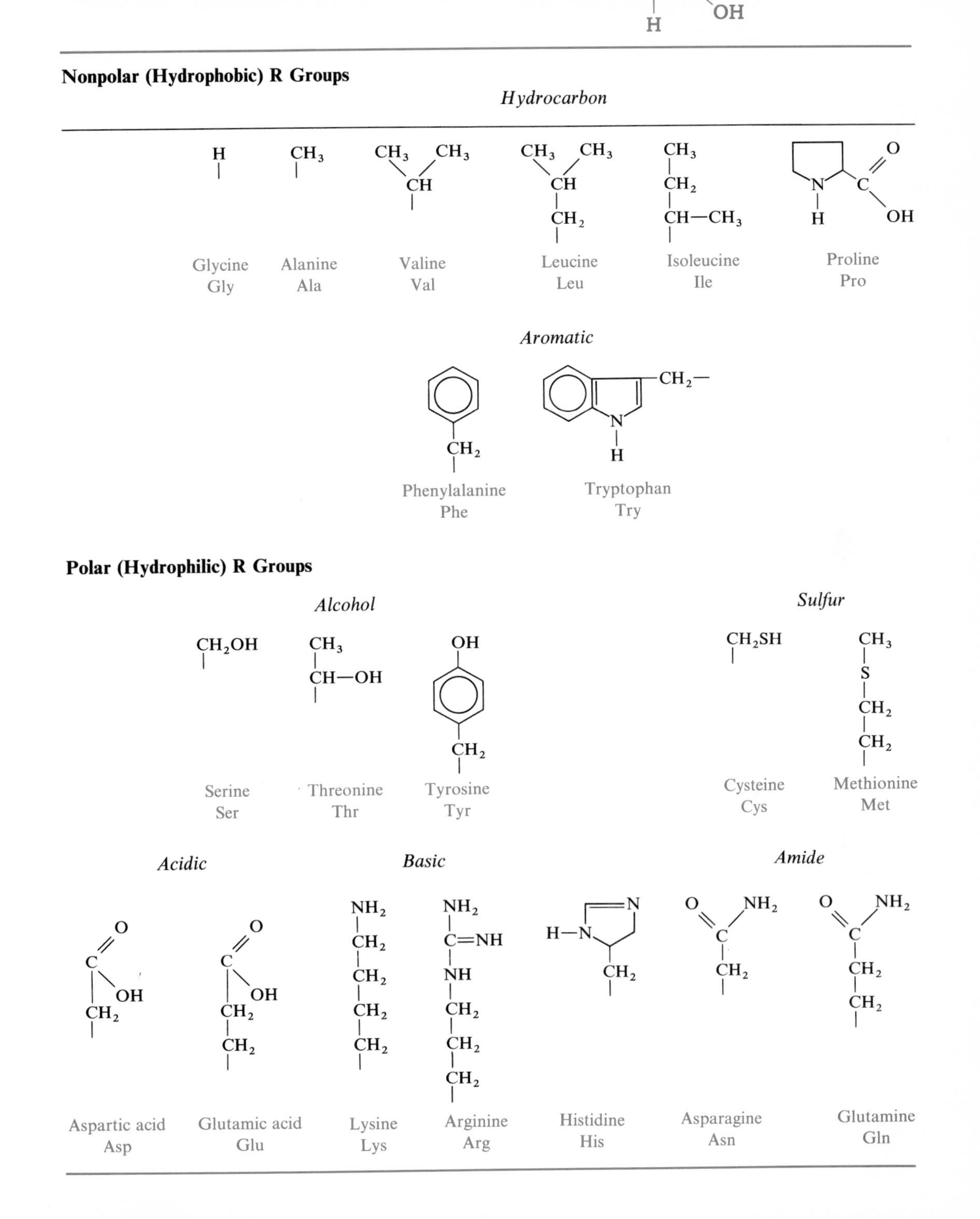

17.3 Essential Amino Acids

OBJECTIVE Distinguish between complete and incomplete proteins.

Proteins, which are constantly being produced in the body for growth and repair, contain amino acids, the building blocks of proteins. There are about 20 amino acids found in the proteins of the body. Those amino acids produced in the body are called *nonessential* amino acids. However, the body is unable to synthesize 10 amino acids; these are called *essential* amino acids:

Essential Amino Acids

Arginine	Methionine
Histidine	Phenylalanine
Isoleucine	Threonine
Leucine	Tryptophan
Lysine	Valine

Histidine appears to be an essential amino acid for infants, but may not be for adults. Arginine may be an essential amino acid even though it is produced in the body. However, the quantities of arginine produced in the body are not sufficient to meet the protein demand for arginine.

complete protein

A *complete* protein supplies all of the necessary quantities of each essential amino acid. An *incomplete* protein is low in one or more of the essential amino acids. In general, proteins from animal sources are complete, whereas proteins from vegetable sources are incomplete, often lacking the amino acids lysine and tryptophan. Table 17.3 gives some examples of complete and incomplete protein sources.

If a diet includes food of animal origin, such as meat, milk, eggs, or cheese, all the essential amino acids will be supplied. But a diet that consists mostly of

Table 17.3 Examples of Complete and Incomplete Protein

Source	Type of Protein	Amino Acid(s) Missing
Animal Protein		
Egg	Complete	None
Milk (cow)	Complete	None
Meat, fish, poultry	Complete	None
Vegetable Protein		
Wheat	Incomplete	Lysine
Corn	Incomplete	Lysine, tryptophan
Rice	Incomplete	Lysine
Beans	Incomplete	Methionine, tryptophan
Peas	Incomplete	Methionine
Almonds	Incomplete	Lysine, tryptophan

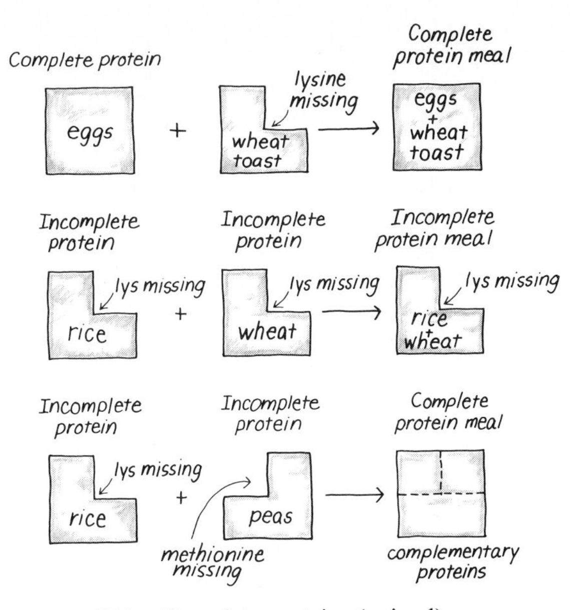

Figure 17.1 Complete proteins (animal) or complementary proteins of vegetable origin will provide a complete protein at a meal.

vegetables may not provide complete protein. For such a vegetable diet to be complete, a large variety of vegetables must be consumed. A complete assortment of amino acids can be obtained by pairing a vegetable protein that is missing in one essential amino acid with a vegetable that contains it. The two vegetable proteins make up a pair called *complementary proteins.* See Figure 17.1.

complementary proteins

When a diet is lacking in essential amino acids, a protein deficiency occurs. In a condition known as *kwashiorkor,* the number of calories in the diet are sufficient, but protein is deficient. Without a sufficient source of protein, a child fails to grow, develops a fatty liver, and the abdomen becomes enlarged as a result of edema.

Example 17.3 Using Table 17.3, indicate whether each of the following combinations of proteins provide a complete or an incomplete protein at a meal:

1. corn and rice
2. corn, eggs, and rice
3. rice and beans

Solution

1. Corn is a vegetable protein that lacks lysine and tryptophan. Rice is a vegetable protein that is missing lysine. Since they both lack lysine, the combination is an incomplete protein.
2. Although corn and rice are lacking one or more amino acids and do not complement each other, eggs are a complete protein and will provide all of the essential amino acids. The combination is a complete protein.
3. Rice is lacking in the amino acid lysine, while beans are low in methionine and tryptophan. Since they are missing different essential amino acids, their combination will provide for all the necessary amino acids, thus giving a complete protein.

17.4 Properties of Amino Acids

OBJECTIVE Write the ionic form of an amino acid as the zwitterion; write the structure of an amino acid in acidic and basic solutions.

Amino acids are solids at room temperatures, have high melting points, and are soluble in water. Such characteristics suggest that amino acids exist in an ionic rather than un-ionized form. An amino acid becomes ionic when the carboxyl group donates a proton to the amino group to form a positive and negative region within the same molecule. A dipolar ion with two opposing ionic regions is called a *zwitterion*:

zwitterion

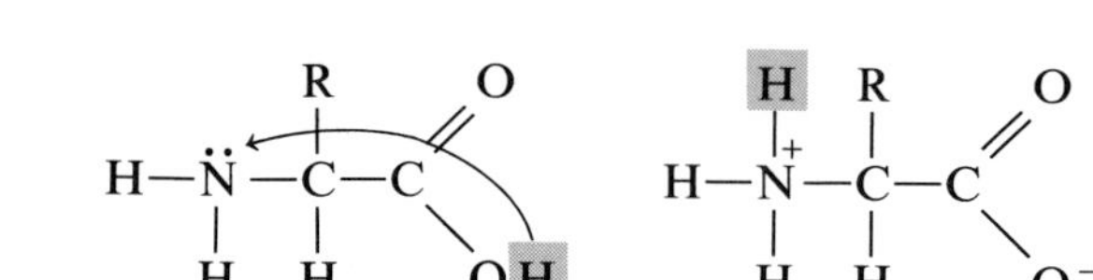

Amino acids in the zwitterion form are shown in the following examples:

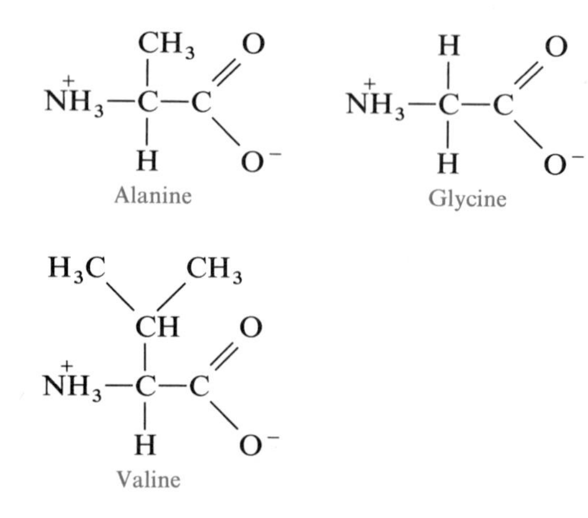

Example 17.4 Draw the zwitterion (dipolar ion) of serine.

Solution: The zwitterion form of serine shows the ionized carboxyl group and amino group:

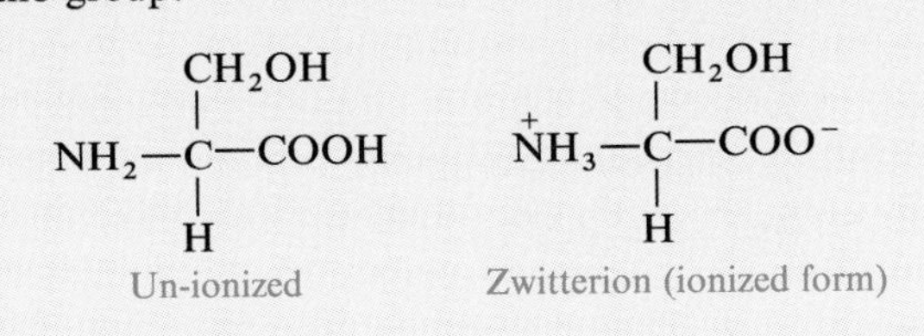

Amphoteric Properties

An *amphoteric* substance acts as an acid when it donates protons; it acts as a base when it accepts protons. Because amino acids exist in the zwitterion form, they have the capability of acting as acids or bases. When placed in an acidic solution (low pH), amino acids act as bases by accepting protons and becoming positively charged. In basic solutions (high pH), amino acids act as acids by donating protons and becoming negatively charged. (See Figure 17.2.)

Amino acids can function as buffers since they are capable of neutralizing small increases of acid or base. Even the amino acids in a protein molecule exhibit this buffer action. Proteins are one of the major buffering systems in the body.

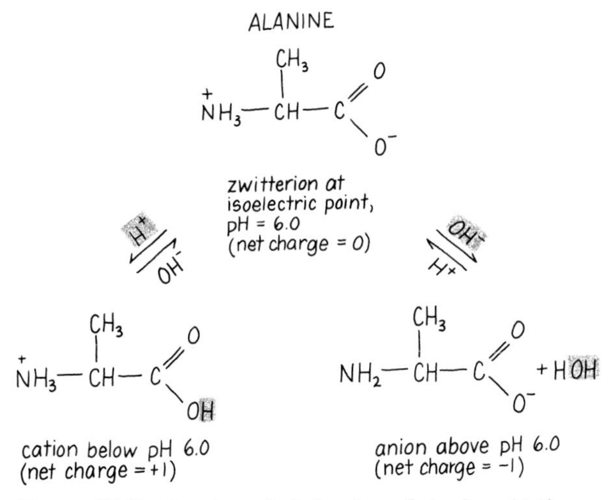

Figure 17.2 Amphoteric behavior of alanine. At the isoelectric point (pH 6.0), alanine is electrically neutral; at pH values below 6.0, it is positively charged; at pH values above 6.0, it is negatively charged.

Example 17.5 Draw the structure of valine in a solution that has a high pH.

Solution: First, we need to look at the zwiterion form of valine:

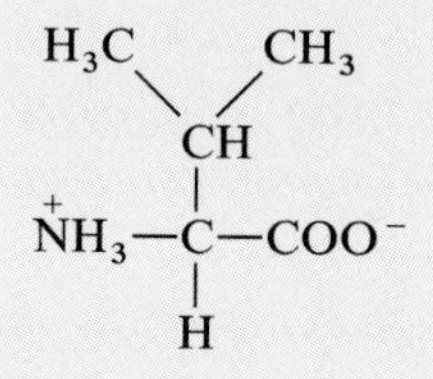

A solution with a high pH contains hydroxide ions (OH^-) that will pick up the proton:

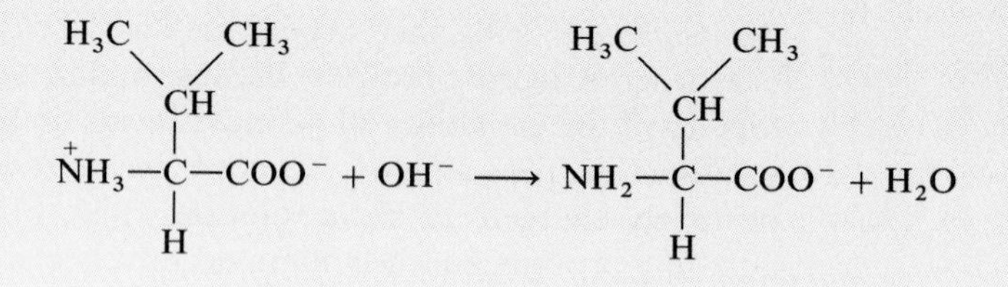

Isoelectric Point

The zwitterion, which is electrically neutral overall, can only exist at a specific pH value. This pH value, called the *isoelectric point*, is different for each amino acid. At the isoelectric point, the amino acid no longer migrates in an electric field. Amino acids with hydrocarbon R groups attain their isoelectric point between a pH of 5.0 and 7.0. The basic amino acids need high pH values to reach their isoelectric point, whereas acidic amino acids need low pH values.

Proteins also have isoelectric points that depend upon the particular amino acids in that protein. At their isoelectric points, proteins become insoluble in water, clump together, and precipitate out of solution. For example, when acid is added to milk, the pH changes until the isoelectric point of casein (milk protein) is reached. The protein becomes insoluble, solidifies, and separates. The production of yogurt and cheese is a result of the change in pH by an acid produced by bacteria added to the milk. Table 17.4 lists isoelectric points for some amino acids and proteins.

Electrophoresis

It is possible to separate amino acids and proteins by using solutions at different pH values. This technique, called *electrophoresis*, is used as an important diagnostic tool in protein analysis in the hospital laboratory. A protein can be identified by finding the pH of its isoelectric point.

Table 17.4 Isoelectric Points of Some Amino Acids and Proteins

Amino Acid	pH	Protein	pH
Aspartic acid (acidic)	3.0	Egg albumin	4.6
Glutamic acid (acidic)	3.2	Casein (milk)	4.6
Serine (polar, neutral)	5.7	Insulin	5.3
Alanine (nonpolar, neutral)	6.0	Hemoglobin	6.8
Lysine (basic)	9.7	Chymotrypsin	9.5
Arginine (basic)	10.8	Lysosyme	11.0

When an amino acid or protein is exposed to pH values below its isoelectric point, it will become positively charged. If the solution has a pH above its isoelectric point, the amino acid will take on a negative charge.

In electrophoresis, a sample of several amino acids is placed on a paper strip or gel between two electrodes. The positively charged ions migrate toward the negative electrode, and the negatively charged ions migrate toward the positive electrodes. Those amino acids or proteins that are at their isoelectric

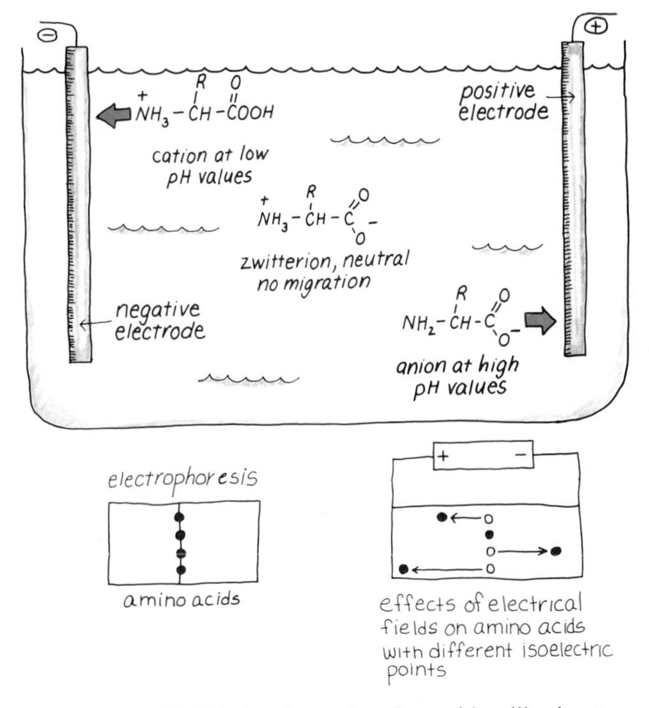

Figure 17.3 Positively charged amino acids will migrate to a negative electrode; negatively charged amino acids will migrate to the positive electrode. Zwitterions, which are neutral, do not migrate.

points are neutral (zwitterions) and will remain stationary during electrophoresis. (See Figure 17.3.) The paper strip is then sprayed with reagents that make amino acids visible. The individual amino acids in the sample can be identified by their direction and rate of migration in the electric field.

17.5 Peptide Bonds

OBJECTIVE Write the structure of a di- or tripeptide from the individual amino acids.

Proteins consist of long chains of amino acids held together by amide linkages called *peptide bonds.* The amide linkage occurs between the carboxyl group of one amino acid and the amino group of the next amino acid. When two amino acids react to form a peptide bond, a molecule of water is removed. The resulting compound is a *dipeptide.* The name of the dipeptide is found by naming the amino acid from the free amino end as an alkyl group and then attaching the name of the amino acid from the free carboxyl end:

dipeptide

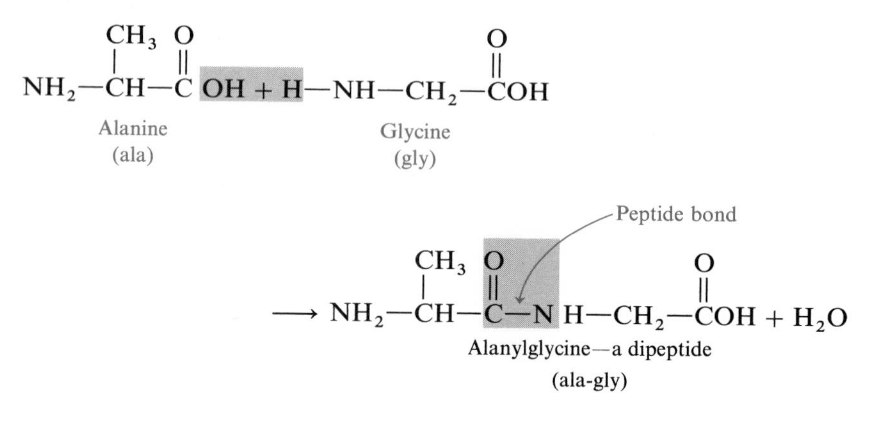

Example 17.6 Write the structure of the dipeptide valylserine.

Solution: The peptide bond forms between the carboxyl group of valine and the amino group of serine; the dipeptide will have the following structure:

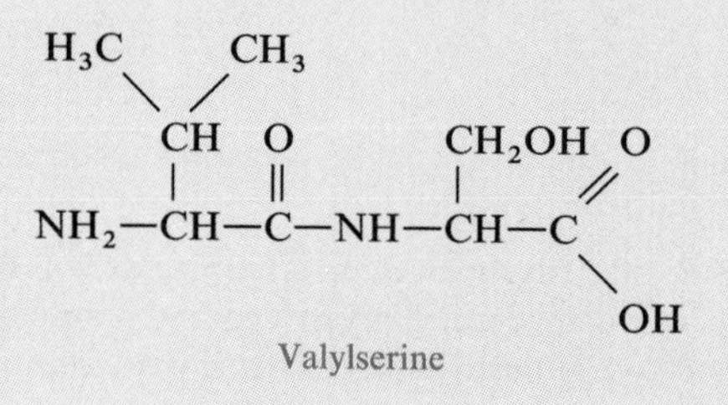

Tripeptides

When three amino acids combine, a *tripeptide* results. Suppose we want to write the tripeptide glycylalanylserine (Gly-Ala-Ser). Glycine is written as the amino end, alanine in the middle, and serine as the free carboxyl end:

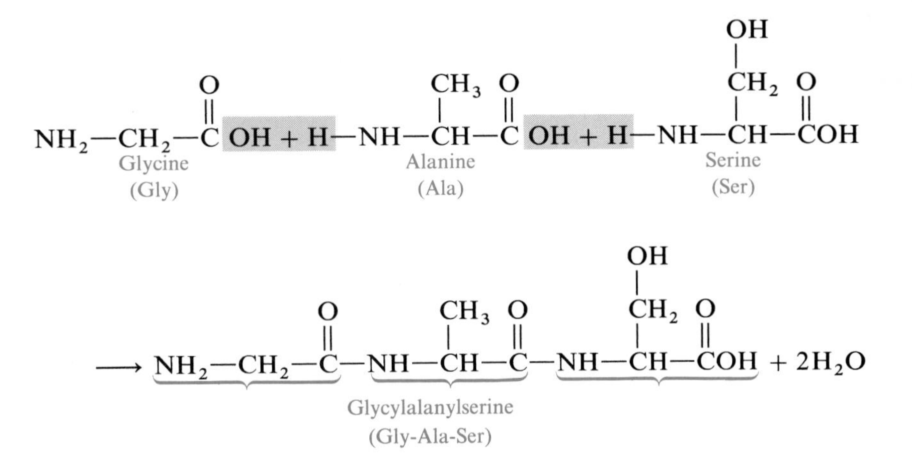

There are a total of six tripeptide combinations possible from the three amino acids glycine, alanine, and serine when one molecule of each is used. (See Table 17.5.) When you consider that a single protein may have from 50 to 400,000 amino acids, the number of possible amino acid arrangements is astronomical.

The amino acid composition of a protein can be determined by hydrolyzing the protein to give the individual amino acids. The amino acid composition of a few proteins is given in Table 17.6.

Example 17.7 Write the structure of the tripeptide Gly-Thr-Asp.

Solution: In the tripeptide, there are peptide bonds between glycine and threonine and between threonine and aspartic acid:

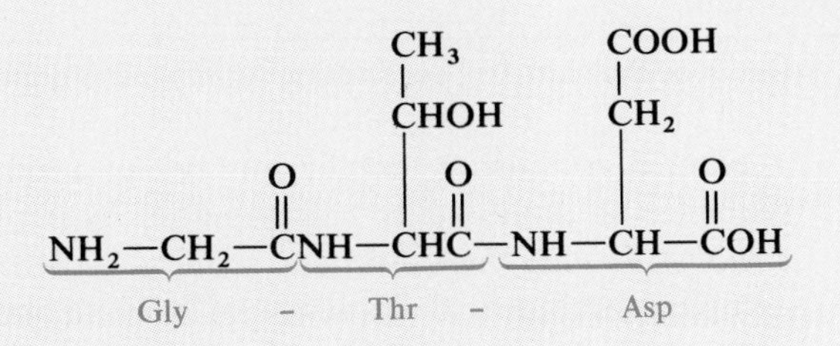

Table 17.5 Tripeptides Possible with Glycine, Alanine, and Serine

Glycylalanylserine	Gly-Ala-Ser
Glycylserylalanine	Gly-Ser-Ala
Alanylserylglycine	Ala-Ser-Gly
Alanylglycylserine	Ala-Gly-Ser
Serylalanylglycine	Ser-Ala-Gly
Serylglycylalanine	Ser-Gly-Ala

Table 17.6 Amino Acid Composition

Amino Acid	Insulin (Beef)	Hemoglobin (Human)	Pepsin
Essential			
Arginine	1	12	2
Histidine	2	38	1
Isoleucine	2	0	27
Leucine	6	72	28
Lysine	1	44	1
Methionine	0	6	5
Phenylalanine	3	30	14
Threonine	3	32	28
Tryptophan	0	6	6
Valine	4	62	21
Nonessential			
Alanine	3	72	18
Asparagine	3	20	3
Aspartic acid	0	30	41
Cysteine	3	6	4
Glutamine	3	8	19
Glutamic acid	4	24	8
Glycine	4	40	38
Proline	1	28	15
Serine	3	32	44

17.6 Protein Structure

OBJECTIVE Describe the primary, secondary, tertiary, and quaternary structural levels of proteins.

primary structure

A protein contains a specific number of amino acids in a certain order. This sequence of amino acids, held together by peptide bonds, is called the *primary structure.* Today the amino acid sequences of many proteins are known. The first complete amino acid sequence of beef insulin was determined in 1953. The primary structure of the 51 amino acids in beef insulin is shown in Figure 17.4.

```
            ┌──S–S──┐
Gly-Ile-Val-Glu-Gln-Cys-Cys-Ser-Cys-Ser-Leu-Tyr
                    S                          Gln
                    S                          Leu
Phe-Val-Asn-Gln-His-Leu-Cys-Gly-Ser-His-Leu-Val-Glu-Ala  Glu
                                               Leu  Asn
                                               Tyr  Tyr
                                               Leu
                                               Val
                                               Cys–S–S–Cys
Thr-Lys-Pro-Tyr-Phe-Phe-Gly-Arg-Glu-Gly              Asn
```

Figure 17.4 Primary structure of beef insulin indicates the order of amino acids held together by peptide bonds.

Secondary Structure

By the 1930s, experimental data indicated that the peptide chains are stabilized by interactions between the amide bonds of the peptide chain. These *secondary structures* may take the form of an alpha helix or a beta-pleated sheet.

In the 1940s, Linus Pauling determined that the amino acid chain in many proteins had a helical or corkscrew shape. This coiled pattern of an amino acid chain is called an *alpha helix* (α-helix). The α-helix is a result of hydrogen bonding between the amide groups along the chain. Hydrogen bonds occur between the hydrogen atom attached to the nitrogen atom of each peptide bond and the carboxyl oxygen of an amino acid in the next turn. One turn of the α-helix contains 3.6 amino acids. The R groups extend out from the helical portion of the α-helix. (See Figure 17.5.)

α-helix

The fibrous proteins that make up hair, wool, skin, nails, feathers, and horns of animals are made of proteins called α-keratins. The α-keratins consist of several α-helix chains. In hair and wool, three or seven α-helix chains wrap or coil together in a regular pattern much like the fibers in a rope. The chains are

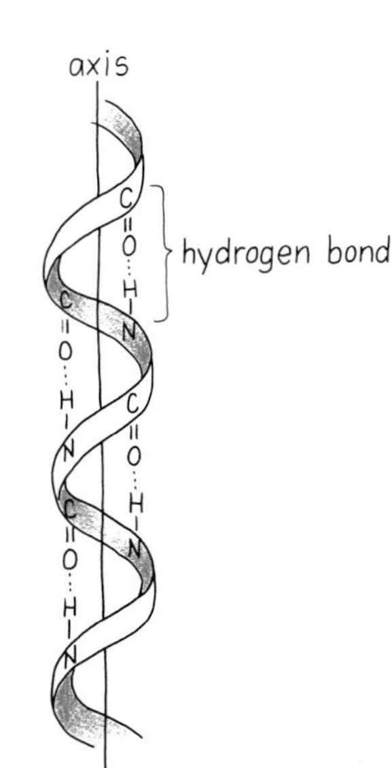

Figure 17.5 The α-helix, a secondary protein structure.

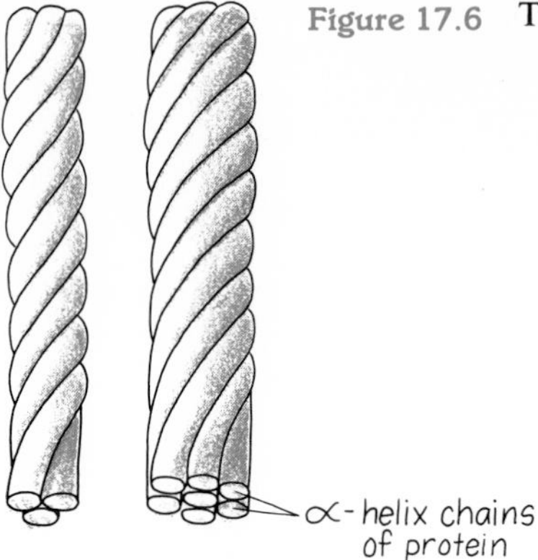

Figure 17.6 The secondary structure shown by fibrous proteins.

held together by disulfide linkages between the chains. When several of these chains combine into a bundle, a hair fiber results. (See Figure 17.6.)

β-pleated sheet

Another type of secondary structure found in proteins such as silk is the *beta-pleated* (β-pleated) *sheet.* In the pleated sheet, several peptide chains are held together in parallel fashion by hydrogen bonding between the amide groups of adjacent chains. The major amino acids in silk with pleated-sheet

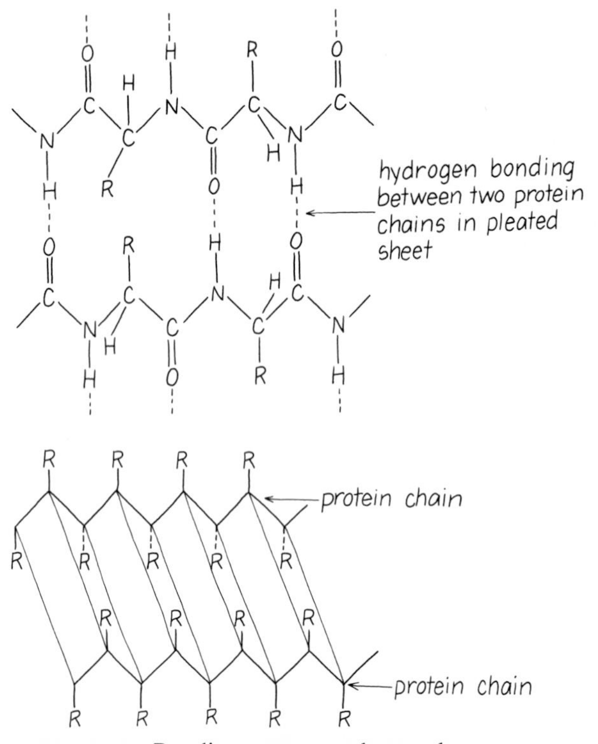

Figure 17.7 Bonding pattern and secondary structure of β-pleated sheet for parallel protein chains.

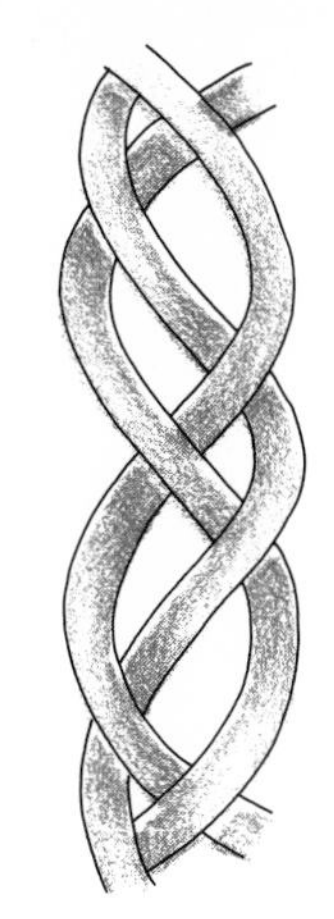

Figure 17.8 The triple helix of collagen (secondary protein structure).

structure are glycine, alanine, and serine. Their R groups extend above and below the plane of the pleated sheet. (See Figure 17.7.)

collagen

Another type of secondary structure is found in collagen, an important protein in connective tissue that makes up skin, tendons, the cornea of the eye, and cartilage. In the human body, collagen makes up as much as 30 percent of all protein. In collagen, three peptide chains are woven together like a braid forming a triple helix. (See Figure 17.8.) Several of the braids combine to form the fibrils of a tendon.

Example 17.8 Indicate the secondary structure (α-helix, β-pleated sheet, triple helix) that each of the following phrases describes:

1. hydrogen bonding that occurs between the oxygen of the carbonyl group and the hydrogen of the amide groups in the same peptide chain
2. hydrogen bonding that occurs between carbonyl groups and amide groups of parallel peptide chains

Solution

1. When hydrogen bonds occur between carbonyl groups and amide groups in the same peptide chain, a coiled peptide chain (α-helix) forms.
2. When hydrogen bonds hold together adjacent peptide chains, a β-pleated sheet structure forms.

Tertiary Structure

For proteins consisting of an α-helix, there is a unique and specific pattern of folding of the protein chain that makes the molecule very compact. Much of the α-helical structure is retained, but now the entire chain folds in upon itself. The tertiary structure, which is typical of globular proteins, is found in egg and

globular protein

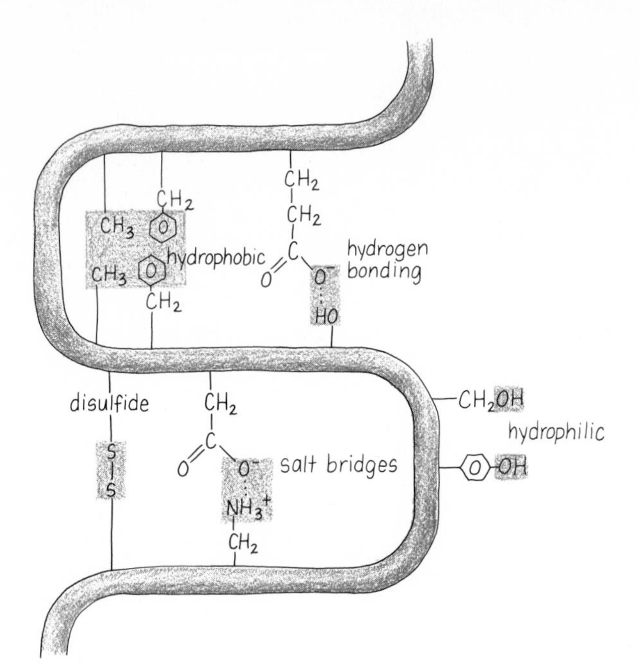

Figure 17.9 Interaction among R groups of amino acids to stabilize the tertiary structure of globular proteins.

serum albumin, hemoglobin, and myoglobin, as well as in enzymes and antibodies.

The polarity of the R groups is a stabilizing factor in maintaining the tertiary structure. The protein chain folds in such a way that the amino acids with nonpolar side groups are directed toward the center of the tertiary structure. The polar and ionic side groups that are hydrophilic are found along the outside portions of the molecule.

cross-linkages
disulfide bond
salt bridges
hydrogen bonding

There are attractions between side groups that provide *cross-linkages* between the folded portions of the protein chain. A common cross-link is a disulfide bond between two cysteine groups. Ionic bonds can occur between two oppositely charged side groups. These ionic bonds, or *salt bridges*, are most likely to exist between an acidic group of aspartic acid or glutamic acid and a basic group of lysine or arginine. *Hydrogen bonds* can occur between an alcohol group and the oxygen of a carbonyl group in another amino acid. (See Figure 17.9.)

Example 17.9 What type of interaction would you expect when the side groups of the following amino acids come close together in a tertiary structure of a protein?

1. cysteine and cysteine
2. glutamic acid and lysine
3. leucine and valine

Solution

1. Since two cysteine molecules contain —SH side groups, a disulfide bond will form:

$$—SH + HS— \longrightarrow —S—S—$$

Disulfide bond

2. The acidic side group of glutamic acid will form an ionic bond with the basic side group of lysine:

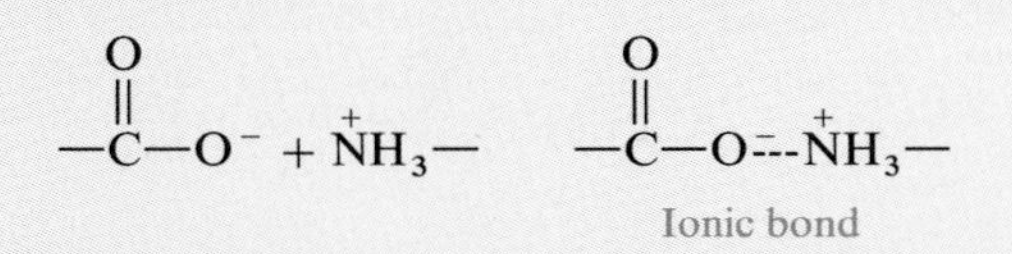

Ionic bond

3. Since valine and leucine have nonpolar side groups, we would expect hydrophobic attraction:

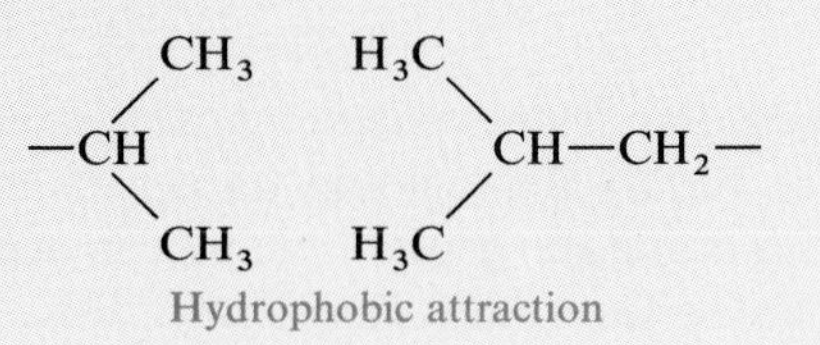

Hydrophobic attraction

myoglobin Myoglobin, which carries oxygen in muscle, consists of 153 amino acids in a single polypeptide chain. The myoglobin protein folds into a tertiary structure to form a very compact molecule. One atom of iron in the molecule is available for the oxygen transport function of the protein. (See Figure 17.10.)

Quaternary Structure

The quaternary structure is an extension of the tertiary structure. When two or more tertiary units combine to form a protein, their arrangement is called a *quaternary structure.* Each tertiary protein is then called a *subunit.*

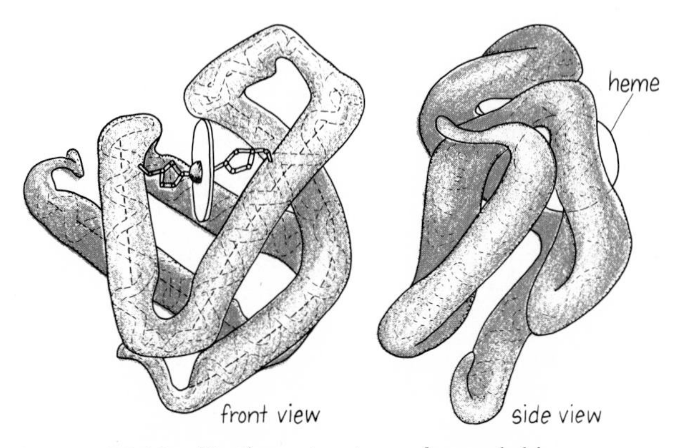

Figure 17.10 Tertiary structure of myoglobin.

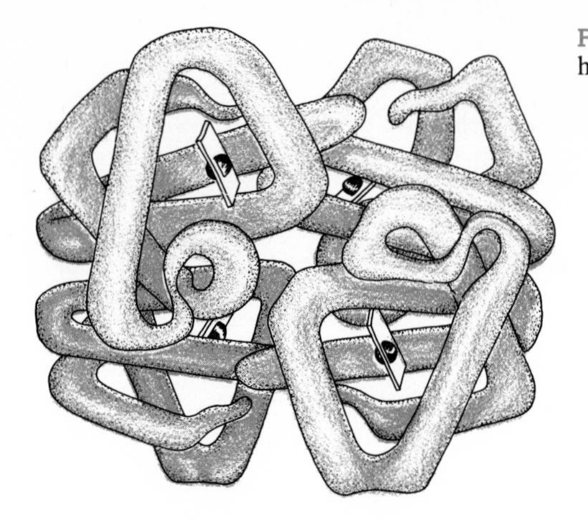

Figure 17.11 Quaternary structure of hemoglobin.

hemoglobin

Hemoglobin is one protein whose quaternary structure is known in detail. (See Figure 17.11.) Hemoglobin transports oxygen and carbon dioxide in the blood. Four separate peptide chains or subunits make up the hemoglobin quaternary structure. All four subunits must be present for the hemoglobin to function in oxygen transport. Each subunit is capable of combining with one oxygen molecule, so the complete hemoglobin structure can transport up to four molecules of oxygen.

It is interesting to note that hemoglobin and myoglobin have similar functions. Hemoglobin carries oxygen in the blood, whereas myoglobin carries oxygen in muscle. However, myoglobin is a single polypeptide chain with a molecular weight of 17,000, about one-fourth the molecular weight of hemoglobin (64,000). In addition, the tertiary structure of the single polypeptide myoglobin is almost identical to the tertiary structure of a single subunit of hemoglobin. Myoglobin transports only one molecule of oxygen, just as each subunit of hemoglobin can carry one oxygen molecule. The similarity in tertiary structures is probably the key to the similarity in function of these two proteins.

Example 17.10 Indicate whether the following bonds are responsible for primary, secondary, tertiary, or quaternary protein structures:

1. disulfide bonds forming cross-links between portions of a protein chain
2. peptide bonds forming a chain of amino acids

Solution

1. Disulfide cross-links between portions of a protein chain help to stabilize the folded shape of a protein called the *tertiary* structure.
2. The peptide bonds that hold the amino acids in a specific sequence determine the first, or *primary*, structure.

17.7 Denaturation of a Protein

OBJECTIVE Discuss denaturation and the factors that cause denaturation of a protein.

Since the biological function of a protein depends upon its three-dimensional shape, a change or alteration in that shape can cause a protein to become biologically ineffective (inactive). A loss of biological activity resulting from a loss of the secondary, tertiary, or quaternary protein structure while retaining the peptide bond structure is called *denaturation* of the protein. More specifically, denaturation involves the breaking of the cross-linkages that maintain the tertiary and secondary levels of structure. We might imagine the compact shape of a protein unfolding to become a loose piece of spaghetti. (See Figure 17.12.)

When denaturation occurs under rather mild conditions, the protein may be restored to its original conformation and biological activity by carefully reversing the conditions. However, most changes in conditions are so drastic that the protein remains denatured. Then the protein strands become insoluble and precipitate out of solution. *Coagulation* of the protein occurs, and the effects of coagulation are irreversible.

Heat and Ultraviolet Light

optimum temperature

The conditions at which proteins function most effectively are called *optimum conditions*. The optimum temperature for most proteins is 37°C (body temperature). Few proteins can remain biologically active above 50°C. Certain bacteria that are *thermophilic*, or "heat-loving," have protein that remains quite stable up to temperatures of 70°C and higher.

Denaturation occurs when increased thermal activity caused by heat or ultraviolet light disrupts some of the hydrogen bonds and attractions between nonpolar side groups that maintain secondary and tertiary structures. Whenever you cook food, you are denaturing protein. You have probably seen the denaturation of an egg. When the egg is placed in boiling water (100°C), the effects of heat are great enough to cause irreversible coagulation of the protein.

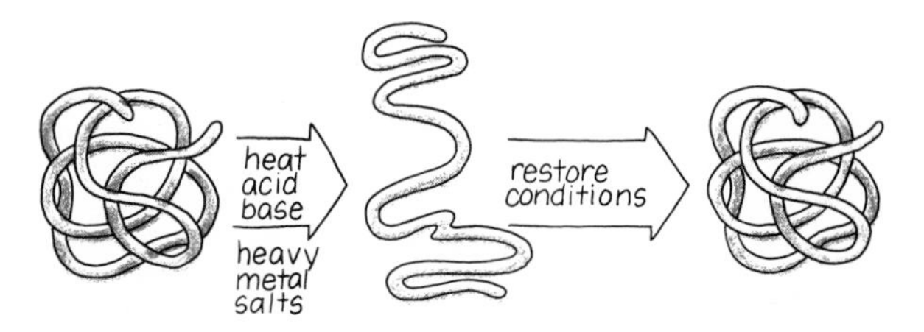

Figure 17.12 Reversible denaturation of a protein. Denaturing agents such as heat, acid, base, and heavy metal salts cause a disruption of the tertiary structure, destroy its globular shape, and render it inactive. If conditions are mild, the denaturation can be reversed if original conditions are restored.

High temperatures are used to disinfect surgical instruments, gowns, and gloves. When these materials are subjected to the temperatures of boiling water or an autoclave, the proteins of any bacteria present will be denatured, thereby rendering the bacteria inactive.

Acids and Bases

Proteins are often biologically active only within narrow ranges of pH. The pH at which a protein such as an enzyme is most active is called its *optimum pH*. For most enzymes, the optimum pH is around the physiological pH of 7.0 to 7.5. Two important exceptions are the digestive enzymes pepsin and trypsin. Pepsin, which must function in the very acidic environment of the stomach, has an optimum pH of 2. Trypsin, which breaks apart peptide bonds in the small intestine, has an optimum pH of 8.

optimum pH

The effect of an acid or base is to add or remove hydrogen ions to the ionic side groups. As a result, ionic bonds (salt bridges) and hydrogen bonds are disrupted. When a protein is placed in a strong acid or base, coagulation may also occur. In cheese production, casein, milk protein, is subjected to acid. The casein coagulates and forms curds, which are then collected and made into cheese. Tannic acid is a weak acid used in burn ointments to cause a coagulation of the proteins at the site of the burn. The coagulated proteins form a protective cover that prevents a further loss of fluids from the burn.

Organic Solvents

Solvents such as ethanol, isopropyl alcohol, and acetone disrupt the hydrogen bonding of proteins by forming their own hydrogen bonds with the protein. The secondary and tertiary levels of structure are disrupted, and coagulation may follow. Such solvents are used as disinfectants. A 70% solution of isopropanol or ethanol can pass through the cell walls of bacteria and cause coagulation of the bacterial proteins within the cell. The use of an alcohol swab for wounds and the practice of immersing thermometers in alcohol solution are examples of efforts to provide aseptic conditions.

Heavy Metal Ions

Salts containing the heavy metal ions Ag^+, Pb^{2+}, and Hg^{2+} cause denaturation of protein. The heavy metal ions react with the disulfide bonds of the tertiary structures and the carboxyl groups of acidic amino acids. The denatured protein is insoluble and precipitates out of solution.

The salts of the heavy metals are sometimes used as antiseptics. In some hospitals, a dilute (1%) solution of $AgNO_3$ is placed in the eyes of newborn babies to kill the bacteria that causes gonorrhea.

If heavy metal ions are ingested, the disruption of the body proteins, especially the stomach proteins, is quite severe. An antidote for their ingestion is

Table 17.7 Substances That Denature Protein

Substance	Effects on Protein Structure
Heat and ultraviolet light	Disrupt hydrogen bonds and hydrophobic attractions; can cause coagulation
Acids and bases	Disrupt hydrogen bonds and ionic bonds
Organic solvents	Disrupt hydrogen bonds
Heavy metal ions	React with disulfide bonds and carboxylate groups; the insoluble combination precipitiates out of solution
Agitation	Stretch globular proteins until they denature and form solids

a high-protein food such as milk, eggs, or cheese. The food protein ties up the heavy metal ions until the stomach can be pumped or an emetic can be given.

Agitation

The agitation or whipping of cream and the beating of egg whites are examples of mechanical agitation that denatures a protein. The violent whipping action causes a stretching of globular proteins which turns egg whites into meringues and whipping cream into a topping.

Table 17.7 summarizes the substances that can denature protein and discusses their effects.

Example 17.11 Explain what happens to the tertiary structure of a protein when it is placed in an acidic solution.

Solution: An acid can cause denaturation of the protein. The acidic side groups will gain hydrogen ions and lose their ability to form ionic bonds or salt bridges. The loss in cross-linkages will cause the tertiary structure to unfold and lose its shape.

17.8 Testing for Amino Acids and Proteins

OBJECTIVE Describe some chemical tests for amino acids and proteins.

Amino acids, peptide bonds, and some side groups of amino acids react with certain reagents to give very distinctive colors.

Biuret Test

In the biuret test, a peptide or protein is treated with biuret reagent ($CuSO_4$ and NaOH). The cupric ion forms a violet color when it reacts with the peptide bonds. No reaction occurs when amino acids or dipeptides are treated with biuret reagent.

Ninhydrin Test

Amino acids with free amino groups produce a blue to purple color when heated with ninhydrin. When amino acids are separated by paper chromatography, the formation of blue spots on the paper after spraying with ninhydrin identifies the position of the individual amino acids. Two amino acids, proline and hydroxyproline, which contain substituted amino groups, give a yellow color.

Xanthoproteic Test

A protein that contains amino acids with cyclic groups such as tryptophan, tyrosine, or phenylalanine gives a yellow color when nitric acid, HNO_3, is added. This is called the *xanthoproteic test*. If you spill nitric acid on your skin, a yellow spot appears because these three amino acids are also present in skin proteins.

Sulfur Test

The sulfur-containing amino acids cysteine and cystine react with lead acetate to form a black precipitate of lead sulfide, PbS. No reaction occurs with methionine another sulfur-containing amino acid.

Glossary

α-helix A secondary level of protein structure consisting of hydrogen-bonded amino acids within a polypeptide chain that forms a coil-like structure.

amino acid The building blocks of proteins, consisting of an amino group, a carboxylic acid group, and an organic group (R) that is different for each type of amino acid.

amphoteric substance A substance that behaves as both an acid and a base.

biuret test A test using cupric hydroxide to give a violet color with peptides and proteins.

complementary proteins A pair of proteins that provides a complete source of essential amino acids.

complete protein A protein source that supplies all the essential amino acids.

denaturation Any process that destroys the secondary and/or tertiary level of protein structure, thereby causing a protein to become biologically inactive.

dipeptide Two amino acids bonded by a peptide bond.

essential amino acids Amino acids that must be supplied by the diet because the body is unable to synthesize the amounts required for protein production.

fibrous proteins Proteins consisting of groups of polypeptide chains (some having α-helixes and others having pleated-sheet or braided structures) that make up hair, wool, skin, nails, and feathers.

globular proteins Proteins such as enzymes and antibodies that acquire a compact shape from a tertiary-level structure.

incomplete protein A source of protein that is low in one or more of the essential amino acids.

isoelectric point The pH value at which the zwitterion form of an amino acid occurs, thereby causing the acid to become insoluble in the solution.

ninhydrin A substance used for the identification of amino acids. Amino acids with free amino groups give a purple color with ninhydrin, and those with substituted amino acids give a yellow color.

nonessential amino acids Amino acids that can be synthesized by the body and are not necessary in the diet.

nonpolar amino acids Amino acids that are not soluble in water because of the presence of a hydrocarbon R group, which makes them nonpolar.

peptide bonds The amide linkages that hold the amino acids together in polypeptides and proteins.

β-pleated sheet A secondary level of protein structure that consists of hydrogen bonds between several parallel polypeptide chains.

polar amino acids Amino acids that are soluble in water because of the presence of an R group that is a hydroxyl group, a thiol group, a carbonyl group, or an amino group.

quaternary structure An extension of the tertiary level of protein structure in which two or more protein units are combined to form an active protein.

sulfur test A test using lead acetate to give insoluble lead sulfide with some sulfur-containing amino acids.

tertiary structure A level of protein structure acquired through the interactions of R groups such as salt bridges and disulfide bonds; these interactions give additional stabilization and a compact shape to the protein.

triple helix A secondary level of protein structure consisting of three polypeptide chains woven together like a braid. Found in collagen.

xanthoproteic test A protein test using nitric acid; this gives a yellow color when amino acids with cyclic R groups are present in the protein.

zwitterion The salt form of an amino acid consisting of two oppositely charged ionic regions that form when the carboxylic acid portion donates its hydrogen ion to the amino group:

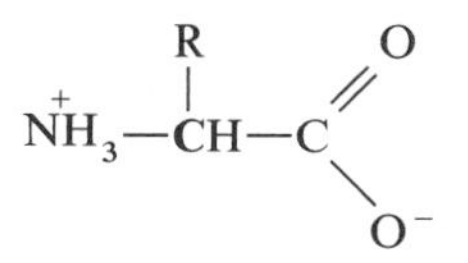

Problems

Functions of Proteins (Objective 17.1)

17.1 Match the following function of proteins with the examples listed below:
(1) catalytic, (2) structural, (3) transport, (4) storage, (5) protective, (6) hormonal.

a. hemoglobin, oxygen carrier in the blood
b. collagen, a major component of tendons and cartilage
c. keratin, a protein found in hair
d. amylase, an enzyme that hydrolyzes starch
e. insulin, a hormone needed for glucose utilization
f. antibodies, proteins that disable foreign proteins
g. casein, milk protein
h. lipase, enzyme that hydrolyzes lipids

Amino Acids (Objective 17.2)

17.2 Describe the features of an amino acid.

17.3 Name the following amino acids:

$COOH = -\overset{\overset{\displaystyle O}{\|}}{C}-OH$

a. $NH_2-\overset{\overset{\displaystyle CH_2OH}{|}}{\underset{\underset{\displaystyle H}{|}}{C}}-COOH$

b. $NH_2-\overset{\overset{\displaystyle (H_3C)(CH_3)CH}{|}}{\underset{\underset{\displaystyle H}{|}}{C}}-COOH$

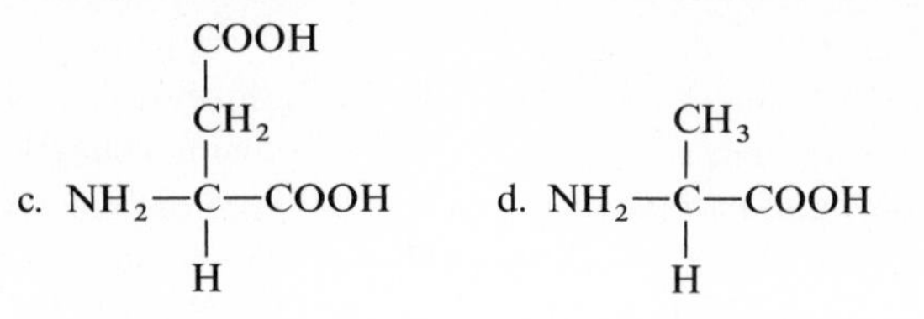

17.4 Indicate whether each side group of the amino acids in Problem 17.3 is hydrophobic or hydrophilic.

17.5 Name the amino acids that contain the following:
a. a sulfur group
b. an alcohol group
c. a basic side group
d. an acidic side group
e. a benzene ring

Essential Amino Acids (Objective 17.3)

17.6 What is an essential amino acid?

17.7 Seeds and vegetables are often deficient in one or more essential amino acids. Using the following table, state whether the following combinations would provide complementary proteins:

Levels for Three Essential Amino Acids

Source	Lysine	Tryptophan	Methionine
Oatmeal	Low	OK	OK
Rice	Low	OK	OK
Garbanzo beans	OK	Low	OK
Lima beans	OK	Low	Low
Cornmeal	Low	OK	OK

a. rice and garbanzo beans
b. lima beans and cornmeal
c. a salad of garbanzo beans and lima beans
d. rice and lima beans
e. rice and oatmeal

17.8 What condition can develop when the diet does not include all of the essential amino acids?

Properties of Amino Acids (Objective 17.4)

17.9 Write the dipolar ion (zwitterion) for each of the following amino acids:
a. glycine
b. phenylalanine
c. glutamine
d. tyrosine
e. methionine
f. leucine

17.10 Leucine has an isoelectric point pH of 6.0. Indicate the ionic structure of leucine at the following pH values:

a. pH 2.0
b. pH 4.0
c. pH 6.0
d. pH 8.0
e. pH 10.0

17.11 A sample containing alanine, aspartic acid, and lysine is placed on a gel for electrophoresis. The isoelectric points are: alanine pH 6.0, aspartic acid pH 3.0, and lysine pH 9.7. A buffer of pH 6.0 is placed on the gel.
a. Which amino acid will migrate toward the positive electrode?
b. Which amino acid will migrate toward the negative electrode?
c. Which amino acid will remain at its original position?

Peptide Bonds (Objective 17.5)

17.12 Write out the structural formulas of the following dipeptides:
a. ala-cys b. ser-ile c. met-gln

17.13 Write out two possible structural formulas and names for the dipeptides from the following:
a. glycine and leucine b. alanine and tyrosine

17.14 Write the structural formulas for the following tripeptides:
a. gly-ala-ala a. met-tyr-phe c. glu-val-ile

17.15 Write the structural formulas for the following tetrapeptides:
a. val-ser-cys-gln b. asp-val-gly-tyr

Protein Structure (Objective 17.6)

17.16 Indicate whether the following statements apply to primary, secondary, tertiary, or quaternary protein structure:
a. Side groups interact to form cross-links such as disulfide bonds or salt bridges.
b. Two protein units combine to form an active protein.
c. Peptide bonds hold amino acids together in a polypeptide.
d. Several peptide chains are held together in a pleated sheet by hydrogen bonds between the chains.
e. Hydrogen bonding between carboxyl groups and amino groups cause a peptide to coil.
f. Hydrophobic R groups seeking a nonpolar environment within the protein chain cause a certain type of shape.
g. Collagen proteins form a three-chain braid.

17.17 Which type of cross-linkage would you expect from the following R groups in a protein chain?
a. two cysteines
b. glutamic acid and lysine
c. serine and aspartic acid
d. two leucines

17.18 A portion of a polypeptide chain contains the following sequence of amino acids

etc.-leu-val-cys-asp

a. Which amino acid in the sequence can form a disulfide cross-link?
b. Which amino acid(s) would most likely be found inside the protein structure? Why?
c. Which amino acid(s) would be found on the outside of the protein? Why?

17.19 What are some differences between the following pairs:

a. secondary and tertiary protein structures
b. essential and nonessential amino acids
c. polar and nonpolar amino acids
d. di- and tripeptides
e. a salt bridge and a disulfide bond
f. fibrous and globular proteins
g. α-helix and β-pleated sheet
h. α-helix and collagen
i. tertiary and quaternary protein structures

Denaturation of a Protein (Objective 17.7)

17.20 An egg placed in boiled water is hardboiled in 3 min. What changes have occurred in the protein structure of the egg? What parts of the protein structure remain unchanged?

17.21 How would the strong acid in the stomach affect the structure of the protein eaten during a meal?

17.22 Prior to an injection, the skin is cleansed with an alcohol swab. What is the effect of the alcohol on any bacteria present on the skin?

17.23 How do each of the following take advantage of the denaturation reactions of proteins?

a. adding $AgNO_3$ to the eyes of newborn infants
b. placing tannic acid on a burn
c. heating milk to 60°C to make yogurt
d. placing surgical instruments in a 120°C autoclave
e. using alcohol on the skin before an injection
f. frying an egg

Testing for Amino Acids and Proteins (Objective 17.8)

17.24 Describe the chemical test you would use to determine the presence of each of the following:

a. tripeptide
b. tyrosine
c. cysteine
d. protein
e. proline

18 Enzymes and Vitamins and the Digestion of Foodstuffs

Objectives

18.1 Draw an energy diagram for a reaction and label the energy levels for reactants and products, the energy of activation with and without a catalyst, and the overall energy of reaction.

18.2 Given the components of an enzyme, state whether it is a simple enzyme or a holoenzyme. Name the enzyme or a class of enzymes that would catalyze a reaction.

18.3 Write an equation for an enzyme-catalyzed reaction and describe the effects of substrate concentration, enzyme concentration, temperature, and pH on the activity of an enzyme.

18.4 Describe the effect of an enzyme inhibitor upon the enzyme activity in competitive and noncompetitive inhibition.

18.5 Given the name of a vitamin, indicate whether it is a water- or fat-soluble vitamin and give its dietary source and biological function.

18.6 Given a foodstuff, state the site of its digestion in the intestinal tract and list the enzymes involved and the products of hydrolysis (digestion).

Scope

Today we know more than 1000 different enzymes, and there is strong evidence that there are more enzymes still to be discovered. Enzymes are a group of proteins that regulate biological reactions within the cells of your body. Enzymes catalyze the reactions that supply you with all the necessary materials and energy for survival. Enzymes cause the fermentation of sugar to alcohol and change milk to cheese. Nerve impulses are conducted and muscles are contracted through enzyme action. Digestive enzymes work within the mouth, stomach, and intestinal tract, breaking down large food particles into smaller molecules that can be absorbed.

Certain kinds of enzymes (conjugated) require the presence of a nonprotein group called a *vitamin.* There was a time when sailors on long sea voyages became ill with scurvy, which is characterized by bleeding, skin hemorrhages, and extreme tenderness in the joints. Eventually, it was discovered that the use of lemons or limes prevents scurvy. The preventive agent in the limes and lemons was found to be vitamin C. Vitamin deficiency can lead to other diseases such as beriberi, a disease that can cause paralysis of the legs, loss of appetite, and constipation. Beriberi is a result of vitamin B_1 (thiamine) deficiency in the diet. The disease is prevented by eating sufficient amounts of unpolished rice, wheat, and vegetables. In vitamin-deficiency diseases such as scurvy and beriberi, the symptoms of the disease disappear when the necessary vitamin is added to the diet.

The foods that your body needs are carbohydrates, fats, and proteins, along with small amounts of vitamins and minerals. However, these foods are useless as nutrients until they go through the process of digestion. Digestion is the chemical breakdown of the foods that you eat.

The process of digestion is primarily one of hydrolysis. Hydrolytic enzymes in the digestive fluids add the components of water to the large polymers, polysaccharides, proteins, and fats. Sugar-digesting enzymes break down polysaccharides into monosaccharides, and fat-digesting enzymes split fats into glycerol and fatty acids. Protein-digesting enzymes split proteins into amino acids. The resulting digestion products are then small enough to be absorbed through the intestinal membranes into the bloodstream. In this way, digestion processes provide nutrients for the maintenance and growth of the cells of your body.

18.1 Energy Diagrams and Catalysts

OBJECTIVE Draw an energy diagram for a reaction and label the energy levels for reactants and products, the energy of activation with and without a catalyst, and the overall energy of reaction.

Most biological reactions would occur too slowly to support life if it were not for the presence of enzymes. In the laboratory, we can break down complex carbohydrates or lipids by using a strong acid or base and by maintaining prolonged high temperatures. Such drastic conditions are not possible in biological systems. Instead, enzymes acting as biological catalysts carry out similar reactions under milder conditions.

Energy Diagrams

In order to understand the function of an enzyme as a catalyst, we need to look at energy diagrams for reactions. For a reaction to take place, molecules of the reactants must first collide. Upon collision, energy is used to break the bonds of the reacting molecules. The products form as new bonds are formed. As the reactants collide and start to react, they reach a high-energy phase, called the *transition state*. The amount of energy needed to reach the transition state is called the *activation energy*. All reactions have an energy of activation, which is defined as the amount of energy that must be available to have a reaction and obtain a product.

transition state

activation energy

The concept of activation energy is analogous to going over a hill to a nearby valley. A certain amount of energy must be expended if we are to climb over the hill successfully. Once we are at the top of the hill, we can effortlessly roll down the other side into the valley. The activation energy is the energy needed to get us from our starting point to the top of the hill.

Futhermore, if we could travel through a tunnel in the hill, we would need less energy to reach the valley. A *catalyst* supplies such an energy tunnel by providing an alternative pathway for the formation of the product. The alternative route occurs at a lower energy, so the activation energy of a catalyzed reaction is lower than that of the uncatalyzed reaction. The catalyst allows the transition to occur more easily between the initial and final energy states with a lower expenditure of energy.

catalyst

Since a catalyst provides a lower energy of activation, more molecules of the reactant acquire the energy needed to reach the transition state and form the product. Since more molecules of the reactant can form the product by way of the new, lower energy pathway, the rate of the reaction increases. (See Figure 18.1.)

a catalyst lowers the energy of activation

If the energy of the products is lower than the initial energy state of the reactants, we say that the reaction is *exothermic*. This means that the energy change for the reaction involves a net release of energy. When the energy of the products is higher than that of the initial reactions, the reaction is *endothermic*, or heat requiring.

exothermic reactions release energy

endothermic reactions absorb energy

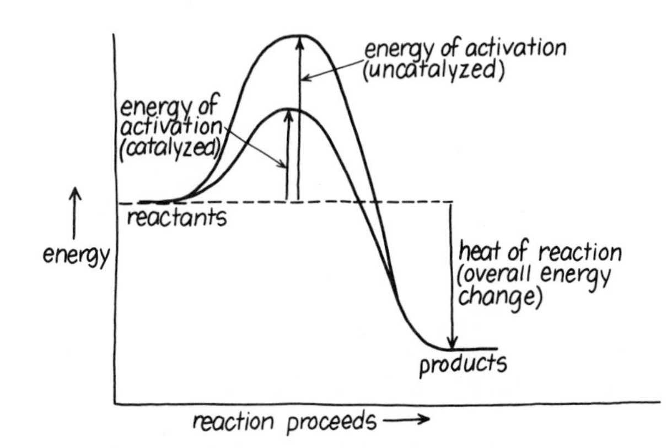

Figure 18.1 Energy diagram for an exothermic reaction with and without a catalyst.

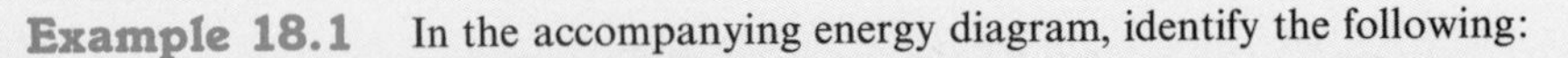

Example 18.1 In the accompanying energy diagram, identify the following:

________ energy level of reactants
________ energy level of products
________ energy of activation (uncatalyzed)
________ energy of activation (catalyzed)
________ overall energy change for the reaction

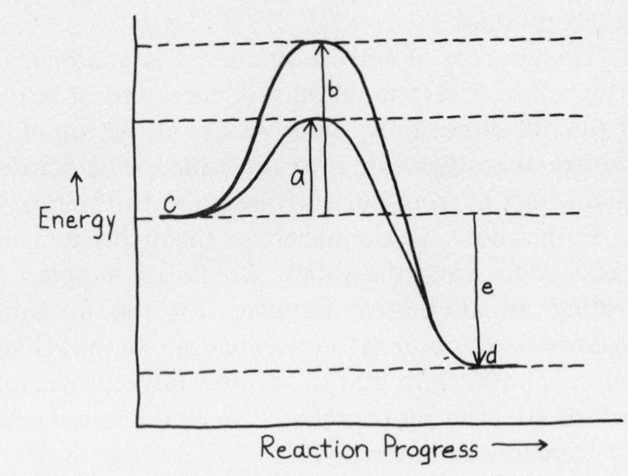

Solution: Arrow (a) represents the energy of activation when a catalyst is present, whereas the energy represented by arrow (b) is the energy of activation for the same reaction when no catalyst is present. The energy level of the reactants is indicated by letter (c). The energy level of the products is indicated by letter (d). The overall change in energy from the reactants to the products is represented by arrow (e). In this case, the energy of the products is less than the energy of the reactants; the reaction is an exothermic reaction, where energy is released.

18.2 Classification of Enzymes

OBJECTIVE Given the components of an enzyme, state whether it is a simple or a holoenzyme. Name the enzyme or a class of enzymes that would catalyze a reaction.

simple enzyme

Many enzymes consist only of protein and are called *simple enzymes*. The biological activity of a simple enzyme is solely a consequence of its protein structure.

cofactor

holoenzyme

apoenzyme

Other enzymes contain a nonprotein portion, called a *cofactor*, which is essential to their biological activity. The protein portion together with the cofactor is called a *holoenzyme*. The inactive protein portion alone is called the *apoenzyme*.

A cofactor may be a *metal ion*, such as Zn^{2+}, Cu^{2+}, Mg^{2+}, Fe^{2+}, or Fe^{3+}. For example, many enzymes that remove hydrogen from biological compounds (dehydrogenases) require the cofactor zinc for their activity. Another type of cofactor is an organic compound, such as a vitamin or hormone, which is called a *coenzyme*.

coenzyme

Apoenzyme +	Cofactor	= Holoenzyme
Protein	[Metal ion or vitamin/hormone (coenzyme)]	

Example 18.2 Indentify the following enzymes as simple or holoenzymes:

1. This enzyme requires Mg^{2+} for activity.
2. This enzyme consists of a protein only.
3. This enzyme contains vitamin B_1.

Solution:

1. An enzyme requiring a cofactor such as the metal ion Mg^{2+} is a holoenzyme.
2. An enzyme consisting only of protein is a simple enzyme.
3. Vitamin B_1, would be an organic cofactor or coenzyme, and the enzyme would be a holoenzyme.

Naming Enzymes

substrate name

The name of an enzyme is often derived from the name of the *substrate* upon which it acts. The substrate is the reactant that undergoes a change during a reaction. Changing the end of the substrate name to *-ase* gives the name of the enzyme. For example, the enzyme that hydrolyzes sucrose is called *sucrase*, and the enzymes that hydrolyze lipids are called *lipases*:

Substrate	*Enzyme*
Sucrose	Sucrase
Lipid	Lipase
Maltose	Maltase

reaction name

Other enzymes are named by the reaction they catalyze. An enzyme that removes CO_2 is called *decarboxylase*. The prefix de- means "to remove":

Reaction	*Enzyme*
Remove CO_2	Decarboxylase
Add H_2O	Hydrase
Hydrolyzes	Hydrolase

trivial name

There are also some trivial names still in use, such as pepsin, rennin, and trypsin, that do not indicate any substrate or reaction:

Function	*Enzyme*
Hydrolyzes proteins	Pepsin, rennin, trypsin

Example 18.3 Predict the name of an enzyme that would catalyze the reaction for the following substrates:

1. lactose
2. urea

Solution

1. The common name of an enzyme is obtained by changing the suffix of the substrate name to *-ase*. The enzyme for lactose would be called *lactase*.
2. The name of the enzyme for urea would be *urease*.

Example 18.4 State the names of the compounds that would be the substrates for the following enzymes whose names are given:

1. maltase
2. peptidase

Solution

1. The name of the substrate would be determined by removing the suffix *-ase* from the name of the enzyme and completing the name of the compound. The substrate for the maltase enzyme would be *maltose*.
2. The substrate of a peptidase is a *peptide*.

Classification of Enzymes

A systematic method of classifying enzymes similar to the IUPAC system for organic compounds was established in 1961. This system classifies enzymes according to six general types of reactions that enzymes catalyze. The major classifications are oxidoreductases, transferases, hydrolases, lyases, isomerases, and ligases.

oxidoreductases

The *oxidoreductases* are the enzymes that catalyze oxidation–reduction reactions. In these types of reactions, electrons or hydrogen are added to or removed from a substrate. For example, a dehydrogenase removes two hydrogen atoms from a substrate:

$$\underset{\text{Ethanol}}{CH_3CH_2OH} \xrightarrow[\text{dehydrogenase}]{\text{Alcohol}} \underset{\text{Acetaldehyde}}{CH_3\overset{\overset{\displaystyle O}{\|}}{C}H}$$

transferases

The *transferases* move a chemical group from one substrate to another. For example, a transaminase catalyzes the transfer of an amino group from an amino acid to a keto acid:

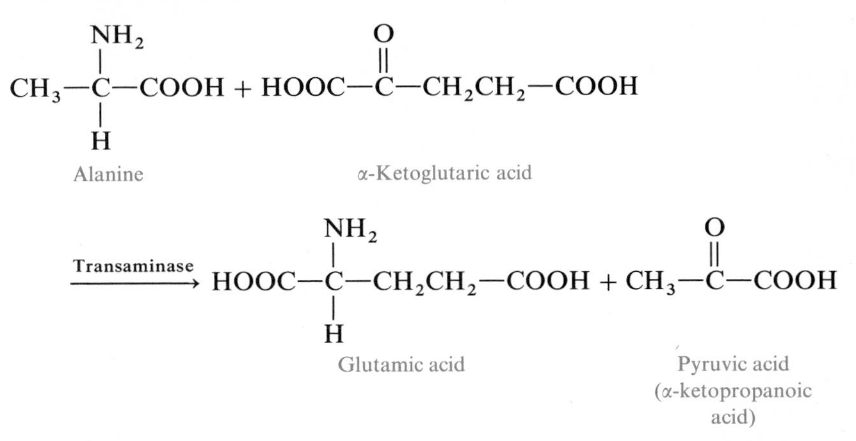

hydrolase

The *hydrolases* split a substrate into smaller molecules by adding water. They include many of the digestive enzymes that hydrolyze (a) amide bonds in proteins (amidases), (b) carbohydrates (carbohydrases), and (c) ester bonds in lipids (lipases):

$$\underset{\text{(Lipid)}}{\text{Triacylglycerol}} + 3H_2O \xrightarrow{\text{Lipase}} 3\text{ Fatty acids} + \text{Glycerol}$$

lyases

The *lyases* add groups such as CO_2, H_2O, and NH_3 to substrates with double bonds, or they remove groups to form double bonds. Examples of lyases include decarboxylases, deaminases, and hydrases:

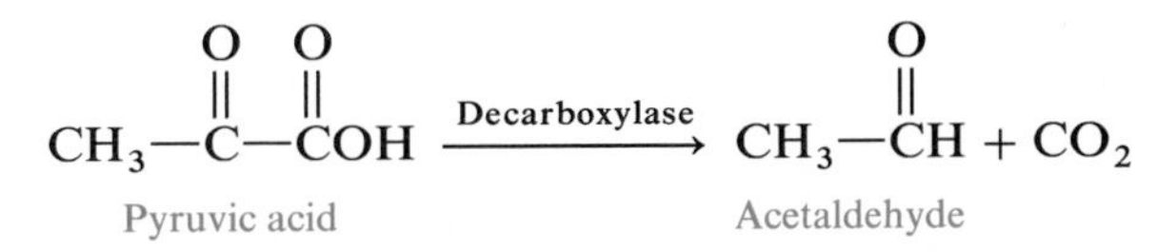

isomerases

The *isomerases* rearrange the molecular structure of a substrate by isomerization:

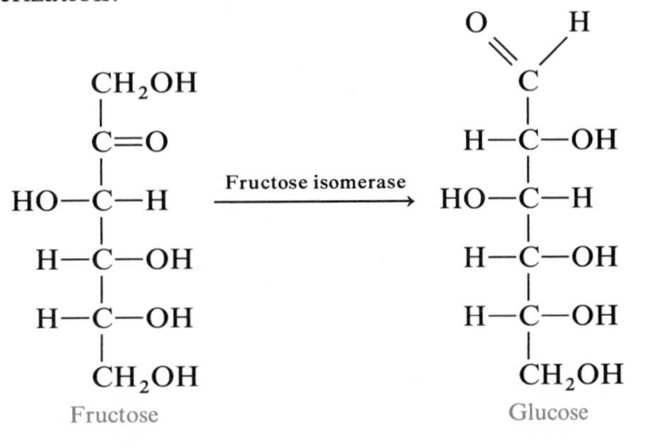

ligases

The *ligases* are enzymes that combine two molecules using an energy source such as ATP. Examples of ligases include synthetases and carboxylases:

$$\underset{\text{Pyruvic acid}}{CH_3\overset{\overset{\displaystyle O}{\|}}{C}-\overset{\overset{\displaystyle O}{\|}}{C}OH} + CO_2 + ATP \xrightarrow{\text{Carboxylase}} \underset{\text{Oxaloacetic acid}}{HO\overset{\overset{\displaystyle O}{\|}}{C}CH_2\overset{\overset{\displaystyle O}{\|}}{C}-\overset{\overset{\displaystyle O}{\|}}{C}OH} + ADP + Pi$$

18.3 Enzymes as Catalysts

OBJECTIVE Write an equation for an enzyme-catalyzed reaction and describe the effects of substrate concentration, enzyme concentration, temperature, and pH on the activity of an enzyme.

substrate

Enzymes are biological catalysts that lower the energy of activation of a chemical reaction in a living cell. Enzymes must first combine with a reactant called the *substrate* (S). Many enzymes demonstrate a strong attraction to just one specific substrate. We say that they are *substrate specific*. For example, the enzyme maltase hydrolyzes only the substrate maltose; sucrase only catalyzes the hydrolysis of sucrose. The recognition of a certain substrate is a result of the three-dimensional structure of the protein portion of the enzyme. Only certain substrates are compatible with the unique tertiary structure of a particular enzyme. Every molecule of a particular enzyme has the same amino acid sequence, the same tertiary structure, and identical substrate specificity.

Enzyme Action

lock-and-key theory

enzyme–substrate complex

The reaction between an enzyme and its substrate can be described by the lock-and-key theory. The enzyme can be thought of as the lock, with the substrate acting as the key. The substrate (S) must have a structure that fits into the enzyme's (E) structure. The combination of the substrate with the enzyme is called the *enzyme–substrate complex* (ES):

$$E + S \rightleftharpoons ES$$

active site

Within the enzyme, there is an area called the *active site*, which comes in contact with the substrate and converts it to product. This active site of an enzyme may consist of just a few of the amino acids that are close together in the spatial orientation of the protein. Finally, the product is released and the enzyme is available to react with another substrate molecule:

$$E + S \rightleftharpoons ES \rightleftharpoons EP \rightleftharpoons E + P$$

For example, the enzyme sucrase is released by the pancreas into the small intestine, where it reacts with the sugar sucrose. The enzyme and the sugar form a sucrase–sucrose complex. Then the active site on the enzyme catalyzes the

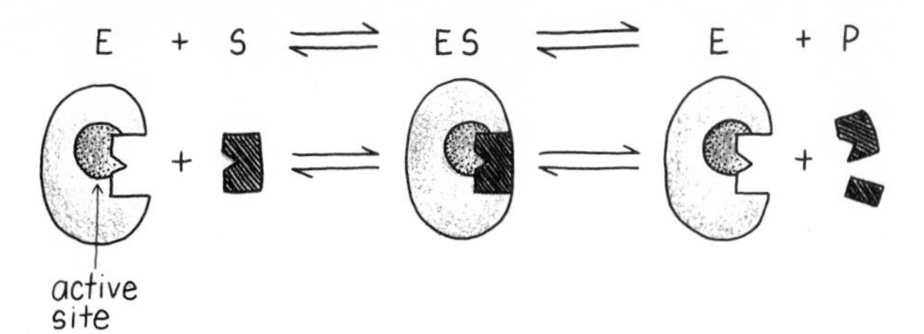

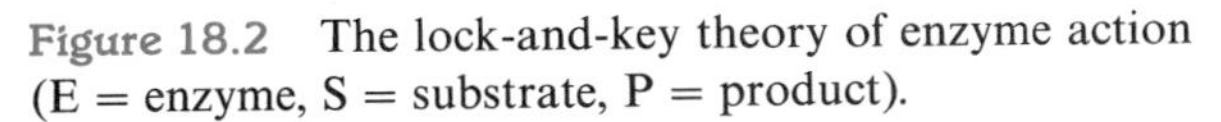

Figure 18.2 The lock-and-key theory of enzyme action (E = enzyme, S = substrate, P = product).

hydrolysis of the sucrose to give the monosaccharide glucose and fructose. The products are then released from the enzyme (see Figure 18.2):

E	+	S	⇌	ES	⇌	E	+	P
Sucrase		Sucrose		Sucrase–sucrose complex		Sucrase		Glucose + Fructose

Example 18.5 The enzyme maltase catalyzes the hydrolysis of the disaccharide maltose. Write a symbolic equation to represent this process.

Solution: In the equation for the hydrolysis of maltose, the enzyme combines with the substrate to form an enzyme–substrate complex. Within this complex, the hydrolysis reaction occurs and the products consisting of two glucose molecules form. The enzyme no longer holds on to the products and the glucose molecules are released. leaving the enzyme available to catalyze additional maltose molecules:

E	+	S	⇌	ES	⇌	E	+	P
Maltase		Maltose		Maltose–maltase complex		Maltase		2 Glucose molecules

Enzyme Activity

substrate concentration

The activity of an enzyme is a measure of the rate of formation of product(s). Several factors influence the rate at which an enzyme converts a substrate to product. An increase in the concentration of substrate (S) causes an increase in the rate of reaction (enzyme activity) until the enzyme is saturated with substrate molecules. Then no further increase in rate can occur even if the concentration of S is increased. (See Figure 18.3).

enzyme concentration

If the concentration of the enzyme (E) is increased, there will be an increase in the rate of reaction. This occurs because the availability of more catalyst means that more substrate molecules can undergo reaction. (See Figure 18.4).

Enzymes show little activity at low temperatures such as 0°C. Low temperatures slow down enzymatic and metabolic activity. As temperatures

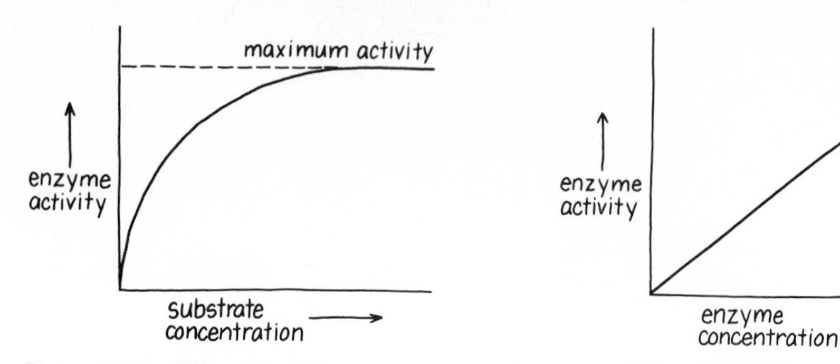

Figure 18.3 Effect of substrate concentration on enzyme activity.

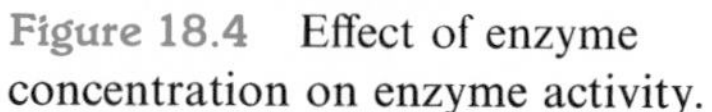

Figure 18.4 Effect of enzyme concentration on enzyme activity.

increase, the rates of enzyme-catalyzed reactions increase until the protein of the enzyme is denatured and rendered inactive. Enzymes usually show maximum activity at *optimum temperature*, which is usually 37°C (body temperature). Increasing the temperature up to 50°C may accelerate enzymatic reactions; however above 50°C, denaturation of the enzyme begins and biological activity decreases rapidly. (See Figure 18.5.)

optimum temperature

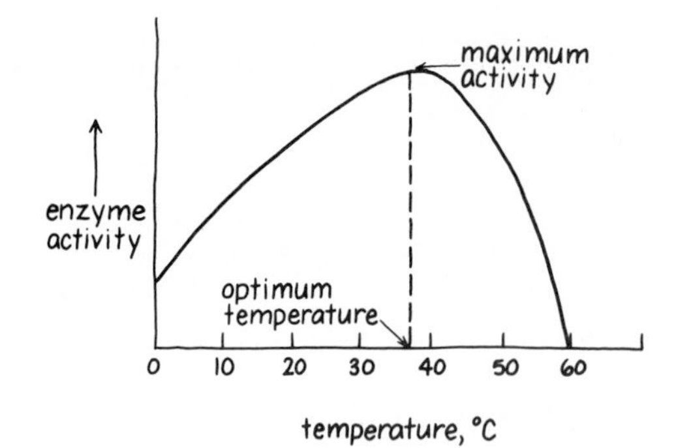

Figure 18.5 Effect of temperature on enzyme activity.

optimum pH

Enzymes also have an *optimum pH* at which they are most active. At high or low pH values, enzymatic activity decreases markedly because denaturation of the protein structure occurs when pH is below or above the optimum pH. In the body, optimum pH is generally the pH of the body fluids. The digestive enzymes in the stomach, such as pepsin, are most active at about pH 1.0–2.0. If the pH in the stomach rises, the digestive enzymes can no longer catalyze the hydrolysis of proteins. (See Figure 18.6.) Table 18.1 lists the optimum pH values for some enzymes.

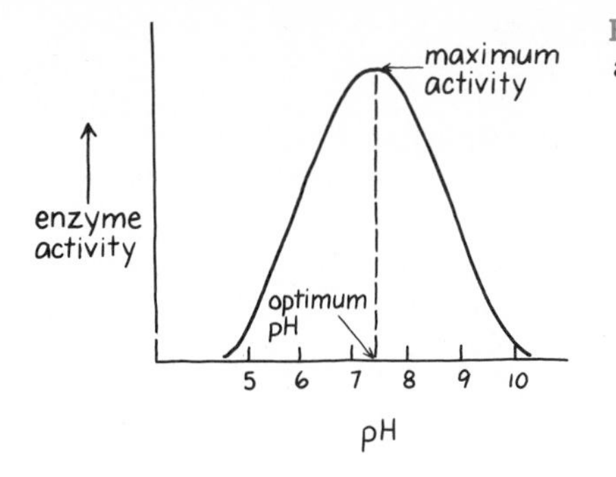

Figure 18.6 Effect of pH on enzyme activity.

Table 18.1 Optimum pH for Some Enzymes

Enzyme	Optimum pH
Arginase	9.7
Trypsin	7.7
Catalase	7.6
Fumarase	6.5
Pyruvate carboxylase	4.8
Pepsin	1.5

Example 18.6 Describe the effects of the following on the activity of the maltase enzyme, which catalyzes the hydrolysis of the substrate maltose:

1. increasing the maltose concentration
2. increasing the maltase concentration
3. decreasing the temperature to 20°C
4. increasing the pH above optimum pH

Solution

1. The activity of the enzyme will increase as the concentration of the substrate maltose increases. The ES complex will form more rapidly, giving a more rapid production of the products. This increase will continue until all the available maltase enzyme is saturated (combined with substrate). Then no further change in activity will occur.
2. Increasing the enzyme concentration increases the rate of formation of the ES complex. Thus the activity of the enzyme increases.
3. Lowering the temperature will slow down the rate of the reaction. Thus the enzyme activity will decrease.
4. If an enzyme is operating at optimum pH, then increasing the pH above optimum pH will be detrimental to the protein structure of the enzyme and may completely denature the enzyme. Thus the enzyme activity will decrease.

HEALTH NOTE

Enzymes as Diagnostic Tools

isoenzymes

Enzymes can occur as isomeric forms, called *isoenzymes*, which are utilized as diagnostic tools in medicine. Isoenzymes have slightly different structures but catalyze the same reactions. Probably the most fully studied group of isoenzymes are those of lactic dehydrogenase (LDH). The active form of the enzyme consists of four subunits. These subunits are of two different types. One is labeled H because it is the predominant subunit present in the LDH enzyme in heart muscle cells. The other is M, the predominant subunit found in other muscle cells. There are five possible combinations of subunits, each having slightly different properties, which allows their separation and identification:

$$M_4 \quad M_3H \quad M_2H_2 \quad MH_3 \quad H_4$$

The LDH enzyme in healthy tissues stays within the cells, so the level of LDH enzyme in the serum of the blood is normally quite low. However, if the cells are damaged, the LDH enzymes spill into the blood, thereby elevating the serum levels of LDH that can be detected. By determining which isoenzyme is present in the blood, it is possible to identify the type of tissue that has been damaged.

The M_4 LDH isoenzyme predominates in liver tissue. Liver disease and consequent damage to the liver may be detected by the elevation of the serum M_4 LDH level. One of the results of myocardial infarction (MI) is necrosis (death) of some of the cells of the heart muscle. This increases the level of H_4 LDH isoenzyme in the serum.

Another enzyme that is used in the assessment of myocardial infarction is glutamate oxaloacetate transaminase (GOT). This enzyme is present in large amounts in the heart muscle, but only a trace level is normally found in the serum as serum

SGOT

GOT or SGOT. When heart damage occurs, the SGOT level rises rapidly 6–12 h after

18.4 Enzyme Inhibition

OBJECTIVE Describe the effect of an enzyme inhibitor upon the enzyme activity in competitive and noncompetitive inhibition.

inhibitors slow or stop enzyme action

Inhibitors are chemical compounds that interfere with the ability of an enzyme to react properly with its substrate. In medicine, certain drugs act as inhibitors by inactivating an enzyme essential to the growth process of bacteria. The growth cycle for bacteria requires enzymes different from those in the host cells (human body). By inhibiting one of the enzymes in the bacteria, bacterial growth can be stopped. Inhibition of growth in viruses has been less successful because viruses use the enzyme systems of the host cells; an inhibitor of a virus also destroys the host cells. There are two types of inhibitors: competitive inhibitors and noncompetitive inhibitors.

the first symptoms of difficulty. In severe cases, the level of SGOT enzyme can reach 20 times the normal value. The SGOT enzyme will return to normal levels in 2–3 days because it is eliminated rapidly from the bloodstream. (See Figure 18.7.)

Acid phosphatase, an enzyme found in the prostate and the intestine, is an indicator of cancer when found in blood serum. When amylase, an enzyme of the pancreas, is found in the blood serum, it can indicate liver disease.

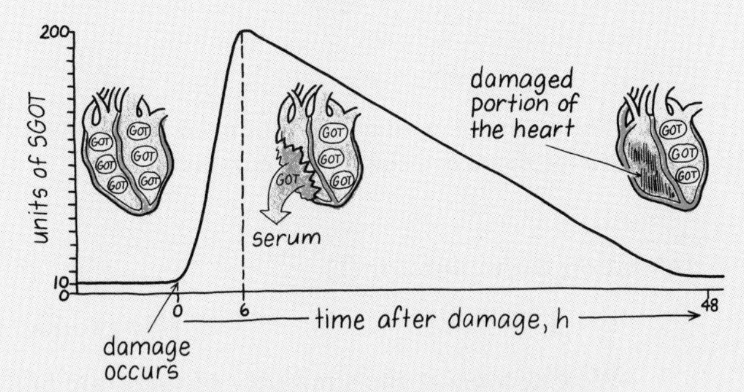

Figure 18.7 Serum glutamate oxaloacetate transaminase (SGOT). In a normal heart, glutamate oxaloacetate transaminase (GOT) is maintained within the cells. Should heart damage and necrosis (death) of heart cells occur, GOT spills into the general circulatory system, where it is then called serum GOT, or SGOT. The amount of SGOT appearing in the blood can indicate the severity of heart damage.

Competitive Inhibitors

A substance that acts as a competitive inhibitor competes with the substrate for the active site on the enzyme. The enzyme mistakes the inhibitor molecule for the substrate because a competitive inhibitor is very similar in structure to the substrate. When the inhibitor substance occupies the active site, the substrate of that enzyme cannot undergo reaction. A competitive inhibition is reversible. If more substrate molecules are added to compete with the inhibitor for the active site, the effects of the inhibition can be reversed. (See Figure 18.8.)

Competitive Inhibition

Reaction with substrate (S):

$$E + S \rightleftharpoons ES \rightleftharpoons E + P \qquad \text{(Product forms)}$$

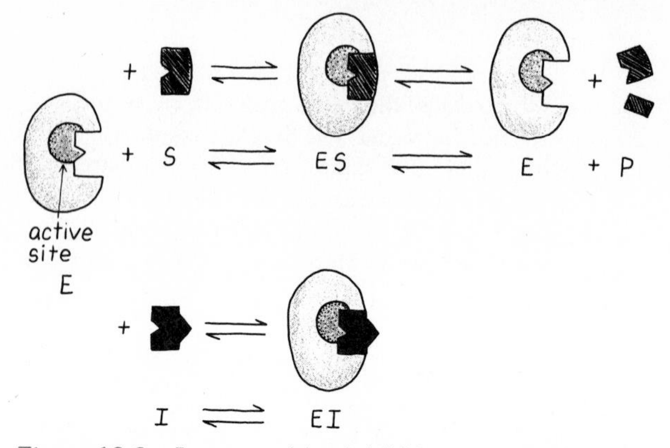

Figure 18.8 In competitive inhibition, an inhibitor (I) competes for the active site on the enzyme (E).

Inhibition with inhibitor (I):

$$E + I \rightleftharpoons EI \qquad \text{(No product forms)}$$

We can look at the similarity of substrate and inhibitor by examining the reaction by the enzyme succinic dehydrogenase. Normally, the substrate is succinic acid, which is oxidized by the enzyme. An inhibitor for the reaction is malonic acid, which has a structure similar to that of succinic acid:

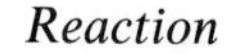

Reaction

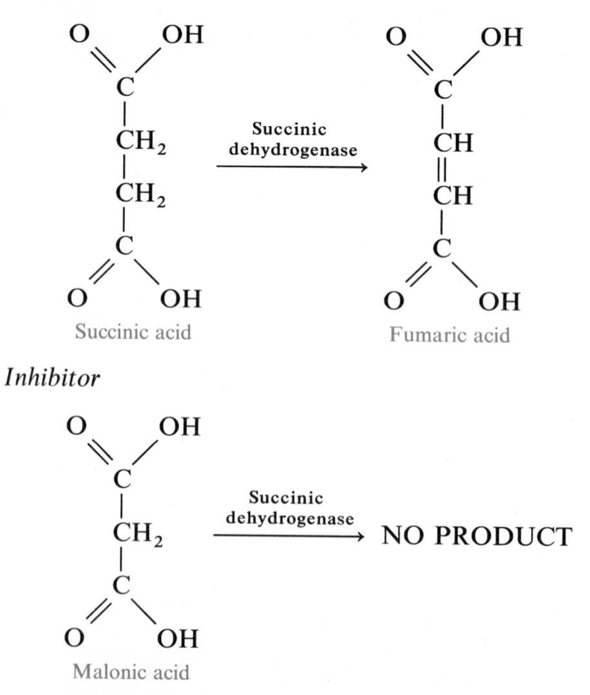

HEALTH NOTE

Competitive Inhibitors Used in Medicine

antimetabolite

When an illness is caused by an invading microorganism such as a bacterium, a competitive inhibitor called an *antimetabolite* may be used to inhibit an enzyme and disrupt the action of that bacterium. A classic example is sulfanilamide, one of the early sulfa drugs used to fight infection. Sulfanilamide competes with an essential substrate (metabolite) in the growth cycle of bacteria. During their growth, many bacteria must synthesize folic acid, and PABA (*p*-aminobenzoic acid) is required for that synthesis. To control the reproduction of these bacteria, sulfanilamide and another antimetabolite, *p*-aminosalicylic acid, are used. Both are chemically similar to PABA and can inhibit the enzyme, tying it up so that little or no enzyme is available for folic acid synthesis. The growth of the bacteria stops.

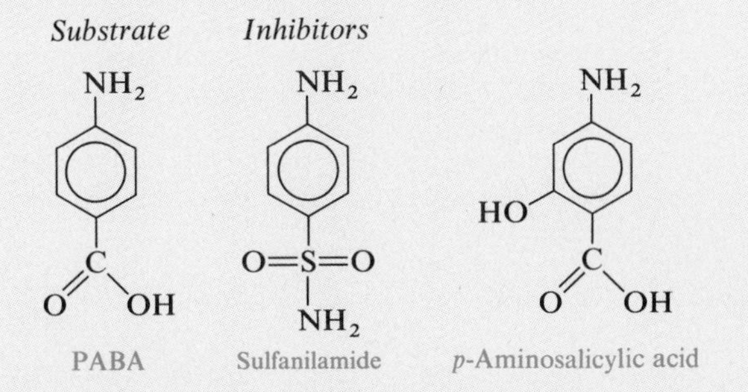

dicumarol

The drug dicumarol is used to control blood clotting through interference with the enzymatic production of prothrombin, an agent essential to the clotting of blood. The actual mode of dicumarol action is not known. Research indicates that half the dicumarol molecule (called *coumarin*) competes with vitamin K for the protein portion of the enzyme responsible for the synthesis of prothrombin. As a result, dicumarol inhibits the blood-clotting process. Such an effect is desired in treating coronary and pulmonary emboli and in preventing the formation of emboli in the treatment of high blood pressure. If hemorrhaging occurs or surgery is required, large amounts of vitamin K are usually given to reverse the effects of the inhibitor.

Noncompetitive Inhibition

A noncompetitive inhibitor is thought to alter the shape of the enzyme and greatly reduce its affinity for the substrate. It is not competing with the substrate for the active site and does not need to resemble the structure of the substrate. The effect of noncompetitive inhibition is not reversed by increasing the substrate concentration. (See Figure 18.9.)

A noncompetitive inhibitor can bind to an enzyme in many ways. If it binds somewhere on the surface of the enzyme, a change occurs in the tertiary structure of the enzyme. This can prevent the substrate from combining with the

HEALTH NOTE

Antibiotics

An *antibiotic* is a type of antimetabolite produced by one microorganism that is poisonous to another microorganism. One of the most commonly used antibiotics is penicillin, which interferes with transpeptidase, an enzyme needed in the formation of bacteria cell walls. Without a complete cell wall, bacteria cannot survive; thus the infection is stopped. (The enzyme transpeptidase is not found in human cell membrane formation.) However, some bacteria have developed a resistance to penicillin by forming an enzyme called penicillinase, which breaks down penicillin. Therefore, researchers have prepared several derivatives of penicillin, such as penicillin G and V, ampicillin, and amoxicillin, to which the bacteria have not yet developed a resistance:

Penicillin | *R Groups for Penicillin Derivatives*

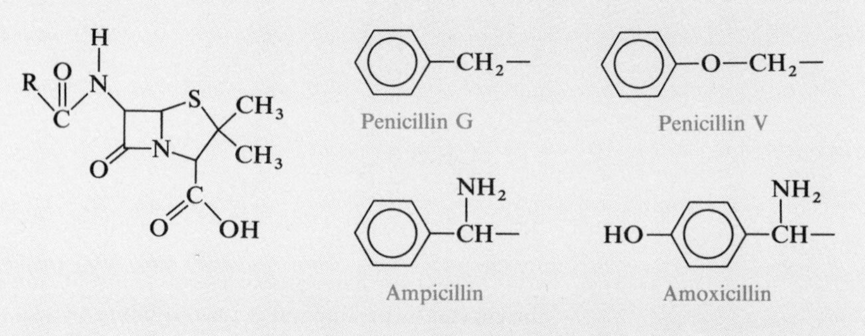

Penicillin G | Penicillin V | Ampicillin | Amoxicillin

Another group of antibiotics, which include tetracycline, streptomycin, and aureomycin, are also used to stop bacterial infections. They inhibit the synthesis of proteins in certain microorganisms. They are also beneficial for fighting infections in people who are allergic to the penicillin type of antibiotic.

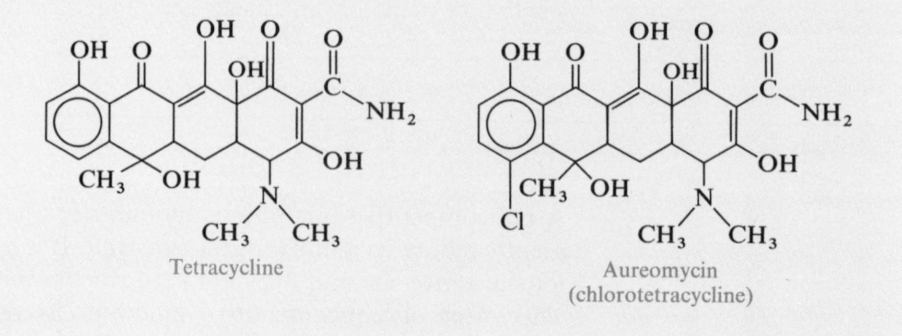

Tetracycline | Aureomycin (chlorotetracycline)

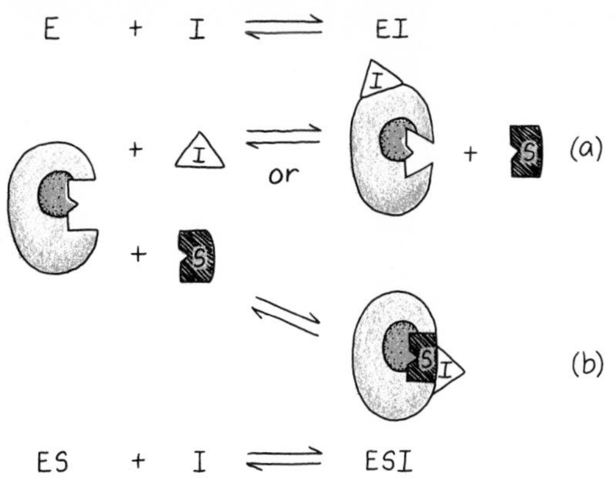

Figure 18.9 In noncompetitive inhibition, the following events take place: (a) The enzyme structure is altered by the inhibitor, preventing substrate attachment; or (b) the substrate is blocked by the inhibitor.

enzyme or make it difficult for the substrate to react with the active site. In that case, both inhibitor and substrate are attached to the enzyme at the same time:

Reaction with substrate: $E + S \rightleftharpoons ES \rightleftharpoons E + P$

Reaction with noncompetitive inhibitor: $E + I \rightleftharpoons EI$ or

$ES + I \rightleftharpoons ESI$

Example 18.7 State the type of inhibition in the following:

1. The inhibitor has a structure that is similar to the structure of the substrate.
2. The inhibitor binds to the surface of the enzyme, changing its structure such that the enzyme cannot accept the substrate.
3. The inhibitor competes with the substrate for the active site.

Solution

1. competitive inhibition
2. noncompetitive inhibition
3. competitive inhibition

18.5 Vitamins and Coenzymes

OBJECTIVE Given the name of a vitamin, indicate whether it is a water- or fat-soluble vitamin and give its dietary source and biological function.

Vitamins are substances that are found naturally in various foods. They are required in the diet because the body is unable to synthesize them. They are also needed regularly because they are metabolized in the body or are eventually excreted. Most vitamins are present only in limited amounts in the body, although the fat-soluble vitamins can be stored by the liver for longer periods of time.

Vitamins are organic compounds that are carried by the bloodstream and are needed in small amounts by the body. They often act as coenzymes and if not present in proper amounts are likely to cause certain illnesses.

The recommended daily requirements for vitamins vary considerably, depending on the size, age, and activity of a person. A large person or one who is very active, as well as growing children, require larger amounts of vitamins.

Table 18.2 Some Important Vitamins and Their Major Sources in the Diet

Vitamin	Recommended Daily Requirement	Major Source
A (retinol)	1.7 mg	Milk, butter, cod liver oil, green and yellow vegetables
B_1 (thiamine)	1.5 mg	Cereal grains, organ meats, yeast, nuts
B_2 (riboflavin)	1.8 mg	Cheese, milk, meats, yeast, liver
B_3 (niacin)	20.0 mg	Yeast, wheat germ, meat
B_6 (pyridoxine)	1.5 mg	Egg yolks, yeast, liver, wheat germ
B_{12} (cyanocobalamin)	1.2 μg	Liver and kidneys, meats
C (ascorbic acid)	75.0 mg	Citrus fruits, green vegetables, tomatoes
Pantothenic acid	10–15 mg	Liver, beef, milk, fish, eggs
Biotin	0.25 mg	Liver, eggs, milk, molasses
Folic acid	0.4 mg	Liver, green leafy vegetables
D (calciferol)	11.0 μg	Fish oils, egg yolks, sunlight
E (tocopherol)	Not known	Many foods, including eggs, milk, fish, green leafy vegetables, wheat germ oil
K (phylloquinone)		Green leafy vegetables, liver, eggs; synthesized in the colon

When a person is sick, vitamins are needed to help fight the illness. During pregnancy, a mother needs increased amounts of vitamins, particularly vitamin D. Table 18.2 lists some important vitamins, as well as their typical daily requirements and their major sources in the diet.

Classification of Vitamins

Vitamins are usually divided into two classes, depending on their solubilities. Those that are soluble in water, the water-soluble vitamins, include vitamins C, B_1 (thiamine), and B_2 (riboflavin). Other vitamins are insoluble in water but

Table 18.3 Water-Soluble and Fat-Soluble Vitamins

Water Soluble	Fat Soluble
Vitamin B_1 (thiamine)	Vitamin A (retinol)
Vitamin B_2 (riboflavin)	Vitamin D (calciferol)
Vitamin B_5 (niacin)	Vitamin E (tocopherol)
Vitamin B_6 (pyridoxine)	Vitamin K (phylloquinone)
Vitamin B_{12} (cyanocobalamin)	
Vitamin C (ascorbic acid)	
Pantothenic acid	
Biotin	
Folic acid	

soluble in organic solvents. These fat-soluble vitamins include vitamins A, E, D, and K. (See Table 18.3.)

Vitamin A (Retinol)

β-carotene

vitamin A is fat soluble

Vitamin A occurs in the body as retinal, the aldehyde form of retinol. The precursors of vitamin A, the β-carotenes, are found in vegetables such as tomatoes, carrots, and sweet potatoes, where they are responsible for the red or yellow colors. The carotenes are changed into vitamin A within the body. Both substances are terpenes, which makes them part of the lipid family. Thus, vitamin A is a fat-soluble vitamin:

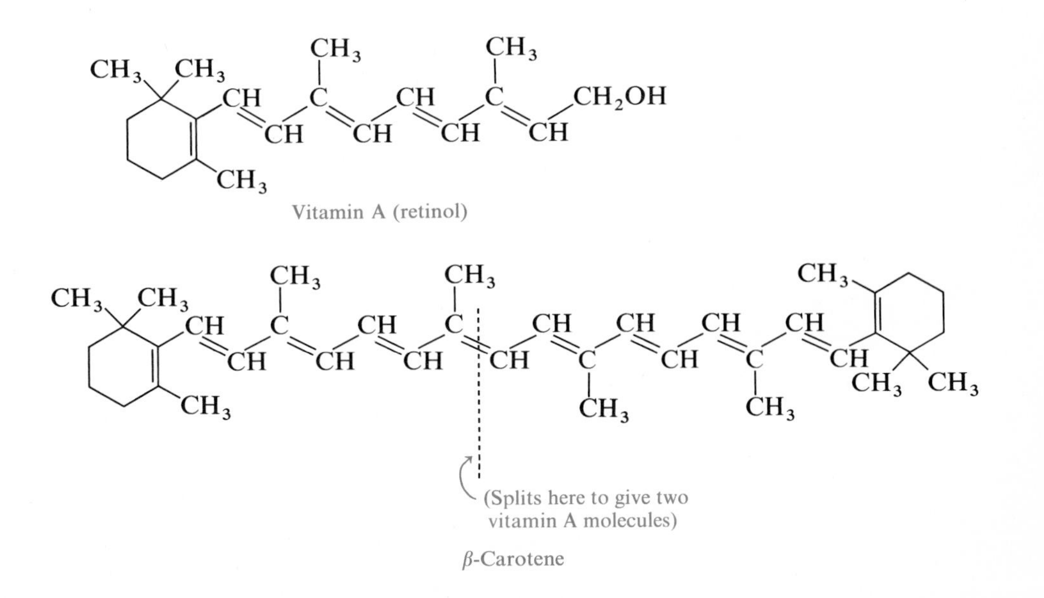

Vitamin A (retinol)

β-Carotene

HEALTH NOTE

Role of Vitamin A in Vision

A major function of vitamin A is the formation of pigments of the retina. Vitamin A is oxidized to form an aldehyde called *retinal.* Retinal combines with a protein (opsin) in the retina to produce *rhodopsin,* also known as *visual purple.* When light strikes rhodopsin, it breaks down into opsin and *trans*-retinal, which is the isomer of the active form, *cis*-retinal. At the same time, the optic nerve is stimulated, and an image is seen. The *trans*-retinal is converted back to *cis*-retinal by an enzyme (isomerase). The *cis*-retinal can now reform the visual purple by combining with opsin:

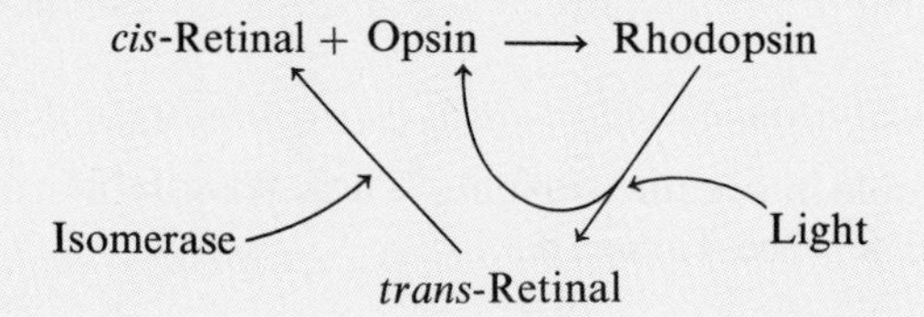

When the quantity of vitamin A in the body is low, the re-formation of rhodopsin is slow. When there is a deficiency of vitamin A, night blindness may occur. The amount of light is too small to stimulate the breakdown of the small amount of rhodopsin present. A deficiency of vitamin A can also cause the skin to become scaly and subject to acne. A lack of vitamin A stunts growth and causes sterility in animals.

hypervitaminosis A

Hypervitaminosis A (too much vitamin A) can occur because vitamin A is a fat-soluble vitamin and thus can be stored for longer periods of time in the body. Such cases have been reported by hunters who have ingested polar bear liver, which is high in vitamin A. Toxicity symptoms include nausea, vomiting, dizziness, and irritability followed by a sloughing of the skin. Persons taking megadosages (over 40,000 units per day) of vitamin A for long periods of time may experience bone pain, hair loss, fissure around the lips, and weight loss.

Vitamin B_1 (Thiamine)

The B vitamins are water soluble

Vitamin B_1, thiamine, is a water-soluble vitamin:

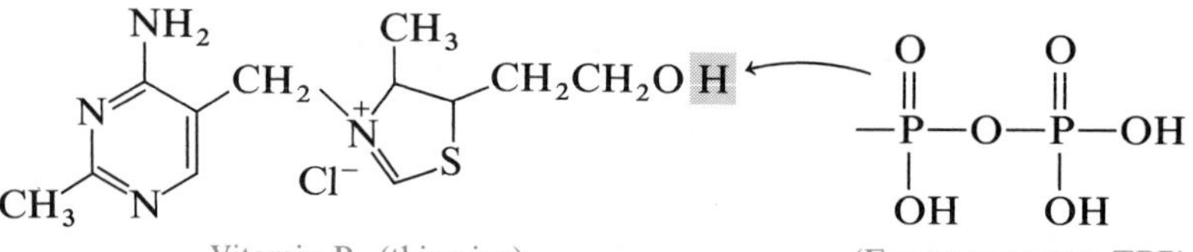

Vitamin B_1 (thiamine)

(Forms coenzyme TPP)

In a phosphate form, thiamine (thiamine pyrophosphate) functions as a coenzyme for a decarboxylase enzyme in decarboxylation reactions of α-keto acids such as pyruvic acid and α-ketoglutaric acid:

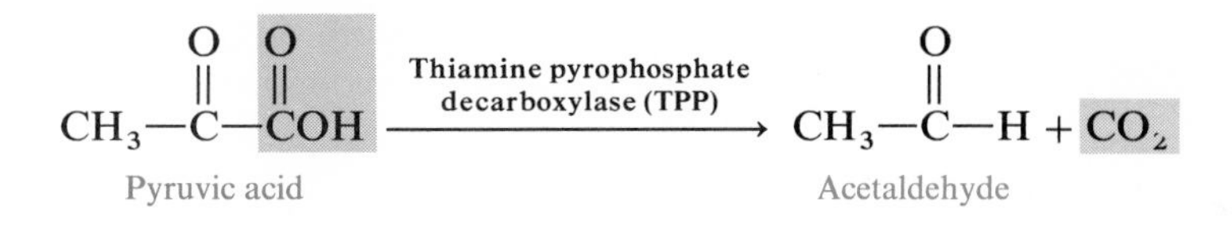

A deficiency of thiamine in the diet decreases the utilization of pyruvic acid and affects the metabolism of carbohydrates. Because the nervous system depends on the metabolism of carbohydrates for most of its energy, thiamine deficiency has a major effect on the nervous system. The protective sheath surrounding the nerve fibers may also degenerate, causing pain and irritation of the nerves, a condition known as *polyneuritis*. In a severe thiamine deficiency, paralysis and heart failure may occur. All these symptoms are collectively referred to as *beriberi*.

beriberi

Vitamin B_2 (Riboflavin)

Vitamin B_2, riboflavin, another B vitamin, is also water soluble:

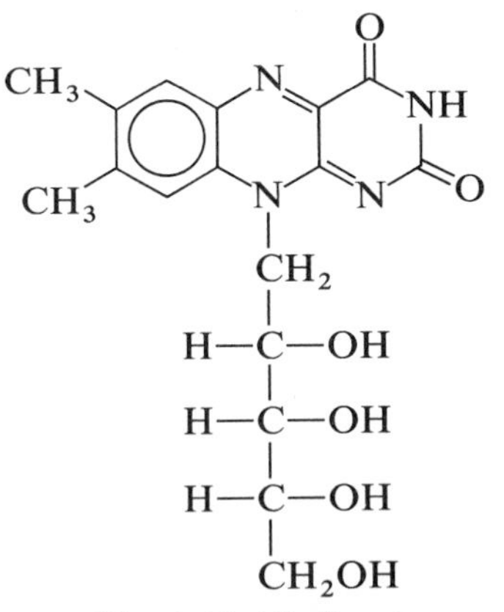

Vitamin B_2 (riboflavin)

Riboflavin combines with phosphoric acid to form two coenzymes that are essential to the cells of the body: flavin mononucleotide (FMN) and flavin adenine dinucleotide (FAD). Both of these coenzymes are involved with enzymes (flavoenzymes) that catalyze oxidation–reduction reactions in which these coenzymes accept or donate hydrogen, i.e., they act as dehydrogenases.

One of the most common symptoms of riboflavin deficiency is inflammation and cracking of the skin at the corners of the mouth. Severe lack of riboflavin does not seem to be a serious problem in humans.

Vitamin B_3 [Niacin (Nicotinic Acid)]

Vitamin B_3 functions in the tissues of the body as a component of the coenzymes nicotinamide adenine dinucleotide (NAD) and nicotinamide adenine dinucleotide phosphate (NADP). Both of these coenzymes are involved with hydrogen transfer during oxidation–reduction reactions in the cells:

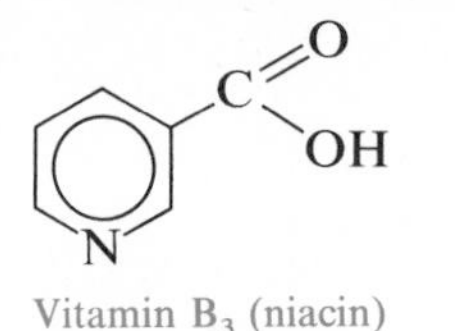

Vitamin B_3 (niacin)

Niacin does not meet the strict definition of a vitamin since it can be synthesized from the amino acid tryptophan in the body. If the levels of tryptophan as well as niacin are low, there will be a deficiency in the vitamin. Such deficiency can occur in people whose diets depend on corn, which lacks the amino acid tryptophan. Lack of niacin in the diet causes a condition called *pellagra*, whose symptoms include muscular weakness, scaliness of the skin, and severe irritations of the mouth and gastrointestinal tract. Pellegra can be prevented by adding whole wheat cereals to the diet.

pellagra

Vitamin B_6 (Pyridoxine)

Pyridoxine functions as a coenzyme in the form of pyridoxal phosphate. It is required by enzymes, such as transaminases and other group-transfer enzymes, that catalyze reactions involving amino acids and the metabolism of proteins. The metabolism of the amino acids tryptophan, glycine, serine, glutamic acid, and cysteine depends on vitamin B_6:

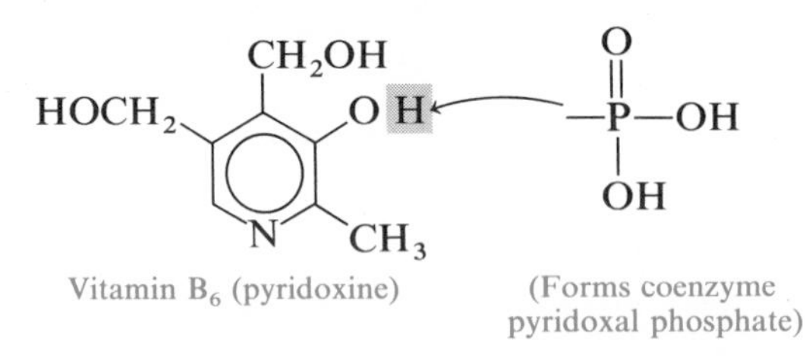

Vitamin B_6 (pyridoxine) (Forms coenzyme pyridoxal phosphate)

A pyridoxine deficiency causes dermatitis, skin irritations, nausea, and vomiting, but the condition is rare since the vitamin is abundant in many foods. However, it has occurred in some children because of the destruction of the vitamin through food processing methods used in the preparation of some infant formulas.

Vitamin B_{12} (Cobalamin)

Cobalamin represents several compounds that contain the element cobalt. Vitamin B_{12} functions primarily as a coenzyme in the conversion of ribonucleotide to deoxyribonucleotides, an important step in the formation of DNA and methionine. It is also important in the growth and maturation of red blood cells:

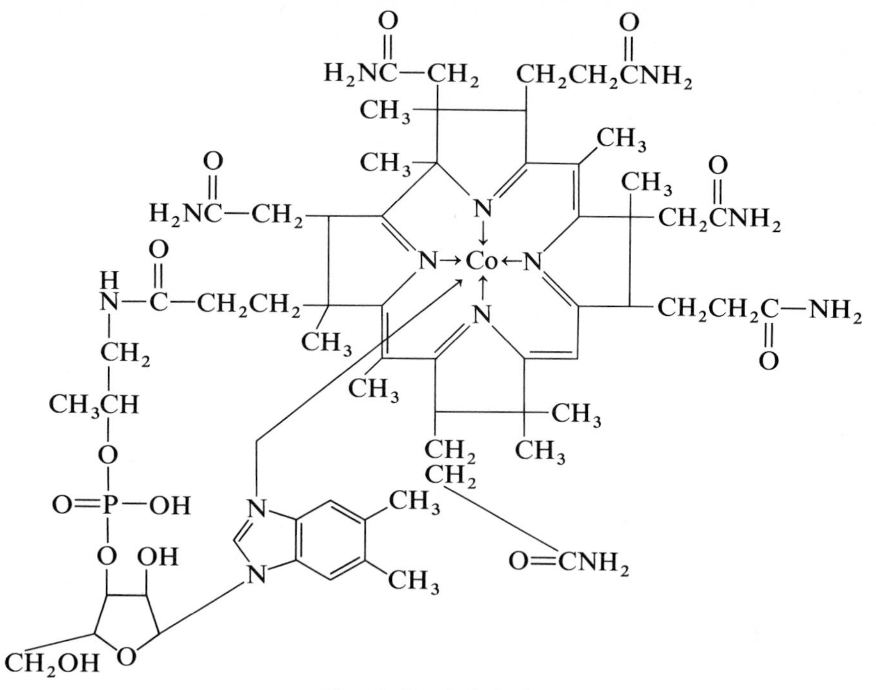

Vitamin B_{12} (cobalamin)

In vitamin B_{12} deficiency, the myelin sheath around the nerves degenerates, causing a loss of peripheral sensation and paralysis. This also occurs in pernicious anemia, where blood cells do not mature because of the intestine's inability to absorb vitamin B_{12} from dietary sources.

Vitamin C (Ascorbic Acid)

vitamin C is water soluble

Although the exact function of vitamin C, ascorbic acid, is not known, it does appear to play a role in the hydroxylation reaction in the synthesis of collagen. Vitamin C is a water-soluble vitamin and is rapidly excreted in the urine. Therefore, vitamin C must be replenished by the diet every day:

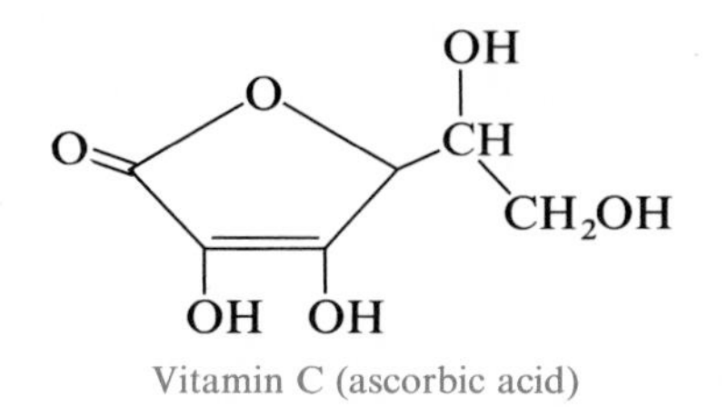

Vitamin C (ascorbic acid)

scurvy

The lack of ascorbic acid brings about a condition called *scurvy*, whose symptoms include slow healing of wounds, hemorrhaging caused by a weakening of the walls of the blood vessels, bleeding gums, an itching of the skin, anemia, and vomiting. This is corrected when vitamin C is added to the diet. There is much controversy over the effects of high dosages (over 1 g per day) of vitamin C on the common cold. However, several studies have shown no significant beneficial effects. There appear to be no toxicity levels of vitamin C, although high dosages may interfere with the absorption of vitamin B_{12}.

Pantothenic Acid

Pantothenic acid is an important component in the structure of coenzyme A, which is involved in enzymatic reactions that transfer acyl groups during the synthesis of fatty acids and in their oxidation. Deficiency symptoms are rare since the vitamin is found in many foods, including liver, beef, fish, milk, and eggs:

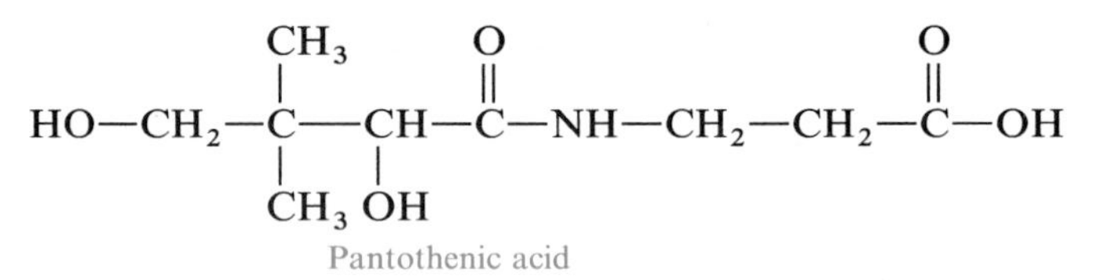

Pantothenic acid

Biotin

Biotin is a vitamin that acts as a coenzyme for enzymes (carboxylases) that combine carbon dioxide with other substances during the synthesis reactions of fatty acids and urea:

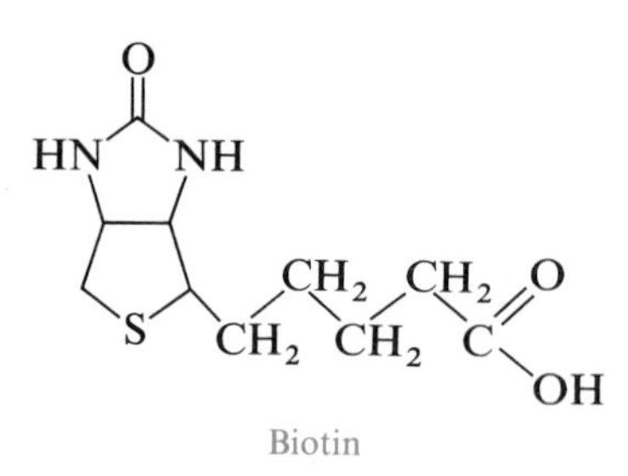

Biotin

Biotin deficiencies are rare because the vitamin is normally produced by intestinal bacteria in sufficient quantities. However, large quantities of egg whites can bind biotin to produce deficiency symptoms of fatigue, depression, nausea, dermatitis, and muscle pains.

Folic Acid

The vitamin folic acid is reduced to form its coenzyme, tetrahydrofolic acid (FH_4), which aids in the transfer reactions of single-carbon groups such as methyl, ($-CH_3$).

Symptoms of folic acid deficiency include diarrhea and anemia. Sources of folic acid include liver and leafy green vegetables:

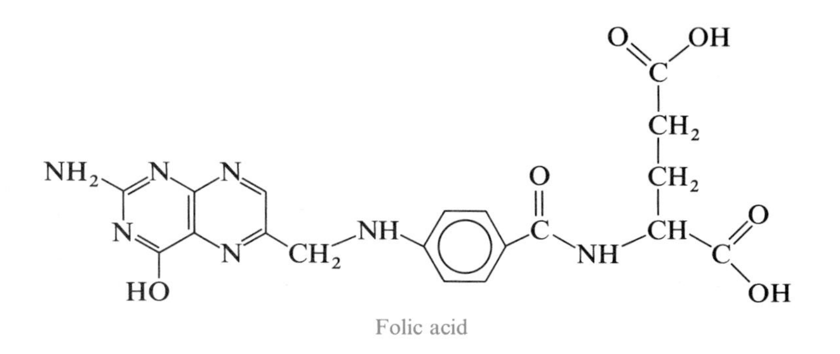

Folic acid

Vitamin D

vitamin D is fat soluble

Vitamin D is a member of the sterol family and is therefore a fat-soluble vitamin. It increases the absorption of calcium from the intestinal tract and regulates the amount of calcium deposited in the bones and teeth:

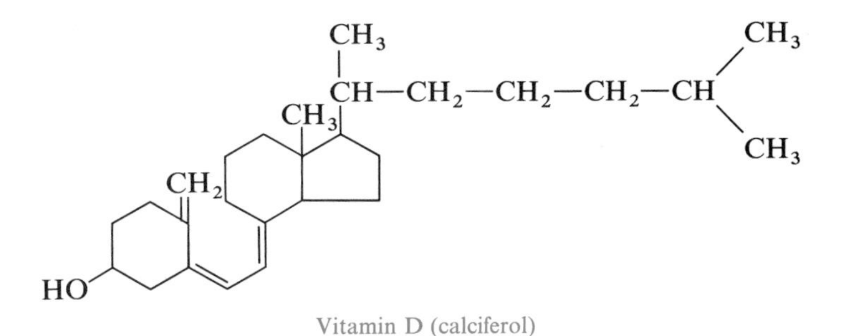

Vitamin D (calciferol)

rickets

If a child lacks vitamin D, *rickets* may develop. In rickets, calcium is not properly absorbed, the blood calcium level drops, and calcium is removed from the bones to raise its concentration in the blood. The bones weaken and bend, causing deformities (rickets) in the child. This happens most often when a child remains out of the sunlight. Sunlight promotes the synthesis of vitamin D in the skin. Rickets is rare in adults, although a lack of vitamin D can cause decalcification, which leads to bones that are brittle and easily broken.

hypervitaminosis D

Since vitamin D is a fat-soluble vitamin, too much vitamin D (hypervitaminosis D) can become a problem, causing weakness, nausea, and diarrhea. Excess absorption of calcium occurs, promoting calcium deposits in the tissues of the body. A person with too much vitamin D is put on a low-calcium diet until the excess vitamin D is used up.

Vitamin E

vitamin E is fat soluble

There are several compounds associated with vitamin E activity, the most common being α-tocopherol:

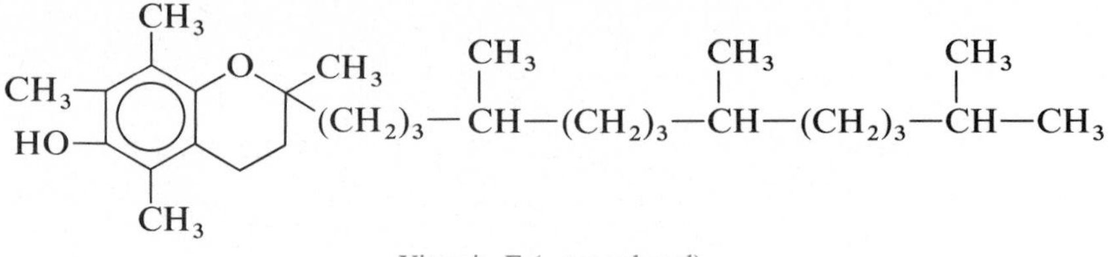

Vitamin E (α-tocopherol)

The tocopherols are a part of the terpene group in the lipid family; vitamin E is fat soluble. The function of vitamin E is not known, but it is thought to protect unsaturated fatty acids from being oxidized. Vitamin E has been shown to prevent sterility in animals, but this effect has not as yet been demonstrated in humans.

Vitamin K

vitamin K is fat soluble

Vitamin K is a member of the terpene group of the lipids and is fat soluble. It is essential to the blood-clotting process:

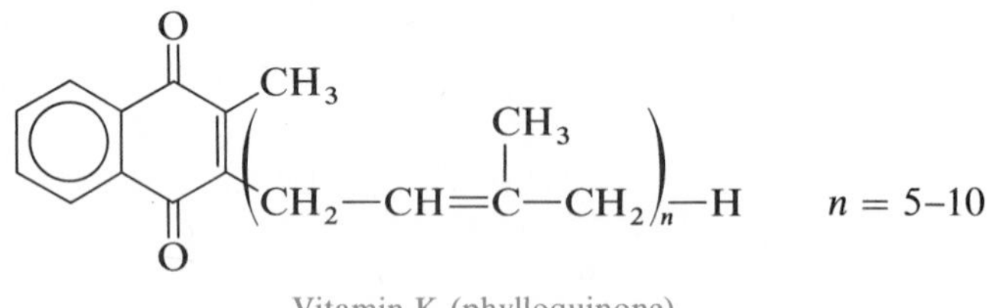

Vitamin K (phylloquinone)

When there is a deficiency of vitamin K in the body, prolonged bleeding will occur. Vitamin K is one vitamin that can be synthesized in the body. Bacteria in the colon produce vitamin K, so a dietary source is not normally needed. However, large amounts of antibiotics destroy these bacteria, and a vitamin K deficiency can occur. High doses of vitamin K may cause a jaundice condition. A person taking an anticoagulant, which prevents blood clotting, may be given vitamin K when it is necessary to cause the blood to clot.

Drugs That Cause Vitamin Depletion

Some drugs can cause malnutrition because they deplete the body of certain vitamins (coenzymes) and minerals (cofactors):

Drug	Nutrient Depleted	Reason
KCl	Vitamin B_{12}	Poor absorption
Phenobarbital	Vitamin D, Ca^{2+}, folic acid	Changes in metabolism
Alcohol	Folic acid, Mg^{2+}, thiamine, protein	Poor absorption, anorexia
Penicillamine	Zn^{2+}, Cu^{2+}, Fe^{2+}, Ca^{2+}, Mg^{2+}	Increased renal excretion
Mineral oil	Fat-soluble vitamins	Poor absorption
Colchicine	Fat-soluble vitamins	Poor absorption
Isoniazid	Pyridoxine	Hydrazide formation

Table 18.4 provides a summary of the vitamins and their functions.

Table 18.4 Summary of Several Vitamins and Their Functions

Vitamin	Function	Effects of Deficiency
Water Soluble		
B_1 (thiamine)	Coenzyme for pyruvic decarboxylase	Beriberi, neuritis, mental disorders
B_2 (riboflavin)	Component of FMN and FAD; forms coenzymes	Decreased production of flavoproteins and hydrogen transport process
B_3 (niacin)	Coenzymes in hydrogen transport, NAD, NADP	Pellagra
B_6 (pyridoxine)	Coenzyme for amino acids and proteins	Dermatitis, weakness, irritability
B_{12} (cobalamin)	Coenzyme in DNA formation	Malformed red blood cells, pernicious anemia
C (ascorbic acid)	Regulates levels of intercellular substances; collagen formation	Scurvy
Pantothenic acid	A component of coenzyme A needed for transferases	Fatigue, nausea, lack of coordination, sleep difficulty
Biotin	Coenzyme for carboxylases	Fatigue, depression, nausea, dermatitis, muscle pains
Folic acid	Coenzyme (tetrahydrofolic acid) in transfers of single-carbon groups	Diarrhea, anemia
Fat Soluble		
A (retinol)	Rhodopsin formation	Night blindness, scaly skin
D (calciferol)	Increases calcium absorption	Rickets, decalcification of bones
E (tocopherol)	Antioxidant	None known
K (phylloquinone)	Blood clotting	Increased bleeding time or no clotting

Example 18.8 Indicate whether each of the following are water or fat soluble:

1. vitamin D, calciferol
2. vitamin B_3, niacin
3. vitamin E, tocopherol
4. vitamin C, ascorbic acid

Solution

1. fat soluble
2. water soluble
3. fat soluble
4. water soluble

Example 18.9 Indicate the vitamin involved in the following:

1. blood clotting
2. absorption of calcium
3. necessary for the coenzyme NAD used in hydrogen transport

Solution

1. Vitamin K is necessary for blood-clotting processes.
2. Vitamin D is needed to increase calcium absorption.
3. Niacin (vitamin B_3) is needed in the formation of NAD, a coenzyme used in hydrogen transfer.

18.6 Digestion of Foodstuffs

OBJECTIVE Given a foodstuff, state the site of its digestion in the intestinal tract and list the enzymes involved and the products of hydrolysis (digestion).

The digestion of food involves mechanical and chemical processes. The mechanical processes include chewing (*mastication*) and the movement of food down the gastrointestinal tract by means of contractions and relaxations of the muscles along the tract (*peristalsis*). Throughout the gastrointestinal tract, various enzymes come in contact with specific substrates and catalyze the chemical process of hydrolysis. Recall that hydrolysis involves the splitting up of large molecules by water until they are small enough for absorption through cell walls.

Digestion of Carbohydrates

amylase

The digestion of starches begins in the mouth with salivary enzymes called *salivary amylases.* The salivary amylases act randomly on the glycosidic linkages of the amylose chains of starch and on some of the branches of amylopectin. The α-1,4 bonds are split, producing smaller oligosaccharides (from three to eight glucose units) and free maltose molecules. The α-1,6 branch points are not split by salivary amylase and remain as fragments called *dextrins.*

The chewed food now forms into a *bolus* (a soft mass of chewed food) to be swallowed. Upon encountering the highly acidic environment of the stomach, the amylase ceases its action. Little carbohydrate digestion occurs while the food is in the stomach.

glucosidase

In the small intestine, the partially digested carbohydrates come in contact with secretions from the pancreas and gallbladder in a slightly basic solution (pH 8). Pancreatic amylase completes the hydrolysis of dextrins and oligosaccharides to give maltose units. Another enzyme, called *glucosidase,* hydrolyzes the α-1,6 linkages from the branches of amylopectin.

In the resulting solution, the disaccharides maltose, sucrose, and lactose encounter their specific enzymes, maltase, sucrase, and lactase, to complete their hydrolysis. Following hydrolysis, the monosaccharides (glucose, fructose, and galactose) are absorbed through the walls of the small intestine. Most of the

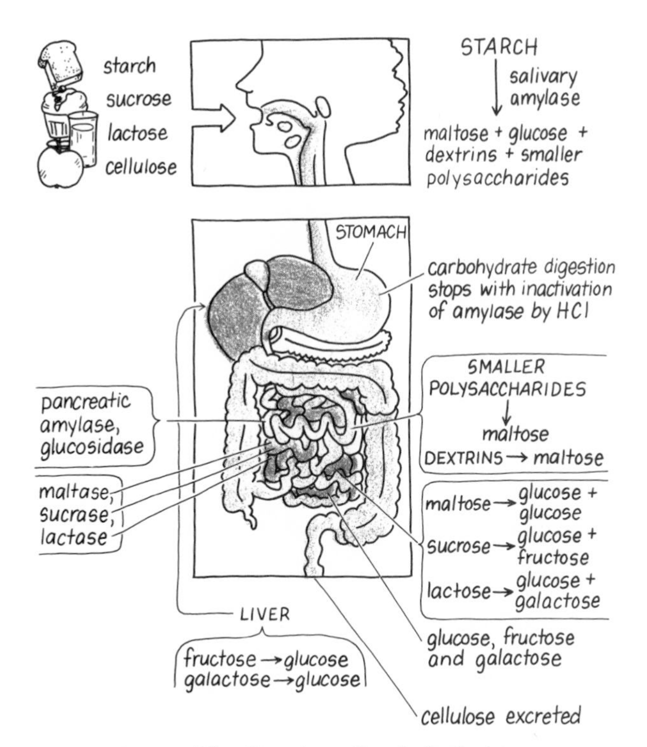

Figure 18.10 The digestion of carbohydrates.

galactose and fructose absorbed will be converted to glucose by the liver and returned to the general circulatory system:

Mouth

Starch
H_2O ↓ Salivary amylase
Pancreatic amylase
↓ Glucosidase

Small Intestine

Maltose	Lactose	Sucrose
H_2O ↓ Maltase	H_2O ↓ Lactase	H_2O ↓ Sucrase
Glucose	Glucose	Glucose
+	+	+
Glucose	Galactose	Fructose

Generally the ordinary diet contains much more starch than sucrose or lactose. As a result, carbohydrate digestion produces about 80 percent glucose, 10 percent galactose, and 10 percent fructose. The diet also contains a large amount of cellulose. However, in humans there are no enzymes in the digestive tract that can hydrolyze cellulose. Cellulose adds fiber to the diet, but it has no nutritive value. (See Figure 18.10.)

Example 18.10 Give the digestion site(s), enzymes needed, and products of digestion for amylose.

Solution: Amylose is a starch whose digestion begins in the mouth by the enzyme salivary amylase. The products consist of smaller chains of polysaccharides and maltose. The digestion of these carbohydrates continues in the small intestine. Pancreatic amylase continues the hydrolysis of the polysaccharides and eventually all of the starch material is hydrolyzed to maltose. The enzyme maltase then hydrolyzes maltose to give two glucose molecules, which are absorbed into the bloodstream.

Digestion of Fats

Fats do not undergo hydrolysis until they reach the small intestine. There, the fats mix with the bile released by the gallbladder into the small intestine. The bile disperses the large globules of fat into small droplets, creating an emulsion. In the emulsified state, the fat has a greater surface area available to the action of digestive enzymes called *lipases*, which are contained in the pancreatic secretions. Lipases are hydrolytic enzymes and digest the fats by converting them to glycerol and fatty acids:

lipase

$$\text{Fat} \xrightarrow{\text{Bile}} \text{Emulsified fat}$$

$$H_2O + \text{Emulsified fat} \xrightarrow[\text{lipase}]{\text{Pancreatic}} \text{Fatty acids} + \text{Glycerol}$$

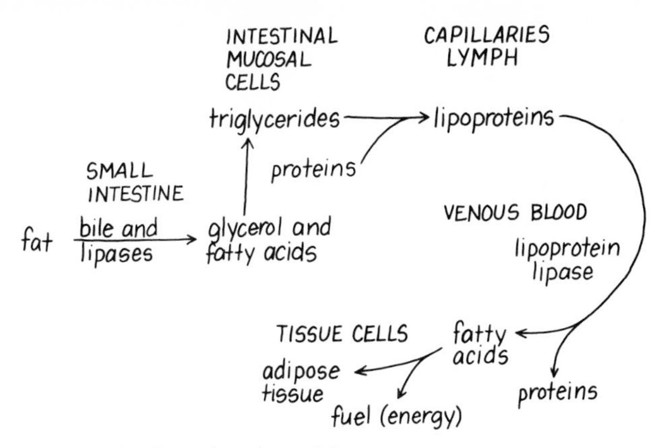

Figure 18.11 Digestion of fats.

The glycerol and fatty acids diffuse to the intestinal mucosal cells and are absorbed. Within the mucosal cells, triglycerides are re-formed from the digestion products of fats. Droplets of fat form in the mucosal cells. These droplets are not very soluble in the aqueous fluid of the body. To aid their movement through the lymph and eventually the blood, the lipids in the mucosal cells are combined with a protein, forming a transport molecule for lipids, called a *lipoprotein.*

lipoprotein

The lipoproteins are hydrophilic and move easily within the lymphatic capillaries. Other lipids, such as cholesterol and phospholipids, are also carried into the lympthatic capillaries as lipoproteins. From the lymph stream, the lipoproteins enter the venous blood. Here, a lipoprotein lipase separates the lipids from the protein and releases the fatty acids from the triglycerides. Thus, the fatty acids have been transported to the cells of the body. They may be used by the cells of energy, or they may be converted to adipose tissue for storage. (See Figure 18.11.)

Example 18.11 Describe the action of lipases on fat digestion.

Solution: No fat digestion occurs until the lipid material reaches the small intestine. Once there, the fat material is emulsified by the action of bile from the gall bladder. Then the lipase enzymes from the pancreas act on the smaller droplets of fat, causing hydrolysis of the ester bonds of the fat and producing glycerol and fatty acids.

Digestion of Proteins

The large protein molecules obtained from the diet are broken down into amino acids that can be absorbed through the intestinal walls into the bloodstream. Protein digestion begins in the stomach. When food is eaten, there is an increase in hydrochloric acid production in the stomach, causing the pH to drop to between 1 and 2. The acidic environment converts an inactive enzyme (proenzyme) *pepsinogen* into *pepsin*, an active enzyme. The hydrolysis action of pepsin splits peptide bonds wherever there is a tyrosine or phenylalanine. Proteins are broken down into smaller peptide chains called *polypeptides*:

pepsin

Stomach

$$\text{Pepsinogen} \xrightarrow{\text{pH } 1-2} \text{Pepsin}$$

$$\text{Proteins} \xrightarrow{\text{Pepsin}} \text{Polypeptides}$$

The polypeptides enter the small intestine where there are peptidases such as trypsin and aminopeptidase, which function in the basic (pH 8) environment of the small intestine. Trypsin splits peptide bonds containing the carboxyl end of lysine or arginine. With different peptidases splitting peptide bonds at different points in the chain, a protein is eventually completely hydrolyzed to amino acids, the end products of protein digestion. The amino acids are absorbed through the intestinal walls into the bloodstream for transport to tissues in the body:

Small Intestine

$$\text{Polypeptides} + H_2O \xrightarrow[\text{pH } 8-9]{\text{Peptidases, trypsin}} \text{Amino acids}$$

Example 18.12 Describe the digestion of protein in the stomach and in the small intestine.

Solution: Protein digestion begins in the stomach, where pepsin hydrolyzes peptide bonds, thus producing smaller proteins or polypeptides. In the small intestine, peptidases such as trypsin and aminopeptidase hydrolyze the polypeptides into amino acids, which can be absorbed into the bloodstream.

Glossary

activation energy The energy needed for reactants to reach the transition state in a chemical reaction.

active site The portion of the enzyme that catalyzes a reaction.

amylase The enzyme present in saliva and pancreatic juice that hydrolyzes α-1,4 bonds in amylose and branches of amylopectin.

antibiotic An antimetabolite produced by a microorganism that acts as a poison to another microorganism.

antimetabolite A compound that inhibits an enzyme-catalyzed reaction.

apoenzyme Protein portion of a conjugated enzyme.

beriberi A disease characterized by neuritis, paralysis, and heart failure because of severe thiamine (vitamin B) deficiency.

bile The secretion from the pancreas that emulsifies fats.

carotenes Compounds found in red and yellow vegetables that are converted by the body to vitamin A.

catalyst A compound or enzyme that usually lowers the energy of activation of a reaction.

coenzyme A cofactor in an enzyme that is an organic compound, often a vitamin.

cofactor The nonprotein portion of an enzyme, such as a metal ion, that is necessary for enzyme activity.

competitive inhibition A compound with a structure similar to that of the substrate that competes for the active site on the enzyme to inhibit enzyme action.

enzyme A type of protein that catalyzes biological reactions.

enzyme–substrate complex An intermediate (transition) state in an enzyme-catalyzed reaction that consists of the enzyme and substrate (ES).

fat-soluble vitamin A vitamin soluble in nonpolar solvents, such as vitamins A, D, E, and K.

holoenzyme A enzyme containing an apoenzyme and a cofactor.

inhibitor Chemical compounds that interfere with enzyme action.

isoenzymes Forms of the same enzyme that vary in structure, depending on the tissue's source of the enzyme.

lock-and-key theory A model of enzyme action that represents enzymes as locks and substrates as the keys.

noncompetitive inhibition Inhibition caused by a substance that alters the surface of an enzyme or blocks the ES complex.

optimum pH The pH at which an enzyme reaches its maximum activity.

pellagra A disease caused by a deficiency of niacin (vitamin B_3) that is characterized by muscular weakness, scaliness of the skin, and irritations of the mouth and gastrointestinal tract.

rhodopsin A compound formed in the retina of the eye from vitamin A and opsin (protein) that is needed for adequate vision at night.

rickets A condition that results from a lack of vitamin D and causes bone deformities.

scurvy A disease caused by vitamin C deficiency that is characterized by slow healing of wounds, hemorrhaging, and anemia.

simple enzymes Enzymes that are only protein.

substrate The substance upon which an enzyme acts in a reaction.

substrate specificity The tendency of an enzyme to react with only specific substrates.

transition state High-energy state necessary for the formation of product in a chemical reaction.

vitamin An organic compound that is needed in small amounts in the body to maintain adequate levels of metabolic activity and must be derived from the diet.

water-soluble vitamins Vitamins soluble in water and fluids of the body; these include vitamins B and C, pantothenic acid, biotin, and folic acid.

Problems

Energy Diagrams and Catalysts (Objective 18.1)

18.1 Consider the following reaction:

$$2H_2 + O_2 \longrightarrow 2H_2O + \text{Energy}$$

a. What are the reactants?
b. What are the products?
c. Is the reaction endothermic or exothermic?
d. Draw an energy diagram to represent the energy of the reactants and products, the energy of activation, and the overall energy.
e. How does a catalyst affect the energy of activation?
f. Why would a catalyst make the reaction go faster?

Classification of Enzymes (Objective 18.2)

18.2 Indicate whether the following phrases describe a simple enzyme or a holoenzyme:
a. an enzyme that contains a sugar
b. an enzyme requiring Zn^{2+} for activity
c. an enzyme that is only protein
d. an enzyme that uses a vitamin coenzyme for activity
e. an enzyme composed of 155 amino acids

18.3 What are the apoenzyme and cofactor of a holoenzyme?

18.4 Name the substrate for the following enzymes:
a. lipase b. peptidase c. maltase d. esterase

18.5 What are the six classes of enzymes?

18.6 Give the class of enzyme that might catalyze the reaction for each of the following substrates:
a. hydrolysis of sucrose
b. hydrolysis of galactose
c. breaking the ester bond of a triacylglycerol
d. addition of oxygen
e. removal of hydrogen
f. isomerization of glucose to fructose

g. $$\underset{\text{Pyruvic acid}}{CH_3{-}\overset{\displaystyle O}{\overset{\|}{C}}{-}\overset{\displaystyle O}{\overset{\|}{C}}{-}OH} \longrightarrow \underset{\text{Acetic acid}}{CH_3{-}\overset{\displaystyle O}{\overset{\|}{C}}{-}OH} + \underset{\text{Carbon dioxide}}{CO_2}$$

h. addition of water to a double bond (hydration)
i. addition of an amino group to a compound (amidation)
j. combination of two molecules to form a larger molecule

Enzymes as Catalysts (Objective 18.3)

18.7 Match the phrases below with the following answers:
(1) cofactor, (2) enzyme–substrate complex, (3) lock-and-key, (4) active site.
a. the temporary combination of an enzyme with the substrate
b. an inorganic compound that is sometimes needed to complete an enzyme
c. the portion of an enzyme where catalytic activity occurs
d. the theory that accounts for the unusual specificity of an enzyme
e. a high-energy intermediate state of enzyme action

18.8 The equation for the lock-and-key theory is as follows:

$$E + S \rightleftharpoons ES \rightleftharpoons EP \rightleftharpoons E + P$$

a. Identify each of the symbols used in the equation.
b. How is the active site different from the whole enzyme structure?
c. After the products have formed, what happens to the enzyme?

18.9 Enzyme activity is affected by several conditions. Explain how each of the following affects the rate of an enzyme-catalyzed reaction:
a. lowering the concentration of the substrate
b. increasing the concentration of the enzyme

c. adjusting the pH to optimum pH
d. adding acid to lower the pH below optimum pH
e. running the reaction 10°C below optimum temperature
f. running the reaction 20°C above optimum temperature

18.10 On the following graph, plot the graph for the rate of an enzyme-catalyzed reaction when the optimum pH is 2; do the same for optimum pH 9:

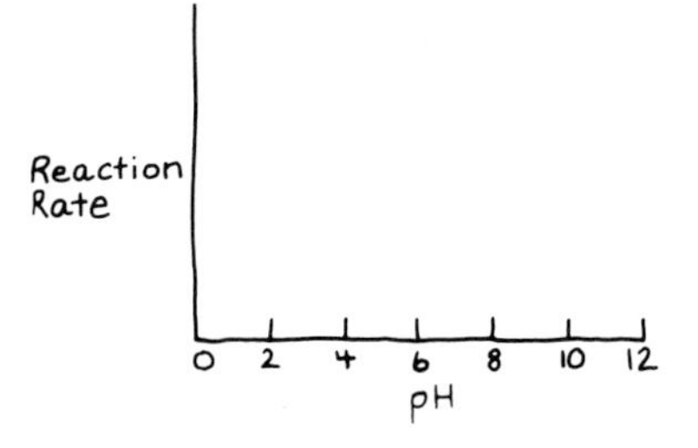

Enzyme Inhibition (Objective 18.4)

18.11 Indicate whether the following describe a competitive or a noncompetitive inhibitor:
a. an inhibitor that has a structure similar to the substrate
b. an inhibitor whose effect cannot be reversed by adding more substrate
c. an inhibitor that competes with the substrate for the active site
d. an inhibitor that changes the shape of the enzyme

18.12 Oxaloacetic acid is an inhibitor of succinic dehydrogenase:

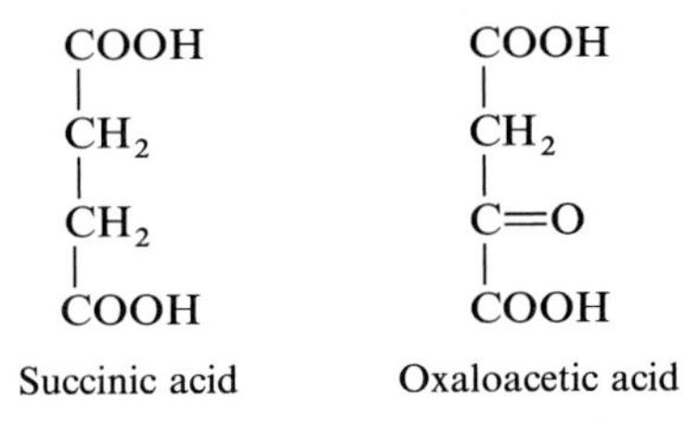

a. Would you expect oxaloacetic acid to be a competitive or a noncompetitive inhibitor? Why?
b. How would you expect the inhibition of the enzyme to occur?
c. How would you reverse the effect of the inhibitor?

Vitamins and Coenzymes (Objective 18.5)

18.13 State whether each of the following statements about vitamins is true (T) of false (F):
a. Vitamins are needed in small amounts by the body.
b. Vitamins are produced by the body.
c. Vitamins must be supplied by the diet.
d. Vitamins often act as coenzymes in metabolic reactions.
e. A vitamin deficiency can cause illness.
f. Vitamins are carried by the bloodstream.
g. Vitamins are proteins.
h. Vitamins are inorganic compounds.

18.14 Give the major dietary sources of each of the following vitamins:
a. vitamin A (retinol)
b. vitamin B_3 (niacin)
c. vitamin D (calciferol)
d. vitamin B_{12} (cyanocobalamin)
e. vitamin C (ascorbic acid)
f. folic acid
g. pantothenic acid
h. vitamin K

18.15 Classify each of the vitamins in problem 18.14 as water soluble or fat soluble. Which type of vitamin is lost more readily in the urine? Which type of vitamin is most likely to be stored in the body?

18.16 Identify the vitamin associated with each of the following functions or conditions:
a. formation of rhodopsin
b. pellagra
c. beriberi
d. decarboxylation of pyruvic acid
e. night blindness
f. scurvy
g. flavin nucleotide (FAD)
h. rickets
i. absorption of calcium from the intestinal tract
j. protection of cells from oxidation
k. involved in blood-clotting processes
l. decalcification of bone
m. poor coagulation time

Digestion of Foodstuffs (Objective 18.6)

18.17 Describe the digestive site(s), enzymes, and products for the following carbohydrates:
a. amylose b. amylopectin c. lactose d. sucrose

18.18 Describe the function of each in fat (triacylglycerol) digestion:
a. bile b. lipases c. lipoprotein

18.19 a. When you begin to ingest food, acid enters the stomach. What would be the purpose of this? Why isn't the acid always in the stomach?
b. Where is protein digestion completed? What are the end products of protein digestion?

Metabolism: Energy Production in the Living Cell

Objectives

19.1 Describe the role of ATP in the metabolic reactions of the cell. Write equations for the formation and the hydrolysis of ATP.

19.2 Describe the role of the electron-transport system and its production of ATP molecules when $NADH/H^+$ or $FADH_2$ enters the system.

19.3 Describe the role of the citric acid cycle. State the number of ATP molecules formed by the oxidation of acetyl coenzyme A.

19.4 Write the equations for the overall reactions of the following: glucose to pyruvic acid; pyruvic acid to acetyl CoA; complete combustion of glucose. Account for the number of ATP molecules produced by each.

19.5 Describe the β-oxidation of fatty acids. Account for the number of ATP molecules produced when a fatty acid is completely oxidized.

19.6 Describe the transamination and oxidative deamination of amino acids for use in the citric acid cycle.

Scope

The amount of food you eat must meet all your metabolic needs. Your body is continually using energy to perform its vast number of functions. When you do any kind of work—bicycling, dancing, or studying—you are using energy. The movement of a muscle, the transmission of an impulse along a nerve, the pumping of your heart, and the contraction of your diaphragm all require energy. The energy to do work is extracted from the nutrients obtained from the food you eat.

Metabolism begins with the products obtained from digestion such as glucose, amino acids, and fatty acids, which have been carried through the bloodstream or lymphatic system and absorbed by the tissues of the body. Within the cells, these substances undergo enzyme-catalyzed reactions that provide energy for the cells as well as the compounds needed for the synthesis of cellular molecules.

19.1 The Role of ATP

OBJECTIVE Describe the role of the ATP in the metabolic reactions of the cell. Write equations for the formation and the hydrolysis of ATP.

catabolic reactions

anabolic reactions

There are several thousand metabolic reactions occurring continually in the cells of the body. However, two major types of reactions are evident. Energy-producing (catabolic) reactions extract energy from the products of digestion through degradation and oxidation reactions. Energy-requiring (anabolic) reactions build new and larger molecules for the cells and carry out the work of the cell. The two types of reactions work together to provide the energy for the activities within the cells:

Catabolism

Large molecules $\longrightarrow$ Smaller molecules + Energy

Anabolism

Smaller molecules + Energy $\longrightarrow$ Large molecules

ATP

The link between the energy-producing reactions and the energy-requiring reactions is a high-energy compound called *adenosine triphosphate* (ATP). ATP is the storage form of the energy made available from the chemical breakdown of substances within the cell. The ATP molecule is composed of adenosine

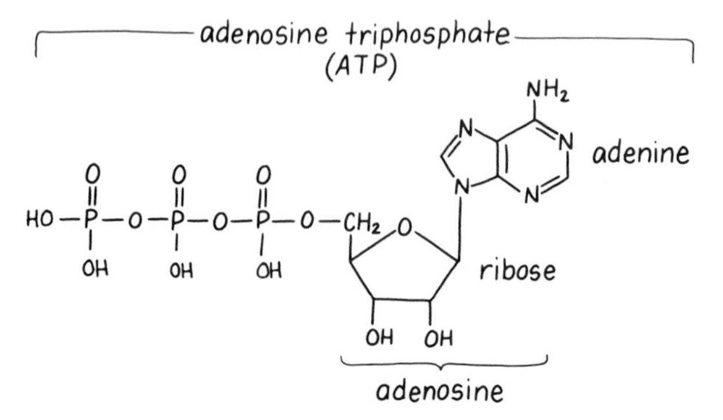

Figure 19.1 Structure of ATP.

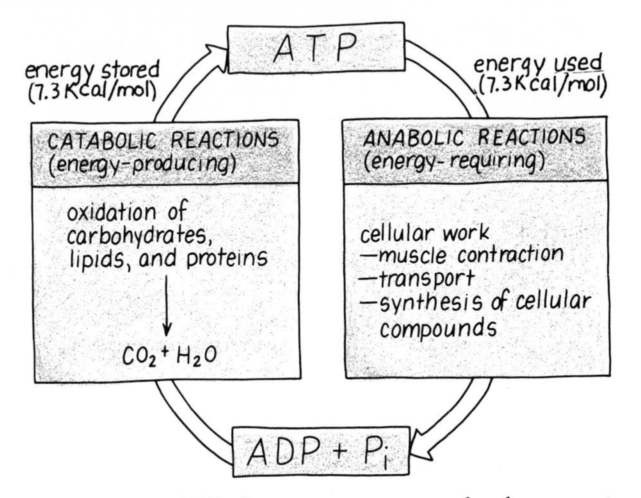

Figure 19.2 ATP, the energy storage molecule, connects the energy-requiring reactions with the energy-producing reactions in the cell.

(adenine and ribose, a pentose sugar) attached to three phosphate groups. (See Figure 19.1.)

The hydrolysis of ATP yields *adenosine diphosphate* (ADP), a phosphate group (P_i), and energy. The energy released by the hydrolysis of ATP is 7.3 kcal/mol:

$$\text{ATP} \longrightarrow \text{ADP} + \text{P}_i + 7.3 \text{ kcal}$$

This is the energy that you use for all of the energy-requiring (anabolic) reactions in the cells. For example, you use ATP energy for the mechanical work of moving your muscles, for transport work when substances are moved across cellular membranes, and for chemical work in the synthesis of larger molecules such as proteins needed to repair cells and to provide enzymes to catalyze reactions (see Figure 19.2):

$$\text{Relaxed muscle} \xrightarrow{\text{ATP} \quad \text{ADP} + \text{P}_i} \text{Contracted muscle}$$

Example 19.1 Write an equation for the hydrolysis of ATP.

Solution: The hydrolysis of ATP produces ADP, P_i, and energy. The energy that was stored in the phosphate bond is released and can be utilized by energy-requiring processes in the cell.

$$\text{ATP} \longrightarrow \text{ADP} + \text{P}_i + 7.3 \text{ kcal}$$

19.2 The Electron-Transport System

OBJECTIVE Describe the role of the electron-transport system and its production of ATP molecules when NADH/H^+ or $FADH_2$ enters the system.

oxidative phosphorylation

The synthesis of ATP occurs in a series of oxidation–reduction reactions called the *electron-transport system.* There are several steps within the electron-transport system that release enough energy to synthesize molecules of ATP. The formation of ATP is called *oxidative phosphorylation*:

$$\text{ADP} + \text{P}_i + 7.3\ \text{kcal} \longrightarrow \text{ATP}$$

mitochondria

The electron-transport system is contained within subcellular structures of the cell called *mitochondria.* Each of the several hundred mitochondria in a cell consist of an outer membrane and an inner membrane. The inner membrane is highly convoluted with many ridgelike folds, along which is a dense network of enzymes that catalyze energy-producing reactions. This is the site for the oxidation of glucose, amino acids, and fatty acids and for the formation of ATP. The mitochondria are considered the power houses of the cell because they contain the machinery for producing energy. (See Figure 19.3.)

metabolite

dehydrogenase

hydrogen acceptors

In metabolic oxidation reactions, a *metabolite* (a substrate in metabolism) is oxidized when two hydrogen atoms (2H) are removed. The enzymes that catalyze the removal of hydrogen atoms are called *dehydrogenases.* Their coenzymes, called *hydrogen acceptors,* pick up or accept the hydrogen atoms. The hydrogen acceptors transport the hydrogen atoms to the electron transport system, where the hydrogen atoms or their electrons pass through several reactions involving many enzymes and coenzymes. The overall process involves the combination of hydrogen from a metabolite with oxygen obtained from respiration to form a molecule of water. The series of reactions in the electron-transport system is energy-producing because the end product, water, has a lower energy than the initial reactants. The energy released is used to form ATP:

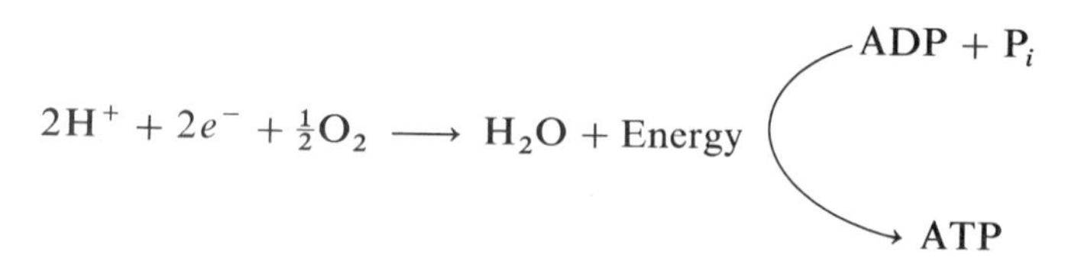

NAD^+, FMN, FAD, and CoQ Coenzymes: The Hydrogen Acceptors

The hydrogen acceptors in the electron-transport system are NAD^+ (nicotinamide adenine dinucleotide), FAD (flavin adenine dinucleotide), FMN (flavin mononucleotide), and CoQ (coenzyme Q).

NAD^+

The NAD^+ coenzyme is derived from niacin (vitamin B_3). (See Figure 19.4.) It is NAD^+ that accepts hydrogen atoms from the initial metabolite

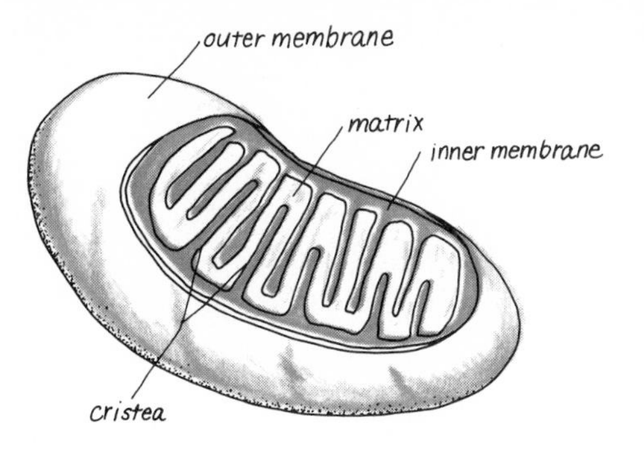

Figure 19.3 Structure of a mitochondrion. There are several hundred mitochondria in a cell containing the enzymes and electron transport system for oxidation reactions and ATP synthesis.

(MH_2). NAD^+ accepts hydrogen atoms when the oxidation involves a carbon–oxygen bond:

$$\underset{\text{Lactic acid}}{CH_3{-}\overset{\overset{\displaystyle OH}{|}}{C}H{-}COOH} + NAD^+ \longrightarrow \underset{\text{Pyruvic acid}}{CH_3{-}\overset{\overset{\displaystyle O}{\|}}{C}{-}COOH} + NADH/H^+$$

In this example, the alcohol group of the lactic acid (the metabolite) is oxidized, NAD^+ accepts the hydrogen atoms; it is reduced. In general, the metabolite may be represented as MH_2 and the reaction is written

$$MH_2 + NAD^+ \longrightarrow NADH/H^+ + M$$

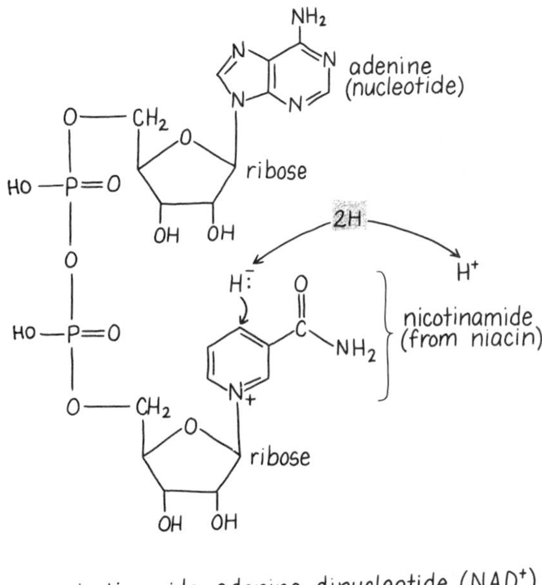

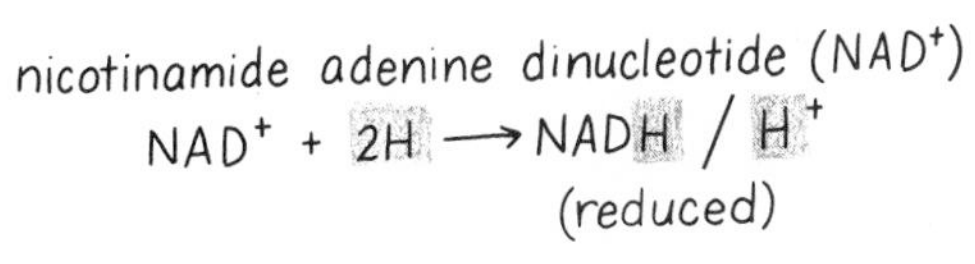

Figure 19.4 Structure of coenzyme NAD^+ and formation of its reduced form $NADH/H^+$.

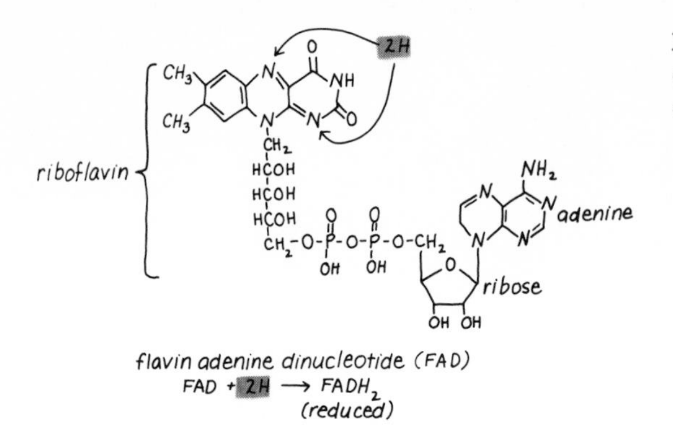

Figure 19.5 Structure of coenzyme FAD and formation of its reduced form $FADH_2$.

Sometimes we represent this kind of reaction using curved arrows to link the process of oxidation and reduction:

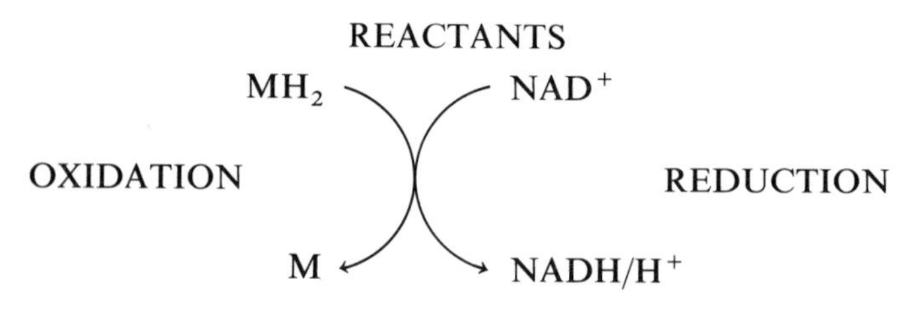

FAD

There is another coenzyme, FAD, derived from riboflavin (vitamin B_2) that also accepts hydrogen atoms directly from metabolites. FAD is the hydrogen acceptor when hydrogen atoms are removed from two carbon atoms to form a carbon–carbon double bond (see Figure 19.5):

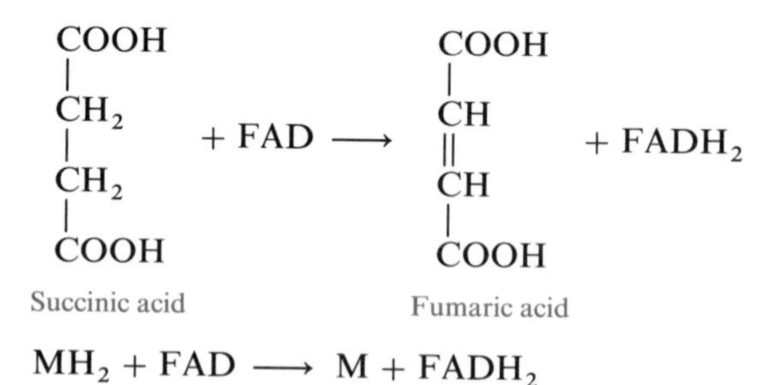

$$MH_2 + FAD \longrightarrow M + FADH_2$$

Therefore, hydrogen atoms from metabolites can be carried to the electron-transport system by one of two hydrogen carriers, $NADH/H^+$ or $FADH_2$.

Example 19.2 The citric acid cycle produces many hydrogen atoms when oxidation reactions occur. What hydrogen acceptors of the electron-transport system can pick up those hydrogen atoms?

Solution: The hydrogen acceptors are NAD^+ and FAD. They can accept hydrogen atoms from the critic acid cycle and transport those hydrogen atoms to the electron-transport system. NAD^+ is used for oxidation of carbon–oxygen bonds; FAD is used for removing hydrogen from carbon–carbon bonds:

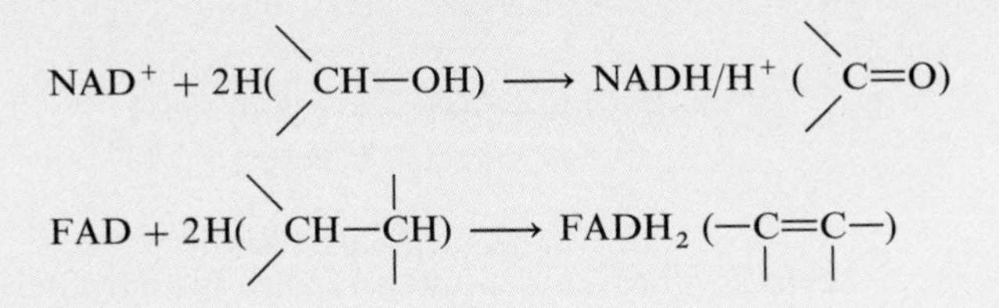

FMN

The next step in the electron-transport system is the transfer of hydrogen from $NADH/H^+$ to another flavin coenzyme called *flavin mononucleotide* (FMN). (See Figure 19.6.) The FMN is reduced and the original hydrogen carrier is reoxidized (NAD^+). Thus, the coenzymes are regenerated to be used again as hydrogen acceptors in oxidation reactions:

$$NADH/H^+ + FMN \longrightarrow FMNH_2 + NAD^+$$

CoQ

The next component in the electron-transport system is coenzyme Q (CoQ), another hydrogen acceptor derived from quinone. (See Figure 19.7.) CoQ accepts the hydrogen atoms from $FMNH_2$ or $FADH_2$, each of which brings hydrogens directly from a metabolite:

$$CoQ + FMNH_2 \longrightarrow CoQH_2 + FMN$$

or

$$CoQ + FADH_2 \longrightarrow COQH_2 + FAD$$

Cytochromes: The Electron Acceptors

$CoQH_2$ separates the hydrogens into protons ($2H^+$) and electrons ($2e^-$). When $CoQH_2$ is reoxidized, the protons are released inside the mitochondria. The electrons are passed on to a group of electron acceptors called *cytochromes*. The

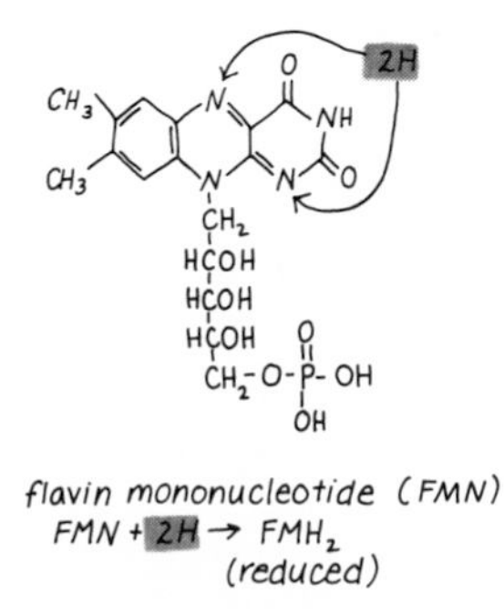

Figure 19.6 Structure of coenzyme FMN and formation of its reduced form $FMNH_2$.

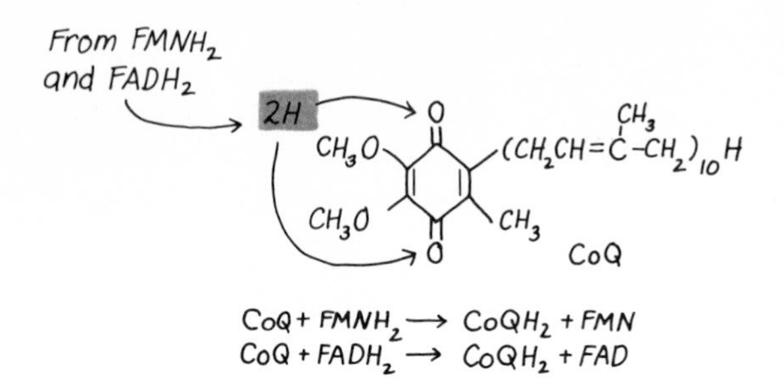

Figure 19.7 Coenzyme Q accepts hydrogens from $FMNH_2$ or $FADH_2$.

cytochromes all contain an iron ion that alternates between the reduced state (Fe^{2+}) and the oxidized state (Fe^{3+}). For every pair of electrons released by $CoQH_2$, two Fe^{3+} ions are needed. The first cytochrome in the series is cytochrome b:

$$\underset{\text{Oxidized form}}{2 \text{ cyt b } (Fe^{3+})} + 2e^- \longrightarrow \underset{\text{Reduced form}}{2 \text{ cyt b } (Fe^{2+})}$$

As the pair of electrons moves from one cytochrome to the next—much like buckets of water in a fire brigade—iron is reduced as electrons are accepted and oxidized as electrons are lost. The different cytochromes are identified by the letters b, c_1, c, a, and a_3. Finally, at the last cytochromes in the series, a and a_3, called *cytochrome oxidase*, the electrons are transferred to molecular oxygen, the protons released earlier are picked up, and a molecule of water is produced:

$$2 \text{ cyt } a_3 \ (Fe^{2+}) + 2H^+ + \tfrac{1}{2}O_2 \longrightarrow H_2O + 2 \text{ cyt } a_3 \ (Fe^{3+})$$

The overall sequence of reactions in the electron-transport system is shown in Figure 19.8.

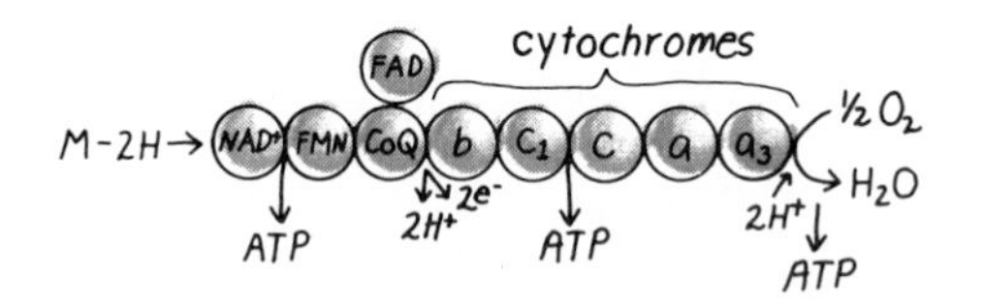

Figure 19.8 Hydrogen and electron acceptors of the electron-transport system in mitochondria.

Example 19.3 Identify the following steps in the electron-transport system as oxidation or reduction:

1. $NAD^+ + 2H \longrightarrow NADH/H^+$
2. $FMNH_2 \longrightarrow FMN + 2H$
3. $2 \text{ cyt a } (Fe^{2+}) \longrightarrow 2 \text{ cyt a } (Fe^{3+})$

Solution

1. The gaining of hydrogen is reduction.
2. The loss of hydrogen is oxidation.
3. The loss of electrons is oxidation.

Example 19.4 Use curved arrows to illustrate the transfer of hydrogen atoms from $FMNH_2$ to CoQ. Indicate the oxidation and reduction reactions.

Solution: The reactants $FMNH_2$ and CoQ are written with the reduced form ($FMNH_2$) on the left. The products, FMN and $CoQH_2$, are written below. Curved arrows connect the oxidation and reduction reactions to indicate that two hydrogen atoms have been transferred:

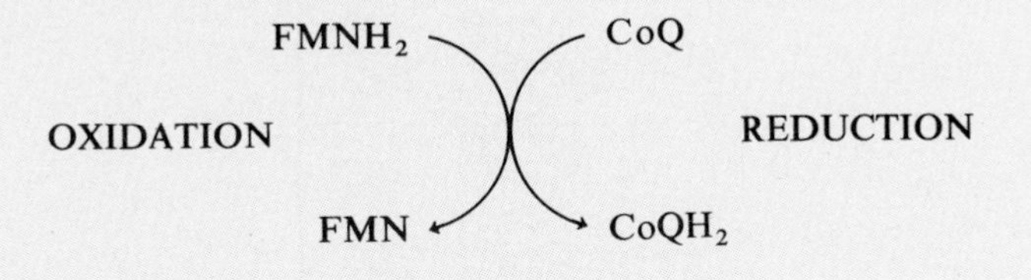

Oxidative Phosphorylation

The electron-transport chain may seem to have little importance—until we point out that three of the reactions in the series yield energy in amounts greater than 7.3 kcal. This is significant because the formation of a mole of ATP requires 7.3 kcal. The combination of the oxidation in the electron-transport system with the synthesis of ATP is called *oxidative phosphorylation*:

oxidative phosphorylation

$$ADP + P_i + 7.3\text{ kcal} \longrightarrow ATP$$

When NAD^+ is the initial hydrogen acceptor, three molecules of ATP are generated. If FAD is the initial hydrogen acceptor, the $FADH_2$ enters the system at a lower energy level and generates only two molecules of ATP:

Initial Hydrogen Acceptor	*ATP Produced*
NAD^+	3
FAD	2

Figure 19.9 illustrates the oxidation–reduction reactions and ATP production in the electron-transport system.

Example 19.5 How many ATP molecules are generated when $NADH/H^+$ enters the electron-transport system?

Solution: NAD^+ is one of the initial hydrogen acceptors for the electron-transport system. When it brings hydrogen atoms in the form of $NADH/H^+$ to the respiratory chain, three ATP molecules can be generated.

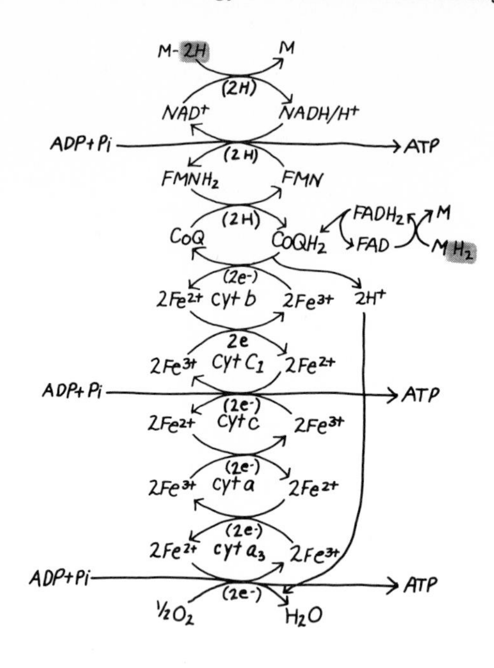

Figure 19.9 Oxidation–reduction reactions and ATP production in the electron-transport system.

19.3 Citric Acid Cycle

OBJECTIVE Describe the role of the citric acid cycle. State the number of ATP molecules formed by the oxidation of acetyl coenzyme A.

The citric acid cycle is a sequence of reactions in which several of the metabolites provide all the hydrogen atoms for the electron-transport system. Its enzymes

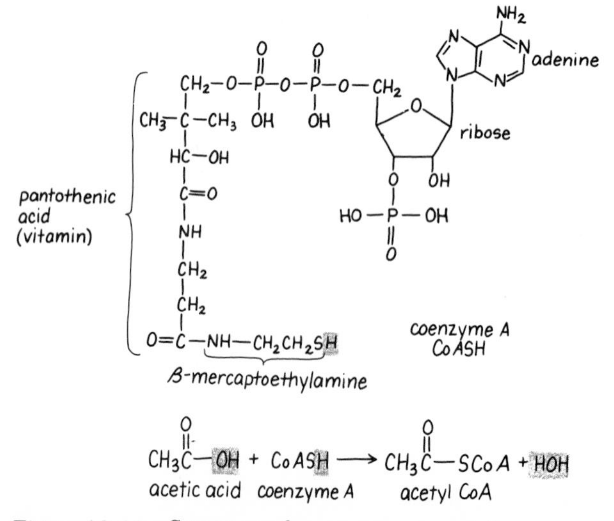

Figure 19.10 Structure of coenzymes A and the formation of acetyl CoA.

and reacting components are located in the mitochondria so that the hydrogen atoms removed by the reactions in the citric acid cycle can be picked up by hydrogen acceptors (NAD^+ or FAD) to be transported to the electron-transport system and the synthesis of ATP.

acetyl coenzyme A

The citric acid cycle begins with a two-carbon molecule called *acetyl coenzyme A* (acetyl CoA). (See Figure 19.10 for the structure of acetyl CoA.) The acetyl portion is obtained from degradation and oxidation reactions of glucose, fatty acids, and proteins, although glucose is the primary source. To be utilized by the cells, this acetyl unit must be attached to coenzyme A, an important intermediate in metabolic reactions. Figure 19.11 illustrates the central role of acetyl CoA and the citric acid cycle in the production of energy (ATP) from food.

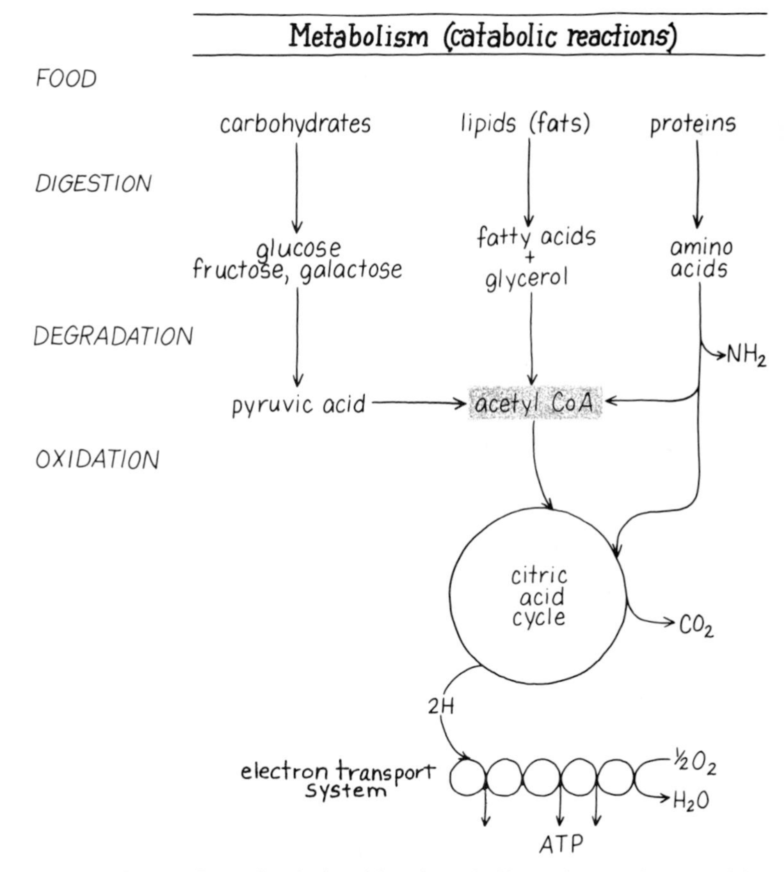

Figure 19.11 Overall relationship of catabolic pathways that provide for the production of ATP from food.

Reactions of the Citric Acid Cycle

As the two-carbon acetyl unit moves through the citric acid cycle, two important types of reactions occur. One is decarboxylation, the removal of a carbon atom as CO_2; the other is dehydrogenation, whereby two hydrogen atoms are removed as a metabolite is oxidized. Each oxidation step is coupled with the reduction of NAD^+ or FAD to take two hydrogen atoms to the electron-transport system. As we discuss the steps of the citric acid cycle, you may wish to refer to the diagram in Figure 19.12 for the structures of the compounds as they undergo reaction.

Step 1: Formation of Citric Acid We can think of the citric acid cycle as beginning with the reaction of acetyl CoA with a four-carbon compound,

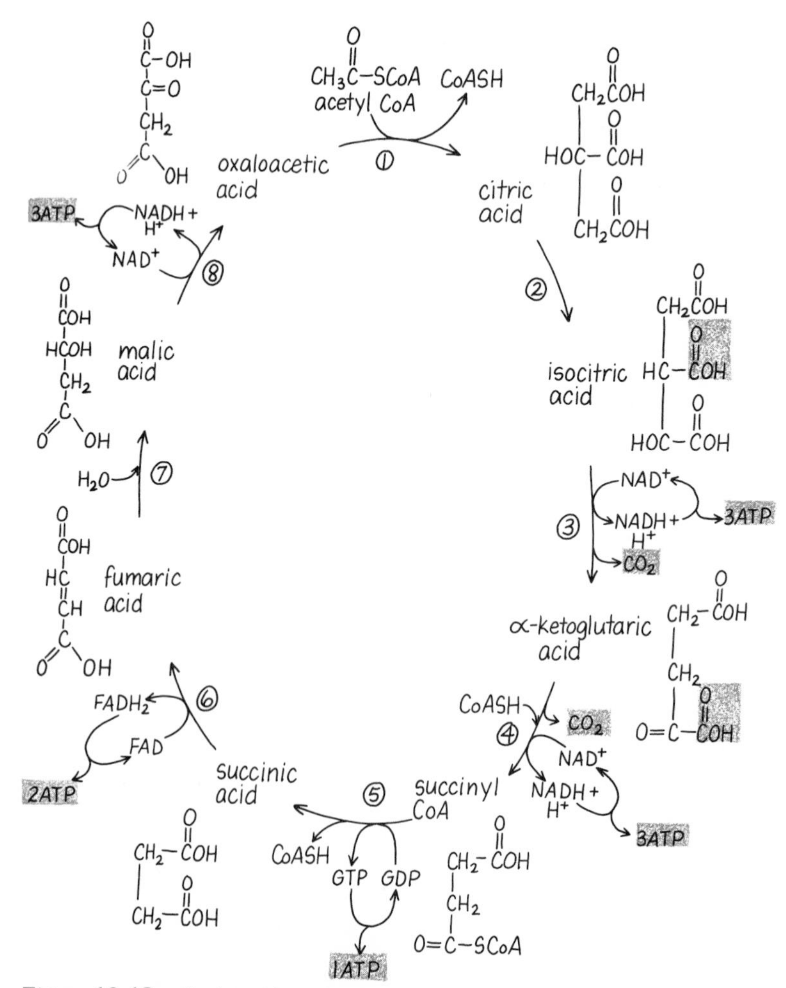

Figure 19.12 Citric acid cycle.

citric acid synthetase

oxaloacetic acid, to produce a six-carbon compound, citric acid. The reaction is catalyzed by an enzyme called *citric acid synthetase*. The coenzyme A that is released can pick up more acetyl units:

$$\text{Oxaloacetic acid} + \text{Acetyl CoA} \xrightarrow{\text{Citric acid synthetase}} \text{Citric acid} + \text{CoASH}$$

isocitric acid

Step 2: Formation of Isocitric Acid In this step, citric acid undergoes isomerization. A dehydration occurs followed by a hydration, both catalyzed by *aconitase*. As a result, the position of the hydroxyl group is changed:

$$\text{Citric acid} \xrightarrow{\text{Aconitase}} \text{Isocitric acid}$$

α-ketoglutaric acid

Step 3: Formation of α-Ketoglutaric Acid At this point in the citric acid cycle, two enzymes called *isocitric acid dehydrogenase* catalyze the first oxidative decarboxylation. First, the hydroxyl group is oxidized to an α-keto group, providing two hydrogens for the hydrogen acceptor NAD^+. Then a carboxylic acid group is removed in a decarboxylation reaction to form CO_2, resulting in a five-carbon compound, α-ketoglutaric acid:

$$\text{Isocitric acid} + NAD^+ \xrightarrow[\text{dehydrogenase}]{\text{Isocitric acid}} \alpha\text{-Ketoglutaric acid} + CO_2 + NADH/H^+$$

succinyl CoA

Step 4: Formation of Succinyl SCoA In this step, a group of enzymes called *α-ketoglutaric acid dehydrogenase complex* catalyzes another oxidative decarboxylation. Another carboxyl group is removed as CO_2, and the molecule is oxidized in the presence of the CoASH. The product is a high-energy compound, succinyl CoA. NAD^+, the coenzyme, is reduced:

$$\alpha\text{-Ketoglutaric acid} + \text{CoASH} + NAD^+ \xrightarrow[\text{dehydrogenase complex}]{\alpha\text{-Ketoglutaric acid}} \text{Succinyl SCoA} + NADH/H^+ + CO_2$$

succinic acid

phosphorylation

Step 5: Hydrolysis of Succinyl SCoA Succinyl coenzyme A is now hydrolyzed to form *succinic acid* and coenzyme A. Sufficient energy is released to add phosphate to GDP (guanosine diphosphate), forming GTP. This is called a *phosphorylation* reaction. The phosphate group from GTP is then used to form ATP. This is the only place in the citric acid cycle that forms ATP directly from energy changes in the substrates of the cycle:

$$\text{Succinyl CoA} + \text{GDP} + P_i \xrightarrow[\text{synthetase}]{\text{Succinyl CoA}} \text{Succinic acid} + \text{GTP} + \text{CoASH}$$

$$\text{GTP} + \text{ADP} \xrightarrow{\text{Phosphokinase}} \text{ATP} + \text{GDP}$$

fumaric acid

Step 6: Formation of Fumaric Acid Succinic acid loses two hydrogens as *succinic acid dehydrogenase* catalyses the synthesis of fumaric acid. The hydrogens are picked up by the coenzyme FAD:

$$\text{Succinic acid} + \text{FAD} \xrightarrow[\text{}]{\text{Succinic acid dehydrogenase}} \text{Fumaric acid} + \text{FADH}_2$$

malic acid

Step 7: Formation of Malic Acid Fumaric acid undergoes hydration as a molecule of water is added to the double bond. The reaction is catalyzed by *fumarase* and produces malic acid:

$$\text{Fumaric acid} + H_2O \xrightarrow{\text{Fumarase}} \text{Malic acid}$$

oxaloacetic acid

Step 8: Formation of Oxaloacetic Acid Malic acid transfers two hydrogens to NAD^+ during oxidation by *malic acid dehydrogenase.* The resulting formation of *oxaloacetic acid* completes one full turn of the citric acid cycle, and the whole process starts again, with another acetyl CoA molecule combining with the new oxaloacetic acid:

$$\text{Malic acid} + NAD^+ \xrightarrow{\text{Malic acid dehydrogenase}} \text{Oxaloacetic acid} + NADH/H^+$$

Example 19.6 Use curved arrows to illustrate the reaction of succinic acid with FAD.

Solution: In step 6, succinic acid is oxidized to fumaric acid. The coenzyme FAD is reduced to $FADH_2$:

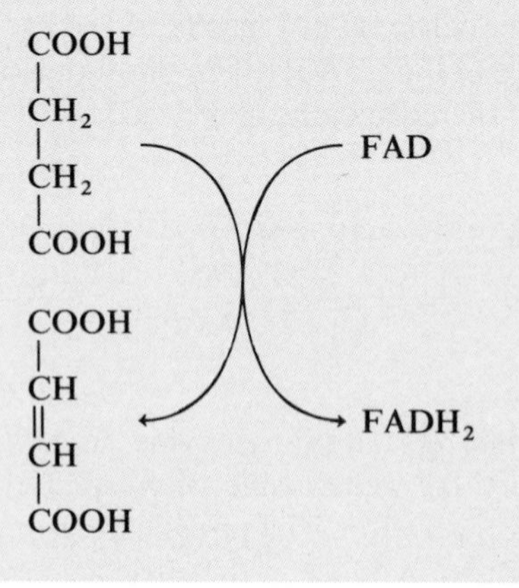

ATP Production by the Citric Acid Cycle

In the citric acid cycle, there are three oxidations (steps 3, 4, and 8) in which hydrogen atoms and NAD^+ produce three molecules of $NADH/H^+$. The FAD acceptor provides one molecule $FADH_2$ in step 6. Each $NADH/H^+$ produces three ATP molecules in the electron-transport system, and each FAD produces two ATP molecules. (At step 5, a direct phosphorylation of ADP by GTP provides one more ATP molecule.) We can now calculate the total ATP energy production by the citric acid cycle:

Reaction	*ATP Produced*
Step 3: $NAD^+ \longrightarrow NADH/H^+$	3
Step 4: $NAD^+ \longrightarrow NADH/H^+$	3
Step 5: $ADP \longrightarrow ATP$	1
Step 6: $FAD \longrightarrow FADH_2$	2
Step 8: $NAD^+ \longrightarrow NADH/H^+$	3
Total (one turn of cycle)	12

The total ATP produced from one molecule of acetyl CoA is 12 molecules of ATP. The overall process for the citric acid cycle can be written as follows:

Citric Acid Cycle

$$\text{Acetyl CoA} \longrightarrow 2CO_2 + 12ATP$$

Example 19.7 How is the oxidative decarboxylation of isocitric acid related to the production of ATP?

Solution: The alcohol group of isocitric acid is oxidized to α-ketoglutaric acid and two hydrogens are removed. They are accepted by NAD^+ to form $NADH/H^+$ and are carried to the electron-transport system with enough energy to synthesize three ATP molecules.

Example 19.8 Write a general equation for the number of ATP molecules produced when one acetyl CoA passes through the citric acid cycle.

Solution: The general equation for acetyl CoA produces two molecules of CO_2 and 12 molecules of ATP in one turn of the citric acid cycle:

$$\text{Acetyl CoA} \longrightarrow 2CO_2 + 12ATP$$

19.4 Glycolysis: Oxidation of Glucose

OBJECTIVE Write the equations for the overall reactions of the following: glucose to pyruvic acid; pyruvic acid to acetyl CoA; complete combustion of glucose. Account for the number of ATP molecules produced by each.

The carbohydrate most prevalent in the blood and tissues of the body is glucose, which is obtained from the polysaccharides and disaccharides in the diet. When

glycogen

glucose is not immediately used by the cells for energy, it can be stored to a limited extent in the liver and muscles as *glycogen*, a polymer of glucose. The formation of glycogen from glucose is called *glycogenesis*. When the levels of glucose in the blood become low, the glycogen is broken down, a process called *glycogenolysis* (splitting of glycogen):

$$\text{Glucose} \underset{\text{Glycogenolysis}}{\overset{\text{Glycogenesis}}{\rightleftharpoons}} \text{Glycogen}$$

Glycolysis

Glucose is the primary fuel used by the cells in the production of energy (ATP). A series of reactions, called *glycolysis*, converts the glucose molecules (six carbon atoms) to two pyruvic acids, which are smaller molecules with three carbon atoms. As we discuss glycolysis, refer to Figure 19.13.

Step 1 Glucose is converted to glucose 6-phosphate, a phosphorylation that utilizes the energy from one ATP:

$$\text{Glucose} + \text{ATP} \xrightarrow{\text{Hexokinase}} \text{Glucose 6-phosphate} + \text{ADP}$$

Step 2 An isomerization of glucose 6-phosphate forms fructose 6-phosphate:

$$\text{Glucose 6-phosphate} \xrightarrow{\text{Phosphoglucoisomerase}} \text{Fructose 6-phosphate}$$

Step 3 Another ATP molecule is used in a second phosphorylation reaction to form fructose 1,6-diphosphate:

$$\text{Fructose 6-phosphate} + \text{ATP} \xrightarrow{\text{Phosphofructokinase}} \text{Fructose 1,6-diphosphate}$$

Step 4 The six-carbon diphosphate molecule splits, producing dihydroxy-acetone phosphate and glyceraldehyde 3-phosphate:

$$\text{Fructose 1,6-diphosphate} \xrightarrow{\text{Aldolase}} \text{Glyceraldehyde 3-phosphate} + \text{Dihydroxyacetone phosphate}$$

Step 5 A triose phosphate isomerase converts dihydroxyacetone to another molecule of glyceraldehyde 3-phosphate:

$$\text{Dihydroxyacetone phosphate} \xrightarrow{\text{Triose phosphate isomerase}} \text{Glyceraldehyde 3-phosphate}$$

This is necessary because dihydroxyacetone cannot be oxidized any further by the tissues, so that we would lose half of the available energy from the glucose

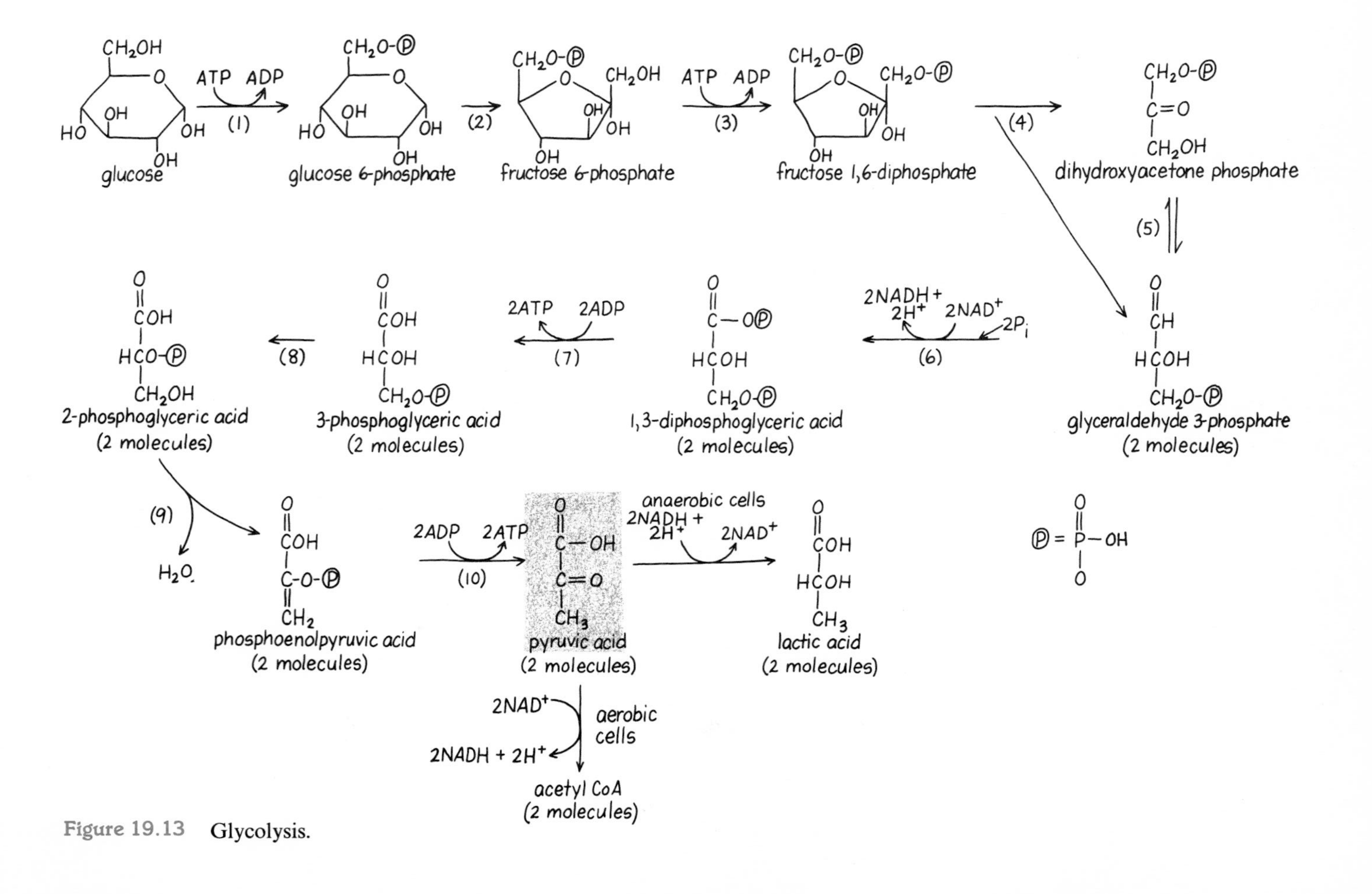

Figure 19.13 Glycolysis.

molecule. At this point in glycolysis, one glucose molecule has been converted to two triose phosphate molecules. The rest of the reactions will occur with two molecules of substrate and enzyme if the initial substrate is glucose.

Step 6 An oxidation of the triose molecules removes two pairs of hydrogens, which are accepted by two molecules of NAD^+. Phosphate is also added to form a high-energy compound:

$$\text{Glyceraldehyde 3-phosphate} + NAD^+ + P_i \xrightarrow{\text{Glyceraldehyde 3-phosphate dehydrogenase}} \text{1,3-Diphosphoglyceric acid} + NADH/H^+$$

Step 7 One of the phosphate groups is transferred to ADP to produce a molecule of ATP. This is a direct substrate–phosphate transfer:

$$\text{1,3-Diphosphoglyceric acid} + \text{ADP} \xrightarrow{\text{Phosphoglycerokinase}} \text{3-Phosphoglyceric acid} + \text{ATP}$$

Step 8 An isomerization reaction forms 2-phosphoglyceric acid:

$$\text{3-Phosphoglyceric acid} \xrightarrow{\text{Phosphoglyceromutase}} \text{2-Phosphoglyceric acid}$$

Step 9 A dehydration occurs in which a molecule of water is removed:

$$\text{2-Phosphoglyceric acid} \xrightarrow{\text{Enolase}} \text{Phosphoenolpyruvic acid (PEP)} + H_2O$$

Step 10 Pyruvic acid is formed by a transfer of the phosphate group to ADP to form another molecule of ATP:

$$\text{Phosphoenolpyruvic acid} + \text{ADP} \longrightarrow \text{Pyruvic acid} + \text{ATP}$$

ATP Production in Glycolysis

ATP energy from glycolysis

The initial reactions in glycolysis (steps 1 and 3) use up two ATP molecules. Later reactions (steps 7 and 10) produce a total of four molecules of ATP. Note that once the six-carbon compound divides into two triose molecules, all production of ATP or $NADH/H^+$ must be doubled:

Glucose to Pyruvic Acid

$$\text{Glucose} + 2\text{ADP} + 2P_i + 2NAD^+ \longrightarrow 2\text{ Pyruvic acid} + 2\text{ATP} + 2H_2O + 2NADH/H^+$$

Additional ATP production is made possible under aerobic (with oxygen) conditions. Then the electron-transport system converts the $NADH/H^+$ into more ATP. Earlier we learned that $NADH/H^+$ provided three molecules of ATP. However, in skeletal muscle cells, glycolysis occurs in the cytoplasm that is separated by a membrane from the mitrochondria. The $NADH/H^+$ must be carried across the membrane before it can be utilized by the electron-transport system. $NADH/H^+$ crosses the membrane, but uses up two ATP molecules in the process. Therefore, the two $NADH/H^+$ molecules produced in glycolysis will produce just four ATP molecules (6ATP − 2ATP to cross the membrane). The ATP production from glycolysis under aerobic conditions is summarized:

Glycolysis (Aerobic Conditions)

Reaction		*ATP Produced*
Step 1, 3:	Phosphorylation of hexoses	−2
Step 6:	$2NAD^+ \longrightarrow NADH/H^+$	+4
Step 7:	$2ADP \longrightarrow 2ATP$	+2
Step 10:	$2ADP \longrightarrow 2ATP$	+2
Glucose $\longrightarrow$ 2 Pyruvic acid		+6

Glycolysis (Aerobic)

Glucose $\longrightarrow$ 2 Pyruvic acid + 6ATP

Anaerobic Fate of Pyruvic Acid

Anaerobic conditions can occur during heavy exercise, when the oxygen stores in the muscles are depleted. The electron-transport system cannot operate in the absence of oxygen, so the $NADH/H^+$ collected during glycolysis is reoxidized by the reduction of pyruvic acid to lactic acid:

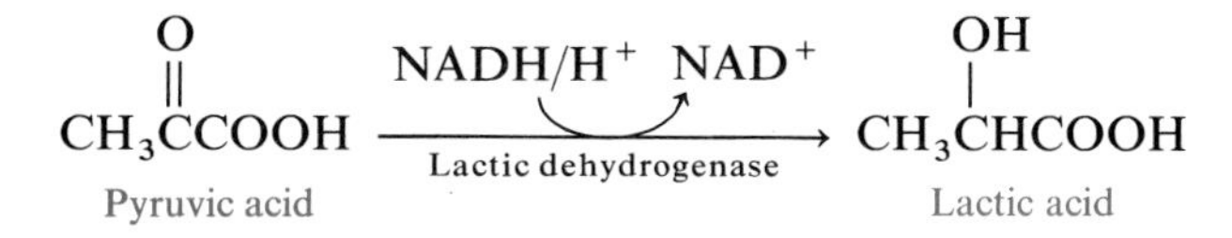

Pyruvic acid Lactic acid

The only ATP production in anaerobic glycolysis occurs through the direct phosphorylation reactions, giving a net total of two ATP molecules:

Glycolysis (Anaerobic)

Glucose $\longrightarrow$ 2 Lactic acid + 2ATP

Under anaerobic conditions, the lactic acid leaves the cells as a waste product, and no further energy is extracted. While muscle cells normally have oxygen present, it may be depleted by rapid exercise. The muscle cells can continue to extract some energy from glucose by the anaerobic pathway. However, much glucose is expended, so the process cannot go on long before exhaustion occurs. The accumulation of lactic acid in the cells causes the muscles to tire rapidly and become sore.

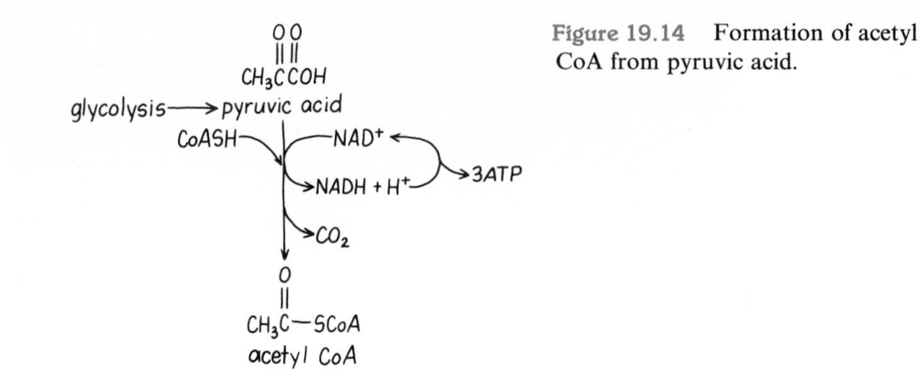

Figure 19.14 Formation of acetyl CoA from pyruvic acid.

Formation of Acetyl CoA from Pyruvic Acid (Aerobic)

When oxygen is available, pyruvic acid is converted to acetyl CoA, which can undergo oxidation in the citric acid cycle. The formation of acetyl CoA from pyruvic acid occurs as an oxidative decarboxylation:

Pyruvic Acid to Acetyl CoA

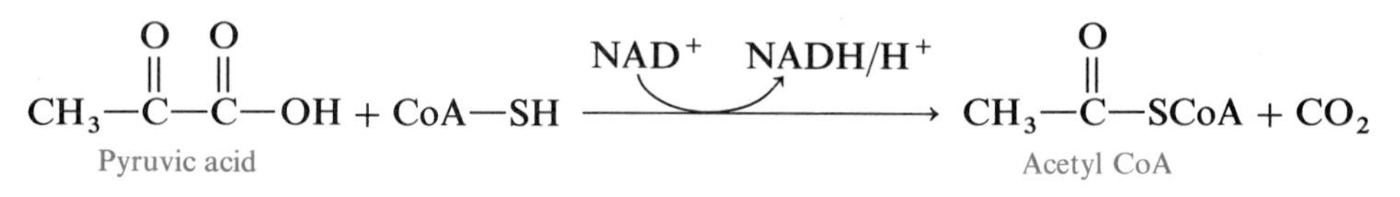

$$CH_3-\overset{O}{\overset{\|}{C}}-\overset{O}{\overset{\|}{C}}-OH + CoA-SH \xrightarrow{NAD^+ \quad NADH/H^+} CH_3-\overset{O}{\overset{\|}{C}}-SCoA + CO_2$$

Pyruvic acid — Acetyl CoA

A multienzyme for this reaction is called *pyruvic acid dehydrogenase complex.* (See Figure 19.14.) The oxidative decarboxylation of pyruvic acid provides $NADH/H^+$ that can be converted to three ATP molecules by the electron-transport system:

$$\text{Pyruvic acid} \xrightarrow{\text{Oxidative phosphorylation}} \text{Acetyl CoA} + CO_2 + 3\text{ATP}$$

Table 19.1 Total ATP Produced by Complete Oxidation of Glucose

Reaction Series	Total ATP
1. Glycolysis (Aerobic):	
Glucose ⟶ 2 Pyruvic acid + 6ATP	6
2. Pyruvic Acid to Acetyl CoA:	
2 (Pyruvic Acid ⟶ Acetyl CoA + CO_2 + 3ATP):	6
3. Citric Acid Cycle and Electron Transport System:	
2 (Acetyl CoA ⟶ $2CO_2$ + 12ATP)	24

Complete Combustion of Glucose

ATP energy from complete oxidation of glucose

We can now calculate the total ATP production for the complete combustion of glucose by considering the ATP production in aerobic glycolysis, the conversion of pyruvic acid to acetyl CoA, and the oxidation of acetyl CoA in the citric acid cycle. (See Table 19.1.) Remember that the degradation of glucose leads to the formation of two molecules of pyruvic acid, which provides two molecules of acetyl CoA for the citric acid cycle. The energy provided by glucose depends, therefore, on the conditions in the cells. This is summarized by the following scheme:

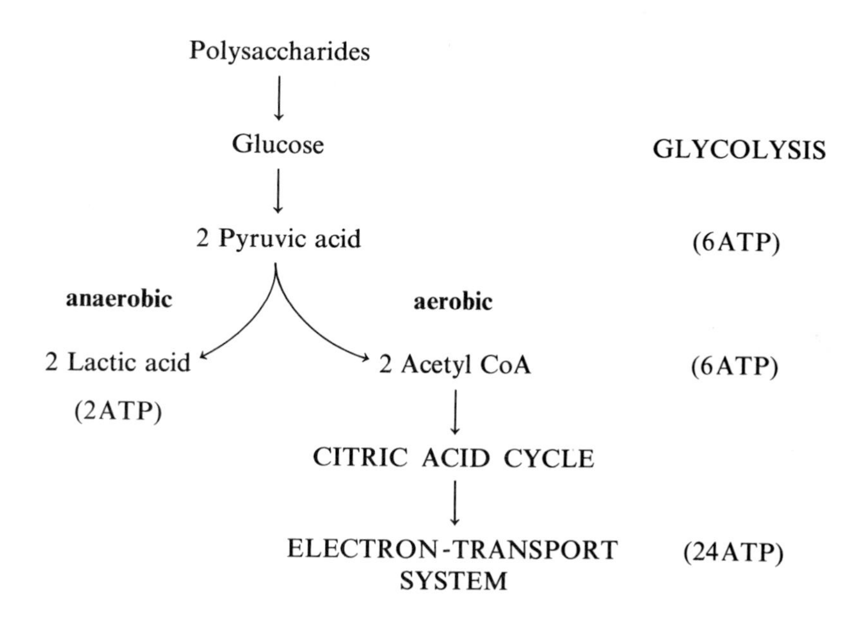

Example 19.9 Indicate the number of molecules of ATP produced by each of the following conversions:

1. The aerobic conversion of glucose to pyruvic acid.
2. The anaerobic conversion of glucose to lactic acid.
3. The complete oxidation of glucose to CO_2.

Solution

1. The conversion of glucose to pyruvic acid produces a total of six ATP molecules.
2. The conversion of glucose to lactic acid produces a total of two ATP molecules.
3. The complete oxidation of glucose produces 36 ATP molecules.

19.5 Oxidation of Fatty Acids

OBJECTIVE Describe the β-oxidation of fatty acids. Account for the number of ATP molecules produced when a fatty acid is completely oxidized.

Body fats (lipids) are the primary form of stored energy. When there are no sources of glucose left, the body begins to utilize its fats as an energy source (9 kcal/g). The triacylglycerols in the adipose cells are hydrolyzed to yield glycerol and fatty acids:

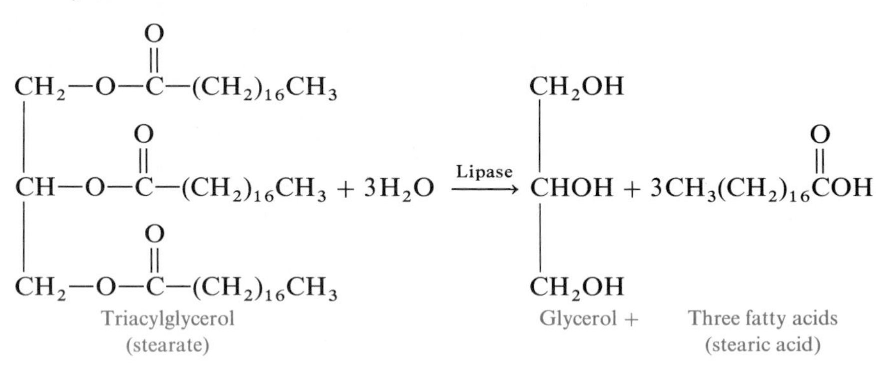

β-Oxidation of Fatty Acids

Fatty acids resulting from the hydrolysis of triacylglycerols undergo oxidation in a series of reactions called β-oxidation. The reactions result in the breaking of the bond between the α-carbon and the β-carbon of each long-chain fatty acid, yielding two-carbon units of acetyl CoA.

Activation of the Fatty Acid

Fatty acid oxidation begins with the activation of a fatty acid by *acyl CoA synthetase*, which uses a molecule of ATP to combine a fatty acid molecule with a molecule of coenzyme A. (See Figure 19.15.) For our example, we will use a 10-carbon fatty acid:

$$\underset{\text{Capric acid (10-carbon atoms)}}{CH_3(CH_2)_6CH_2{-}CH_2{-}\overset{\overset{\displaystyle O}{\|}}{C}{-}OH} + CoA{-}SH \xrightarrow[\text{Acyl CoA synthetase}]{\text{ATP} \quad \text{AMP} + \text{PP}_i}$$

$$\underset{\text{Activated capryl CoA}}{CH_3(CH_2)_6CH_2{-}CH_2{-}\overset{\overset{\displaystyle O}{\|}}{C}{-}S{-}CoA} + H_2O$$

β-Oxidation Spiral

The diagram for the β-oxidation series of reactions is shown in Figure 19.15.

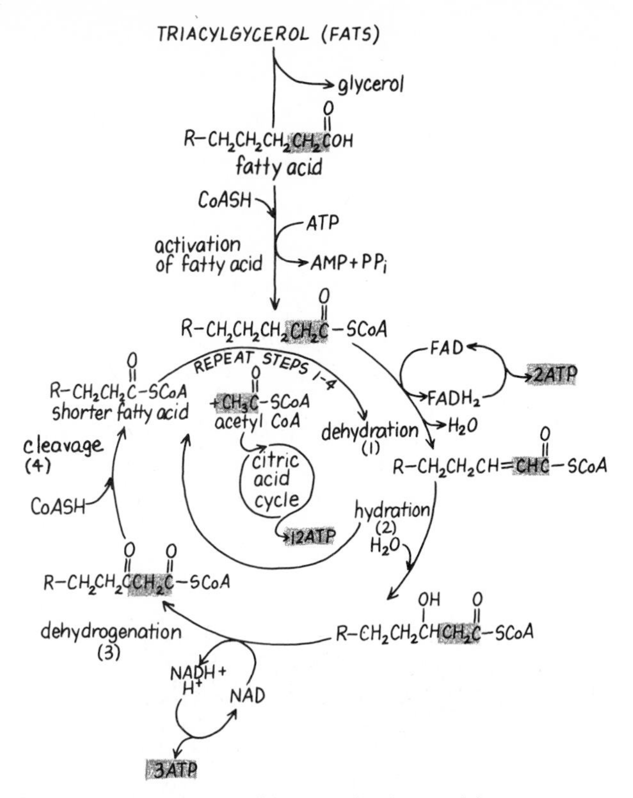

Figure 19.15 Beta oxidation of a fatty acid.

oxidation **Step 1** *Oxidation* occurs within the fatty acid at the α- and β-carbon atoms, with FAD accepting the hydrogen:

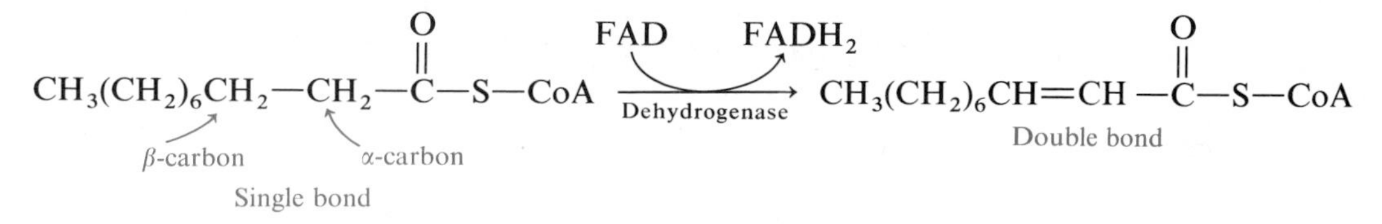

hydration **Step 2** Water is added to the double bond in a *hydration* reaction catalyzed by a hydrolase to give a β-hydroxy compound:

$$CH_3(CH_2)_6CH{=}CH{-}\overset{\overset{\displaystyle O}{\|}}{C}{-}S{-}CoA + H_2O \xrightarrow[\text{Hydrolase}]{} CH_3(CH_2)_6\overset{\overset{\displaystyle OH}{|}}{C}H{-}CH_2{-}\overset{\overset{\displaystyle O}{\|}}{C}{-}S{-}CoA$$

(Alcohol)

oxidation **Step 3** The β-hydroxy group is oxidized by a dehydrogenase enzyme using the NAD^+ hydrogen acceptor to form a β-keto compound:

$$CH_3(CH_2)_6\underset{|}{\overset{OH}{C}}H-CH_2-\overset{O}{\overset{\|}{C}}-S-CoA \xrightarrow[\text{Dehydrogenase}]{NAD^+ \quad NADH/H^+} CH_3(CH_2)_6\overset{O}{\overset{\|}{C}}-CH_2-\overset{O}{\overset{\|}{C}}-S-CoA$$

Ketone

cleavage **Step 4** A two-carbon unit of acetyl CoA is split from the 10-carbon molecule catalyzed by a thiolase enzyme. The remaining long chain from the fatty acid has decreased in length by two carbon atoms and is now an eight-carbon fatty acid CoA:

$$CH_3(CH_2)_6\overset{O}{\overset{\|}{C}}-CH_2-\overset{O}{\overset{\|}{C}}-S-CoA \xrightarrow{\text{Ketothiolase}} CH_3-\overset{O}{\overset{\|}{C}}-SCoA + CH_3(CH_2)_6\overset{O}{\overset{\|}{C}}-S-CoA$$

+ HSCoA

Acetyl CoA → Citric acid

Caprylyl CoA (8 carbon atoms) → Repeat β-oxidation

The fatty acid chain, which is now two carbon atoms shorter, will repeat the β-oxidation spiral (steps 1–4) until the entire chain has been degraded to units of acetyl CoA. The units of acetyl CoA (the end products of β-oxidation) enter the citric acid cycle for further oxidation:

β-Oxidation (Four Times Through Spiral)

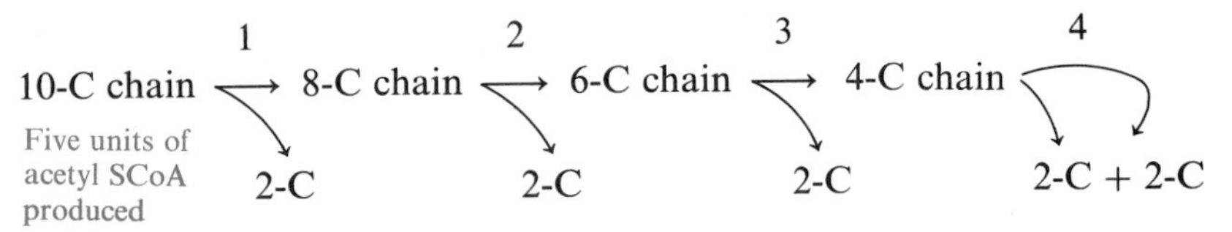

The number of acetyl CoA molecules can be determined by counting the number of carbon atoms in the chain and dividing by two. The number of times that the fatty acid will go through the β-oxidation spiral is one less than the number of acetyl CoA molecules. In our example, capric acid, which has 10 carbon atoms, will produce a total of five acetyl CoA units upon complete oxidation. It will go through the β-oxidation sequence four times (one less than the number of acetyl CoA molecules):

Capric Acid	
Number of carbon atoms	10
Number of acetyl CoA molecules	$10/2 = 5$
Number of times through cycle	$5 - 1 = 4$

Energy Production in β-Oxidation

We can calculate the amount of energy available from a typical fatty acid. Using the 10-carbon fatty acid as our example, we have determined that it will produce five acetyl CoA molecules and go through the β-oxidation spiral four times.

Each turn of the β-oxidation spiral produces one $FADH_2$ molecule and one $NADH/H^+$ molecule, which together produce a total of five ATP molecules via the electron-transport system. In our example, there are four turns of β-oxidation, giving a total of 20 ATP molecules. In addition, the acetyl CoA units will produce ATP when they enter the citric acid cycle. Since each turn of the citric acid cycle produces 12 ATP molecules, our five acetyl CoA molecules will produce a total of 60 molecules of ATP. Also, since the initial activation step had to cleave two phosphate bonds, we count the energy expenditure as being equal to two molecules of ATP. Therefore, two ATP molecules must be subtracted from the total energy output.

Energy Produced by a 10-Carbon Fatty Acid

Acetyl CoA/Citric Acid Cycle:

$$5\ \text{Acetyl CoA} \times \frac{12\,\text{ATP}}{1\,\text{Acetyl CoA}} = 60\,\text{ATP}$$

β-Oxidation (4 cycles):

$$4\text{FADH}_2 \times \frac{2\,\text{ATP}}{1\,\text{FADH}_2} = 8\,\text{ATP}$$

$$4\text{NADH/H}^+ \times \frac{3\,\text{ATP}}{1\,\text{NADH/H}^+} = 12\,\text{ATP}$$

Activation of fatty acid = − 2ATP

Total from 10-C fatty acid: 78ATP

Example 19.10 How much ATP could be produced from β-oxidation and the citric acid cycle for a 14-carbon fatty acid?

Solution: A 14-carbon fatty acid would produce seven acetyl CoA units (14/2 = 7) and undergo β-oxidation six times (7 − 1):

Acetyl CoA = 14/2 = 7

Turns of β-oxidation spiral = 7 − 1 = 6

Citric Acid Cycle:

7 Acetyl CoA × 12ATP/mol = 84ATP

β-Oxidation of a 14-C Fatty Acid:

$6\text{FADH}_2 \times 2\text{ATP/mol} = 12\text{ATP}$

$6\text{NADH/H}^+ \times 3\text{ATP/mol} = 18\text{ATP}$

Activation of fatty acid = −2ATP

Total from 14-C fatty acid: 112ATP

HEALTH NOTE

Formation of Ketone Bodies

When the body breaks down large quantities of body fat to meet energy needs normally satisfied by carbohydrates, an accumulation of acetyl CoA occurs. As a result, the citric acid cycle becomes saturated with acetyl CoA, and the excess follows a pathway in which ketone (acetone) bodies are produced. Ketone bodies appear in the urine and blood when a person is on a rigid diet, is starving, or has a disease, such as diabetes, in which carbohydrates cannot be utilized.

One such ketone body, acetoacetic acid, is formed from the accumulation of acetyl CoA from the β-oxidation of fatty acids. It is converted by a series of reactions to acetone and β-hydroxybutyric acid:

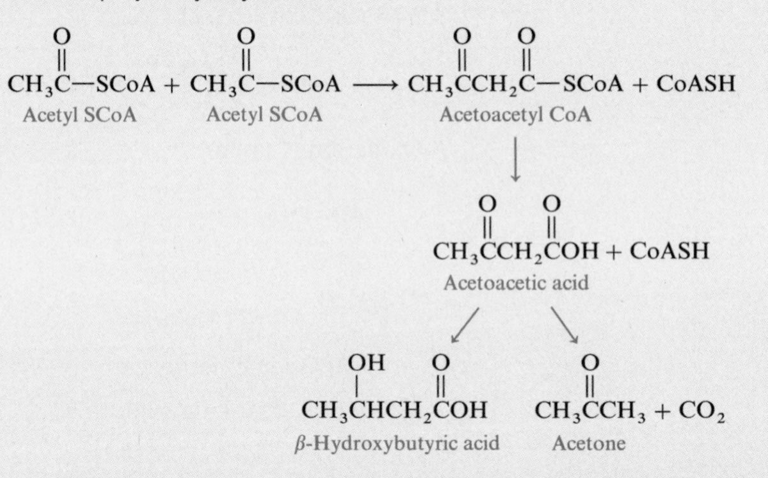

Since many of the ketone bodies are acids, the pH of the blood is lowered; low pH can lead to acidosis and dehydration. A high level of ketone bodies in the blood is called *ketosis.* When ketone bodies are eliminated from the body, they carry sodium cations with them, depleting the body's sodium supply. The loss of sodium electrolyte results in a large output of urine, which creates a strong sensation of thirst. These symptoms are seen in cases of untreated diabetes. The fruity odor of acetone can be detected on the breath of a person with ketosis.

19.6 Oxidation of Amino Acids

OBJECTIVE Describe the transamination and oxidative deamination of amino acids for use in the citric acid cycle.

Proteins in the diet and in the body can be hydrolyzed to yield amino acids. The amino acids move through intestinal cell membranes and cell walls to become part of the amino acid pool from which they can be used to build needed proteins, or be degraded (metabolized) if they need to be used as an energy source.

transamination

In order to use the carbon structure of amino acids for energy, the amino group must first be removed. Amino acids can transfer amino groups through a process called *transamination* or *oxidative deamination.* When transamination occurs, the amino group is transferred from the amino acid to the keto group of an α-ketoacid. Usually, the α-ketoacid is α-ketoglutaric acid, a metabolite in the citric acid cycle:

$$\alpha\text{-Amino acid} + \alpha\text{-Ketoglutaric acid} \xrightarrow{\text{Glutamate transaminase}} \alpha\text{-Keto acid} + \text{Glutamic acid}$$

We can write the transamination reaction using alanine as the α-amino acid:

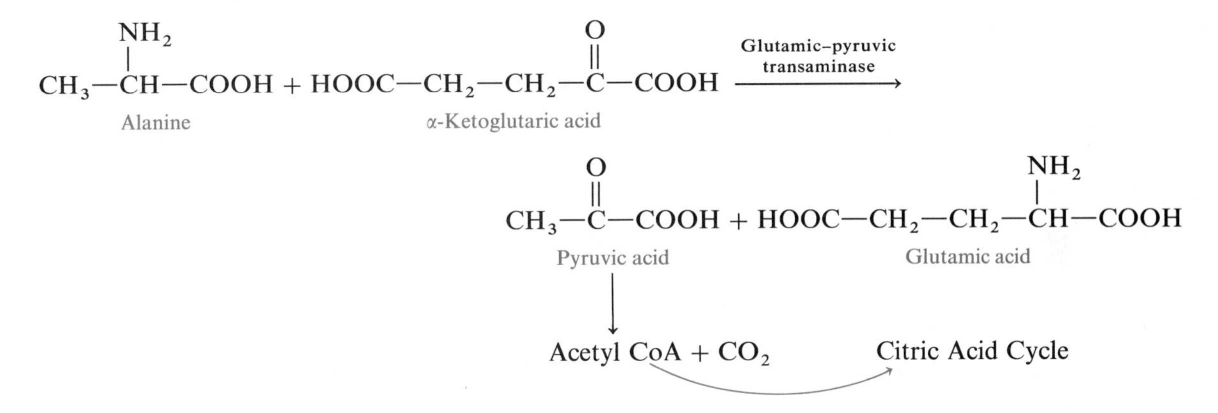

oxidative deamination

In oxidative deamination, the amino acid is converted into an α-keto acid, with the amino group converted to ammonia (NH_3), which is then converted to urea for excretion through the kidneys:

$$\underset{\text{Alanine}}{CH_3CH(NH_2)\text{—}C(=O)OH} + NAD^+ + H_2O \xrightarrow{\text{Alanine oxidase}} \underset{\text{Pyruvic acid}}{CH_3C(=O)\text{—}C(=O)OH} + \underset{\text{Ammonia}}{NH_3} + NADH/H^+$$

When the body needs to use amino acids in the production of energy, transamination and oxidative deamination converts them to α-keto acids that can enter the citric acid cycle at various points. (See Figure 19.16.)

In starving conditions, the body relies on energy from amino acids, indicating that fat reserves have been used up. However, this eventually causes a depletion of body protein and destruction of body tissues. Certain diets, particularly high-protein diets, overload the system with amino acids whose deamination will cause an abundance of ketone bodies, resulting in ketosis.

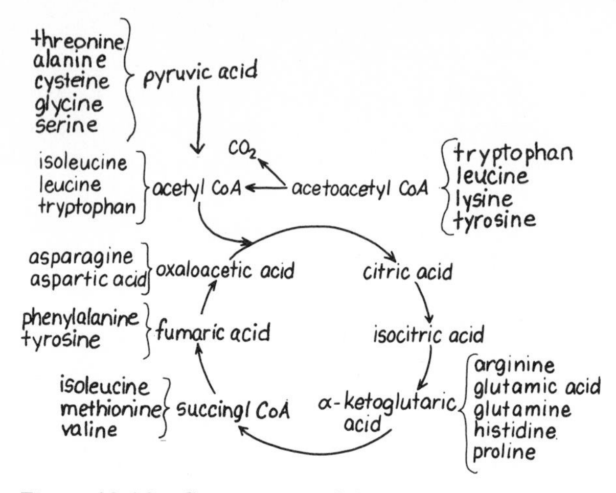

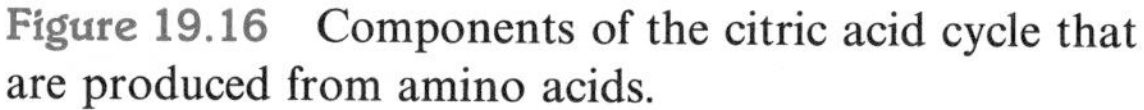

Figure 19.16 Components of the citric acid cycle that are produced from amino acids.

Example 19.11 Serine can be converted to pyruvic acid by a series of reactions including deamination. How does this reaction represent a way in which amino acids can be utilized as an energy source?

Solution: Amino acids can be converted to compounds that are a part of the energy-producing citric acid cycle. After the removal of the amino group of serine (deamination), it is converted to pyruvic acid. Pyruvic acid undergoes oxidative decarboxylation to give acetyl CoA, which can enter the citric acid cycle to produce 12 ATP molecules in one turn.

Energy-Consuming Processes

In the beginning of this chapter we mentioned that metabolic reactions consisted of energy-producing (catabolic) and energy-requiring reactions (anabolic). Until now we have concentrated on those major reactions that involve the production of energy by the cells of the body. To complete the concept of total metabolism, we need to mention the ways in which energy is consumed. As we saw, the oxidative pathways degrade large molecules to small molecules that can be used for energy production via the citric acid cycle. These molecules can also follow routes that lead to the synthesis of larger molecules in the cell. Such synthesis is energy consuming and depends on a supply of ATP.

For example, pyruvic acid can be converted to glycogen for storage in a series of reactions similar to a reversal of glycolysis. The β-oxidation of fatty acids, the oxidative deamination of amino acids, and the oxidation of glucose all can produce acetyl CoA, which in turn can enter several synthesis reactions. For instance, the two-carbon units of acetyl CoA can join together step by step to

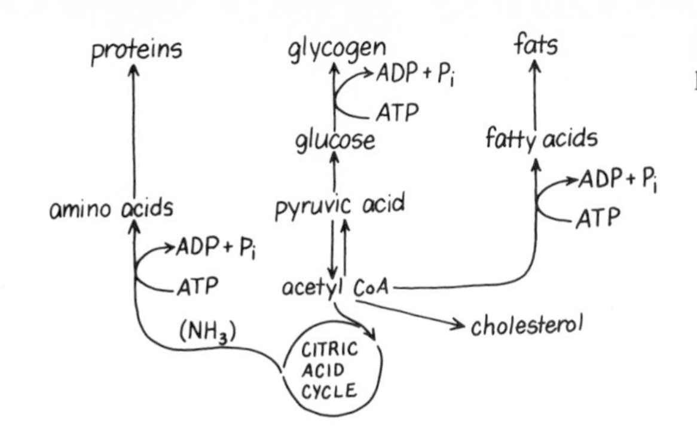

Figure 19.17 Energy-requiring pathways (anabolic).

form long-chain fatty acids. When there is an abundance of carbohydrates in the diet, the citric acid cycle produces sufficient energy so that the acetyl CoA that is not needed is moved into a pathway leading to the synthesis of fatty acids and then to fats (triacylglycerols) that can be stored without limitation by the body. Acetyl CoA is also a precursor in the synthesis of cholesterol.

The α-keto acids of the citric acid cycle, pyruvic acid, and acetyl CoA may be used in the synthesis of several amino acids (nonessential) by transamination reactions. With the production of these amino acids and the essential amino acids obtained from the diet, the cells build the necessary proteins for structure and activity. (See Figure 19.17.)

Glossary

acetyl CoA An end product of glucose, fatty acid, and protein metabolic pathways that can enter the citric acid for oxidation reactions leading to the production of ATP. Acetyl CoA is a two-carbon acetyl unit combined with coenzyme A.

ADP Adenosine diphosphate, a compound consisting of adenosine (adenine and a ribose sugar) and two phosphate groups; ADP forms when a phosphate group is cleaved from ATP, releasing 7.3 kcal of energy.

aerobic An oxygen-containing environment in the cells.

anaerobic A cellular environment without oxygen.

ATP Adenosine triphosphate, a compound consisting of adenosine (adenine and a ribose sugar) and three phosphate groups; ATP is formed in the mitochondria to store energy for the cells.

β-oxidation spiral A series of reactions that cleave acetyl CoA units from fatty acids to provide energy for the production of ATP.

citric acid cycle A series of oxidation reactions in the mitochrondria that convert acetyl CoA to CO_2 and provide hydrogen atoms for energy production by the electron-transport system.

coenzyme Q An acceptor of hydrogen atoms from $FMNH_2$ or $FADH_2$ in the electron-transport system. When $CoQH_2$ reoxidizes, a pair of electrons ($2e^-$) passes on to the cytochromes and two protons ($2H^+$) are released into the mitochondria.

cytochromes A group of electron acceptors in the electron transport system that use the oxidation states of iron (Fe^{3+}, Fe^{2+}) to transfer electrons to oxygen to form water.

electron-transport system A series of reactions occurring in the mitochondria in which hydrogen atoms or electrons are transferred among coenzymes and cytochromes to oxygen to produce water and energy in the form of two or three ATP molecules.

FAD A hydrogen carrier (flavin adenine dinucleotide) that accepts hydrogen atoms from carbon–carbon bonds and carries them to the electron-transport system, where they yield two ATP molecules.

FMN A hydrogen acceptor (flavin mononucleotide) in the electron-transport system.

glycogenesis The chemical process of converting glucose to glycogen for storage.

glycogenolysis The hydrolysis of glycogen to glucose.

glycolysis The oxidation reactions of glucose that yield two pyruvic acid molecules under aerobic conditions and two lactic acid molecules under anaerobic conditions, producing ATP for the cell.

metabolite A compound that is part of a series of metabolic reactions.

mitochondria Structures within a cell that contain enzymes and coenzymes needed for oxidation reactions; mitochondria produce energy.

NAD^+ A hydrogen acceptor that accepts hydrogen atoms directly from carbon–oxygen oxidations and carries them to the electron-transport system to yield three ATP molecules.

oxidation The loss of two hydrogen atoms or the loss of two electrons in a chemical reaction.

oxidative deamination Reactions that convert the amino group of an amino acid to ammonia to produce α-keto acids that can enter the citric acid cycle for energy production.

transamination The transfer of an amino group from an α-amino acid to an α-keto acid to produce intermediates in the citric acid cycle that can be oxidized for the production of energy in the cell.

Problems

The Role of ATP (Objective 19.1)

19.1 Write an equation to represent the hydrolysis of ATP.

19.2 What is the role of ATP in the cells?

The Electron-Transport System (Objective 19.2)

19.3
a. What is oxidative phosphorylation?
b. How much energy is needed to form a mole of ATP from a mole of ADP?
c. Write an equation for the formation of ATP from ADP.

19.4
a. What are the hydrogen acceptors in the electron-transport system?
b. Which coenzymes accept hydrogen directly from metabolites?
c. Write an equation to illustrate the transfer of hydrogen atoms from a metabolite MH_2 to FAD; to NAD^+.

19.5 Write an equation to illustrate the following:
a. oxidation of $NADH/H^+$
b. reduction of CoQ
c. oxidation of $FMNH_2$

19.6 Complete the blanks in the following series of reactions in the electron-transport system:

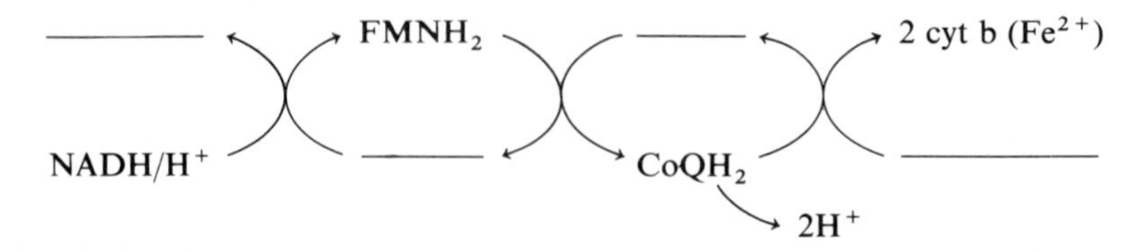

19.7 What is the function of the cytochromes in the electron-transport system?

19.8 Indicate whether the following represent oxidation or reduction reactions:
a. 2 cyt c $(Fe^{3+}) + 2e^- \longrightarrow$ 2 cyt c (Fe^{2+})
b. 2 cyt a $(Fe^{2+}) \longrightarrow$ 2 cyt a $(Fe^{3+}) + 2e^-$

19.9 a. What is the end product of the electron transport system?
b. How many ATP molecules are produced when $NADH/H^+$ is oxidized?
c. How many ATP molecules are produced when $FADH_2$ is oxidized?

Citric Acid Cycle (Objective 19.3)

19.10 a. What is the function of acetyl CoA in the citric acid cycle?
b. What are the six-carbon compounds in the citric acid cycle?
c. What are the five-carbon compounds?
d. What are the four-carbon compounds?
e. How is the number of carbon atoms of the compounds in the citric acid cycle decreased?
f. What are the end products of the citric acid cycle?

19.11 A dehydrogenase is used to catalyze the oxidation of isocitric acid, malic acid, and succinic acid. Why is NAD^+ used for the oxidation of isocitric acid and malic acid, whereas FAD is used for the oxidation of succinic acid?

19.12 How is the citric acid cycle responsible for the production of ATP by the electron transport system?

19.13 Write the equation for the hydrogen or P_i acceptor for each of the following reactions; state the number of ATP molecules produced:

	Acceptor	*ATP*
a. Isocitric acid ⟶ α-Ketoglutaric acid	$2H + NAD^+ \longrightarrow NADH/H^+$	3 ATP
b. α-Ketoglutaric acid ⟶ Succinyl CoA	______	______
c. Succinyl CoA ⟶ Succinic acid	______	______
d. Succinic acid ⟶ Fumaric acid	______	______
e. Malic acid ⟶ Oxaloacetic acid	______	______

19.14 Write the overall equation for the citric acid cycle, including the number of ATP molecules produced in one completion of the cycle.

19.15 How many ATP molecules (or moles) would be produced when the following numbers of acetyl CoA molecules (or moles) enter the citric acid cycle?
a. five molecules of acetyl CoA
b. two moles of acetyl CoA
c. 25 molecules of acetyl CoA
d. 10 moles of acetyl CoA

Oxidation of Glucose (Objective 19.4)

19.16 a. What is the starting reactant for glycolysis?
b. What is the end product of glycolysis under aerobic conditions?
c. What is the end product of glycolysis under anaerobic conditions?

19.17 The first three steps of glycolysis involve the phosphorylation of glucose.
a. How does ATP function in these initial steps?
b. How many ATP molecules are used?

19.18 a. What is the coenzyme used in glycolysis?
b. What happens to the two $NADH/H^+$ molecules produced in glycolysis under aerobic conditions?
c. How are the two $NADH/H^+$ molecules reoxidized under anaerobic conditions?
d. How does direct phosphorylation account for the production of ATP in glycolysis?

19.19 The $NADH/H^+$ in glycolysis is produced in the cytoplasm and must be transported across the mitochondrial membranes to the electron-transport system. How does this condition affect the ATP produced by $NADH/H^+$ in glycolysis?

19.20 a. Under aerobic conditions, what happens to the pyruvic acid produced by glycolysis?
b. What is the energy yield in ATP molecules when pyruvic acid undergoes oxidative decarboxylation?
c. Write the overall equation for the oxidative decarboxylation of pyruvic acid?

19.21 Indicate the number of ATP molecules associated with each of the following conversions under aerobic conditions:

a. $\text{Glucose} \xrightarrow{\text{Glycolysis}} \text{2 Pyruvic acid}$

b. $\text{2 Pyruvic acid} \xrightarrow[\text{decarboxylation}]{\text{Oxidative}} \text{2 acetyl CoA} + 2CO_2$

c. $\text{2 Acetyl CoA} \xrightarrow{\text{Citric acid cycle}} 4CO_2$

d. $\text{Glucose} + 6O_2 \longrightarrow 6CO_2 + 6H_2O$

Oxidation of Fatty Acids (Objective 19.5)

19.22 What is required to prepare a fatty acid for β-oxidation?

19.23 a. What are the four major reactions in β-oxidation?
b. In which steps are dehydrogenase enzymes used?
c. Why are different hydrogen acceptors used?
d. What is the ATP yield from the two oxidation steps in β-oxidation?

19.24 Palmitic acid, $CH_3(CH_2)_{14}COOH$, is a common fatty acid in butter, tallow, and human fat. The activated form of palmitic acid can be written

$$CH_3(CH_2)_{12}{-}CH_2{-}CH_2{-}\overset{\overset{\displaystyle O}{\|}}{C}{-}SCoA$$

a. Indicate the α- and β-carbon atoms in the compound.
b. Write equations for palmitic acid going through the first turn of the β-oxidation spiral.

19.25 For palmitic acid (Problem 19.24), give the following information for complete oxidation:
a. What is the total number of carbon atoms in palmitic acid?
b. How many acetyl CoA units will be produced when β-oxidation is complete?
c. How many turns of the β-oxidation spiral are needed to completely oxidize palmitic acid?

d. Account for the ATP yield by completing the following list:

_______ Acetyl CoA units ⟶ Citric acid cycle (and electron-transport system) _______ ATP

_______ $FADH_2$ produced ⟶ Electron transport _______ ATP

_______ $NADH/H^+$ produced ⟶ Electron transport _______ ATP

Activation step −2ATP

Total _______ ATP

Oxidation of Amino Acids (Objective 19.6)

19.26 Write a transamination reaction for the following amino acids and α-ketoglutaric acid, $HOOC{-}CH_2{-}CH_2{-}\overset{\overset{\displaystyle O}{\|}}{C}{-}COOH$:

a. $CH_3{-}\underset{}{\overset{\overset{\displaystyle NH_2}{|}}{CH}}{-}COOH$
Alanine

b. $HOOC{-}CH_2{-}\overset{\overset{\displaystyle NH_2}{|}}{CH}{-}COOH$
Aspartic acid

19.27 What steps in metabolism can the α-keto acids produced in Problem 19.26 enter?

19.28 How are amino acids converted to α-keto acids by oxidative deamination?

19.29 Write a reaction for oxidative deamination of the following amino acids:

a. $CH_3{-}\overset{\overset{\displaystyle NH_2}{|}}{CH}{-}COOH$
Alanine

b. $HOOC{-}CH_2{-}\overset{\overset{\displaystyle NH_2}{|}}{CH}{-}COOH$
Aspartic acid

19.30 How does acetyl CoA play a central role in both catabolic and anabolic pathways?

20 Chemistry of Heredity: DNA and RNA and Hormonal Action

Objectives

20.1 Given the name of a nucleotide, describe its nitrogenous base (type of purine or pyrimidine), sugar, and phosphate groups.

20.2 Use complementary base pairing to explain the interaction between base pairs in a DNA molecule; draw or describe a portion of a DNA helix.

20.3 Describe the replication process of DNA.

20.4 Describe the transcription and translation processes in protein synthesis by writing codons for RNA, anticodons for tRNA, and the resulting amino acid sequences.

20.5 Differentiate between the reaction control systems of feedback control, enzyme induction, and enzyme repression.

20.6 Describe a hormone and two methods of hormonal action.

20.7 Describe the function(s) of the hormones released by the anterior and posterior pituitary glands.

20.8 State the function(s) of the hormones released by the thyroid gland, parathyroid glands, pancreas, adrenal cortex, gonads, and hypothalamus.

Scope

In the preceding chapters we discussed three groups of substances—carbohydrates, proteins, and lipids—that are essential to the survival of living organisms. There is one more group of compounds that must be included in this discussion of life: the nucleic acids.

Nucleic acids play a critical role in the control and direction of cellular growth and reproduction. The nucleic acid DNA—the genetic material in the nucleus of the cell—contains all the information needed for the development of a complete living system. The way you grow, your hair, your eyes, your total physical appearance, and the internal activities of the cells in your body are all dictated by the set of directions held in the nucleus of each of your cells. DNA of the nucleus controls the functioning of the cell by controlling the synthesis of enzymes (proteins). Another group of nucleic acids, RNA, is responsible for transmitting the information from the nucleus to the ribosomes, where protein synthesis occurs. Errors in the transmittal of information can lead to faulty or nonfunctioning enzymes. Genetic diseases are a result of such mutations.

20.1 Components of the Nucleic Acids

OBJECTIVE Given the name of a nucleotide, describe its nitrogenous base (type of purine or pyrimidine), sugar, and phosphate groups.

DNA

RNA

Nucleic acids are very large molecules found in and produced by the nucleus of a cell. There are two major types: *deoxyribonucleic acid* (*DNA*) and *ribonucleic acid* (*RNA*). DNA, along with some protein, constitutes the material of the chromosomes, the threadlike structures observed in the nucleus during cellular division. In humans, every cell contains 23 pairs of chromosomes, or 46 chromosomes in all (except egg and sperm cells, each of which contains half that number, 23 chromosomes total). Structural characteristics and cellular components develop from directions given by the DNA of the chromosomes. Sets of directions, or genes, for a particular characteristic or enzyme are found within the chromosomes. (See Figure 20.1.)

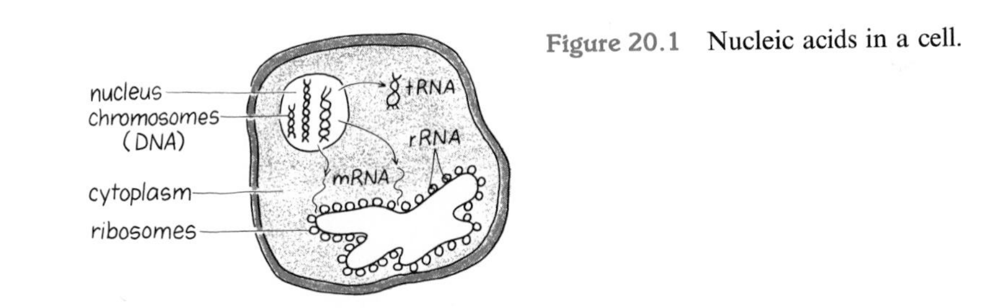

Figure 20.1 Nucleic acids in a cell.

The second type of nucleic acid, RNA, is produced in the nucleus by the DNA. However, the RNA transmits information to the cell by migrating into the cytoplasm where it participates in protein synthesis. There are three types of RNA in a cell. Table 20.1 lists the types of nucleic acids found in the cells.

Table 20.1 Nucleic Acids in the Cells

Nucleic Acid	Function
DNA (deoxyribonucleic acid)	Holds the genetic information for the cell
RNA (ribonucleic acid)	
mRNA (messenger RNA)	An RNA that carries information for the construction of proteins from the DNA to the ribosomes in the cytoplasm
rRNA (ribosomal RNA)	An RNA that makes up 60 percent of the ribosomes and that along with proteins is involved in the process of synthesizing proteins for the cell. However, the role of the rRNA in this process is not yet understood
tRNA (transfer RNA)	A smaller RNA that participates in the collection of amino acids from the cytoplasm for protein synthesis at the ribosomes

Purines and Pyrimidines

One of the subunits of nucleic acids is a nitrogenous base that is a nitrogen-containing cyclic compound having one or two rings. The *purines* adenine (A) and guanine (G) have the double-ring structures; the *pyrimidines* cytosine (C), thymine (T), and uracil (U) have the single-ring structures. (See Figure 20.2.)

Sugars and Phosphate in Nucleic Acids (Nucleotides)

There are two kinds of pentose sugars in nucleic acids, a ribose sugar and a deoxyribose sugar. DNA molecules contain only deoxyribose (*deoxy-* means "without oxygen"); RNA molecules contain only ribose. (See Figure 20.3.)

When a sugar is combined with one of the nitrogenous bases by a glycosidic link, a nucleoside forms. When a phosphate is added, a nucleotide, the building block of all nucleic acids, is formed. (See Figure 20.4.)

purines

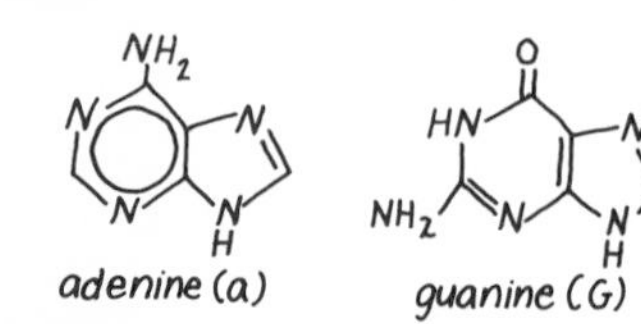

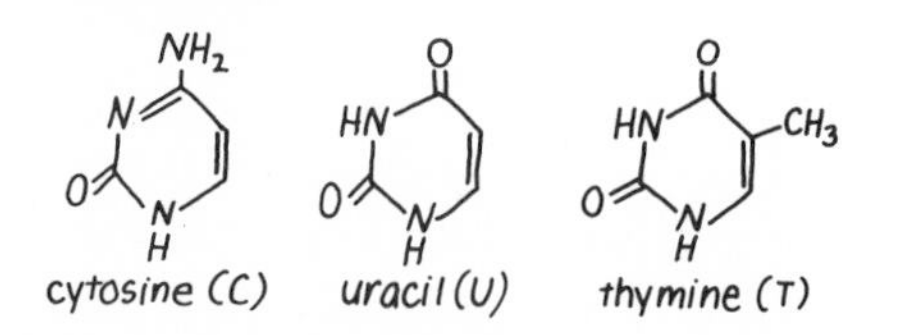

Figure 20.2 Purine and pyrimidine bases found in nucleic acids.

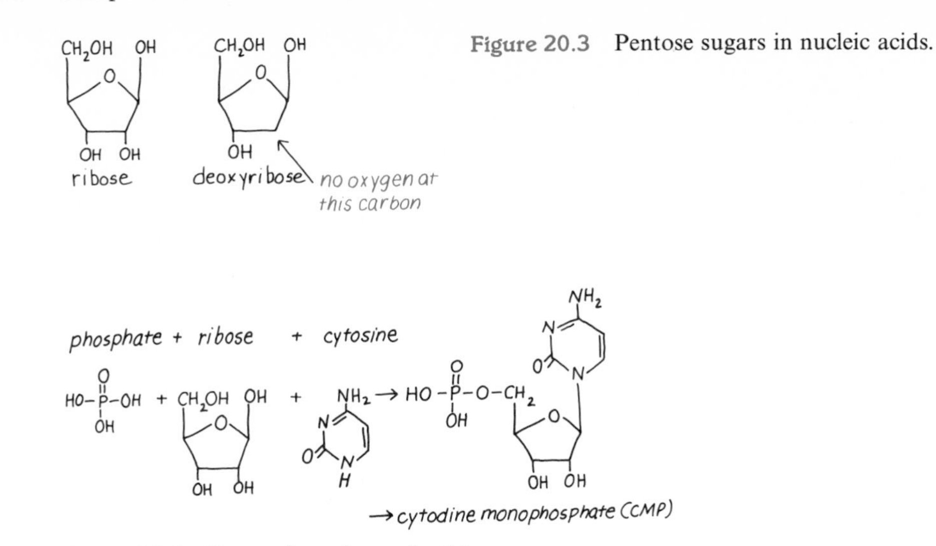

Figure 20.3 Pentose sugars in nucleic acids.

Figure 20.4 Formation of a nucleotide.

naming nucleotides

Nucleotides of RNA formed with ribose are named by changing the end of the base name to *-osine* or *-idine* and adding the word *monophosphate*. Nucleotides of DNA formed with deoxyribose add the prefix *deoxy-* to the name. The names of the nucleotides and their abbreviations are given in Table 20.2.

Table 20.2 Nucleotides of DNA and RNA

DNA Nucleotides		RNA Nucleotides	
Phosphate: HO—P(=O)(OH)—O—CH2 attached to Deoxyribose (OH); Purine or pyrimidine (A, G, C, or T)		Phosphate: HO—P(=O)(OH)—O—CH2 attached to Ribose (OH OH); Purine or pyrimidine (A, G, C, or U)	
Base	Nucleotide	Base	Nucleotide
Adenine (A)	Deoxyadenosine monophosphate (dAMP)	Adenine (A)	Adenosine monophosphate (AMP)
Guanine (G)	Deoxyguanosine monophosphate (dGMP)	Guanine (G)	Guanosine monophosphate (GMP)
Cytosine (C)	Deoxycytidine monophosphate (dCMP)	Cytosine (C)	Cytidine monophosphate (CMP)
Thymine (T)	Deoxythymidine monophosphate (dTMP)	Uracil (U)	Uridine monophosphate (UMP)

Example 20.1 State the nitrogen base, sugar, and phosphate groups in each of the following nucleotides and indicate whether they are part of DNA or RNA:

1. deoxyguanosine monophosphate
2. AMP

Solution

1. In general, nucleotides consist of a nitrogenous base, a sugar, and a phosphate group. In this nucleotide, there is guanine, a nitrogenous base, deoxyribose, a sugar, and a phosphate group. The nucleotide deoxyguanosine monophosphate would be found in DNA.
2. AMP is the abbreviation for adenosine monophosphate, which consists of adenosine (adenine and a ribose sugar) and a phosphate group. It would be found in RNA.

Example 20.2 Indicate whether each of the following is found in DNA and/or RNA:

1. ribose
2. adenine
3. uracil

Solution

1. Ribose is the sugar found in RNA.
2. Adenine is a nitrogenous base that is found in both DNA and RNA.
3. Uracil is a nitrogenous base found only in RNA.

20.2 Structure of Nucleic Acids

OBJECTIVE Use complementary base pairing to explain the interaction between base pairs in a DNA molecule; draw or describe a portion of a DNA helix.

primary structure

A nucleic acid is a polymer made up of an assortment of the four nucleotides in much the same way that the 20 amino acids form polymers of protein. The primary structure of a nucleic acid consists of nucleotides joined by bonds between a sugar of one nucleotide and the phosphate of the next. Specifically, the phosphate group of one nucleotide bonds with carbon 5 in the sugar of another nucleotide. The bonds between the nucleotides form an alternating pattern of sugar–phosphate–sugar–phosphate and so on, which can be thought of as the backbone of the nucleic acids with a nitrogenous base attached to each of the sugar groups. (See Figure 20.5.)

The structure of DNA was elucidated in 1953 by J. D. Watson, an American biologist, and F. H. C. Crick, a British biophysicist—an achievement that earned them a Nobel Prize. In the Watson–Crick model of DNA, the nucleotides are arranged to form two continuous chains of repeating sugar and phosphate units with a nitrogenous base attached to each sugar. The two chains

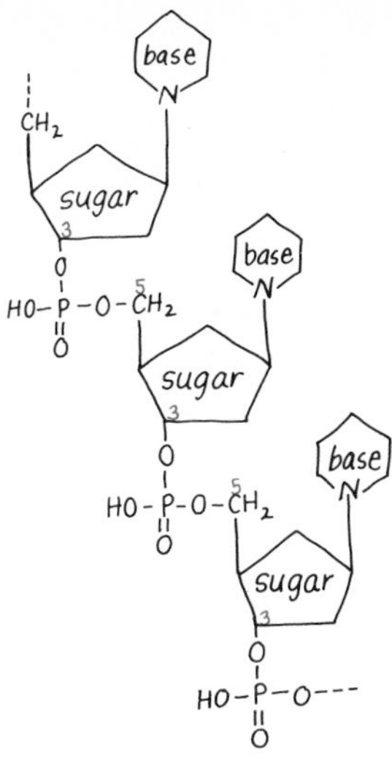

Figure 20.5 A portion of a nucleic acid (DNA) in which alternating sugar-phosphate groups form a backbone with a purine or pyrimidine base attached on each sugar group.

of nucleic acid in DNA are held together by hydrogen bonds between each pair of bases, one from each chain.

Complementary Base Pairing

In DNA, only certain pairs of nitrogenous bases are possible. The bases of adenine and thymine are always paired and held together by *two* hydrogen bonds. This may be abbreviated as follows:

A-T, A: :T or T-A, T: :A

Cytosine and guanine are paired by *three* hydrogen bonds to give

G-C, G: : :C or C-G, C: : :G

See Figure 20.6 for the specific forms of the hydrogen bonding.

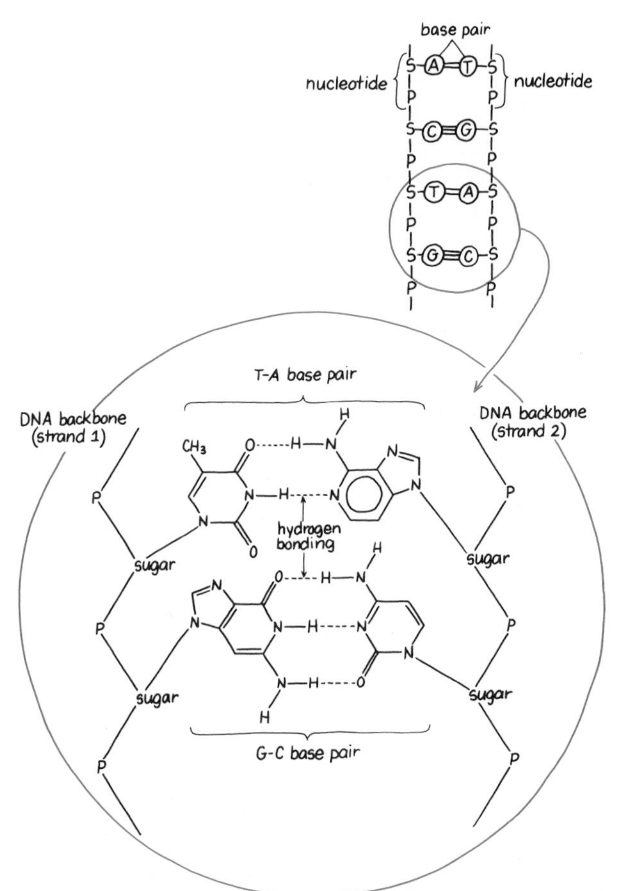

Figure 20.6 Hydrogen bonding between base pair in DNA.

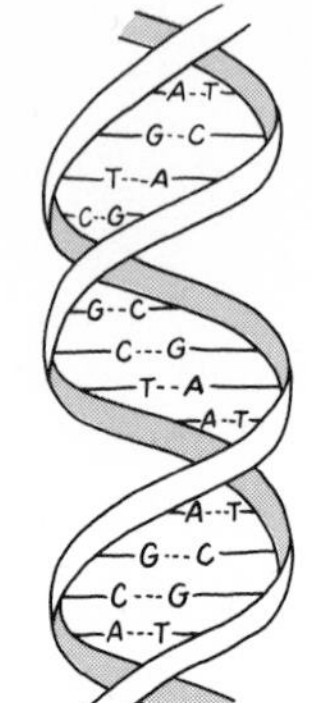

Figure 20.7 Watson–Crick model of DNA in a double-stranded *a* helix.

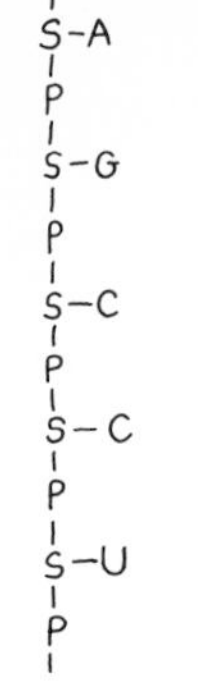

Figure 20.8 Representation of RNA.

This formation of selective base pairs is called *complementary base pairing*. In DNA, adenine is always bonded to thymine, and guanine is always bonded to cytosine. Complementary base pairing is of great importance; it is the key to cell replication, the transfer of hereditary information, and the survival of a living system.

DNA: An α-Helix

secondary structure

Finally, the double strand of DNA with its complementary base pairs is coiled upon itself, forming an α-helix, the secondary structure of DNA. The double helix winds like a spiral staircase with the sugar–phosphate units along the railing and the base pairs as the steps. All DNA molecules contain the same components, but each molecule differs in the order and number of nucleotides along the two strands. (See Figure 20.7.)

RNA structure

The RNA molecules are single strands of nucleotides consisting of a single sugar–phosphate backbone with each ribose attached to adenine, cytosine, guanine, or uracil. (See Figure 20.8.)

Example 20.3 One strand of DNA contains the nitrogenous bases in the order ACGAT. Complete this segment of the DNA molecule by supplying the complementary bases for the second strand. Describe the secondary structure of DNA.

Solution: The second strand contains bases that are complementary to the given sequence. The base adenine (A) always pairs with thymine (T); guanine (G) always pairs with cytosine (C). The backbone of the second strand of base is a sugar–phosphate chain. The secondary structure of the DNA indicates that this double strand of nucleic acid is coiled into a spiral shape called an α-helix.

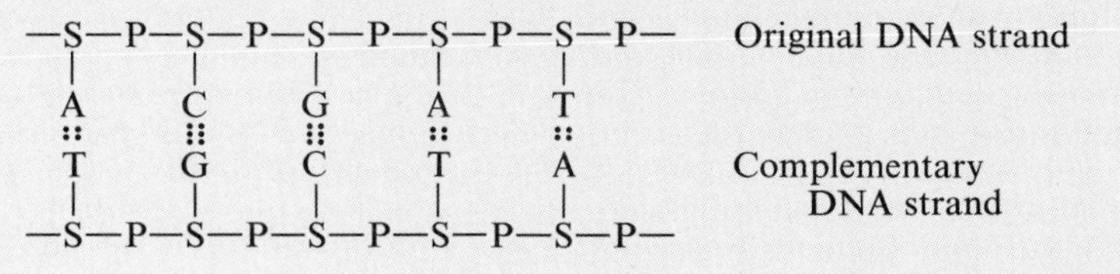

20.3 DNA Replication

OBJECTIVE Describe the replication process of DNA.

One criterion of a living system is the ability of its cells to reproduce themselves. When a cell divides during the process of *mitosis*, two new cells, called *daughter cells*, are produced. These are exact copies of the parent cell. Each daughter cell grows in exactly the same way and produces exactly the same proteins as the parent cell. The information for this duplication or replication is contained in the DNA molecules of the chromosomes. We might think of DNA as a master blueprint. Each new cell must receive an exact copy of that same blueprint. The way in which exact copies of DNA are produced during every cellular division is called *DNA replication*.

DNA replication

Central to the process of replication is the concept of complementary base pairing. Recall that the base pairs in DNA molecules are held together by hydrogen bonds. These hydrogen bonds are rather weak compared to covalent and ionic bonds and can be pulled apart. DNA replication begins when an enzyme causes one end of a DNA molecule to start unwinding and thereby causes the base pairs to separate. Within the nucleus are many single nucleotides that now align their bases with a complementary base on the separated portion of the DNA strands. New hydrogen bonds form, and a new sugar–phosphate backbone builds as each nucleotide attaches to its complementary base. A new half of a DNA strand is built on each old strand of the original DNA. (See Figure 20.9.)

Finally, the entire DNA molecule becomes unwound, and each strand is paired with new nucleotides. Two double strands of DNA have thus been produced, each an exact copy of the original DNA. Hence, replication has occurred. The new DNA molecules are separated into two new daughter cells.

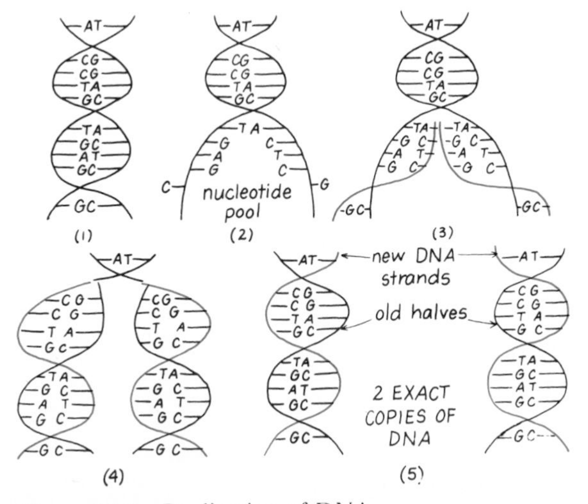

Figure 20.9 Replication of DNA.

The ability to form DNA duplicates with 100 percent accuracy is due to the specificity of complementary base pairing requirements.

Example 20.4 How is a DNA molecule copied during replication?

Solution: The double strand of DNA separates. Each half bonds to new nucleotides according to complementary base pairing (A to T and G to C only). As each nucleotide joins the duplicating chains, its sugar-phosphate groups join the newly forming backbone. When this process is completed, two identical molecules of DNA have formed.

20.4 Protein Synthesis

OBJECTIVE Describe the transcription and translation processes in protein synthesis by writing codons for RNA, anticodons for tRNA, and the resulting amino acid sequences.

transcription

DNA controls protein synthesis by the production of messenger RNA (mRNA) molecules within the nucleus—a process called *transcription.* A different mRNA is produced for each protein to be synthesized. A portion of the DNA that corresponds to an mRNA molecule loosens, and one of the strands is used as a pattern (template) for the mRNA synthesis. Since the message for the building of a specific protein is encoded in this portion of the DNA molecule, it is important that the sequence of bases be retained. This is accomplished through complementary base pairing. The bases in the single-stranded portion of the DNA form a single-stranded RNA by pairing with their complementary bases. Wherever adenine (A) appears in the DNA sequence, a uracil (U) appears in the RNA. (In RNA molecules, uracil replaces thymine.) An enzyme called *RNA polymerase* joins all of the nucleotides to complete the formation of an mRNA. (See Figure 20.10.)

After mRNA has formed, it migrates out of the nucleus into the cytoplasm, where it attaches to the ribosomes. If we think of the ribosomes as factories for protein synthesis, then we might think of mRNA as the mold for a particular protein. There is a different mRNA produced by a portion of a DNA molecule for every kind of protein in the cell. Thus, among the DNA of the 23 pairs of chromosomes is a sequence of bases that serves as a template for an mRNA of every protein that is produced in the cell.

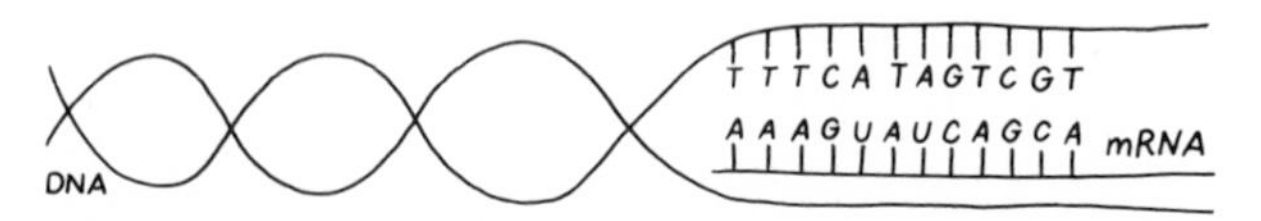

Figure 20.10 Transcription: formation of mRNA along a template strand of DNA.

Surrounding the ribosomes in the cytoplasm are all the amino acids needed to build a protein molecule. There are also transfer RNA (tRNA) molecules for each different amino acid. A tRNA attaches to a particular amino acid and directs that amino acid to the mRNA at the ribosome.

Triplets of the Genetic Code

Research has shown that there is a relationship, called a *triplet*, between every three nitrogenous bases and the particular amino acids making up the protein. Early work on protein synthesis found that an mRNA consisting of repeating uracil(U)-containing nucleotides, polyuracil, produces a polypeptide that contains repeating amino acid units of only a single amino acid, phenylalanine. In the polyuracil mRNA, the only possible triplet is UUU, which directs only the amino acid phenylalanine into protein formation. The triplet of bases in the mRNA is called a *codon*. Each different triplet calls for a certain amino acid. Table 20.3 lists the codons in the genetic code. Several codons can represent the same amino acid.

codon

Translation

The amino acids have no way of recognizing the sequence of bases in the mRNA, but the transfer RNA molecules do. The tRNA has a single polynucleotide chain that folds upon itself, allowing base pairing to occur *within* the molecule. At a

Table 20.3 RNA Codons for Amino Acids in the Genetic Code

First Base	Second Base: U	C	A	G	Third Base
U	UUU Phe	UCU Set	UAU Tyr	UGU Cys	U
	UUC Phe	UCC Ser	UAC Tyr	UGC Cys	C
	UUA Leu	UCA Ser	UAA stop	UGA stop	A
	UUG Leu	UCG Ser	UAG stop	UGG Trp	G
C	CUU Leu	CCU Pro	CAU His	CGU Arg	U
	CUC Leu	CCC Pro	CAC His	CGC Arg	C
	CUA Leu	CCA Pro	CAA Gln	CGA Arg	A
	CUG Val	CCG Pro	CAG Gln	CGG Arg	G
A	AUU Ile	ACU Thr	AAU Asn	AGU Ser	U
	AUC Ile	ACC Thr	AAC Asn	AGC Ser	C
	AUA Ile	ACA Thr	AAA Lys	AGA Arg	A
	AUG Met	ACG Thr	AAG Lys	AGG Arg	G
G	GUU Val	GCU Ala	GAU Asp	GGU Gly	U
	GUC Val	GCC Ala	GAC Asp	GGC Gly	C
	GUA Val	GCA Ala	GAA Glu	GGA Gly	A
	GUG Val	GCG Ala	GAG Glu	GGG Gly	G

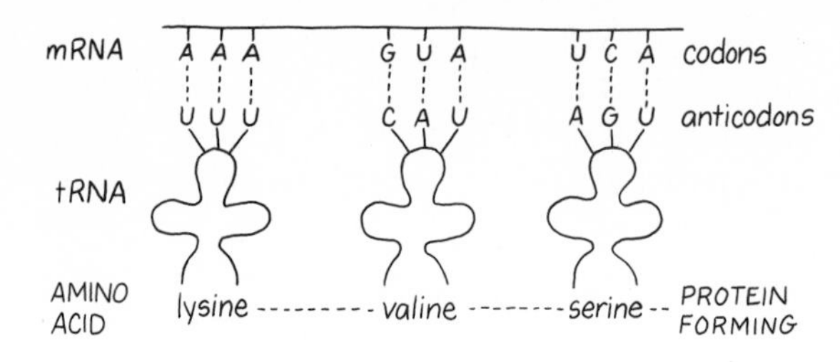

Figure 20.11 Translation: The triplet code (codon) of the mRNA is paired with the triplet (anticodon) of the tRNA to bring the appropriate amino acid into the protein chain.

middle point in the chain, a loop with three unpaired bases, a triplet, forms. This triplet on the tRNA, called an *anticodon*, will complement a triplet (codon) on the mRNA. This process is called *translation*. The amino acid corresponding to that triplet is attached to the opposite end of the tRNA molecule. (See Figure 20.11.)

anticodon

Example 20.5 The nitrogenous bases in a portion of a DNA strand follow this order: C-G-A-T-C-A-G-G-T. Complete the following:

1. the corresponding portion of mRNA
2. the codons of that portion of mRNA
3. the amino acid sequence

Solution

1. The corresponding portion of mRNA consists of complementary bases with uracil (U) pairing with adenine (A):

DNA portion C-G-A-T-C-A-G-G-T

: : : : : : : : :

nRNA portion G-C-U-A-G-U-C-C-A

2. The codons are the triplets of bases in the mRNA:

G-C-U A-G-U C-C-A

3. Each of the codons will code for specific amino acids:

G-C-U is the codon for alanine
A-G-U is the codon for serine
C-C-A is the codon for proline

The amino acid sequence for this portion of the DNA is Ala-Ser-Pro.

Protein Formation

As each tRNA attaches to the mRNA by the pairing of triplets, one amino acid is placed in line with another amino acid so that a peptide bond can form. The tRNA is released and returns to the cytoplasm to pick up another amino acid. As peptide bonds continue to knit on each amino acid brought to the mRNA, a

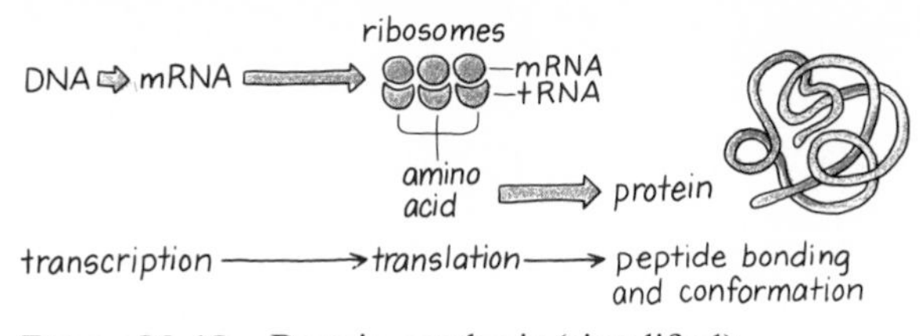

Figure 20.12 Protein synthesis (simplified).

protein chain grows. Each codon in the mRNA attracts one particular amino acid by way of a tRNA molecule. The sequence of the mRNA prescribes the order of amino acids for the entire protein. When the last amino acid has been added, the protein leaves the ribosome. At this point, the protein has achieved only a primary-level structure, a sequence of amino acids held together by peptide bonds. As secondary, tertiary, and possibly quaternary structures occur, the protein becomes biologically active and participates in the metabolic reactions of the cell. (See Figures 20.12 and 20.13.)

Mutations

Mutations are errors in the transmission of the correct base sequence of a DNA. Some mutations are known to result from radiation, such as gamma rays and x-rays, and from chemical agents. If a change in the DNA occurs in a cell other than a reproductive cell, that alteration will be limited to that cell alone and any offspring of that cell. If the DNA error occurs in the reproductive cells of the eggs or sperm, then all the cells produced in a new individual will contain the same error in every DNA. If the error greatly affects the balance of reactions of the new cells, the new cells may not survive.

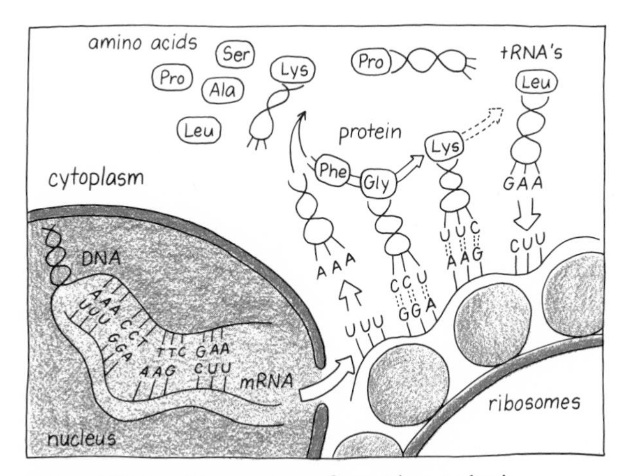

Figure 20.13 The process of protein synthesis.

HEALTH NOTE

Genetic Diseases Resulting from Mutations

sickle-cell anemia

In the genetic disease *sickle-cell anemia*, the red blood cells have a sicklelike shape. (See Figure 20.14.) The sickle shape is caused by a change in the structure of the protein hemoglobin. The structural change is associated with a substitution of valine for glutamic acid in one of the chains of the hemoglobin molecule:

Normal cell: -Val-His-Leu-Thr-Pro-Glu-Glu-Lys- (Correct amino acid order)

↓

Sickled cell: -Val-His-Leu-Thr-Pro-Val-Glu-Lys- (Incorrect amino acid order)

The sicklelike shape of the red blood cells interferes with their ability to transport adequate quantities of oxygen. The sickled cells are removed from circulation more rapidly than normal red blood cells, causing a condition of anemia and low oxygen tension in the tissue. The sickled cells also form aggregates that plug up the capillaries and cause much pain and critically low oxygen levels in the affected tissues.

PKU

Phenylketonuria, PKU, results when DNA cannot produce the proper enzyme to catalyze the conversion of phenylalanine to the amino acid tyrosine. The phenylalanine that cannot react accumulates in the body. Another pathway, by which phenylalanine is converted to phenylpyruvic acid, is activated. In infants, the production of phenylpyruvic acid leads to rapid mental impairment. Brain damage

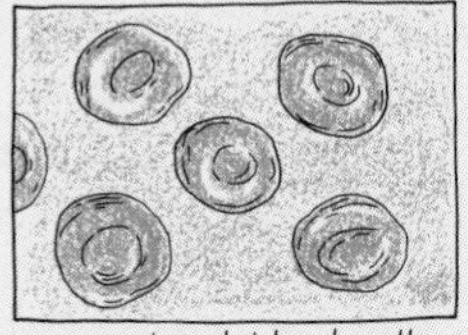

Figure 20.14 Comparison of the shape of normal red blood cells with sickled red blood cells.

can be averted through a test for phenylpyruvic acid that is now routinely performed on infants at birth in most areas in the United States. If phenylpyruvic acid is detected, a diet change is ordered. The special diet avoids food containing phenylalanine, thus preventing the buildup and damaging effects of phenylpyruvic acid. Normal growth and development ensue.

galactosemia

Galactosemia can also be treated by a diet change. In this genetic disease there is a low level or a total absence of the enzyme necessary for cellular utilization of galactose. If an afflicted infant is fed milk containing lactose (milk sugar), the compound galactose-1-phosphate accumulates in the cells. The mental and physical impairments of galactosemia can be avoided if the disease is detected at birth or soon after and the infant is switched to a galactose-free diet. If the galactose-1-phosphate does accumulate in the cells, the effects of the disease become irreversible. Families with known galactosemia should have children tested at birth for the disease so that a diet change can be made immediately if necessary.

Several genetic errors in metabolism resulting from enzyme deficiency can now be detected through prenatal diagnosis. Table 20.4 lists some of these as well as the areas in which metabolism is affected.

Table 20.4 Genetic Diseases Detectable Through Prenatal Diagnosis

Lipid Metabolism	
Fabry's disease	Neimann-Pick disease
Gaucher's disease	Tay-Sachs disease
Carbohydrate Metabolism	
Galactosemia	Glycogen storage diseases
Glucose-6-phosphate dehydrogenase deficiency	Mannosidosis
Amino Acid Metabolism	
Argininosuccinicaciduria	Histidinemia
Cystathionine synthetase deficiency (homocystinuria)	Maple syrup urine disease
Cystinosis	
Miscellaneous	
Adenosine deaminase deficiency	Thalassemia
Hypophosphatasia	Xeroderma pigmentosa
Sickle cell anemia	

In many cases, individuals do survive with genetic errors (*genetic diseases*). The effects of the genetic disease may be manifested in physical malformations or may be apparent only in metabolic misfunctions. Occasionally a mutation may improve the condition of the organism and improve its chances of survival. This is the manner in which variation and evolutionary change occur in a living system. However, most mutations are harmful and affect the well-being of an organism.

Let us trace the effect of a change in the DNA to the eventual effect on the synthesis of a protein. Consider a triplet of bases in a DNA such as CCC, which produces the codon in mRNA of GGG. A tRNA would place the amino acid glycine in the protein that forms. Now, suppose there is a change in one of the bases in the triplet of the DNA: Perhaps an adenine is substituted for the middle cytosine, giving CAC. The codon for the mRNA now becomes GUG and no longer codes for the same amino acid. This codon calls for the amino acid valine, which would enter the protein chain instead of the glycine. This alteration in the primary amino acid sequence of a protein leads to incorrect secondary, tertiary, and quaternary structures. If such an alteration affects the shape of the enzyme so it cannot form the ES complex or changes the amino acid sequence of the active site, an inactive enzyme may result. Such an enzyme can be lethal to the cell because it disrupts a metabolic pathway. Accumulations of substances in the cells resulting from a noncatalyzed step can eventually poison the cell.

Recombinant DNA

Over the last 15 years, geneticists have been cutting, splicing, and rejoining DNAs from different genes to form new or synthetic DNA, called *recombinant DNA*. Most of this experimentation has been done with bacteria called *Escherichia coli* (*E. coli*), which has a simple chromosome and several small cyclic DNA particles called *plasmids*. It is the plasmids with their few genes that have been most helpful in the field of recombinant DNA.

E. coli
plasmids

The plasmids are obtained from the *E. coli* cells by soaking the cells in a detergent solution that breaks them open. The plasmids are then separated from the rest of the cell debris. Enzymes called *restriction enzymes* splice the plasmids at two sites so that a section of DNA is removed. The splicing is done in such a way that the ends of the remaining DNA portion are "sticky" and therefore reactive. When the same enzymes are combined with a foreign DNA (DNA not normally found in the cell), they remove a DNA section having "sticky" ends complementary to those found in the plasmid portion. Another enzyme, called *DNA ligase*, joins the two DNA sections to form a new plasmid of recombinant DNA which now contains the DNA section from the foreign DNA. The new synthetic plasmids are recombined, with some *E. coli* cells becoming a part of the cell again. As the cell divides, the recombinant DNA plasmids replicate to make more copies of the new recombinant DNA. The protein that was synthesized by the foreign DNA can now be produced by the *E. coli* cells. (See Figure 20.15.)

foreign DNA

Concern over recombinant DNA research includes its potential use in producing toxic strains of bacteria for which there would be no biological

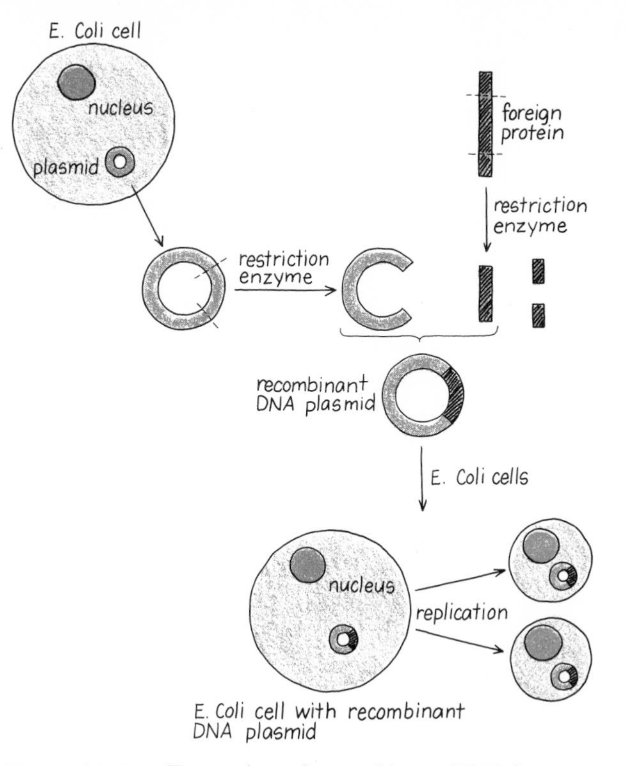

Figure 20.15 Formation of recombinant DNA by plasmid in *E. coli* cell.

defenses. Guidelines to control such possibilities have been established by the National Institutes of Health. In one control measure, the *E. coli* strain used for research is made deficient in an amino acid, so that it becomes dependent on laboratory conditions for survival.

On the other hand, there is much excitement over the medical and industrial uses of recombinant DNA technology. Scientists are beginning to produce enzymes, hormones, antibodies, clotting factors, and other compounds that some people do not produce in their cells as a result of genetic diseases. Recently, the DNA that synthesizes human insulin has been successfully incorporated into *E. coli* plasmids. In the near future, diabetics will have available an insulin that is an exact copy of human insulin. Other substances necessary to save lives have been difficult and time consuming to obtain in sufficient quantities for the number of patients requiring them. Now recombinant DNA techniques are making their production possible. Already substances such as interferon, a protein that aids in fighting viral infections and possibly cancer, and urokinase, an enzyme that dissolves blood clots, have been successfully produced by recombinant DNA technology.

HEALTH NOTE

Viruses

Viruses, which are composed of nucleic acids (DNA or RNA) contained in a protein coat, make up a family of infectious agents that can invade living cells. Viruses are responsible for many diseases, including the common cold, influenza, measles, mumps, meningitis, and polio.

A virus invades a cell by attaching to the surface and then releasing the viral DNA or RNA particles into the cell. Using the machinery within the host cell, the viral particles direct the synthesis of viral nucleic acid and the protein coat. The RNA virus can direct nucleic acid synthesis by using the viral RNA as its own template or by forming a DNA (reverse transcription), which then directs the formation of copies of the viral RNA.

With both types of viruses, many new copies of the virus are produced. The host cell eventually bursts, thereby releasing the viral particles, which go on to infect more cells. A viral infection has set in. As of now, we do not have the means of preventing or curing several of the known viral diseases, including the common cold. We also know that some viruses cause tumors and cancers in laboratory mice and monkeys. It is speculated that viruses may be involved in certain forms of human cancers such as lymphomas and leukemias, but more research is needed to confirm this hypothesis.

20.5 Cellular Control

OBJECTIVE Differentiate between the reaction control systems of feedback control, enzyme induction, and enzyme repression.

If a cell is to maximize the energy available to it, it must operate efficiently. Materials are produced in a cell as they are needed and are removed when not needed. When a substance is produced in the cell or enters the cell, the enzyme (protein) required to catalyze a reaction for that substance is produced via protein synthesis. In other words, proteins are not randomly synthesized at the ribosomes. Rather, the DNA turns off and turns on, through reaction control systems, the production of the particular mRNA that directs the synthesis of the protein needed.

Feedback Control

The first control system, *feedback control*, affects the activity of the enzyme itself. Because metabolic reactions involve several reaction steps, the product of the final step serves as a control for the rate of the reaction. If too much end product is produced, the excess can inhibit the activity of an enzyme involved in the early steps of the sequence. If one enzyme in the series is tied up, the production of the end product will be slowed or shut down until a need for it arises or the level again drops.

Enzyme Induction

The second control system is called *enzyme induction* and occurs at the level of protein synthesis. The French biologists François Jacob and Jacques Monod developed a model of enzyme induction in which the part of the DNA that produces mRNA is considered a structural gene and is joined to a gene called an *operator*. The unit formed is called *operon* and is controlled by a *regulator gene*.

operon
regulator gene

We can think of the operator gene as a switch. When the operator switch is on, the structural gene is producing mRNA for protein synthesis. When the operator is off, the structural gene is repressed (not active) and does not produce mRNA. There may be one or several structural genes under the control of one operator gene. We would expect the structural genes under one operator to be related in such a way that they produce a series of enzymes for a particular metabolic pathway. When that pathway is not operating, those structural genes would be turned off by the operator. No enzymes of that pathway would be produced.

The regulator gene is responsible for turning the operator on and off. It produces some type of protein that attaches to the operator and turns it off. If a substrate that needs those enzymes now enters the cell, some of the substrate combines with the protein from the regulator gene. The substrate reacts with the regulator and "ties" it up, so that it cannot turn off the operator gene. With the operator gene on, mRNA is produced to direct the needed protein synthesis. The substrate has induced the production of its own enzymes: Enzyme induction has taken place. When the substrate has undergone metabolic reactions and its concentration drops in the cell, the regulator protein is free from substrate once again and proceeds to block and turn off the operator gene. (See Figure 20.16.)

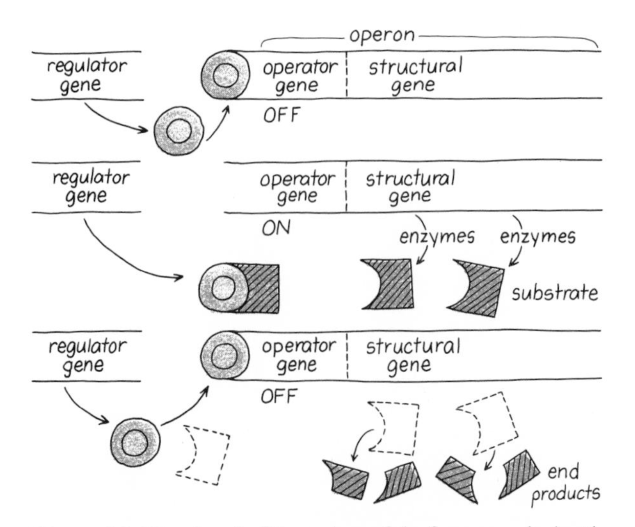

Figure 20.16 Jacob–Monod model of enzyme induction.

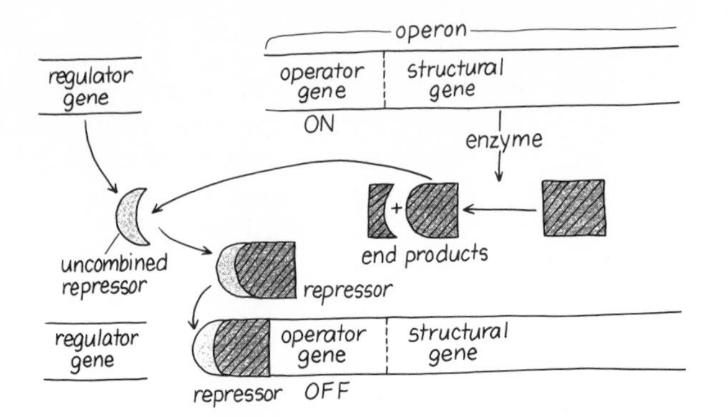

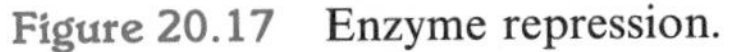
Figure 20.17 Enzyme repression.

Enzyme Repression

Jacob and Monod also introduced a model for the third control system, *enzyme repression.* The end product of a certain enzyme reaction regulates the synthesis of that enzyme. As long as the amount of end product is low, the structural gene produces mRNA for continued enzyme synthesis. But when the end product exceeds the level required by the cell, some of the end product combines with a protein from the regulator gene. This combined unit, a *repressor*, attaches to the operator gene and blocks the synthesis of protein by turning off or repressing the operator gene. (See Figure 20.17.)

repressor

In enzyme induction, the *substrate level* controls enzyme (protein) synthesis; in enzyme repression, the *end product* regulates protein synthesis.

Example 20.6 Indicate whether the following statements are true or false:

1. Feedback control operates at the enzyme-activity level.
2. In enzyme repression, the regulator protein is free from the operator gene when there is no end product of the reaction enzyme.
3. In enzyme induction, the substrate combines with a protein from the regulator gene to unblock the operator gene and turn on the structural gene.

Solution

1. True. In feedback control, the first enzyme in a sequence is inhibited when the end product of the sequence is produced in sufficient quantity for the needs of the cell.
2. True. As long as the amount of end product is low, the gene produces mRNA. When the end product is too high, it combines with the regulator gene to form a repressor that attaches to the operator gene to block the synthesis of mRNA and therefore stops the production of that particular protein.
3. True. A substrate attaches to the regulator gene and prevents the regulator gene from turning off the operator gene. The operator gene continues to produce the mRNA for the enzyme needed. When the substrate is used up, it no longer attaches to the regulator gene. Then the regulator gene causes the operator to shut down.

20.6 Hormonal Action

OBJECTIVE Describe a hormone and two methods of hormonal action.

The word *hormone* comes from the Greek "to arouse" or "to excite." Hormonal action is the stimulation of some process somewhere in the body other than where the hormone is produced. In the absence of a hormone, a process operates at a lower level or is inactive.

Hormones are chemical substances that may be amino acids, peptides, or steroids. These substances are formed and released by a gland to help regulate a cellular process somewhere else in the body. Hormones serve as a kind of communication system that operates alongside the nervous system. However, the nervous system acts rapidly, whereas the hormonal system sends very slow messages with long-lasting effects.

Example 20.7 Indicate whether the following statements are true or false:

1. Hormones must be provided in the diet.
2. Hormones may be amino acids, peptides, or steroids.

Solution

1. False. Hormones are produced by glands in the body.
2. True. Hormones are chemical substances that may be amino acids, peptides, or steroids.

Cyclic AMP System

The organs affected by a particular hormone are called the *target organs*; the cells within those organs reacting to the hormone are the *target cells*. Hormones affect the level of activity in these target cells by one of two known mechanisms. Some hormones activate a cyclic AMP system in the target cells, which then causes a certain response in that cell. Other kinds of hormones activate the genes of the target cells. As a result, the genes in these target cells increase the production of certain proteins. These proteins, usually enzymes, then bring about an increase in the rate of certain cellular activities.

The cyclic AMP system is activated by hormones that are polypeptides. The hormone changes the shape of the enzyme adenyl cyclase, which then converts ATP to cyclic AMP. (See Figure 20.18.) The cyclic AMP converts glycogen to glucose to give the cell an increase in energy and therefore an increase in activity. Sometimes the cyclic AMP also appears to activate an enzyme, which then increases the rate of certain metabolic reactions. (See Figure 20.19.)

We might say that cyclic AMP acts as a secondary messenger, causing reactions to occur inside the cell. It activates enzymes, alters the permeability of

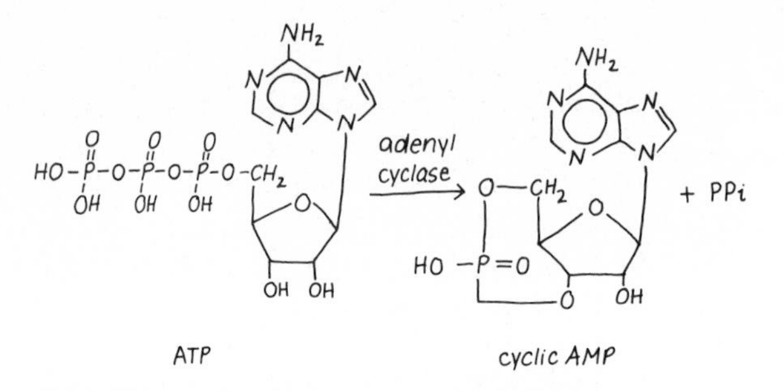

Figure 20.18 Formation of cyclic AMP.

the cell, and causes muscles to contract or relax. For example, when thyroid cells are stimulated by thyroid-stimulating hormone, thyroid hormones are produced. In the kidney, cyclic AMP increases the cellular permeability to water.

Hormonal Effects on Gene Action

The other major type of hormonal action affects protein synthesis by the genes. The steroid hormones produced by the adrenal cortex, ovaries, and testes appear to act in this way. The steroid hormone enters the cytoplasm of the cell, where it combines with a receptor molecule. This hormone–receptor complex binds with the DNA of the chromosomes. This DNA then increases the synthesis of mRNA, which leads to an increase in the synthesis of certain proteins by the cell. These proteins are usually enzymes whose increased concentration increases the rate of specific reactions within the target cell.

Aldosterone appears to act this way. The appearance of aldosterone in the tubules is accompanied by an increase in the number of protein molecules. The additional protein increases sodium reabsorption and the release of potassium into the tubules. Figure 20.20 illustrates the action of the steroid hormones.

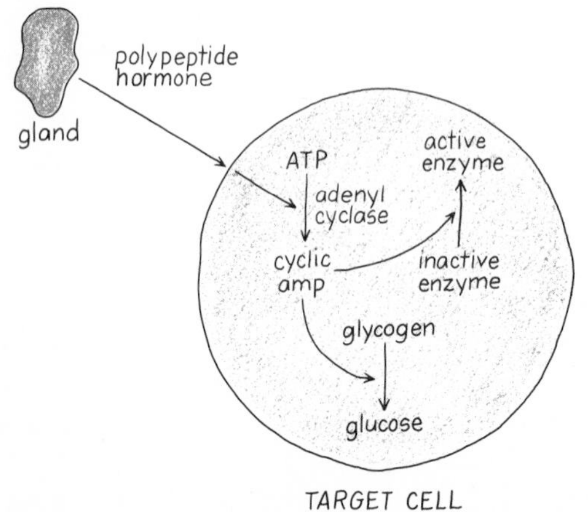

Figure 20.19 Mechanism of action of a polypeptide hormone.

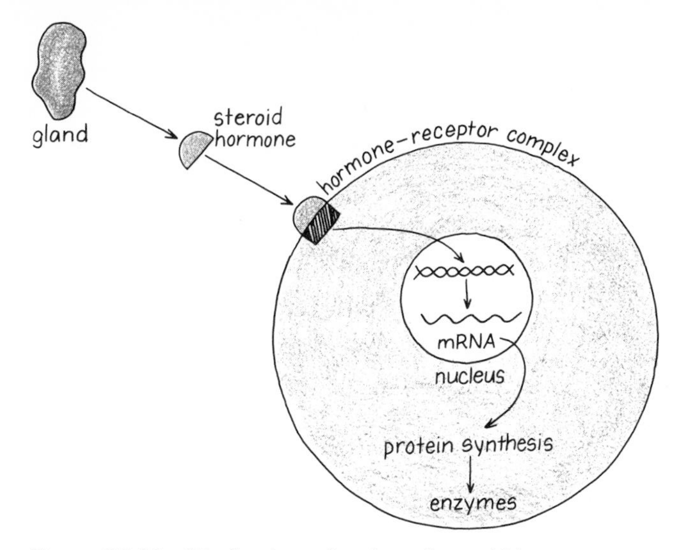

Figure 20.20 Mechanism of action of steroid hormones.

Example 20.8 State whether each of the following would be part of the cyclic AMP system of hormonal action or the stimulation of protein synthesis by the genes:

1. a polypeptide hormone
2. activation of the enzyme adenyl cyclase
3. a steroid hormone

Solution

1. Hormones that are polypeptides activate the cyclic AMP.
2. The enzyme adenyl cyclase converts ATP to cyclic AMP.
3. Steroid hormones affect protein synthesis by the genes.

20.7 Hormones of the Pituitary Gland

OBJECTIVE Describe the function of the hormones released by the anterior and posterior pituitary glands.

Hormones are produced in small amounts by several glands of the body. The names and locations of these glands are illustrated in Figure 20.21. The pituitary glands produce several hormones called *trophic* hormones. These trophic hormones stimulate the actions of other glands, the target organs. When the target organs are stimulated, they secrete other substances, which are usually additional hormones. When these substances reach sufficient levels in the bloodstream, the pituitary gland reduces the output of trophic hormone. In this way, the pituitary gland controls the levels of the various hormones produced by

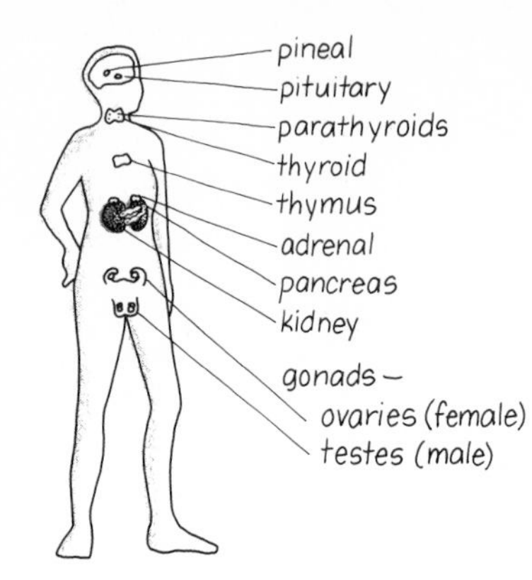

Figure 20.21 The names and locations of hormone-producing glands in the body.

the target organs. Because the pituitary has this regulatory function, it is considered the master controller of the hormonal activities in the body.

vasopressin
oxytocin

The pituitary gland is really two glands, the anterior pituitary and the posterior pituitary. The anterior pituitary gland produces the bulk of the pituitary hormones. These include hormones that stimulate growth in the body and the activities of the thyroid gland, the gonads, and the adrenal gland. The posterior pituitary releases the hormones *vasopressin* and *oxytocin*, both pep-

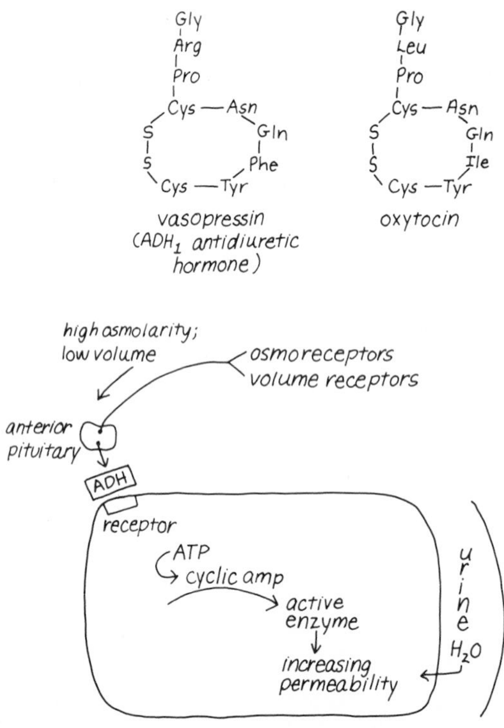

Figure 20.22 Increasing osmotic pressure or low volume of plasma triggers release of ADH from anterior pituitary, which increases permeability of tubule cells and returns more water to plasma.

tides of nine amino acids. Vasopressin affects urine formation in the kidneys, and oxytocin affects muscular contraction in the uterus. (See Figure 20.22.)

growth hormone

The growth hormone (GH) is a protein molecule consisting of 188 amino acids. It stimulates an increase in the rate of protein synthesis in all cells of the body. This increases the rate of mitotic divisions and the size of the cells, and growth occurs. If the amount of *growth hormone* is too low, the growth rate during childhood is slowed to such an extent that as an adult the person's stature is that of a dwarf. An overproduction of growth hormone in childhood can cause gigantism, a condition in which a person grows to a very large size. When an adult produces too much growth hormone, the joints and cartilage portions of the body become enlarged, causing a condition called *acromegaly*. Such people have enlarged portions of the face, hands, and feet. Sometimes this is caused by a pituitary tumor that can be treated with surgery or radiation.

LTH

The hormone prolactin, also called *luteotropin* (LTH), stimulates the production of milk by the mammary glands. During pregnancy, this hormone promotes mammary gland development and, when birth occurs, stimulates milk production as well. The hormones of the pituitary glands are listed in Table 20.5 along with their target organs and major effects.

Table 20.5 Some Hormones of the Pituitary Glands

Hormone	Target Organ	Effects
Anterior Pituitary		
Growth hormone (GH)	Body tissues	Increases rate of protein synthesis; promotes growth of cells
Prolactin (LTH)	Mammary glands	Stimulates milk production
Follicle-stimulating hormone (FSH)	Ovaries (female)	Stimulates follicle growth
Luteinizing hormone (LH)	Testes (male)	Stimulates production of testosterone
	Ovaries (female)	Stimulates formation of corpus luteum and production of progesterone
Adrenocorticotrophic hormone (ACTH)	Adrenal cortex	Stimulates production of corticosteroids
Thyroid-stimulating hormone (TSH)	Thyroid gland	Stimulates production of thyroid hormones
Posterior Pituitary		
Vasopressin (antidiuretic hormone, ADH)	Kidneys	Stimulates reabsorption of water by kidneys; decreases urine formation
	Smooth muscle	Constricts blood vessels; raises blood pressure
Oxytocin	Uterus; mammary glands	Stimulates muscle contraction of uterus
		Stimulates let-down reflex of milk in mammary glands

Example 20.9 Complete the following list for hormones of the pituitary glands:

Hormone	*Function*
1. Thyroid-stimulating hormone	____________________
2. ____________________	Promotes body growth
3. Vasopressin	____________________

Solution

1. The thyroid-stimulating hormone stimulates the thyroid to increase its production of thyroid hormones such as thyroxine. An increase in thyroid hormones will increase the rates of metabolic reactions in the body.
2. The hormone responsible for body growth is the growth hormone (GH) produced in the pituitary gland. It stimulates an increase in the rate of protein synthesis necessary for growth of the body.
3. Vasopressin is a hormone produced in the posterior pituitary that stimulates the reabsorption of water by the kidneys when the osmotic pressure of the blood is elevated.

20.8 More Hormones

OBJECTIVE State the function(s) of the hormones released by the thyroid gland, parathyroid glands, pancreas, adrenal cortex, gonads, and hypothalamus.

Thyroid

thyroxine

The thyroid gland is shaped like a butterfly and is located in the front of the lower neck. When the pituitary releases thyroid-stimulating hormone, the thyroid increases its production of hormones such as *thyroxine*. The release of thyroid hormones increases the rates of metabolic reactions in the tissues of the body. Food is utilized at a faster rate, more protein is synthesized, and the growth rate is accelerated. The process is self-regulating. When the levels of thyroid hormones become high in the blood, the pituitary decreases the output of thyroid-stimulating hormones, thereby slowing down the activity of the thyroid. When the levels of thyroid hormones become too low, the pituitary increases the output of TSH, increasing the thyroid activity.

The element *iodine* plays an important role in the hormones of the thyroid. Iodine is present as ions and also as part of the thyroid hormones. The hormone thyroxine is an iodine-containing derivative of the amino acid tyrosine. About

1 mg of iodine is needed each week to produce the necessary quantities of thyroxine:

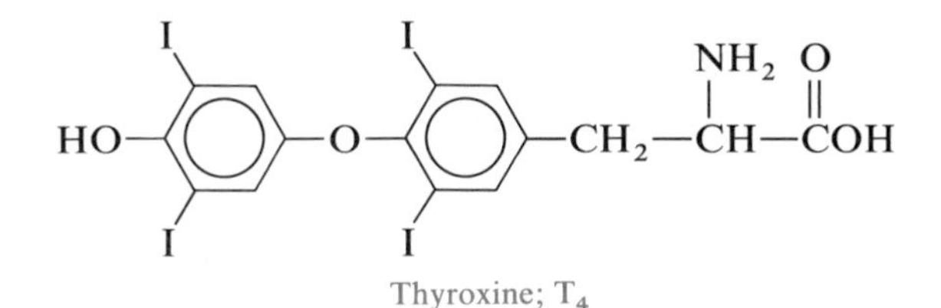

Thyroxine; T_4

Sometimes the diet is deficient in iodine, so that the thyroid cannot synthesize enough thyroid hormone. The pituitary tries to respond to this low level by increasing the output of TSH. Eventually this leads to an overproduction of TSH, causing the thyroid to increase in size, and a *goiter* develops. The increased size of the thyroid is an attempt to compensate for the low levels of thyroid hormone. The occurrence of goiters in the United States and many parts of the world has been greatly reduced by the addition of iodine to salt.

Other conditions can also occur as a result of a thyroid hormone deficiency. If an infant suffers iodine deficiency, growth is limited to the size of an 8-year-old. The result is dwarfism. Even treatment with thyroid hormones cannot completely reverse this lack of development.

Too much thyroid activity also creates problems. A hyperthyroid condition can lead to *Grave's disease.* In this illness the person is very excitable, has a big appetite but loses weight, exhibits a rapid heart rate, is flushed, and has protruding eyes. The disease can occur when there is a tumor of the thyroid or when thyroid function is excessive. Treatment includes the surgical removal of a portion of the hyperactive gland or destruction of a part of the gland by using radioactive iodine, ^{131}I.

Parathyroid

parathormone

The *parathyroid glands* consist of four small oval glands that are located within the thyroid gland. They secrete a hormone called *parathormone*, a polypeptide, which regulates the levels of calcium and phosphate in the body fluids. When the level of calcium is low because of a lack of calcium in the diet or a deficiency of vitamin D, the parathyroid glands increase the output of parathormone. A small amount of calcium is transferred from the bone to increase the calcium concentration in the body fluids. If only a small amount of calcium is removed, the bone is not damaged. Overactive parathyroid glands, which produce too much parathormone, can cause the removal of too much calcium from the bones. The result is the formation of pockets or cavities within the bone, making the bone very fragile.

Pancreas

insulin

The pancreas produces enzymes for digestion within the small intestine and two important hormones, insulin and glucagon. *Insulin* is a small protein consisting of two amino acid chains; it increases the rate of carbohydrate metabolism in the body, lowers the glucose concentration in the blood, and promotes the formation of glycogen by the tissues.

When carbohydrates are broken down in the intestine, the glucose formed enters the blood and increases the blood glucose levels. Cells in the pancreas secrete insulin to increase the rate at which glucose is transported into the cells of the body tissues. If the pancreas does not produce sufficient insulin, the blood glucose level remains high, resulting in a condition called *hyperglycemia*. This condition often occurs in diabetes mellitus, where blood glucose is high and can be detected in the urine (glucosuria). Acidosis usually accompanies this disease because of the formation of ketone bodies. The treatment of hyperglycemia is a daily injection of insulin and a diet low in carbohydrates.

glucagon

Glucagon, a hormone that is also produced by the pancreas, has an effect opposite that of insulin; the secretion of glucagon increases the blood level of glucose. Glucagon is a small peptide consisting of 29 amino acids; it stimulates the phosphorylase activity in the liver. This increases the rate at which glycogen is broken down, thereby releasing glucose, which enters the bloodstream. Glucagon protects the body from too low a blood glucose level, a condition called *hypoglycemia*. A blood glucose level that is too low can seriously impair the energy available to the brain. Glucagon is also used to treat an overdose of insulin by a diabetic. (See Figure 20.23.)

Adrenal Gland

The adrenal gland, located near the kidneys, is composed of two parts: the adrenal medulla and the adrenal cortex. The adrenal cortex secretes a group of hormones called the *adrenocorticosteroids* or *corticosteroids*. All these hormones

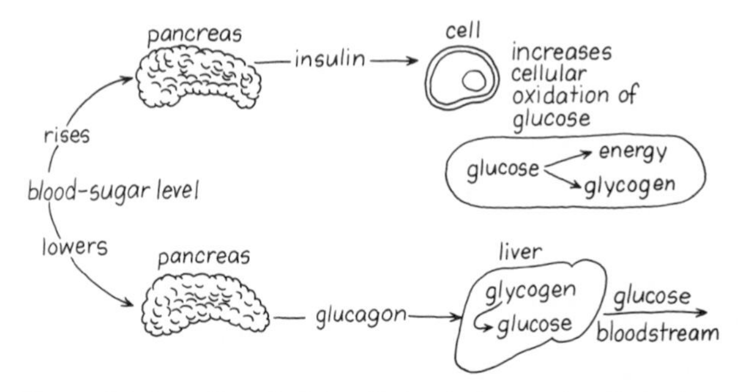

Figure 20.23 Regulation of blood glucose levels by hormones of the pancreas.

are related to the steroid nucleus. The three groups of hormones secreted by the adrenal cortex are the

1. *mineralocorticoids*, which regulate the levels of the electrolytes in body fluids, such as aldosterone;
2. *glucocorticoids*, which regulate carbohydrate and protein metabolism and increase blood glucose levels, such as cortisol; and
3. *androgens*, which affect the development of the secondary sex characteristics, such as estrone, estradiol, and testosterone.

aldosterone

Let us look at a few of these hormones. Aldosterone is the major mineralocorticoid:

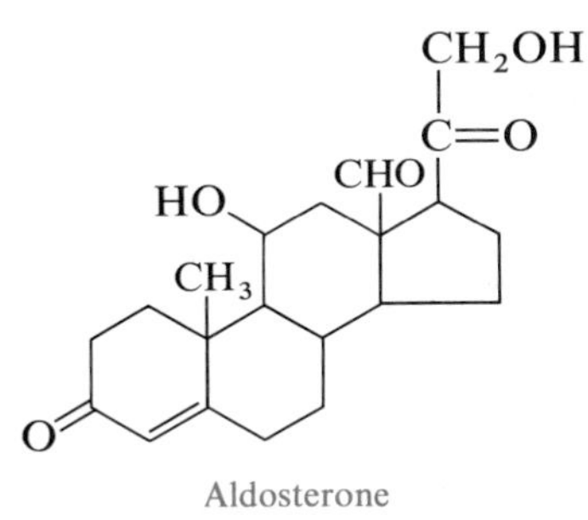

Aldosterone

The most important function of aldosterone and the other mineralocorticoids is to increase the rate of sodium reabsorption by the kidney tubules. If aldosterone is secreted at a high rate, almost all the sodium is reabsorbed, and very little is found in the urine. If aldosterone secretions are too low, large amounts of sodium appear in the urine. A loss of large amounts of sodium is accompanied by a loss of water from the body fluids, which leads to dehydration. When the adrenal cortex is inactive, the condition called *Addison's disease* results. The symptoms are poor kidney function, weight loss, low levels of glucose in the blood, and a severe drop in sodium levels in the blood and body fluids.

cortisol

The major glucocorticoid is *cortisol*, whose major function is to provide an emergency energy source. When cortisol levels increase under stress, amino acids and fats are made available for energy and synthesis to other compounds. Cortisol facilitates the synthesis of glucose from amino acids in two ways. Cortisol increases the rate at which amino acids are transported to the liver cells. In the liver, cortisol also stimulates the production of the enzymes that are needed to synthesize glucose from amino acids. When the level of cortisol is too low, the blood glucose level falls, and the amount of glycogen stored in the liver drops.

Gonads

The hormones produced by the gonads are responsible for the development of the sexual organs, growth, and metabolism. Most male hormones are produced in the testes; a few are produced in the adrenal glands. At puberty, the testosterone

stimulating trophic hormones such as FSH from the pituitary trigger the production of testosterone by the testes. This androgen, along with androgens from the adrenal gland, directs the development of the male sex organs, sperm maturation, and secondary male characteristics.

estrogen
progesterone

Several female characteristics depend on the two hormones estrogen and progesterone. The trophic hormones FSH and LH from the pituitary stimulate the production of the female hormones by the ovaries. With the production of estrogens and progesterone, body changes occur, and ovulation and menstruation begin.

In the female, the mature follicles in the ovaries produce estradiol, the major estrogen. When estrogens are secreted, female sex characteristics develop: the uterus increases in size, fat is deposited in the breasts, the pelvis broadens, and body growth ceases.

The hormone progesterone is released by the corpus luteum when a mature ovum is released into the uterus. The most important function of progesterone is to produce changes in the uterus necessary for implantation of a fertilized ovum. If the ovum is not fertilized, the progesterone and estrogen levels drop sharply, and menstruation follows. A summary of the hormones we have discussed is in Table 20.6.

Hypothalamus and Hormone Control

The feedback control mechanisms between the pituitary and other glands are controlled by the secretions of the hypothalamus glands, located near the anterior pituitary. The hypothalamus receives signals from all parts of the nervous system via nerve endings located within the gland. The hypothalamus responds to nutrient levels, electrolyte levels, glucose levels, and the amount of water in the blood. All this information enables the hypothalamus to control the secretions of the anterior pituitary. The feedback system is summarized in the following scheme:

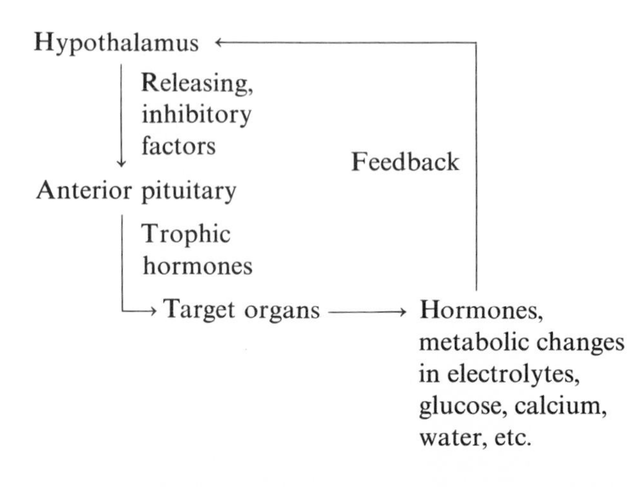

Table 20.6 Summary of Hormones and Functions

Gland	Stimulating Hormone or Agent	Hormone Produced	Effect
Thyroid	TSH	Thyroxine	Stimulates metabolic reactions
Adrenal cortex	ACTH	Corticoids Aldosterone Cortisol Testosterone	Regulates carbohydrate metabolism, electrolyte levels, glycogen
Parathyroid	Calcium levels	Parathormone	Regulates levels of calcium and phosphate in the blood
Pancreas	Glucose levels	Insulin	Decreases blood glucose level
		Glucagon	Increases blood glucose level
Ovaries	FSH, LH	Estrogens	Stimulates secondary sex characteristics
		Progesterone	Prepares uterus for fertilized ovum
Testes	FSH	Testosterone	Stimulates secondary sex characteristics

The neurons within the hypothalamus produce hormones called *releasing* and *inhibitory factors.* These factors control the secretions by the pituitary. There is a releasing factor for every hormone of the anterior pituitary, and for some there is a known inhibitory factor. A list of several factors produced by the hypothalamus is in Table 20.7.

Many of the releasing and inhibitory factors are small peptide molecules. The thyroid-stimulating hormone releasing factor TRF is a tripeptide consisting of these amino acids, Glu-His-Pro. The luteinizing hormone releasing factor LRF is a peptide of 10 amino acids, and the growth hormone inhibiting factor consists of 14 amino acids. Not all factors have been identified, but they are also expected to be small peptide molecules.

Table 20.7 Releasing and Inhibitory Factors from the Hypothalamus

Factor	Effect
Thyroid-stimulating hormone releasing factor (TRF)	Causes release of thyroid-stimulating hormone (TSH)
Corticotropin releasing factor (CRF)	Causes release of ACTH
Growth hormone releasing factor (GRF)	Causes release of growth hormone
Growth hormone inhibitory factor (GIF)	Inhibits release of growth hormone
Luteinizing hormone releasing factor (LRF)	Causes release of follicle-stimulating hormone
Prolactin inhibitory factor (PIF)	Causes inhibition of prolactin secretion
Prolactin releasing factor (PRE)	Causes release of prolactin

Example 20.10 Indicate the function of the following hormones:

1. insulin
2. parathormone

Solution

1. Insulin is a hormone that controls blood glucose levels by increasing carbohydrate metabolism.
2. Parathormone regulates calcium and phosphate levels in body fluids.

Example 20.11 How does thyroid-releasing factor affect the production of thyroxine by the thyroid gland?

Solution: The thyroid-releasing factor produced by the hypothalamus stimulates the anterior pituitary to release thyroid-stimulating hormone. The TSH in turn stimulates the thyroid gland to produce and release thyroxine to regulate carbohydrate metabolism.

Glossary

alpha helix The secondary structural feature of the nucleotide chains of nucleic acid, whereby the chains form a coil.

androgens Steroid hormones, produced by the adrenal cortex, that are responsible for the development of the secondary sex characteristics.

anticodon The triplet of bases on a tRNA that is the complement of the codon triplet on the mRNA.

chromosome A structure, usually linear, containing a series of genes and made of DNA, found in the nucleus of the cell

codon A sequence of three bases in DNA that specifies the incorporation of a certain amino acid into a protein chain or the beginning or ending of transcription.

complementary base pairing The formation of specific combinations of base pairs. In DNA, adenine is always paired with thymine, whereas guanine is always paired with cytosine. In the formation of RNA, the adenine of DNA is paired with uracil in the single strand of RNA.

cyclic AMP A cyclic compound, formed from ATP by the presence of a hormone, that carries out the hormonal action in a cell by increasing the energy in the cell or by activating specific enzymes in the cell.

DNA Deoxyribonucleic acid—the genetic material of all cells found in the chromosomes of the nucleus.

double helix The three-dimensional geometric structure of DNA depicted as a "double staircase."

gene A specific portion of DNA that codes for a certain product, such as a protein.

genetic code The relationship between codons of DNA, anticodons of RNA, and the amino acids of the resulting protein.

genetic disease The manifestation of mutations in an individual by way of physical malformations or metabolic misfunctions.

glucocorticoids Steroid hormones, produced by the adrenal cortex, that regulate carbohydrate and protein metabolism.

hormones Compounds, synthesized in the body, that are needed in small amounts to stimulate metabolic processes elsewhere in the body.

hyperglycemia An elevated blood glucose level resulting from a lack of insulin secretion by the pancreas.

mRNA Messenger ribonucleic acid—an RNA, produced in the nucleus by DNA, that carries information to the ribosomes for the construction of a protein.

mineralocorticoids Steroid hormones, secreted by the adrenal cortex, that regulate the electrolyte levels in the body fluids.

mutation A change in the genetic code that affects the formation and function of a protein in the cell.

nitrogenous base Nitrogen-containing compounds, called purines and pyrimidines, that are found in DNA and RNA: adenine (A), thymine (T), cytosine (C), guanine (G), and uracil (U).

nucleic acids Large molecules, composed of nucleotide monomers, that form the DNA of the chromosomes found in the nucleus of the cells and the RNAs of mRNA, tRNA, and rRNA.

nucleotide A single monomer unit of a nucleic acid consisting of a nitrogenous base (purine or pyrimidine), a pentose sugar (ribose or deoxyribose), and a phosphate group.

operon A group of genes whose transcription is controlled by the same regulator protein.

plasmid One of the small units of DNA found in the cytoplasm of cells. It contains only a few genes and is used in recombinant DNA research.

primary structure The order of nucleotides in a nucleic acid.

purine Double-ringed nitrogen-containing compounds, called *adenine* and *guanine*, that are found in nucleic acids.

pyrimidine Single-ringed nitrogen-containing compounds of cytosine, uracil, and thymine found in nucleic acids.

recombinant DNA The formation of a synthetic DNA for a cell by replacing a portion of the DNA in a plasmid by a DNA of a foreign DNA. When reincorporated into *E. coli* cells, the new DNA undergoes replication and directs the synthesis of protein previously not produced in that cell.

releasing factors Hormones, produced by the hypothalamus, that stimulate or inhibit the secretion of the trophic hormones by the pituitary glands.

replication The process of duplicating DNA through formation of new strands by base pairing with the old strands of DNA.

repressor A chemical substance that interacts with the operon to inhibit the transcription of mRNA.

secondary structure The coiling of a nucleotide chain to form an α-helix.

target organ The specific organ affected by a hormone.

transcription The transfer of genetic information from the DNA base sequence to the base sequence of RNA through complementary pairing.

translation The transfer of genetic information, as a triplet of bases, from mRNA to an amino acid of a forming protein.

trophic hormones Hormones, produced by the pituitary gland, that stimulate the production of hormones in the target organs and glands of the body.

virus A group of infectious agents composed of nucleic acids and contained in a protein coat.

Problems

Components of the Nucleic Acids (Objective 20.1)

20.1 Give the abbreviation for each of the following nucleic acids:
a. deoxyribonucleic acid
b. transfer ribonucleic acid
c. messenger ribonucleic acid
d. ribosomal ribonucleic acid

20.2 State the type of nucleic acid (DNA, RNA, or both) in which each of the following is found:
a. uracil b. ribose c. thymine d. cytosine

20.3 Give the name of the nucleotide, the specific subunits, and the nucleic acid source of the following:
a. CMP b. dAMP c. dTMP d. UMP e. dGMP

Structure of Nucleic Acids (Objective 20.2)

20.4 Draw a model of a DNA having one strand with the following base sequences:

a. A-G-T-G-C-A c. A-A-A-A-A
b. C-T-G-T-A-T-A d. G-C-G-C

20.5 In what ways are the DNA molecules of all the chromosomes similar? How do they differ?

DNA Replication (Objective 20.3)

20.6 How is the replication process of DNA controlled to assure exact copies of DNA in dividing cells?

20.7 Diagram the replication of the following portion of a DNA molecule:

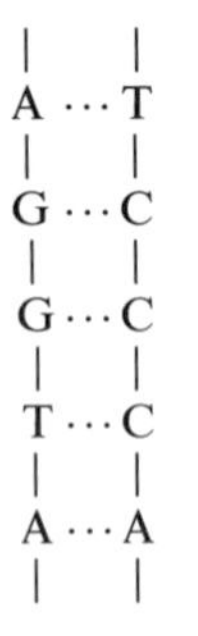

Protein Synthesis (Objective 20.4)

20.8 Consider a portion of the template strand of a DNA having the following base sequence:

G-G-C-T-T-C-C-A-A-G-A-G

a. Write the corresponding portion of the mRNA it will produce.
b. Indicate the codons on the mRNA (reading from left to right).
c. Write the anticodon for each tRNA.
d. Indicate the amino acid sequence in the resulting protein.

20.9 Repeat the questions in Problem 20.8 for the following DNA template:

A-A-A-A-T-G-C-C-C

20.10 In a mutation, the order of amino acids may be altered.

a. How does this happen?
b. What is the effect on the resulting protein formation and function?

20.11 How can a gene from a foreign DNA be incorporated in a cell?

Cellular Control (Objective 20.5)

20.12 In feedback control, how does the end product of a sequence of reactions shut down its synthesis?

20.13 How does a substrate induce the synthesis of specific proteins?

20.14 How does an end product repress the synthesis of enzymes in its own pathway?

Hormone Action (Objective 20.6)

20.15 a. What kinds of chemical compounds can act as hormones?
b. How does the body acquire hormones?
c. What are target organs?

20.16 Indicate whether the following descriptions are: (1) true of polypeptide hormonal action, (2) true of steroid hormonal action, (3) true of both polypeptide and steroid hormonal action.
a. changes the form of the enzyme adenyl cyclase
b. forms a hormone–receptor complex
c. converts ATP to cyclic AMP
d. binds with DNA
e. increases the synthesis of RNA
f. provides energy by increasing the rate of glycogen breakdown to glucose
g. activates an inactive enzyme
h. increases the amount of protein in the cell

Hormones of the Pituitary Gland (Objective 20.7)

20.17 Supply the names of the anterior pituitary and posterior pituitary hormone(s) responsible for the following function or conditions:
a. promotes body growth
b. stimulates the thyroid gland
c. stimulates reabsorption of water by the kidneys
d. stimulates production of corticosteroids
e. gigantism
f. production of milk by mammary glands

More Hormones (Objective 20.8)

20.18 Name the hormone or component that stimulates the following organs and the hormone or action that is thereby produced:

Organ	*Stimulating Hormone or Component*	*Resulting Hormone or Action*
Thyroid	________	________
Adrenal cortex	________	________
Ovaries	________	________
Testes	________	________
Parathyroid	________	________
Pancreas	________	________

20.19 a. Indicate whether each of the following conditions is the result of too much or too little hormone.
b. Name the hormone responsible for the condition.

Condition	*Too Much or Too Little Hormone*	*Hormone Responsible*
Goiter	______________	__________
Diabetes mellitus	______________	__________
Dwarfism	______________	__________
Addison's disease	______________	__________
Acromegaly	______________	__________
Grave's disease	______________	__________
Immature sexual development in females	______________	__________

20.20 What would be the function of the following releasing factors?
a. growth hormone releasing factor
b. corticotropin releasing factor
c. luteinizing hormone releasing factor

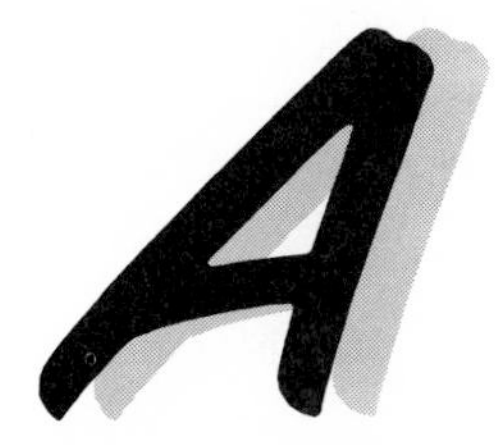

Appendix: Math and the Calculator

In the sciences, we make many measurements such as the length of a bacterium, the mass of a liquid, or the temperature of a patient. We need to know how to use measured numbers in calculations and how to report final answers properly.

A.1 Significant Figures

The results of a measurement are never exact. Several students determined the height of a patient, with the following results:

172.7 cm   172.6 cm   172.4 cm   172.8 cm

All of the students agreed upon the first three figures in the measurement, but not the last. The last figure is an estimation, and its value depends upon the measuring device and the ability of the student to measure. In general, the last figure in a measurement is an estimation. The above measurements all have four significant figures. *Significant figures* refers to all the numbers recorded in a measurement, including the last estimated figure.

Rules for Counting Significant Figures

For a measured number, the following points apply:

1. *All nonzero numbers are significant.* The value 8.2 cm has two significant figures, and 125.78 g has five significant figures.

2. *Zeros within the measurement are significant.* The value 50.2 mm has three significant figures, and 23.08 g has four significant figures.
3. *Zeros that follow other digits in a number with a decimal point are significant.* The value 22.50 has four significant figures, and 0.200 has three significant figures.
4. *Zeros appearing in front of the number are not significant figures.* The value 0.0075 has two significant figures, whereas 0.0004 has one significant figure.

Summary of Rules for Significant Figures

Significant figures are:	Examples	Number of Significant Figures
a. all nonzero digits	122.35 cm	5
	4.5 g	2
b. zeros occurring in the middle of a number or occurring at the end of a number that has a decimal point	205 mm	3
	5.082 kg	4
	21.40 g	4
	1.2500 μm	5
Nonsignificant zeros occur:		
a. in front of a number	0.055 m	2
	0.0004 lb	1
b. in large numbers that have no decimal point	55,000	2
	1,065,000	4

Example A.1 State the number of significant figures in each of the following measured quantities:

1. 24.2 cm
2. 40.1 kg
3. 0.00025
4. 450,000 km
5. 2.440 g

Solution

1. three
2. three
3. two
4. two
5. four

A.2 Rounding Off Calculator Results

When we use measured numbers in calculations, it is sometimes necessary to round off the calculator answer. Rounding off allows us to express an answer to the same degree of precision as our initial measurements. For example, dividing the measured mass of 7.2 g by a measured volume of 3.8 mL gives a calculator answer 1.8947368 g/mL. Since our measurements have only two significant figures, we need to round off the calculator results to 1.9 g/mL, which has two significant figures. Although there is some variation in rounding-off rules, we will use the ones given below.

Rules for Rounding Off

1. If the first digit to be dropped is less than 5, it and all following digits are simply dropped from the number.
2. If the first digit to be dropped is 5 or greater, the preceding digit is increased by 1.

		3 Sig. Figs.	*2 Sig. Figs.*
Example:	8.4234 rounds off to	8.42	8.4
Example:	14.784 rounds off to	14.8	15

Example A.2 Round off each of the following numbers to the number of significant figures indicated:

1. 35.7823 (to 3 figures)
2. 0.002624 (to 2 figures)
3. 381.268 (to 4 figures)
4. 0.578 (to 1 figure)
5. 2.5734 (to 3 figures)

Solution

1. 35.8
2. 0.0026
3. 381.3
4. 0.6
5. 2.57

A.3 Calculations with Measured Numbers

A calculator will permit you to make calculations much faster than you could do using longhand. However, calculators cannot think for you. It is up to you to enter the numbers correctly, push the right buttons, and adjust the calculator answer to give a proper answer. For many typical calculators, you can process the number and mathematical operations just as they are written. Be sure to enter a decimal point when it appears in a number.

$$\frac{2 \times 6}{4} = 3$$

To do this problem on a calculator, we press the number keys and then the operation keys. In this case, we might press the keys in the following order:

2	×	6	÷	4 =	3
	Multiplied by		Divided by		(Answer in display area)

The number of digits that you keep in the final answer depends on the significant figures in your measured numbers. In addition or subtraction, the final answer can have the same number of digits as the measurement with the *fewest* digits after the decimal point:

Add:

2.045	
34.11	
396.4	(Only one digit after the decimal point)
432.555	(Calculator answer)
432.6	(Rounded-off answer with one digit past the decimal point)

Subtract:

84.4674	
−3.56	(Two digits after the decimal point)
80.9074	(Calculator answer)
80.91	(Rounded-off answer with two digits past the decimal point)

When the calculations involve multiplication and division, the final answer must have the same number of digits as the measurement with the *fewest* significant figures. The location of the decimal point is not a consideration.

Multiply: 24.5×16.25

$$24.5 \times 16.25 = 398.125 \text{ (Calculator answer)}$$

$$= 398 \text{ (Rounded-off answer to three significant figures)}$$

The measurement 24.5 has the fewest number of significant figures, three. Therefore, the calculator answer is rounded off to three significant figures.

Divide: $\dfrac{45.75}{2.8}$

$$\frac{45.75}{2.8} = 16.339285 \text{ (Calculator answer)}$$

$$= 16 \text{ (Rounded-off answer to two significant figures)}$$

In the division problem, the term with the fewest significant figures is 2.8. Therefore, the calculator answer is rounded off to give a final answer with two significant figures.

Example A.3 Perform the following addition and subtraction calculations and give final answers with the correct number of significant figures:

1. 5.25 cm + 27.8 cm + 0.235 cm
2. 104.45 mL + 0.838 mL + 46 mL
3. 53.24 g − 14.8 g
4. 154.45 mL − 8.2 mL

Solution

1. 33.3 cm
2. 151 mL
3. 38.4 g
4. 146.3 mL

Example A.4 Perform the following multiplication and division calculations and give final answers with the correct number of significant figures:

1. 56.8×0.37
2. 34.05×0.108
3. $\frac{71.4}{11}$
4. $\frac{2.075}{8.42}$

Solution

1. 21
2. 3.68
3. 6.5
4. 0.246

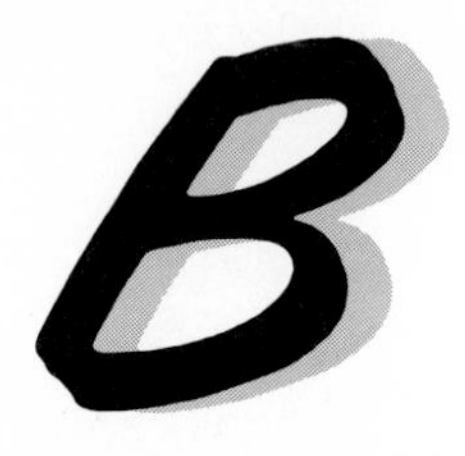

Appendix: Exponential Notation

B.1 Exponents

The exponent of a number tells how many times that number is to be multiplied by itself. When exponential notation is used, the exponent is written as a raised number and is placed on the right-hand side of the number to be multiplied:

$$3^2 \qquad 10^3 \qquad \left(\frac{1}{10}\right)^2$$

To find the value of these expressions, multiply the number itself by the number of times indicated by the exponent:

$$3^2 = 3 \times 3 = 9$$

$$10^3 = 10 \times 10 \times 10 = 1000$$

$$\left(\frac{1}{10}\right)^2 = \frac{1}{10} \times \frac{1}{10} = \frac{1}{100} \quad \text{or} \quad 0.01$$

Suppose we have a number written as 3×10^2. To find its value, follow this procedure:

$$3 \times 10^2 = 3 \times (10 \times 10) = 3 \times 100 = 300$$

When a number is multiplied by 10 or by a power of 10 where the exponent is positive, move the decimal point to the right:

$2.45 \times 10^1 = 24.5$ (Decimal moved one place)
$2.45 \times 10^2 = 245$ (Decimal moved two places)
$2.45 \times 10^3 = 2450$ (Decimal moved three places)
$4 \times 10^3 = 4000$ (Decimal moved three places)
$2.14 \times 10^5 = 214000$ (Decimal moved five places)

When the exponent is negative, divide the number of times indicated by the exponent:

$$2^{-2} = \frac{1}{2 \times 2} = \frac{1}{4}$$

$$10^{-3} = \frac{1}{10 \times 10 \times 10} = \frac{1}{1000}$$

When using negative powers of 10, we are dividing by 10. The decimal place moves to the left:

$$2 \times 10^{-2} = \frac{2}{10 \times 10} = 002. = 0.02$$ (Decimal moved two places)

$$8.5 \times 10^{-3} = \frac{8.5}{10 \times 10 \times 10} = 0008.5 = 0.0085$$ (Decimal moved three places)

B.2 Converting a Number to Scientific Notation

Scientific notation requires that the decimal point be moved to the right of the first digit in the number. (For example, 2 is the first digit in the number 2000.) We indicate the number of places moved by the exponent of 10.

Let us look at the scientific notation for the number 2000. To place the decimal point to the right of the digit 2, we need to move the decimal point three places. This would be the same as multiplying by 1000, which we express as an exponent of 10, or 10^3.

$$2000. = 2 \times 10^3$$

To convert the number 0.00019 to scientific notation, we move the decimal point four places and obtain the number 1.9. This is the same as dividing by 10 four times, so our exponent of 10 is -4:

$$0.00019 = 1.9 \times 10^{-4}$$

Write in scientified notation:

3000	3000.	3×10^3
425	425.	4.25×10^2
520,000	520,000.	5.2×10^5
0.125	0.125	1.25×10^{-1}
0.00858	0.00858	8.58×10^{-3}
0.000005	0.000005	5×10^{-6}

Example B.1 Write the value of

1. 3.5×10^4
2. 5.5×10^{-2}

Write in scientific notation:

3. 30,000,000
4. 0.00128

Solution

1. 35,000
2. 0.055
3. 3×10^7
4. 1.28×10^{-3}

Chapter 1

1.1 a. meter–length b. gram–mass c. liter–volume d. meter–length

1.2 a. mg b. dL c. km d. kg

1.3 a. centimeter b. millimeter c. deciliter d. kilogram

1.4 a. 0.01 (1/100) b. 1000 c. 0.001 (1/1000) d. 0.1 (1/10)

1.5 a. deci- b. deka- c. kilo- d. centi-

1.6 a. milli-, centi-, kilo-
b. micro-, milli-, centi-
c. milli-, deci-, deka-, mega-
d. mg, cg, dg, g, kg
e. mm, dm, m, hm, km
f. mL, cL, L, kL

1.7 a. 100 cm b. 1000 m c. 0.001 m d. 1000 mL e. 10 dL f. 0.1 L
g. 0.001 kg h. 1000 mg

1.8 a. 1 yd = 3 ft, 3 ft/1 yd, 1 yd/3 ft b. 1 min = 60 sec; 1 min/60 sec, 60 sec/1 min
c. $1 = 4 quarters, $1/4 quarters, 4 quarters/$1
d. 1 gal = 4 qt; 1 gal/4 qt, 4 qt/1 gal
e. 1 mile = 5280 ft, 1 mile/5280 ft, 5280 ft/1 mile
f. 1 week = 7 days; 7 day/1 week, 1 week/7 day

1.9 a. 100 cm = 1 m, 1 m/100 cm, 100 cm/1 m
b. 1000 mg = 1 g, 1 g/1000 mg, 1000 mg/1 g

c. 1 cm = 10 mm, 1 cm/10 mm, 10 mm/1 cm
d. 1 L = 1000 mL, 1 L/1000 mL, 1000 mL/1 L
e. 1 dL = 100 mL, 1 dL/100 mL, 100 mL/1 dL
f. 1000 g = 1 kg, 1 kg/1000 g, 1000 g/1 kg

1.10 a. 2.54 cm = 1 in., 1 in./2.54 cm, 2.54 cm/1 in.
b. 2.2 lb = 1 kg, 1 kg/2.2 lb, 2.2 lb/1 kg
c. 946 mL = 1 qt, 1 qt/946 mL, 946 mL/1 qt
d. 454 g = 1 lb, 1 lb/454 g, 454 g/1 lb
e. 1.06 qt = 1 L, 1 L/1.06 qt, 1.06 qt/1 L

1.11 a. 8 yd b. 900 sec c. 0.75 gal d. 14 games e. 13,000 ft

1.12 a. 1.75 m b. 5.5 L c. 50 g d. 0.8 g e. 85 mL f. 2.84 g

1.13 a. 710 mL b. 75.0 kg c. 495 mm d. 20 gal e. 114 g f. 2.5 pt g. 1500 mm h. 11 kg

1.14 a. 1500 mg b. 3 tablets c. 66 gal d. 182 mg e. 8 tablets f. 0.6 mL g. 19 dL

1.15 a. 1.2 g/mL b. 4.4 g/mL c. 1.55 g/mL d. 1.28 g/mL e. 1.05 g/mL

1.16 a. 210 g b. 575 g c. 1700 g d. 20 mL e. 250 lb f. 26.8 mL

1.17 a. 1.030 b. 0.85 g/mL c. 430 g d. 1.13 e. 353 mL f. 0.13 L

1.18 a. 96.8°F b. −4°F c. 302°F d. 15.6°C e. 43.3°C f. −31.7°C

1.19 a. 333 K b. 253 K c. 266 K d. 267°C e. −53°C f. 1341°F

1.20 41.1°C

1.21 Yes. The temperature is 101.7°F

1.22 62.8°C

1.23 13 K = −436°F, 683 K = 770°F

Chapter 2

2.1 a. Cu b. Si c. K d. Co e. Fe f. Ba g. Pb h. Ne i. 0 j. Li k. S l. Al m. He n. B o. H

2.2 a. carbon b. chlorine c. iodine d. phosphorus e. silver f. fluorine g. argon h. zinc i. magnesium j. sodium k. potassium l. nickel m. mercury n. calcium o. bromine

2.3 a. nonmetal b. metal c. metal d. nonmetal e. metal f. nonmetal g. nonmetal h. nonmetal i. metal j. metal

2.4 a. family b. period c. family d. period e. family f. family

2.5 a. metal b. metal c. metal d. nonmetal e. metal f. nonmetal g. metal h. metal

2.6 a. electron b. proton c. electron d. neutron e. electron f. neutron

2.7 a. atomic number 15, mass number 31 b. atomic number 35, mass number 80 c. atomic number 11, mass number 23 d. atomic number 26, mass number 56

2.8

	Atomic Number	Mass Number	Protons	Neutrons	Electrons	Name	Symbol
a.	13	27	13	14	13	Aluminum	Al
b.	12	24	12	12	12	Magnesium	Mg
c.	6	13	6	7	6	Carbon	C
d.	16	31	16	15	16	Sulfur	S
e.	16	34	16	18	16	Sulfur	S
f.	20	42	20	22	20	Calcium	Ca

2.9 a. 13 protons, 14 neutrons and 13 electrons
b. 24 protons, 28 neutrons and 24 electrons
c. 16 protons, 18 neutrons and 16 electrons
d. 26 protons, 30 neutrons and 26 electrons

2.10 a. $^{44}_{20}Ca$ b. $^{59}_{28}Ni$ c. $^{24}_{11}Na$ d. $^{80}_{35}Br$

2.11 a. $^{32}_{16}S$ $^{33}_{16}S$ $^{34}_{16}S$ $^{36}_{16}S$
b. They all have the same number of protons and electrons. (They all have the same atomic number.)
c. They have different numbers of neutrons- different mass numbers.

2.12 a. 16.0 b. 14.0 c. 55.8 d. 1.0 e. 24.3 f. 35.5 g. 23.0 h. 31.0

2.13 a. 2, 4 b. 2, 8, 8 c. 2, 8, 3 d. 2, 8, 6 e. 2, 8, 8, 1 f. 2, 8, 5 g. 2, 5 h. 2, 8

2.14 a. Li b. Mg c. H d. Cl e. 0

2.15 a. absorb b. emit c. high-energy, low-energy

2.16 a. B 2, 3 Al 2, 8, 3 b. 3 c. III-A d. the highest energy level—the outer shell

2.17 a. $2e^-$, II-A b. $7e^-$, VII-A c. $6e^-$, VI-A d. $5e^-$, V-A e. $1e^-$, I-A f. $8e^-$, VIII-A g. $4e^-$, IV-A h. $8e^-$, VIII-A

Chapter 3

3.1 a. :Ṡ̤· b. ·Ṅ̤· c. Ċa· d. Na˙ e. K˙ f. ·Ċ̤· g. :Ö̤· h. :F̤̈· i. Li· j. :C̤̈l·

3.2 a. IIIA b. VIA c. IA d. VA e. VIIA

3.3 a. 5 b. 6 c. 7 d. 1 e. 6 f. 1 g. 2 h. 3 i. 7 j. 7

3.4 a. 2, 8; stable—no loss or gain of electrons
b. 2, 6; gain electrons
c. 2, 1; lose electrons
d. 2, 8, 8; stable—no loss or gain of electrons
e. 2, 8, 5; gain electrons

3.5 a. nonmetal b. metal c. metal d. nonmetal e. nonmetal f. metal g. metal h. metal i. metal j. nonmetal

3.6 a. loses two b. gains three c. gains one d. loses one e. loses three f. gains two g. gains two h. gains one i. loses one j. gains three

3.7 a. 1− b. 2+ c. 3− d. 1+ e. 0

3.8 a. Cl^- b. Mg^{2+} c. K^+ d. O^{2-} e. Al^{3+} f. F^- g. Ca^{2+} h. K^+ i. Na^+ j. Li^+

3.9 a. sodium b. magnesium c. fluoride d. chloride e. oxide f. aluminum g. calcium h. potassium

3.10 a. 2, 8 b. 2, 8 c. 2, 8, 8 d. 2, 8, 8

3.11 a. iron(II); or ferrous b. silver c. copper(II); or cupric d. copper(I); or cuprous e. iron(III); or ferric f. zinc

3.12 a. K^+ b. Cu^{2+} c. Fe^{3+} d. Fe^{2+} e. Ag^+ f. Zn^{2+}

3.13 a. sulfate b. carbonate c. phosphate d. nitrate e. hydroxide f. sulfite

3.14 a. HCO_3^- b. NH_4^+ c. PO_4^{3-} d. NO_2^- e. SO_3^{2-} f. OH^-

3.15 a. $K\cdot \curvearrowright \cdot\ddot{\underset{\cdot\cdot}{Cl}}: \longrightarrow K^+:\ddot{\underset{\cdot\cdot}{Cl}}:^- \longrightarrow KCl$

b. $\dot{Mg}\cdot$ $\cdot\ddot{\underset{\cdot\cdot}{Cl}}:$ $\cdot\ddot{\underset{\cdot\cdot}{Cl}}:$ $\longrightarrow Mg^{2+}\ 2[:\ddot{\underset{\cdot\cdot}{Cl}}:^-] \longrightarrow MgCl_2$

c. $Na\cdot$ $Na\cdot$ $Na\cdot$ $\cdot\dot{\underset{\cdot}{N}}\cdot \longrightarrow 3Na^+\ :\ddot{\underset{\cdot\cdot}{N}}:^{3-} \longrightarrow Na_3N$

d. $\dot{Mg}\cdot \curvearrowright \cdot\dot{\underset{\cdot\cdot}{S}}: \longrightarrow Mg^{2+}\ :\ddot{\underset{\cdot\cdot}{S}}:^{2-} \longrightarrow MgS$

e. $\cdot\dot{Al}\cdot$ $\cdot\ddot{\underset{\cdot\cdot}{Cl}}:$ $\cdot\ddot{\underset{\cdot\cdot}{Cl}}:$ $\cdot\ddot{\underset{\cdot\cdot}{Cl}}:$ $\longrightarrow Al^{3+}\ 3[:\ddot{\underset{\cdot\cdot}{Cl}}:^-] \longrightarrow AlCl_3$

3.16 a. Na_2O b. $FeCl_3$ c. $BaCl_2$ d. CuO e. KI f. $ZnCl_2$ g. Al_2S_3 h. Li_2S i. Fe_2O_3 j. Ag_3N

3.17 a. Na_2S b. Al_2O_3 c. $FeCl_2$ d. Cu_2S e. $CaCl_2$ f. $BaBr_2$ g. LiCl h. Zn_3P_2 i. Fe_2O_3 j. Ag_3N

3.18 a. aluminum oxide b. calcium chloride c. sodium oxide d. magnesium nitride e. sodium sulfide f. potassium phosphide g. magnesium oxide h. lithium bromide

3.19 a. Mg^{2+} b. Fe^{2+} c. Cu^+ d. Ag^+ e. Fe^{3+}

3.20 a. iron(II) chloride; or ferrous chloride
b. copper(II) oxide; or cupric oxide
c. iron(III) sulfide; or ferric sulfide
d. copper(I) chloride; or cuprous chloride
e. silver phosphide
f. sodium sulfide
g. zinc fluoride
h. aluminum phosphide

3.21 a. $MgCl_2$ b. Na_2S c. Cu_2O d. Zn_3P_2 e. Ba_3N_2 f. Fe_2O_3 g. BaF_2 h. $AlCl_3$ i. Ag_2S j. $CuCl_2$

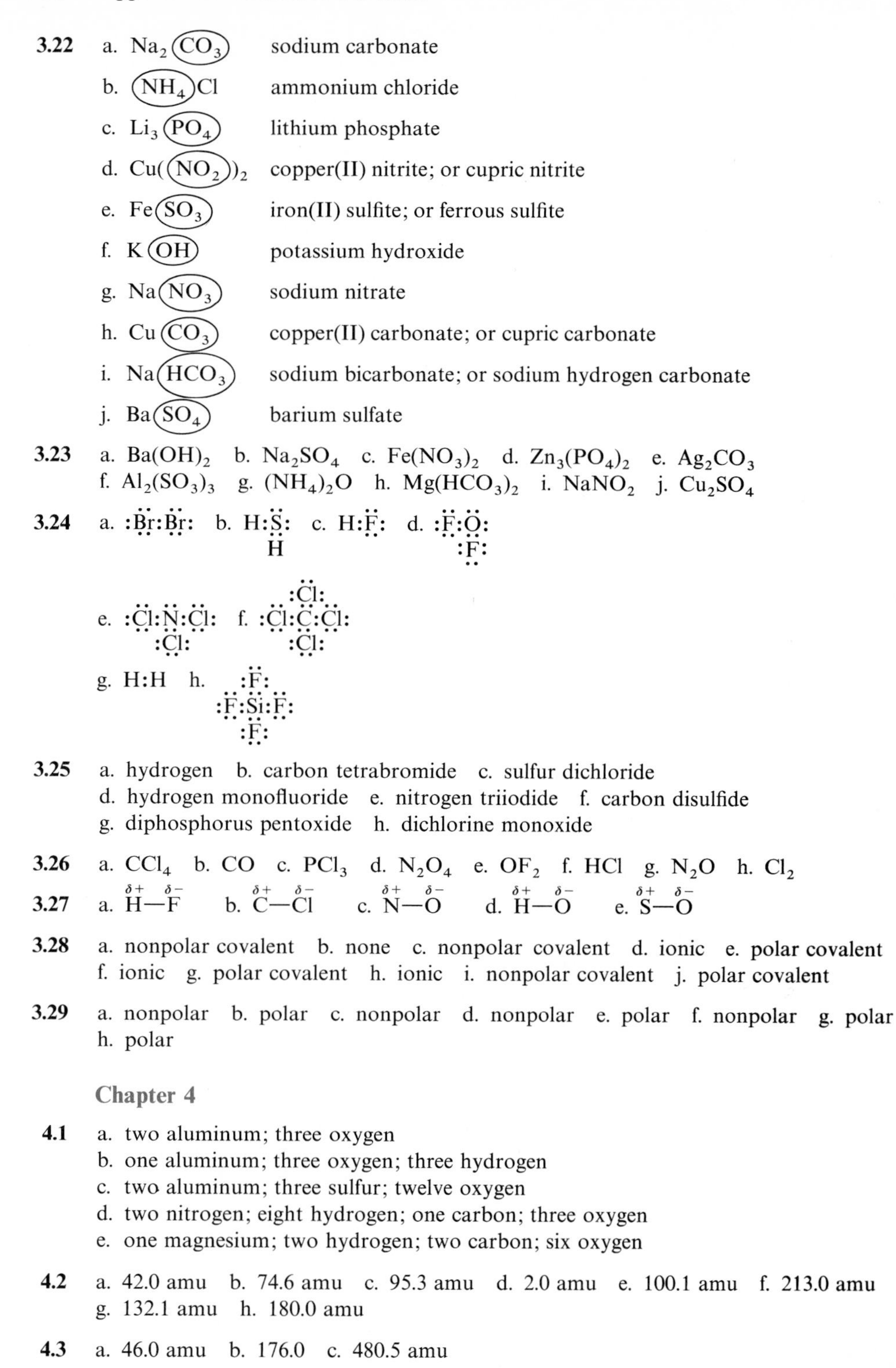

3.22
a. $Na_2(CO_3)$ sodium carbonate
b. $(NH_4)Cl$ ammonium chloride
c. $Li_3(PO_4)$ lithium phosphate
d. $Cu((NO_2))_2$ copper(II) nitrite; or cupric nitrite
e. $Fe(SO_3)$ iron(II) sulfite; or ferrous sulfite
f. $K(OH)$ potassium hydroxide
g. $Na(NO_3)$ sodium nitrate
h. $Cu(CO_3)$ copper(II) carbonate; or cupric carbonate
i. $Na(HCO_3)$ sodium bicarbonate; or sodium hydrogen carbonate
j. $Ba(SO_4)$ barium sulfate

3.23 a. $Ba(OH)_2$ b. Na_2SO_4 c. $Fe(NO_3)_2$ d. $Zn_3(PO_4)_2$ e. Ag_2CO_3
f. $Al_2(SO_3)_3$ g. $(NH_4)_2O$ h. $Mg(HCO_3)_2$ i. $NaNO_2$ j. Cu_2SO_4

3.24 a. :B̤̈r:B̤̈r: b. H:S̤̈: (with H below S) c. H:F̤̈: d. :F̤̈:Ö: (with :F̤: below O)

e. :C̤̈l:N̤̈:C̤̈l: (with :C̤l: below N) f. :C̤̈l:C̤̈:C̤̈l: (with :C̈l: above and :C̤l: below C)

g. H:H h. :F̤̈:S̤̈i:F̤̈: (with :F̈: above and :F̤: below Si)

3.25 a. hydrogen b. carbon tetrabromide c. sulfur dichloride
d. hydrogen monofluoride e. nitrogen triiodide f. carbon disulfide
g. diphosphorus pentoxide h. dichlorine monoxide

3.26 a. CCl_4 b. CO c. PCl_3 d. N_2O_4 e. OF_2 f. HCl g. N_2O h. Cl_2

3.27 a. $\overset{\delta+}{H}—\overset{\delta-}{F}$ b. $\overset{\delta+}{C}—\overset{\delta-}{Cl}$ c. $\overset{\delta+}{N}—\overset{\delta-}{O}$ d. $\overset{\delta+}{H}—\overset{\delta-}{O}$ e. $\overset{\delta+}{S}—\overset{\delta-}{O}$

3.28 a. nonpolar covalent b. none c. nonpolar covalent d. ionic e. polar covalent
f. ionic g. polar covalent h. ionic i. nonpolar covalent j. polar covalent

3.29 a. nonpolar b. polar c. nonpolar d. nonpolar e. polar f. nonpolar g. polar
h. polar

Chapter 4

4.1 a. two aluminum; three oxygen
b. one aluminum; three oxygen; three hydrogen
c. two aluminum; three sulfur; twelve oxygen
d. two nitrogen; eight hydrogen; one carbon; three oxygen
e. one magnesium; two hydrogen; two carbon; six oxygen

4.2 a. 42.0 amu b. 74.6 amu c. 95.3 amu d. 2.0 amu e. 100.1 amu f. 213.0 amu
g. 132.1 amu h. 180.0 amu

4.3 a. 46.0 amu b. 176.0 c. 480.5 amu

4.4 a. 23.0 g/mol b. 35.5 g/mol c. 207.2 g/mol d. 55.8 g/mol e. 24.3 g/mol f. 126.9 g/mol

4.5 a. 42.0 g/mol b. 80.0 g/mol c. 84.0 g/mol d. 322.0 g/mol e. 78.0 g/mol f. 430.0 g/mol

4.6 a. 80 g b. 3.9 g c. 270 g d. 54 g e. 59 g f. 0.54 g g. 39 g h. 78 g

4.7 a. 8.5 g b. 124 g c. 66 g d. 180 g e. 350 g f. 15 g

4.8 a. 600 g b. 11 g c. 11 g d. 7.1 g

4.9 a. 0.463 mol b. 0.630 mol c. 0.833 mol d. 2.7 mol e. 12 mol f. 0.0534 mol

4.10 a. 0.0869 mol b. 0.178 mol c. 336 mol

4.11 a. 0.50 mol b. 10 mol c. 2.00 mol d. 1.6 mol e. 2.00 mol f. 0.00107 mol

4.12 a. 12 mol b. 2300 mol c. 0.0087 mol

4.13 a. physical b. physical c. physical d. chemical e. physical f. chemical g. chemical

4.14
a. Two molecules of nitrogen monoxide react with one molecule of oxygen to produce two molecules of nitrogen dioxide.
b. Two molecules of dihydrogen sulfide react with three molecules of oxygen to produce two molecules of sulfur dioxide and two molecules of water.
c. One molecule of ethanol reacts with three molecules of oxygen to produce two molecules of carbon dioxide and three molecules of water.

4.15
a. $4NH_3 + 3O_2 \longrightarrow 2N_2 + 6H_2O$
b. $4Fe + 3O_2 \longrightarrow 2Fe_2O_3$
c. $2C_3H_8O + 9O_2 \longrightarrow 6CO_2 + 8H_2O$

4.16
a. 2Na, 2Cl
b. 2N, 8H, 4O
c. 4P, 16O, 12H

4.17
a. $N_2 + O_2 \longrightarrow 2NO$
b. $2HgO \longrightarrow 2Hg + O_2$
c. $4Fe + 3O_2 \longrightarrow 2Fe_2O_3$
d. $2Na + Cl_2 \longrightarrow 2NaCl$
e. $2Cu_2O + O_2 \longrightarrow 4CuO$

4.18
a. $2Al + 3Br_2 \longrightarrow 2AlBr_3$
b. $P_4 + 5O_2 \longrightarrow P_4O_{10}$
c. $C_3H_8 + 5O_2 \longrightarrow 3CO_2 + 4H_2O$
d. $Sb_2S_3 + 6HCl \longrightarrow 2SbCl_3 + 3H_2S$
e. $Fe_2O_3 + 3C \longrightarrow 2Fe + 3CO$

4.19
a. $Mg + 2AgNO_3 \longrightarrow Mg(NO_3)_2 + 2Ag$
b. $CuCO_3 \longrightarrow CuO + CO_2$
c. $CaCO_3 \longrightarrow CaO + CO_2$
d. $2Al + 3CuSO_4 \longrightarrow 3Cu + Al_2(SO_4)_3$
e. $Pb(NO_3)_2 + 2NaCl \longrightarrow PbCl_2 + 2NaNO_3$

4.20 a. $Zn + H_2SO_4 \longrightarrow ZnSO_4 + H_2$
b. $Al_2(SO_4)_3 + 6KOH \longrightarrow 2Al(OH)_3 + 3K_2SO_4$
c. $K_2SO_4 + BaCl_2 \longrightarrow BaSO_4 + 2KCl$

4.21 a. One mole NaCl reacts with one mole $AgNO_3$ to produce one mole AgCl and one mole $NaNO_3$.
b. Four moles Al react with three moles oxygen to produce one mole Al_2O_3.

4.22 a. $N_2 + O_2 = 60.0$ g; $2 \times NO = 60.0$ g
b. $CaCO_3 = 100.1$ g; $CaO + CO_2 = 100.1$ g
c. $2 \times SO_2 = 128.2$ g $\quad 2 \times SO_3 = 160.2$ g
$O_2 = 32.0$ g
160.2 g

4.23 a. 1.0 mol S b. 4.0 mol Cu_2S c. 200 g Cu_2S

4.24 a. 3.0 mol H_2 b. 12 g H_2 c. 68 g NH_3

4.25 a. 6 mol O_2 b. 140 g N_2 c. 80 g O_2 d. 0.12 lb H_2O

4.26 a. 20 mol O_2 b. 39.3 g H_2O c. 107 g O_2

4.27 a. 5.56 mol CO_2 b. 123 g ethanol

Chapter 5

5.1

Nuclear Symbol	Mass Number	Number of Protons	Number of Neutrons
	51	24	27
$^{60}_{27}Co$			33
$^{59}_{26}Fe$	59		
	131	53	78
$^{32}_{15}P$		15	

5.2 $^{75}_{35}Br$ $^{76}_{35}Br$ $^{77}_{35}Br$ $^{80}_{35}Br$ $^{82}_{35}Br$ $^{83}_{35}Br$

5.3 a. $^{4}_{2}\alpha$ or $^{4}_{2}He$ b. $^{1}_{0}n$ c. $^{0}_{-1}\beta$ or $^{0}_{-1}e$ d. $^{1}_{1}p$ or $^{1}_{1}H$ e. $^{0}_{0}\gamma$

5.4 a. β b. α c. n d. γ e. p

5.5 a. Certain types of substances placed between the radioactive source and the body will absorb some or all the radiation, thereby providing shielding. Clothing and skin shield a person from alpha particles. Heavy clothing provides shielding from beta particles while lead or concrete is needed to provide sufficient shielding from gamma rays.
b. The less time a person is exposed to a radioactive source, the less radiation that person receives. Therefore, the amount of time spent near a radioactive source should be kept to a minimum.
c. Increasing the distance that a person stands from a radioactive source decreases the intensity of the radiation received. By doubling the distance, the intensity drops to one fourth of its previous value.

5.6 a. ${}^{208}_{84}Po \longrightarrow {}^{4}_{2}\alpha + {}^{204}_{82}Pb$
b. ${}^{232}_{90}Th \longrightarrow {}^{4}_{2}\alpha + {}^{228}_{88}Ra$
c. ${}^{251}_{102}No \longrightarrow {}^{4}_{2}\alpha + {}^{247}_{100}Fm$

5.7 a. ${}^{25}_{11}Na \longrightarrow {}^{0}_{-1}\beta + {}^{25}_{12}Mg$
b. ${}^{59}_{26}Fe \longrightarrow {}^{0}_{-1}\beta + {}^{59}_{27}Co$
c. ${}^{92}_{38}Sr \longrightarrow {}^{0}_{-1}\beta + {}^{92}_{39}Y$

5.8 a. ${}^{28}_{14}Si$ b. ${}^{87}_{36}Kr$ c. ${}^{0}_{-1}\beta$ d. ${}^{238}_{92}U$ e. ${}^{4}_{2}\alpha$ f. ${}^{207}_{81}Tl$ g. ${}^{35}_{17}Cl$

5.9 a. ${}^{10}_{4}Be$ b. ${}^{0}_{-1}\beta$ c. ${}^{27}_{13}Al$ d. ${}^{4}_{2}\alpha$ e. ${}^{239}_{92}U$ f. ${}^{14}_{7}N$

5.10 Radiation from a radioactive source in the body strikes a detection tube within a scanner. The radiation causes the formation of ion pairs within the gas of the detection tube. These charged particles cause an electric current which is detected and changed to a flash of light that exposes a photographic plate to give an image of the organ.

5.11 a. Since the elements Ca and P are part of bone, their radioactive isotopes will be attracted to bone. The radioactive Sr also seeks out bone since Sr acts much like Ca. Once the radioactive isotopes are in the bone, their radiation can be used to reduce a bone lesion or bone tumor.
b. ${}^{32}P$ seeks out bone where its radiation will decrease some of the bone marrow that produces red cells.
c. ${}^{131}I$ seeks the thyroid where its radiation will decrease the number of thyroid cells that produce thyroid hormone.

5.12 a. 200.0 mg ${}^{59}Fe$ remain b. 100.0 mg ${}^{59}Fe$ remain c. 50.0 mg ${}^{59}Fe$ remain d. 12.5 mg ${}^{59}Fe$ remain e. 88 d = 2 half-lives; 100.0 mg ${}^{59}Fe$ remain f. 132 d = 3 half-lives; 50.0 mg ${}^{59}Fe$ remain

5.13 a. 84 d = 3 half-lives; 6.3 μg remain
b. 156 hr; 234 hr
c. 360 d = 6 half-lives; 3.1 mg remain
d. 36 hr = 6 half-lives; 7.8 mg ${}^{99m}Tc$ remain

5.14 a. fission b. fusion c. fission d. fusion e. fusion f. fusion g. fusion h. both i. fission

5.15 a. 294 μCi b. 1 mCi${}^{99m}Tc$

Chapter 6

6.1 a. potential b. kinetic c. potential d. potential e. potential f. potential g. kinetic h. kinetic

6.2 a. 1000 cal b. 4700 J c. 48.4 cal

6.3 40 kcal

6.4 12,000 cal b. 94000 J c. 0.71 kcal

6.5 a. liquid b. gas c. solid d. gas e. gas

6.6

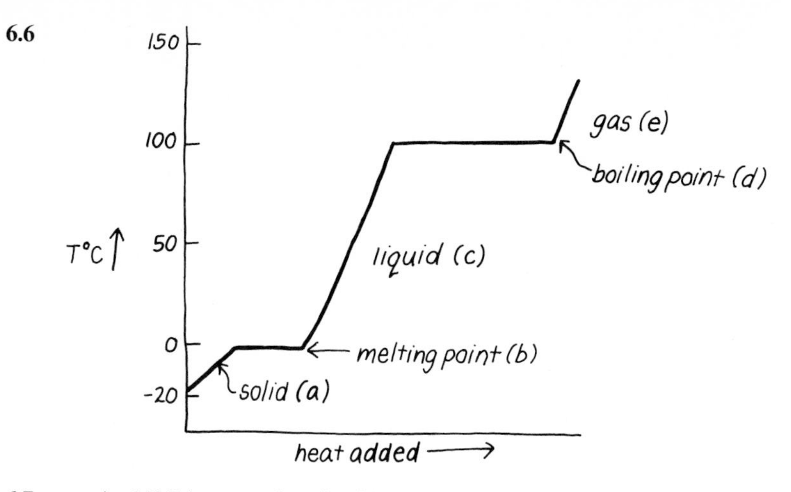

6.7 a. At 15°C benzene is a liquid.
b. The curve has a plateau at 5°C as benzene freezes.
c. At 60°C benzene is a liquid; at 90°C it is a gas.
d. Liquid and gas are both present at 80°C.

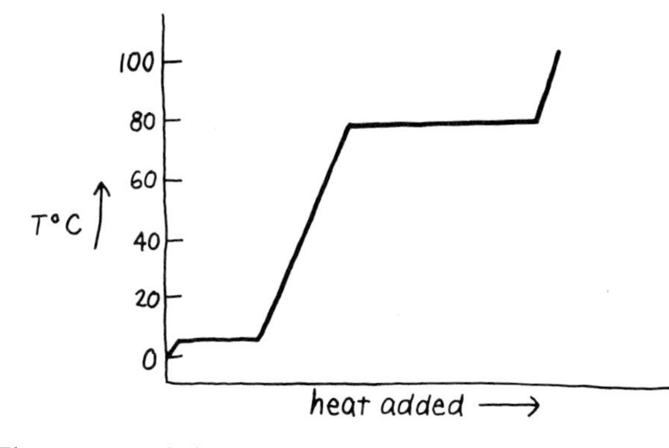

6.8 The steam cools from 110°C to 100°C; at 100°C it will condense to a liquid. It will cool to 0°C, and then freeze. The solid will then cool to −10°C.

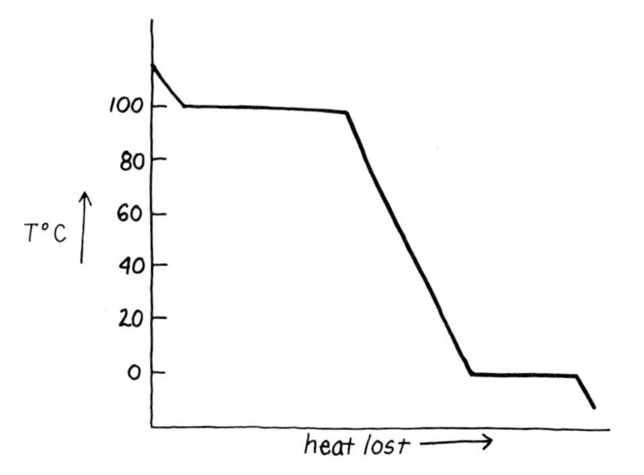

6.9 a. melting point b. evaporation c. melting d. boiling

6.10 a. 1200 cal b. 25,100 J c. 7250 cal

6.11 20 kcal

6.12 a. 27 kcal b. 6100 cal c. 144 kcal

6.13 21,000 cal

6.14 360 kcal

6.15 500 kcal

6.16 a. 5 kcal b. 210 kcal c. 25 kcal

6.17 a. 68 kcal b. 18 g carbohydrate c. 275 kcal d. 37 g fat

6.18 2100 kcal

6.19 925 kcal

6.20 a. 38 g protein, 113 g carbohydrate, 44 g fat
b. 68 g protein, 203 g carbohydrate, 80 g fat
c. 98 g protein, 293 g carbohydrate, 116 g fat

Chapter 7

7.1 a. Gases are composed of very small particles (atoms or molecules).
b. Particles of a gas move in all directions in straight lines.
c. The particles of a gas are far apart.
d. There are no attractive forces between gas particles.
e. Gas particles move at high speeds: they move faster at higher temperature.

7.2 Items a, d, and e describe gas pressure.

7.3 a. 1520 torr b. 29.4 psi c. 1520 mmHg

7.4 a. 1.05 atm b. 0.943 atm c. 0.974 atm

7.5 a. boiling point b. vapor pressure c. atmospheric pressure d. boiling point
e. boiling point

7.6 a. Boiling would not occur.
b. Boiling would occur.
c. Boiling would occur.
d. Boiling would not occur.

7.7 a. The atmospheric pressure at the top of Mount Whitney is less than 760 torr, so boiling occurs at a temperature lower than 100°C.
b. The pressure inside a pressure cooker is greater than 760 torr, so boiling occurs above 100°C and the food cooks quicker.
c. Water boils at 120°C at a pressure of 2.0 atm.

7.8 Volume, pressure, temperature, and amount (moles)

7.9 a. temperature b. pressure c. amount d. temperature e. volume f. volume

7.10 a. increases b. increases c. decreases

7.11 a. 300 torr b. 771 torr c. 2.97 atm

7.12 a. 1.25 L b. 606 mL c. 300 L

7.13 a. expiration b. inspiration c. expiration d. inspiration e. inspiration

7.14 a. increases b. decreases c. decreases

7.15 a. 774 mmHg b. 4.14 atm c. 12.8 atm

7.16 a. 819°C b. −229°C

7.17 a. increases b. decreases c. decreases

7.18 a. 195 mL b. 603 mL c. 104 mL

7.19 a. 410°C b. −218°C c. −102°C

7.20 a. 25,000 torr b. 0.302 atm c. 1600 L d. 175 mL

7.21 a. 5.0 L b. 1320 mL c. 4.0 mol SO_2 lost

7.22 a. 2.0 mol O_2 b. 0.18 mol CO_2 c. 10.1 g Ne d. 0.0714 mol H_2

7.23 a. 56 L b. 9.0 L c. 22.4 L

7.24 a. 4.9 atm b. 56 L c. 0.487 mol O_2

7.25 a. 650 torr b. 920 torr c. 505 torr d. He 300 torr, O_2 900 torr

Chapter 8

8.1 a. NaCl solute, water solvent
b. water solute, ethanol solvent
c. O_2 solute, N_2 solvent
d. mercury solute, silver solvent
e. sugar solute, water solvent

8.2 a. water b. water c. hexane d. hexane e. hexane f. water

8.3 a. saturated b. unsaturated c. saturated d. unsaturated e. saturated

8.4 a. unsaturated b. saturated c. unsaturated d. unsaturated e. unsaturated

8.5 a. unsaturated b. saturated

8.6 a. The pressure of CO_2 above the solution decreases when the bottle is opened. That causes a decrease in the solubility of the $CO_2(g)$ in the liquid, bubbles of CO_2 form and leave the solution.
b. Heating the water will make the $Cl_2(g)$ less soluble. Bubbles of $Cl_2(g)$ form and leave the solution.

8.7 a. increase b. decrease c. decrease d. increase e. increase

8.8 a. surface tension b. oxygen atom c. hydrogen bonding d. hydration
e. hydrogen atom

8.9 a. $KCl(s) \xrightarrow{H_2O} K^+(aq) + Cl^-(aq)$

b. $Na_2SO_4(s) \xrightarrow{H_2O} 2Na^+(aq) + SO_4^{2-}(aq)$

c. $CaCl_2(s) \xrightarrow{H_2O} Ca^{2+}(aq) + 2Cl^-(aq)$

d. $LiNO_3(s) \xrightarrow{H_2O} Li^+(aq) + NO_3^-(aq)$

e. $K_3PO_4(s) \xrightarrow{H_2O} 3K^+(aq) + PO_4^{3-}(aq)$

f. $Ba(NO_3)_2(s) \xrightarrow{H_2O} Ba^{2+}(aq) + 2NO_3^-(aq)$

8.10 a. endothermic b. endothermic c. exothermic d. endothermic

8.11 a. $AgNO_3(s) + 5.4\ kcal/mol \xrightarrow{H_2O} Ag^+(aq) + NO_3^-(aq)$

b. $KBr(s) + 4.8\ kcal/mol \xrightarrow{H_2O} K^+(aq) + Br^-(aq)$

c. $LiBr(s) \xrightarrow{H_2O} Li^+(aq) + Br^-(aq) + 11.7\ kcal/mol$

d. $CsF(s) \xrightarrow{H_2O} Cs^+(aq) + F^-(aq) + 8.8\ kcal/mol$

8.12 a. ions only b. molecules only c. molecules and some ions d. ions only e. molecules only

8.13 a. strong b. weak c. non d. non e. weak

Chapter 9

9.1 a. 2% sucrose b. 5% KCl c. 15% Na_2SO_4 d. 6% glucose e. 20% $CaCl_2$

9.2 a. 2.5 g KCl b. 4.0 g NaCl c. 200 g NH_4Cl d. 25 g glucose

9.3 100 g glucose

9.4 a. 20 g mannitol b. 480 g mannitol

9.5 a. 3000 mL b. 4000 mL c. 2500 mL d. 25 mL

9.6 2.0 L

9.7 a. 100 mL b. 1000 mL c. 800 mL

9.8 a. 0.5% KCl b. 4% mannitol c. 5% NaCl

9.9 a. 2.0 *M* KOH b. 0.50 *M* glucose c. 10 *M* NaOH d. 2.5 *M* NaCl

9.10 a. 1.0 *M* HCl b. 1.0 *M* NaOH c. 3.56 *M* glucose

9.11 a. 3.0 mol NaCl b. 10 mol $CaCl_2$ c. 0.80 mol glucose d. 0.50 mol sucrose

9.12 a. 40 g NaOH b. 600 g KCl c. 29 g NaCl d. 48 g NaOH e. 135 g glucose

9.13 a. 1.0 L b. 10 L c. 2 L d. 0.50 L

9.14 3.54 g Na^+, 5.47 g Cl^-

9.15 155 meq/L

9.16 a. true b. colloidal c. colloidal d. suspension e. suspension f. true g. suspension

9.17 a. the 5% starch solution
b. from the 1% starch solution into the 5% starch solution
c. the 5% starch solution

9.18 a. the 2% albumin solution
b. from the 0.1% albumin solution into the 2% albumin solution
c. the 2% albumin solution

9.19 a. 1 osmol b. 2 osmol c. 3 osmol d. 1 osmol

9.20 a. 1% glucose and 0.05% NaCl are hypotonic.
b. 4% NaCl and 10% glucose are hypertonic.
c. 5% glucose and 0.9% NaCl are isotonic.
d. 4% NaCl and 10% glucose would cause crenation.
e. 1% glucose and 0.05% NaCl would cause hemolysis.
f. 5% glucose and 0.9% NaCl would not cause any change.

9.21 a. isotonic b. hypertonic c. hypertonic d. hypotonic e. isotonic

9.22 a. NaCl b. KCl and glucose c. urea and NaCl

9.23 The dialysate is prepared with a high potassium level so the potassium will move from the dialysate into the patient's blood. The low levels of sodium and urea in the dialysate will cause these compounds to move out of the patient's blood into the dialysate.

Chapter 10

10.1 a. $HCl \xrightarrow{H_2O} H^+(aq) + Cl^-(aq)$

$HCl + H_2O \longrightarrow H_3O^+(aq) + Cl^-(aq)$

b. $HNO_3 \xrightarrow{H_2O} H^+(aq) + NO_3^-(aq)$

$HNO_3 + H_2O \longrightarrow H_3O^+(aq) + NO_3^-(aq)$

10.2 a. $H_3BO_3 \underset{}{\overset{H_2O}{\rightleftharpoons}} 3H^+(aq) + BO_3^{3-}(aq)$
b. $H_2CO_3 \underset{}{\overset{H_2O}{\rightleftharpoons}} 2H^+(aq) + CO_3^{2-}(aq)$

10.3 a. weak b. strong c. weak d. weak

10.4 a. $LiOH \xrightarrow{H_2O} Li^+(aq) + OH^-(aq)$

b. $Mg(OH)_2 \xrightarrow{H_2O} Mg^{2+}(aq) + 2OH^-(aq)$

c. $KOH \xrightarrow{H_2O} K^+(aq) + OH^-(aq)$

10.5 $NH_3 + H_2O \rightleftharpoons NH_4^+(aq) + OH^-(aq)$

10.6 a. acid b. base c. base d. base e. acid

10.7 a. $[H^+] = 1 \times 10^{-7}\ M$; $[OH^-] = 1 \times 10^{-7}\ M$
b. $[H^+][OH^-] = 1 \times 10^{-14}$

10.8 a. $1 \times 10^{-3}\ M$ b. $1 \times 10^{-9}\ M$ c. $1 \times 10^{-13}\ M$ d. $1 \times 10^{-2}\ M$

10.9 a. 1×10^{-3} b. 1×10^{-12} *M* c. 1×10^{-10} d. 1×10^{-6} *M*

10.10 a. basic b. acidic c. basic d. acidic e. acidic f. neutral g. acidic h. basic i. basic j. basic

10.11 a. 0.4, 4.5, 6.8, 13.0 b. 1.6, 2.3, 7.1, 8.5, 11.7 c. 2.9, 3.3, 4.4, 9.8, 14.0 d. 7.4, 8.8, 9.7, 11.4, 13.4

10.12 a. acidic b. neutral c. basic d. acidic e. acidic f. acidic g. basic h. neutral i. acidic j. basic

10.13 a. 4 b. 7 c. 11 d. 4 e. 2 f. 2 g. 10 h. 7 i. 4 j. 11

10.14

$[H^+]$	$[OH^-]$	pH	Acidic, Basic, Neutral
1×10^{-8} *M*	1×10^{-6} *M*	8	basic
1×10^{-2} *M*	1×10^{-12} *M*	2	acidic
1×10^{-5} *M*	1×10^{-9} *M*	5	acidic
1×10^{-10} *M*	1×10^{-4} *M*	10	basic
1×10^{-7} *M*	1×10^{-7} *M*	7	neutral

10.15 a. $HCl + KOH \longrightarrow KCl + H_2O$
b. $2HNO_3 + Ca(OH)_2 \longrightarrow Ca(NO_3)_2 + 2H_2O$
c. $H_3PO_4 + 3LiOH \longrightarrow Li_3PO_4 + 3H_2O$
d. $3HBr + Al(OH)_3 \longrightarrow AlBr_3 + 3H_2O$

10.16 a. $2NaOH + H_2SO_4 \longrightarrow Na_2SO_4 + 2H_2O$
b. $KOH + HCl \longrightarrow KCl + H_2O$
c. $Ca(OH)_2 + H_2SO_4 \longrightarrow CaSO_4 + 2H_2O$
d. $NaOH + HBr \longrightarrow NaBr + H_2O$
e. $H_3PO_4 + 3NaOH \longrightarrow Na_3PO_4 + 3H_2O$
f. $2Al(OH)_3 + 3H_2SO_4 \longrightarrow Al_2(SO_4)_3 + 6H_2O$

10.17 a. $2KOH + H_2SO_4 \longrightarrow K_2SO_4 + 2H_2O$
b. $LiOH + HNO_3 \longrightarrow LiNO_3 + H_2O$
c. $3Ca(OH)_2 + 2H_3PO_4 \longrightarrow Ca_3(PO_4)_2 + 6H_2O$

10.18 a. 2.2 *M* HCl b. 0.76 *M* H_2SO_4 c. 0.53 *M* H_3PO_4

10.19 Item c and e are buffers systems. Item (c) contains a weak acid (HAc) and its sodium salt, NaAC. Item (e) contains the weak acid (H_2CO_3) and its sodium salt, $NaHCO_3$.

10.20 a. Maintain the pH of a system.
b. Provide additional negative ions.
c. Ac^- $Ac^- + H^+ \longrightarrow HAc$
d. HAc $HAc + OH^- \longrightarrow H_2O + Ac^-$

Chapter 11

11.1 a. inorganic b. organic c. organic d. organic

11.2

```
      H  H  H  H  H
      |  |  |  |  |
a. H—C—C—C—C—C—H
      |  |  |  |  |
      H  H  H  H  H
```

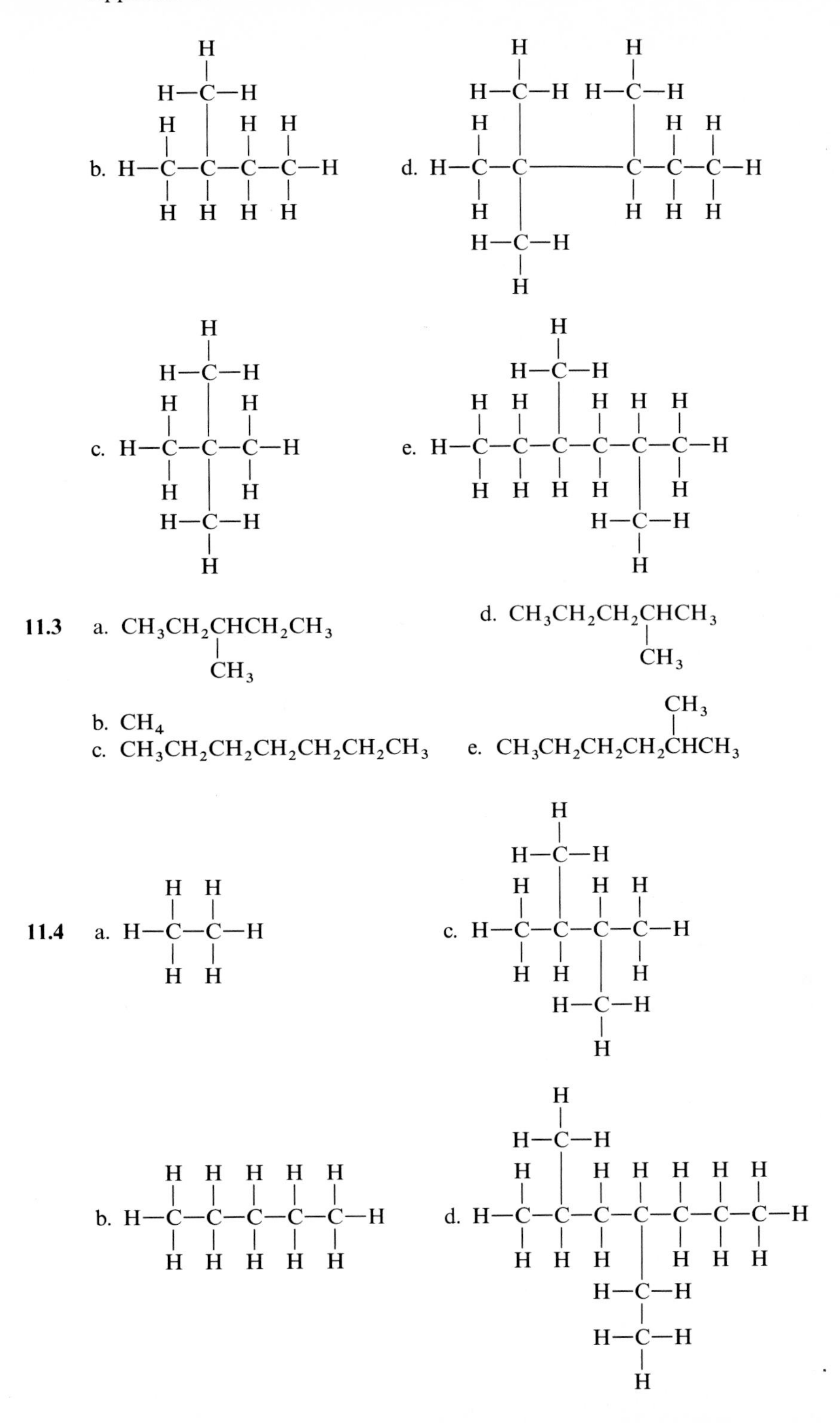
b.
d.
c.
e.
11.3
a. CH3CH2CHCH2CH3
CH3
d. CH3CH2CH2CHCH3
CH3
b. CH4
c. CH3CH2CH2CH2CH2CH2CH3
CH3
e. CH3CH2CH2CH2CHCH3
11.4
a.
c.
b.
d.

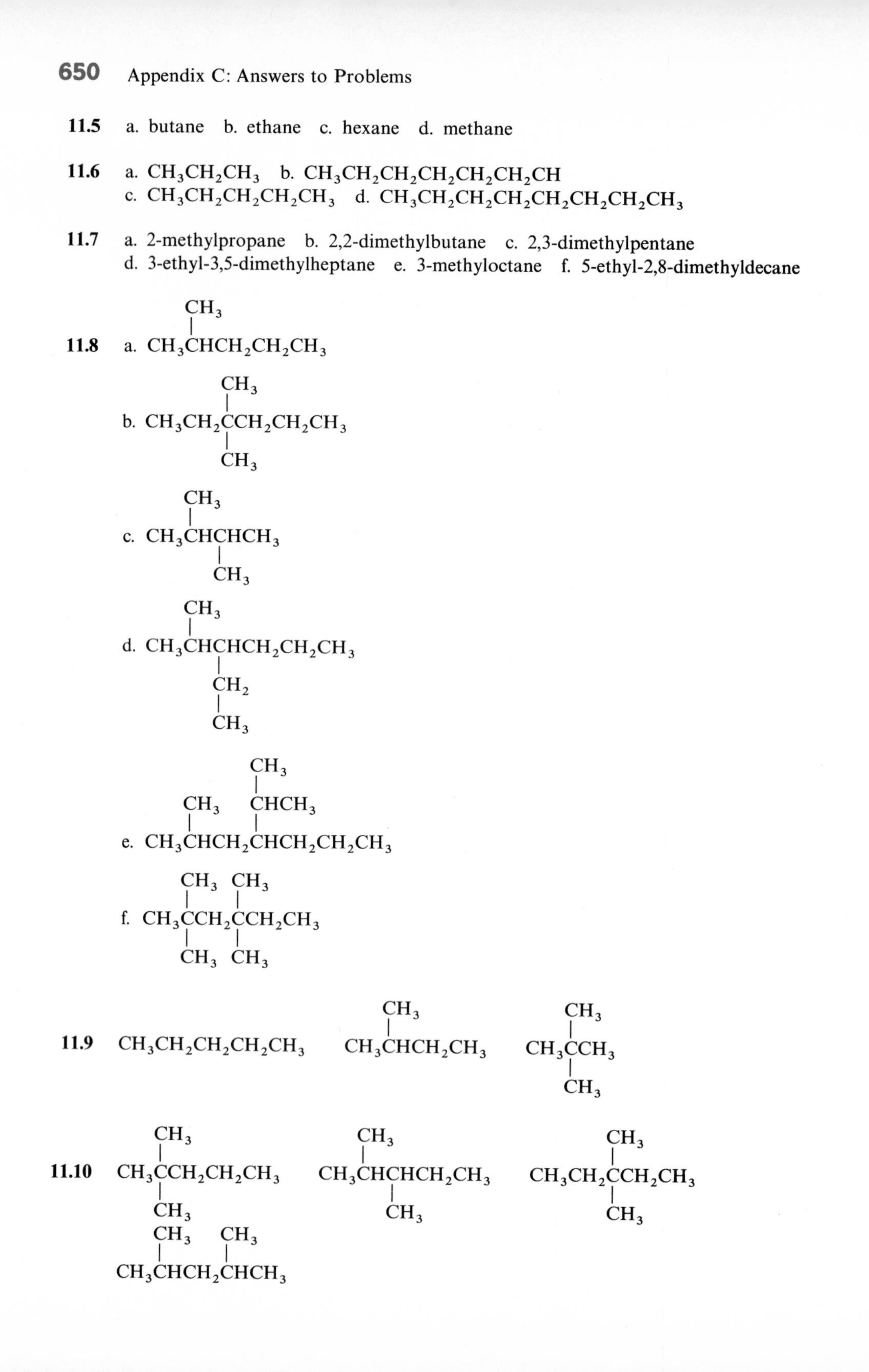

11.5 a. butane b. ethane c. hexane d. methane

11.6 a. $CH_3CH_2CH_3$ b. $CH_3CH_2CH_2CH_2CH_2CH_2CH$
c. $CH_3CH_2CH_2CH_2CH_3$ d. $CH_3CH_2CH_2CH_2CH_2CH_2CH_2CH_3$

11.7 a. 2-methylpropane b. 2,2-dimethylbutane c. 2,3-dimethylpentane
d. 3-ethyl-3,5-dimethylheptane e. 3-methyloctane f. 5-ethyl-2,8-dimethyldecane

11.8

a.
CH_3
|
$CH_3CHCH_2CH_2CH_3$

b.
CH_3
|
$CH_3CH_2CCH_2CH_2CH_3$
|
CH_3

c.
CH_3
|
$CH_3CHCHCH_3$
|
CH_3

d.
CH_3
|
$CH_3CHCHCH_2CH_2CH_3$
|
CH_2
|
CH_3

e.
CH_3
|
CH_3 $CHCH_3$
| |
$CH_3CHCH_2CHCH_2CH_2CH_3$

f.
CH_3 CH_3
| |
$CH_3CCH_2CCH_2CH_3$
| |
CH_3 CH_3

11.9

$CH_3CH_2CH_2CH_2CH_3$

CH_3
|
$CH_3CHCH_2CH_3$

CH_3
|
CH_3CCH_3
|
CH_3

11.10

CH_3
|
$CH_3CCH_2CH_2CH_3$
|
CH_3

CH_3
|
$CH_3CHCHCH_2CH_3$
|
CH_3

CH_3
|
$CH_3CH_2CCH_2CH_3$
|
CH_3

CH_3 CH_3
| |
$CH_3CHCH_2CHCH_3$

11.11 a. isomers b. isomers c. identical d. identical

11.12 a. cyclobutane b. cyclopentane c. methylcyclohexane
d. 1,2,3-trimethylcyclohexane e. 2-ethyl-1,1,3-trimethylcyclopentane
f. 1-ethyl-1-methylcyclopentane

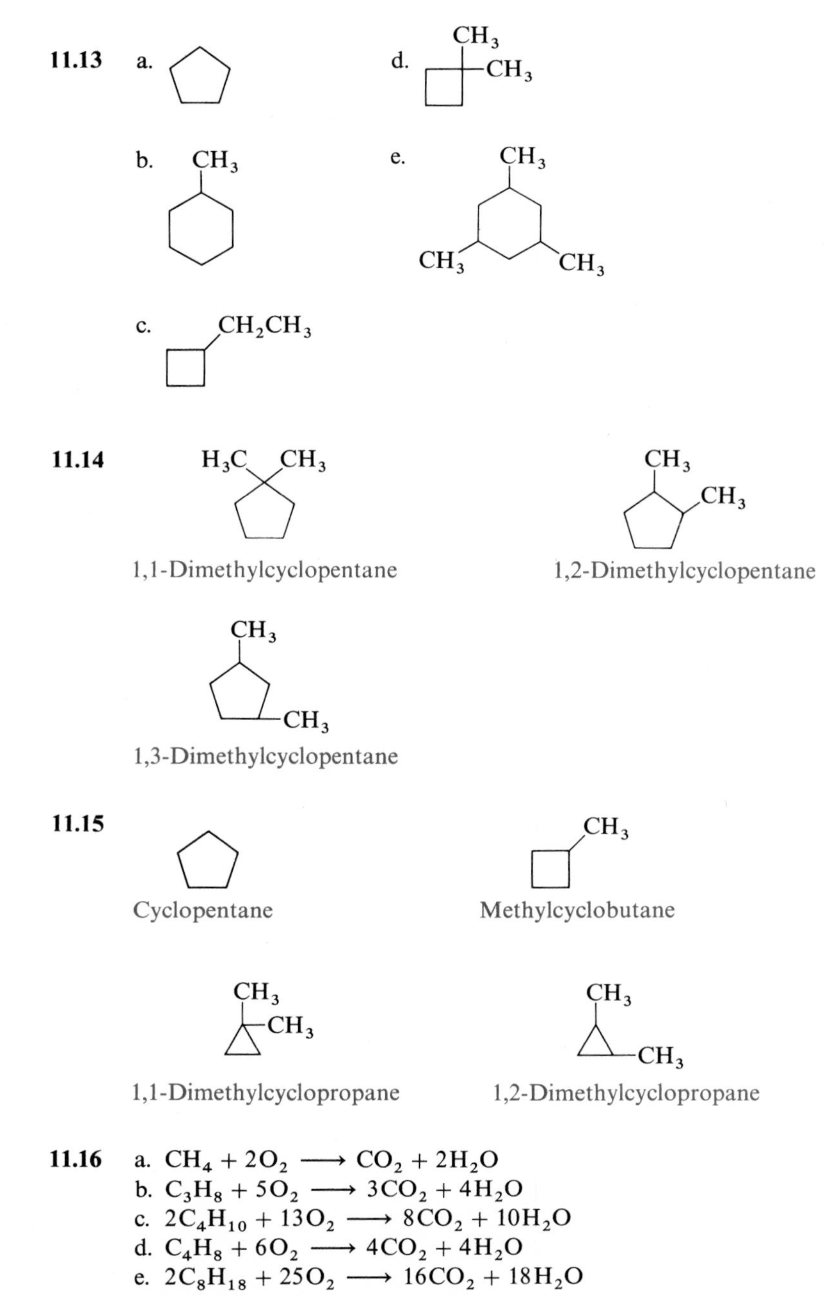

11.16 a. $CH_4 + 2O_2 \longrightarrow CO_2 + 2H_2O$
b. $C_3H_8 + 5O_2 \longrightarrow 3CO_2 + 4H_2O$
c. $2C_4H_{10} + 13O_2 \longrightarrow 8CO_2 + 10H_2O$
d. $C_4H_8 + 6O_2 \longrightarrow 4CO_2 + 4H_2O$
e. $2C_8H_{18} + 25O_2 \longrightarrow 16CO_2 + 18H_2O$

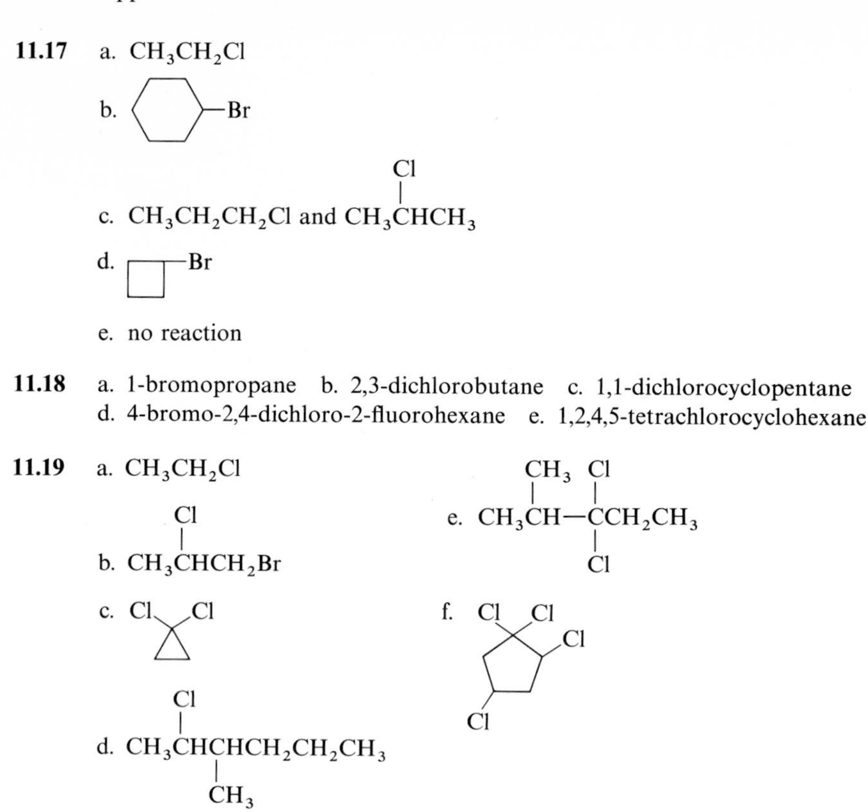

11.17 a. CH_3CH_2Cl

b. (cyclohexane ring)—Br

c. $CH_3CH_2CH_2Cl$ and $CH_3CHClCH_3$

d. (cyclobutane ring)—Br

e. no reaction

11.18 a. 1-bromopropane b. 2,3-dichlorobutane c. 1,1-dichlorocyclopentane d. 4-bromo-2,4-dichloro-2-fluorohexane e. 1,2,4,5-tetrachlorocyclohexane

11.19 a. CH_3CH_2Cl

b. $CH_3CHClCH_2Br$

c. (1,1-dichlorocyclopropane)

d. $CH_3CHClCH(CH_3)CH_2CH_2CH_3$

e. $CH_3CH(CH_3)—CCl_2CH_2CH_3$

f. (1,1,2,4-tetrachlorocyclopentane)

Chapter 12

12.1 Cyclohexane is a six-carbon cycloalkane. It is a saturated alkane that is not very reactive, although it will undergo substitution reactions and combustion. Cyclohexene is a six-carbon cycloalkene. It is unsaturated and much more reactive undergoing addition reactions at the site of the double bond.

12.2 a. ethene; ethylene b. 2-pentene c. 3-methyl-1-pentene d. 3-methylcyclopentene e. 2-methyl-1-pentene f. 2,3-dimethylcyclobutene

12.3 a. $CH_3CH{=}CH_2$

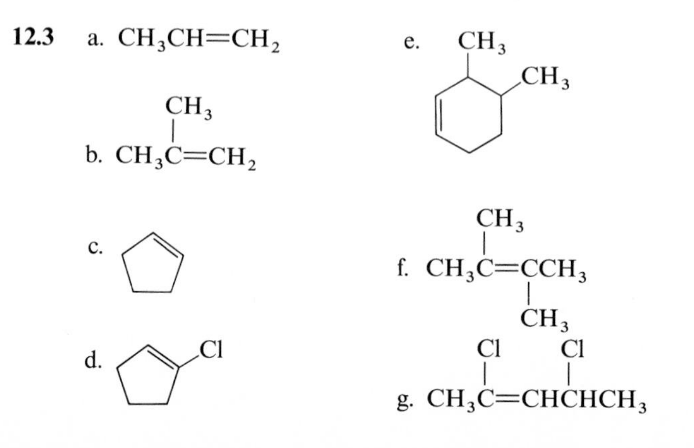

b. $CH_3C(CH_3){=}CH_2$

c. (cyclopentene)

d. (1-chlorocyclopentene)

e. (3,4-dimethylcyclohexene)

f. $CH_3C(CH_3){=}C(CH_3)CH_3$

g. $CH_3CCl{=}CHCHClCH_3$

12.4

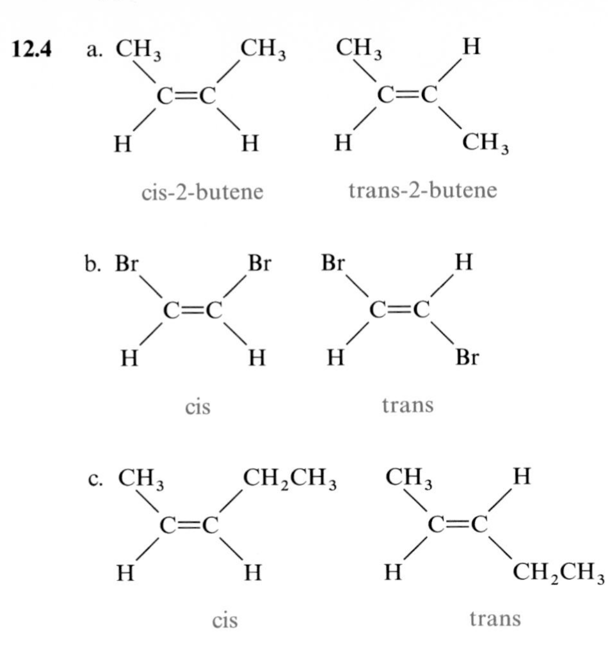

12.5 a. ethyne; acetylene b. 2-hexyne c. 4,5-dichloro-1-heptyne d. 4-methyl-2-hexyne

12.6

a. $CH_3C{\equiv}CCH_3$

b. $CH{\equiv}CH$

c. $CH_3C{\equiv}CC(Cl)_2CH_3$ (Cl above and below the fourth C: $CH_3C{\equiv}C\overset{Cl}{\underset{Cl}{C}}CH_3$)

12.7

a. CH_3CH_2Br

b. $CH_3\overset{Br}{C}H{-}\overset{Br}{C}H\underset{CH_3}{C}HCH_3$

c. $CH_3CH_2CH_2CH_3$

d. cyclopropane–Cl

e. $CH_3\underset{CH_3}{\overset{Cl}{C}}{-}\overset{Cl}{C}HCH_2CH_3$

f. cyclobutane–CH_3

g. $CH_3\underset{Cl}{\overset{CH_3}{C}}CH_3$

12.8

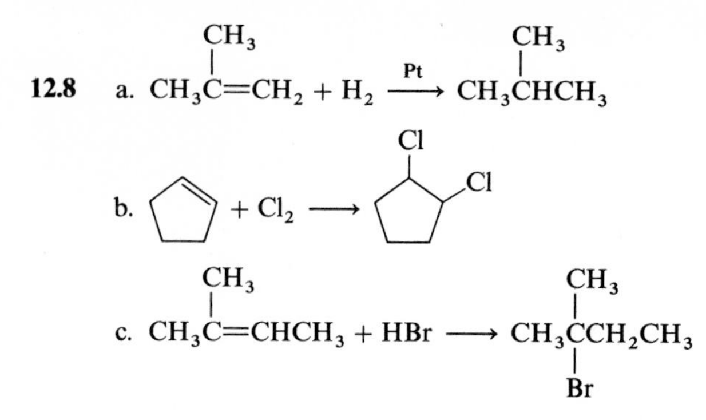

12.9 Benzene is an aromatic compound that is made stable by the unique arrangement of electrons within the six-carbon ring. Cyclohexene is a very reactive compound because of the unsaturated site (double bond) within its six-carbon ring. Cyclohexane is a saturated compound that contains only single bonds between the carbon atoms in its ring.

12.10 a. benzene b. methylbenzene; toluene c. 1,2-dichlorobenzene; *o*-dichlorobenzene d. 1,3-dimethylbenzene; m-xylene e. ethylbenzene f. 3-chloro-1-methylbenzene; m-chlorotoluene g. 1,2-dimethylbenzene; o-xylene h. 3,4-dichlorotoluene

12.11

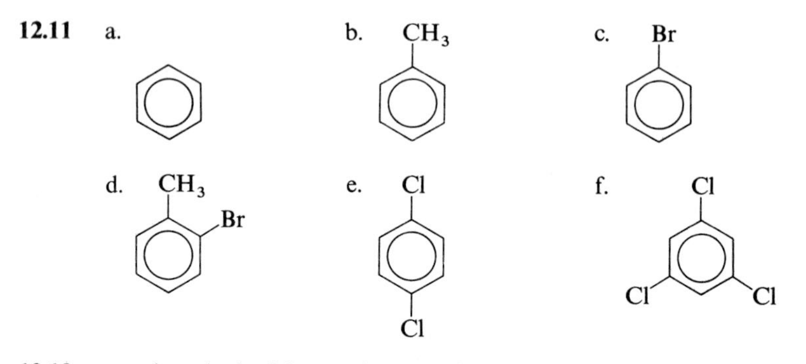

12.12 a. ethanol b. 2-butanol c. cyclohexanol d. 3-methyl-1-butanol e. 2-methylcyclopentanol f. 1-methylcyclobutanol g. 2,3-dimethyl-1-pentanol h. 3-methyl-3-pentanol

12.13 a. $CH_3CH_2CH_2OH$

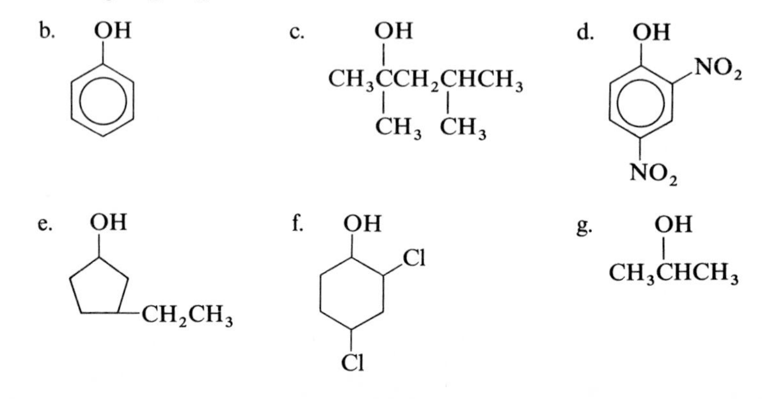

12.14 a. methanethiol b. ethanethiol c. 2-butanethiol d. 2-methyl-1-propanethiol

12.15 a. secondary b. tertiary c. primary d. secondary e. secondary f. primary g. primary h. tertiary

12.16 a. $CH_3CH_2CH(OH)CH_3$

b. cyclohexyl–OH

c. $CH_3CH(OH)CH_3$

d. $C_6H_5CH(OH)CH_3$

e. 1-methylcyclohexene (CH_3)

12.17 a. $CH_3CH{=}CH_2$

b. $CH_3CH(CH_3)CH_2CH_2OH$

c. $CH_3CH(CH_3)CH{=}CH_2$

d. cyclobutyl–OH

e. C_6H_5–$CH{=}CH_2$

12.18 a. cyclopentene $+ H_2O \xrightarrow{H^+}$ cyclopentyl–OH

b. $CH_2{=}C(CH_3)CH_2CH_3 + H_2O \xrightarrow{H^+} CH_3C(CH_3)(OH)CH_2CH_3$

c. cyclobutyl–OH $\longrightarrow$ cyclobutene $+ H_2O$

d. $CH_3CH(OH)CH_2CH(CH_3)CH_3 \xrightarrow{H^+} CH_3CH{=}CHCH(CH_3)CH_3 + H_2O$

12.19 a. ethyl methyl ether b. cyclopentyl methyl ether c. methyl propyl ether d. diethyl ether e. dimethyl ether f. cyclobutyl ethyl ether

12.20 a. $CH_3OCH_2CH_2CH_3$

b. CH_3CH_2O–cyclopentyl

c. CH_3OCH_3

d. $CH_3CH_2OCH_2CH_3$

12.21 a. CH_3OCH_3 b. cyclopentyl–O–cyclopentyl c. $CH_3CH_2OCH_2CH_3$

12.22
a. alkane 2,2,5-trimethylhexane
b. haloalkane 2,2,4-trichlorohexane
c. alcohol 4-methyl-2-pentanol
d. alkene 1-butene
e. aromatic 2-bromo-1,4-dichlorobenzene
f. alkyne 2-pentyne
g. aromatic phenol
h. ether ethyl methyl ether
i. thiol ethanethiol, ethyl mercaptan
j. alcohol 3-methylcyclopentanol

12.23

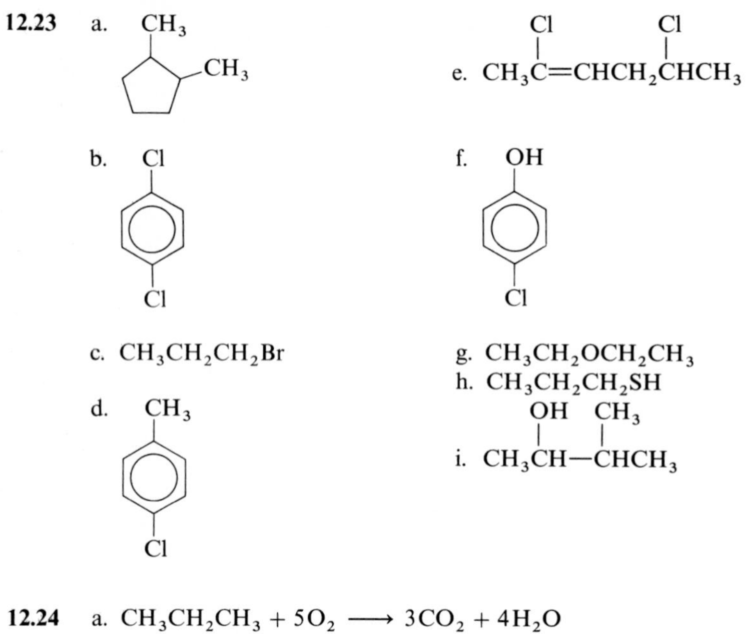

12.24

a. $CH_3CH_2CH_3 + 5O_2 \longrightarrow 3CO_2 + 4H_2O$

b. (1-methylcyclohexene)$-CH_3 + H_2 \xrightarrow{Pt}$ (cyclohexane)$-CH_3$

c. $CH_3\overset{\overset{\displaystyle CH_3}{|}}{C}{=}CHCH_3 + Cl_2 \longrightarrow CH_3\overset{\overset{\displaystyle CH_3}{|}}{\underset{\underset{\displaystyle Cl}{|}}{C}}{-}\underset{\underset{\displaystyle Cl}{|}}{C}HCH_3$

d. $CH_3\overset{\overset{\displaystyle CH_3}{|}}{C}{=}CHCH_2CH_3 + HCl \longrightarrow CH_3\overset{\overset{\displaystyle CH_3}{|}}{\underset{\underset{\displaystyle Cl}{|}}{C}}CH_2CH_2CH_3$

e. (1-methylcyclopentene, CH_3) $+ H_2O \xrightarrow{H^+}$ (1-methylcyclopentanol, H_3C OH)

Chapter 13

13.1 a. ethanal; acetaldehyde b. methanal; formaldehyde c. pentanal d. benzaldehyde e. butanal, butyraldehyde f. 2-methylbutanal

13.2

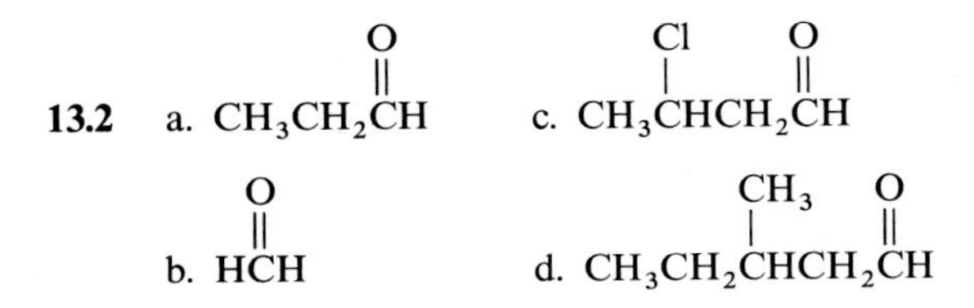

13.3 a. butanone; methyl ethyl ketone b. 3-hexanone; ethyl propyl ketone c. cyclohexanone d. 4-methyl-2-pentanone e. 2-methylcyclopentanone

13.4

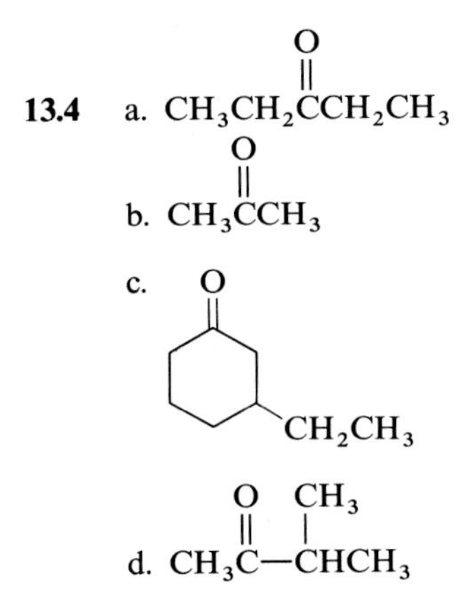

13.5 a. aldehyde; propanal, propionaldehyde
b. alcohol; 2-propanol, isopropyl alcohol
c. alcohol; 1-propanol, n-propyl alcohol
d. ketone; 2-propanone, dimethyl ketone, acetone

13.6

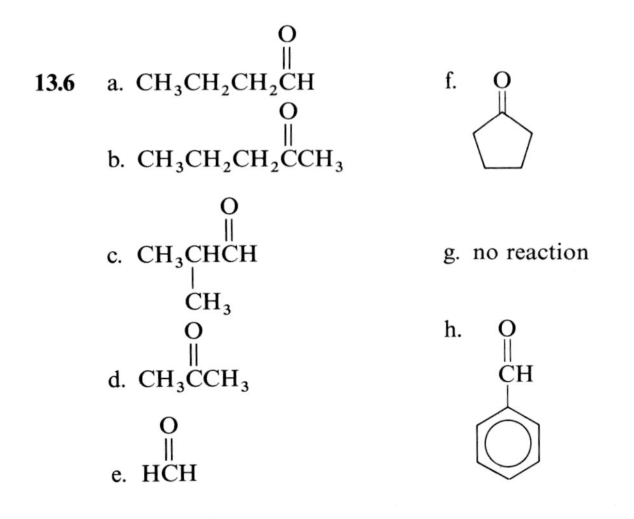

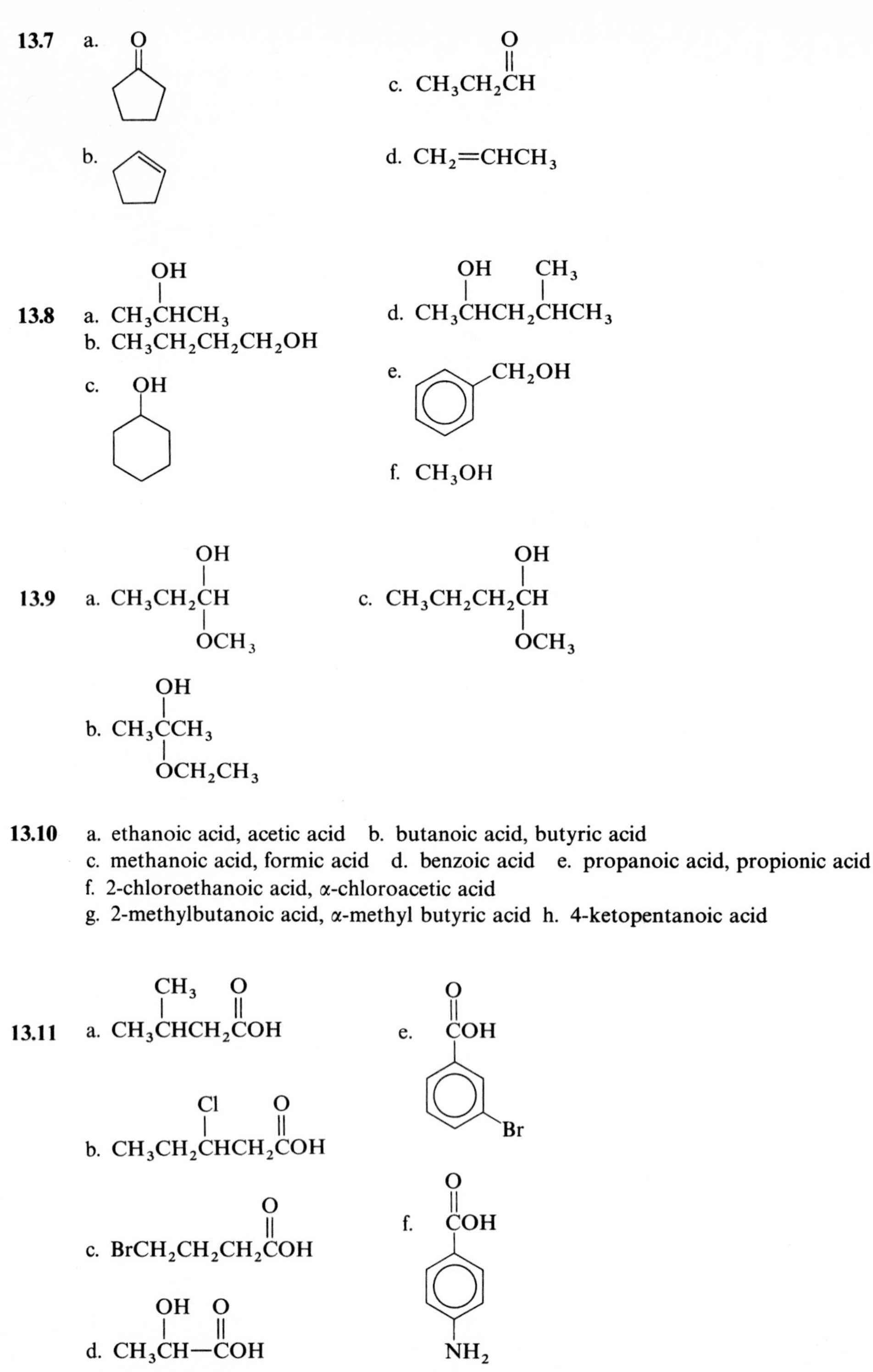
13.7
a. O
b.
c. O CH3CH2CH
d. CH2=CHCH3
13.8
a. OH CH3CHCH3
b. CH3CH2CH2CH2OH
c. OH
d. OH CH3 CH3CHCH2CHCH3
e. CH2OH
f. CH3OH
13.9
a. OH CH3CH2CH OCH3
b. OH CH3CCH3 OCH2CH3
c. OH CH3CH2CH2CH OCH3
13.10
a. ethanoic acid, acetic acid b. butanoic acid, butyric acid
c. methanoic acid, formic acid d. benzoic acid e. propanoic acid, propionic acid
f. 2-chloroethanoic acid, α-chloroacetic acid
g. 2-methylbutanoic acid, α-methyl butyric acid h. 4-ketopentanoic acid
13.11
a. CH3 O CH3CHCH2COH
b. Cl O CH3CH2CHCH2COH
c. O BrCH2CH2CH2COH
d. OH O CH3CH—COH
e. O COH Br
f. O COH NH2

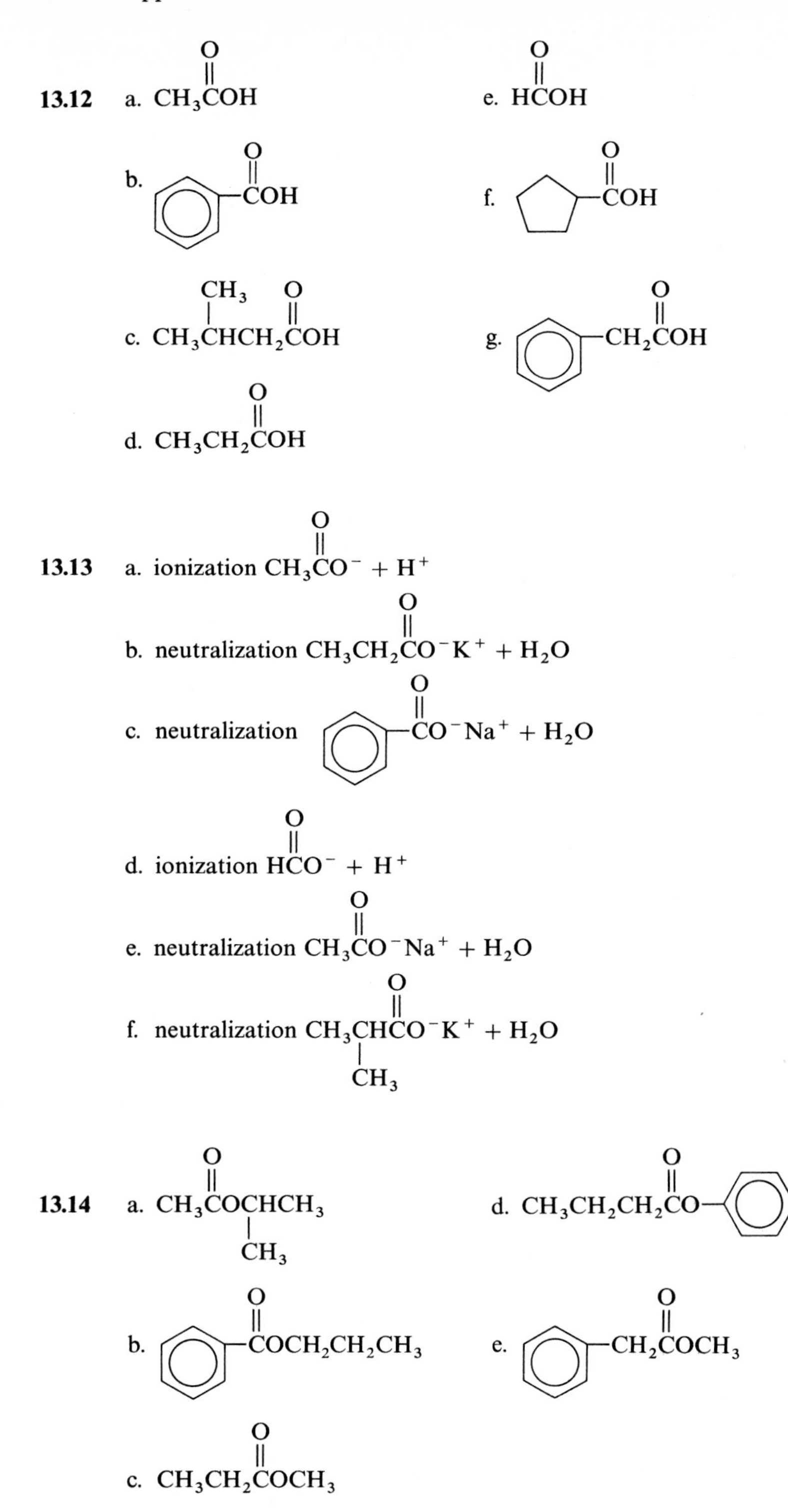

13.12

a. CH_3COOH

b. C_6H_5COOH

c. $CH_3CH(CH_3)CH_2COOH$

d. CH_3CH_2COOH

e. $HCOOH$

f. C_5H_9COOH

g. $C_6H_5CH_2COOH$

13.13

a. ionization $CH_3COO^- + H^+$

b. neutralization $CH_3CH_2COO^-K^+ + H_2O$

c. neutralization $C_6H_5COO^-Na^+ + H_2O$

d. ionization $HCOO^- + H^+$

e. neutralization $CH_3COO^-Na^+ + H_2O$

f. neutralization $CH_3CH(CH_3)COO^-K^+ + H_2O$

13.14

a. $CH_3COOCH(CH_3)CH_3$

b. $C_6H_5COOCH_2CH_2CH_3$

c. $CH_3CH_2COOCH_3$

d. $CH_3CH_2CH_2COOC_6H_5$

e. $C_6H_5CH_2COOCH_3$

13.15 a. methyl ethanoate; methyl acetate b. ethyl ethanoate, ethyl acetate c. propyl methanoate, propyl formate d. methyl benzoate e. ethyl pentatnoate f. phenyl ethanoate; phenyl acetate g. 1-butyl butanoate, n-butyl butyrate h. ethyl benzoate

13.16 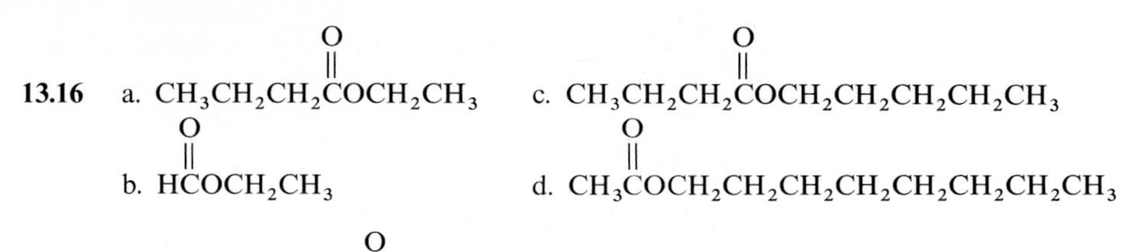

a. $CH_3CH_2CH_2\overset{O}{\overset{\|}{C}}OCH_2CH_3$ c. $CH_3CH_2CH_2\overset{O}{\overset{\|}{C}}OCH_2CH_2CH_2CH_2CH_3$

b. $H\overset{O}{\overset{\|}{C}}OCH_2CH_3$ d. $CH_3\overset{O}{\overset{\|}{C}}OCH_2CH_2CH_2CH_2CH_2CH_2CH_2CH_3$

13.17 a. saponification; $H\overset{O}{\overset{\|}{C}}O^-Na^+ + CH_3CH_2CH_2OH$

b. saponification; $CH_3CH_2\overset{O}{\overset{\|}{C}}O^-Na^+ +$ C_6H_5—OH

c. saponification: C_6H_5—$\overset{O}{\overset{\|}{C}}O^-K^+ +$ C_6H_5—OH

d. hydrolysis; $CH_3CH_2CH_2\overset{O}{\overset{\|}{C}}OH + CH_3CH_2OH$

e. saponification; $CH_3\overset{O}{\overset{\|}{C}}O^-Na^+ + CH_3OH$

f. hydrolysis; C_6H_5—$\overset{O}{\overset{\|}{C}}OH + CH_3\underset{CH_3}{\underset{|}{C}}HCH_2OH$

Chapter 14

14.1 a. primary b. secondary c. primary d. tertiary e. secondary f. secondary

14.2 a. ethylamine b. butyl amine c. aniline d. diethylamine e. trimethylamine f. diethylmethylamine g. N-methyl aniline h. N-ethyl-N-methyl aniline

14.3 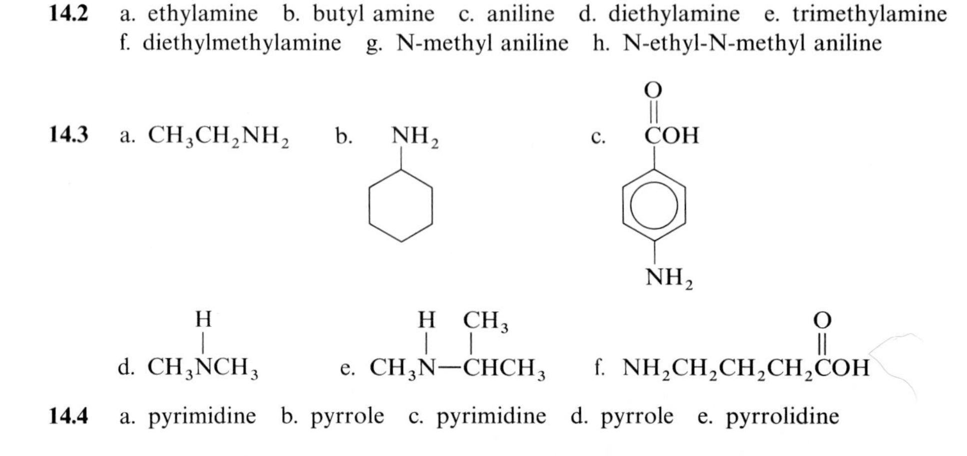

a. $CH_3CH_2NH_2$ b. NH_2 (on cyclohexane ring) c. $\overset{O}{\overset{\|}{C}}OH$ / NH_2 (para on benzene ring)

d. $CH_3\overset{H}{\overset{|}{N}}CH_3$ e. $CH_3\overset{H}{\overset{|}{N}}—\overset{CH_3}{\overset{|}{C}}HCH_3$ f. $NH_2CH_2CH_2CH_2\overset{O}{\overset{\|}{C}}OH$

14.4 a. pyrimidine b. pyrrole c. pyrimidine d. pyrrole e. pyrrolidine

14.5 a. neutralization; $CH_3\overset{+}{N}H_3NO_3^-$

b. ionization; $-NH_3^+ + OH^-$

c. ionization; $CH_3NH_3^+ + OH^-$

d. neutralization; $CH_3CH_2NH_3^+Cl^-$

e. ionization; $(CH_3)_2NH_2^+ + OH^-$

f. neutralization; $(CH_3CH_2)_3NH^+Cl^-$

g. neutralization; $NH_3^+Cl^-$

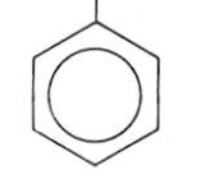

h. ionization; $CH_3CH_2CH_2NH_3^+ + OH^-$

14.6 a. ethanamide, acetamide b. butanamide, butyramide c. benzamide

d. N,N-dimethylethanamide; N,N-dimethylacetamide

e. N-ethylmethanamide, N-ethylformamide

f. N-methylethanamide, N-methylacetamide

g. N-ethyl-N-methylbutanamide, N-ethyl-N-methylbutyramide

14.7 a. $CH_3\overset{O}{\overset{\|}{C}}NH_2$

b. $CH_3CH_2CH_2CH_2\overset{O}{\overset{\|}{C}}-NH_2$

c. $CH_3\overset{CH_3}{\overset{|}{C}}HCH_2\overset{O}{\overset{\|}{C}}NH_2$

d. $CH_3CH_2CH_2\overset{O}{\overset{\|}{C}}-NHCH_3$

e. $CH_3\overset{O}{\overset{\|}{C}}-\overset{CH_3}{\overset{|}{N}}-CH_3$

f. $ClCH_2CH_2\overset{O}{\overset{\|}{C}}NH_2$

14.8 a. 1. amine 2. ester 3. amine b. 1. carboxylic acid 2. amine c. 1. amide

d. 1. carboxylic acid 2. amine 3. amide 4. ester

14.9

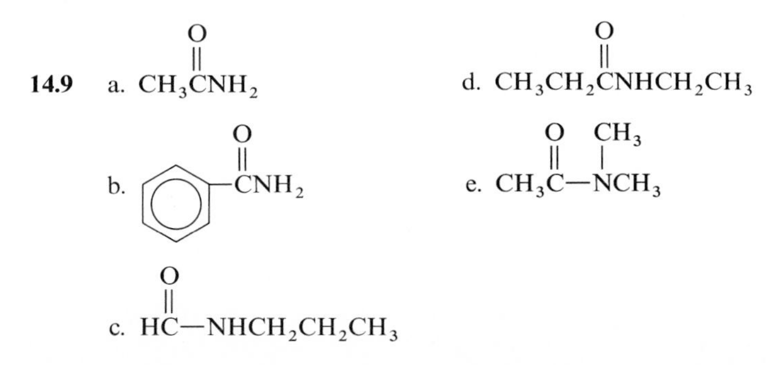

a. $CH_3\overset{O}{\overset{\|}{C}}NH_2$

b. $-\overset{O}{\overset{\|}{C}}NH_2$

c. $H\overset{O}{\overset{\|}{C}}-NHCH_2CH_2CH_3$

d. $CH_3CH_2\overset{O}{\overset{\|}{C}}NHCH_2CH_3$

e. $CH_3\overset{O}{\overset{\|}{C}}-\overset{CH_3}{\overset{|}{N}}CH_3$

14.10

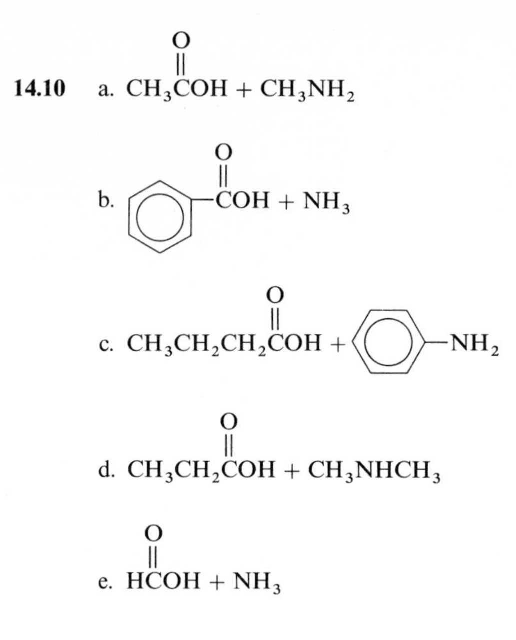

Chapter 15

15.1 a. ketohexose b. aldopentose c. ketotriose d. aldopentose e. aldohexose

15.2 Items b and e contain a chiral carbon atom.

15.3 a. L b. D c. L d. D e. D

15.4

a.
CHO
HCOH
HOCH
HOCH
CH_2OH

b.
CHO
HCOH
HCOH
HOCH
HOCH
CH_2OH

c.
CH_2OH
C=O
HCOH
HOCH
HCOH
CH_2OH

d.
CHO
HCOH
HOCH
HCOH
HOCH
CH_2OH

e.
CHO
HCOH
HCOH
HCOH
CH_2OH

15.5

15.6

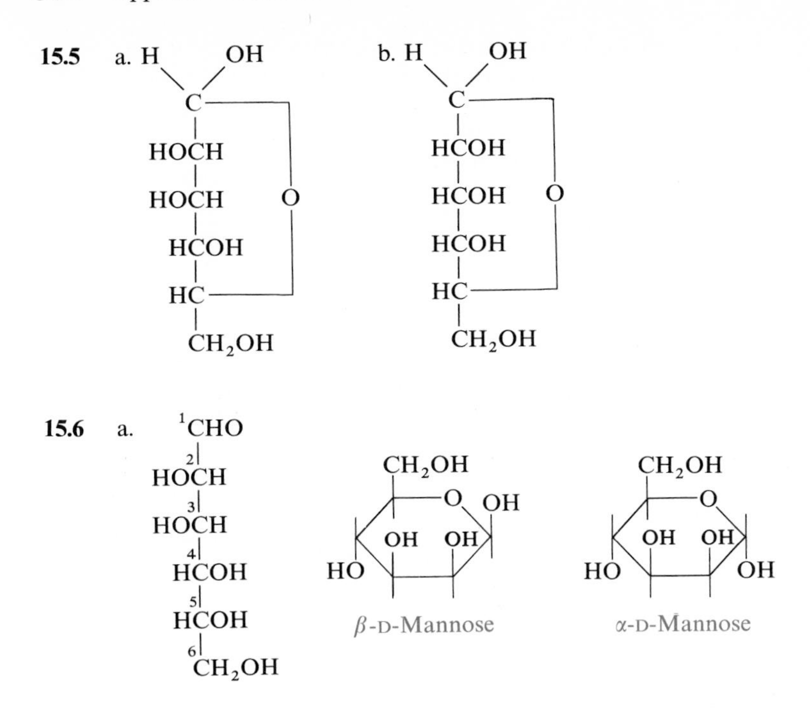

β-D-Mannose

α-D-Mannose

15.7 a. α-D-Fructose b. α-D-glucose c. β-D-Galactose
d. β-D-Glucose e. β-D-Fructose

15.8

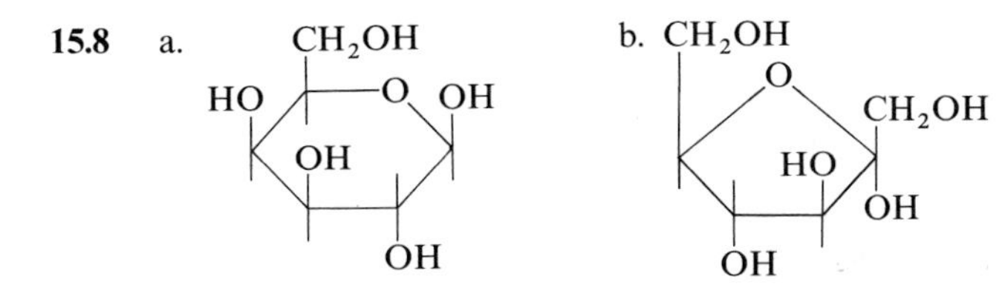

15.9

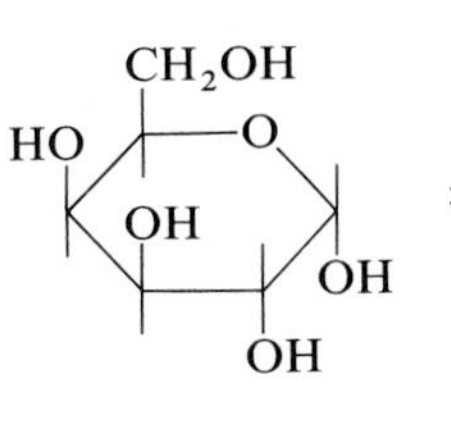

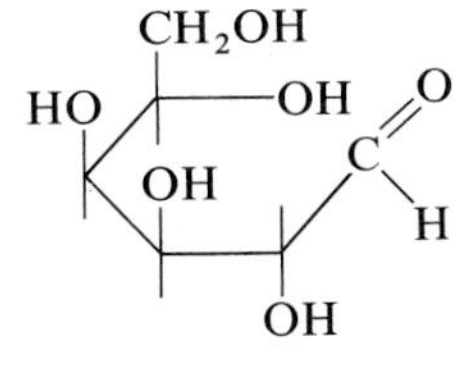

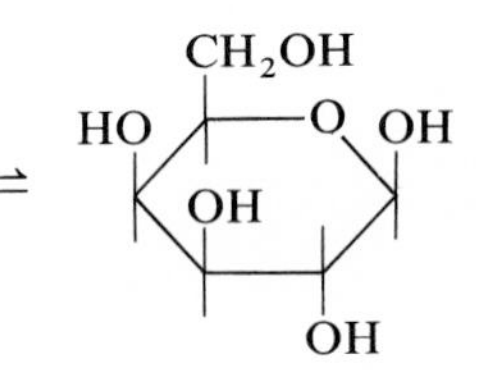

15.10

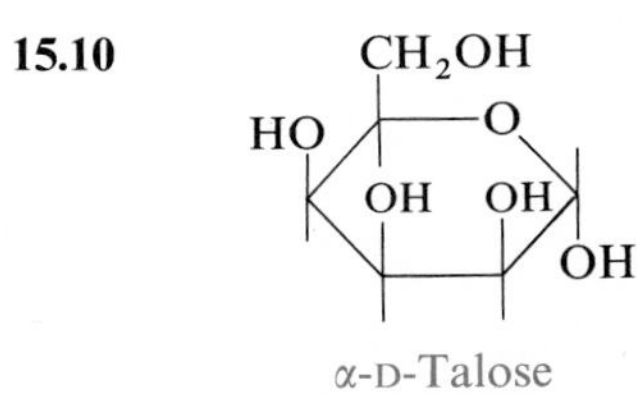

α-D-Talose

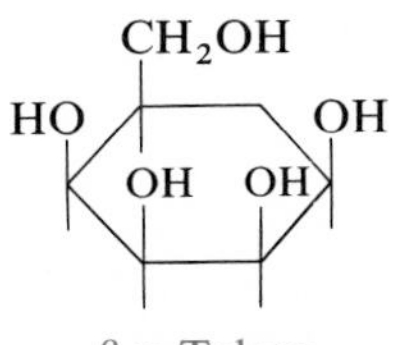

β-D-Talose

15.11 a. fruits, vegetables, corn syrup, honey b. fruit juices, honey

15.12 a. β-galactose and β-glucose; β-1,4; β-lactose
b. α-glucose and β-glucose; α-1,4; β-maltose
c. α-glucose and α-glucose; α-1,4, α-maltose
d. α-glucose and β-fructose; α-1, β-2; sucrose

15.13 amylose; α-glucose; straight chain; α-1,4 bonds amylopectin, α-glucose; branched; α-1,4 and α-1,6 bonds glycogen; α-glucose, branched; α-1,4 and α-1,6 bonds cellulose; β-glucose; straight chain; β-1,4

15.14 a. cellulose b. amylopectin or glycogen c. amylose

15.15 a. $Cu(OH)_2$ is added to sugar solution. A free aldehyde group on a reducing sugar reduces the Cu^{2+} to Cu^{+} which forms a reddish-orange color. Positive tests occurs with reducing sugars which include all monosaccharides and disaccharides except sucrose.
b. When yeast is added to certain sugars, they ferment to ethanol and CO_2. Glucose and fructose undergo fermentation as well as maltose and sucrose.
c. Iodine reacts with polysaccharides to give a blue-black color with amylose and reddish-purple and brown colors with other polysaccharides.

15.16

	Iodine	Fermentation	Benedict's
Galactose	none	none	positive
Maltose	none	positive	positive
Sucrose	none	positive	none
Starch	positive	none	none

15.17 Sugar A is lactose. Sugar B is sucrose.

15.18 a. sucrose b. maltose c. sucrose d. starch e. glycogen f. cellulose g. galactose h. galactose i. maltose j. cellulose

Chapter 16

16.1 Saturated fats contain single carbon to carbon bonds, whereas unsaturated fats contain one or more double bonds. Both are long chain carboxylic acids with 16–18 carbon atoms when present in biological systems.

16.2 Saturated fatty acids have higher melting points than unsaturated fatty acids. More saturated fats are found in animal fats, whereas vegetable oils contain more unsaturated fats. A saturated fatty acid molecule has a regular, linear structure such that molecules can fit close together allowing attractions between the molecules. Unsaturated fatty acids have an angular shape which results from a cis double bond in the carbon chain. The bent shapes cannot align in a regular pattern and therefore do not allow the molecular attractions found in the saturated fatty acids. Thus, less energy and a lower melting point will separate(melt) unsaturated fatty acids.

16.3 a. $CH_3(CH_2)_{12}\overset{\overset{\displaystyle O}{\|}}{C}OH$

b. $CH_3(CH_2)_{14}\overset{\overset{\displaystyle O}{\|}}{C}OH$

c. $CH_3(CH_2)_{16}\overset{\overset{\displaystyle O}{\|}}{C}OH$

16.4 linoleic acid a. unsaturated b. vegetable c. liquid
palmitic acid a. saturated b. animal c. solid

16.5 $CH_3(CH_2)_{14}\overset{\overset{\displaystyle O}{\|}}{C}O(CH_2)_{29}CH_3$ beeswax

16.6 a. CH_2OH
|
$CHOH$
|
CH_2OH

b. $CH_2O\overset{\overset{\displaystyle O}{\|}}{C}(CH_2)_{16}CH_3$
|
$CHO\overset{\overset{\displaystyle O}{\|}}{C}(CH_2)_{16}CH_3$
|
$CH_2O\overset{\overset{\displaystyle O}{\|}}{C}(CH_2)_{16}CH_3$

16.7 palmitic; linoleic; oleic

16.8 $CH_2O\overset{\overset{\displaystyle O}{\|}}{C}(CH_2)_{14}CH_3$
|
$CHO\overset{\overset{\displaystyle O}{\|}}{C}(CH_2)_{14}CH_3$
|
$CH_2O\overset{\overset{\displaystyle O}{\|}}{C}(CH_2)_7CH{=}CH(CH_2)_7CH_3$

$CH_2O\overset{\overset{\displaystyle O}{\|}}{C}(CH_2)_{14}CH_3$
|
$CHO\overset{\overset{\displaystyle O}{\|}}{C}(CH_2)_7CH{=}CH(CH_2)_7CH_3$
|
$CH_2O\overset{\overset{\displaystyle O}{\|}}{C}(CH_2)_{14}CH_3$

16.9 Triolein contains three unsaturated fatty acids while tristearin contain only saturated stearic acid.

16.10 a.

$$\begin{array}{l} CH_2OC(=O)(CH_2)_{12}CH_3 \\ | \\ CHOC(=O)(CH_2)_{12}CH_3 + 3H_2O \xrightarrow{H^+} \\ | \\ CH_2OC(=O)(CH_2)_{12}CH_3 \end{array} \quad \begin{array}{l} CH_2OH \\ | \\ CHOH + 3HOC(=O)(CH_2)_{12}CH_3 \\ | \\ CH_2OH \end{array}$$

$$\begin{array}{l} CH_2OC(=O)(CH_2)_7CH{=}CH(CH_2)_7CH_3 \\ | \\ CHOC(=O)(CH_2)_7CH{=}CH(CH_2)_7CH_3 + 3H_2O \xrightarrow{H^+} \\ | \\ CH_2OC(=O)(CH_2)_7CH{=}CH(CH_2)_7CH_3 \end{array}$$

$$\begin{array}{l} CH_2OH \\ | \\ CHOH + 3HOC(=O)(CH_2)_7CH{=}CH(CH_2)_7CH_3 \\ | \\ CH_2OH \end{array}$$

16.11 a.

$$\begin{array}{l} CH_2OC(=O)(CH_2)_{14}CH_3 \\ | \\ CHOC(=O)(CH_2)_{14}CH_3 + 3NaOH \longrightarrow \\ | \\ CH_2OC(=O)(CH_2)_{14}CH_3 \end{array} \quad \begin{array}{l} CH_2OH \\ | \\ CHOH + 3Na^+ \; ^-OC(=O)(CH_2)_{14}CH_3 \\ | \\ CH_2OH \end{array}$$

b.

$$\begin{array}{l} CH_2OC(=O)(CH_2)_7CH{=}CH(CH_2)_7CH_3 \\ | \\ CHOC(=O)(CH_2)_7CH{=}CH(CH_2)_7CH_3 + 3NaOH \longrightarrow \\ | \\ CH_2OC(=O)(CH_2)_7CH{=}CH(CH_2)_7CH_3 \end{array}$$

$$\begin{array}{l} CH_2OH \\ | \\ CHOH + 3Na^+ \; ^-OC(=O)(CH_2)_7CH{=}CH(CH_2)_7CH_3 \\ | \\ CH_2OH \end{array}$$

16.12

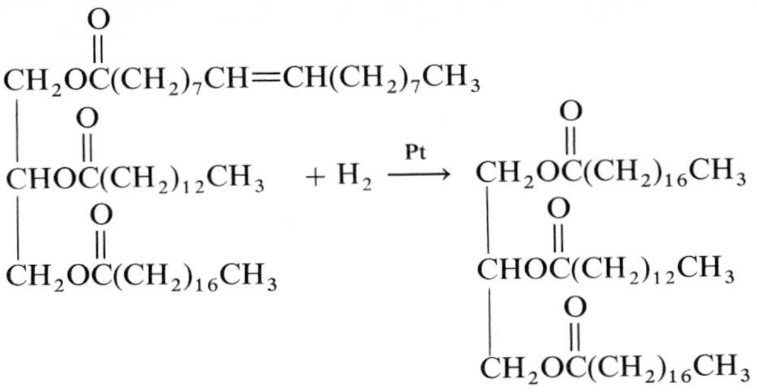

$$\begin{array}{l} \overset{O}{\overset{\|}{}} \\ CH_2OC(CH_2)_7CH{=}CH(CH_2)_7CH_3 \\ |\quad O \\ |\quad \| \\ CHOC(CH_2)_{12}CH_3 \\ |\quad O \\ |\quad \| \\ CH_2OC(CH_2)_{16}CH_3 \end{array} + H_2 \xrightarrow{Pt} \begin{array}{l} \quad O \\ \quad \| \\ CH_2OC(CH_2)_{16}CH_3 \\ |\quad O \\ |\quad \| \\ CHOC(CH_2)_{12}CH_3 \\ |\quad O \\ |\quad \| \\ CH_2OC(CH_2)_{16}CH_3 \end{array}$$

16.13 A fat that reacts with a small amount of iodine has a low iodine number. This means that there are only a few unsaturated sites in the fat. A fat with many double bonds will absorb a greater amount of iodine.

16.14 soybean oil, peanut oil, butter, coconut oil

16.15 a. Triolein b. The fat with two molecules linoleic acid and one molecule oleic acid.

16.16 items b and c

16.17 One of the fatty acids has been replaced by a phosphate group that is attached to an amino alcohol.

16.18 The phosphate group and amino alcohol make up the polar part. The two fatty acids are the nonpolar part.

16.19

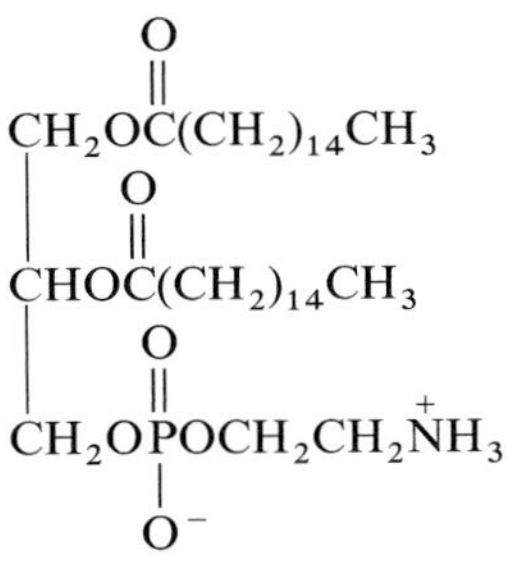

$$\begin{array}{l} \quad O \\ \quad \| \\ CH_2OC(CH_2)_{14}CH_3 \\ |\quad O \\ |\quad \| \\ CHOC(CH_2)_{14}CH_3 \\ |\quad O \\ |\quad \| \\ CH_2OPOCH_2CH_2\overset{+}{N}H_3 \\ \quad\;\; | \\ \quad\;\; O^- \end{array}$$

16.20 Lecithin contains the amino alcohol choline whereas a cephalin contains serine or ethanolamine.

16.21 Sphingolipids use the alcohol sphinogosine in place of glycerol.

16.22 The phospholipids are arranged in a double row with the polar part which is hydrophilic on the outside and the nonpolar hydrophobic fatty acids in the middle.

16.23 a. steroid nucleus b. isoprene

16.24 a. steroid b. steroid c. steroid d. terpene e. terpene f. steroid

16.25 a. Contains a steroid nucleus.
b. Most of the side groups are identical to estradiol except for the addition of the triple bond.

16.26 a. 2 b. 6 c. 3 d. 5 e. 4 f. 1

16.27 a. long chain alcohol and fatty acid b. steroid nucleus
c. two fatty acids, glycerol, phosphate and amino alcohol
d. glycerol and three fatty acids e. isoprene f. salts of fatty acids
g. sphingosine, fatty acids, and a monosaccharide h. steroid nucleus i. isoprene

Chapter 17

17.1 a. 3 b. 2 c. 2 d. 1 e. 6 f. 5 g. 4 h. 1

17.2 All amino acids contain an amino group and a carboxylic acid group. The distinguishing feature is a side R group which may be one hydrogen atom or several carbon atoms. Sometimes an oxygen atom or a sulfur atom may be present in the side group.

17.3 a. serine b. valine c. aspartic acid d. alanine

17.4 a. hydrophilic b. hydrophobic c. hydrophilic d. hydrophobic

17.5 a. cysteine, methionine b. serine, tyrosine, threonine c. lysine, arginine, histidine
d. aspartic acid, glutamic acid e. phenylalanine, tryptophan, tyrosine

17.6 An essential amino acid is not synthesized in the body and must be provided by the diet.

17.7 a. yes b. yes c. no d. yes e. no

17.8 Kwashiorkor

17.9

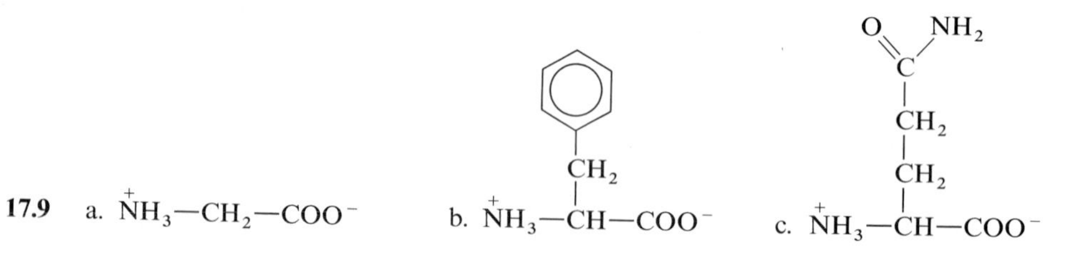

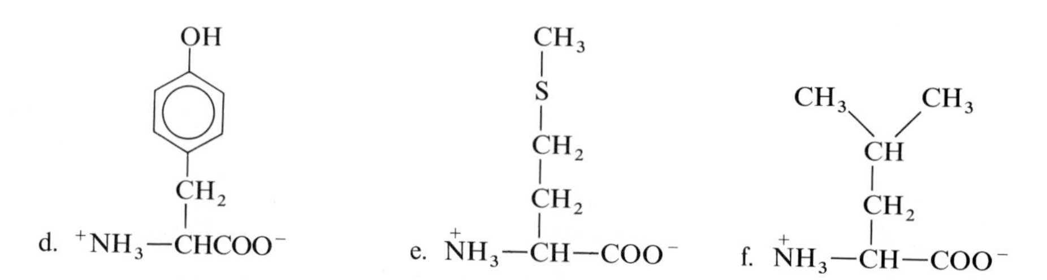

17.10 a and b. c.

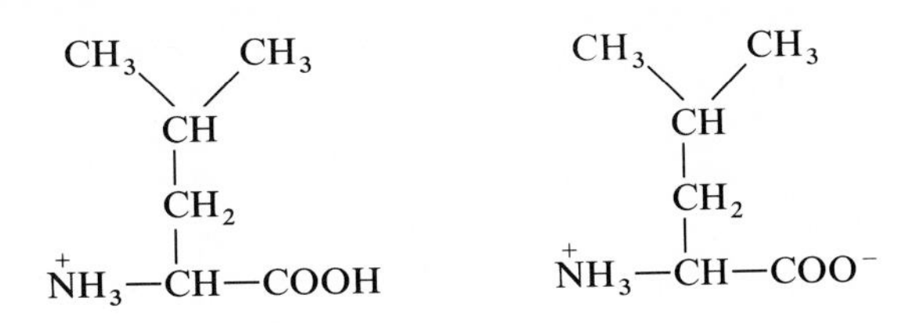

d and e.

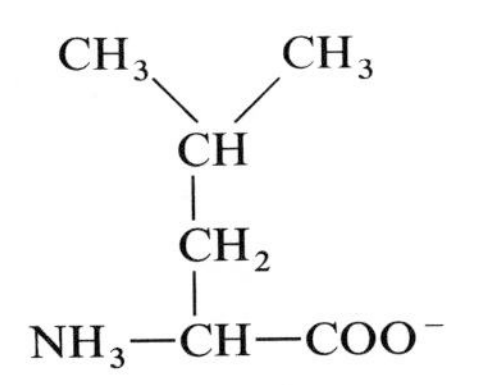

17.11 a. aspartic acid b. lysine c. alanine

17.12

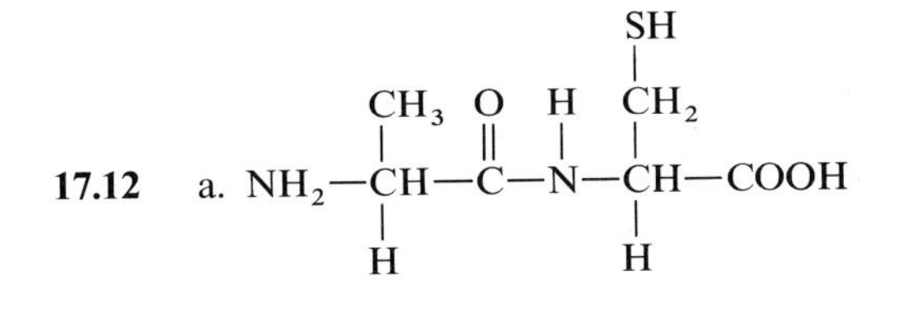

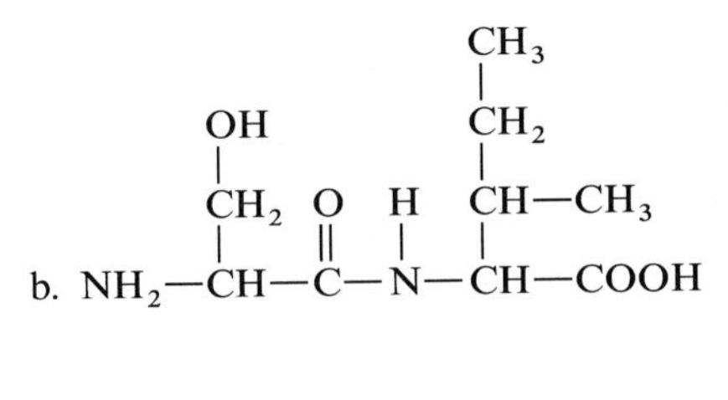

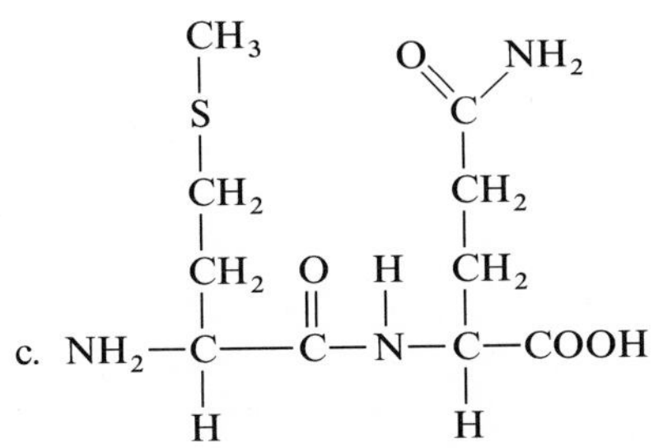

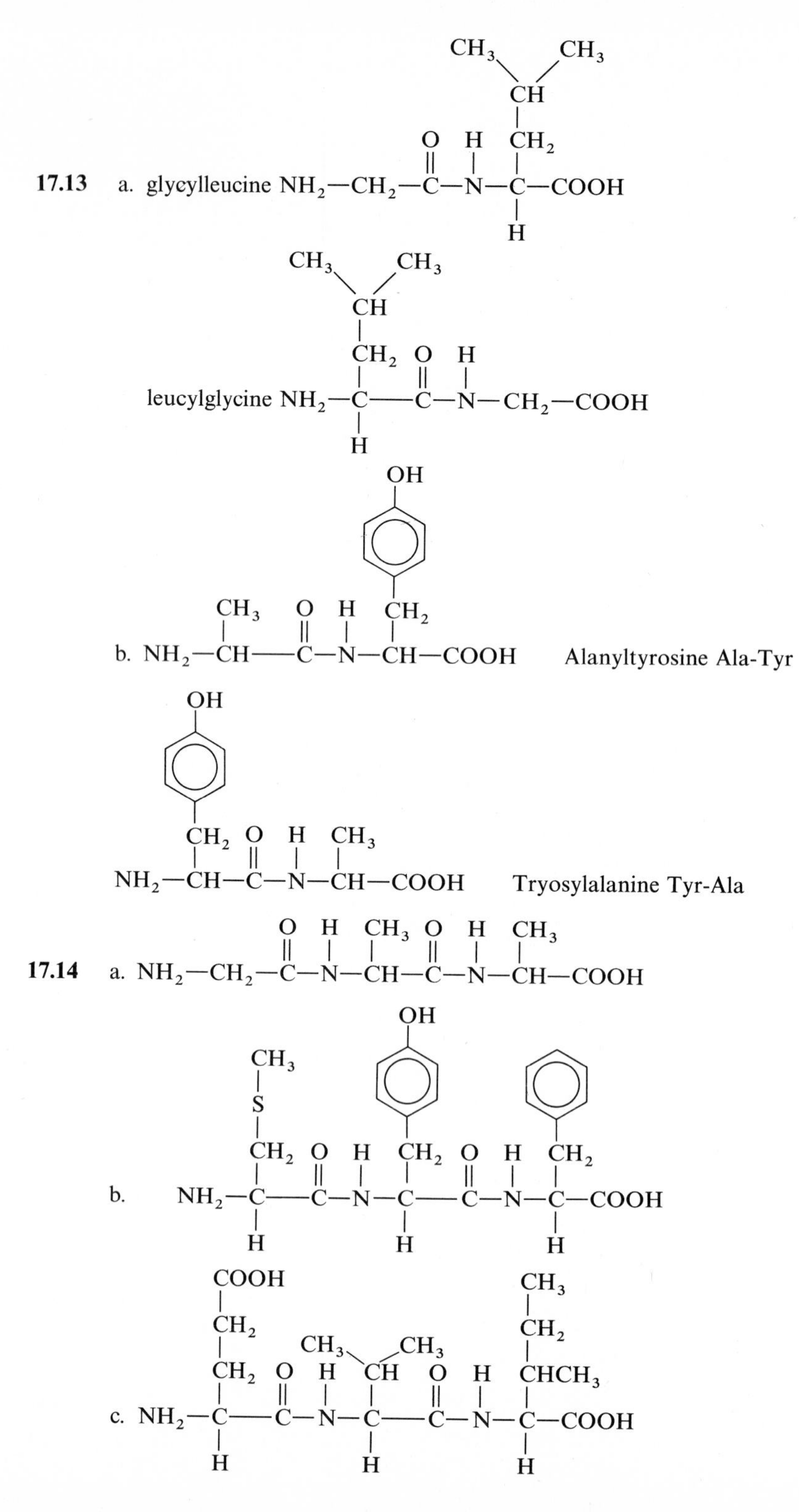
17.13
a. glycylleucine
leucylglycine
b.
Alanyltyrosine Ala-Tyr
Tryosylalanine Tyr-Ala
17.14
a.
b.
c.

17.15 a.

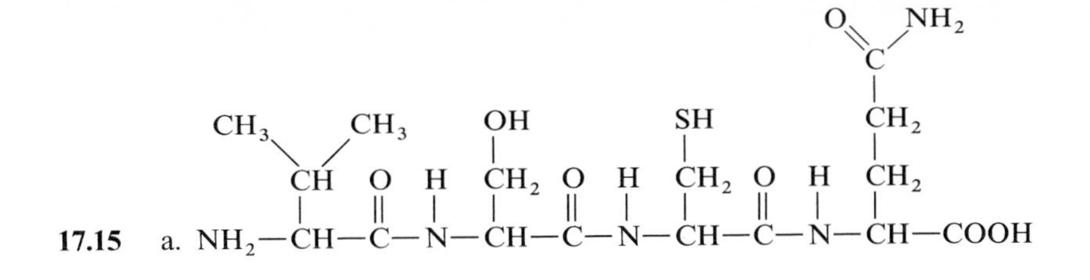

b.

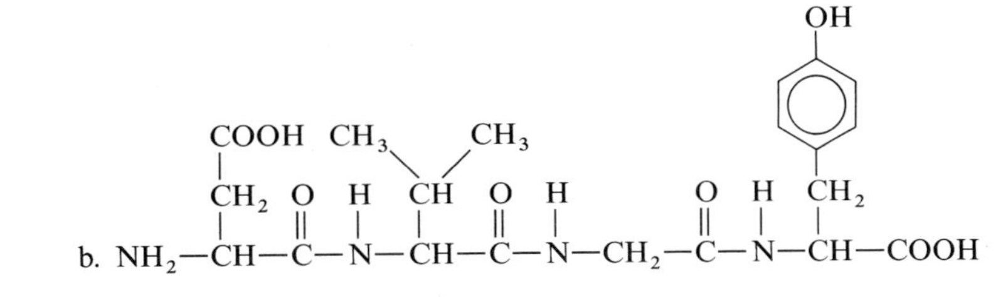

17.16 a. 3° b. 4° c. 1° d. 2° e. 2° f. 3° g. 2°

17.17 a. disulfide —S—S— b. $-\overset{\displaystyle O}{\overset{\|}{C}}O^{-} \cdots\cdots {}^{+}H_3N-$

c. hydrogen bonding $-CH_2OH \cdots\cdots {}^{-}O-\overset{\displaystyle O}{\overset{\|}{C}}-$ d. hydrophobic

17.18 a. cysteine b. leucine and valine, they are hydrophobic
c. aspartic acid, it is hydrophilic.

17.19

a. Secondary protein structures are the result of hydrogen bonding within a polypeptide chain to form an α helix or between polypeptide chains to give pleated-sheet or braided structures of fibrous proteins. Tertiary structures are formed when further interactions occur between the R groups to give the compact shapes of the globular proteins.

b. Essential amino acids must be acquired through the diet, while nonessential amino acids may be synthesized in the body.

c. Polar amino acids contain R groups that are hydrophilic due to the presence of a polar group such as a sulfur, hydroxy, acidic, or basic group, while nonpolar amino acids contain R groups that are primarily hydrocarbon in nature and therefore hydrophobic.

d. A dipeptide contain two amino acids with one peptide bond; a tripeptide has three amino acids attached by two peptide bonds.

e. A salt bridge is a tertiary feature that results from the ionic attraction between acidic and basic side groups. A disulfide bond is a tertiary feature that occurs when two cysteine side groups bond.

f. A fibrous protein uses β-pleated sheet or a collagen braid to form structural components in the body such as muscle, hair, nails and tendons. Globular proteins found in the cells such as antibodies and enzymes have a compact three-dimensional shape.

g. In an α-helix, the hydrogen bonding occurs between the amino acids in a single polypeptide chain. In the β-pleated sheet, the hydrogen bonding occurs between aligned polypeptide chains.

h. An α-helix is the secondary structure that forms the spiral chain in a polypeptide. The secondary structure of collagen consists of 3 polypeptide chains.
i. The tertiary structure is a three-dimensional shape of a single protein caused by side group attractions. The quaternary structure combines two or more of the tertiary structures to make an active protein.

17.20 Heat breaks apart the hydrogen bonds causing a loss of the tertiary and secondary structures. The protein will retain the primary structure of the amino acids.

17.21 Acid causes the disruption of ionic bonds and hydrogen bonds; tertiary structure breaks down.

17.22 Alcohol disrupts hydrogen bonds causing a loss of the tertiary and secondary structures of the proteins in the bacteria on the skin.

17.23 a. Heavy metal salt is used to destroy the protein of bacteria in the eyes of newborns.
b. Tannic acid is an alkaloidal substance that will react with proteins in the burn area to form an insoluble crust which protects the damaged area.
c. The increased temperature causes coagulation of milk protein to give a solid product.
d. Heat is used to sterilize the instruments by destroying the protein of bacteria present.
e. The alcohol is used to destroy bacteria on the skin prior to an injection.
f. The protein of the egg is coagulated by high temperatures.

17.24 a. A tripeptide will give a violet color in the biuret test.
b. Tyrosine will give a blue color with ninhydrin and a yellow color in the xanthoproteic test.
c. Cysteine will give a blue color with ninhydrin and a black precipitate (PbS) in the sulfur test.
d. A protein will give a violet color in the biuret test.
e. Proline will give a yellow color with ninhydrin.

Chapter 18

18.1 a. $2H_2 + O_2$ b. $2H_2O$ and energy c. exothermic

d.

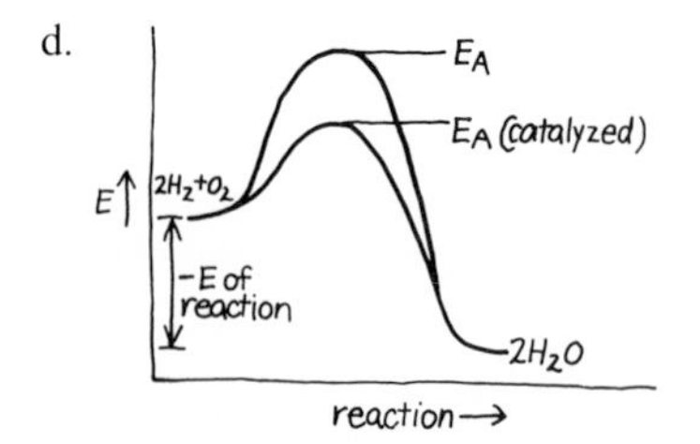

e. lowers the energy of activation
f. More molecules will have sufficient energy to react; the rate of the reaction increases.

18.2 a. holoenzyme b. holoenzyme c. simple enzyme d. holoenzyme
e. simple enzyme

18.3 The apoenzyme is the protein portion. The cofactor is an additional component such as a metal ion, vitamin or hormone required for activity by some enzymes.

18.4 a. lipid b. peptide c. maltose d. ester

18.5 oxidoreductase; transferase, hydrolase, lyase, ligase, isomerase

18.6 a. hydrolase b. hydrolase c. hydrolase d. oxidoreductase e. oxidoreductase f. isomerase g. lyase h. lyase i. lyase j. ligase

18.7 a. 2 b. 1 c. 4 d. 3 e. 2

18.8
a. E = enzyme S = substrate ES = enzyme-substrate complex
EP = enzyme-product complex P = product
b. The active site is a small portion of the enzyme at which the reaction of the substrate molecule takes place.
c. After the product is released from the enzyme, the enzyme is available to react with another molecule of substrate.

18.9
a. Less substrate will slow down the rate of reaction.
b. More enzyme will increase the rate of the reaction.
c. Maximum activity will occur at optimum pH; the rate of reaction will increase.
d. The rate of reaction will decrease, and reaction may stop if the enzyme is completely denatured.
e. Activity of the enzyme will be decreased (assuming that the optimum temperature is 37°C).
f. Temperatures above optimum temperature will slow or stop the reaction.

18.10

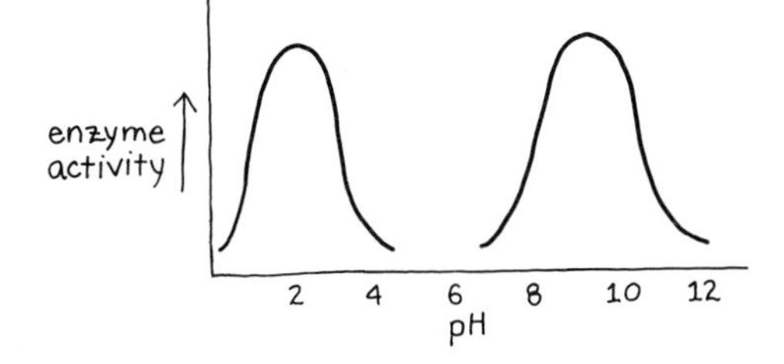

18.11 a. competetive b. noncompetetive c. competetive d. noncompetetive

18.12
a. competetive; structure is similar
b. compete for the active site
c. increase the concentration of succinic acid

18.13 a. T b. F c. T d. T e. T f. T g. F h. F

18.14
a. milk butter, codliver oil, green and yellow vegetables
b. yeast, wheat germ, meat
c. fish oils, egg yolk, sunlight
d. liver, kidney, meats
e. citrus fruits, green vegetables, tomatoes
f. liver, green leafy vegetables
g. liver, beef, milk, fish, eggs
h. liver, eggs, green leafy vegetables, synthesized in colon

18.15 a. fat b. water c. fat d. water e. water f. water g. water h. fat

18.16 a. A b. B_3 (niacin) c. B_1 (thiamine) d. B_1 (thiamine) e. A f. C (ascorbic acid) g. B_2 (riboflavin) h. D i. D j. E k. K l. D m. K

18.17
a. Amylose begins digestion in the mouth by salivary amylase to produce smaller polysaccharides, maltose, and glucose. In the small intestine, pancreatic amylase completes the breakdown of amylose to form maltose and glucose. Maltose is hydrolyzed by maltase to give glucose molecules.
b. Amylopectin begins its digestion in the mouth by salivary amylase, which breaks only the α-1,4 bonds to product dextrins, maltose, and glucose. In the small intestine, digestion is completed by pancreatic amylase and pancreatic glucosidase (for the α-1,6 linkages) to form maltose and glucose. Maltose is hydrolyzed by maltase to give glucose molecules.
c. Lactose is hydrolyzed in the small intestine by lactase to form galactose and glucose.
d. Sucrose is hydrolyzed in the small intestine by sucrase to form fructose and glucose.

18.18
a. Bile from the gallbladder emulsifies the fat particles to form smaller fat globules.
b. Lipases are enzymes that hydrolyze the ester bonds in triacylglycerols to produce glycerol and fatty acids.
c. Lipoproteins are polar and therefore more soluble in the aqueous lymph and bloodstream. They serve as the transport form of lipids in the body.

18.19
a. The anticipation of food causes the secretion of HCl into the stomach, where it activates the protein-hydrolyzing enzyme pepsin. If pepsin were continually active, it would attack the protein lining of the stomach when there was no food present, causing ulcerations of the stomach.
b. Protein digestion is completed in the small intestine by additional peptidases to form amino acids.

Chapter 19

19.1 $ATP \longrightarrow ADP + P_i + 7.3\ \text{kcal}$

19.2 ATP is a storage form of energy in cells. Its hydrolysis provides energy for energy-requiring processes in the cells such as moving substances across cell membranes and synthesizing cellular substances.

19.3
a. Oxidative phosphorylation occurs with phosphorus is added to a molecule of adenosine diphosphate.
b. 7.3 kcal
c. $ADP + P_i + 7.3\ \text{kcal} \longrightarrow ATP$

19.4
a. NAD^+, FAD, FMN, CoA
b. NAD^+, FAD
c. $MH_2 + FAD \longrightarrow M + FADH_2$
$MH_2 + NAD^+ \longrightarrow M + NADH/H^+$

19.5
a. $NADH/H^+ \longrightarrow NAD^+ + 2H$
b. $CoQ + 2H \longrightarrow CoQH_2$
c. $FMNH_2 \longrightarrow FMN + 2H$

19.6

NAD ⟷ $FMNH_2$ ⟷ CoQ ⟷ 2 cyt $b(Fe^{2+})$

$NADH/H^+$ — FMN — $CoQH_2$ — 2 cyt $b(Fe^{3+})$

$CoQH_2 \rightarrow 2H^+$

19.7 The cytochromes transfer electrons obtained from coenzyme Q to oxygen to form H_2O.

19.8 a. reduction b. oxidation

19.9 a. H_2O and ATP b. 3 c. 2

19.10 a. Acetyl CoA is the end product of several metabolic pathways that oxidize food stuffs. It serves as the fuel for the citric acid cycle in the production of energy.
b. citric acid, isocitric acid
c. α-ketoglutaric acid
d. succinic acid, fumaric acid, malic acid, oxaloacetic acid
e. decarboxylation of isocitric acid and α-ketoglutaric acid
f. CO_2 and ATP

19.11 The hydrogen acceptor NAD^+ is used when hydrogen atoms are removed from atoms in a carbon–oxygen bond. FAD is used when hydrogen atoms are removed from the atoms in a carbon–carbon bond.

19.12 Oxidation reactions of compounds in the citric acid cycle provide the hydrogen atoms for the electron transport system where ATP is produced.

19.13 a. $2H + NAD^+ \longrightarrow NADH/H^+$ 3ATP
b. $2H + NAD^+ \longrightarrow NADH/H^+$ 3ATP
c. $GDP + P_i \longrightarrow GTP$ 1ATP
d. $2H + FAD \longrightarrow FADH_2$ 2ATP
e. $2H + NAD^+ \longrightarrow NADH/H^+$ 3ATP

19.14 Acetyl CoA $\longrightarrow 2CO_2 + 12ATP$

19.15 a. 60 molecules ATP b. 24 mol ATP c. 300 molecules ATP d. 120 mol ATP

19.16 a. glucose b. pyruvic acid c. lactic acid

19.17 a. ATP is used first to activate glucose. b. 2

19.18 a. NAD^+
b. In skeletal muscle, $NADH/H^+$ must cross the membrane from the cytoplasm into the mitochondria at the expense of 2 ATP.
c. by reducing pyruvic acid to lactic acid

$$NADH/H^+ + CH_3\overset{\overset{\displaystyle O}{\|}}{C}COOH \longrightarrow CH_3\overset{\overset{\displaystyle OH}{|}}{C}HCOOH + NAD^+$$

d. In two of the steps, phosphate is transferred directly to ADP. $ADP + P_i \longrightarrow ATP$.

19.19 Two ATP are used up in transporting the $NADH/H^+$ across the membrane into the mitochondria. The net production of ATP in glycolysis is 4 ATP (6 ATP − 2 ATP).

19.20 a. enters the citric acid cycle
b. 3ATP
c. Pyruvic acid + HSCoA $\longrightarrow$ Acetyl-SCoA + CO_2 + 3ATP

19.21 a. 6 b. 2(3) = 6 c. 2(12) = 24 d. 36

19.22 Activation of a fatty acid requires 2 ATP and coenzyme A.

19.23 a. (1) oxidation, (2) hydration, (3) oxidation, (4) cleavage
b. oxidation steps
c. FAD removes hydrogen atoms from a carbon to carbon bond in the first oxidation step. In the second oxidation, NAD^+ is used because hydrogen atoms are removed from a carbon to oxygen bond.
d. 5 ATP

19.24 a. $CH_3(CH_2)_{12}-\underset{\beta}{CH_2}-\underset{\alpha}{CH_2}-\overset{O}{\overset{\|}{C}}-CoA$

b. $CH_3(CH_2)_{12}-CH_2-CH_2-\overset{O}{\overset{\|}{C}}-CoA \xrightarrow[\text{dehydrogenase}]{\text{FAD} \quad \text{FADH}_2}$

$CH_3(CH_2)_{12}-CH{=}CH-\overset{O}{\overset{\|}{C}}-CoA$

$CH_3(CH_2)_{12}-CH{=}CH-\overset{O}{\overset{\|}{C}}-CoA + H_2O \xrightarrow{\text{hydrolase}}$

$CH_3(CH_2)_{12}-\overset{OH}{\overset{|}{C}H}-CH_2-\overset{O}{\overset{\|}{C}}-CoA$

$CH_3(CH_2)_{12}-\overset{OH}{\overset{|}{C}H}-CH_2-\overset{O}{\overset{\|}{C}}-CoA \xrightarrow[\text{dehydrogenase}]{\text{NAD}^+ \quad \text{NADH/H}^+}$

$CH_3(CH_2)_{12}-\overset{O}{\overset{\|}{C}}-CH_2-\overset{O}{\overset{\|}{C}}-CoA$

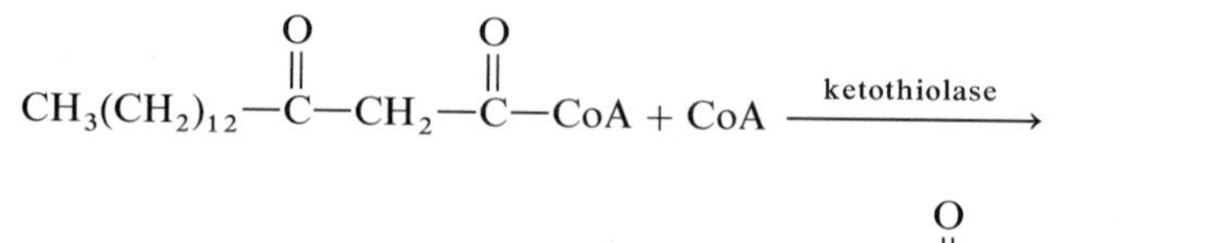

$CH_3(CH_2)_{12}-\overset{O}{\overset{\|}{C}}-CoA + CH_3-\overset{O}{\overset{\|}{C}}-CoA$

19.25 a. 16 b. $16/2 = 8$ c. $8 - 1 = 7$ turns
d. 8 acetyl CoA × 12 = 96 ATP
$FADH_2$ 7 × 2 = 14 ATP
$NADH/H^+$ 7 × 3 = 21 ATP
−2 ATP (activation)
129 ATP

19.26

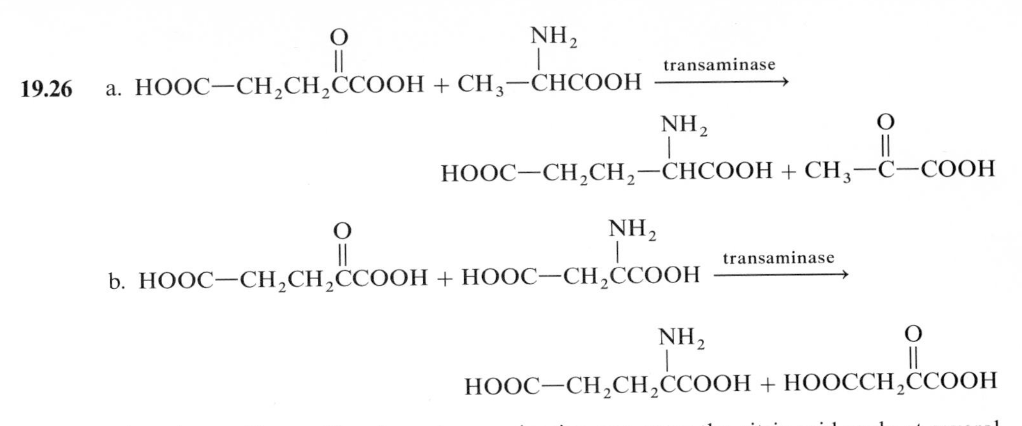

19.27 The α-ketoacids resulting from transamination can enter the citric acid cycle at several points for energy production.

19.28 In oxidative deamination, the amine portion of an amino acid is converted to ammonia and replaced by an α-keto group.

19.29 a. $CH_3CH(NH_2)COOH + NAD^+ + H_2O \longrightarrow CH_3C(=O)COOH + NH_3 + NADH/H^+$

b. $HOOCCH_2CH(NH_2)COOH + NAD^+ + H_2O \longrightarrow HOOCCH_2C(=O)COOH + NH_3 + NADH/H^+$

19.30 All foodstuffs can be degraded to form acetyl CoA, which can then be used for energy needs or as the starting compound for the synthesis of many other necessary compounds in the cell.

Chapter 20

20.1 a. DNA b. tRNA c. mRNA d. rRNA

20.2 a. RNA b. RNA c. RNA d. both

20.3
a. cytidine monophosphate, cytosine, phosphate, ribose, RNA
b. deoxyadenosine monophosphate, adenosine, phosphate, deoxyribose, DNA
c. deoxythymidine monophosphate, thymine, phosphate, deoxyribose, DNA
d. uridine monophosphate, uracil, phosphate, ribose, RNA
e. deoxyguanosine monophosphate, guanine, phosphate, deoxyribose, DNA

20.4

a. —A—G—T—G—C—A—
   —T—C—A—C—G—T—

b. —C—T—G—T—A—T—A—
   —G—A—C—A—T—A—T—

c. —A—A—A—A—A
   —T—T—T—T—T—

d. —G—C—G—C—
   —C—G—C—G—

20.5 All DNA molecules are made up of nucleotides containing adenine, thymine, cytosine, and guanine along with a backbone of phosphate and deoxyribose sugar. In all DNA molecules, A is always paired with T, and G is paired with C to form the double-helix structure of DNA. DNA molecules differ in the number of A-T and C-G pairs and in the order in which these pairs occur along the DNA helix.

20.6 An enzyme splits the A-T and G-C pairs along the double DNA helix. Each nucleotide in the original half bonds with a nucleotide available in the cytoplasm to produce a new half that is complementary to the original. The complementary base-pairing requirement ensures that the same order of nucleotides is retained in each portion to produce exact duplications of the original DNA.

20.7

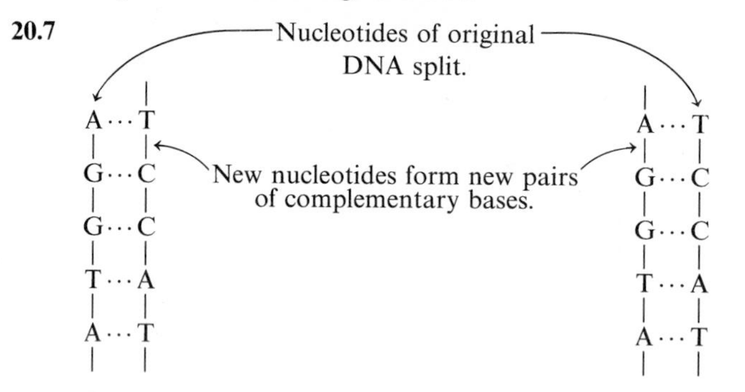

Two duplicates of the original DNA portion are produced.

20.8 One DNA strand: -G-G-C-T-T-C-C-A-A-G-A-G-

a. mRNA: -C-C-G-A-A-G-G-U-U-C-U-C-
b. Codons: -CCG-AAG-GUU-CUC-
c. Anticodon (tRNA): -GGC-UUC-CAA-GAG-
d. Amino acids in protein: Pro-Lys-Val-Leu

20.9 One DNA strand: -A-A-A-A-T-G-C-C-C-

a. mRNA: -U-U-U-U-A-C-G-G-G-
b. Codons: -UUU-UAC-GGG-
c. Anticodons (tRNA): -AAA-AUG-CCC-
d. Amino acids: Phe-Tyr-Gly

20.10

a. The order of amino acids may be changed when the order of nucleotides in the DNA is altered. The change in the triplet code results in the placement of a different amino acid.
b. If the amino acid order of a protein changes, the shape of the protein is altered when tertiary structure occurs. Since proteins such as enzymes use structure to identify specific substrates, a metabolic reaction with a certain substrate may be affected. Such metabolic disturbances are called *genetic disease*.

20.11 A gene from a foreign DNA can replace a section of a plasmid in a cell by recombinant DNA methods. Restriction enzymes separate a piece of DNA from a plasmid and a piece of DNA from a foreign DNA. The piece from the foreign DNA combines with the plasmid, giving a cell the instructions for the synthesis of a new protein.

20.12 The end product ties up an enzyme in the early part of the pathway, thereby shutting down the reactions and therefore the formation of that particular substance. As its level drops, the enzyme is no longer inhibited, and the reactions leading to that end product begin.

20.13 When a substrate enters a cell, certain enzymes are required for the reactions of that substrate. In enzyme induction, the substrate combines with a protein from the regulator gene. The regulator gene can no longer keep the operator gene from producing mRNA that will eventually lead to the production of enzymes needed to metabolize the substrate. When the substrate has undergone reaction, it is removed from the regulator protein. The regulator protein reacts with the operator gene and enzyme synthesis is stopped.

20.14 In enzyme repression, the buildup of an end product causes some of the end product to react with protein from the regulator gene. This unit, called a repressor, attaches to the operator gene and prevents the synthesis of proteins involved with the formation of the end product that has accumulated in the cell.

20.15 a. Hormones can be peptides, or steroids.
b. Hormones are produced by the glands of the body.
c. Target organs are the sites of action for particular hormones.

20.16 a. 1 b. 2 c. 1 d. 2 e. 2 f. 1 g. 3 h. 2

20.17 a. growth hormone b. thyroid-stimulating hormone c. vasopressin [antidiuretic hormone (ADH)] d. adrenocorticotrophic hormone (ACTH) e. growth hormone f. prolactin

20.18

Organ	Stimulating Hormone or Component	Resulting Hormone
Thyroid	Thyroid-stimulating hormone (TSH)	Thyroxine
Adrenal cortex	Adrenocorticotrophic hormone (ACTH)	Corticosteroids
Ovaries	Follicle-stimulating hormone (FSH)	Stimulates follicle growth; estrogens
	Luteinizing hormone (LH)	Stimulates formation of corpus luteum; progesterone
Testes	Luteinizing hormone (LH)	Stimulates production of testosterone
Parathyroid	Ca levels in blood	Parathormone
Pancreas	Glucose	Insulin, glucagon

20.19

Condition	Amount of Hormone	Hormone Responsible
Goiter	Too little	Thyroid-stimulating hormone
Diabetes mellitus	Too little	Insulin
Dwarfism	Too little	Growth hormone
Addison's disease	Too little	Aldosterone
Acromegaly	Too much	Growth hormone
Grave's disease	Too much	Thyroxine
Immature sexual development in females	Too little	Follicle-stimulating (luteinizing) hormone

20.20 a. Stimulates the production of growth hormones by the pituitary gland.
b. Stimulates the production of adrenocorticotrophic hormones by the pituitary gland.
c. Stimulates the production of follicle-stimulating hormone (luteinizing hormone).

Index